From The PAST to The FUTURE

The Legacy of Lev Lipatov

Recommended Titles in Related Topics

From My Vast Repertoire ...
Guido Altarelli's Legacy
edited by Stefano Forte, Aharon Levy and Giovanni Ridolfi
ISBN: 978-981-3238-04-6

Gribov-90 Memorial Volume: Field Theory, Symmetry, and Related Topics Proceedings of the Memorial Workshop Devoted to the 90th Birthday of V N Gribov
edited by Yu L Dokshitzer, P Lévai, Á Lukács and J Nyíri
ISBN: 978-981-123-839-0

From The PAST to The FUTURE

The Legacy of Lev Lipatov

editors

Joachim Bartels
Hamburg University, Germany

Victor Fadin
Russian Academy of Sciences, Russia

Evgeny Levin
Tel Aviv University, Israel

Aharon Levy
Tel Aviv University, Israel

Victor Kim
National Research Center "Kurchatov Institute", Russia &
Saint Petersburg Polytechnic University, Russia

Agustin Sabio-Vera
Universidad Autonoma de Madrid, Spain

World Scientific

NEW JERSEY • LONDON • SINGAPORE • BEIJING • SHANGHAI • HONG KONG • TAIPEI • CHENNAI • TOKYO

Published by

World Scientific Publishing Co. Pte. Ltd.
5 Toh Tuck Link, Singapore 596224
USA office: 27 Warren Street, Suite 401-402, Hackensack, NJ 07601
UK office: 57 Shelton Street, Covent Garden, London WC2H 9HE

Library of Congress Control Number: 2021049422

British Library Cataloguing-in-Publication Data
A catalogue record for this book is available from the British Library.

FROM THE PAST TO THE FUTURE
The Legacy of Lev Lipatov

ISBN 978-981-123-111-7 (hardcover)
ISBN 978-981-123-112-4 (ebook for institutions)
ISBN 978-981-123-113-1 (ebook for individuals)

For any available supplementary material, please visit
https://www.worldscientific.com/worldscibooks/10.1142/12127#t=suppl

Typeset by Stallion Press
Email: enquiries@stallionpress.com

Printed in Singapore

https://doi.org/10.1142/9789811231124_fm

Foreword

Lev Lipatov died in September 2017, while participating in a Physics conference. This was a shock for all of us who had known him for so many years.

In 1963, Lev had joined the department of V. N. Gribov in Leningrad, and this was the time when one of us (Genya Levin) became a life-long close friend. After completing the first dissertation (candidate of science) it took only a few years when Lev, after a seminar talk in 1970 of V. N. Gribov on the parton model and its derivation from quantum field theory, stood up and by saying 'Vladimir Naumovich you are wrong with your procedure to find and sum the logs' demonstrated his coming scientific leadership. In the following years he developed his fundamental contributions to deep inelastic scattering (DGLAP) and to the Regge limit in quantum field theory (BFKL). During this time, one of us (Victor Fadin) became his closest collaborator and friend. Since 1997 Lev was the director of the theory division of the Petersburg Nuclear Physics Institute (now NRC Kurchatov Institute), and he held this position until his death. One of us (Victor Kim) is a member of this institute.

It was only towards the end of the 1980's when members of Gribov's department were able to go abroad and build up international connections. However, their pioneering work had become known much earlier, and had started to attract visitors to come to Leningrad/Gatchina, in particular leading scientists from the US. One of us (Jochen Bartels), right after his PhD in the early 1970's had the privilege to visit Gribov's group and to learn the basics of their research: it was during this visit that he first met Lev and learned about his ideas. Later on in 1990, when HERA started the deep inelastic scattering (DIS) experiments, researchers at DESY (and at Hamburg University) had understood that the expertise of Russian theorists around Lev (in particular Lev himself, E. Levin, M. Ryskin, and V. Fadin) would be indispensable for deepening our understanding of QCD and of DIS physics and analyzing HERA data. As a result, during the following years Lev became a regular visitor of DESY where he initiated and stimulated research work, including also young theorists. During this time, experimentalists, e.g. one of us (Aharon Levy), came into closer contacts with Lev's results and their implications for understanding experimental data.

During Lev's numerous visits to research institutes and conferences, he built up contacts with younger scientists. As an example, one of us (Agustin Sabio Vera), as a postdoc, gave a talk in a conference, and caught Lev's attention ('Lev stood up and said out loud that all Agustin had said was completely wrong'). This tough discussion was the beginning of a long lasting journey into a much deeper understanding of quantum field theory and, later on, resulted in numerous common papers with Lev.

This book 'From the Past to the Future — The Legacy of Lev Lipatov' is meant to honor Lev Lipatov, the wide spectrum of his scientific work, the broad class of collaborators, and, last but not least, his friendly personality. As an introduction, the book contains three articles about Lev as a person, written by family members, by a very close friend, Genya Levin, and by another outstanding scientist, Al Mueller.

The main part then consists of a collection of 18 articles written by several scientists who closely knew and/or collaborated with Lev. Each article refers to a different scientific topic.

It summarizes parts of Lev's scientific achievements, presents new results which are based upon Lev's work, and gives an outlook on possible future developments. Lev's theoretical work has had an impact also on phenomenology and experimental high energy physics; the book therefore also includes several articles on these experimental aspects.

Jochen Bartels, Victor Fadin, Victor Kim,
Eugene Levin, Aharon Levy, and Agustin Sabio Vera

https://doi.org/10.1142/9789811231124_fm

Contents

https://doi.org/10.1142/9789811231124_0001

Chapter 1

Family Memories: Husband, Father and Grandfather

Elvira, Irina and Ekaterina Lipatov and grandchildren

1.1 Introduction

These are the memories of the family of the renowned theoretical physicist, the academician of Russian Academy of Science, Lev Lipatov. He was the smartest and the kindest man of his time with overflowing energy which sufficed not only to develop science but also to crucially contribute to the bringing up of daughters, grandchildren and great grandchildren. The Lipatov family infinitely loved, loves, and will love him: the husband, the father and the grandfather. Bearing this in mind, everyone from the family spared no time and efforts to write these notes. Everyone remembers the best of this extraordinary man who meticulously shared everything to everyone who takes an interest. Lev liked to tell that: "Everyone has to build a house, plant a tree and leave behind a son, but theoretical physicists have only daughters, and I have two of them". Two daughters, their mom and their children wrote these notes. Enjoy reading!

1.2 Wife Elvira

It was in 1961. This number is magical and invertible,[1] for me even fortunate: I met Lev (see Fig. 1.1). It was on a student party in Pesochnaya near Leningrad on the "dacha" of our fellow student Valerij Nefedov. It was November 7, the great holiday of the Bolshevik upheaval, but for us it was only the happy occasion to get to know each other: I, Elvira Logvin (from Siberia), and Lev Lipatov (from Leningrad). We were very young (21 years) and cheerful; we were dancing, singing, kissing and did not know what life had in store for us. In 6 months, we had married and graduated from Leningrad University (see Fig. 1.1).

Lev made his diploma work with Michail Braun, who had three promising students — Alexander Vasiliev, Eugene Levin and Lev Lipatov — and had to choose one of these "whales" for the post-graduate work. His choice was Alexander Vasiliev. Lev was distributed to the Electro-Mechanical Institute. As often being repeated by A. Vasiliev, Lev was an original thinker. He was not only a distinguished physicist but also a good father of two daughters and a very good chess- and (especially) draught-player. In school and early student years, he had to choose between a science and a chess career. Once, he played with Boris Spassky, often solved chess problems and liked blitz-game very much.

Later, science became the main matter and he used chess only as an exercise for the brain (every morning he began with the computer chess-game). So, he went to the third-grade institute and had to work there for two years by law (some kind of serve-labour). Fortunately, we heard about Gribov's seminars in the Physico-Technical Institute and it turned out well. Lev was discharged from electro-mechanic

Fig. 1.1 1962: the newlyweds.

[1]Elvira noted that 1961 has symmetry under rotation by 180 degrees

after a year and with some attempts he entered a post-graduate study with V. N. Gribov. It was the beginning of his scientific work, collaborating with V. Gribov, V. Gorshkov, E. Levin and later E. Kuraev, V. Fadin, et cetera.

Let me cite a paragraph from the memories of S. G. Sherman: "... When Lev came to the Physical Technical Institute, he became immediately noticeable. Viktor Gorshkov (who died last year), with whom Lev talked a lot at seminars and on the sidelines, told a friend: "In our department, there appeared one who creates an order throughout everything". A little bit later, when they were engaged in a common work, he remarked: "In darkness and off-road, where we are sticking and tearing tires, Lev goes like a tank!...".

During the defense of Lev's doctoral dissertation in Novosibirsk, an opponent, Professor Sudakov, gave a brilliant review and said a catch phrase: "When you listen to Bach's music, you want to take your hat off!". The scientific results of Lev speak for themselves.

I want to dwell on his contacts with foreign colleagues. It was in the 70's. We were living behind the 'iron curtain'. How to break it? I think, it was E. Levin, from whom we got an idea that the necessary first step was Poland. And for Lev, it turned out well: the sip of air out of doors, some seminars at Jagiellonian University (Krakov), acquaintance with Lech Schimanowsky. For the beginning, it was not bad. But then Lev said: "Hen is not a bird, Poland is not a foreign country".[2] The next refuge in those times was Finland. A Finnish collaborator of Lev had expressed fears of the Soviet Union. So, the second option also worked. Lev had permission (from the Gatchina party committee) and gave several seminars in Helsinki. Other options were not so simple. I remember several of Lev's foreign colleagues from the 80's, Jochen Bartels and Henry Abarbanel among them. The latter once said (having in mind the Soviet system): "I would like to understand what this thoroughly rotten system sustains on".

To travel to Western countries, it was necessary to obtain permission from the foreign department of the Gatchina district committee, where the main violin was Comrade Lisina, the head of the foreign policy department. To obtain her permission, it was necessary to come up with a convincing explanation to the decisions made at plenary sessions of the CPSU (Communist Party of the Soviet Union). Lev's explanations were not always sound. At the end, they stopped to release him. It was stated: "How can you be the head of the theoretical department without being a member of the party?". The trip was delayed.

I remember the year our troops entered Afghanistan. For several months, Lev waited for travel permissions to Italy and had already lost hope, when suddenly, on December 24, the long-awaited permission came with air tickets for December 25, when the entire Western World celebrates Christmas. Lev, who had flown to Italy, unsuccessfully tried for a long time to enter through the tightly closed doors of the

[2]In Russian this phrase is a verse.

Institute in Trieste. He was saved by a colleague from Novosibirsk who happened to be there (Weinstein).

But 'perestroika' (in the 90's) soon began, and a broad road to the West opened before Lev, where he could fruitfully cooperate with his colleagues without asking permission from incompetent party leaders. Lev was a pedestrian. He did not have a driver's license and he considered as a waste of time getting it. Sometimes he even walked from the airport, amusing oncoming motorists, as it once happened in St. Louis. When we were in Argonne, we walked to the supermarket by the High-Way until the neighbors offered us a ride. And at Los Angeles, Lev once wandered into the territory of a military base and noticed only on his way back, that entrance was strictly prohibited. As he walked, Lev continued to solve equations and sometimes it ended badly. So, in Odessa, at Privoz (market), a 'diplomat' (flat small bag for the papers) placed on the ground with an article prepared for printing was dragged from under his feet. In moscow, the same thing happened in front of Viktor Fadin. In Krakow, a wallet with documents and money was pulled from his pocket. In Madrid, I managed to save the wallet (they threw it on the metro to his feet). And in Paris, in the last year of his life, I failed to save his wallet. True, this time the wallet was without important documents. Apparently, swindlers saw detachment in the face of Lev, who was reflecting on the structure of the universe. Conferences, seminars, lectures... this was the life in which Lev felt "on the horseback", never left out and always participated with enthusiasm.

"We are happy people; we have lived to see the time when the existence of gravitational waves was experimentally confirmed," he said to his grandchildren. Lev would be even happier if he lived to 2019 when new particles, Λ_b and Σ_b baryons, were discovered in CERN. I think that it would inspire him to write another equation.

Returning to the student's party of 1961, I must say that the community of our friends, so called "kontora", had accompanied us for our entire lives. Lev was very glad to see our friends on each of our birthdays. The number of friends grew smaller with time: some friends have gone far away (E. Levin), others were dead (A. Vasilev, A. Sovestnov, M. Dyakonov). But every year the rest of us were gathered on December 19 (my birthday) and on May 2 (Lev's birthday). Lev considers this company as an 'unusual phenomenon' and uniquely endearing. Our friend V. Startseva made a remarkable film, in which Lev talked about it.

About Lev as a father, I give the word to our daughters Irina and Ekaterina.

1.3 The eldest daughter Irina

My first memory of childhood is how Dad had read a book to me. In the evenings, my father read fairy tales to me, very expressively, with different voices (see Fig. 1.2). Dad loved poetry, especially Lermontov. He knew many poems by heart; in his youth he even wrote poems by himself.

Fig. 1.2 Lev and his eldest daughter Irina in 1964.

Dad himself began to read early and always read a lot. He said that in the kindergarten he read books in front of the children, at a time when the educators were busy with something else. He taught me to read early, and I was very proud when I was also put to read in front of a group of kindergarten children.

Dad taught us to play chess. He played chess with me without a queen, but I always won. One of Dad's students recalled how he was playing blindfold chess in the back of a truck transporting first-year students to agricultural work. After that, the student decided that Dad was a genius.

When I was a teenager, Dad created with me a prototype of a learning computer capable of playing a game, by cell, something in between a nod-cross (the game was called tic-tac-toe) and mini-checkers. We glued together a few dozens of matchboxes: each paper was glued with a composition of chips located on nine squares. Playing with this 'computer', it was necessary after every move to get the marble out of the box on which the combination on the board was pasted. As a result, the wrong development of the game was gradually stopped by the withdrawal of the marble from the 'wrong' box, which led to the defeat. This device, as a result, was trained to play correctly. Dad was very excited about the process.

In general, Dad until the end of his life (I could write "until old age", but this is not true: Dad was never old) remained to be a person who was fond of and interested in everything around him, just like children who were not so bored as long as they could. Probably, it was this quality that helped him to find completely unexpected ways in solving amazing problems. Often these solutions came to him in a dream.

We lived in Leningrad near the large Piskarevsky park. In the summer of 1969, walking there with a newborn baby lying in a pram (she was my little sister Katya), my father met with a moose. Having majestically stepped out from behind the trees, it carefully examined dad with the pram and then majestically moved away. Daddy really enjoyed telling everyone this episode.

The largest civilian cemetery in the world is located on the border of Piskarevsky park. More than one million people who died from starvation in the siege of Leningrad during the World War II are buried there. Our family owes its existence to the village Nizhne Nashchekino, that is 80 kilometers away from Tambov. Dad's father, my grandfather Nikolai Mikhailovich, a military veterinarian, served near Leningrad. In August 1941, having been on leave for three days, waving a Nagant (according to a family legend), he put on the train his wife Maria Pavlovna Terekhina, his mother-in-law Pelageja and his two sons: Leonid 11 years old and Leva one year old.

So, my great-grandmother, grandmother, uncle and Dad left just before the siege to their relatives, first in Tambov, then in the village near Tambov, where Dad's parents were coming from. Great grandmother Pelageja starved to death because she gave her food to Lev and Leonid. We all survived because she died. My father was dystrophic, but he survived. Dad loved this village for all his life, although of course, he did not remember it, and was there only a few times at a conscious age.

Dad loved his parents very much. Grandma Masha (Maria Pavlovna Teryokhina) and grandfather Kolya (Nikolai Mikhailovich Lipatov), like most of my ancestors and relatives, had an extraordinary desire to learn. Despite hunger and hardship, they graduated and trained their sons. Dad became a theoretical physicist, while his elder brother, Leonid, was a sea captain, went several times around the world, wrote memoirs about his travels (see Fig. 1.3).

Grandmother was a teacher of geography; grandfather began after the war to teach military chemistry. Everyone read a lot. Dad was named in honor of Lev Nikolaevich Tolstoy, and his elder brother in honor of Leonid Nikolaevich Andreev.

Fig. 1.3 The Lipatovs in 1949: Maria (mother), Leonid (brother), Lev and Nikolaj (father) in Leningrad.

Fig. 1.4 1972: Irina (eldest daughter), Ekaterina (youngest daughter) and Lev.

Being a grown-up father, he was also close to his parents, and tried to visit them on weekends with us. He loved his mother very much. When she died in 1973, Dad cried for several days.

Our Dad walked with us and taught us to swim and ride a bike. On weekends, he woke up early (in general, he was a morning person, a 'lark'): he went to bed early, fell asleep quickly, and got up easily in the early morning. At about 8 o'clock on a Sunday winter morning, he felt very uneasy since the day was so good, while all others were still sleeping. Two daughters were pushing apart, climbing, and the whole family went with skis on a long journey by bus and train to ride from the mountains (see Fig. 1.4).

Dad always worked. In the morning, waking up, we asked: "Dad, are you writing?" "Yes, I am writing" answered Dad from the big room, "I am writing." In the evening, when we went to bed, we again asked: "Dad, are you writing?". And we received the same answer.

Dad did not know how to rest; he always was busy, either with his hands or with his head. When he and his co-author, Eduard Kuraev, were sawing wood in the forest, right next to the fence of our dacha, they talked about amazing things. A new fantastic theory of the world was born! The forest never heard such a talk. One day,

Dad and Kuraev decided to build a bench. Of course, the easiest way for the two theorists of science was to cut a bench out of a huge log. Pundits, enthusiastically talking about their own matter of interest, started to cut this thick log of wood. Having reached the core of the log, the saw suddenly began to squeal. It turned out that the saw's teeth stumbled upon ... a bullet! In places, where our dacha is located, the terrible battles in the WWII took place. In the first years, working on the site, we found a lot of cartridges and bullets. They managed sawing the log, and the gigantic bench pleased us for many years, standing in the place of honor on the site.

Thinking about his scientific problems, Dad sometimes made mistakes in everyday life. When we obtained the telephone line in the mid-1970s, Dad, hearing the bell and coming up to the door, said:"Hello?". But picking up the phone, he shouted: "Who is there?". Once, he flew away to England for receiving an award in casual pants, along with the mantle (gown), and with the quadrangular academic cap. He packed his luggage without our Mom. When Mom came home, she was horrified to find father's formal pants remaining in place. The prize was awarded to him after all.

When I was three or four years old, Dad, thinking about his scientific problems, forgot me at the store. Somehow I reached our house otherwise I would not be writing these lines now. More interestingly, my sister was forgotten immediately after birth in the hospital. Dad always wanted to have a son. He even chose a name: Peter — in honor of Peter the Great, the founder of St. Petersburg. When the second "non-Peter" was born, instead of 'Peter' the historical Petersburg name 'Catherine' was given to her by Dad... Dad met Mom with the newborn Ekaterina in the lobby of the maternity hospital where the nurse took out the children wrapped in a blanket with a bow. Parents, joyful, hugged, kissed and ... went. "And the baby, baby!" — the nurse shouted after them.

When we grew up, our parents began to go hiking with us. I remember rafting on a canoe on the river Wolf. We had the right camping equipment: a tent, kettles, saws, sleeping bags. The kayak, disassembled, was placed in two huge blue backpacks. The backpacks were placed on a wobbly trolley of four rails and two shaky wheels, and dad dragged the whole construction to the bus stop and into the train. In a double kayak, my sister and I were always sitting between my parents. My sister was a preschool child, and I was in the elementary school. In the bow and stern of the kayak, backpacks with luggage were stuffed, the ropes from the steering wheel were put on Daddy's big toes, so he taxied. Both Mom and Dad were rowing. Every hike was a real adventure. The water ran along the inflated sides; the sun was blinding through its reflection off the surface of the water. Sometimes the river was littered with fallen trees, which had to be cut to free the way. On the stops, a bonfire was lit under a pot — Dad chopped firewood while we carried dead wood, picked mushrooms and berries. At the bottom of the tent, the spruce fir twigs were first laid out and smelled wonderfully. How interesting it was all! Every trip was a real adventure (See Fig. 1.5).

Fig. 1.5 1972: Lev, Irina and Ekaterina.

And once, we almost drowned in Vuoksa. We came out of the ducts into the wide water, and the light kayak began to be carried away by a suddenly rising wind into the open sea. The rising waves flowed over the edge of the overloaded canoe. The wind and the waves roared, my mother shouted: "Help!", for some reason my sister laughed, and my Dad rowed like Ilya of Murom.[3] I remember how the muscles of his back walked. It was very scary. Thanks to father's strength, we somehow landed on the shore. After that, Mom went with us further on foot, and Dad dragged the kayak with things along the coast.

Dad loved every move. He especially loved to walk. Dad's grandfather, Mikhail Afanasyevich Lipatov, who was a member of the Second State Duma in 1907 (from the Labor Party of the Tambov Gubernia), was also very fond of walking. In the memoirs of his son, and of my grandfather, it is stated that he could walk 100 kilometers a day. It seems to me that this generation, whose childhood fell on the war, is distinguished by incredible energy and vitality. Alas, the less adapted and less persistent fellow citizens did not survive those years, and those who remained

[3]Ilya of Murom is a folk hero of ancient Kievan Rus' — a bogatyr (akin to knight-errant) and a character of many bylinas, East Slavic medieval epic poems.

lived for themselves and for them. This internal energy was also manifested in the absence of gray hair. Dad turned gray only after seventy years.

Dad loved to sing, although he did this not correctly enough. Our guests at the end of the evening always sang their favorite songs of their student youth. The guests came often, fellow students kept together, there were many married couples, children appeared after graduating from the university almost at the same time; everyone was a friend. On the weekend, everyone went to visit each other: to the Vasilievs, to the Levins, to the Boutikovs. Children climbed under the holiday table to tickle all sorts of interesting legs. The children said that my Dad was the kindest.

In the evenings on weekdays, after dinner, Dad told us in the kitchen all sorts of amazing, fantastic stories from the field of physical science. About black holes, about the Universe, about infinity and eternity.

In the evening on the weekend in the country, everyone sat down to play 'King' — a card game of our mild preference. It is worth mentioning that Dad always won because he remembered all the cards: those that came out and those that we had.

Dad's ability to find solutions is beautifully illustrated by the story of the Rubik's Cube. Dad brought it from Hungary in the early 1980s, when it first appeared. We all played with pleasure with our family. Once I saw him lying on the couch (he generally liked to work lying down, on the floor or on the couch), and next to it was the cube, and my Dad wrote his formulae on a white sheet of paper with his fine-grained small hand. Then, without taking his eyes off the sheet, Dad took the cube and with a few movements of his hands ordered its sides in the correct colors. It seemed to me that the formulae on the sheet were written in small letters and numbers, less than one page. I was then a student at the Physics Department of the Leningrad State University. With admiration for how Dad mastered this Rubik's cube, I took the puzzle to the physics school hostel with a story about how to solve it: on a piece of paper. After a while, in giggles and murmurs, classmates brought me a correctly assembled cube. Then it turned out that the students simply disassembled the Rubik's cube and assembled it correctly. So, theorists and experimenters approached the problem differently.

In 1983, Dad obtained the place for the summer cottages to be built. The first summer, we lived in the tent, sawed wood and uprooted stumps. Dad assembled the house from the panels, along with his friends, theoretical physicist Vasilyev, Startsev, Fadin, Levin. Dad loved the country house which was built by his own hands, so that even in winter, without paths, they traveled there with our Mom up to the waist in the snow, where they would cut firewood, heat the stove, maintain the heat in the house, feed the birds and admire the winter views all day long.

The foundation under the house was poured by Dad himself. The solution was prepared without a special mixer, with just a shovel in a basin. The amount of work is difficult to imagine. The foundation had the depth of more than one and a half metres. Dad dug also the basement to the same depth — just with a shovel under the whole house (6×6 meters). It was to store things. All gardening neighbors were extremely grateful to us because their plots were thus perfectly drained. All the water from there merged under our house. One summer, during which I lived in this cottage with my baby son (it was 1985, in the first year after the house was built), a lot of frogs croaked under the floor in a fun and cozy way. It was impossible to store things in the basement. Then we dug for a long time a variety of ditches around the house, even deeper than the basement itself, and laid the drainage pipes there. The basement was somehow drained, and my Dad was very happy when we kept in there the pots with the soup for storage. Sometimes we did this just to please him. Then, a quarter of a century later, when we finally covered the basement with sand, hiring labor migrants as workers, Dad bitterly regretted it. By the way, four dump trucks of sand entered the basement; the workers worked for a whole week.

Here, I tell another story about the family couch, which is the best example of describing the energy of my father and his friends. For a long time, our family did not have a car. At the cottage, that we had rented for many years in Sosnovo, there was a sofa. Dad decided to transport it to Stroganovo. And he did it together with Eduard Kuraev. This is hard to believe because the distance between the old cottage and the new one was more than two hundred kilometers. First, with the sofa on the bus, they got to the train station in Sosnovo; then by train from Sosnovo to the Finland Station; by tram across the city to the Baltic Station; by train to Stroganovo; finally, with a sofa in their hands they walked through the forest one and a half kilometers to the dacha! "But," said Dad, "we could always put the sofa on the ground and take a break!"

I guess one of the consequences of a hungry military father's childhood was the desire that our dacha should 'work' (his peasant roots could play some role here as well). When Dad and Mom were young, they were digging at the country house, planting and growing vegetables. Then they began more and more often in summer to leave abroad for job contracts, and I sowed everything with grass to make lawns. Now in the country we have the large lawn instead of vegetables. The only thing I could not win was the greenhouse. Whatever happened in the world of physics, in the early days of May including his birthday, one could find my Dad at the dacha, where he and Mom, sometimes with the help of friends, sometimes alone, covered the greenhouse with a polyethylene. Cucumbers and tomatoes grew beautifully in the greenhouse.

"To build a house, to plant a tree, to raise a son." Dad built not only his house in science, but also the real one, at the place of our country house. In the same place of the country, he planted not only one, but nearly fifteen trees. Twelve of the trees planted by our parents are apple trees. Where all decent apple trees have periods of rest, our apple trees do not. In the fall, I give all my friends and acquaintances two or three hundred kilograms of apples, depending on the crop of the year. Dad did not allow us to clean the apple trees, even if they were partially dry. In the corner of the site, in the place of the finally destroyed greenhouse, I wanted to build a gazebo. But there grew an apple tree, which Dad forbade us from touching. The builder of the gazebo turned out to be a creative person; he gladly built a gazebo that surrounded this apple tree on all sides, without destroying a single branch.

Fig. 1.6 2009. Lev's daughters and grandchildren: Kirill, Sergej with their mother Ekaterina; Irina with her children Anton and Antonina.

Dad had no sons, but two daughters were born: Irina and Ekaterina. Irina has two children: Anton and Antonina. Ekaterina has three sons: Sergey, Georgy and Kirill. The daughters and grandchildren love physics and maths very much (see Fig. 1.6).

Dad loved children. First me and my sister, then his grandchildren. Then the great-grandchildren. When I brought my own grandchildren to the dacha, my father enjoyed their company, talked to them, and invited them to play chess.

The great-grandfather could manage to hold his last great-granddaughter Verochka smiling in his arms in June 2017 in Berlin (see Fig. 1.7).

What is surprising: for some reason, all newborns in the new generations of our family resemble the grandfather and great-grandfather Lev in the first months of their life. I have not noticed it alone.

Dad had four grandchildren (Anton, Sergej, Georgij, Kirill) one granddaughter (Antonina), two great-grandchildren (Grigorij, Michail) and one great-granddaughter (Vera).

In his last year, Dad loved telling us that we all lived in a happy time. That we were all happy people. Because in our time, the existence of gravitational waves had been proven. Dad really was a happy man. Not only because he lived in a happy time. For all his life he was engaged in an interesting business, and he had a loving and beloved family.

He was always, in every instance, very interesting. Perhaps, equally true: every instance was interesting to him.

Fig. 1.7 2017. Lev with his great-granddaughter Verochka in Berlin.

Fig. 1.8 Lev and wife Elvira in 2007.

1.4 The youngest daughter Ekaterina

О милых спутниках, которые наш свет
Своим присутствием для нас животворили,
Не говори с тоской: их нет;
Но с благодарностию: были.
В.А. Жуковский "Воспоминание"

About dear friends, who our light
By their presence livened up with might,
Do not weep: they are not here;
But say with thanks, that they were near.
V.A. Zhukovsky "Memories"[4]

Fig. 1.9 Lev with his youngest daughter Ekaterina (Kate).

On May 2, 1940 in Leningrad, a boy was born into the family of a war veterinary surgeon and a teacher, who would later be my Dad. Grand-Dad Nikolay liked Russian literature. Therefore, when his two sons were born, he gave them the names Lev (to my father) and Leonid (to my uncle) in honour of the great Russian writers Lev Nikolaevich Tolstoy and Leonid Nikolaevich Andreev. When Lev went to school, he first studied in the school number 284, then in number 281 of Leningrad (now Lyceum; the later President of the Russian Federation V.V. Putin studied there). He finished the school with a silver medal (4 for Russian language, 5 for all other subjects[5]).

In 1957 he entered the Physics Department of the Leningrad State University, where he graduated in 1962. At first he worked at the Institute of Electrical Engineers on the Neva river (it was associated with the factory "Elektrosila", where they produced turbines). He was working there for about 1.5 years, then he entered the postgraduate study at the Institute of Physics and Technology of the Academy of Sciences of the USSR (enrolled on December 1, 1963), completed it and defended

[4]*translated by Ekaterina*
[5]Students are graded on a 5-step scale, ranging from 1 (unacceptable) to 5 (excellent)

Fig. 1.10 Lev with his father, N.M. Lipatov and his eldest brother, L.N. Lipatov, (sea captain).

his candidate dissertation in 1968. In 1963 my sister was born, in 1969 I was born. My granny, Dad's mother, died in 1973. It was immensely stressful for Dad, but as it often happened, there soon came a quiet time. In the next year (1974) Dad defended his Doctor of Science dissertation. He was 34 years old, and it was said among scientists that it was an extraordinary feat to obtain a Doctor of Science degree at such a young age. I was then 5 years old during which my father often went with me for a walk, thinking about his work. His dissertation was devoted to me.

My memories of my father began just at that age, 5–6 years old. I fragmentarily remember with pleasure the birthdays of my parents' university friend A. Vasiliev in Peterhof, where Mom and Dad brought my sister and me every year. Many girls gathered there: daughters of the Vasilyev, the Startsev, the Levin, and often other university friends came with their children. On all those evenings, Father joked and laughed in the company of his former university mates, and he was a very sociable man. Once someone asked about somebody, who was present there: "What bad can you say about him?". Startsev answered: "The goat". For a moment there was silence. Then, a joke: "The little goat". Dad laughed.

In another case, during the 80's, Dad was nominated for the first time as a corresponding member of the Russian Academy of Science. There was an article in the newspaper titled "Leningrad Truth" about him, as a great scientist. A few years before, the wife of his friend Startsev had decided to begin painting pictures. She painted them them so well that some of her pictures were shown in a small exhibition, together with the works of another novice painter. Soon there was an article in a small newspaper in Gatchina about this event with photos of Startzev and the other painter and his pictures. She said to my Dad: "You and I are the most

Fig. 1.11 Portrait of Lev, drawn by the artist, Devyatov for the newspaper: "Leningrad Truth", where the article about Lev was published.

famous people in our company: they have written about us in the newspapers!". Father answered, laughing, that they had written about him long ago. Dad had the good sense of humour.

He especially liked A. Vasilev, who was his best friend.

Dad took me and my sister, since the age of 5 years old, to the House of Scientists in the Forest to become familiar with the English language. He spoke English well and thought his children should study English too. Before New Year's day, my father and I went to the fair to choose a present for my mother. He bought a New Year tree and I dolled it up with beautiful colored balls and New Year toys.

Throughout my childhood, I had been coughing persistently. Mummy treated me with nasty wet mustard plasters and jars. This treatment routine became a bit milder as my father read to me tales about the Old Man-"Lesovichok" and "The Adventures of a doll Kate". Dad read expressively.

My sister once decided to get a dog, a black terrier. Mummy loved animals and was not against a dog in our flat, but Dad was not happy with this idea. It was understandable: me, my mother, my father and my sister, lived then in a tiny three-room apartment with 2 interconnecting rooms. The corridor was occupied by a massive bicycle, the wheel of which was rotated in such a way that everybody, entering the room, inevitably stumbled across. Nobody, I think, could say whether anybody ever went on this bicycle or not. The prospect of stumbling upon it when entering the room and now having a dog did not really please my father. My sister, knowing that Dad treated me, the youngest daughter, more favorably, persuaded me to talk to Dad when he was drinking (in such instances, he became kinder). I successfully coped with the task entrusted to me: the bike evaporated to somewhere, and in its place a huge black terrier Martin loudly snored. By the way, in contrast to the absolutely useless (in urban conditions) bicycle, Martin was useful: my sister cut its hair, someone spun its wool, and my grandmother knitted from it warm socks.

Dad went to his job twice a week: on Mondays at the Physico-Technical Institue and on Thursdays to Gatchina at the Leningrad (Petersburg) Nuclear Physics Institute. The carpet on the floor in the living room of our apartment was covered with white sheets of paper with many formulas, which had been written by Dad. He liked to create his scientific works, lying on the floor; he had no desk. When I was schooling, Dad cooked a meal every day to feed me after school. In the evenings, before my mother came back from her work, I, my father and my sister together peeled potatoes for supper. Dad put us, his daughters, to bed early in the evening, at 8 o'clock, and then often looked into the bedroom to see if we were sleeping. When we were left alone, I jumped into my sister's bed, where she told me various interesting stories about the Bermuda Triangle, about black holes, etc. My sister and I piled up the blanket on my bed in such a way that when Dad had suddenly looked into the bedroom, he got the illusion that I was sleeping under that blanket, covering my head.

Dad loved his relatives very much. He often traveled to the city of Tambov, in the suburbs of which the village of Nizhnee Nashchekino is located, where all the Lipatovs came from. By that time (in the 60's-70's) many of his relatives were living in Tambov. My Dad's cousin's nephew Sergey Ehlakov remembers. "It was in the summer of 1968 or 1969. I finished my 8th or 9th school class then. I had to enter the university. Uncle Lev, as a real physicist, brought me a gift of 'Elementary Textbook of Physics' edited by G.S. Landsberg. The textbook, though elementary, was designed for a university course. I never managed to master it... After this meeting we were sent to the river, to sunbathe. For some reason no one went with us. We sunbathed across the river, on the opposite bank. Lev Nikolaevich told me about quarks and quasars, apparently trying to raise my interest in this topic. For me, his stories were unusual and very interesting. Though I was regularly reading the popular science and life magazine in those days, everything I heard was a great discovery for me".

In the summers my parents rented a cottage in Sosnovo on the isthmus 'Karelsky' near Leningrad on the bank of Deaf Lake. Its landlady was a judge. It was a large two-floor house, on the one side of which the Deaf Lake splashed, and on the other side a steep little mountain emerged. We occupied the entire first floor with three small rooms; a wobbly wooden bridge led from the exit of the middle room to the slope, on which a small dilapidated building containing a kitchen with a gas stove was located. To the left of the kitchen there was an object that I, being a little girl, was terribly afraid of: the ants could not think of anything more clever than to arrange an anthill, almost on the path to the outhouse. To the right of the kitchen there were birch trees, and there was a small vegetable-bed where, thanks to our efforts, peas and strawberries were growing. Dad liked agricultural work very much and was glad to receive harvest. In his childhood, that was in war time, he had been starving, therefore he liked to grow vegetables and fruits by himself.

My sister and I lived at this dacha every summer with my grandmother, who came from a town Kustanay to look after us. Dad and Mom came only for the weekend. Dad came before Mom and brought books for me and my sister from the library of the Physico-Technical Institute. I loved to read and was always looking forward to my father's arrival. On early mornings of Sundays, we went from Sosnovo to a place called "The Big Stone". There were a lot of mushrooms. Dad collected everything; he was sorry to leave something in the woods, so the rotten, huge russules, as well as suspicious mushrooms, the edibility of which were in doubt, also fell into his basket.

My father also fished. There were many fishes in Deaf Lake, especially roaches and perches. Sometimes the owners gave us a boat, and my parents and I went on exciting trips across the lake, with stops on the uninhabited islands, on which there were many charming lilies of the valley during the month of June. On the weekends of July, we went to a huge sandy beach. After sunning we went to swim. At that time I could not swim very well, so my father took me in his arms, carried me to the depths, and I jumped from his arms and swam to the shore. I remember how Dad taught me to ride a bike at this dacha, and then I went on the bicycle to a nearby settlement to buy milk.

Later, when I was in the seventh form, Dad got a piece of land in horticulture "Proton" in Stroganovo near Siverskaya. Father always tried to help my mother in house-keeping: peeling potatoes, mushrooms, cutting salads before guests were coming. On the dacha in Stroganovo he gathered and dismantled a warmer for vegetables. Dad brought a cart of manure (for that warmer and a vegetable-bed with strawberries), set potatoes, filed beams and broke them for stove.

In 1984 he and a classmate of my sister built in the country a two-storey shield house, in which later his four grandsons and granddaughters were growing up. Two great-grandchildren also liked to come to this cottage. Some grandchildren and great-grandchildren had other holiday homes belonging to other relatives, but the cottage in Stroganovo, built by their grandfather was a favorite one.

Dad had another idea to build a bath on the site. He began to attach it to the barn (rather, a shed for storing firewood), but after a few years of work, he gave up this endeavor (it seems, after the 5th tier of logs), as he lost enthusiasm for the construction business. After 10 years, at the initiative of the sisters 'the barn' and 'bath' were dismantled, and in their place (handy Foundation) a handsome guest house and a new barn have been built, sheathed with siding. All this was not built by Dad, but by builders hired by my sister (payed with Dad's money). With his own money the main house was completed: to the sun room in front of the house a gallery of unknown purpose was attached; there was also the right glass wall, which ran almost along the border with the neighboring piece of land. At the end of the gallery near the forest there was a kitchen and another glass sun room, from which they made a door to the backyard. Dad for a long time resisted the appearance of this door. He imagined detachments of thieves who would find it convenient for a robbery. Yet he had to give in to the opinion of most relatives. From the kitchen they built a spiral staircase to the second floor, where the parents were living (Dad wasn't happy with these stairs either: there was another one at the far end of the house, so why change anything? He was conservative).

Every year on May 2, 20 friends and relatives gathered at this dacha for Dad's birthday. On the 75-th anniversary of my father the grandchildren wrote with large colored chalk on the biggest stone: "Grandfather 75 years." I think my dad was very pleased. Dad loved to ski in winter. Our garden plot was located next to the forest. If you go through it on the path after half an hour, you can walk to the Creek, behind which begins the Christmas swamp; moving further in the same direction, through the swamp, you can walk to the estate of Nabokov. Once my father and I went there during the winter on skis, but did not reach the goal because my ski broke in the middle of the swamp.

My father was quite a well-read man, and often told interesting stories about writers and their works. For example, he said that S.T. Aksakov, when writing the second part of the "Family chronicle" trilogy: "Children's years of Bagrov's grandson", based the description of the Blizzard on the "Captain's daughter" of Pushkin. Following the advice of Dad, I read the books by a significant representative of the "village pros" Valentin Rasputin ("Live and remember", "Money for Mary", "Farewell to the Mother"); the trilogy "Walking in pain" of A. Tolstoy and the historical novel "Evil spirit" by V.S. Pikul. I did not like the last novel very much because of the historically incorrect represention of the moral image of the Holy Emperor Nicholas II, his relatives and those, who were close to them. I told Dad this; he objected that, perhaps, at the time when Pikul wrote his novel there was still not enough information to objectively describe the members of the Royal family. My father liked "country prose", of course, because he had spent his childhood in rural areas.

Dad loved his grandchildren very much (see Fig. 1.13). With his second grandson, my eldest son Sergei, I stayed at home until he turned two (due to a lack of availability in kindergarten schooling), and I was studying at the Institute. Dad's last grandson, Cyril, my younger son, rode in a stroller during his infancy on Vasilevsky island, shaking the stroller with all his might. Cyril had a nasty habit of crying and pitching which led him into a silent reverie. My Dad helped his third grandson, my middle son George, to solve problems in physics and linear algebra, when he studied in his first year at the Physics Department of Moscow State University, and shared with George about nuclear decay, nuclear fusion, TOKAMAK, and interesting stories from the life of scientists. Once we were talking about the soviet scientist Gamow, who emigrated to the United States. Dad explained to little George the essence of alpha-beta-gamma theory and said that Gamow was the first scientist who suggested what happened in the first seconds after the Big Bang. With Dad, everyone was always interested; he was a very versatile

Fig. 1.12 Lev in Hamburg.

person. When grandchildren grew up, Dad took them with him abroad and to Kolomna for school holidays.

In 1997, my father became a corresponding member of the RAS in the Nuclear Physics Department. In 2011, he was elected academician of RAS in the Department of Physical Sciences. Since the 80's, my father often went abroad with my mother on business trips: he was in Spain, Israel, Chile, Australia, China, Vietnam, America, Italy, Portugal and Greece in Crete. He worked for long in France and Germany, (see Fig. 1.12).

In 1991, I was with my eldest son for a week visiting my parents in the German city of Siegen, where my father worked at the university on the invitation of Professor Hans Damen. Then we went to Munich to Pinacoteca to see the "Vier Apostel" of Duerer, then to Padua and Venice, where we walked along the canals and were in the Palace of the Dodges. In 1993, my eldest son and I were in Zeuthen, a suburb of Berlin, where the Pope gave a course of lectures in the branch of the University of Berlin. In 2008, I was visiting my parents in Hamburg, while my father worked at DESY, where his theoretical predictions were confirmed at the HERA accelerator. My sister once asked my dad: "What will you do, if your theory does not work in practice?". He said: "I have an emergency theory". During this trip, my parents and I went to the Hamburg Zoo and to the Botanical garden, in the Chinese garden of which we came across a nice gazebo, where we sat down with pleasure. Now, as I open my album of family photos, I feel profoundly sad. Oh, how I want to get back into this happy past and wander through this beautiful garden with my father and mother!

Fig. 1.13 Lev and his wife, Elvira with her youngest daughter Ekaterina and her two sons in Hamburg 2008: (from left to write): grandson George, daughter Ekaterina, wife Elvira, Lev and grandson Cyrill.

In 2009, Dad was invited to lecture in Israel, in Tel-Aviv for 3 months. My mother and father lived there in the "Maxim" hotel on the Mediterranean Sea. Every morning, as he loved to swim, Dad went to the sea. My parents and I went on an excursion to Jerusalem, visited the vicinity of Haifa, Caesarea. In the same year I moved from St. Petersburg to Moscow with my husband and younger sons. The parents came to visit us twice a year (Dad participated in the meetings of the Academy of Sciences). We lived then in the southern outskirts of the capital, near the Moscow ring road, near the metro Zyablikovo. The Lenin Hills were half an hour of driving away; after one of the arrivals of our parents, all of us together went to walk in the Lenin Hills. Dad joked: "Let's go to Lenin's places!" We went to Lublin Park; one of the houses located in it was a cottage rented by Fyodor Dostoevsky in 1866. Here he wrote the 2nd part of the novel "Crime and punishment" and the story "Eternal husband". Dad went with me and Mom to exhibitions in art galleries, and he loved the impressionists. Together we watched films about artists who belonged to this direction.

In the summer of 2017, I sold my apartment in St. Petersburg and with the financial help of Dad I bought another one in Moscow. We now live with the whole family in the exclusive district of the capital, from which I go to work quicker than before. My younger and middle sons now get to schools faster. From the height of the 17-th floor, it is pleasant to admire the panorama of modern Moscow houses with multi-colored balconies. Mom often comes to visit and likes to watch the stars from the balcony of our apartment. It's all thanks to Dad.

My father used to say: "We live in happy times, we are happy people, because we have discovered gravitational waves". Dad was, without a doubt, a happy man. One of his grandchildren, my middle son George, after listening to such speeches, became interested in astrophysics. In his first year at the department of physics in MGU, he chose astrophysics among several subjects as the study of his choice. Now, in his second year, he is reading recently published articles in English about the registration of the gravitational waves in the collision of neutron stars: it is obvious that he decided to become a happy man in full. I also collected an article about gravitational waves and found it so interesting that I read for a few days almost without sleep. And I felt happy too. We are all happy.

To conclude my recollections, I would like to return to the beginning and note that the fates of my father and his brother were somewhat similar to the fates of the great Russian writers, whose double namesakes they were. My uncle, Leonid Nikolaevich, died suddenly from heart disease, as well as Leonid Nikolaevich Andreev. The ways of life of my father and Leo Tolstoy were also similar: both became famous, one among writers, the other among physicists; they also had a lot of similarity in their outlook. The end of their roads also looked alike: Tolstoy before his death secretly left Yasnaya Polyana; Dad also left home and went to Dubna, refusing to have anyone from his family accompanying him. He went to a scientific conference, where he stayed forever. It was as Minkin, Director of the Physics Institute where my father worked, said "He like a soldier died in the line of duty".

I am grateful to my father for what he was in my life and made it so good, bright and beautiful that it is impossible to express in any words.

1.5 Grandchildren

1.5.1 *Anton*

My grandfather Lev Lipatov was a tremendously advanced scientist who always wanted me to become a physicist. In fact, I realized that I might have an option to become something else only when I was already a master student at the Physics Department of Saint Petersburg State University — the same department where my parents and grandparents met and studied. The subject of my study was solid state physics, which my grandfather described as 'too simple'.

My life later moved to an engineering and entrepreneurial direction. At one point I wanted to make use of all the knowledge and skills gathered from physics and met grandfather and his colleague Vitalii Velizhanin in Berlin to have a deep dive into the current state of research. These meetings reassured me that the days, when everything will be clear in physics, are yet to come.

I loved my grandfather very much. Besides being one of the smartest people of his time, he was also one of the kindest. His enthusiasm and fun he had explaining the most complicated matters to everyone ready to listen will be forever an example to follow. My whole family is one or another way connected to physics and maybe someday I will get back to it, and mankind will get its long anticipated cold fusion reactor.

1.5.2 *Antonina*

You know, my grandfather was a great person with a great soul and a great mind. I am so happy to be part of such an astonishing family. He always told us mysterious things about science and space right at the dinner table. When I was a child, I thought that he never had time for me, but now I don't think so. My grandfather could find a place in his schedule even to go to my little physics course in school and give seven students a lecture about his work, and that was fantastic. He brought me to his trips in different countries — God knows how much I could see and know in that time. I could even eat and talk with Higgs or Antonio Zichichi (who is president of the part of Sicily), can you imagine? However, I grew up in his society, and it was so obvious that almost everyone around me is one of the best physicists in the world. He built the fortress, protecting his family from another world, where we could study and mature. Thank you, grandfather; I wouldn't be that person who I am now without you.

https://doi.org/10.1142/9789811231124_0002

Chapter 2

Lev Lipatov: My Friend and Renowned Physicist

Eugene Levin

2.1 Introduction

Prof. Lev Lipatov (see Fig. 2.1) died on September 4, 2017 during a conference in Dubna. Physicists all around the globe, who are engaged in high energy physics, and especially Quantum Chromodynamics, know or feel that this area of physics would be quite different without his contributions, and the future progress will be slower without his ability to solve difficult problems. I believe that this book will show you what we lost as far as physics is concerned. The goal of my presentation is quite different: I wish to share with you the atmosphere and the flavour of everyday life in Gribov's theory department, where Lev matured as an independent researcher and wrote all his important papers. Actually, his entire life was in this department, where he started as Gribov's Ph.D. student, and finished as the head of the department.

Fig. 2.1 Quam bene vivas refert non quam diu (It is how well you live that matters, not how long).

If you wish, my paper deciphers the main credo, which we were taught by Prof. Gribov and by the physicists of his generation: *"Physics first"*[1]

2.2 The beginning

Let me start from the beginning. The first time I met Lev was in July 1957. It was a rare sunny day in Sankt Petersburg (Leningrad at that time) in the country, which does not exist anymore in the Soviet Union. We did not pay too much attention to the beauty of the day, since we had to write the entry exam in physics to the Physics Department of St. Peterburg University. Nowadays, it is difficult to believe that physics at that time was so popular in Russia, that every mother wished her daughter would marry a physicist. The competition was very strong: ten people for one place. Therefore, the five problems that we had to solve, were rather difficult, especially one. When I finished, I went around asking about the answer to the most difficult problem. The first three gave me an answer, which differed from mine. The fourth was Lev, who had the same answer as I had. Three days later, we found ourselves among the people who passed the exam (see Fig. 2.2).

Starting from that day, we were in touch on an everyday basis for the rest of our lives. We went to the same group for the general course of physics, together decided to choose theory of elementary particles, and joined the theory group at

[1]It should be pronounced as 'ladies first' in a chivalrous code. Meaning is obvious: we think first about physics. It is so natural as to let a lady go first.

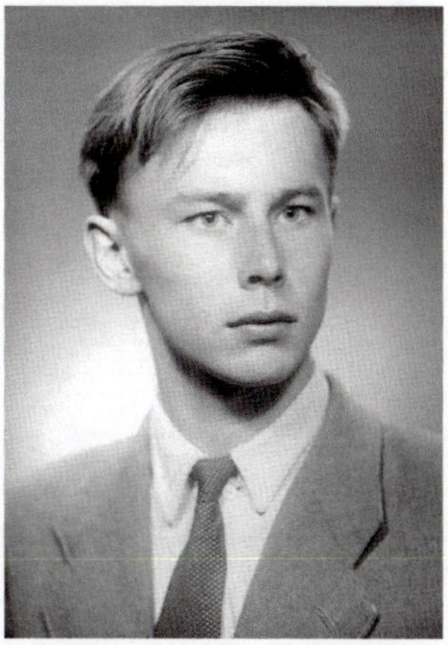

Fig. 2.2 A man comes into this world with clenched fists and a cry; he seems to be saying the whole world will be mine — Lev Lipatov, student of the physics department of St. Petersburg University.

Fig. 2.3 And he leaves with open hands, as if he says, I don't take anything with me — one of the last pictures of Lev that I have.

the university. We created a net of common friends in the university and found our wives among them.

Our friends (see Fig 2.4) deserve more than several lines in this paper. They shared with us all our ups and downs, supported us without questioning whether our behavior was correct or not. I know Lev was proud that we have not lost even one friend in 60 years after university. Even now, as I am writing this paper, sitting at home in Israel, I hear my wife discussing important matters on children and grandchildren with her friends in St. Petersburg, over Skype.

Fig. 2.4 We and our girls: theory group of the physics department of St. Petersburg university, 25 years after graduation. Lev and I are in the first row.

During the university years, the mathematical capabilities of Lev were fully recognized by us. Actually, he wanted to get a Master's degree in mathematical physics. Fortunately for theoretical physics, he changed his plans and studied high energy physics. I believe he realized that high energy physics was full of difficult unsolved problems. One episode was very instructive, which showed that we were very well motivated kids. Prof. Yu.V. Novozhilov, who taught the course in quantum field theory, chose to follow the variational approach of J. Schwinger to quantum mechanics and quantum field theory. Frankly speaking, it was very difficult for us and when we found the book of Bethe, Hoffman and Schweber: "Mesons and Fields", that had been recently translated to Russian, we decided to ask the young star of the theoretical department Prof. Mikhail Braun to organize an informal seminar in which we read and discussed this book. I remember that three of us, Alexander Vassiliev, Lev Lipatov and I, were brave enough to ask Prof. Braun for his help. He agreed and most of my friends, including Lev and myself, consider him as our first teacher. This paper is a good opportunity for me to express in public how we were and are thankful to him. I believe that after this event Lev changed his mind and joined us in our attempts to learn quantum field theory.

Impressed by the personality of Prof. Braun, we decided to write the diploma paper with him. This was a mistake. Neither Lev nor I remembered the subject of this paper, which we did together. It was never published. The only result from this work was that we learned some basics of the Feyman diagrams approach, and found out that the only place in St. Petersburg, where the high energy physics was on a good international level, was the Gribov department in the Ioffe Physico-Technical Institute. After getting our masters degrees, we were allocated to the research institutes in St. Petersburg, in which we were involved in quite different physics. However, we were shy but stubborn kids, and in 1963 we started to take the so called Landau exams in theoretical physics. Our goal was to get a position in the Gribov department, but this was only possible after passing the set of exams. For us, since we were fresh from the university, Gribov reduced the set of exams we needed to pass: classical mechanics, classical electrodynamics and quantum mechanics, based on the Landau and Lifshitz book, and quantum electrodynamics from the book of Akhiezer and Berestetsky. It was a hard time for our young families, working at night after a day in our institutes, but we passed all exams, solving difficult problems. Prof. Gribov took us as his only Ph.D. students. Gribov's third Ph.D. student, L. Frankfurt, also passed the exam procedure. This was the end of the beginning for the three of us.

2.3 Gribov's department and high energy physics seminar

Let me start with two citations. The first one is from Wikipedia: "Vladimir Naumovich Gribov was a prominent Russian theoretical physicist, who worked on high-energy physics, quantum field theory and the Regge theory of the strong interactions. His best known contributions are the Pomeron, the DGLAP equations, and the Gribov copies." The second is from the paper of Dokshitzer and Kharzeev [Dokshitzer and Kharzeev (2004)]: "Suffice it to mention that his name is attached to many key notions of the theory arsenal: Gribov-Froissart projection and the Gribov vacuum pole (Pomeron), Gribov factorization, Reggeon Calculus and Reggeon Field Theory (RFT), Gribov diffusion, the AGK (Abramovsky-Gribov-Kancheli) cutting rules, the Gribov bremsstrahlung theorem, Gribov-Glauber theory of relativistic multiple scattering, Gribov-Lipatov evolution equations, Gribov copies and the horizon, etc." The dry language of these citations shows that we were lucky to get as our supervisor the outstanding physicist, but they do not describe the Gribov personality, his drive for physics, his intuition and ability to create a picture for complicated physics problems. This is not a paper about Gribov, but his influence on Lipatov and us is difficult to overestimate. Gribov, whom we know and respect, is described by Dokshitzer in [Dokshitzer (1998)] and I can only certify that he gave a portrait of our teacher, which is close to reality.

We came, when Gribov had published his famous papers on the Pomeron and Gribov-Froissart projection, but other items, listed above, were done in our time and with our participation.

However, our first problem was to survive, when we came to the department after passing the Landau exams. Prof. Gribov summoned the three of us and gave the following instructions. You have two duties: to come and participate in any seminar of the department, and to answer all questions that an experimentalist would ask you. You are free to do whatever you want, but do not think that I will find a problem for you. You need to find it by yourselves. However, you have a privilege: you can ask me any questions and I will answer. Therefore, we understood that Prof. Gribov used the principle of deep water in his pedagogical approach. You know that there is a method to teach kids to swim: to hurl them in the deep water. They say that some of them will start to swim.

We survived, but many did not. Lev and I swore that if we would survive and have our own students, they will never suffer as we did. My former students, S. Bondarenko and A. Prygarin, and the young researcher in UTFSM, M. Siddikov, with whom Lev wrote his last papers, have no idea that the patient and attentive attitude of Lev to them was formed in the distant year of 1964, due to our experience.

In 1963, when we joined the department, Gribov's department in the Physico-Technical Institute was a unique phenomenon having about 15–20 men, who were young (all, except Ilia Mironovich Shmushkevich, were younger or of the same age as Gribov, who was 33 when we came), ambitious, creative and well educated. All of them were doing the same physics as Gribov. We actually worked at home having two visiting days on Mondays and Thursdays. On Monday was the Gribov seminar, which I will discuss below, but on Thursday the informal discussions went from the morning till 6 p.m. We had only one common room[2] in the institute and the typical scene on Thursday was Gribov relating what he did or thought during the days since the previous meeting on Monday (see Fig. 2.5). We all were listening and trying to understand and argue against. For us, youngsters, it was difficult to digest what he was talking about, and even much more difficult to take part in the discussions.

Fortunately, there were three of us and each of us worked hard. Our internal chatting, with attempts to understand the new ideas and accumulate them, was the main factor that helped us a lot. The second factor was that Gribov was not strict with his promise not to meddle in our search for the subject: Frankfurt and I got the idea from him to look at hadron scattering in the additive quark model, using the analogy with deuteron-deuteron collisions; Lev was assigned to his team for reconsidering the double log approximation in quantum electrodynamics.

This department had a regular seminar, which was as unique as the department itself. After leaving Russia in 1990, I travelled a lot, but had not seen anything which is close to the Gribov seminar. The goal of the seminar was not to learn what the invited speaker did, but to understand the physics problem and find the solution to this problem. Sometimes with the help of a speaker, sometimes without his help. It was assumed, that: (i) the speaker was a good physicist, since he was invited to give a talk; (ii) he wanted to convince us that he made a breakthrough, although he could likely be wrong; (iii) the speaker was responsible for everything, that he used and should prove every point of his talk[3]; (iv) the speaker must answer any questions related to his talk. The big shots and the young students were equal in giving the talk and in asking the questions. Your past merits cannot save you from deserved criticism. The seminar starts at 10 a.m. on Monday. The speaker was afforded five minutes, during which he should not be interrupted. The duration of the seminars

[2]Actually, this fact: one room, was the official argument, for why we had only two visiting days.
[3]If somebody would ask him to prove the Pythagoras theorem, the speaker should prove it.

Fig. 2.5 Magister dixit: Prof. B.H. Gribov in our common room is telling us, what he understood. The best picture that I found which reflects his personality.

were not limited. I did not remember a seminar that ended before 2 p.m. but very often after lunch break it continued to 6 p.m. I remember several cases, when we gathered together on the next day to continue the discussions.

The typical talk at the seminar was as follows. At 10 a.m. Gribov introduces the speaker. The speaker talks for five minutes. After that Gribov jumps up saying first, the speaker is completely wrong, and second, the correct answer should be like this. The description of the answer followed, taking 30 minutes. The weak speakers started to be nervous, but the ones who trusted themselves, countered, "you, Prof. Gribov, are completely wrong and that you have not taken into account of this and that". Other people entered the discussion. Sometimes the discussion became so heated that everybody forgot about the speaker. Normally, after two hours of discussions, the common opinion was to let the speaker resume, but to emphasize the points that had been raised in the discussions.

It was a painful experience for anybody who talked at that seminar, an ordeal not only for your knowledge and ability in physics, but for your character. We know this from our own experience, since everyone had to report on his paper at the seminar before sending it for publication. However, after the talk, when the dust had settled, you realized that the friendly fight at the seminar was very useful for you: you were able to look at the problem, that you solved in the paper, from a different angle, which would help you both in writing the paper and in the future work.

After 1985 we started to go abroad and many of us found jobs in the West. We found ourselves quite capable to discuss and defend our ideas against the experts, but were very badly prepared to give a talk for a more general public: for the physicists, who were doing the high energy physics, but whose interests were rather far away from ours. Prof. Lev Lipatov suffered from this problem to the end of his days. His talks were very useful and enjoyable for a couple of dozen of experts (most of them are the authors of this book), but were rather a black box for the others. He actually recognized this problem and worked hard to solve it. Our bad luck, that he left us before solving it.

My last comment in this section is about our second duty: to answer all questions of an experimentalist. The origin of this demand lies in the strong belief that physics is an experimental science. Nature is so beautiful that we do not need a substitute: the formalized theory, let's say the string theory, as an example. Our teachers demanded of us that we know the mathematics, which is a tool for solving the problem, but the main task of physicists is to correctly formulate the physics problem, which will be beautiful and solvable only because it is part of nature. Lev Lipatov spent most of his life solving theoretical problems, but he evaluated very highly his attempt with D. Ross and H. Kowalski (see the paper in this book) to describe the HERA data. He told me that, finally, he felt himself as a normal physicist, who was investigating nature, not only writing abstract formulae.

2.4 1966–1974: Double logs in QED (The 'log academy' and its president)

As I have mentioned, Gribov added Lev to his team, which reconsidered and developed the double log approximation for QED [Dokshitzer (1998); Frolov *et al.* (1966); Gribov *et al.* (1967); Gorshkov *et al.* (1968b,a)]. It was a lucky chance for Lev. His mathematical talent, which was known to his friends, found a fertile ground for improvement and development. Frankly speaking, at that time, we looked at him a bit from above. Indeed, we were involved in such important issues like additive quark model, finite energy sum rules, duality models, Reggeon calculus, parton and multiperipheral models, etc. Field theory was considered as the past. Ten years later the situation reversed itself.

We all started to re-learn field theory with the advent of QCD. Lev Lipatov was better prepared than everybody else, even better than Gribov himself, as you will see from the next section. I cannot express in a proper form how much I, personally, learned from Lev about QCD and the secrets of the perturbative approach to it. Lev's contributions to team work was recognized and he defended his first dissertation (candidate of science) in 1968 on the basis of

[Gorshkov *et al.* (1968b,a, 1969); Lipatov and Frolov (1969)].[4] For a long time Lev continued to work in calculating different processes in QED (see [Anselm *et al.* (1969, 1970); Lipatov (1971b,a); Lipatov and Frolov (1971); Gorshkov *et al.* (1971); Kuraev and Lipatov (1972b,a); Kurayev *et al.* (1973); Kuraev *et al.* (1973, 1974a); Kuraev and Lipatov (1974); Kuraev *et al.* (1974b)]). I put the titles of the papers so that you can see the diversity of Lev's interests. In calculating these QED processes, Lev became a great expert in finding and summing the log contributions in perturbative QED Feynman diagrams.

The double log approach consists of two ingredients. First, in each QED Feyman diagram of the order $\alpha^n_{e.m.}$ the contribution, which is proportional to $(\alpha_{e.m} \ln s \ln t)^n$, has to be calculated. Second, all these contributions have to be summed as an analytical function, mostly by writing the equation which sums all these contributions. In brief

$$\text{DLA} : A(s,t) = \sum_n C_n \left(\alpha_{e.m} \ln s \ln t\right)^n \tag{2.1}$$

for $\alpha_{e.m} \ln s \ll 1$ and $\alpha_{e.m} \ln t \ll 1$ while $\alpha_{e.m} \ln s \ln t \sim 1$.

The technique, which is used, includes the Sudakov decomposition:

$$k_\mu = \alpha p_{1,\mu} + \beta\, p_{2,\mu} + \vec{k}_\perp, \tag{2.2}$$

where p_1 and p_2 are the momenta of the colliding particles, and $k_{\perp,\mu} p_{1,\mu} = 0$ as well as $k_{\perp,\mu} p_{2,\mu} = 0$. Lev was not only very skillful in finding log integrations in α, β and $k_\perp$, but he introduced the analyticity and unitarity, which were often discussed in our department, in perturbative calculations. It was new at that time.

We recognized his superiority and joked that he created a 'log academy', where he was the president and E. Kuraev and V. Fadin were the academicians. Humor is humor, but I was proud when in 1982 they incorporated me as the corresponding member of this academy.

Working on his QED papers, Lev found and created his own team: E. Kuraev (from Khar'kov) and V. Fadin (from Novosibirsk). Below you would see how this team would work. Edik Kuraev had left us and I wish to remind that he came to our department from Khar'kov and quickly understood that this was his chance to learn physics. He worked as a night guard in St. Petersburg to earn money and stayed with us for a long time. As you can see from the list of references, Lev did a lot of work with him on QED. He also initiated the Gribov's lecture on strong interactions, which he printed as a Khar'kov preprint. He is the letter K in the celebrated BFKL Pomeron, which we will discuss below. He was not well known, mostly because he could not speak English, but he was the key member of the Lipatov's team. It would make me happy, if you consider these several words as a tribute to him.

[4] I believe that these papers give a good insight into the comparison between the Russian Candidate of Science degree and the Ph.D. one. I think that in the best Russian institutions the C.Sc. degree was more or less the same as the Ph.D.

2.5 1971–1974: The DGLAP evolution equation

In 1969–1970 four papers of S. D. Drell, D. J. Levy and T. M. Yan were published in Phys. Rev. [Drell *et al.* (1969, 1970a,b); Yan and Drell (1970)], in which they demonstrated that the main features of the parton model can be derived from quantum field theory assuming that all transverse momenta are cut off at some value, which does not depend on energy. These papers impressed Prof. Gribov, and he started to discuss what happens with the deep inelastic processes if we do not introduce a momentum cutoff. Every Monday and Thursday we heard about his ideas on this subject and, finally, a seminar was announced where he would present his findings.

The seminar proceeded in the usual fashion. Gribov explained why he thought that the correct integration over transverse momenta would result in log contributions ($\ln Q^2$, where Q^2 is the photon virtuality in inclusive $\gamma^* p$ scattering). He proposed to calculate the deep inelastic structure function as[5]

$$F(x, Q^2) = \sum_n C_n(x)(\alpha_S \ln Q^2)^n, \tag{2.3}$$

assuming $\alpha_S \ln Q^2 \sim 1$ while $\alpha_s \ll 1$ and $\alpha_s \ln(1/x) \ll 1$.

He started to tell us how to sum all these logs, but in five minutes Lev Lipatov raised his hand and very politely said: "Vladimir Naumovich, you are wrong with your procedure to find and sum the logs"; and in ten minutes he explained why and what we needed to do to find the sum. When he finished everyone understood two things: first, Prof. Gribov was, indeed, wrong; and second, Lev Lipatov would be the head of this department after Gribov. Those ten minutes changed the standing of Lev from the expert in QED calculations, which were not the primary interest of the group, to the main expert in quantum field theory.

Frankly speaking, Gribov tried to answer Lev to defend his point of view. It lasted about 40 minutes. Finally, he gave up and said: "Lev, you are the part of all this; help me to answer."

It goes without saying that this seminar was the shortest that I remember. A few months later, a new seminar was held, in which Lev Lipatov was the speaker. It was also very short, since Gribov was mostly supportive. Lev told us how to sum logs of transverse momenta and derived the equation for the deep inelastic function from which we see the violation of the Bjorken scaling behaviour due to these log contributions [Gribov and Lipatov (1972a,b)]. The estimates were made in vector and pseudo-scalar meson theories, which were on the market at that time, but all essential elements: gauge invariance, violation of the Bjorken scaling and so on, were found. Frankly speaking, when QCD appeared, the only item that we needed was to calculate the gluon $\to$ gluon + gluon splitting function. However, both Gribov and Lev were very busy with different problems: Gribov was doing his copies, which was

[5] The modern notations are used in Eq. 2.3: $x = Q^2/s$ where Q is the photon virtuality, s is the energy squared. α_s is QCD coupling which in the paper of Gribov and Lipatov is denoted by g^2.

really more important than the correct evolution equation. Lev responded to my suggestion, that he could calculate the gluon splitting and write the QCD equation in two days, and he did not have a good student for this simple job, but, "I am close to understanding what is the Pomeron in gauge theories". Therefore, the correct evolution equations were waiting for a good student. Finally, Alexej Anselm, gave this problem to his good student, Yura Dokshitzer, who wrote the QCD evolution equation [Dokshitzer (1977)]. But in 1977, it was done independently by Altarelli and Parisi [Altarelli and Parisi (1977)].

I would like to quote here the citation from Wikipedia about QCD evolution equations: "DGLAP (Dokshitzer-Gribov-Lipatov-Altarelli-Parisi) are the authors who first wrote the QCD evolution equation of the same name. DGLAP was first published in the western world by Altarelli and Parisi in 1977 [Altarelli and Parisi (1977)], hence DGLAP and its specializations are sometimes still called Altarelli-Parisi equations. Only later did it become known that an equivalent formula had been published in Russia by Dokshitzer also in 1977 [Dokshitzer (1977)], and by Gribov and Lipatov already in 1972 [Gribov and Lipatov (1972a,b)]".

I do not think that this citation correctly reflects the situation. Certainly, experts knew about the Gribov-Lipatov approach before 1977. It is enough to mention that one of the principle papers, that sets the formalism of the operator product expansion for the deep inelastic function (see N. H. Christ, B. Hasslacher and A. H. Mueller, [Christ *et al.* (1972)]) is based on the Gribov-Lipatov equations. I think that situation is different, people in the West did not know the next paper of Lipatov [Lipatov (1975)], in which he explained how the partons appear in the gauge theory. About partons they learned from Altarelli and Parisi, and this fact explains why the Altarelli-Parisi paper was so popular.

In 1974 he was the first among us to obtain the degree of Doctor of Science[6] for these equations.

In 2015 the 'High Energy and Particle Physics Prize', for an outstanding contribution to High Energy Physics, had been awarded to James D. Bjorken for his prediction of scaling behaviour in the structure of the proton that led to a new understanding of the strong interaction, and to Guido Altarelli, Yuri L. Dokshitzer, Lev Lipatov, and Giorgio Parisi for developing a probabilistic field theory framework for the dynamics of quarks and gluons, enabling a quantitative understanding of high-energy collisions involving hadrons.

I believe that the above citation nicely summarizes the achievement of Lev and the importance of his work for the high energy community. It goes without saying that Lev Lipatov got international recognition after this work, and his papers jumped over the iron curtain barriers and became known in the West. However, as

[6]The second degree in Russia, which is similar to habilitation in Germany. You cannot get a full professor position without this degree. Lipatov, once more, gave a clear example what it meant to be Doctor of Science in Russia at that time.

I have mentioned, his main contribution to DIS [Lipatov (1975)] remained unknown till the 80's, when we started to travel abroad.

2.6 1975–1990: The Pomeron in QCD (BFKL Pomeron)

The Pomeron structure was Lev's entire life's work. The high energy asymptotic behaviour of the scattering amplitude, which in the Regge approach was related to the exchange of the Pomeron, was discussed in the Gribov department every visiting day and was considered as the key principle question of high energy physics. Thinking about this distant past, frankly speaking, I think it was, to a large extent, the relic of the non field theory (FT) approach, which was the standard in the 60's. The idea of how to approach a high energy amplitude in the FT was suggested in [Gribov *et al.* (1971)] by Gribov, Lipatov and Frolov in QED, but the following simple equation clarified this approach in QCD:

$$\text{LL(s)A}: \quad A(x,t) = \sum C_n(t)\Big(\bar{\alpha}_S \ln(s)\Big)^n; \quad \bar{\alpha}_S \ln(t) \ll 1;\ \bar{\alpha}_S \ll 1; \tag{2.4}$$

where t is the momentum transferred squared.

Over four years Fadin, Kuraev and Lipatov[7] developed the leading log(s) approximation for non-abelian gauge theories from calculating the diagrams in the first orders of α_S^n to the solution of the general equation [Fadin *et al.* (1975); Lipatov (1976b); Kuraev *et al.* (1976, 1977)]. I believe that the numerous features of this equation will be the subject of most papers in this book. Therefore, I am not going to discuss them. I wish only to give you a flavour of the hot discussion in our department on this equation. Nowadays, most physicists think about the BFKL equation as the evolution of the DIS structure function in the region of small x. Lev, when he discussed this equation, did not see this application. For him it was an attempt to solve the Pomeron problem and to answer the question, what is the structure of the Pomeron in the gauge field theories. Figure 2.6 shows the main ingredients of this structure. The left figure shows that the FKL production of n gluons at high energy can be described as the ladder diagram, and has the following expression:

$$M_{2\to \text{n}} = sT_{A,A'}\frac{s_{0,1}^{\alpha_G(q_1)}}{q_1^2}\Gamma_L(k_{1,T})\frac{s_{1,2}^{\alpha_G(q_2)}}{q_2^2}\Gamma_L(k_{2,T})\cdots\Gamma_L(k_{n-2,T})\frac{s_{n-2,n-1}^{\alpha_G(q_{n-1})}}{q_{n-1}^2}T_{BB'}, \tag{2.5}$$

where q_i are the transverse momenta of the vertical reggeized gluons, $\alpha_G(q_i)$ is the spin of these gluons, and Γ_L is the Lipatov vertex which differs from the triple

[7] In our department the names of authors follow the alphabetical order. Since the main papers were written in Russian, one needs to know that the letter F is one of the last letter in the Russian alphabet, while in English it is one of the first. I follow the English order, which is commonly accepted for the BFKL Pomeron.

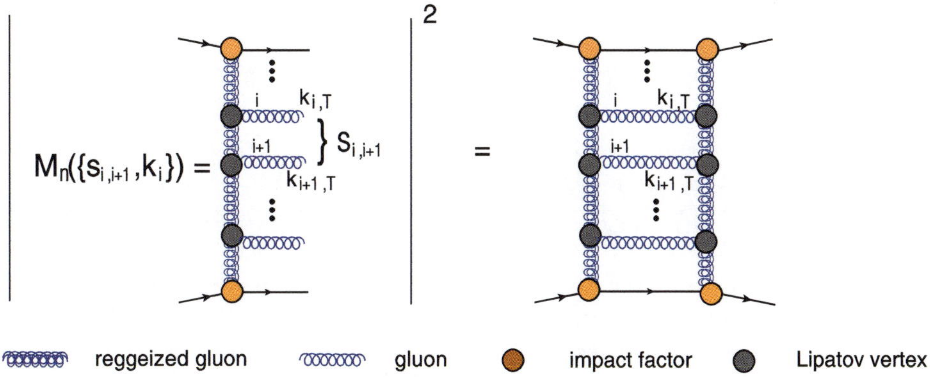

Fig. 2.6 BFKL Pomeron and its main ingredients.

gluon vertex in the QCD Lagrangian. The contribution to the scattering amplitude is equal to

$$\mathrm{Im}A(s,t) = \frac{1}{s}\sum_{n=2}^{\infty} |M_{2\rightarrow \mathrm{n}}|^2. \tag{2.6}$$

These two equations show that FKL gave the s-channel picture for the Pomeron exchange and, actually, stated that the maximum contribution comes from the gluon production in the typical multi-reggeon kinematics at high energies. The first surprise is that the same structure satisfies the t-channel unitarity. In the framework of Regge approach, we believed that the t-channel unitarity and the crossing symmetry are more important than the s-channel unitarity, which will come automatically, when we will build the theory which holds unitarity in the t-channel. Why surprise? Because it turns out that due to the exchange of a gluon, whose spin depends on its transverse momentum (reggeized gluon), the emitted gluons do not follow the Gribov diffusion and do not lead to the amplitude which depends on t. In principle, we knew that there is such a general possibility, that the asymptotic behaviour of the amplitude will be determined by the standing singularities, but nobody expected that QCD, the long awaited microscopic theory, will produce the result which is opposite to what we have believed at least ten years before its advent.

The reggeization of the gluon, together with the structure of the Lipatov vertex, lead to a different kind of the diffusion process: diffusion in logs of transverse momentum. It means that, in the process of evolution, the transverse momenta of the produced gluons acquire larger as well as smaller transverse momenta. Therefore, from the early days of the BFKL Pomeron we knew that at high energy we have to understand what happens to the QCD Pomeron: calculated in LL(s)A of perturbative QCD it generates gluons with small transverse momenta, which have to be treated in non-perturbative QCD. If you take into account that the intercept of the BFKL Pomeron turns out to be larger than 1, in contradiction to the Froissart theorem, I hope that you understand how important the BFKL Pomeron was and is in high energy physics.

Now I need to clarify the letter B in the BFKL Pomeron. As you can see from [Balitsky (1978)] this letter stands for Balitsky, who was the lucky guy. Lev did not want to repeat his mistake, when he refused to develop the Gribov-Lipatov equation for the case of QCD. Fortunately, he had a good student, Ian Balitsky, and he gave him a problem: to rewrite the FKL equations which they derived in Yang-Mills SU(2) theory for the more trivial case of gluons with zero mass. In this paper, they not only rewrote the equation for the only interesting case, but also found the exact solution for the total cross section.

Lev considered the reggeization of the gluon in QCD as a most important result. He thought that the major theoretical problem was to build the effective theory for QCD at high energies, which will include the new degrees of freedom: the reggeized gluons. He suggested the effective action (see [Lipatov (2017)]) which, he hoped, would solve all problems and give the effective theory for high energy QCD. He was so firm in his belief, that he thought that CGC and the colour dipole approaches were just wrong, because in the framework of these approaches the new phenomena of the gluon reggeization appears as a trivial property of the normalization of the dipole wave function. Frankly speaking, we only once discussed this problem, when he learned that I like both the CGC-dipole approach and the fact that the reggeization looks so simple. I feel that it is good that I am in Israel and not in Russia, for keeping our relationship on a positive level. I was really afraid that our long term friendship would be broken. All my arguments, that simplification is the main driving force of physics and that he was in the beginning of the coordinate approach, which seriously clarifies the BFKL Pomeron structure, were in vain. After Yura Kovchegov and I wrote the book [Kovchegov and Levin (2012)] on high energy QCD, he was very unhappy. I was able to placate him, saying that we presented his point of view and also that he had a book, where he argued his point of view in full [Ioffe *et al.* (2010)]. He agreed, saying that he wrote a book not for kindergarten kids. However, I must admit, I have not found courage to tell him that I wrote a paper, which shows how trivial the reggeization is in the CGC approach [Altinoluk *et al.* (2010)].

The twist of fate is indeed the fact that Lev first wrote the BFKL Pomeron Green function in the coordinate representation and found the solution to the BFKL equation in the general case [Lipatov (1986)]. Writing this note I looked and found that this paper has less than a thousand citations. I was upset, since this paper is the most essential paper in the whole business of the BFKL equations.

2.7 Lipatnik and the beginning of the Lipatov school

As I have mentioned, our duty was to answer all questions of the experimentalists. However, as it is well known, experimentalists have so many problems, besides physics, that they very often have no time to think about physics. For physicists at the universities this problem shows up less, since they have a common seminar, where both theoreticians and experimentalists discuss their results and related physics. In the research institute, as it was in the PNPI, the seminars were separate. More than this, before 1972 theoretical high energy physics was concentrated in the Ioffe Physical-Technical Institute in S. Petersburg (Leningrad at that time) while the experimental laboratories were placed in the small town of Gatchina which is 44 km to the south of the city. After 1972 the separate institute, the Petersburg Nuclear Physics Institute, was organized, and we started to go to Gatchina, but only on Thursdays, as the seminar fell on Mondays in the city. So, you see, that the communications were hampered. Fortunately, for all physicists of the institute, the experimentalists found the solution, organizing a regular Winter school. The idea was simple and brilliant: in February-March we go for a week outside of the town for skiing and physics. The theoreticians would teach the experimentalists what had happened in high energy physics during the past year, and they would be able to ask whatever they want. The scheme worked perfectly and the 53rd school was held on 2–8 March 2019.

The first school was in 1966, but we started to come to the school a bit later. It happened very fast that the school became popular in the Soviet Union and many physicists from other institutes started to come to the school. Naturally, it was quickly transformed into a kind of informal conference or, better to say, workshop: the place where we, the theoretical physicists, could discuss our problems with the guests from all over Soviet Union. However, the task to teach remained and a large number of talks, devoted to the different aspects of nuclear and high energy physics, were read and gave us the possibility to be educated besides the problems that we worked on.

Lev Lipatov transformed this school into the place, where he presented his ideas and advocated his way of doing physics. He did this in a very unusual and genius way: he organized his own seminar, which we called Lipatnik. It was outside of the official program and it was held at a specific time: after supper and the night cultural program. The duration was indefinite and very often he finished the seminar at 3 a.m. The main difference of this seminar from others was that Lipatov did not discuss the ideas and the way to realize these ideas: he discussed just how to work out the ideas with an emphasis on how to calculate.

The participants felt more like students in the lecture but he was so skillful in giving his presentation that everyone took part in the process. The benefits of such approach were obvious and the seminar was very popular. Actually, Lev had only one Ph.D. student: Ian Balitsky, but with these seminars he taught a lot of students and young researchers. I believe this seminar was the foundation of the so called 'Lipatov school'.

As far as Winter schools are concerned, Lev felt that it was his duty to take part in preparing and conducting the school, I put in this note four photos of him. In the first two you see him in 2005 and 2006. Just to give you a flavour of both of his participation and of the problems that were discussed in the school, I put below the titles of his talks at the 39th school:

1. Kuraev E.A., Cherednikov I. (JINR, Dubna), Antonov E.N., Lipatov L.N. (PNPI) — Feynman diagrams for the effective field theory at high energies in QCD;
2. Lipatov L.N. (PNPI) — Solution of the Balitsky-Kovchegov equation in three-dimensional Yang-Mills theory;
3. Lipatov L.N., Volkov G., Velizhanin V.N. (PNPI), Sabio Vera A. (ITP, Hamburg, Germany) — Calabi-Yau spaces, applications, and classification.

I hope you see the diversity of the problem and the serious attitude that Lev had at these schools.

In Fig. 2.8 you can see Lev during his last school in 2017. Figure 2.8(a) shows him in front of the chart, written in Russian: The 51st Winter School of Petersburg Nuclear Physics Institute, February 27–March 4, 2017. In Fig. 2.8(b) Lev and Victor Fadin were discussing something during this school.

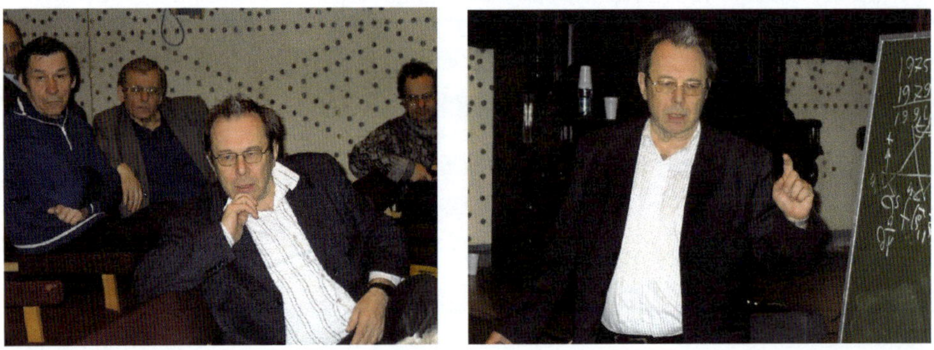

(a) (b)

Fig. 2.7 Lev Lipatov at the Winter school in 2005 (a) and in 2006 (b). He was masterful at both listening and discussing.

(a) (b)

Fig. 2.8 Lev Lipatov at his last school in 2017.

The title of his last talk was: "Scattering amplitudes in QCD and gravity". He made the Winter School a place where every young theoretical physicist wished to participate. I have sent as many as possible of my students to him. All of them came back saying that this school was very useful for them. Therefore, Prof. Lev Lipatov provided a considerable contribution to Israeli physics as well as to the Russian one.

2.8 The BFKL Pomeron and me

As I have alluded, I was doing mainstream physics before the BFKL Pomeron time: the additive quark model (with L. Frankfurt), the reggeon calculus (with V. Gribov and A. Migdal) and duality models and hadronic strings (with V. Kudryavtsev). Starting from 1967, I had the possibility to work with Misha Ryskin on the processes of multiparticle generation: multiperipheral and parton models for the Reggeon (mostly Pomeron) structure. Misha is a physicist of Lipatov's scale, and to work with him was a god's gift to me. By 1974 we were disappointed by the theoretical background of the parton model and were searching for the theory that would solve the problems that we had accumulated. Prof. Gribov advised us to look at the new QCD theory for this, saying that this theory looked like a good candidate for nature. We learned very fast the basics of the DLAP evolution equations but, fortunately, we also quickly realized that the BFKL approach was our ticket to the new QCD physics. We found that the structure of the parton cascade of the BFKL approach was very similar to our familiar parton model and we learned how to treat the differences.

Our first QCD papers [Ryskin (1980); Levin and Ryskin (1980)] were actually a breakthrough: we found that for the inclusive production in the BFKL cascade we have the k_T factorization formula. We based these papers on Eq. 2.5 (the first diagram of Fig. 2.6) and wrote the formulae for the inclusive process $h + h \to G(\gamma^*) + X$ in the form

$$\frac{d^2\sigma}{dyd^2p_T} \propto \int d^2k_T \frac{d\sigma\left(GG \to G(\gamma^*)\right)}{d^2p_T}\, \phi^G_h\left(Y - y; \vec{p}_T - \vec{k}_T\right)\, \phi^G_h(y;\, \vec{k}_T), \qquad (2.7)$$

where γ^* denotes the virtual gamma quantum and G the gluons. ϕ is the probability to find a gluon in a hadron which is the solution to the BFKL equation.

Being newcomers to the field, we did not realize the importance of this formula or the fact that we were the first to write it. The papers, even with their English translations, did not make any impact, and the high energy community was mostly not aware of our results. Lev, who was the expert, was never interested in the multiparticle production, and he could not help us. We understood all these issues only ten years later and published the paper [Levin *et al.* (1991)] which has everything, including the correct terminology. Slow as we might be, nevertheless, we were among the first people who suggested the k_T-factorization — but not the very first. *Sic transit gloria mundi.*

Lev, when I told him that his equation is suited for the estimates of the inclusive production at large values of transverse momenta, was very pleased and supportive. His demeanor encouraged me to tell him that his equation gave not the structure of the Pomeron, but the generalization of the DGLAP equation for the evolution of DIS at small values of Bjorken x. For the first ten minutes he told me how stupid I was, but for another ten minutes he decided maybe we were right. Our chat finished with Lev promising me a bottle of vodka, since he finally realized that we blew a new life into his beloved BFKL equation. We shared this bottle discussing how beautiful the BFKL equation is.

Lev was also very supportive when we made the first attempt to include the BFKL evolution of the gluon density in the GLAP equation (see our review [Gribov *et al.* (1983)]) and to take into account the running QCD coupling. However, he was lukewarm with our attempt to include the interaction of partons into the BFKL parton cascade by introducing the interaction of several parton showers. These ideas were taken from our previous experience with the Pomeron calculus and from the proof, which we gave, that the interaction of the BFKL Pomerons can be treated in a similar fashion as it is done in the Gribov Pomeron calculus.

He certainly knew the two major problems with the BFKL Pomeron: power-like increase with energy, and the influence of the confinement of quarks and gluon on the energy evolution. We showed that the interaction of the BFKL Pomeron leads to the saturation of the gluon density, which stops increasing, and to the transverse momentum in the energy evolution, which does not decrease. The typical transverse momentum increases, making the approach better and better from the point of the perturbative estimates. In spite of his numerous attempts to find a hole in our proof, he failed. However, his intuition did not trust our approach up to the end of his days. As I have mentioned, he hoped to solve these problems using the effective action, which he suggested. In simple words, he wanted to include the parton annihilation directly in the BFKL kernel.

Actually, Misha and I knew that the non-linear evolution equation that we suggested did not lead to the parton cascade that satisfies the s-channel unitarity at any values of rapidity of gluons. It is written clearly in our paper [Gribov *et al.* (1983)]. However, we believe that the equation is only valid in the vicinity of the new dimensional scale: the saturation momentum where it correctly introduces the gluon interaction. Lev thought that we needed to solve exactly and for the entire region of the rapidity.

I have a history of my understanding of his ideas. In the beginning, as I have mentioned, I believed that our approach and the non-linear equation which we suggested is only valid in the vicinity of the saturation momentum. Inside the saturation domain we cannot trust the perturbative approach to QCD, if not QCD itself. We suggested just to assume that the density reaches some value, as the practical way to introduce the unsolved confinement into our evolution.

When Balitsky [Balitsky (1996, 1999)] and Kovchegov [Kovchegov (1999)] followed the ideas of Al Mueller [Mueller (1994, 1995)], re-wrote our equation in the coordinate representation and clarified the physical meaning of ϕ, relating it to the dipole scattering amplitude, they managed to convince me that the resulting equation is valid in the wide kinematic region. This claim was in direct contradiction with Lev's expectation, but since it was done in the dipole approach, Lev refused even to discuss this issue with me.

However, later Al Mueller and A. Shoshi [Mueller and Shoshi (2004)] showed that the non-linear equation, even in the form that Balitsky and Kovchegov derived, does not provide s-channel unitarity for colourless dipoles with any value of its rapidity, reviving our old arguments. Starting from 2004, I went back in my disbelief that this equation is correct for any values of energy. In a recent paper [Kovner *et al.* (2016)] we found that this equation contradicts s-channel unitarity and thus is sick. At the moment, I do not know how the contradictions, that we found in [Kovner *et al.* (2016)], are interrelated with Mueller's and Shoshi's arguments, but I hope to find the generalization of this equation with the help of Alex Kovner and Misha Lublinsky and, perhaps, that Lev's ideas will help us in this adventure. The simple model, which has the same form and the same defects as the Balitsky-Kovchegov equation, and, which we solved in the paper, showed that the solution will be non-trivial. It is interesting to note that this paper has no citations and in a sense appears to be similar to our papers of 1980. I consider this as an indication that something interesting is awaiting us in the future.

Looking at the picture in Fig. 2.9, I imagine that Lev is instructing me, how to solve the problem. Actually, I know what he would say: "as I told you many times, you need to use my effective action". Perhaps he was right, but I will try something less complicated.

Having so different points of view on the non-linear evolution, we could not find a subject for joint work for a long time. However, rather recently, we did have one [Levin *et al.* (2014)]. The problem, which we addressed in this paper, has its roots in 1980. During our first talk at the Gribov seminar, we were attacked

Fig. 2.9 Lev Lipatov discussing with me his effective action in 2013.

with the question, what happens with our non-linear corrections at large impact parameters, and how they can fix the problem that the BFKL Pomeron has the power-like decrease at large b. Prof. V. Anisovich was a leader of these attacks. Since we knew this problem, we were able only to tell them that we propose our equation for the gluon densities at $t = 0$ and proved that we can take the integration over the momentum transferred by the BFKL Pomerons, introducing only one new phenomenological parameter: the correlation length of gluons in the parton cascade. Certainly, we were aware that this is not a satisfactory answer. We made several attempts to introduce a new non-perturbative soft parameter to describe the behaviour at large b (see [Levin and Ryskin (1989, 1990); Levin and Tapia (2013); Levin (2015)] but we failed to find an unambiguous, theoretical way to introduce a new non-perturbative parameter with a clear physical meaning into the kernel of the BFKL equation. The non-linear Balitsky-Kovchegov equation in the coordinate representation is written at fixed impact parameter. Both Ian and Yura considered this as an essential difference compared to our approach. When I heard this for the first time, I just laughed: kids did not understand what pandora's box they opened. Actually, I agree with them that the two equations are different, but ours is better, since it shows a way to avoid the problem of the large b behaviour, at least for the deep inelastic structure functions. Indeed, very soon after, A. Kovner and U. Wiedemann [Kovner and Wiedemann (2002a,b, 2003)] made this secret public, and the large b dependence is the most important unsolved problem which prevents the CGC approach from being an effective theory for QCD at high energies.

So, it is quite natural that both Lev and I tried to join our efforts to think about this. In our paper with Lev, we suggested to consider not QCD, but the gauge theory with the same colour structure as QCD with Higgs mechanism for the mass of gluons. This theory has no problem with the behaviour of the scattering amplitude at large b, but, at first sight, we changed the eigenfunctions and eigenvalues of the Pomeron in this theory in comparison with QCD. Then, we found that the intercept of the Pomeron is the same as the intercept of the BFKL Pomeron in QCD. We consider this fact as an indication that introducing corrections from confinement of quarks and gluons into the kernel of the BFKL equation will not change the main prediction of the CGC approach. On the other hand, Victor Fadin poured a bucket of cold water on us, by claiming that this result is trivial. I am still waiting for his paper in which he will prove that all possible modifications of the BFKL kernel will give the same spectrum.

2.9 The best papers

I believe that, without an exaggeration, we can say that all our understanding of high energy scattering in QCD is based on Lev Lipatov and his team's papers. Perhaps, they will live longer than his other contributions, but he himself as well as many other physicists, considered that his best papers are the ones on high order of the perturbative series [Lipatov (1976a, 1977b,a); Lipatov *et al.* (1979); Bukhvostov and Lipatov (1977)]. Indeed, the method to calculate the asymptotic behaviour even in the BFKL kinematic region, was suggested in the paper with Frolov and Gribov, and Lev always remembered that he was a developer but not a creator. The problem of the behaviour of the large order of the perturbative estimates and the divergence of the perturbative series was started by him, was formulated and solved. Not being an expert in this field, I took the following lines from the obituary of the theory division of PNPI faculty and staff:

"Lev Nikolaevich Lipatov excelled in research of the perturbative Quantum Field Theory. He had an amazing ability to find elegant and mathematically rigorous solutions of prohibitively difficult problems. He had suggested and developed an effective method for analysis of the asymptotic behavior of the perturbation theory series in Quantum Field Theory. His works on this subject now acquired the status of the theoretical physics classics. They now form the basis of our knowledge about the behavior of the perturbation theory series and their singularities in the complex coupling constant plane."

In these papers Lev showed that he was more than the expert in high energy QCD. He became an internationally renowned theoretical physicist with deep knowledge in all methods of modern Theoretical Physics. He became a friend of the Chernogolovka theory group of Sasha Polyakov and, when most of this group moved to Paris and other places in the West, he frequently visited them, discussing problems that are far from high energy scattering.

2.10 1990–2017: Climbing the ladder of success

In 1990 I left Russia and in 1996 I realized the dream of my teacher, Prof. V. Gribov, and became a member of the particle physics department of Tel Aviv University. I remember, as if it was yesterday, in 1964 or 1965, Prof. Gribov came to the common room saying: "Guys, now the center of the theoretical physics moves from Caltech and Harvard to Tel Aviv. Yuval Ne'eman organizes a high energy group at the new university there". He did not say anything else, but sighed heavily. Of course, he exaggerated a bit, but every time when my student comes to my office, opening the door with his foot, and telling me, "Professor, you are stupid; everything that you told me yesterday, is just rubbish", I feel that Tel Aviv is a good place for doing physics with a bright future ahead. Hence, I could watch only from the side, how my friend went from one success to another, but an onlooker sees most of the game. Lev started to travel abroad very frequently and we met both at conferences and at different institutions (see Fig. 2.10 and Fig. 2.11). It goes without saying that I used every possibility to invite him to Tel Aviv and to conferences in Israel, where we can talk in the very informal atmosphere of my home (see Fig. 2.12).

Fig. 2.10 Lev Lipatov and me at Diffraction 2014., Primosten, Croatia.

Fig. 2.11 Lev Lipatov, his wife Elvira, and me at Diffraction 2012., Lanzarote, Canary Islands, Spain.

During teaching, I have tried to use Lev for training my students. I remember him saying: " I cannot teach students how to think, but I can teach even a child how to calculate". I tried, for every student of mine, to organize either the possibility to work with him, or to meet him and discuss physics during a period of time for

Fig. 2.12 Lev Lipatov and his wife, Elvira, with me in my place at Be'er Yakov, Israel.

as long as possible. I used to send my students to DESY, to Prof. J. Bartels, since I know that Lev visits him on a more or less permanent basis. For three of my students the cooperation with Lev resulted in joint papers: nine for Alex Prygarin, three for Andrey Kormilitzin and two for Sergey Bondarenko. After my retirement, Bondarenko and Prygarin found money for Lev's visits, giving me a chance to see my friend (see Fig. 2.12). By the way, the last of Lev's papers were written in collaboration with them.

In almost three decades Lev generated several breakthrough successes, but one of them: the calculation to the BFKL equation in the next-to-leading order [Fadin and Lipatov (1998)], impressed me enormously. I believe that Victor Fadin and he accomplished the feat of performing the needed calculations. It took them almost ten years to find the answer, and I am sure that they did this only because Jochen Bartels helped them in organizing the regular visits to Hamburg and convincing them to continue these time consuming calculations. I think that the high energy community should be very thankful to Jochen. The NLO kernel for the BFKL equation led to a deeper understanding of the interrelation between BFKL and DGLAP evolutions [Salam (1998); Ciafaloni *et al.* (1999)]. On the other hand, the fact that the NLO corrections turn out to be large, resolve the contradiction with the attempt to describe the experimental data in the framework of the BFKL evolution or/and in the framework of the CGC/saturation approach. Indeed, the two essential parameters that determine the high energy scattering are the BFKL Pomeron intercept, which is approximately equal to $2.8\alpha_S$ and the energy behaviour of the new dimensional scale: the saturation momentum $Q_s^2 \propto \exp\left(4.88\,\alpha_S \ln(\frac{1}{x})\right)$. Both of them show the increase in the leading order BFKL approach, which cannot be reconciled with the available experimental data. So, the large NLO corrections appear to be the only way out, even now as well as two decades ago.

I tried to follow other achievements of my friend, especially his proof that the BFKL Hamiltonian is equivalent to an integrable Heisenberg spin model which introduced the concept of integrability into the high energy physics. He developed further his effective action approach and his lecture on this subject constituted the last words that I heard from him [Lipatov (2017)]. For a couple of years, I was very enthusiastic about AdS/CFT duality and discussed a lot with Lev his duality between BFKL Pomeron and the graviton. However, I felt disappointed and cheated, since this picture, in which the BFKL Pomeron mostly contributes to the real part of the scattering amplitude, contradicts everything that we have learned, both experimentally and theoretically, about the processes of multi-particle generation. Fortunately, physics is an experimental science, and I firmly believe that the LHC data on soft interactions killed these ideas as they do not show any significant increase of the diffraction dissociation cross sections with the growth of energy.

It goes without saying that his scientific achievements were widely recognized, resulting in his advancing in rank. In 1990 he became Professor of Physics, and from 1996 to the end of his days he was the head of the Gribov department at the Petersburg Nuclear Physics Institute. In 1997 he was elected as a corresponding member of the Russian Academy of Science. Lev Lipatov was a great patriot of Russia. His dream from the student years was to become the Russian Academician. This dream was accomplished in 2011, when he became a full member of the Russian Academy of Science.

I wish to make a lyrical digression in my narration about his life. I just want to tell my friends from the West that Russia is a very complicated country, but please think about Prof. Lipatov and not about politicians when you discuss Russians. Actually, Lev was against my move to Tel Aviv and argued with several reasons why I needed to stay in Russia. I failed to convince him that it was difficult to live, knowing that you needed to be the best to survive. It is funny, but he had only one argument in favor. He told me that he envied me, because I will have extraordinary bright students in Tel Aviv. Indeed, in Russia the Jewish students were the top of the class. To my surprise, he was wrong in his expectation; the students in Tel Aviv turned out to be the same as Russian students — even they were lazy like Russians. So antisemitism has a positive impact.

Since the differences in our opinions were not of the scale of the gluon reggeization, we disregarded these differences and Lev was a frequent guest in Tel Aviv. In 2009–2010 he was a scholar at Sackler Institute of Advance Studies and spent almost one year with us. He told me that he likes Israel, since it is not a

Fig. 2.13 2001: Lev Lipatov is getting the Pomeranchuk prize from Mikhail Danilov, the director of ITEP.

boring country. During the last three or four years he has come to Israel, to Ariel University, in order to work with my former students, allowing me the pleasure to talk with him.

I think it suffices to list the prizes and awards that were bestowed upon him, to illustrate what we have lost:

1. 1993 — the Research Award of the Alexander-von-Humboldt Foundation. He spent more than one year at DESY.
2. 2001 — the I. Ya. Pomeranchuk Prize (2001) for the BFKL Pomeron (see Fig. 2.13);
3. 2006–2009: the Marie Curie Excellence Chair of the European Community, hosted by Hamburg University;
4. 2009–2010: Scholar at Sackler Institute of Advance Studies (Tel Aviv);
5. 2015 — the European Physical Society High Energy and Particle Physics Prize "for developing a probabilistic field theory framework for the dynamics of quarks and gluons, enabling a quantitative understanding of high-energy collisions involving hadrons", or in simple words, for the DGLAP evolution equation.

2.11 Life in numbers

The content of this section is just the table below which speaks for itself without any comment.

Number of papers	160
Number of citations	26 169
Citations per paper:	170.9
Papers with citation 3000–4000	3 ([Gribov and Lipatov (1972a)], [Kuraev *et al.* (1977); Balitsky (1978)])
Papers with citation 1000–3000	6
Papers with citation 500–1000	1
Famous papers (250–499)	8
Very well-known papers (100–249)	18
Well-known papers (50–99)	21
Reviews	6
Books	3
Number of co-authors for ≥3 papers	36
Total number of co-authors:	86

Fig. 2.14 2015: DGLAP (without Gribov): Dokshitzer, Lipatov, Altarelli and Parisi, are getting the European Physical Society High Energy and Particle Physics Prize.

I believe that these numbers tell us more about the main contributions that Lev Lipatov made to high energy physics, than the prizes and awards.

2.12 My several words

Every day, when I start to work, I find myself as if in Plato's cave — your faces, my dear friends: Sasha Vassiliev, Alyosha Kaidalov, Lev Lipatov, Odette Benary, Uri Maor, Yona Oren, are on the wall in front of me. Thank you for sharing your life with me, thank you for your support and generosity, thank you for the intellectual

Fig. 2.15 "Vitam regit fortuna, non sapientia" ("Fortune, not wisdom, rules lives") (*Cicero*).

pleasure of our discussions, which adorned my life. I wish to report to you that I am still doing physics, but, I must admit, only nine hours per day. I still enjoy doing physics, I think, that we made a correct decision, choosing physics as the way of life. At least I have not felt bored for even a minute of my life.

Every time when I put my formulae on the sheet of paper, I feel myself as a young guy from Gribov's department who is hurrying up to meet you and discuss what I have understood. All our hopes, all our ideas, all our dreams are still with me and I am trying to share them with young physicists.

Dear Lev, the renowned physicist and my friend for 60 years, my memory will not fade; I will remember you, our discussions, our approaches to life and physics. You will be always as in Fig. 2.15: smiling, active, always in motion and always ahead of all of us.

Bibliography

Altarelli, G. and Parisi, G. (1977). Asymptotic freedom in parton language, *Nucl. Phys. B* **126**, pp. 298–318, doi:10.1016/0550-3213(77)90384-4.

Altinoluk, T., Kovner, A., and Levin, E. (2010). Inside looking out: Probing JIMWLK wave functions with BFKL calculations, *Phys. Rev. D* **82**, p. 074016, doi:10.1103/PhysRevD.82.074016, arXiv:1004.5113 [hep-ph].

Anselm, A., Lipatov, L., and Winbow, G. (1969). Three-particle production amplitude at high ernergies and momentum transfers, *Pisma Zh. Eksp. Teor. Fiz.* **10**, pp. 499–504.

Anselm, A., Lipatov, L., and Winbow, G. (1970). Amplitude for three-particle production at high energies and large momentum transfers, *Nuovo Cim. A* **69**, pp. 57–79, doi:10.1007/BF02728771.

Balitsky, I. (1978). Effective electric charge and asymptotic freedom, *Yad. Fiz.* **27**, pp. 1091–1097.

Balitsky, I. (1996). Operator expansion for high-energy scattering, *Nucl. Phys. B* **463**, pp. 99–160, doi:10.1016/0550-3213(95)00638-9, arXiv:hep-ph/9509348.

Balitsky, I. (1999). Factorization and high-energy effective action, *Phys. Rev. D* **60**, p. 014020, doi:10.1103/PhysRevD.60.014020, arXiv:hep-ph/9812311.

Bukhvostov, A. and Lipatov, L. (1977). High orders of the perturbation theory in scalar electrodynamics, *Phys. Lett. B* **70**, pp. 48–50, doi:10.1016/0370-2693(77)90341-0.

Christ, N. H., Hasslacher, B., and Mueller, A. H. (1972). Light cone behavior of perturbation theory, *Phys. Rev. D* **6**, p. 3543, doi:10.1103/PhysRevD.6.3543.

Ciafaloni, M., Colferai, D., and Salam, G. (1999). Renormalization group improved small x equation, *Phys. Rev. D* **60**, p. 114036, doi:10.1103/PhysRevD.60.114036, arXiv:hep-ph/9905566.

Dokshitzer, Y. L. (1977). Calculation of the structure functions for deep inelastic scattering and e+ e− annihilation by perturbation theory in quantum chromodynamics, *Sov. Phys. JETP* **46**, pp. 641–653.

Dokshitzer, Y. L. (1998). V.N. Gribov: 1930–1997, in *3rd Workshop on Continuous Advances in QCD (QCD 98)*, pp. xv–xxi, arXiv:physics/9801025.

Dokshitzer, Y. L. and Kharzeev, D. E. (2004). The Gribov conception of quantum chromodynamics, *Ann. Rev. Nucl. Part. Sci.* **54**, pp. 487–524, doi:10.1146/annurev.nucl.54.070103.181224, arXiv:hep-ph/0404216.

Drell, S., Levy, D. J., and Yan, T.-M. (1969). A theory of deep inelastic lepton-nucleon scattering and lepton pair annihilation processes. 1, *Phys. Rev.* **187**, pp. 2159–2171, doi:10.1103/PhysRev.187.2159.

Drell, S., Levy, D. J., and Yan, T.-M. (1970a). A theory of deep inelastic lepton nucleon scattering and lepton pair annihilation processes. 2. Deep inelastic electron scattering, *Phys. Rev. D* **1**, pp. 1035–1068, doi:10.1103/PhysRevD.1.1035.

Drell, S., Levy, D. J., and Yan, T.-M. (1970b). A theory of deep inelastic lepton-nucleon scattering and lepton pair annihilation processes. 3. Deep inelastic electron-positron annihilation, *Phys. Rev. D* **1**, pp. 1617–1639, doi:10.1103/PhysRevD.1.1617.

Fadin, V. S., Kuraev, E., and Lipatov, L. (1975). On the pomeranchuk singularity in asymptotically free theories, *Phys. Lett. B* **60**, pp. 50–52, doi:10.1016/0370-2693(75)90524-9.

Fadin, V. S. and Lipatov, L. (1998). BFKL pomeron in the next-to-leading approximation, *Phys. Lett. B* **429**, pp. 127–134, doi:10.1016/S0370-2693(98)00473-0, arXiv:hep-ph/9802290.

Frolov, G., Gorshkov, V., and Gribov, V. (1966). Backward Compton scattering at high-energies, *Phys. Lett.* **22**, pp. 662–664, doi:10.1016/0031-9163(66)90697-4.

Gorshkov, V., Gribov, V., Lipatov, L., and Frolov, G. (1968a). Backward electron — positron scattering at high-energies, *Sov. J. Nucl. Phys.* **6**, p. 262.

Gorshkov, V., Gribov, V., Lipatov, L., and Frolov, G. (1968b). Doubly logarithmic asymptotic behavior in quantum electrodynamics, *Sov. J. Nucl. Phys.* **6**, p. 95.

Gorshkov, V., Kuraev, E., Lipatov, L., and Nesterov, M. (1971). Interference between coulomb and strong interactions at high energies, *Zh. Eksp. Teor. Fiz.* **60**, pp. 1211–1215.

Gorshkov, V., Lipatov, L., and Nesterov, M. (1969). On j-plane singularities of e-minus muon-minus scattering amplitudes, *Yad. Fiz.* **9**, pp. 1221–1231.

Gribov, L., Levin, E., and Ryskin, M. (1983). Semihard Processes in QCD, *Phys. Rept.* **100**, pp. 1–150, doi:10.1016/0370-1573(83)90022-4.

Gribov, V., Gorshkov, V., and Frolov, G. (1967). Double logarithmic asymptotics of the large angle Compton scattering, *Annals Phys.* **43**, pp. 201–227, doi:10.1016/0003-4916(67)90149-2.

Gribov, V. and Lipatov, L. (1972a). Deep inelastic e p scattering in perturbation theory, *Sov. J. Nucl. Phys.* **15**, pp. 438–450.

Gribov, V. and Lipatov, L. (1972b). e+ e- pair annihilation and deep inelastic e p scattering in perturbation theory, *Sov. J. Nucl. Phys.* **15**, pp. 675–684.

Gribov, V., Lipatov, L., and Frolov, G. (1971). The leading singularity in the j plane in quantum electrodynamics, *Sov. J. Nucl. Phys.* **12**, p. 543.

Ioffe, B. L., Fadin, V. S., and Lipatov, L. N. (2010). *Quantum chromodynamics: Perturbative and nonperturbative aspects*, Vol. 30 (Cambridge Univ. Press), ISBN 978-1-107-42475-3, 978-0-521-63148-8, 978-0-511-71744-4, doi:10.1017/CBO9780511711817.

Kovchegov, Y. V. (1999). Small x F(2) structure function of a nucleus including multiple pomeron exchanges, *Phys. Rev. D* **60**, p. 034008, doi:10.1103/PhysRevD.60.034008, arXiv:hep-ph/9901281.

Kovchegov, Y. V. and Levin, E. (2012). *Quantum chromodynamics at high energy*, Vol. 33 (Cambridge University Press), ISBN 978-0-521-11257-4, 978-1-139-55768-9, doi:10.1017/CBO9781139022187.

Kovner, A., Levin, E., and Lublinsky, M. (2016). QCD unitarity constraints on Reggeon Field Theory, *JHEP* **08**, p. 031, doi:10.1007/JHEP08(2016)031, arXiv:1605.03251 [hep-ph].

Kovner, A. and Wiedemann, U. A. (2002a). Nonlinear QCD evolution: Saturation without unitarization, *Phys. Rev. D* **66**, p. 051502, doi:10.1103/PhysRevD.66.051502, arXiv:hep-ph/0112140.

Kovner, A. and Wiedemann, U. A. (2002b). Perturbative saturation and the soft pomeron, *Phys. Rev. D* **66**, p. 034031, doi:10.1103/PhysRevD.66.034031, arXiv:hep-ph/0204277.

Kovner, A. and Wiedemann, U. A. (2003). No Froissart bound from gluon saturation, *Phys. Lett. B* **551**, pp. 311–316, doi:10.1016/S0370-2693(02)02907-6, arXiv:hep-ph/0207335.

Kuraev, E. and Lipatov, L. (1972a). Electron and muonic production in e- e- and e+ e- colliding beams, *Yad. Fiz.* **16**, pp. 1060–1077.

Kuraev, E. and Lipatov, L. (1972b). Total cross-section for electron and muon pair production in colliding electron beams, *Pisma Zh. Eksp. Teor. Fiz.* **15**, pp. 229–232.

Kuraev, E. and Lipatov, L. (1974). Bremsstrahlung mechanism of e+ e- and mu+ mu- pair production in electron colliding beams, *Yad. Fiz.* **20**, pp. 112–121.

Kuraev, E., Lipatov, L., and Fadin, V. S. (1977). The Pomeranchuk Singularity in Nonabelian Gauge Theories, *Sov. Phys. JETP* **45**, pp. 199–204.

Kuraev, E., Lipatov, L., and Merenkov, N. (1973). High energy behavior of total cross-sections for e+ e- scattering and the slope at q-squared = 0 of the Dirac form-factor, *Phys. Lett. B* **47**, pp. 33–35, doi:10.1016/0370-2693(73)90562-5.

Kuraev, E., Lipatov, L., Merenkov, N., Fadin, V. S., and Khoze, V. A. (1974a). Double bremsstrahlung in the same direction in e+ e- colliding beams, *Sov. J. Nucl. Phys.* **19**, pp. 164–167.

Kuraev, E., Lipatov, L., and Strikman, M. (1974b). On the relation between various inclusive processes in quantum electrodynamics, *Zh. Eksp. Teor. Fiz.* **66**, pp. 838–848.

Kuraev, E. A., Lipatov, L. N., and Fadin, V. S. (1976). Multi — Reggeon processes in the Yang-Mills theory, *Sov. Phys. JETP* **44**, pp. 443–450.

Kurayev, E., Lipatov, L., Merenkov, N., and Fadin, V. (1973). Radiative corrections to bremsstrahlung radiation at high energies, *Zh. Eksp. Teor. Fiz.* **65**, pp. 2155–2160.

Levin, E. (2015). Large **b** behaviour in the CGC/saturation approach: BFKL equation with pion loops, *Phys. Rev. D* **91**, 5, p. 054007, arXiv:1412.0893 [hep-ph].

Levin, E., Lipatov, L., and Siddikov, M. (2014). BFKL Pomeron with massive gluons, *Phys. Rev. D* **89**, 7, p. 074002, doi:10.1103/PhysRevD.89.074002, arXiv:1401.4671 [hep-ph].

Levin, E. and Ryskin, M. (1980). Quantum chromodynamics prediction for heavy lepton pair production in the pionization region (in russian), *Yad. Fiz.* **32**, pp. 802–816.

Levin, E. and Ryskin, M. (1989). The shrinkage of the diffraction peak of the bare Pomeron in QCD, *Sov. J. Nucl. Phys.* **50**, pp. 881–887, doi:10.1007/BF01554471.

Levin, E. and Ryskin, M. (1990). High-energy hadron collisions in QCD, *Phys. Rept.* **189**, pp. 267–382, doi:10.1016/0370-1573(90)90016-U.

Levin, E., Ryskin, M., Shabelski, Y., and Shuvaev, A. (1991). Heavy quark production in semihard nucleon interactions, *Sov. J. Nucl. Phys.* **53**, p. 657.

Levin, E. and Tapia, S. (2013). BFKL Pomeron: Modeling confinement, *JHEP* **07**, p. 183, doi:10.1007/JHEP07(2013)183, arXiv:1304.8022 [hep-ph].

Lipatov, L. (1971a). Bremsstrahlung in backward e+ e- scattering as a multiregge process, *Yad. Fiz.* **14**, pp. 160–177.

Lipatov, L. (1971b). Bremsstrahlung in e+ e- backward scattering at high energies. *Yad. Fiz.* **14**, pp. 396–407.

Lipatov, L. (1975). The parton model and perturbation theory, *Sov. J. Nucl. Phys.* **20**, pp. 94–102.

Lipatov, L. (1976a). Calculation of the Gell-Mann-Low function in a scalar field theory with strong nonlinearity, *Sov. Phys. JETP* **44**, p. 1055.

Lipatov, L. (1976b). Reggeization of the vector meson and the vacuum singularity in nonabelian gauge theories, *Sov. J. Nucl. Phys.* **23**, pp. 338–345.

Lipatov, L. (1977a). Divergence of perturbation series and pseudoparticles, *JETP Lett.* **25**, pp. 104–107.

Lipatov, L. (1977b). Divergence of the perturbation theory series and the quasiclassical theory, *Sov. Phys. JETP* **45**, pp. 216–223.

Lipatov, L. (1986). The Bare Pomeron in quantum chromodynamics, *Sov. Phys. JETP* **63**, pp. 904–912.

Lipatov, L. (2017). Effective actions for high energy scattering in QCD and in gravity, *EPJ Web Conf.* **164**, p. 02002, doi:10.1051/epjconf/201716402002.

Lipatov, L., Bukhvostov, A., and Malkov, E. (1979). Large order estimates for perturbation theory of a Yang-Mills field coupled to a scalar field, *Phys. Rev. D* **19**, pp. 2974–2983, doi:10.1103/PhysRevD.19.2974.

Lipatov, L. and Frolov, G. (1969). Some processes in quantum electrodynamics at high energies and small scattering angles, *Pisma Zh. Eksp. Teor. Fiz.* **10**, pp. 399–402.

Lipatov, L. and Frolov, G. (1971). Some processes in quantum electrodynamics at high energies. *Yad. Fiz.* **13**, pp. 588–599.

Mueller, A. and Shoshi, A. (2004). Small x physics beyond the Kovchegov equation, *Nucl. Phys. B* **692**, pp. 175–208, doi:10.1016/j.nuclphysb.2004.05.016, arXiv:hep-ph/0402193.

Mueller, A. H. (1994). Soft gluons in the infinite momentum wave function and the BFKL Pomeron, *Nucl. Phys. B* **415**, pp. 373–385, doi:10.1016/0550-3213(94)90116-3.

Mueller, A. H. (1995). Unitarity and the BFKL Pomeron, *Nucl. Phys. B* **437**, pp. 107–126, doi:10.1016/0550-3213(94)00480-3, arXiv:hep-ph/9408245.

Ryskin, M. (1980). High transverse momentum hadron production in the pionization region and vacuum singularity in QCD (in russian), *Yad. Fiz.* **32**, pp. 259–274.

Salam, G. (1998). A resummation of large subleading corrections at small x, *JHEP* **07**, p. 019, doi:10.1088/1126-6708/1998/07/019, arXiv:hep-ph/9806482.

Yan, T.-M. and Drell, S. (1970). A theory of deep inelastic lepton-nucleon scattering and lepton pair annihilation processes. 4.Deep inelastic neutrino scattering, *Phys. Rev. D* **1**, p. 2402, doi:10.1103/PhysRevD.1.2402.

https://doi.org/10.1142/9789811231124_0003

Chapter 3

Lev Lipatov: Some Personal Reminiscences

A. H. Mueller

Department of Physics
Columbia University
New York, NY 10027

Lev Lipatov and I were about the same age. He was a little less than a year younger than me. For most of his career we had similar interests in physics. In this short note I would like to recount my interest in his work and later describe meeting and interacting with him. I also want to talk about the different physics cultures we came from and how I learned from the papers of he and his colleagues a style of doing physics which at the beginning was foreign to me, but which in the end I found very natural and attractive.

I was maturing as a theoretical particle physicist in the late 1960's and early 1970's. By this time the S-matrix theory of strong interactions which arose in the late 1950's and early 1960's was already showing signs that it didn't possess enough information to be a genuine theory of hadronic physics. It was a time when there was no theory of the strong interactions which seemed really promising. It was becoming fashionable to study Feynman graphs to try and get ideas about how high energy scattering amplitudes might behave. A big motivation for this was to try to carry over T. Regges results in potential theory to a realistic theory. Φ^3 theory was widely used in these studies. Regge poles and, later, Regge cuts were found as well as many interesting results on the analytic properties of scattering amplitudes. A good summary of the state of the art in the middle of the 1960's is given in the book The Analytic S-Matrix [Eden *et al.* (1966)]. However, it was clear that Φ^3 was not an interesting theory in itself. In the weak coupling limit cross sections decreased rapidly with energy while it was known experimentally that it was not the case for, say, proton-proton scattering.

At the very end of the 1960's groups in the Soviet Union (V.N. Gribov, L.N. Lipatov and G.V. Frolov) and in the United States (H. Cheng and T.T. Wu) began to calculate high energy elastic scattering in quantum electrodynamics (QED). These were very difficult calculations, even in a leading logarithmic approximation. The two groups found the same answer [Gribov *et al.* (1971); Frolov *et al.* (1970); Cheng and Wu (1970)], a cross section which at weak coupling grows slowly as a small power of the energy, a behavior which would be essentially reproduced later by Lipatov and his collaborators in QCD. This growth in energy would have to be tamed if a realistic answer were to emerge. I was very interested in these calculations and studied them in great detail. The thing that struck me the most was how hard Cheng and Wu had worked to get their result and how simple the procedure of Gribov, Lipatov and Frolov was. This was an important lesson for me. Cheng and Wu had to work very hard because they were being very complete, not making approximations until the end of the calculation, while Gribov, Lipatov and Frolov seemed to know, beforehand, exactly what region of phase space was important and what approximations could be made from the start of the calculation. In the US this was an unusual way to proceed. We were taught to do calculations as completely as possible in order not to miss any contribution, and only at the end to make approximation. The approximations made by Gribov, Lipatov and Frolov were systematically done and no contribution to the leading logarithmic result was lost.

I don't believe this was a style invented by the Gribov group, after all the early vertex and self-energy calculations done in QED by L.D. Landau, A.A. Abrikosov and I.M. Khalatnikov [Landau *et al.* (1954)] and the double logarithmic vertex function done later by V.V. Sudakov [Sudakov (1956)] followed this same procedure of first understanding where the dominant contributions come from and then focusing on the specific graphs in the dominant regions of phase space. While the Gribov school may not have invented this procedure Lipatov and his colleagues certainly perfected it.

While I very much appreciated the Gribov, Lipatov and Frolov calculation it was not clear what to do with it. There were attempts to tame the energy growth by various, ad hoc, procedures of unitarization, but none of these procedures were convincing. Later, the idea that cross sections could grow over a limited energy region became the basis of a successful phenomenology by A. Donnachie and P. Landshoff [Donnachie and Landshoff (1992)] and when the Balitsky-Kovchegov (BK) equation came along, much later, one learned that such a temporary growth of cross sections with energy was even natural.

In 1972 an impressive and influential paper by V.N. Gribov and L.N. Lipatov appeared in which they calculated the leading logarithmic limit of deep inelastic electron-proton scattering (DIS). (DIS was the most active field of research in particle physics at the time. Experiments had rescued Lagrangian field theory from the ill repute that it had suffered at the hands of both S-matrix theory and axiomatic field theory.) They did their calculation in several fermionic theories including QED. A few years later Lev added the partonic interpretation [Lipatov (1975)] to their calculation and the partonic interpretation was extended to QCD by Yu. L. Dokshitzer [Dokshitzer (1977)] and by G. Altarelli and G. Parisi [Altarelli and Parisi (1977)]. I was just coming to Columbia when the Gribov-Lipatov paper came out and I, along with N. Christ and B. Hasslacher, were inspired to cast their calculation into the new technology of the operator product expansion and the renormalization group [Christ *et al.* (1972)]. When asymptotic freedom was discovered in QCD in 1973 by H.D. Politzer [Politzer (1973)] and by D.J. Gross and F. Wilczek [Gross and Wilczek (1973, 1974)] the Gribov-Lipatov calculation could easily be taken over to QCD by this new technology and the era of precise tests of QCD was ready to emerge. The Gribov-Lipatov calculation was masterful. Understanding what the contribution would be and from what regions of phase space it would come made the calculation almost understandable to particle theorists of the time. To me, personally, it further impressed on me the efficiency of the Russian style of doing complex diagrammatic calculations.

Between 1976 and 1978 the Balitsky, Fadin, Kuraev and Lipatov (BFKL) Pomeron was born. The original calculation of V.S. Fadin, E.A. Kuraev and L.N. Lipatov [Fadin *et al.* (1975); Kuraev *et al.* (1976)] followed a semi S-matrix method of calculation, using a multiperipheral approach where the exchanged objects were reggeized gluons and a new (Lipatov) vertex was introduced. A little later Lev and his student Ian Balitsky [Balitsky and Lipatov (1978)] followed a more conventional Feynman diagram approach. The two methods agreed and the leading logarithmic result for high energy scattering in QCD was very similar to that found earlier in QED, a cross section that grows as a small (at weak coupling) power of the energy. In 1986 Lev wrote a thorough summary [Lipatov (1986)] of what was known about the leading logarithmic BFKL Pomeron.

I met Lev for the first time in 1983 at an official US-USSR conference in Yerevan. After the conference I spent a week in Leningrad and saw him again at the Leningrad Nuclear Physics Institute (LNPI) in Gatchina. This was a very special experience for me. Never before had I seen, and met, so many outstanding theorists in one institute focused on strong interaction physics. This was the theory group assembled by V.N. Gribov, although Gribov had gone to Moscow a few years earlier for personal reasons. Theorists at the LNPI typically only came to the institute one day a week as they mostly lived in Leningrad and the lack of office space in Gatchina and the difficult commute made it more natural to assemble in small groups in each others apartments and do physics there most of the days of the week. The sociology of the LNPI group was fascinating. Theorists had formed working groups of 2-3 people. There working groups were very stable in time and each group focused on certain types of problems of which they were often the worlds expert. In my 1983 visit I had the opportunity to meet with a number of these groups with which I had some overlapping interest. Because of my interest in similar topics and my experience with the Russian style of physics I found it not too difficult to reasonably understand and appreciate the very high level of work being done in Leningrad. Over the next ten years I came back to Leningrad almost every year to continue interactions with this exceptional group of theorists.

At the time of my first visit to Leningrad I recognized that Lev was not part of any of these small groups within the LNPI. Levs directions were somewhat more technical than those of most of his colleagues. However, he did do most of his work in collaboration, often with E.A. Kuraev (Kharkov) and V.S. Fadin (Novosibirsk). Their technical abilities and inclinations were well matched to those of Lev.

Lev's competiveness was not so open as that of most of his Leningrad colleagues. He was certainly sensitive to the regard with which his work was held, and he had strong feelings about the work of the others, especially when it touched the issues of high energy scattering. However, I have never seen Lev openly attack another physicist when he disagreed, even strongly, with that person's work. I believe Lev was too kind to make such an attack, and it did not fit his rather reserved personality to attack someone else. This was quite in contrast to many of his colleagues, and not uniquely at the LNPI, who felt obliged to point out the foolishness present in much of the work being done in theoretical physics.

Lev was the recognized master of calculating high energy scattering in QCD. He had no real competitors in this area, and his goal was to turn the BFKL Pomeron into a unitarity obeying realistic description of high energy hard scattering. However, it was two younger theorists who would take the next essential steps in this direction, Ian Balitsky an ex-LNPI theorist and a student of Lev and Yuri Kovchegov, an ex-student of mine, who had a really deep insight into high energy scattering just a few years after his PhD. Balitsky's and Kovchegov's works would come in the late 1990's. Ian wrote a set of coupled equations [Balitsky (1996, 1999)] for the scattering of objects ranging from a simple dipole to very complicated partonic structures. The equations implicitly had unitarity constraints and resembled Schwinger-Dyson equations familiar from QED. The fact that there seemed to be no natural parameter limiting the complexity of the coupling of the equations suggested that the stochasticity in the problem was very deep and important. However, Yuri recognized that if one scatters a dipole off a large nucleus the self-averaging over the independent parts of the nucleus, along with the large number of colors approximation, reduced this coupled system of equations to a simple nonlinear equation, the Balitsky-Kovchegov (BK) equation [Kovchegov (1999, 2000)], for the dipole scattering amplitude. (An earlier nonlinear equation had been found by L.V. Gribov, E.M. Levin and M.G. Ryskin [Gribov *et al.* (1983)], and this equation was phenomenologically useful and furnished much motivation for searching for nonlinear extensions to the BFKL equation. However, the GLR equation was more a generalization of the DGLAP equation than of the BFKL equation.) When used on a non self-averaging target, like a proton, the BK equation remains an interesting and useful mean field equation which agrees with the BFKL equation at moderate energy and tames the BFKL growth, in a unitarity preserving way, at high energies. The stochasticity present in the fully coupled equations is, I believe, conceptually understood but has yet to be transformed into a usable calculational system. I dont think Lev was ever happy with the BK equation. It was not the way he had imagined that high energy scattering would be described.

I saw Lev regularly between 1983 and around 2000 in Russia, at conferences and during a few visits he made to Columbia. After 2000 Lev's physics direction turned more formal. He became very interested in problems involving conformal symmetry and AdS/CFT duality. I still saw him quite regularly at DESY during the first decade of the 2000s, however, by this time our physics interests were too different for useful conversations. I believe his last truly important contribution to the phenomenology of high energy reactions was the next-to-leading order BFKL kernel calculation that he and Victor Fadin did [Fadin and Lipatov (1998)]. Their result was later confirmed by I. Balitsky and G. Chirilli [Balitsky and Chirilli (2008)] using the simpler but more general technique of dipole scattering which also extended the Fadin-Lipatov calculation to a next-to-leading order result for the BK kernel.

I have always liked and had great admiration for Lev Lipatov. He was unquestionably one of the greatest technical theoretical physicists of the last half century. It has been a pleasure and a privilege to have him as a colleague and friend for so many years.

Bibliography

Altarelli, G. and Parisi, G. (1977). Asymptotic freedom in parton language, *Nucl. Phys. B* **126**, pp. 298–318, doi:10.1016/0550-3213(77)90384-4.

Balitsky, I. (1996). Operator expansion for high-energy scattering, *Nucl. Phys. B* **463**, pp. 99–160, doi:10.1016/0550-3213(95)00638-9, arXiv:hep-ph/9509348.

Balitsky, I. (1999). Factorization and high-energy effective action, *Phys. Rev. D* **60**, p. 014020, doi:10.1103/PhysRevD.60.014020, arXiv:hep-ph/9812311.

Balitsky, I. and Chirilli, G. A. (2008). Next-to-leading order evolution of color dipoles, *Phys. Rev. D* **77**, p. 014019, doi:10.1103/PhysRevD.77.014019, arXiv:0710.4330 [hep-ph].

Balitsky, I. and Lipatov, L. (1978). The Pomeranchuk singularity in quantum chromodynamics, *Sov. J. Nucl. Phys.* **28**, pp. 822–829.

Cheng, H. and Wu, T. T. (1970). Logarithmic factors in the high-energy behavior of quantum electrodynamics, *Phys. Rev. D* **1**, pp. 2775–2794, doi:10.1103/PhysRevD.1.2775.

Christ, N. H., Hasslacher, B., and Mueller, A. H. (1972). Light cone behavior of perturbation theory, *Phys. Rev. D* **6**, p. 3543, doi:10.1103/PhysRevD.6.3543.

Dokshitzer, Y. L. (1977). Calculation of the structure functions for deep inelastic scattering and e+ e− annihilation by perturbation theory in quantum chromodynamics, *Sov. Phys. JETP* **46**, pp. 641–653.

Donnachie, A. and Landshoff, P. (1992). Total cross-sections, *Phys. Lett. B* **296**, pp. 227–232, doi:10.1016/0370-2693(92)90832-O, arXiv:hep-ph/9209205.

Eden, R. J., Landshoff, P. V., Olive, D. I., and Polkinghorne, J. C. (1966). *The Analytic S-matrix* (Cambridge Univ. Press, Cambridge).

Fadin, V. S., Kuraev, E., and Lipatov, L. (1975). On the Pomeranchuk singularity in asymptotically free theories, *Phys. Lett. B* **60**, pp. 50–52, doi:10.1016/0370-2693(75)90524-9.

Fadin, V. S. and Lipatov, L. (1998). BFKL pomeron in the next-to-leading approximation, *Phys. Lett. B* **429**, pp. 127–134, doi:10.1016/S0370-2693(98)00473-0, arXiv:hep-ph/9802290.

Frolov, G., Gribov, V., and Lipatov, L. (1970). On Regge poles in quantum electrodynamics, *Phys. Lett. B* **31**, pp. 34–35, doi:10.1016/0370-2693(70)90013-4.

Gribov, L., Levin, E., and Ryskin, M. (1983). Semihard processes in QCD, *Phys. Rept.* **100**, pp. 1–150, doi:10.1016/0370-1573(83)90022-4.

Gribov, V., Lipatov, L., and Frolov, G. (1971). The leading singularity in the j plane in quantum electrodynamics, *Sov. J. Nucl. Phys.* **12**, p. 543.

Gross, D. and Wilczek, F. (1974). Asymptotically free gauge theories. 2, *Phys. Rev. D* **9**, pp. 980–993, doi:10.1103/PhysRevD.9.980.

Gross, D. J. and Wilczek, F. (1973). Ultraviolet behavior of nonabelian gauge theories, *Phys. Rev. Lett.* **30**, pp. 1343–1346, doi:10.1103/PhysRevLett.30.1343.

Kovchegov, Y. V. (1999). Small x F(2) structure function of a nucleus including multiple pomeron exchanges, *Phys. Rev. D* **60**, p. 034008, doi:10.1103/PhysRevD.60.034008, arXiv:hep-ph/9901281.

Kovchegov, Y. V. (2000). Unitarization of the BFKL pomeron on a nucleus, *Phys. Rev. D* **61**, p. 074018, doi:10.1103/PhysRevD.61.074018, arXiv:hep-ph/9905214.

Kuraev, E. A., Lipatov, L. N., and Fadin, V. S. (1976). Multi-reggeon processes in the Yang-Mills theory, *Sov. Phys. JETP* **44**, pp. 443–450.

Landau, L., Abrikosov, A., and Khalatnikov, I. (1954). An asymptotic expression for the photon green function in quantum electrodynamics, *Dokl. Akad. Nauk SSSR* **95**, 6, pp. 1177–1180, doi:10.1016/b978-0-08-010586-4.50085-7.

Lipatov, L. (1975). The parton model and perturbation theory, *Sov. J. Nucl. Phys.* **20**, pp. 94–102.

Lipatov, L. (1986). The Bare Pomeron in quantum chromodynamics, *Sov. Phys. JETP* **63**, pp. 904–912.

Politzer, H. (1973). Reliable perturbative results for strong interactions? *Phys. Rev. Lett.* **30**, pp. 1346–1349, doi:10.1103/PhysRevLett.30.1346.

Sudakov, V. (1956). Vertex parts at very high-energies in quantum electrodynamics, *Sov. Phys. JETP* **3**, pp. 65–71.

https://doi.org/10.1142/9789811231124_0004

Chapter 4

BFKL — Past and Future

V. S. Fadin*,†
*Budker Institute of Nuclear Physics of SD RAS, 630090 Novosibirsk Russia
†Novosibirsk State University, 630090 Novosibirsk, Russia

Abstract

This paper contains my recollection of the creation and development of the so-called BFKL approach and my ideas about the ways of its further development.

4.1 BFKL history

The BFKL story goes back to the middle of 70-th of the last century, when Lev started [Lipatov (1976)] to study the high energy behaviour of non-Abelian gauge theories [Yang and Mills (1954)]. To avoid infrared singularities, the investigation was performed in the theories with Higgs mechanism of mass generation [Higgs (1966); Kibble (1967); Englert and Brout (1964)] preserving renormalizability. More definitely, the model conserving global isotopic invariance was considered, with local gauge symmetry SU(2) and isotopic doublet of scalar fields. The paper [Lipatov (1976)] was devoted to check of Reggeization of the gauge boson and investigation of the vacuum singularity in this theory.

4.1.1 *Historical background*

Recall that the notion of Reggeization of elementary particles in perturbation theory was introduced and study of this phenomenon was started in the series of papers by Gell-Mann and co-authors [Gell-Mann and Goldberger (1962); Gromov *et al.* (2015b); Kazakov and Kotikov (1988); Gell-Mann *et al.* (1964b)]. In terms of the relativistic partial wave amplitude $A_j(t)$, analytically continued to complex j values, the Reggeization was defined as the disappearance, due to radiative corrections, of the non-analytic terms of the Born approximation on account of one-particle exchanges in the t channel. In other words, Reggeization of an elementary particle with spin j_0 and mass m means that at large s and fixed t Born amplitudes with exchange of this particle in the t channel acquire a factor $s^{j(t)-j_0}$, with $j(m^2) = j_0$, as a result of radiative corrections. Gell-Mann and his collaborators established the presence of this phenomenon for backward Compton scattering in QED with heavy photon and thus showed the fermion Reggeization in this theory. They also investigated scalar electrodynamics and concluded that a scalar is not Reggeized. In [Gell-Mann *et al.* (1964a)] they formulated necessary conditions for Reggeization on the Born scattering amplitudes of the theory. These conditions were generalized later in [Mandelstam (1965)]. In [Mandelstam (1964)] it was shown that they are not fulfilled in the vector channel of massive QED. Non-Reggeization of the heavy photon was confirmed by direct two-loop calculations in [Chester (1965)].

Note that the term "Reggeization" can be understood in different ways. If we assume that it means only the existence of the Regge trajectory on which the particle lies, then the scalar interacting with the photon is Reggeized [Cheng and Lo (1975)]. We use this term in a stronger sense: it means not only the existence of the Regge trajectory of the particle, but that the entire amplitude is given by the Reggeon with this trajectory.

In non-Abelian gauge theories the Reggeization problem was considered in Refs. [Abers *et al.* (1970); Dicus *et al.* (1971); Grisaru *et al.* (1973b,a); Grisaru (1977)]. In particular, it was shown in [Grisaru *et al.* (1973b)], [Grisaru *et al.* (1973a)] that in the theories with the Higgs mechanism of mass generation (unlike the theories with mass terms in original Lagrangian [Dicus *et al.* (1971)]) the necessary conditions for Reggeization [Gell-Mann *et al.* (1964a,b); Mandelstam (1965)] are fulfilled for the vector channel. Fulfilment of these conditions allowed to find [Grisaru *et al.* (1973b)] the trajectory of the vector meson using the t-channel unitarity:

$$j_V(t) = 1 + \frac{8g^2}{\pi}(t - m_V^2)\int_{4m_V^2}^{\infty} \frac{dt'}{t'-t}\frac{\rho(t')}{t' - 4m_V^2}, \tag{4.1}$$

where

$$\rho(t) = \frac{1}{32\pi}\sqrt{\frac{t-4m_V^2}{t}}. \tag{4.2}$$

However, this was only an indication of the possibility of the Reggeization, and not at all a proof of it. Therefore, it was very important to check the Reggeization by direct calculations, and the Lev's first aim was to do it.

The second aim was to investigate the vacuum singularity, i.e. asymptotics of amplitudes with vacuum quantum numbers in the t-channel and positive signature. In the Regge-Gribov theory of complex angular momenta this singularity originally was introduced (with intercept equal to one) [Gribov (1962)], [Chew and Frautschi (1961)] to provide constant cross sections at asymptotically high energies. Because of its fundamental role, this singularity has received a special name: it was called Pomeron after I. Ya. Pomeranchuk.

This singularity was investigated in QED [Gribov *et al.* (1971); Frolov *et al.* (1970); Cheng and Wu (1970)] (Lev was one of the main investigators), and it was shown that in the LLA it is a fixed branch point to the right of 1 (at $j = 1+\frac{11}{32}\pi\alpha^2$), i.e. the LLA violates the Froissart bound [Froissart (1961)]. Investigation of the Pomeron in the non-Abelian gauge theories was especially interesting because of vector boson Reggeization, unlike photon.

4.1.2 *Dispersive approach and the first results*

It was important to check the Reggeization by direct calculations, and it was done in [Lipatov (1976)] in two loops with the leading logarithmic accuracy (LLA), when in each order of perturbation theory only terms with the highest powers of $\ln s$ are kept.

For the calculations Lev used the dispersive approach based on the general properties of the theory: analyticity, unitarity and renormalizability. To my mind, it was the first application of the dispersive approach to the non-Abelian gauge theories. This approach turned out very successful, because it permits to escape consideration of great number of Feynman diagrams and to work only with physical particles in the unitarity relations thus avoiding the use of ghosts by Faddeev-Popov. Now it is widely used, unfortunately, without any reference to Lev.

Note that the term LLA was used to indicate that only terms with the highest powers of the logarithm of c.m.s. energy $\sqrt{s}$ are being held in amplitude discontinuities, not in the amplitudes themselves. Total amplitudes are obtained from their s-channel discontinuities in this approximation by the substitutions

$$is\ln^n s \to -\frac{s}{2\pi(n+1)}\left(\ln^{n+1}(-s-i\epsilon) \pm \ln^n(s-i\epsilon)\right), \tag{4.3}$$

where $+(-)$ sign for negative (positive) signature (symmetry with respect to the replacement $s \leftrightarrow u \simeq -s$). Therefore the degrees of logarithms in the real parts of the amplitudes with a negative signature are one more than the degree in the imaginary parts, whereas for the amplitudes with a positive signature, the reverse is true, and in the LLA they are purely imaginary. It is very important, because thanks to this in the LLA (and in the next to it, as it will be discussed below) only amplitudes with negative signature must be kept on the right side of the unitarity relations

$$\mathrm{Im}_s \mathcal{A}_{AB}^{A'B'} = \frac{1}{2} \sum_{n=0}^{\infty} \sum_{\{f\}} \int \mathcal{A}_{AB}^{\tilde{A}\tilde{B}+n} \left(\mathcal{A}_{A'B'}^{\tilde{A}\tilde{B}+n} \right)^* d\Phi_{\tilde{A}\tilde{B}+n}, \tag{4.4}$$

where $\sum_{\{f\}}$ means sum over discrete quantum numbers of intermediate particles, $\mathcal{A}_{AB}^{\tilde{A}\tilde{B}+n}$ and $\left(\mathcal{A}_{A'B'}^{\tilde{A}\tilde{B}+n}\right)$ are the amplitudes of n-particle production and $d\Phi_{\tilde{A}\tilde{B}+n}$ is the corresponding phase space element.

As it is seen from (4.4), calculation of elastic amplitudes in the dispersive approach requires knowledge of inelastic ones. It was recognized by Lev that the inelastic amplitudes are necessary in special kinematics. In the LLA it is so called multi-Regge kinematics (MRK), where produced particles have limited (not growing with s) transverse momenta and are strongly ordered in rapidity. For two-loop calculations one-particle production amplitudes in the Born approximation are necessary. Using the t-channel unitarity, Lev obtained a simple factorized form of these amplitudes with famous now Lipatov's vertex for production of vector mesons.

The main result of Ref. [Lipatov (1976)] was that in the LLA two-loop radiative corrections to elastic scattering amplitudes with the gauge boson quantum numbers in the t-channel and negative signature have the Regge form. As for the vacuum singularity, it was shown only that it has not a pole form. More definite conclusions about the nature of this singularity on the basis of two-loop calculations can not be drawn.

4.1.3 *Pre-BFKL*

Next important steps in the investigation of the high energy behaviour of non-Abelian gauge theories was done in [Fadin *et al.* (1975); Kuraev *et al.* (1976, 1977)]. Two-particle production amplitudes were found in the Born approximation and the one-loop corrections to the one-particle production amplitudes were calculated. It turned out that the first ones have the factorized form with the same vertices as the one-particle production amplitudes and that the corrections to the last ones have the Regge form. Three-loop corrections to elastic amplitudes calculated using these results proved the vector boson Reggeization in this order. Details of derivation of all these results and their generalization on colours group $SU(N)$ with arbitrary N were described in [Kuraev *et al.* (1976)].

An extremely important step made on this basis was the hypothesis of vector meson Reggeization. It was assumed that in the LLA not only elastic, but also inelastic amplitudes in the MRK with quantum numbers of vector bosons and negative signatures in all cross-channels are given by the Regge pole contributions in all orders of perturbation theory. Since only such amplitudes are important in the unitarity relations (4.4), it is possible to express partial waves $(A_{AB}^{A'B'})_{\omega}^{T}(t)$ $(\omega = j - 1)$ of amplitudes of the process $AB \to A'B'$ with the global

$SU(2)$ "isospin" T

$$(A_{AB}^{A'B'})_{\omega}^{T}(t) = \int_{s_0}^{\infty} \frac{ds}{s} s^{-\omega-1} \mathrm{Im}_s (A_{AB}^{A'B'})^{T}(s,t),$$

$$(A_{AB}^{A'B'})^{T}(s,t) = \frac{s}{2\pi i} \int_{\delta-i\infty}^{\delta+i\infty} \frac{d\omega}{\sin(\pi\omega)} \times \left((-1)^T \left(\frac{s}{s_0}\right)^{\omega} - \left(\frac{-s}{s_0}\right)^{\omega} \right) (A_{AB}^{A'B'})_{\omega}^{T}(t), \quad (4.5)$$

through partial waves of amplitude $(F_{B'B})_{\omega}^{T}(k_\perp, q_\perp - k_\perp)$ of Reggeon scattering on the particle B

$$(A_{AB}^{A'B'})_{\omega}^{T}(t) = \Phi_{A'A}^{T} \int \frac{d^2k'_\perp (F_{B'B})_{\omega}^{T}(k'_\perp, q_\perp - k'_\perp)}{(k'^2_\perp - m^2)((q-k')^2_\perp - m^2)} \quad (4.6)$$

and to write for them the equation [Fadin *et al.* (1975)]:

$$[2 + \omega - \alpha(k_\perp^2) - \alpha((q-k)_\perp^2)](F_{B'B})_{\omega}^{T}(k_\perp, q_\perp - k_\perp) = \Phi_{B'B}^{T} + \frac{g^2}{(2\pi)^3} \times \int \frac{d^2k'_\perp (F_{B'B})_{\omega}^{T}(k'_\perp, q_\perp - k'_\perp)}{(k'^2_\perp - m^2)((q-k')^2_\perp - m^2)} \Big[A_T(q_\perp^2) - \left(2 - \frac{1}{2}T(T+1)\right) \times \frac{(k_\perp^2 - m^2)((q-k')_\perp^2 - m^2) + (k'^2_\perp - m^2)((q-k)_\perp^2 - m^2)}{(k-k')_\perp^2 - m^2} \Big], \quad (4.7)$$

where $t = q^2 = q_\perp^2$, $\alpha(t)$ is the vector meson trajectory,

$$\alpha(t) = 1 + \frac{g^2}{(2\pi)^3}(t - m^2) \int \frac{d^2k_\perp}{(k_\perp^2 - m^2)((q-k)_\perp^2 - m^2)}, \quad t = q^2 = q_\perp^2, \quad (4.8)$$

$\Phi_{A'A}^{T}$ and $\Phi_{B'B}^{T}$ are the impact factor for the $A \to A'$ and $B \to B'$ transitions,

$$A_T(q^2) = \frac{m^2}{2} + \left(2 - \frac{1}{2}T(T+1)\right)\left(t - \frac{1}{2}m^2\right). \quad (4.9)$$

It is not difficult to see that $\alpha(t)$ is equal to $j_V(t)$ defined in (4.8).

It was shown in [Fadin *et al.* (1975)] that there is bootstrap in the vector boson channel, corresponding to $T = 1$: the equation (4.7) gives in the j-plane only the same Regge pole which was assumed. Indeed, the solution of Eq. (4.7) for $T = 1$ is

$$(F_{B'B})_{\omega}^{T}(k_\perp, q_\perp - k_\perp)\Big|_{T=1} = \frac{\Phi_{B'B}^{T}}{\omega + 1 - \alpha(q^2)}, \quad (4.10)$$

$$(A_{AB}^{A'B'})_{\omega}^{T}(t)\Big|_{T=1} = \Phi_{A'A}^{T}\Phi_{B'B}^{T}\Big|_{T=1} \frac{(\alpha(q^2) - 1)}{g^2(t - m^2)(\omega + 1 - \alpha(q^2))}, \quad (4.11)$$

and since $\Phi^T_{P'P}\Big|_{T=1} = g\Gamma_{P'P}$, where $\Gamma_{P'P}$ is the vertex of reggeon-particle interaction,

$$(A^{A'B'}_{AB})(s,t)\Big|_{T=1} = \Gamma_{A'A}\frac{\left((-u)^{\alpha(q^2)} - (-s)^{\alpha(q^2)}\right)}{(t-m^2)}\Gamma_{B'B}, \quad u \approx -s, \tag{4.12}$$

accordance with the Reggeization hypothesis. Of course, this could in no way be considered as a proof of a hypothesis, because the hypothesis refers not only to elastic amplitudes. Nevertheless, it was the first step to the proof. Later the bootstrap conditions for inelastic amplitudes were formulated and the proof of the hypothesis was held on their basis [Balitsky *et al.* (1979)].

So, a primary Reggeon in this approach (subsequently named BFKL) is the Reggeized gauge boson. The Pomeron, which corresponds to the rightmost j-plane singularity of the partial wave with $T = 0$ in (4.7) and determines the high energy behaviour of cross sections, appears as a compound state of two Reggeized gauge bosons. It was shown [Fadin *et al.* (1975)] that in the Pomeron channel the leading j-plane singularity turns out to be a square root branch point at

$$j = 1 + N\frac{g^2}{\pi}\ln 2, \tag{4.13}$$

for the gauge group $SU(N)$. Therefore, the Froissart bound [Froissart (1961)] is violated as well as in QED [Frolov *et al.* (1970)], [Cheng and Wu (1970)], although the mechanism of this violation differs from QED: in the non-Abelian theories cross sections for production of any fixed number of particles decrease with energy due to the vector boson Reggeization, and the total cross section increases as power of s only due to increasing number of opening channels. The reason of the violation is that the s-channel unitarity is not fulfilled in the LLA.

Another important observation made in [Fadin *et al.* (1975)] was the growth of the typical transverse momenta with increasing s. It was conjectured that due to this growth and the asymptotic freedom only Regge poles may appear to the right from the point $j = 1$.

Detailed consideration of the Pomeron channel was done in [Kuraev *et al.* (1977)].

4.1.4 *Appearance of the BFKL and its development*

Appearance of the BFKL should be attributed to the end of the 1970s, when Lev turned to QCD and with his student Yan Balitsky applied the methods developed for non-Abelian theories with broken symmetry in [Lipatov (1976)], [Fadin *et al.* (1975); Kuraev *et al.* (1976, 1977)] to scattering of colourless particles in QCD. In 1978, a famous paper [Balitsky and Lipatov (1978)] was published. The results of [Fadin *et al.* (1975)] (see (4.7), (4.8)) can be applied in the massless limit (theory without spontaneous breaking of symmetry) only with some regularization (in the following dimensional regularization will be used). The reason is that they were obtained for scattering of coloured particles (we will call them partons). The key difference of colourless particles from partons is that due to the gauge invariance impact factors of colourless particles, describing their interaction with Reggeized gluons, vanish at zero gluon momenta. It was shown in [Balitsky and Lipatov (1978)] that in this case the transition to massless theory is not difficult: the equation (4.7) with $\Phi^T_{B'B}$ corresponding to the impact factor of a colourless particle is free from the infrared singularities and therefore can be used in QCD.

From now on, we will talk about gauge theories with unbroken symmetry, although, as it follows from the above, the approach can be used in theories with broken symmetry as well. If not specified, QCD will be kept in mind, although for generality, the gauge group will be taken $SU(N_c)$, with the number of colours N_c.

An approximate theory can only be considered consistent when it is possible to calculate corrections to the approximation. Development of the BFKL approach in the next-to-leading logarithmic approximation (NLLA) was started in the late 1980s [Fadin and Lipatov (1989)]. In the NLLA, the same scheme of derivation of the BFKL equation as in the LLA is applicable. Again only amplitudes with negative signature must be kept on the right side of the unitarity relations (4.4), because leading terms in amplitudes with positive signature are imaginary and have one less degree of $\ln s$. Moreover, only real parts of the amplitudes with negative signature are necessary because their imaginary parts are suppressed by one power of $\ln s$ compared to their real parts. It was supposed that these amplitudes have the Regge pole form, i.e. are expressed in terms of the gluon trajectory and the Reggeon vertices. Therefore it was necessary to calculate one-loop corrections to the gluon trajectory and the vertices used in the LLA. Also, instead of one gluon two-gluon and quark-antiquark jets can be produced in the NLLA. Therefore it was necessary to calculate new Reggeon vertices for the jet production.

The calculations required a lot of effort and time. They started with the two-gluon jet production vertex [Fadin and Lipatov (1989)]. A lot of work was done and a lot of papers were published (see [Fadin and Lipatov (1989); Fadin and Fiore (1992); Fadin and Lipatov (1993); Fadin *et al.* (1994b,a); Fadin (1995c,b); Fadin *et al.* (1995); Fadin (1995a); Fadin *et al.* (1996b); Fadin and Lipatov (1996); Kotsky and Fadin (1996); Fadin *et al.* (1996a,c, 1997, 1998)]) before everything necessary for obtaining the NLO BFKL kernel in a colourless (Pomeron) channel at zero momentum transfer became available. Then this kernel and his eigenvalues were found [Fadin and Lipatov (1998)] (see also [Ciafaloni and Camici (1998)]). It is necessary to notice here that although talking about BFKL kernel (or BFKL equation) they usually mean exactly the kernel for forward scattering, although the approach is also applicable to scattering with the transfer of both momentum and colour. To date, a lot of work has been done and a number of important results on the development of the BFKL as applied to such processes have been obtained.

4.2 Basics of the BFKL approach

4.2.1 *The gluon Reggeization*

The BFKL approach is based on a remarkable property of QCD — gluon Reggeization, which gives a very powerful tool for the description of high energy processes. The gluon Reggeization determines the form of QCD amplitudes at large energies and limited transverse momenta. Due to the Reggeization, dominant amplitudes have a simple factorized form in the next-to-leading logarithmic approximation (NLLA) as well as in the LLA. The Reggeization allows to express an infinite number of amplitudes through several effective vertices and gluon trajectory.

In the Regge kinematic region $s \simeq -u \to \infty$, t fixed, amplitudes of the process $A + B \to A' + B'$ with a colour octet t-channel exchange and negative signature can be depicted by the diagram of Fig. 4.1 and have the Regge form

$$\mathcal{A}_{AB}^{A'B'} = \Gamma^c_{A'A}\left[\left(\frac{-s}{-t}\right)^{j(t)} - \left(\frac{s}{-t}\right)^{j(t)}\right]\Gamma^c_{B'B}, \tag{4.14}$$

where $\Gamma^c_{P'P}$ are energy-independent particle-particle-Reggeon (PPR) vertices (or scattering vertices), c is a colour index, $j(t) = 1 + \omega(t)$ is the Reggeized gluon trajectory.

As it is seen from (4.14), in the leading order (LO) PPR vertices are determined by the Born amplitudes, so that their calculation is trivial **assuming the gluon Reggeization, i.e. the form (4.14)**. Expressions for them are also quite simple. In the helicity basis they have the same form for all partons (quarks and gluons in QCD):

$$\Gamma^c_{P'P} = gT^c_{P'P}\delta_{\lambda_{P'}\lambda_P}, \tag{4.15}$$

where $T^c_{P'P}$ are matrix elements of the colour group generators in the corresponding representations and λ are parton helicities. Except for a common coefficient the vertices (4.15) can be written down without calculation, because they are given by

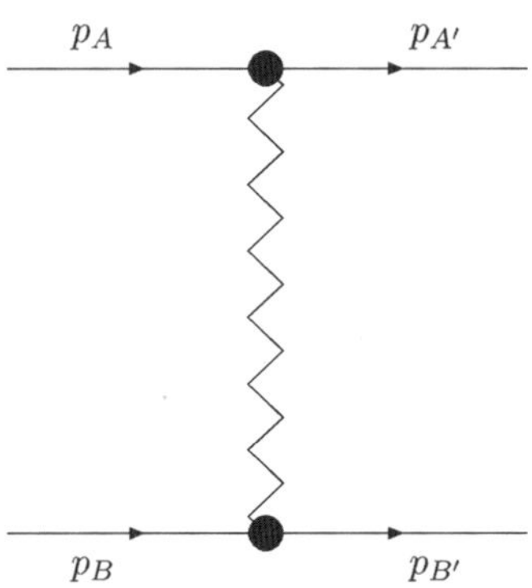

Fig. 4.1 Schematic representation of the process $A + B \to A' + B'$ with colour octet in the t channel and negative signature. The zigzag line represents Reggeized gluon.

forward matrix elements of the conserved current. Note that Eq. (4.15) implies a definite choice of the relative phase of spin wave functions of particles P' and P. Evidently, the phase is zero at $t = 0$. In (4.15) the s-channel helicity conservation is exhibited explicitly. Note that for gluons and for massive quarks it is valid only in the LO.

Again, as it is seen from (4.14), the LLA (one-loop) trajectory is determined by the s-channel discontinuity of any amplitude. It gives

$$\omega(t) = \frac{g^2 N_c t}{2(2\pi)^{D-1}} \int \frac{d^{D-2}q_1}{\vec{q}_1^{\,2}(\vec{q}-\vec{q}_1)^2} = -g^2 \frac{N_c \Gamma(1-\epsilon)}{(4\pi)^{D/2}} \frac{\Gamma^2(\epsilon)}{\Gamma(2\epsilon)} (\vec{q}^{\,2})^{\epsilon}$$
$$\simeq -g^2 \frac{N_c \Gamma(1-\epsilon)}{(4\pi)^{2+\epsilon}} \frac{2}{\epsilon} (\vec{q}^{\,2})^{\epsilon}. \tag{4.16}$$

In (4.16) and below the vector sign means transverse to the p_A, p_B plane components, $D = 4 - 2\epsilon$ is the space-time dimension taken different from 4 to regularize infrared divergencies, which are inevitable in parton amplitudes. Of course, using the dimensional regularization in (4.8) at $m = 0$ one obtains $\alpha(t) = 1 + \omega(t)$, where $\omega(t)$ is given by (4.16) at $N_c = 2$.

The Reggeization means also definite (multi-Regge) form of production amplitudes in the multi-Regge kinematics (MRK). MRK is the kinematics where all particles have limited (not growing with s) transverse momenta and are combined into jets with limited invariant mass of each jet and large (growing with s) invariant masses of any pair of the jets. This kinematics gives dominant contributions to cross sections of QCD processes at high energy $\sqrt{s}$. In each order of perturbation theory dominant (having the largest $\ln s$ degrees) are amplitudes with gluon quantum numbers and a negative signature. For the amplitude $\mathcal{A}_{2\to n+2}$ of the process $A + B \to A' + J_1 + \cdots + J_n + B'$ of production of n jets with momenta $k_1, k_2, \ldots k_n$ the MRK means

$$s \gg s_i \gg |t_i| \simeq \vec{q}_i^{\,2}, \quad s \simeq \frac{\prod_{i=1}^{n+1} s_i}{\prod_{i=1}^{n}(\vec{k}_i^{\,2} + k_i^2)}, \tag{4.17}$$

where

$$s = (p_A + p_B)^2, \quad s_i = (k_{i-1} + k_i)^2, \quad i = 1, \ldots n+1,$$
$$k_0 \equiv P_{A'}, \; k_{n+1} \equiv P_{B'}, \tag{4.18}$$

$$q_1 = p_A - p'_A, \quad q_{j+1} = q_j - k_j, \quad j = 1, \ldots n, \; q_{n+1} = p_{B'} - p_B. \tag{4.19}$$

In this region the amplitudes $\mathcal{A}_{2\to n+2}$ can be represented by Fig. 4.2.

Multi-particle amplitudes have a complicated analytical structure. They are not simple even in MRK (see, for instance, [Bartels (1975a,b, 1980); Fadin and Lipatov (1993)]). Fortunately, only real parts of these amplitudes are used in the BFKL approach in the NLLA as well as in the LLA. The reason is that imaginary parts of amplitudes with negative signature are suppressed by one power of $\ln s$ compared to their real parts, so that account of them in the right part of the

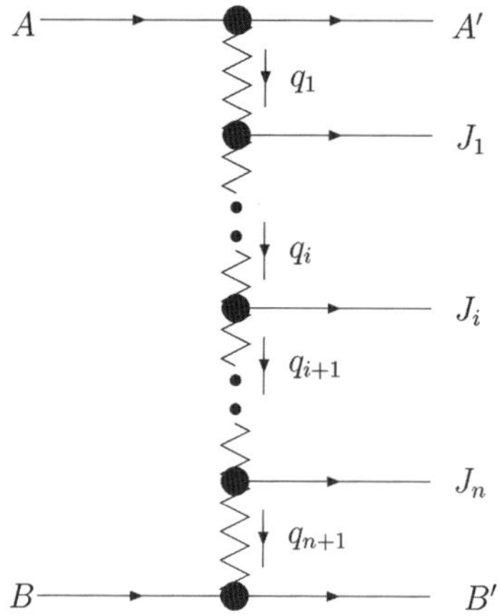

Fig. 4.2 Schematic representation of the process $A + B \to A' + J_1 + \cdots + J_n + B'$ in MRK. The zig-zag lines represent Reggeized gluon exchange; the black circles denote the Reggeon vertices; q_i are Reggeon momenta.

unitarity relations (4.4) means loss of two powers of $\ln s$ (as well as account of amplitudes with positive signature). Restricting ourselves to the real parts we can write

$$\Re \mathcal{A}_{2\to n+2} = 2s\Gamma^{R_1}_{A'A} \left(\prod_{i=1}^{n} \frac{1}{t_i} \left(\frac{s_i}{|\vec{k}_{i-1}||\vec{k}_i|} \right)^{\omega(t_i)} \gamma^{G_i}_{R_i R_{i+1}} \right)$$
$$\frac{1}{t_{n+1}} \left(\frac{s_{n+1}}{|k_n||\vec{q}_{n+1}|} \right)^{\omega(t_{n+1})} \Gamma^{R_{n+1}}_{B'B}. \tag{4.20}$$

Here $\Gamma^R_{A'A}$ and $\Gamma^R_{B'B}$ are the same scattering vertices as in (4.14) and $\gamma^{J_i}_{R_i R_{i+1}}$ are the Reggeon-Reggeon-Jet (RRJ) vertices (or production vertices), i.e. the effective vertices for production of jets J_i with momenta $k_i = q_i - q_{i+1}$ in collisions of Reggeons with momenta q_i and $-q_{i+1}$.

In the LLA only gluons can be produced, $J = G$,

$$\gamma^G_{R_1 R_2} = gT^c_{c_1 c_2} e^*_\mu(k) C^\mu(q_2, q_1), \tag{4.21}$$

where $k = q_1 - q_2$, $e_\mu(k)$ and c are the gluon momentum, polarization vector and colour index, c_1 and c_2 are the colour indices of the Reggeons R_1 and R_1 respectively, and $C^\mu(q_1, q_2)$ is the famous Lipatov's vertex [Lipatov (1976)]:

$$\begin{aligned} C_\mu(q_2, q_1) &= -q_{1\mu} - q_{2\mu} + p_{1\mu}\left(\frac{q_1^2}{kp_1} + 2\frac{kp_2}{p_1p_2}\right) - p_{2\mu}\left(\frac{q_2^2}{kp_2} + 2\frac{kp_1}{p_1p_2}\right) \\ &= -q_{1\perp\mu} - q_{2\perp\mu} - \frac{p_{1\mu}}{2(kp_1)}\left(k_\perp^2 - 2q_{1\perp}^2\right) \\ &\quad + \frac{p_{2\mu}}{2(kp_2)}\left(k_\perp^2 - 2q_{2\perp}^2\right). \end{aligned} \tag{4.22}$$

The Reggeon vertices and the gluon trajectory are known now in the next-to-leading order (NLO), that means the one-loop approximation for the vertices and the two-loop approximation for the trajectory. It is just the accuracy which is required for the derivation of the BFKL equation in the NLLA. Validity of the forms (4.14) and (4.20) is proved now in all orders of perturbation theory in the coupling constant g both in the LLA [Balitsky *et al.* (1979)] and in the NLLA [Fadin *et al.* (2006); Kozlov *et al.* (2011, 2012b,a); Fadin *et al.* (2015)].

In the LLA only gluons can be produced. In the NLA one has to consider not only the amplitudes (4.14), (4.20), but also amplitudes obtained from them by replacement of one of final particles by a jet containing a couple of particles with fixed (of order of transverse momenta) invariant mass.

4.2.2 *The scheme of derivation of the BFKL equation*

The Reggeization provides a simple derivation of the BFKL equation both in the LLA and NLLA. Two-to-two scattering amplitudes with all possible quantum numbers in the t–channel are calculated using the amplitudes (4.20) in the s-channel unitarity relations (4.4) and analyticity. The s-channel discontinuities of the amplitude $\mathcal{A}_{AB}^{A'B'}$ for the high energy process $A + B \to A' + B'$ at fixed momentum transfer $\sqrt{-t}$ may be presented by Fig. 4.3, where $q_1 + q_2 = q_1' + q_2' = q = p_A - p_{A'} = p_{B'} - p_B$ and symbolically written as the convolution

$$\Phi_{A'A} \otimes G \otimes \Phi_{B'B}$$

where the impact factors $\Phi_{A'A}$ and $\Phi_{B'B}$ describe transitions $A \to A'$ and $B \to B'$ due to interactions with Reggeized gluons, G is the Green's function for two interacting Reggeized gluons with an operator form

$$\hat{\mathcal{G}} = e^{Y\hat{\mathcal{K}}}, \tag{4.23}$$

where $Y = \ln(s/s_0)$, s_0 is an energy scale, $\hat{\mathcal{K}}$ is the BFKL kernel. The impact factors and the BFKL kernel are expressed in terms of the Reggeon vertices and trajectory. Energy dependence of scattering amplitudes is determined by the BFKL kernel,

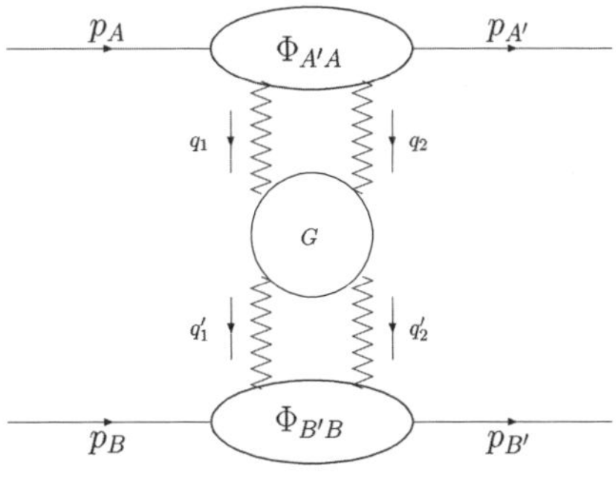

Fig. 4.3 Schematic representation of the s-channel discontinuities of amplitudes $A+B \to A'+B'$.

which is universal (process independent). The kernel

$$\hat{\mathcal{K}} = \hat{\omega}_1 + \hat{\omega}_2 + \hat{\mathcal{K}}_r \tag{4.24}$$

is expressed through the Regge trajectories $\hat{\omega}_1$ and $\hat{\omega}_2$ of two gluons and the "real part" $\hat{\mathcal{K}}_r$ describing production of particles in their interaction:

$$\hat{\mathcal{K}}_r = \hat{\mathcal{K}}_G + \hat{\mathcal{K}}_{Q\bar{Q}} + \hat{\mathcal{K}}_{GG}. \tag{4.25}$$

In the LLA only $\hat{\mathcal{K}}_G$ must be kept, because only gluons can be produced; in the NLLA production of quark-antiquark ($Q\bar{Q}$) and gluon (GG) pairs is also possible. The impact factors $\Phi_{A'A}$ and $\Phi_{B'B}$ describing transitions $A \to A'$ and $B \to B'$ depend on properties of scattering particles. All energy dependence is contained in the Green's function G for two interacting Reggeized gluons.

While the contribution of the trajectories to the kernel is diagonal in colour, the contribution associated with the production of real particles can change colour states. Using the projection operators $\hat{\mathcal{P}}^R$ on the representation R of the colour group, one can perform the decomposition

$$\hat{\mathcal{K}} = \sum_R \hat{\mathcal{P}}^R \hat{\mathcal{K}}^R. \tag{4.26}$$

In QCD, possible representations are $\mathbf{1}$, $\mathbf{8_a}$, $\mathbf{8_s}$, $\mathbf{10}$, $\mathbf{10^*}$ and $\mathbf{27}$. In the following, we will consider only the singlet representation, corresponding to scattering of colourless particles, and omit the index R.

4.2.3 *Forward BFKL kernel*

Separate contributions to the kernel (4.24) are singular. They were calculated in the dimensional regularization, so that in [Fadin and Lipatov (1998)] the kernel also was written at the space-time dimension $D = 4 - 2\epsilon$, different from the physical one. To work with this kernel is rather difficult. But the kernel is infrared safe, i.e. singularities in various parts cancel. It makes possible to perform the cancellation of the most singular terms and to write the kernel in the form [Fadin *et al.* (2009)] (here the Reggeon momenta are q and l)

$$\begin{aligned}
\mathcal{K}(\vec{q},\vec{l}) &= \frac{\alpha_s(\mu^2)N_c}{2\pi^2}\left[\frac{2}{(\vec{q}-\vec{l})^2} - \delta(\vec{q}-\vec{l})\int\frac{d\vec{k}\vec{q}^{\,2}}{(\vec{q}-\vec{k})^2\vec{k}^{\,2}}\right] \\
&\quad\times\left[1+\frac{\alpha_s(\mu^2)N_c}{4\pi}\left(\frac{67}{9}-2\zeta(2)-\frac{10}{9}\frac{n_f}{N_c}\right)\right] \\
&\quad+\frac{\alpha_s^2(\mu^2)N_c^2}{4\pi^3}\left[\frac{1}{(\vec{q}-\vec{l})^2}\left(\frac{\beta_0}{N_c}\ln\left(\frac{\mu^2}{(\vec{q}-\vec{l})^2}\right)-\ln^2\left(\frac{\vec{q}^{\,2}}{\vec{l}^{\,2}}\right)\right)+f_1(\vec{q},\vec{l})\right. \\
&\quad\left.+f_2(\vec{q},\vec{l})+\delta(\vec{q}-\vec{l})\left(\frac{\beta_0}{2N_c}\int\frac{d\vec{k}\vec{q}^{\,2}}{(\vec{q}-\vec{k})^2\vec{k}^2}\ln\left(\frac{(\vec{q}-\vec{k})^2\vec{k}^2}{\mu^2\vec{q}^{\,2}}\right)+6\pi\zeta(3)\right)\right],
\end{aligned} \tag{4.27}$$

where where $\beta_0 = 11N_c/3 - 2n_f/3$ is the first coefficient of the β-function,

$$f_1(\vec{q}_1, \vec{q}_2) = -\frac{2(\vec{q}_1^{\,2} - \vec{q}_2^{\,2})}{\vec{k}^2(\vec{q}_1 + \vec{q}_2)^2}\left(\frac{1}{2}\ln\left(\frac{\vec{q}_1^{\,2}}{\vec{q}_2^{\,2}}\right)\ln\left(\frac{\vec{q}_1^{\,2}\vec{q}_2^{\,2}\vec{k}^4}{(\vec{q}_1^{\,2} + \vec{q}_2^{\,2})^4}\right) + Li_2\left(-\frac{\vec{q}_1^{\,2}}{\vec{q}_2^{\,2}}\right)\right.$$
$$\left. - Li_2\left(-\frac{\vec{q}_2^{\,2}}{\vec{q}_1^{\,2}}\right)\right) - \left(1 - (\vec{q}_1^{\,2} - \vec{q}_2^{\,2})^2/\vec{k}^2(\vec{q}_1 + \vec{q}_2)^2\right)$$
$$\times \left(\int_0^1 - \int_1^\infty\right)\frac{dz\,\ln\frac{(z\vec{q}_1)^2}{(\vec{q}_2)^2}}{(\vec{q}_2 - z\vec{q}_1)^2}, \tag{4.28}$$

$$f_2(\vec{q}_1, \vec{q}_2) = -\left(1 + \frac{n_f}{N_c^3}\right)\frac{2\vec{q}_1^{\,2}\vec{q}_2^{\,2} - 3(\vec{q}_1\vec{q}_2)^2}{16\vec{q}_1^{\,2}\vec{q}_2^{\,2}}\left(\frac{2}{\vec{q}_2^{\,2}} + \frac{2}{\vec{q}_1^{\,2}} + \left(\frac{1}{\vec{q}_2^{\,2}} - \frac{1}{\vec{q}_1^{\,2}}\right)\ln\frac{\vec{q}_1^{\,2}}{\vec{q}_2^{\,2}}\right)$$
$$-\left(3 + \left(1 + \frac{n_f}{N_c^3}\right)\left(1 - \frac{(\vec{q}_1^{\,2} + \vec{q}_2^{\,2})^2}{8\vec{q}_1^{\,2}\vec{q}_2^{\,2}} - \frac{2\vec{q}_1^{\,2}\vec{q}_2^{\,2} - 3\vec{q}_1^{\,4} - 3\vec{q}_2^{\,4}}{16\vec{q}_1^{\,4}\vec{q}_2^{\,4}}(\vec{q}_1\vec{q}_2)^2\right)\right)$$
$$\times \int_0^\infty \frac{dx\,\ln\left|\frac{1+x}{1-x}\right|}{\vec{q}_1^{\,2} + x^2\vec{q}_2^{\,2}}, \tag{4.29}$$

where $\vec{k} = \vec{q}_1 - \vec{q}_2$.

This representation greatly simplifies the calculation of the eigenvalues of the kernel. Strictly speaking, because of the charge renormalization the eigenfunctions of the LO kernel $\left(\vec{l}^{\,2}\right)^{\gamma-1}$, which at $\gamma = 1/2 + i\nu$, $-\infty < \nu < \infty$ form a complete set for the kernel averaged over angles, are not any more eigenfunctions of the NLO kernel. But with the NLO accuracy one can write

$$\alpha_s(\mu^2) = \alpha_s(\vec{q}^{\,2})\left(1 + \beta_0\frac{\alpha_s(\vec{q}^{\,2})}{4\pi}\ln\left(\frac{\vec{q}^{\,2}}{\mu^2}\right)\right), \tag{4.30}$$

and obtain

$$\begin{aligned} \int d^2l\mathcal{K}(\vec{q}, \vec{l})\left(\frac{\vec{l}^{\,2}}{\vec{q}^{\,2}}\right)^{\gamma-1} &= \omega(\vec{q}^{\,2}, \gamma) = \frac{\alpha_s(\vec{q}^{\,2})N_c}{\pi}\chi(\gamma), \\ \chi(\gamma) &= \chi_B(\gamma) + \frac{\alpha_s N_c}{\pi}\chi^{(1)}(\gamma). \end{aligned} \tag{4.31}$$

Here $\chi_B(\gamma)$ gives the LO eigenvalues,

$$\chi_B(\gamma) = 2\psi(1) - \psi(\gamma) - \psi(1-\gamma) \tag{4.32}$$

and the correction $\chi^{(1)}(\gamma)$ is:

$$\chi^{(1)}(\gamma) = -\frac{1}{4}\left[\frac{\beta_0}{2N_c}\left(\chi_B^2(\gamma)(\gamma) + \chi_B'(\gamma) - 6\zeta(3)\right)\right.$$
$$+\frac{\pi^2\cos(\pi\gamma)}{\sin^2(\pi\gamma)(1-2\gamma)}\left(3 + \left(1+\frac{n_f}{N_c^3}\right)\frac{2+3\gamma(1-\gamma)}{(3-2\gamma)(1+2\gamma)}\right)$$
$$\left. -\left(\frac{67}{9} - \frac{\pi^2}{3} - \frac{10}{9}\frac{n_f}{N_c}\right)\chi_B(\gamma) + \chi_B''(\gamma)) - \frac{\pi^3}{\sin(\pi\gamma)} + 4\phi(\gamma)\right], \tag{4.33}$$

where

$$\phi(\gamma) = -\int_0^1 \frac{dx}{1+x}\left(x^{\gamma-1} + x^{-\gamma}\right)\int_x^1 \frac{dt}{t}\ln(1-t)$$
$$= \sum_{n=0}^{\infty}(-1)^n\left[\frac{\psi(n+1+\gamma)-\psi(1)}{(n+\gamma)^2} + \frac{\psi(n+2-\gamma)-\psi(1)}{(n+1-\gamma)^2}\right]. \tag{4.34}$$

This result was firstly obtained in [Fadin and Lipatov (1998)].

The representation (4.27) is useful also for finding all eigenvalues of the kernel. Defining them as

$$\int d^2l K(\vec{q},\vec{l})\left(\frac{\vec{l}^{\,2}}{\vec{q}^{\,2}}\right)^{\gamma-1} e^{in(\phi_l-\phi_q)} = \frac{\alpha_s(\vec{q}^{\,2})\,N_c}{\pi}\left(\chi_B(\gamma,n) + \frac{\alpha_s N_c}{\pi}\chi^{(1)}(\gamma,n)\right), \tag{4.35}$$

where

$$\chi_B(\gamma,n) = 2\psi(1) - \psi\left(\gamma + \frac{|n|}{2}\right) - \psi\left(1-\gamma+\frac{|n|}{2}\right), \tag{4.36}$$

we come to the correction obtained in [Kotikov and Lipatov (2000)]:

$$4\chi^{(1)}(\gamma,n) = -\frac{\beta_0}{2N_c}\left(\chi_B^2(\gamma,n) + \chi_B'(\gamma,n)\right) + 6\zeta(3) - \chi_B''(\gamma,n)$$
$$+\left(\frac{67}{9} - \frac{\pi^2}{3} - \frac{10}{9}\frac{n_f}{N_c}\right)\chi_B(\gamma,n) - 2\Phi(n,\gamma)$$
$$-2\Phi(n,1-\gamma) + F(n,\gamma), \tag{4.37}$$

where

$$F(n,\gamma) = \frac{\pi^2\cos(\pi\gamma)}{\sin^2(\pi\gamma)(1-2\gamma)}\left[\frac{\gamma(1-\gamma)(\delta_{n,2}+\delta_{n,-2})}{2(3-2\gamma)(1+2\gamma)}\right.$$
$$\left.\times\left(1+\frac{n_f}{N_c^3}\right) - \left(\frac{3\gamma(1-\gamma)+2}{(3-2\gamma)(1+2\gamma)}\left(1+\frac{n_f}{N_c^3}\right)+3\right)\delta_{n,0}\right], \tag{4.38}$$

and

$$\Phi(n,\gamma) = \int_0^1 \frac{dt}{1+t} t^{\gamma-1+n/2} \left\{ \frac{\pi^2}{12} - \frac{1}{2}\psi'\left(\frac{n+1}{2}\right) - Li_2(t) \right.$$
$$- Li_2(-t) - \left(\psi(n+1) - \psi(1) + \ln(1+t) + \sum_{k=1}^{\infty} \frac{(-t)^k}{k+n} \right) \ln t$$
$$\left. - \sum_{k=1}^{\infty} \frac{t^k}{(k+n)^2} \left[1 - (-1)^k\right] \right\}. \tag{4.39}$$

4.3 Investigation of the BFKL properties

In the wide range of problems that Lev dealt with, the BFKL approach was one of the first. Lev many times turned to it, and his analysis was always distinguished by its depth and originality.

4.3.1 *Conformal invariance*

In 1985 he discovered that for scattering of colourless particles the LO BFKL equation can be solved in a general form not only for the forward case, but for general momentum transfer [Lipatov (1986)]. This striking fact is related to the remarkable property of the BFKL equation discovered in [Lipatov (1986)]: for scattering of colourless particles it can be written in the special representation (which was called later Möbius representation [Bartels *et al.* (2005)]), which is invariant with respect to conformal (Möbius) transformations in the impact parameter space $\vec{r}_1, \vec{r}_2$ of the Reggeized gluons. In this representation the BFKL kernel $\hat{\mathcal{K}}_M$ operates in the space of functions having the colour transparency property (turning into zero at $\vec{r}_1 = \vec{r}_2$). Note that the colourlessness means the Pomeron exchange (while the reverse statement is not correct) and that it is necessary for writing the BFKL equation in the Möbius representation. The colourlessness of colliding particles provides the gauge invariance of impact factors, i.e. their vanishing at zero Reggeized gluon momenta. Together with an analogous property of the BFKL kernel, it gives a possibility to restrict the space of functions where the kernel acts by the functions having the colour transparency.

The conformal invariance of the BFKL equation in the Möbius representation is extremely important because it permits to classify all solutions of the homogeneous equation and to find their complete set. Another significant property is the holomorphic separability of BFKL kernel in this representation. Actually, this property was used already in [Lipatov (1986)]; it was explicitly exhibited and used later [Lipatov (1990, 1993)] for solving the Bartels-Kwiecinski-Praszalowichz (BKP) equation — the generalization of the BFKL equation to the case of many Reggeized gluons in the t–channel [Bartels (1980); Kwiecinski and Praszalowicz (1980)]. The conformal invariance permits to obtain eigenfunctions of the kernel $\hat{\mathcal{K}}_M$ from eigenfunctions for forward scattering by transformations of the Möbius group (inversion and shift). In [Lipatov (1986)] they were chosen in the form

$$E_{n,\nu}(\vec{r}_{10},\vec{r}_{20}) = (-1)^n \left(\frac{z_{12}}{z_{10}z_{20}}\right)^{\frac{1}{2}+i\nu-\frac{n}{2}} \left(\frac{z_{12}^*}{z_{10}^*z_{20}^*}\right)^{\frac{1}{2}+i\nu+\frac{n}{2}}, \tag{4.40}$$

where $z = x+iy$ for $\vec{r} = (x,y)$, the vector $\vec{r}_0$ is introduced for indexing of the wave functions, $\vec{r}_{ij} = \vec{r}_i - \vec{r}_j$, n is integer and ν is real. The eigenvalues coincide with the eigenvalues of the forward kernel

$$\omega_B(\gamma,n) = \frac{\alpha_s N_c}{\pi}\chi_B(\gamma,n), \gamma = 1/2 + i\nu, \tag{4.41}$$

$\chi_B(\gamma,n)$ is given in (4.36). One can see it noticing that the eigenvalues do not depend on $\vec{r}_0$ and that integration over $\vec{r}_0$ gives

$$\int d\vec{r}_0 E_{n,\nu}(\vec{r}_{10},\vec{r}_{20}) = C_{n,\nu}(z_{12})^{\frac{1}{2}-i\nu+\frac{n}{2}}(z_{12}^*)^{\frac{1}{2}-i\nu-\frac{n}{2}} = C_{n,\nu}\, e^{in\phi_{12}}\vec{r}_{12}^{\,2(1-\gamma)}, \tag{4.42}$$

where ϕ_{12} is the azimuthal angle of the vector $\vec{r}_{12}$. Then the problem is reduced to the calculations of the integrals for the forward case. The form of the integral (4.42) can be easily obtained using the change of the integration variables $z_{10} = z_{12}z$; the coefficient $C_{n,\nu}$ is irrelevant for the calculation of the eigenvalues. In fact, it was found in [Lipatov (1986)]:

$$C_{n,\nu} = \frac{\pi 2^{4i\nu}}{|n|-2i\nu}\frac{\Gamma(\kappa)}{\Gamma(\kappa^*)}\frac{\Gamma(\kappa^*-1/2)}{\Gamma(\kappa-1/2)}, \quad \kappa = \frac{1}{2} - i\nu + \frac{|n|}{2}. \tag{4.43}$$

One can easily show that

$$z_{12}^2\frac{\partial}{\partial z_1}\frac{\partial}{\partial z_2}E_{n,\nu} = \lambda_{n,\nu}E_{n,\nu}, \quad z_{12}^{*2}\frac{\partial}{\partial z_1^*}\frac{\partial}{\partial z_2^*}E_{n,\nu} = \lambda_{n,-\nu}E_{n,\nu},$$
$$\lambda_{n,\nu} = \frac{1}{4} - \left(\frac{n}{2} - i\nu\right)^2. \tag{4.44}$$

The completeness condition derived in [Lipatov (1986)] has the form

$$4\sum_{n=-\infty}^{+\infty}\int_{-\infty}^{+\infty} d\nu \int d^2r_0 \frac{n^2+4\nu^2}{\vec{r}_{12}^{\,2}\vec{r}_{1'2'}^{\,2}} E_{n,\nu}(\vec{r}_{10},\vec{r}_{20})E_{n,\nu}^*(\vec{r}_{1'0},\vec{r}_{2'0})$$
$$= (2\pi)^4\delta(\vec{r}_{11'})\delta(\vec{r}_{22'}). \tag{4.45}$$

Using (4.44) and (4.45) one obtains for the Green's function in the impact parameter space

$$\langle\vec{r}_1,\vec{r}_2|\frac{1}{\omega-\hat{\mathcal{K}}_M}\frac{1}{\hat{\vec{q}}_1^{\,2}\hat{\vec{q}}_2^{\,2}}|\vec{r}_1{}',\vec{r}_2{}'\rangle = 4\sum_{n=-\infty}^{+\infty}\int_{-\infty}^{+\infty} d\nu \int d^2r_0$$
$$\times \frac{(n^2+4\nu^2)E_{n,\nu}(\vec{r}_{10},\vec{r}_{20})E_{n,\nu}^*(\vec{r}_{1'0},\vec{r}_{2'0})}{[(n+1)^2+4\nu^2][(n-1)^2+4\nu^2](\omega-\omega(\gamma,n))}. \tag{4.46}$$

For $n = \pm 1$ the integral over ν here must be taken in the sense of its principal value.

Later an explicit expression for the kernel $\mathcal{K}_{\mathcal{M}}$ in the impact parameter space was obtained [Fadin *et al.* (2007b)]:

$$\langle \vec{r}_1, \vec{r}_2 | \hat{\mathcal{K}}_M | \vec{r}_1{}', \vec{r}_2{}' \rangle = \frac{\alpha_s N_c}{2\pi^2} \int d\vec{r}_0 \frac{\vec{r}_{12}^{\,2}}{\vec{r}_{10}^{\,2} \vec{r}_{20}^{\,2}} (\delta(\vec{r}_{11'})\delta(\vec{r}_{02'})$$

$$+ \delta(\vec{r}_{22'})\delta(\vec{r}_{01'}) - \delta(\vec{r}_{11'})\delta(\vec{r}_{22'})), \tag{4.47}$$

where the subscript M denotes the Möbius form. It is worth to note that the ultraviolet singularities of separate terms in (4.47) cancel in their sum with account of the dipole property of the target impact factors.

The transformations of the Möbius group in the two-dimensional space $\vec{r} = (x, y)$ can be written as

$$z \to \frac{az + b}{cz + d}, \tag{4.48}$$

where $z = x + iy$, a, b, c, d are complex numbers, with $ad - bc \neq 0$. Under these transformations, one has

$$\begin{aligned} z_1 - z_2 &\to \frac{z_1 - z_2}{(cz_1 + d)(cz_2 + d)}(ad - bc), \\ dzdz^* &\to dzdz^* \frac{|ad - bc|^2}{|(cz + d)^2|^2}, \end{aligned} \tag{4.49}$$

so that the conformal invariance of (4.47) is evident.

It turns out that the Möbius form (4.47) coincides with the kernel of the colour dipole model [Nikolaev and Zakharov (1994); Nikolaev *et al.* (1994); Mueller (1994); Mueller and Patel (1994)] formulated in the coordinate space. It means that the colour dipole kernel could have been written by Lev as early as 1985. It was not done only by chance, because he used the operator form of the kernel instead of the explicit expression for it in the impact parameter space.

Investigation of the conformal properties of the BFKL kernel in the NLO [Fadin *et al.* (2007c,a); Fadin and Fiore (2008); Fadin *et al.* (2010)] showed later that the ambiguity of the NLO kernel allows to present the kernel in the form where the conformal invariance is violated only by renormalization.

In the same paper [Lipatov (1986)] Lev performed another great study: account of running of the coupling constant. Strictly speaking, this account oversteps the limits of the LLA. Nevertheless, it is not unreasonable to take it into account just in the LO BFKL kernel in order to understand qualitative effects of the running. One important effect discovered by Lev is the conversion of the cut $[0, \omega_P^B]$ into an infinite series of moving poles with limiting point at $\omega = 0$. The Pomeron trajectories found in [Lipatov (1986)] are represented in the form

$$\omega_k(t) = \frac{c}{k+\eta(q^2)+1/4}, \quad k=0,\pm 1,\pm 2\ldots, \tag{4.50}$$

where c is a calculable constant. The function $\eta(q^2)$ is determined by large distances, but is limited by the inequalities $-1 \le 4\eta \le 1$ at small $|t|$. In the region of large $|t|$ ($\alpha_s(|t|) \ll 1$) the family of poles is approximated by the moving cut with the branch point

$$\omega(t) = \frac{4N_c}{\pi} \ln 2\alpha_s(|t|). \tag{4.51}$$

4.3.2 *Relation with DGLAP*

The relation between the eigenvalues of BFKL kernel $\frac{\alpha_s N_c}{\pi}\chi(\gamma)$ and the anomalous dimension γ_ω of the twist-2 operators with $j = 1+\omega$ was first discovered in the LLA in [Jaroszewicz (1982)], where it has the form:

$$\omega = \frac{\alpha_s N_c}{\pi}\chi(\gamma_\omega). \tag{4.52}$$

It is a remarkable result because it means that the BFKL approach permits to obtain the resummation of the most singular at $j = 1$ terms in the anomalous dimension of the twist-two operators.

It was noticed in [Fadin and Lipatov (1998)] that in the NLLA this relation must be changed. The reason is that energy scale s_0 becomes significant in the NNLLA, and this scale is different from DGLAP. In the NLA $s_0 = Q_A Q_B$, where Q_A and Q_B are typical momenta for the impact factors Φ_A and Φ_B. Therefore the deep-inelastic moments are defined as

$$M_\omega(Q^2) = \frac{1}{4(2\pi)^2}\int \frac{d\vec{q}}{\vec{q}^{\,2}} \int \frac{d\vec{l}}{\vec{l}^{\,2}} \Phi_A(\vec{q}) e^{\omega \ln(Q/Q_B)} G_\omega(\vec{q},\vec{l})\Phi_B(\vec{l}). \tag{4.53}$$

As the result, for the anomalous dimension $\gamma_\omega(\alpha_s)$ one has

$$\gamma_\omega = \frac{\omega}{2} + \chi^{-1}\left(\frac{\pi\omega}{N_c\alpha_s}\right), \tag{4.54}$$

where χ^{-1} is the inverse function. In other words, the anomalous dimensions $\gamma_\omega(\alpha_s)$ of the twist-2 operators near the point $\omega = 0$ are determined from the solution of the equation

$$\begin{aligned}
\omega &= \frac{\alpha_s N_c}{\pi}\chi\left(\gamma - \frac{\omega}{2}\right) \simeq \frac{\alpha_s N_c}{\pi}\chi(\gamma) - \frac{\alpha_s^2 N_c^2}{2\pi^2}\chi(\gamma)\chi'(\gamma) \\
&\simeq \frac{\alpha_s N_c}{\pi}\frac{1}{\gamma} - \frac{\alpha_s^2 N_c^2}{4\pi^2}\left(\frac{11+2n_f/N_c^3}{3\,\gamma^2} + \frac{n_f(10+13/N_c^2)}{9\gamma N_c}\right. \\
&\quad \left. + \frac{395}{27} - 2\zeta(3) - \frac{11}{3}\frac{\pi^2}{6} + \frac{n_f}{N_c^3}\left(\frac{71}{27} - \frac{\pi^2}{9}\right) + \mathcal{O}(\gamma)\right)
\end{aligned} \tag{4.55}$$

for $\gamma \to 0$. This equation was derived in [Fadin and Lipatov (1998)] and used to reproduce the known results and predict the higher loop correction for $\omega \to 0$:

$$\gamma \simeq \frac{\alpha_s N_c}{\pi}\left(\frac{1}{\omega}-\frac{11}{12}-\frac{n_f}{6N_c^3}\right)-\left(\frac{\alpha_s}{\pi}\right)^2\frac{n_f N_c}{6\omega}\left(\frac{5}{3}+\frac{13}{6N_c^2}\right)$$
$$-\frac{1}{4\omega^2}\left(\frac{\alpha_s N_c}{\pi}\right)^3\left(\frac{395}{27}-2\zeta(3)-\frac{11}{3}\frac{\pi^2}{6}+\frac{n_f}{N_c^3}\left(\frac{71}{27}-\frac{\pi^2}{9}\right)\right). \quad (4.56)$$

The result for the three-loop correction was confirmed later in [Vogt *et al.* (2004)], [Kotikov *et al.* (2004)].

4.3.3 *BFKL in N=4 SUSY*

Impressive and surprising results connected with BFKL were obtained by Lev in maximally extended supersymmetric Yang-Mills theory (N=4 SYM). The eigenvalues of the NLO BFKL kernel were calculated in [Kotikov and Lipatov (2000)]. They turned out to be much simpler than in QCD: instead of (4.37) the one-loop correction takes the form

$$4\chi^{(1)}(\gamma,n) = -2\Phi(n,\gamma)-2\Phi(n,1-\gamma)+6\zeta(3)+\psi''\left(\gamma+\frac{n}{2}\right)+\psi''\left(\gamma+\frac{n}{2}\right). \quad (4.57)$$

A remarkable fact is the disappearance of non analytical in conformal spin n terms. Moreover, all functions entering in (4.57) have the property of maximal transcendentality [Kotikov *et al.* (2004)]. The maximal transcendentality of an expression means by definition, that the special functions and numbers with lower complexities do not contribute to it. By definition $\psi(x)$ has the transcendentality equal to 1, the transcendentalities of $\psi^{(n)}$ and $\zeta(n+1)$ are $n+1$ and the additional poles in the sum over k increase the transcendentality of the function $\Phi(M)$ up to 3. Lev put forward a remarkable hypothesis about growing with order of perturbation theory maximal transcendentality of the eigenvalues of the BFKL kernel and the anomalous dimensions of twist-2 operators in $N = 4$ SUSY. This hypothesis is not disproved and widely used now. Another of Lev's remarkable hypothesis is the hypothesis that in $N = 4$ SUSY the Pomeron coincides with the Reggeized graviton, which gives a possibility to calculate its intercept at large coupling constants. Maximal transcendentality together with integrability allow one to find the anomalous dimensions of twist-2 operators in this model up to 4 loops.

4.4 Future development

Evidently, the next step in the development of the BFKL approach should be the next-to-next-to-leading logarithmic approximation (NNLLA). Unfortunately, in QCD the BFKL equation in the next-to-next-to-leading approximation is not yet obtained, although in maximally extended supersymmetric Yang-Mills theory with large number of colours (planar N=4 SYM) impressive results have been obtained [Gromov *et al.* (2015b); Velizhanin (2015); Caron-Huot and Herranen (2018); Alfimov *et al.* (2018)].

4.4.1 *NNLLA BFKL in planar N=4 SYM*

In [Gromov *et al.* (2015b)] the next-to-next-to-leading order (NNLO) corrections to the BFKL Pomeron eigenvalues for conformal spin $n = 0$ were obtained. The method of calculation was quite different from the one used in the BFKL approach, in which, first of all, the kernel of the BFKL equation is calculated using the unitarity and analyticity. It was based on the integrability of the planar N=4 SYM and on the observation of L. N. Lipatov and A.V. Kotikov [Kotikov and Lipatov (2000)] that BFKL eigenvalues and anomalous dimensions of twist-2 operators are connected in planar N=4 SYM by analytic continuation. The problem of calculation of the anomalous dimensions was solved by the quantum spectral curve method (QSC) [Gromov *et al.* (2014)], [Gromov *et al.* (2015a)] giving a finite set of Riemann-Hilbert equations for exact spectrum of planar N=4 SYM theory. Applicability of this method to calculation of the LO BFKL Pomeron eigenvalues was demonstrated in [Alfimov *et al.* (2015)]. Another significant assumption, besides the applicability of the QSC, was the possibility to write the relation between the spin S of the twist-two operator $Tr(ZD_+^S Z)$ and its full conformal dimension $\Delta = 2 + S + \gamma$, γ is the anomalous dimension, in the form

$$j(\Delta) = 2 + S(\Delta) = 1 + \sum_{n=1}^{\infty} \left(\frac{g^2 N_c}{(4\pi)^2}\right)^n \left[F_n\left(\frac{\Delta-1}{2}\right) + F_n\left(\frac{-\Delta-1}{2}\right)\right], \quad (4.58)$$

where $F_n(x)$ are simple linear combinations of the nested harmonic sums with the same transcendentality n

$$S_{a_1,a_2,\ldots,a_n}(x) = \sum_{y=1}^{x} \frac{(\text{sign}\,(a_1))^y}{y^{|a_1|}} S_{a_2,\ldots,a_n}(y) \quad S(x) = 1. \quad (4.59)$$

The possibility of such representation for the LO and NLO eigenvalues with

$$F_1 = -4S_1, \frac{F_2}{4} = -\frac{3}{2}\zeta_3 + \pi^2 \ln 2 + \frac{\pi^2}{3} S_1 + 2S_3 + \pi^2 S_{-1} - 4S_{-2,1} \quad (4.60)$$

was shown in [Costa *et al.* (2012)]. The three-loop result obtained in [Gromov *et al.* (2015b)] is:

$$\begin{aligned}\frac{F_3(x)}{256} = & -\frac{5S_{-5}}{8} - \frac{S_{-4,1}}{2} + \frac{S_1 S_{-3,1}}{2} + \frac{S_{-3,2}}{2} - \frac{5S_2 S_{-2,1}}{4} + \frac{S_{-4}S_1}{4} + \frac{S_{-3}S_2}{8} \\ & + \frac{3S_{3,-2}}{4} - \frac{3S_{-3,1,1}}{2} - S_1 S_{-2,1,1} + S_{2,-2,1} + 3S_{-2,1,1,1} - \frac{3S_{-2}S_3}{4} - \frac{S_5}{8} \\ & + \frac{S_{-2}S_1 S_2}{4} + \pi^2 \left[\frac{S_{-2,1}}{8} - \frac{7S_{-3}}{48} - \frac{S_{-2}S_1}{12} + \frac{S_1 S_2}{48}\right] - \pi^4 \left[\frac{2S_{-1}}{45} - \frac{S_1}{96}\right]\end{aligned}$$

$$+\zeta_3\left[-\frac{7S_{-1,1}}{4}+\frac{7S_{-2}}{8}+\frac{7S_{-1}S_1}{4}-\frac{S_2}{16}\right]$$
$$+\left[2\mathrm{Li}_4\left(\frac{1}{2}\right)-\frac{\pi^2\log^2 2}{12}+\frac{\log^4 2}{12}\right](S_{-1}-S_1)$$
$$+\frac{\log^5 2}{60}-\frac{\pi^2\log^3 2}{36}-\frac{2\pi^4\log 2}{45}-\frac{\pi^2\zeta_3}{24}+\frac{49\zeta_5}{32}-2\mathrm{Li}_5\left(\tfrac{1}{2}\right). \quad (4.61)$$

The eigenvalues $\omega(\gamma) = j(i\nu)-1$ at $\gamma = \frac{1}{2}+i\nu$ correspond to complex values of Δ and require analytical continuation of the harmonic sums. Corresponding prescriptions can be found in [Kotikov and Velizhanin (2005); Kazakov and Kotikov (1988); Lopez and Yndurain (1981); Blumlein (2009)].

The approach used in [Velizhanin (2015)] was also different from the BFKL approach. The NNLO corrections to the Pomeron eigenvalues for conformal spin $n = 0$ were obtained from the constraints, coming from the six-loop anomalous dimension of twist-2 operators and large-gamma limit. The results agree with [Gromov *et al.* (2015b)].

The calculations of the three-loop corrections in [Caron-Huot and Herranen (2018)] is closer to the BFKL approach since the corrections are calculated firstly to the BFKL kernel (in the Pomeron channel). But the method of the calculation is quite different from the dispersive method used in the BFKL approach. Moreover, the kernel is calculated not in the momentum representation, but in the impact parameter space. The method used is based on established in [Caron-Huot (2018)] connection of the BFKL logarithms with so-called non-global logarithms [Dasgupta and Salam (2001)], [Banfi *et al.* (2002)] the physics of soft wide-angle radiation.

In [Alfimov *et al.* (2018)] a general non-perturbative framework for the BFKL spectrum of planar N = 4 SYM was developed. As well as in [Gromov *et al.* (2015b)], it is based on the QSC method, but the spectrum was studied at arbitrary values of conformal spin n.

4.4.2 *NNLLA in the BFKL approach*

The results described above were obtained in $N = 4$ SUSY model with unique properties and concern only the BFKL kernel. Recall that amplitudes are given by convolution of impact factors of scattering particles and the Green's function, which is determined by the kernel, so that to calculate the amplitudes, it is necessary to know not only the kernel, but also the impact factors. The BFKL approach gives an algorithm for calculating both components. Besides this, the amplitudes of both elastic and inelastic processes are calculated in the original (dispersive) BFKL approach. Therefore, its development in the NNLLA is very desirable. And finally, it is highly desirable to develop this approach in QCD. Unfortunately, this development is facing great difficulties.

Remind that original scheme of derivation of the BFKL equation looks as follows. Elastic scattering amplitudes are calculated using the s-channel unitarity and analyticity. In the LLA and in the NLLA, only amplitudes having the pole Regge form contribute to unitarity relations. It permits to present the s-channel discontinuities of elastic processes as the convolutions of impact factors of scattering particles and the Green's function for two interacting Reggeized gluons. All these components are expressed in terms of the Reggeon vertices and trajectory.

If this scheme were applicable, for derivation of the BFKL equation in the NNLLA would be sufficient to calculate three-loop corrections to the trajectory, two-loop corrections to the vertex of one gluon production, one-loop corrections to the vertices of two gluon and quark-antiquark production and to find in the Born approximation new vertices: for three gluon and quark-antiquark-gluon production.

However, this scheme is based on the pole Regge form of amplitudes which are used in the unitarity relations (4.4). It can not be used the NNLLA. In this approximation two large logarithms can be lost in the product of two amplitudes in the unitarity relations (4.4). It can be done losing either one logarithm in each of the amplitudes or both logarithms in one of the amplitudes.

In the first case it becomes necessary to take into account the imaginary parts of amplitudes with negative signature as well as amplitudes with positive signature, which both are suppressed by one degree of the logarithm. Note that the simple factorized form (4.20) is valid only for the real part of the amplitudes and must be corrected to account for the imaginary parts, which creates certain difficulties. But more difficulties are connected with the second case. In this case one of the amplitudes in (4.4) must be taken in the LLA and the other in the NNLLA. Since the amplitudes in the LLA are real, only real parts of the NNLLA amplitudes are important in this case. But even for these parts the pole Regge form becomes inapplicable because of the contributions of the three-Reggeon cuts which appear in this approximation.

The first observation of the violation the pole Regge form was done [Del Duca and Glover (2001)] in the high-energy limit of the results of direct two-loop calculations of the two-loop amplitudes for gg, gq and qq scattering. Then, the terms breaking the pole Regge form in two and three-loop amplitudes of elastic scattering were found in [Del Duca *et al.* (2014, 2013, 2015)] using the techniques of infrared factorization.

It is necessary to say that, in general, breaking of the pole Regge form is not a surprise. It is well known that Regge poles in the complex angular momenta plane generate Regge cuts. Moreover, in amplitudes with positive signature the Regge cuts appear already in the LLA. In particular, the BFKL Pomeron is the two-Reggeon cut in the complex angular momentum plane. But in amplitudes with negative signature Regge cuts appear only in the NNLLA. It is natural to expect that the observed violation of the pole Regge form can be explained by their contributions.

Indeed, all known cases of breaking of the pole Regge form are now explained by the three-reggeon cuts [Fadin (2017); Caron-Huot *et al.* (2017)]. Unfortunately, the approaches and the explanations used in these papers are different. Their results coincide in three loops but may diverge in more loops. It requires further investigation.

Consideration of three-reggeon cuts in many-particle amplitudes is an even more complicated problem.

4.4.3 *Conclusion*

L.N. Lipatov stood at the origins of the BFKL approach and played a prominent role in its development. He left a huge legacy to his disciples and followers. Unfortunately, now they are deprived of his highest skill, and the depth and originality of his ideas.

Bibliography

Abers, E., Keller, R., and Teplitz, V. (1970). Reggeization of the spinor in yang-mills theory, *Phys. Rev. D* **2**, pp. 1757–1759, doi:10.1103/PhysRevD.2.1757.

Alfimov, M., Gromov, N., and Kazakov, V. (2015). QCD Pomeron from AdS/CFT quantum spectral curve, *JHEP* **07**, p. 164, doi:10.1007/JHEP07(2015)164, arXiv:1408.2530 [hep-th].

Alfimov, M., Gromov, N., and Sizov, G. (2018). BFKL spectrum of $\mathcal{N} = 4$: Non-zero conformal spin, *JHEP* **07**, p. 181, doi:10.1007/JHEP07(2018)181, arXiv:1802.06908 [hep-th].

Balitsky, I. and Lipatov, L. (1978). The Pomeranchuk singularity in quantum chromodynamics, *Sov. J. Nucl. Phys.* **28**, pp. 822–829.

Balitsky, I., Lipatov, L., and Fadin, V. S. (1979). *Regge Processes In Nonabelian Gauge Theories* (Materials of,IV Winter School of LNPI, pp. 109–149, Leningrad).

Banfi, A., Marchesini, G., and Smye, G. (2002). Away from jet energy flow, *JHEP* **08**, p. 006, doi:10.1088/1126-6708/2002/08/006, arXiv:hep-ph/0206076.

Bartels, J. (1975a). A Reggeon calculus for the production amplitude. 1. *Phys. Rev. D* **11**, 10, p. 2977, doi:10.1103/PhysRevD.11.2977.

Bartels, J. (1975b). A reggeon calculus for the production amplitude II, *Phys. Rev. D* **11**, 10, p. 2989, doi:10.1103/PhysRevD.11.2989.

Bartels, J. (1980). High-energy behavior in a nonabelian gauge theory (II): First corrections to $T_{n\to m}$ beyond the leading $\ln s$ approximation, *Nucl. Phys. B* **175**, pp. 365–401, doi:10.1016/0550-3213(80)90019-X.

Bartels, J., Lipatov, L., and Vacca, G. (2005). Interactions of reggeized gluons in the Mobius representation, *Nucl. Phys. B* **706**, pp. 391–410, doi:10.1016/j.nuclphysb.2004.10.008, arXiv:hep-ph/0404110.

Blumlein, J. (2009). Structural relations of harmonic sums and mellin transforms up to weight w = 5, *Comput. Phys. Commun.* **180**, pp. 2218–2249, doi:10.1016/j.cpc.2009.07.004, arXiv:0901.3106 [hep-ph].

Caron-Huot, S. (2018). Resummation of non-global logarithms and the BFKL equation, *JHEP* **03**, p. 036, doi:10.1007/JHEP03(2018)036, arXiv:1501.03754 [hep-ph].

Caron-Huot, S., Gardi, E., and Vernazza, L. (2017). Two-parton scattering in the high-energy limit, *JHEP* **06**, p. 016, doi:10.1007/JHEP06(2017)016, arXiv:1701.05241 [hep-ph].

Caron-Huot, S. and Herranen, M. (2018). High-energy evolution to three loops, *JHEP* **02**, p. 058, doi:10.1007/JHEP02(2018)058, arXiv:1604.07417 [hep-ph].

Cheng, H. and Lo, C. C. (1975). Elementary scalar mesons and regge poles, *Phys. Lett. B* **57**, pp. 177–180, doi:10.1016/0370-2693(75)90018-0.

Cheng, H. and Wu, T. T. (1970). Logarithmic factors in the high-energy behavior of quantum electrodynamics, *Phys. Rev. D* **1**, pp. 2775–2794, doi:10.1103/PhysRevD.1.2775.

Chester, A. N. (1965). Non-Reggeization of the vector meson, *Phys. Rev. B.* **140**, p. 85.

Chew, G. and Frautschi, S. C. (1961). Principle of equivalence for all strongly interacting particles within the S matrix framework, *Phys. Rev. Lett.* **7**, pp. 394–397, doi:10.1103/PhysRevLett.7.394.

Ciafaloni, M. and Camici, G. (1998). Energy scale(s) and next-to-leading BFKL equation, *Phys. Lett. B* **430**, pp. 349–354, doi:10.1016/S0370-2693(98)00551-6, arXiv:hep-ph/9803389.

Costa, M. S., Goncalves, V., and Penedones, J. (2012). Conformal Regge theory, *JHEP* **12**, p. 091, doi:10.1007/JHEP12(2012)091, arXiv:1209.4355 [hep-th].

Dasgupta, M. and Salam, G. (2001). Resummation of nonglobal QCD observables, *Phys. Lett. B* **512**, pp. 323–330, doi:10.1016/S0370-2693(01)00725-0, arXiv:hep-ph/0104277.

Del Duca, V., Falcioni, G., Magnea, L., and Vernazza, L. (2013). Beyond Reggeization for two- and three-loop QCD amplitudes, *PoS* **RADCOR2013**, p. 046, doi:10.22323/1.197.0046, arXiv:1312.5098 [hep-ph].

Del Duca, V., Falcioni, G., Magnea, L., and Vernazza, L. (2014). High-energy QCD amplitudes at two loops and beyond, *Phys. Lett. B* **732**, pp. 233–240, doi:10.1016/j.physletb.2014.03.033, arXiv:1311.0304 [hep-ph].

Del Duca, V., Falcioni, G., Magnea, L., and Vernazza, L. (2015). Analyzing high-energy factorization beyond next-to-leading logarithmic accuracy, *JHEP* **02**, p. 029, doi:10.1007/JHEP02(2015)029, arXiv:1409.8330 [hep-ph].

Del Duca, V. and Glover, E. (2001). The High-energy limit of QCD at two loops, *JHEP* **10**, p. 035, doi:10.1088/1126-6708/2001/10/035, arXiv:hep-ph/0109028.

Dicus, D., Freedman, D., and Teplitz, V. (1971). Regge pole content of chiral and gauge field theories, *Phys. Rev. D* **4**, pp. 2320–2330, doi:10.1103/PhysRevD.4.2320.

Englert, F. and Brout, R. (1964). Broken symmetry and the mass of gauge vector mesons, *Phys. Rev. Lett.* **13**, pp. 321–323, doi:10.1103/PhysRevLett.13.321.

Fadin, V. (1995a). The gluon trajectory in the next to leading approximation, in *10th Workshop on Photon-photon Collisions (PHOTON '95)*, pp. 181–186.

Fadin, V. (2017). Particularities of the NNLLA BFKL, *AIP Conf. Proc.* **1819**, 1, p. 060003, doi:10.1063/1.4977159, arXiv:1612.04481 [hep-ph].

Fadin, V. and Fiore, R. (2008). The dipole form of the BFKL kernel in supersymmetric Yang-Mills theories, *Phys. Lett. B* **661**, pp. 139–144, doi:10.1016/j.physletb.2008.01.046, arXiv:0712.3901 [hep-ph].

Fadin, V., Fiore, R., and Grabovsky, A. (2009). On the discrepancy of the low-x evolution kernels, *Nucl. Phys. B* **820**, pp. 334–363, doi:10.1016/j.nuclphysb.2009.05.025, arXiv:0904.0702 [hep-ph].

Fadin, V., Fiore, R., and Grabovsky, A. (2010). Matching of the low-x evolution kernels, *Nucl. Phys. B* **831**, pp. 248–261, doi:10.1016/j.nuclphysb.2010.01.017, arXiv:0911.5617 [hep-ph].

Fadin, V., Fiore, R., Grabovsky, A., and Papa, A. (2007a). The dipole form of the gluon part of the BFKL kernel, *Nucl. Phys. B* **784**, pp. 49–71, doi:10.1016/j.nuclphysb.2007.06.002, arXiv:0705.1885 [hep-ph].

Fadin, V., Fiore, R., Kozlov, M., and Reznichenko, A. (2006). Proof of the multi-Regge form of QCD amplitudes with gluon exchanges in the NLA, *Phys. Lett. B* **639**, pp. 74–81, doi:10.1016/j.physletb.2006.03.031, arXiv:hep-ph/0602006.

Fadin, V., Kotsky, M., and Lipatov, L. (1997). One-loop correction to the BFKL kernel from two gluon production, *Phys. Lett. B* **415**, pp. 97–103, doi:10.1016/S0370-2693(97)01210-0.

Fadin, V., Kozlov, M., and Reznichenko, A. (2015). Gluon reggeization in Yang-Mills theories, *Phys. Rev. D* **92**, 8, p. 085044, doi:10.1103/PhysRevD.92.085044, arXiv:1507.00823 [hep-th].

Fadin, V. S. (1995b). Regge trajectory of a gluon in the two loop approximation, *JETP Lett.* **61**, pp. 346–350.

Fadin, V. S. (1995c). Two loop correction to the gluon trajectory in QCD, *Phys. Atom. Nucl.* **58**, pp. 1762–1766.

Fadin, V. S. and Fiore, R. (1992). Quark contribution to the gluon-gluon — reggeon vertex in QCD, *Phys. Lett. B* **294**, pp. 286–292, doi:10.1016/0370-2693(92)90696-2.

Fadin, V. S., Fiore, R., Flachi, A., and Kotsky, M. (1998). Quark - anti-quark contribution to the BFKL kernel, *Phys. Lett. B* **422**, pp. 287–293, doi:10.1016/S0370-2693(98)00044-6, arXiv:hep-ph/9711427.

Fadin, V. S., Fiore, R., and Kotsky, M. (1996a). Gluon Regge trajectory in the two loop approximation, *Phys. Lett. B* **387**, pp. 593–602, doi:10.1016/0370-2693(96)01054-4, arXiv:hep-ph/9605357.

Fadin, V. S., Fiore, R., and Papa, A. (2007b). On the coordinate representation of NLO BFKL, *Nucl. Phys. B* **769**, pp. 108–123, doi:10.1016/j.nuclphysb.2007.01.030, arXiv:hep-ph/0612284.

Fadin, V. S., Fiore, R., and Papa, A. (2007c). The dipole form of the quark part of the BFKL kernel, *Phys. Lett. B* **647**, pp. 179–184, doi:10.1016/j.physletb.2007.02.022, arXiv:hep-ph/0701075.

Fadin, V. S., Fiore, R., and Quartarolo, A. (1994a). Quark contribution to the reggeon — reggeon — gluon vertex in QCD, *Phys. Rev. D* **50**, pp. 5893–5901, doi:10.1103/PhysRevD.50.5893, arXiv:hep-th/9405127.

Fadin, V. S., Fiore, R., and Quartarolo, A. (1994b). Radiative corrections to quark quark reggeon vertex in QCD, *Phys. Rev. D* **50**, pp. 2265–2276, doi:10.1103/PhysRevD.50.2265, arXiv:hep-ph/9310252.

Fadin, V. S., Fiore, R., and Quartarolo, A. (1996b). Reggeization of quark quark scattering amplitude in QCD, *Phys. Rev. D* **53**, pp. 2729–2741, doi:10.1103/PhysRevD.53.2729, arXiv:hep-ph/9506432.

Fadin, V. S., Kotsky, M., and Fiore, R. (1995). Gluon reggeization in QCD in the next-to-leading order, *Phys. Lett. B* **359**, pp. 181–188, doi:10.1016/0370-2693(95)01016-J.

Fadin, V. S., Kotsky, M., and Lipatov, L. (1996c). Gluon pair production in the quasimulti — Regge kinematics, arXiv:hep-ph/9704267.

Fadin, V. S., Kuraev, E., and Lipatov, L. (1975). On the Pomeranchuk singularity in asymptotically free theories, *Phys. Lett. B* **60**, pp. 50–52, doi:10.1016/0370-2693(75)90524-9.

Fadin, V. S. and Lipatov, L. (1989). High-energy production of gluons in a quasimultiregge kinematics, *JETP Lett.* **49**, p. 352.

Fadin, V. S. and Lipatov, L. (1993). Radiative corrections to QCD scattering amplitudes in a multi — Regge kinematics, *Nucl. Phys. B* **406**, pp. 259–292, doi:10.1016/0550-3213(93)90168-O.

Fadin, V. S. and Lipatov, L. (1996). Next-to-leading corrections to the BFKL equation from the gluon and quark production, *Nucl. Phys. B* **477**, pp. 767–808, doi:10.1016/0550-3213(96)00334-3, arXiv:hep-ph/9602287.

Fadin, V. S. and Lipatov, L. (1998). BFKL pomeron in the next-to-leading approximation, *Phys. Lett. B* **429**, pp. 127–134, doi:10.1016/S0370-2693(98)00473-0, arXiv:hep-ph/9802290.

Froissart, M. (1961). Asymptotic behavior and subtractions in the Mandelstam representation, *Phys. Rev.* **123**, pp. 1053–1057, doi:10.1103/PhysRev.123.1053.

Frolov, G., Gribov, V., and Lipatov, L. (1970). On Regge poles in quantum electrodynamics, *Phys. Lett. B* **31**, pp. 34–35, doi:10.1016/0370-2693(70)90013-4.

Gell-Mann, M. and Goldberger, M. (1962). Elementary particles of conventional field theory as Regge poles, *Phys. Rev. Lett.* **9**, 6, pp. 275–277, doi:10.1103/PhysRevLett.9.275.

Gell-Mann, M., Goldberger, M., Low, F., Marx, E., and Zachariasen, F. (1964a). Elementary particles of conventional field theory as Regge poles. III, *Phys. Rev.* **133**, 1B, pp. B145–B160, doi:10.1103/PhysRev.133.B145.

Gell-Mann, M., Goldberger, M., Low, F., Singh, V., and Zachariasen, F. (1964b). Elementary particles of conventional field theory as Regge poles. IV, *Phys. Rev.* **133**, pp. B161–B174, doi:10.1103/PhysRev.133.B161.

Gribov, V. (1962). Partial waves with complex orbital angular momenta and the asymptotic behavior of the scattering amplitude, *Sov. Phys. JETP* **14**, p. 1395.

Gribov, V., Lipatov, L., and Frolov, G. (1971). The leading singularity in the j plane in quantum electrodynamics, *Sov. J. Nucl. Phys.* **12**, p. 543.

Grisaru, M. T. (1977). Reggeization of scalars and dual model structure of Yang-Mills theory, *Phys. Rev. D* **16**, p. 1962, doi:10.1103/PhysRevD.16.1962.

Grisaru, M. T., Schnitzer, H., and Tsao, H.-S. (1973a). Reggeization of elementary particles in renormalizable gauge theories — vectors and spinors, *Phys. Rev. D* **8**, pp. 4498–4509, doi:10.1103/PhysRevD.8.4498.

Grisaru, M. T., Schnitzer, H., and Tsao, H.-S. (1973b). Reggeization of yang-mills gauge mesons in theories with a spontaneously broken symmetry, *Phys. Rev. Lett.* **30**, pp. 811–814, doi:10.1103/PhysRevLett.30.811.

Gromov, N., Kazakov, V., Leurent, S., and Volin, D. (2014). Quantum spectral curve for planar $\mathcal{N} = 4$ super-yang-mills theory, *Phys. Rev. Lett.* **112**, 1, p. 011602, doi:10.1103/PhysRevLett.112.011602, arXiv:1305.1939 [hep-th].

Gromov, N., Kazakov, V., Leurent, S., and Volin, D. (2015a). Quantum spectral curve for arbitrary state/operator in AdS_5/CFT_4, *JHEP* **09**, p. 187, doi:10.1007/JHEP09(2015)187, arXiv:1405.4857 [hep-th].

Gromov, N., Levkovich-Maslyuk, F., and Sizov, G. (2015b). Pomeron eigenvalue at three loops in $\mathcal{N} = 4$ supersymmetric yang-mills theory, *Phys. Rev. Lett.* **115**, 25, p. 251601, doi:10.1103/PhysRevLett.115.251601, arXiv:1507.04010 [hep-th].

Higgs, P. W. (1966). Spontaneous symmetry breakdown without massless bosons, *Phys. Rev.* **145**, pp. 1156–1163, doi:10.1103/PhysRev.145.1156.

Jaroszewicz, T. (1982). Gluonic Regge singularities and anomalous dimensions in QCD, *Phys. Lett. B* **116**, pp. 291–294, doi:10.1016/0370-2693(82)90345-8.

Kazakov, D. and Kotikov, A. (1988). Total $\alpha^{-}s$ Correction to deep inelastic scattering cross-section ratio, R = σ^{-}l / σ^{-}t in QCD. Calculation of longitudinal structure function, *Nucl. Phys. B* **307**, p. 721, doi:10.1016/0550-3213(88)90106-X, [Erratum: Nucl.Phys.B 345, 299 (1990)].

Kibble, T. (1967). Symmetry breaking in nonAbelian gauge theories, *Phys. Rev.* **155**, pp. 1554–1561, doi:10.1103/PhysRev.155.1554.

Kotikov, A. and Lipatov, L. (2000). NLO corrections to the BFKL equation in QCD and in supersymmetric gauge theories, *Nucl. Phys. B* **582**, pp. 19–43, doi:10.1016/S0550-3213(00)00329-1, arXiv:hep-ph/0004008.

Kotikov, A., Lipatov, L., Onishchenko, A., and Velizhanin, V. (2004). Three loop universal anomalous dimension of the Wilson operators in $N = 4$ SUSY Yang-Mills model, *Phys. Lett. B* **595**, pp. 521–529, doi:10.1016/j.physletb.2004.05.078, arXiv:hep-th/0404092, [Erratum: Phys.Lett.B 632, 754–756 (2006)].

Kotikov, A. and Velizhanin, V. (2005). Analytic continuation of the Mellin moments of deep inelastic structure functions, arXiv:hep-ph/0501274.

Kotsky, M. and Fadin, V. S. (1996). Reggeization of the amplitude of gluon-gluon scattering, *Phys. Atom. Nucl.* **59**, pp. 1035–1045.

Kozlov, M., Reznichenko, A., and Fadin, V. (2011). Check of the gluon-reggeization condition in the next-to-leading order: Quark part, *Phys. Atom. Nucl.* **74**, pp. 758–770, doi:10.1134/S1063778811050152.

Kozlov, M., Reznichenko, A., and Fadin, V. (2012a). Check of the gluon-Reggeization condition in the next-to-leading order: Gluon part, *Phys. Atom. Nucl.* **75**, pp. 493–506, doi:10.1134/S1063778812030106.

Kozlov, M., Reznichenko, A., and Fadin, V. (2012b). Impact factor for gluon production in multi-Regge kinematics in the next-to-leading order, *Phys. Atom. Nucl.* **75**, pp. 850–865, doi:10.1134/S1063778812040072.

Kuraev, E., Lipatov, L., and Fadin, V. S. (1977). The Pomeranchuk singularity in nonabelian gauge theories, *Sov. Phys. JETP* **45**, pp. 199–204.

Kuraev, E. A., Lipatov, L. N., and Fadin, V. S. (1976). Multi — Reggeon processes in the yang-mills theory, *Sov. Phys. JETP* **44**, pp. 443–450.

Kwiecinski, J. and Praszalowicz, M. (1980). Three gluon integral equation and odd c singlet Regge singularities in QCD, *Phys. Lett. B* **94**, pp. 413–416, doi:10.1016/0370-2693(80)90909-0.

Lipatov, L. (1976). Reggeization of the vector meson and the vacuum singularity in nonabelian gauge theories, *Sov. J. Nucl. Phys.* **23**, pp. 338–345.

Lipatov, L. (1986). The bare Pomeron in quantum chromodynamics, *Sov. Phys. JETP* **63**, pp. 904–912.

Lipatov, L. (1990). Pomeron and odderon in QCD and a two-dimensional conformal field theory, *Phys. Lett. B* **251**, pp. 284–287, doi:10.1016/0370-2693(90)90937-2.

Lipatov, L. (1993). High-energy asymptotics of multicolor QCD and two-dimensional conformal field theories, *Phys. Lett. B* **309**, pp. 394–396, doi:10.1016/0370-2693(93)90951-D.

Lopez, C. and Yndurain, F. (1981). Behavior at $x = 0, 1$, sum rules and parametrizations for structure functions beyond the leading order, *Nucl. Phys. B* **183**, pp. 157–181, doi:10.1016/0550-3213(81)90551-4.

Mandelstam, S. (1964). Non-regge terms in the vector channelof the vector-spinor theory, in *12th International Conference on High Energy Physics*, Vol. 1, pp. 376–377.

Mandelstam, S. (1965). Non-Regge terms in the vector-spinor theory, *Phys. Rev.* **137**, pp. B949–B954, doi:10.1103/PhysRev.137.B949.

Mueller, A. H. (1994). Soft gluons in the infinite momentum wave function and the BFKL pomeron, *Nucl. Phys. B* **415**, pp. 373–385, doi:10.1016/0550-3213(94)90116-3.

Mueller, A. H. and Patel, B. (1994). Single and double BFKL pomeron exchange and a dipole picture of high-energy hard processes, *Nucl. Phys. B* **425**, pp. 471–488, doi:10.1016/0550-3213(94)90284-4, arXiv:hep-ph/9403256.

Nikolaev, N. N. and Zakharov, B. (1994). The Triple pomeron regime and the structure function of the pomeron in the diffractive deep inelastic scattering at very small x, *Z. Phys. C* **64**, pp. 631–652, doi:10.1007/BF01957772, arXiv:hep-ph/9306230.

Nikolaev, N. N., Zakharov, B., and Zoller, V. (1994). The s channel approach to Lipatov's pomeron and hadronic cross-sections, *JETP Lett.* **59**, pp. 6–12, arXiv:hep-ph/9312268.

Velizhanin, V. (2015). BFKL pomeron in the next-to-next-to-leading approximation in the planar N=4 SYM theory, arXiv:1508.02857 [hep-th].

Vogt, A., Moch, S., and Vermaseren, J. (2004). The three-loop splitting functions in QCD: The singlet case, *Nucl. Phys. B* **691**, pp. 129–181, doi:10.1016/j.nuclphysb.2004.04.024, arXiv:hep-ph/0404111.

Yang, C.-N. and Mills, R. L. (1954). Conservation of isotopic spin and isotopic gauge invariance, *Phys. Rev.* **96**, pp. 191–195, doi:10.1103/PhysRev.96.191.

https://doi.org/10.1142/9789811231124_0005

Chapter 5

Basics of DGLAP

Yuri Dokshitzer

CNRS, LPTHE, University Pierre et Marie Curie,
UMR 7589, Paris, France
PNPI Kurchatov Institute, St. Petersburg,
Gatchina, Russia

Abstract

We discuss QCD parton picture putting emphasis on the link between space- and time-like evolution.

Although this may seem a paradox, all exact science
is dominated by the idea of approximation.
Bertrand Russell

5.1 Non-scientific preface

Appearance may be deceiving. Lev Lipatov, mathematical physics guru and one of the most knowledgeable and creative theorists of our times, was once considered not a very promising postdoc whose perspectives as a permanent member of the Gribov theory department were not certain.

Gribov was presenting his study of deep inelastic scattering at an internal group seminar when Lipatov saw a flaw in his arguments and attacked. Speaker got stuck; the seminar stalled.

— Lev, you are already the co-author. Why don't you help me out?
Lev's future got sealed.

"Evolution" is the second Lipatov's name. He was the first to combine the words "partons" and "QFT" in a single context when QCD was barely around [Lipatov (1974)].

5.2 Partons and quantum field theory

We will discuss basic physics of QCD quark–gluon multiplication processes, emphasising relations between, and peculiarities of, space- and time-like parton cascades, as well as possibility of a fruitful cooperation between QCD and maximally super-symmetric $\mathcal{N}=4$ Yang–Mills theory.

5.2.1 *Puzzle of DIS*

Deep inelastic lepton-hadron scattering (DIS) is a classical example of a hard process that has become the first laboratory for perturbative QCD.

Here, the momentum q with a large negative virtuality q^2 ($Q^2 = |q^2|$) is transferred from an incident electron (muon, neutrino) to the target proton breaking up into the final multihadron system. Introducing an invariant energy $s = 2(Pq)$ between the exchange photon (Z^0, $W^{\pm}$) q and the proton with 4-momentum P ($P^2 = M_p^2$), one writes the invariant mass of the produced hadron system that measures the inelasticity of the process as

$$W^2 \equiv (q+P)^2 - M_p^2 = q^2 + 2(Pq) = s(1-x), \quad x \equiv \frac{Q^2}{2(Pq)} \leq 1.$$

The cross section of the process depends on two variables: the *hardness* Q^2 and Bjorken x. For the case of *elastic* lepton-proton scattering one has $x \equiv 1$ and it is natural to write the cross section as

$$\frac{d\sigma_{el}}{dQ^2[dx]} = \frac{d\sigma_{\text{Ruth}}}{dQ^2} \cdot f_{el}^2(Q^2) \cdot [\delta(1-x)]. \tag{5.1a}$$

Here $\sigma_{\text{Ruth}} \propto \alpha^2/Q^4$ is the standard Rutherford cross section for e.m. scattering off a point charge and f_{el} stands for the elastic proton form factor.

For inclusive *inelastic* cross section one can write an analogous expression by introducing an inelastic proton "form factor" which now depends on both the momentum transfer Q^2 and the inelasticity parameter x:

$$\frac{d\sigma_{in}}{dQ^2\,dx} = \frac{d\sigma_{\mathrm{Ruth}}}{dQ^2} \cdot f_{in}^2(x, Q^2). \tag{5.1b}$$

What kind of Q^2-behavior of the form factors (5.1) could one expect in the Bjorken limit $Q^2 \to \infty$? Non-relativistic quantum mechanics taught us how the Q^2-behaviour of the electromagnetic form factor could be related to the charge distribution inside a proton:

$$f_{el}(Q^2) = \int d^3r \, \rho(\vec{r}) \exp\left\{ i\vec{Q}\vec{r} \right\}.$$

For a point charge $\rho(\vec{r}) = \delta^3(\vec{r})$, it is obvious that $f \equiv 1$. On the contrary, for a smooth charge distribution $f(Q^2)$ falls with increasing Q^2, the faster the smoother ρ is. Experimentally, the elastic *e-p* cross section does decrease with Q^2 *much faster* that the Rutherford one ($f_{el}(Q^2)$ decays as a large power of Q^2).

Does this imply that $\rho(\vec{r})$ is indeed regular so that there is no well-localised, point charge inside a proton? If this were the case, the *inelastic* form factor would decay as well in the Bjorken limit: a tiny photon with the characteristic size $\sim 1/Q \to 0$ would penetrate through a "smooth" proton like a knife through butter, inducing neither elastic nor inelastic interactions.

However, as was first observed at SLAC in the late sixties, for a fixed x, f_{in}^2 stays practically constant with Q^2. That is, the inelastic cross section (with a given inelasticity) is similar to the Rutherford cross section (Bjorken scaling). It looks *as if* there was a point-like scattering in the guts of it, but in a rather strange way: it results in inelastic break-up dominating over the elastic channel.

Quite a paradoxical picture emerged; Feynman-Bjorken partons came to the rescue.

Imagine that it is not the proton itself that is a point-charge-bearer, but some other guys (quark-partons) inside it. If those constituents were *tightly* bound to one another, the elastic channel would be bigger than, or comparable with, the inelastic one. An excitation of the parton that takes an impact would be transferred, with the help of rigid links between partons, to the proton as a whole, leading to the elastic scattering or to the formation of a quasi-elastic finite-mass system ($N\pi$, $\Delta\pi$ or so), $1 - x \ll 1$.

To match the experimental pattern, $f_{el}^2(Q^2) \ll f_{in}^2(Q^2)$, one has instead to view the parton ensemble as a *loosely* bound system of quasi-free particles. Only under such circumstances does knocking off one of the partons inside the proton inevitably lead to deep inelastic breakup of the latter, with a negligible chance of reshuffling the excitation among the struck parton and the rest of the target.

Within this picture the inelastic scattering off the proton actually proceeds via elastic lepton scattering off the quasi-real quark-parton k with the longitudinal momentum fraction x:

$$k \approx \beta P, \quad (k+q)^2 \simeq 2kq + q^2 = 2Pq\left(\beta - \frac{-q^2}{2Pq}\right) \simeq 0 \implies \beta = x.$$

The parton model, forged to explain the DIS phenomenon, was intrinsically paradoxical. In the sixties and seventies of the past century there was no other way of discussing particle interactions but within the field-theoretical framework. But all renormalisable, 4-dimensional quantum field theories (QFTs) known by then had one feature in common: an effective interaction strength (the running coupling $g^2(Q^2)$) *increasing* with the scale of the hard process Q^2. Actually, this feature was widely believed to be a general law of nature. At the same time, in order to be in accord with the parton model, we need to have it the other way around: we would rather have parton–parton interaction *weakened* at small distances (large Q^2).

Only with the advent of non-Abelian QFTs (and QCD in particular) that exhibit an anti-intuitive asymptotic-freedom behaviour of the coupling, the concept of partons was to become more than a mere phenomenological model.

5.2.2 *Violation of Bjorken scaling in QFT*

It was soon realised that the Bjorken scaling — hypothesis that the inelastic form factor $f_{in}(x, Q^2)$ in (5.1b) tends to a finite distribution $f_{in}(x)$ in the $Q^2 \to \infty$ limit — cannot hold in QFT.

The point is, particle virtualities/transverse momenta in QFT are not limited, as the orthodox parton model once assumed. In a DIS process transverse momenta of *partons* — quarks and gluons — are broadly distributed,

$$dw \propto \frac{\alpha_s}{\pi}\frac{dk_\perp^2}{k_\perp^2}, \quad \text{up to} \quad k_\perp^2 \sim Q^2.$$

As a result, the number of particles that enter the game turns out to be large in spite of the smallness of the coupling:

$$\int dw \propto \int^{Q^2} \frac{\alpha_s}{\pi}\frac{dk_\perp^2}{k_\perp^2} \sim \frac{\alpha_s}{\pi}\ln Q^2 = \mathcal{O}(1).$$

Such *collinear* enhancement is typical for theories with dimensionless coupling — so called *logarithmic* QFTs.

In other words, a QFT particle is not point-like. It is surrounded by its proper field coat — a coherent virtual cloud consisting of gluons and "sea" $q\bar{q}$ pairs. Its visible content depends on the *resolution power* of the probe $\lambda = 1/Q = 1/\sqrt{-q^2}$.

DIS. A hard probe applied to such a *dressed* parton breaks coherence of the cloud. Constituents of these field fluctuations are then released as particles accompanying the hard interaction. The harder the hit, the larger an intensity of bremsstrahlung and, therefore, the fraction of the energy-momentum of the dressed parton that the bremsstrahlung quanta typically carry away.

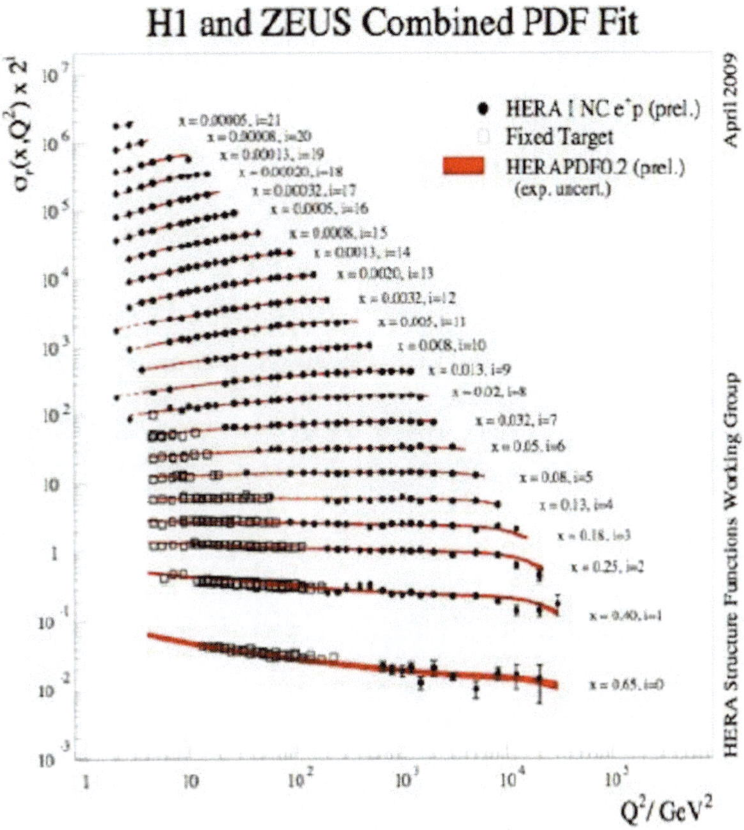

Fig. 5.1 HERA data on DIS spanning five orders of magnitude in Q^2 and x.

Thus we should expect, in particular, that the probability that a "bare" core quark carries a large fraction of the energy of its dressed parent will decrease with increase of Q^2. At the same time, the number of quark–partons with small momentum fraction x that may take an impact should grow with Q^2. Precise HERA measurements shown in Fig. 5.1 do support these expectations [Altarelli (2009)].

Jets. In DIS we look for a "bare" quark inside a target dressed one. In e^+e^- hadron annihilation at large energy $s = Q^2$ the chain of events is reversed. Here we produce instead a bare quark with energy $Q/2$, which then "dresses up". In the process of restoring its proper field-coat our parton produces a controllable amount of bremsstrahlung radiation which leads to formation of a hadron jet. Having done so, in the end of the day it becomes a constituent of one of the hadrons that hit the detector. Typically, this is the leading hadron. However, the fraction x_F of the initial energy $Q/2$ that is left to the leader depends on the amount of accompanying radiation and, therefore, on Q^2 (the larger, the smaller).

What makes the annihilation channel particularly interesting, is that the CERN LEP experiments have reached such level of sophistication as to provide us with a

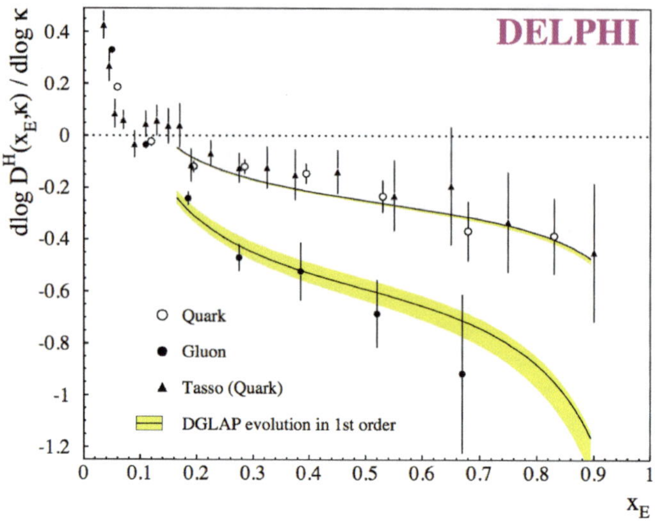

Fig. 5.2 Scaling violation rates in inclusive hadron distributions from gluon and quark jets [Abreu et (2000)].

near-to-perfect separation between quark- and gluon-initiated jets (the latter being extracted from heavy-quark-tagged three-jet events).

In Fig. 5.2 a comparison is shown of scaling violation rates in the hadron spectra from gluon and quark jets, as a function of the hardness scale κ that characterizes a given jet [Abreu et (2000)]. For large values of $x_F \simeq 1$ the ratio of the logarithmic derivatives is predicted to be close to that of the gluon and quark "colour charges", $C_A/C_F = 9/4$. Experimentally, the ratio was measured to be

$$\frac{C_A}{C_F} = 2.23 \pm 0.09_{\text{stat.}} \pm 0.06_{\text{syst.}}. \tag{5.2}$$

In fact, the same rules (and the same formulae) apply to the scaling violation pattern in e^+e^- fragmentation functions (time-like parton evolution) as to that in the DIS parton distributions (space-like evolution). The logarithmic scaling violation pattern in DIS and e^+e^- is well established experimentally and meticulously follows the QCD predictions based on the parton evolution picture.

The first systematic study of DIS and e^+e^- annihilation in a QFT framework was undertaken by V. Gribov and L. Lipatov [Gribov and Lipatov (1972)]. They have studied hard scattering and annihilation processes in two renormalisable QFT models: 1) that of fermions coupled with a pseudoscalar field and 2) a QED-like model of fermions interacting with a vector field. Aiming at an all-order perturbative analysis, Gribov and Lipatov have developed the Leading Logarithmic Approximation (LLA) in which one insists on picking up, in each order of perturbation theory, a logarithmic enhancement factor $g^2 \to g^2 \log Q^2$, and came up with all-order-resummed expressions for DIS structure functions and e^+e^- fragmentation functions.

Two years later Lev Lipatov has presented their results explicitly in terms of parton distributions with a variable ultraviolet cutoff. Corresponding *evolution equations* described a probabilistic Markov chain of independent $1 \to 2$ parton splitting processes [Lipatov (1974)].

Gribov–Lipatov findings for DIS were reformulated in terms of the Operator Product Expansion (OPE) in [Christ, Hasslacher and Mueller (1972)]. With advent of QCD, it is within the OPE framework that the first realistic analysis of DIS structure functions has been performed [Gross and Wilczek (1973)].

The QCD parton picture, extending that of Gribov & Lipatov, has been formulated in [Altarelli and Parisi (1977); Dokshitzer (1977)].

5.2.3 *Hard cross sections in perturbation theory*

The universal nature of the parton dynamics goes under the name of *factorisation* of collinear ("mass") singularities. Physically, it is due to the fact that quark–gluon multiplication processes happen at *much larger* space–time distances than the hard interaction itself. It is this separation that makes it possible to describe quark–gluon cascades in terms of independent classical *parton splitting processes.* They succeed one another in a cleverly chosen *evolution time*, $t \sim \ln Q^2$, whose flow "counts" basic parton splittings that occur at well separated, strongly ordered space–time scales. The perturbative structure of the cross section of a given process (p) characterised by the hardness scale Q^2 can be cast, symbolically, as a product (convolution) of three factors:

$$\sigma_h^{(p)}(t) \propto C^{(p)}[\alpha_s(t)] \otimes \exp\left(\int_{t_0}^{t} d\tau\, \gamma[\alpha_s(\tau)]\right) \otimes w_h(t_0), \quad t \sim \ln Q^2. \tag{5.3}$$

Here the functions $C^{(p)}[\alpha_s]$ (hard cross section; coefficient function) and $\gamma[\alpha_s]$ (parton evolution Hamiltonian; anomalous dimension matrix) are perturbative objects that one analyses in terms of the α_s expansion. The last factor w_h embeds non-perturbative information about parton structure of the participating hadron(s) h, be it a target hadron in the initial state (parton distribution) or a hadron triggered in the final state (fragmentation function).

A borderline between perturbative and non-perturbative ingredients in (5.3) is fictitious; it is set arbitrarily by choosing the launching hardness scale $t_0 \sim \ln Q_0^2$. This is however not the only arbitrariness present in the representation (5.3). Namely, beyond the leading approximation (one loop; $\gamma = \mathcal{O}(\alpha_s)$), the separation between the C and $\exp(\gamma)$ factors becomes scheme dependent. Here one talks about *factorisation scheme* dependence.

Another negotiable object is the expansion parameter α_s itself whose definition depends on the ultraviolet renormalisation procedure (*renormalisation scheme* dependence). The famous MS-bar scheme based on the dimensional regularisation is a precisely prescribed procedure for eliminating ultraviolet divergences. It took over as the best suited scheme for carrying out laborious high order calculations.

It is the MS-bar scheme that was used to calculate the "parton Hamiltonian" $\gamma[\alpha_s]$ beyond the first loop: in the next-to-leading, $\mathcal{O}(\alpha_s^2)$ [Curci, Furmanski and Petronzio (1980)], and next-to-next-to-leading accuracy, $\mathcal{O}(\alpha_s^3)$ [Moch, Vermaseren and Vogt (2004); Vogt, Moch and Vermaseren (2004)].

The physical answer does not depend on which schemes (factorisation, renormalization) one chooses to construct the expansion. In reality, this would have been the case if we had hold of the *full* answer. However, building up an order-by-order perturbative expansions one faces a problem: perturbative series in QFT are, in general, at best *asymptotic.*

This means that starting from some order, $n > n_{\rm crit} = {\rm const}_{(p)}/g^2$, the series for any observable (p) inevitably ceases to represent the answer. For QED where $n_{\rm crit} \sim 100$ this is an academic problem. In QCD, on the contrary, the perturbative expansion may offer no better than a reasonable quantitative *estimate* based on the first few orders of the perturbation theory.[1]

This being understood, it becomes legitimate, and mandatory, to play with perturbative series trying to recast a formal α_s expansion in the most relevant way, as close to the physics of the problem as possible.

5.3 QCD DIS minutes

Typical QCD graphs for DIS amplitudes are shown in Fig. 5.3.

For moderate x-values (say, $x > 0.1$) the process is dominated by lepton scattering off a valence quark in the proton. The scattering cross section has a standard energy behavior $\sigma \propto x^{-2(J_{\rm ex}-1)}$, where $J_{\rm ex}$ is the spin of the exchanged particle in the t-channel. It is the quark with $J_{\rm ex} = \frac{1}{2}$ in the left picture of Fig. 5.3, so that the valence contribution to the cross section decreases at small x as $\sigma \propto x$.

For high-energy scattering, $x \ll 1$, the Bethe-Heitler mechanism takes over which corresponds to the t-channel *gluon* exchange: $J_{\rm ex} = 1$, $\sigma \propto x^0 = {\rm const}$ (modulo logarithms).

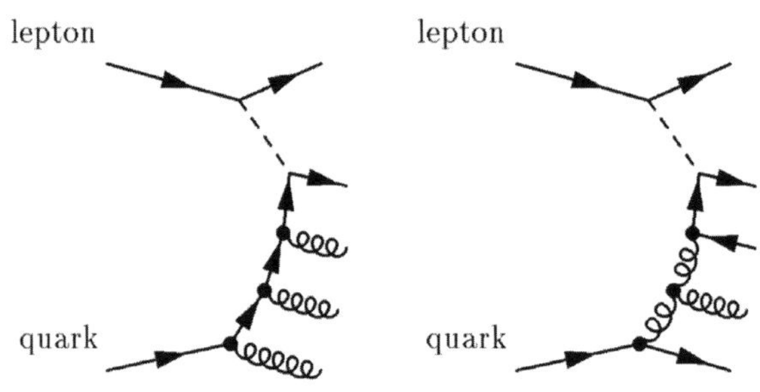

Fig. 5.3 Valence (left) and Bethe-Heitler mechanism (right) of DIS.

[1] An intrinsic uncertainty of such PT estimate can often be linked with genuine non-perturbative effects.

To contribute to the LLA, the transverse momenta of produced partons should be strongly ordered, increasing up the "ladder": $k_{1\perp}^2 \ll \ldots \ll k_{n\perp}^2 \ll Q^2$. (At the level of Feynman *amplitudes* the ladder diagrams dominate, provided a special physical gauge is chosen for gluon fields.)

$$\frac{1}{n!}\left[\frac{\alpha_s}{\pi}\ln\frac{Q^2}{\mu^2}\right]^n$$

$$= \left[\frac{\alpha_s}{\pi}\right]^n \int^{Q^2} \frac{dk_{\perp n}^2}{k_{\perp n}^2} \int^{k_{\perp n}^2} \frac{dk_{\perp n-1}^2}{k_{\perp n-1}^2} \cdots \int^{k_{\perp 3}^2} \frac{dk_{\perp 2}^2}{k_{\perp 2}^2} \int_{\mu^2}^{k_{\perp 2}^2} \frac{dk_{\perp 1}^2}{k_{\perp 1}^2}, \tag{5.4}$$

with $k_{\perp i}^2$ the squared transverse momenta of produced partons.

The expression (5.4) becomes senseless (diverges) in the zero-quark-mass limit, $\mu \to 0$. Well, when you see a nasty thing happen beyond your reach, you can do no better than make good use of it. This "mass singularity" occurs in the lower limit of the $k_\perp$ integration of the very first (and only!) parton branch. By dragging this misbehaving integral to the left one can rewrite (5.4) as

$$\begin{aligned}(5.4)^{[n]}(Q^2,\mu^2) &= \frac{\alpha_s}{\pi}\int_{\mu^2}^{Q^2}\frac{dk_{\perp 1}^2}{k_{\perp 1}^2}\cdot\left[\frac{\alpha_s}{\pi}\right]^{n-1}\int^{Q^2}\frac{dk_{\perp n}^2}{k_{\perp n}^2} \\ &\quad \times \int^{k_{\perp n}^2}\frac{dk_{\perp n-1}^2}{k_{\perp n-1}^2}\cdots\int_{k_{\perp 1}^2}^{k_{\perp 3}^2}\frac{dk_{\perp 2}^2}{k_{\perp 2}^2} \\ &= \frac{\alpha_s}{\pi}\int_{\mu^2}^{Q^2}\frac{dk_{\perp 1}^2}{k_{\perp 1}^2}\cdot(5.4)^{[n-1]}(Q^2,k_{\perp 1}^2).\end{aligned}$$

Here we have combined the internal $(n-1)$ integrals into the same expression that corresponds to the previous order in α_s-expansion and has a new lower limit $k_{\perp 1}^2$ substituted for the original μ^2. Now, we can localize the μ-dependence by evaluating the logarithmic derivative:

$$\mu^2\frac{\partial}{\partial\mu^2}(5.4)^{[n]} = -\frac{\alpha_s}{\pi}\cdot(5.4)^{[n-1]}.$$

This equation relates the nth order of the PT expansion to the previous one. To put this symbolic relation at work one first has to recall the satellite x-dependence.

By extracting the first step one may look upon the rest as DIS off a new "target" — the parton with transverse momentum $k_{\perp 1}^2$ and a finite fraction z of the initial longitudinal momentum P. As a result, there appears an additional integration with the probability of the first splitting, $\Phi(z)$, and the differential equation for the resummed parton distribution $D^{(\text{LLA})}$ takes the form[2]

$$\frac{\partial}{\partial\ln\mu^2}D^{(\text{LLA})}\left(x,Q^2,\mu^2\right) = -\int_x^1\frac{dz}{z}\Phi(z)\frac{\alpha_s}{2\pi}\cdot D^{(\text{LLA})}\left(\frac{x}{z},Q^2,\mu^2\right). \tag{5.5a}$$

[2]Soft divergence at $z \to 1$ gets regulated by inclusion of virtual corrections/Sudakov form factor effects, for review see, e.g. [Dokshitzer, Diakonov and Troian (1980)]

Since a logarithm (like a stick) has two ends, differentiation over the overall hardness scale Q^2 would do the same job, the result being the *evolution equation* in a familiar form:

$$\frac{\partial}{\partial \ln Q^2} D^{(\mathrm{LLA})}(x, Q^2) = \frac{\alpha_s}{2\pi} \Phi(x) \otimes D^{(\mathrm{LLA})}(x, Q^2), \tag{5.5b}$$

where the symbol $\otimes$ stands for convolution in the x-space.

5.4 Probabilistic picture of LLA parton evolution

5.4.1 *Space-like parton evolution (DIS)*

The decay phase space for the *space-like* evolution determining the DIS structure functions is

$$dw^{A \to B+C} = \frac{dk_\perp^2}{k_\perp^2} \frac{\alpha_s(k_\perp^2)}{2\pi} \frac{dz}{z} \Phi_A^{BC(S)}(z), \tag{5.6a}$$

with z the longitudinal momentum fraction carried by the parton B. The functions Φ play the role of "Hamiltonian" (x-dependent kernels) of the evolution equation (5.5) for the LLA parton distributions.

In the DIS environment the initial parton A with a negative (space-like) virtuality decays into $B[z]$ with the large space-like virtual momentum $|k_B^2| \gg |k_A^2|$ and a positive virtuality (time-like) $C[1-z]$. The parton C generates a subjet of secondary partons ($\to$ hadrons) in the final state. Since no details of the final-state structure are measured in an inclusive process, the invariant mass of the subjet C is integrated over. In the dominant integration region $k_C^2 \ll |k_B^2|$ the parton C looks *quasi-real*, compared with the hard scale of $|k_B^2|$. The same is true for the initial parton A.

Splitting can be viewed as a large momentum-transfer process of scattering (turnover) of a "real" target parton A into a "real" C on the external field mediated by high-virtuality B. At the next step of evolution it is B's turn to play a role of a next target $B \equiv A'$, "real" with respect to yet deeper probe $|k_B'^2| \gg |k_B^2|$, and so on.

Successive parton decays with step-by-step *increasing* space-like virtualities (transverse momenta) constitute the picture of parton wave-function fluctuations inside the proton. The sequence proceeds until the overall hardness scale Q^2 is reached.

5.4.2 *Time-like parton cascades*

A similar picture emerges for the time-like branching processes determining the internal structure of jets produced, for example, in e^+e^- annihilation. Here the flow of *hardness* is opposite to that in DIS: evolution starts from a highly virtual quark with positive virtuality, originating from the e.m. vertex, while time-like virtualities of its products ("quasi-real" with respect to predecessors; "high-virtuality" with respect to offspring) degrade.

In the time-like case the longitudinal phase space is symmetric in offspring parton energy fractions, and the differential decay probability reads

$$dw^{A\to B+C} = \frac{d\Theta^2}{\Theta^2}\frac{\alpha_s(k_\perp^2)}{2\pi}dz\Phi_A^{BC(T)}(z). \tag{5.6b}$$

5.4.3 *Relating DIS and e^+e^-*

In the DIS case, the large virtual momentum q transferred from the incident lepton to the target nucleon with momentum P is *space-like*, $q^2 < 0$. The *inelasticity* of the process is conveniently characterised by the *Bjorken* variable $x_B = -q^2/2(Pq)$. On the other side, inclusive fragmentation of an e^+e^- pair with the total momentum q (large *positive* invariant mass squared q^2) into a final state hadron with momentum P is characterised by the *Feynman* variable $x_F = 2(Pq)/q^2$ (hadron energy fraction in the e^+e^- cms.). The fact that Bjorken and Feynman variables are indicated by the same letter is certainly not accidental. In both channels $0 \le x \le 1$ though these variables are actually reciprocal, $x_F \iff 1/x_B$, rather than identical:

$$x_B = \frac{-q^2}{2(Pq)}, \quad x_F = \frac{2(Pq)}{q^2}. \tag{5.7}$$

Apart from the difference in the hadron momentum P belonging to the initial state in the DIS and final state in the e^+e^- case, the Feynman diagrams for the two processes are just the same. In particular, *mass singularities* that emerge when some parton momentum become collinear to P are therefore also the same. That is why, for the two processes a similar parton interpretation emerges in terms of QCD evolution equations, and the space- and time-like evolution anomalous dimensions turn out to be related.

For discussion of the relations between splitting functions it is convenient to strip off colour factors and introduce

$$\begin{aligned} \Phi_F^{FG}(z) &= C_F V_F^{FG}(z), \quad \Phi_F^{GF}(z) = C_F V_F^{GF}(z), \\ \Phi_G^{FF}(z) &= T_R V_G^{FF}(z), \quad \Phi_G^{GG}(z) = N_c V_G^{GG}(z). \end{aligned} \tag{5.8}$$

Here C_F and N_c are the familiar quark and gluon "color charges", while T_R is a scientific name for one half:

$$\mathrm{Tr}(t^a t^b) = \sum_{i,k=1}^{N_c} t^a_{ik} t^b_{ki} \equiv T_R \delta^{ab} = \tfrac{1}{2}\delta^{ab}.$$

Drell–Levy–Yan relation elucidates a deep connection between the channels [Drell, Levy and Yan (1970)]. It follows directly from the comparison of Feynman diagrams for space- and time-like particle multiplication processes. It states that the

e^+e^- splitting functions can be obtained from those of DIS by replacing $x_B \to 1/x_F$ (modulo a kinematical factor):

$$V_A^{BC(T)}(x) = (-1)^{2(J_A - J_B)+1} x \cdot V_B^{AC(S)}(x^{-1}) \tag{5.9}$$

(with J_A the spin of the parton A). Beyond the leading logarithmic approximation (LLA) the DLY relation (5.9) was being used to actually *determine* the time-like splitting functions from their space-like counterparts. True in any QFT,[3] it reflects the crossing and allows one to link the cross sections in the two channels by *analytic continuation* from $x < 1$ to $x > 1$. One obtains the time-like anomalous dimensions, $\gamma^{(T)}(x)$, by analytically continuing $\gamma^{(S)}(x)$ into the unphysical region $x > 1$, and then replacing $x \to 1/x < 1$. The continuation path crosses the singular point $x = 1$ which needs a special care to be taken in defining certain complex logarithms in "arithmetic" sense, $\ln(1-x) \Longrightarrow |\ln(1-x)|$, see [Bukhvostov, Lipatov and Popov (1974); Dokshitzer (1977)]; beyond the first loop, see [Stratmann and Vogelsang (1997); Blümlein, Ravindran and van Neerven (2000)] and references therein.

Gribov–Lipatov relation states that the flow of *energy* (longitudinal momentum) in space- and time-like cascades is governed, in the LLA, by the same splitting functions. This identity is one of many wonderful symmetries that our "parton Hamiltonian" obeys:

$$\Phi_A^{BC(S)}(z) = \Phi_A^{BC(T)}(z) \equiv \Phi_A^{BC}(z) \implies \Phi_A^B(z). \tag{5.10}$$

When studying inclusive characteristics of parton cascades, one traces a single thread of successive parton splittings, $A \to B \to \ldots$. Having this in mind, we can drop the label that marks partons C whose fate does not concern us, $\Phi_A^{BC}(z) \implies \Phi_A^B(z)$ ($\equiv P_{BA}$ in the Altarelli–Parisi notation [Altarelli and Parisi (1977)]).

5.4.4 *LLA parton splitting probabilities*

QCD splitting functions in the notation (5.8) read

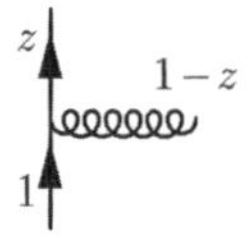

$$V_F^F(z) = \frac{1+z^2}{1-z}, \tag{5.11a}$$

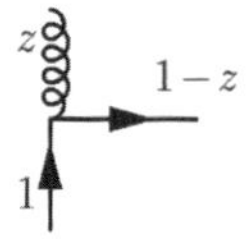

$$V_F^G(z) = \frac{1+(1-z)^2}{z}, \tag{5.11b}$$

[3]When applied to *physical cross sections*, not necessarily to scheme dependent anomalous dimensions [Blümlein, Ravindran and van Neerven (2000)].

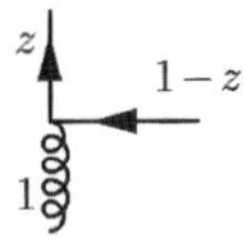

$$V_F^G(z) = z^2 + (1-z)^2\,, \tag{5.11c}$$

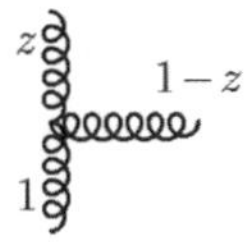

$$V_G^G(z) = 2\left[z(1-z) + \frac{1-z}{z} + \frac{z}{1-z}\right]. \tag{5.11d}$$

5.4.5 *Internal symmetries of parton dynamics*

The LLA parton splitting functions have the following remarkable symmetry properties.

Parton Exchange stems from an obvious relation between probabilities to find decay products with complementary momentum fractions:

$$V_A^{B(C)}(z) \;=\; V_A^{C(B)}(1-z). \tag{5.12a}$$

Gribov–Lipatov Reciprocity emerges when one combines the Gribov–Lipatov (5.10) and Drell–Levy–Yan relations (5.9):

$$V_B^A(z) = (-1)^{2J_A+2J_B-1} z V_A^B\left(\frac{1}{z}\right). \tag{5.12b}$$

These two relations do not leave much freedom for splitting functions. In fact, one could borrow V_F^F from QED textbooks, reconstruct V_F^G by exchanging the decay products (5.12a), and then obtain V_G^F using the crossing (5.12b). This is the way to generate all three splitting functions relevant for QED (5.11a)–(5.11c). The last gluon-gluon splitting function (5.11d) stays invariant under both tranformations (5.12a) and (5.12b). The more surprising is the fact that the gluon self-interaction kernel actually could have been obtained "from QED".

Super-Symmetry Relation

$$V_F^F(z) + V_F^G(z) = V_G^F(z) + V_G^G(z). \tag{5.12c}$$

This relation exploits the existence of the supersymmetric QFT closely related to real QCD. In the SUSY-QCD, "quark" and "gluon" belong to the same (adjoint) representation of the color group, so that all color factors become identical $C_F = C_A = T_R$ (cf. (5.8)). Bearing this in mind one can interpret (5.12c) as an equality between the *total* probabilities of "quark" and "gluon" decays. The fact that it holds identically in z means that there is an infinite number of non-trivial conservation laws in this theory!

However, even this is not the end of the story.

Conformal Invariance leads to a number of relations (involving derivatives) between splitting functions, the simplest of which reads [Bukhvostov, Lipatov and Popov (1974)]

$$\left(z\frac{d}{dz}-2\right)V_G^F(z)=\left(z\frac{d}{dz}+1\right)V_F^G(z). \tag{5.12d}$$

The general character of the symmetry properties makes them practically useful when studying subleading effects in parton distributions where one faces technically difficult high order calculations.

5.5 Evolution time: Coherence

An attentive reader has noticed that we wrote the phase space differently: for the space-like case (5.6a) in terms of transverse momentum $k_\perp$ and for the time-like evolution (5.6b) via the decay angle Θ. Logarithmic differentials by themselves are identical, since $k_\perp^2$ and Θ^2 are proportional for fixed z. We have made this distinction to stress an important difference between probabilistic interpretation of DIS and e^+e^- evolution: different *evolution times.*

Within the LLA framework it does not matter which of evolution parameters $\ln k_{\perp i}^2$, $\ln\Theta_i^2$ or $\ln|k_i^2|$ (with k^2 the total parton virtuality) would do a better job: these choices differ by *subleading* terms formally negligible from the LLA point of view. A mismatch is of the order of

$$\sim \alpha_s \ln^2 x. \tag{5.13}$$

It becomes significant, however, and should be "resummed" in all orders when numerically small values of Bjorken x are concerned.[4]

For space-like cascades, the $k_\perp$-ordering proves to be the one that takes care of potentially disturbing corrections (5.13) in all orders, and in this sense becomes a preferable choice for constructing the probabilistic scheme for single-inclusive parton distributions in DIS.

It is instructive to see how does this come about.

5.5.1 *Collinear logarithms and fluctuation time*

Let us introduce two light-like vectors p_1^μ, p_2^μ and write down the Sudakov (light-cone) decomposition of momenta:

$$k^\mu = \beta p_1^\mu + \alpha p_2^\mu + k_\perp{}^\mu. \tag{5.14}$$

[4]The word "numerically" stands here as a warning for not confusing this kinematical region with a "parametrically" small x, such that $\alpha_s \ln 1/x \sim 1$ — the Regge region — where essentially different physics comes into play.

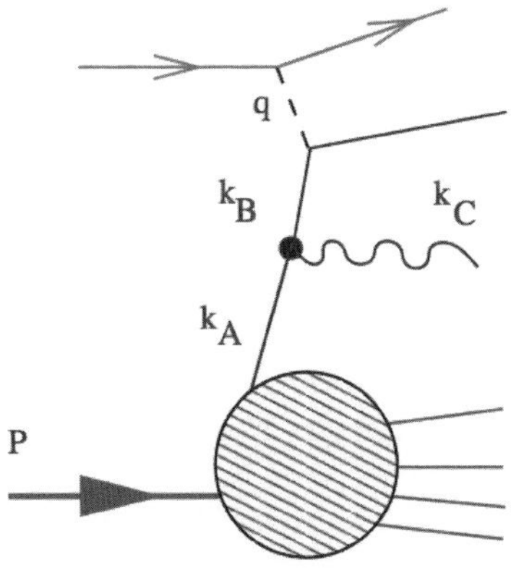

Fig. 5.4 Kinematics of parton splitting in DIS.

For $k_1^\mu + k_2^\mu + k_3^\mu = 0$ the general relation holds:

$$\frac{k_1^2}{\beta_1} + \frac{k_2^2}{\beta_2} + \frac{k_3^2}{\beta_3} = \frac{\beta_1\beta_2}{\beta_3}\left(\frac{\vec{k}_{1\perp}}{\beta_1} - \frac{\vec{k}_{2\perp}}{\beta_2}\right)^2. \tag{5.15}$$

Each parton cell involves a space-like parton A decaying into $B[z] + C[1-z]$, see Fig. 5.4. Relation (5.15) applied to this splitting gives

$$\frac{-k_B^2}{z} = \frac{-k_A^2}{1} + \frac{k_C^2}{1-z} + \frac{k_\perp^2}{z(1-z)}, \tag{5.16}$$

with z the longitudinal momentum fraction — the ratio of the Sudakov light-cone variables β. (We have chosen the direction of p_1 such that $\vec{k}_{A\perp} = 0$, so that $\vec{k}_{B\perp} = -\vec{k}_{C\perp} \equiv \vec{k}_\perp$ is the relative transverse momentum in the splitting.)

Since the 4-momenta of A and B are space-like, all terms in (5.16) are positive. B being an intermediate virtual state, k_B^2 enters in the Feynman denominators in the matrix element. The collinear-log contribution arises upon integration over $k_\perp^2$, over the region where the last term dominates in the r.h.s. of (5.16), that is from the region

$$\frac{|k_B^2|}{z} \simeq \frac{k_\perp^2}{z(1-z)} \gg |k_A^2|, \frac{k_C^2}{1-z}. \tag{5.17}$$

The physical origin of this strong inequality becomes transparent in terms of lifetimes of virtual states

$$\tau_i = \frac{k_i^0}{|k_i^2|}, \tag{5.18}$$

namely $\tau_B \ll \tau_A,\ \tau_C$. This shows that the LLA contributions originate from the sequence of branchings well separated in the *fluctuation time* (5.18). Invoking the local-scattering analogy (recall $A \to C$ on the "external field" B), we can say that the classical picture naturally implies "fast scattering": probing time τ_B much smaller than the lifetime(s) of the "target" before (τ_A) and after the scattering occurs (τ_C).

Assembling a "ladder" of successive parton splittings $i = 1, 2 \ldots, n$, and tracing the space-like parton state, one may expect that the nth-order LLA contribution $\sim [\alpha_s \ln(Q^2/\mu^2)]^n$ should come from the time-ordered kinematics

$$\frac{P}{\mu^2} \gg \tau_1 \gg \tau_2 \gg \ldots \gg \tau_n \gg \frac{xP}{Q^2}. \tag{5.19}$$

Equation (5.17) relates virtuality and transverse momentum of the "t-channel" parton after the ith splitting with the relative longitudinal momentum (β-) fraction z_i:

$$\left|k_i^2\right| \simeq \frac{k_{i\perp}^2}{1 - z_i} \approx k_{i\perp}^2. \tag{5.20}$$

The latter approximation is made by remembering that, because of cancellation between real and virtual contributions in inclusive parton distributions (DIS structure functions), the soft s-channel radiation $1 - z_i \ll 1$ does not matter (as long as we stay away from the quasi-elastic kinematics, $1 - x \ll 1$, where it does).

We are now in a position to compare different orderings:

$$\text{Fluctuation times} \Longrightarrow k_{i\perp}^2 \gg z_i \cdot k_{i-1\perp}^2, \tag{5.21a}$$

$$\text{Transverse momenta} \Longrightarrow k_{i\perp}^2 \gg k_{i-1\perp}^2. \tag{5.21b}$$

For $x = z_1 \cdot z_2 \cdot \ldots \cdot z_n \sim 1$ prescriptions (5.21) are essentially equivalent since each decay fraction stays finite, $z_i \sim 1$, and may be neglected in the logarithmic integral over $k_{i\perp}^2$.

For small x, however, the Bethe-Heitler mechanism of DIS off sea quarks dominates (see Fig. 5.3b). The multi-gluon "ladders" provide longitudinally enhanced contributions $\propto \ln z_i$, which combine with the $\ln z_i$ factors from the collinear-integration phase space to produce, at the end of the day, a large DL-enhanced mismatch $[\alpha_s \ln^2 x]^{n-2}$ between the different options (5.21).

Which ordering is correct? The first prescription (5.21a) is more liberal as it allows for *disordered* transverse momentum configuration:

$$z_i \cdot k_{i-1\perp}^2 \ll k_{i\perp}^2 \ll k_{i-1\perp}^2, \tag{5.22}$$

This interval where the $k_\perp$-ordering is violated, may be quite broad for $z_i \ll 1$. The truth is, this region does not contribute to the answer, so that the $k_\perp$-ordering (5.21b) proves to be the correct one. The reason is quantum mechanical coherence.

5.5.2 *Forward inelastic diffraction and $k_\perp$ ordering (DIS)*

Consider a two-step process shown by the first graph in Fig. 5.5. Let the second decay be *soft* in the t-channel direction, that is $z_2 = \beta_2/\beta_1 \ll 1$. (The first one can

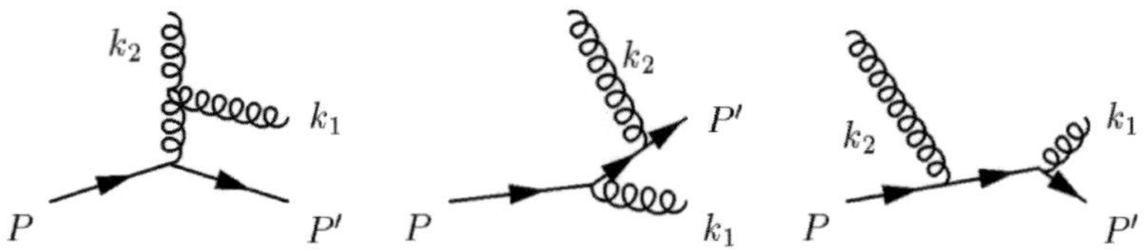

Fig. 5.5 In the "wrong" kinematics $k_{2\perp} < k_{1\perp}$, the sum of the two space-like evolution amplitudes cancels against the final state time-like decay.

be either soft, $z_1 = \beta_1/\beta_0 \ll 1$ or hard, $z_1 \sim 1$.) In the kinematical region (5.22),

$$z_2 \cdot k_{1\perp}^2 \ll k_{2\perp}^2 \ll k_{1\perp}^2, \tag{5.23}$$

the *time-ordering* is still intact, which means that the momentum k_2 is transferred fast as compared with the lifetime of the first fluctuation $P \to P' + k_1$.

Since β_2 is small, the process can be viewed as inelastic relativistic scattering $P \to P' + k_1$ in the external gluon field. The transverse size of the field is $\rho_\perp \sim k_{2\perp}^{-1}$. The characteristic size of the fluctuation $P' + k_1$, according to (5.23), is smaller: $\Delta r_{01\perp} \sim k_{1\perp}^{-1} \ll \rho_\perp$. We thus have a *compact* state propagating through the field that is smooth at distances of the order of the size of the system. In such circumstances the field cannot resolve the internal structure of the fluctuation. Components of the fluctuation, partons P' and k_1 in the first two graphs of Fig. 5.5, scatter *coherently* with the total amplitude identical and opposite in sign to that for scattering of the initial state P (the last graph): inelastic breakup does not occur.

This physical phenomenon is quite general. In our context the cancellation between the amplitudes of Fig. 5.5 in the region (5.23) is a direct consequence of the conservation of the colour current.

5.5.3 *Bremsstrahlung cones and Angular Ordering (Jets)*

Thus we conclude that in the DIS environment, due to quantum-mechanical coherence, the *transverse momentum* ordering (5.24a) proved to be the one that takes care of potentially disturbing corrections (5.13) in all orders, and therefore becomes the preferable choice for constructing the probabilistic scheme for space-like parton cascades (DIS structure functions).

Meantime, for the the time-like cascades it turns out to be the *relative angle* Θ between the offspring partons (rather than the transverse momentum) which has to be kept ordered and decreasing along the evolutionary decay chain, from the hard production vertex towards the registered inclusive hadron:

$$dt_- = d\ln k_\perp^2 \quad \text{(space-like)}, \tag{5.24a}$$

$$dt_+ = d\ln \frac{k_\perp^2}{\beta^2} = d\ln\tan\frac{\Theta}{2} \quad \text{(time-like)}. \tag{5.24b}$$

Since $k_\perp/\beta P = k_\perp/k_+ = 2\tan\Theta/2$, the variable (5.24b) corresponds indeed to the ordering in angles (AO).

Now the process starts from a large scale q^2 (cms annihilation energy) and results in triggering a final particle h with momentum P. Here the order of events is opposite to (5.19): a parton of the generation $(i+1)$ lives *longer* than its parent (i):

$$\frac{P}{xq^2} \ll \tau_1 \ll \tau_2 \ll \ldots \ll \tau_n \ll \frac{P}{\mu^2}; \quad x = x_{\text{Feynman}} \equiv \frac{2Pq}{q^2}, \tag{5.25}$$

where we have used that the energy of the initial quark stemming from the $\gamma^* \to q\bar{q}$ vertex is $q_0/2 = P/x_F$.

Let us examine the mismatch between the fluctuation time and relative angle orderings. For $z \ll 1$ we have $|k^2| \simeq k_\perp^2$ and the comparison goes as follows:

$$e^+e^- \begin{cases} \text{time ordering}: & \tau_i = \dfrac{\beta_i P}{k_{\perp i}^2} \tau_{i+1} = \dfrac{\beta_{i+1} P}{k_{\perp i+1}^2}, \\ \text{angular ordering}: & \Theta_i = \dfrac{k_{\perp i}}{\beta_i P} \Theta_{i+1} = \dfrac{k_{\perp i+1}}{\beta_{i+1} P}; \end{cases}$$

$$\text{mismatch} \Longrightarrow \Theta_i^2 < \Theta_{i+1}^2 < \frac{\Theta_i^2}{z}. \tag{5.26}$$

When the energy of the offspring parton $i+1$ is relatively small, $z \equiv \beta_{i+1}/\beta_i \ll 1$, (soft gluon), the mismatch interval (5.26) widens and gives rise to explosive $(\alpha_s \ln^2 x)^n$ contributions. The truth is, this angular-disordered region when a soft gluon is emitted at angle Θ_{i+1} *exceeding* the relative angle between its parents, Θ_i, must be excluded from the parton multiplication picture. Once again, the reason is quantum-mechanical coherence, as it were the case of space-like evolution that we discussed above.

Disregarding quantum mechanics has had a dramatic consequence: before the AO was established [Ermolaev and Fadin (1981); Mueller (1981)], the rate of particle multiplicity growth in jets was expected to be much faster, with the multiplicity anomalous dimension γ_0,

$$\langle N \rangle \sim (Q^2)^{\gamma_0},$$

larger by a spurious factor $\sqrt{2}$ than the true one.

Even a more spectacular consequence of the AO manifested itself in the x-dependence of the parton density, $xD(x)$ at $x \ll 1$. In was predicted by the orthodox parton model to be a constant in both channels — Feynman plateau.

In DIS, due to QCD parton cascading, $xD(x)$ steadily increases with x decreasing, the faster the larger Q^2. In the time-like cascades (internal structure of jets) the pattern turns out to be very different. Defying naive probabilistic intuition, the softest particles do not multiply at all. The density of partons with $x \sim Q_0/Q$ stays constant while that of their more energetic companions increases with the hardness of the process Q. The particles that multiply most efficiently have "intermediate" energies: $\ln E_{\max}/Q_0 \sim 0.3 - 0.35 \ln Q/Q_0$. A direct consequence of AO, this phenomenon was labeled *hump-backed plateau*. The corresponding

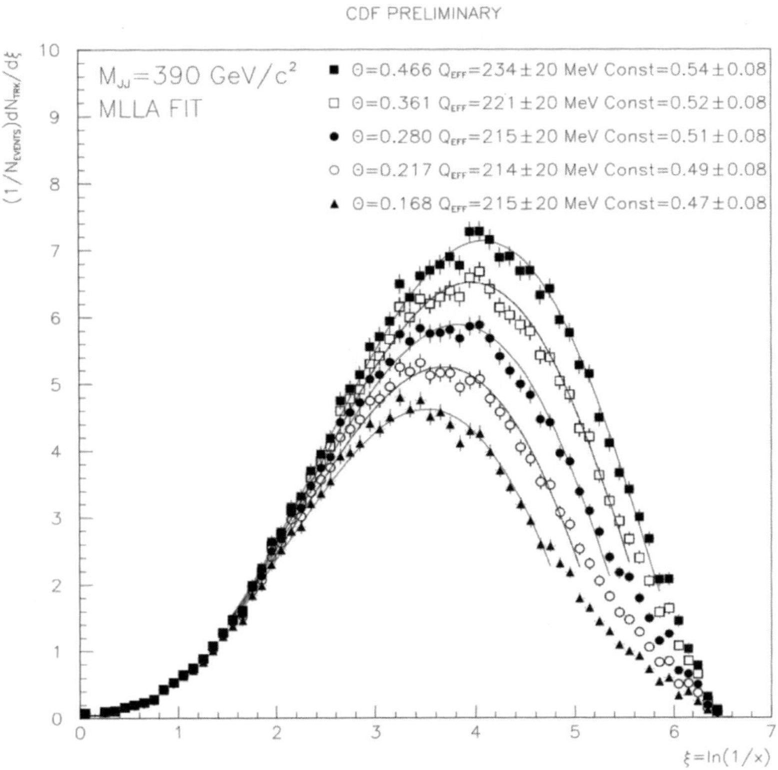

Fig. 5.6 Inclusive distribution of all charged hadrons in CDF jets with restricted opening angle compared with the MLLA gluon spectrum [Goulianos (1997)].

theoretical framework (DGLAP incorporating AO) is known as *modified* leading-log approximation (MLLA), see [Dokshitzer, Khoze, Mueller and Troian (1988)] and references therein. First scrutinised at LEP (ALEPH, OPAL), the predicted shape of the inclusive distribution has been quantitatively verified at SLC and KEK e^+e^- machines, as well as at HERA and Tevatron where hadron jets originate not from bare quarks dug up from the vacuum by a highly virtual photon/Z^0 but from hard partons kicked out from initial hadron(s).

For spectacular CDF checks of analytic QCD predictions see [Korytov (1997); Safonov (2000)]. One of such checks is shown in Fig. 5.6.

AO in QED. So, what is the reason for radiation at angles exceeding the scattering angle to be suppressed? Let us examine semi-classically how the radiation process occurs. We start from good old photon bremsstrahlung.

A physical electron is a charge surrounded by its proper Coulomb field. In quantum language the Lorentz-contracted Coulomb-disk attached to a relativistic particle may be treated as consisting of photons virtually emitted and, in due time, reabsorbed by the core charge. Such virtual emission and absorption processes form a coherent state which we call a physical electron ("dressed" particle).

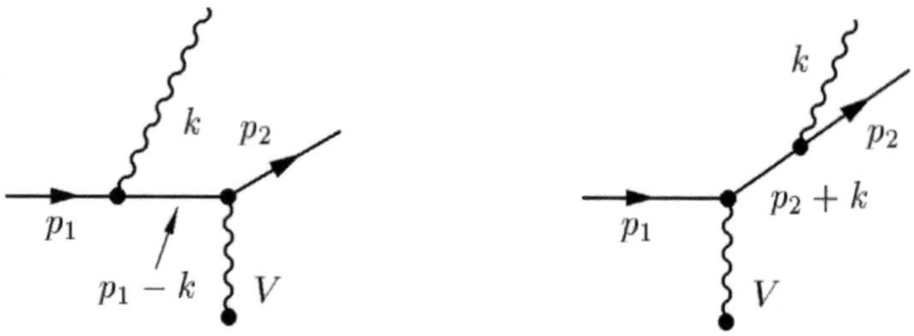

Fig. 5.7 Photon Bremsstrahlung diagrams for scattering off an external field.

This coherence is partially destroyed when the charge experiences an impact. As a result, a part of intrinsic field fluctuations gets released in the form of real photon radiation: the bremsstrahlung cone in the direction of the initial momentum develops. On the other hand, the deflected charge now leaves the interaction region as a "half-dressed" object with its proper field-coat lacking some field components (eventually those that were lost at the first stage). In the process of regenerating the new Coulomb-disk adjusted to the final-momentum direction, an extra radiation takes place giving rise to the second bremsstrahlung cone.

Now we need to be more specific to find out which momentum components of the electromagnetic coat do actually take leave.

A typical time interval between emission and reabsorption of the photon k by the initial electron p_1 may be estimated as the Lorentz-dilated lifetime of the virtual intermediate electron state $(p_1 - k)$ (see the left graph in Fig. 5.7),

$$t_{\rm fluct} \sim \frac{E_1}{|m^2 - (p_1 - k)^2|} = \frac{E_1}{2p_1 k} \sim \frac{1}{\omega\Theta^2} \simeq \frac{\omega}{k_\perp^2}. \tag{5.27}$$

Here we restricted ourselves to small radiation angles, $k_\perp \approx \omega\Theta \ll k_{||} \approx \omega$.

The fluctuation time (5.27) may become macroscopically large for small photon energies ω. It enters as a characteristic parameter in a number of radiation related processes and is often referred to as *formation time.*

Now imagine that within this interval the core charge was kicked by some external interaction and has changed direction by some Θ_s. Whether the photon will be reabsorbed or not depends on the position of the scattered charge with respect to the point where the photon was expecting to meet it "at the end of the day". That is, we need to compare the spatial displacement of the core charge $\Delta\vec{r}$ with the characteristic size of the to-be-emitted photon field, $\lambda_{||} \sim \omega^{-1}$, $\lambda_\perp \sim {k_\perp}^{-1}$:

$$\Delta r_{||} \sim \left|v_{2||} - v_{1||}\right| \cdot t_{\rm fluct} \sim \Theta_s^2 \cdot \frac{1}{\omega\Theta^2} = \left(\frac{\Theta_s}{\Theta}\right)^2 \lambda_{||} \Leftrightarrow \lambda_{||}; \tag{5.28}$$

$$\Delta r_\perp \sim c\Theta_s \cdot t_{\rm fluct} \sim \Theta_s \cdot \frac{1}{\omega\Theta^2} = \left(\frac{\Theta_s}{\Theta}\right) \lambda_\perp \Leftrightarrow \lambda_\perp. \tag{5.29}$$

For large scattering angles, $\Theta_s \sim 1$, two full-size bremsstrahlung cones are present. For numerically small $\Theta_s \ll 1$, however, it is only photons with $\Theta \lesssim \Theta_s$ that can notice the charge being displaced and thus the coherence of the state being disturbed. Therefore only radiation at angles smaller than the scattering angle actually emerges. The other field components have too large a wavelength and are easily reabsorbed *as if* there were no scattering at all.

So what counts is a change of the current, which is sharp enough to be noticed by the "to-be-emitted" quantum within the characteristic formation/field-fluctuation time (5.27) of the latter. Radiation at large angles has too short a formation time to become aware of the acceleration of the charge. No scattering — no radiation.

QCD: Coherent large-angle radiation. Let us substitute quarks for electrons and replace the colourless external field V by gluon exchange in Fig. 5.7. Then the change in the current becomes twofold: the quark not only changes direction but gets "repainted" because of the t-channel colour exchange. As a result, the amplitudes of soft gluon radiation off initial and final quarks at large angles $\Theta > \Theta_s$ no longer cancel. They combine instead into radiation *as if* off the t-channel gluon

$$T^a T^b - T^b T^a = i f_{abc} T^c.$$

Since the commutation relation is universal, the density of the large-angle radiation is always proportional to the gluon colour charge, N_c, irrespective of the nature of the projectile (quark, gluon, any other coloured object).

We considered bremsstrahlung radiation accompanying QED/QCD *scattering*. Exactly the same arguments apply to the dual process of *production* of two opposite charges (decay of a neutral object, vacuum pair production, etc.). The only difference is that now one has to take for $\Delta\vec{r}$ not a displacement between the initial and the final charges, but the actual distance between produced particles (spatial size of the dipole), to compare it with the radiation wavelength.

This results in the Angular Ordering prescription (5.24b) for parton multiplication inside jets (time-like cascades).

5.6 Gribov–Lipatov reciprocity, innovative bookkeeping and SUSY

Let us look at the anomalous dimension matrix $\tilde{\gamma}[x, \alpha_s]$ in the Mellin moment space:

$$\gamma(N) = \mathbf{M}\left[\tilde{\gamma}(x)\right] \equiv \int_0^1 \frac{dx}{x} x^N \tilde{\gamma}(x). \tag{5.30}$$

The Gribov-Lipatov Reciprocity (GLR) (5.12b),

$$\tilde{\gamma}(x) = \mp x \tilde{\gamma}\left(x^{-1}\right), \tag{5.31}$$

then translates into the symmetry $N \to -(N+1)$. In other words, in order to respect the GLR, the anomalous dimension $\gamma(N)$ has to be a function of a single variable $N(N+1)$. This is nothing but the Casimir operator of the collinear $SO(1,2)$ subgroup of the $SO(2,4)$ conformal group [Braun, Korchemsky and Mueller (2003)].

A group symmetry nature of the GL reciprocity was advocated already in the early 70's when the issue was first raised [Ferrara, Gatto and Parisi (1973); Etim (1973)]. Indeed, since x is a light-cone momentum fraction, $y = \ln x$ represents parton (pseudo)rapidity: $k_0 = m_\perp \cosh y$, $k_z = m_\perp \sinh y$. Therefore in the Mellin transformation,

$$\int \frac{dx}{x} x^N = \int dy\, e^{yN} \Longrightarrow \int d\phi\, e^{i\phi N}, \quad \phi = iy,$$

an analogy between the Lorenz boost parameter y and an imaginary rotation angle allows one to look upon the conjugate variable N as "angular momentum", with $\mathbf{J}^2 = N(N+1)$ the corresponding Casimir operator.

True in the LLA (first loop for γ), the relation (5.31), or, alternatively, an identity of space- and time-like evolution kernels (5.10), are known to break already in the second loop [Curci, Furmanski and Petronzio (1980); Furmanski and Petronzio (1980)].

We are going to look deeper into the origin of this breaking [Dokshitzer and Marchesini (2007)].

5.7 Rescuing Gribov–Lipatov reciprocity

We saw that the choice of the variables (5.24) is a clever dynamical move that takes into full consideration soft gluon coherence and prevents explosively large terms (5.13) from appearing in higher loop anomalous dimensions. Interestingly, not only is the fluctuation time ordering physically *wrong* both for space- and time-like cascades, it happens to be *equally, symmetrically* wrong: the τ-ordering positions itself right in the middle between the two "clever" ones:

$$k_\perp^2 \Longrightarrow \frac{k_\perp^2}{z} \Longrightarrow \frac{k_\perp^2}{z^2}.$$

What if we decide to play a fool and stubbornly stick to the "wrong" τ-ordering?

The upper limits of virtuality integrals are

$$\left.\begin{array}{ll} \text{DIS}: & \left|k_i^2\right| \ll \dfrac{\beta_i}{\beta_{i+1}} \left|k_{i+1}^2\right| = z^{-1} \cdot \left|k_{i+1}^2\right| \\[2ex] e^+e^-: & k_{i+1}^2 \ll \dfrac{\beta_{i+1}}{\beta_i} k_i^2 = z \cdot k_i^2 \end{array}\right\} \quad z = \frac{\beta_{i+1}}{\beta_i} \le 1. \tag{5.32}$$

It is different placing of the z factor that causes violation of the GLR beyond the first loop. Moreover, it is likely to be the *one and only* source of this breaking!

Choosing $\kappa^2 = \left|k^2\right|$ as an integration variable and assembling parton evolution sequences, for the probability $D(x, Q^2)$ to find a parton with *virtuality* integrated up to a given Q^2 we obtain (omitting a trivial Born term)

$$D(x, Q^2) = \int_x^1 \frac{dz}{z} \int^{Q^2} \frac{d\kappa^2}{\kappa^2} P[z, \alpha_s] D\left(\frac{x}{z}, z^\sigma \kappa^2\right),$$
$$\text{with} \quad \sigma = \pm 1 \text{ for the t/s channel.} \tag{5.33}$$

Here $P[x, \alpha_s]$ is a new evolution kernel that is (supposed to be) the same for space- and time-like channels. With the one-loop accuracy (LLA), one would drop the z^σ factor as a subleading correction and return to the Gribov–Lipatov identity of anomalous dimensions (5.10):

$$\gamma_-(N) = \gamma_+(N) = \mathbf{M}\left[\frac{\alpha_s}{2\pi}\Phi(x)\right] \equiv \mathcal{P}^{(1)}(N). \tag{5.34}$$

In general, the presence of the factor z^σ in the second argument of (5.33) makes the evolution equation *non-local* in longitudinal (x) and transverse variables (Q^2) and breaks the identification between the splitting functions and anomalous dimensions. What it offers instead is a *link* between the two channels by means of the *universal* reciprocity respecting parton splitting matrix $\mathcal{P}$, the same for t and s evolutions. In spite of the fact that the new *evolution kernel* in (5.33) does not correspond to any clever choice of the evolution variable (explosive $\alpha_s \ln^2 x$ terms being present in both channels), this universality can be exploited to relate the DIS and e^+e^- anomalous dimensions.

Differentiating (5.33) and taking Mellin moments one obtains

$$\gamma_\sigma(N) D_N = \int_0^1 \frac{dz}{z} z^N P[z, \alpha_s] z^{\sigma \partial_{\ln Q^2}} D_N. \tag{5.35a}$$

This integral can be formally evaluated using the Taylor expansion trick,

$$\gamma_\sigma(N) = (D_N)^{-1} \mathcal{P}(N + \sigma \partial_{\ln Q^2}) D_N, \tag{5.35b}$$

expressing the anomalous dimension through the Mellin image of the splitting function that depends on the *differential operator* as an argument, $N \to N + \sigma\partial_{\ln Q^2}$. The derivative acts upon $D_N = D_N(Q^2)$ producing, by definition, $\gamma(N) \cdot D_N$. In high orders it will also act on the *running coupling* the anomalous dimension depends on, $\gamma = \gamma(N, \alpha_s)$. The latter action gives rise to terms proportional to the β-function. Such terms are scheme dependent as they can be reshuffled between the exponent and the coefficient function $C[\alpha_s]$ in (5.3). Neglecting for the time being these contributions by treating α_s as a constant, (5.35) reduces to a compact functional equation

$$\gamma_\sigma(N) = \mathcal{P}\left(N + \sigma\gamma_\sigma(N)\right). \tag{5.36}$$

Since $\gamma = \mathcal{O}(\alpha_s)$, we can expand the argument of the splitting function perturbatively,

$$\gamma_\sigma = \mathcal{P} + \dot{\mathcal{P}} \cdot \sigma\gamma_\sigma + \tfrac{1}{2}\ddot{\mathcal{P}} \cdot \gamma_\sigma^2 + \mathcal{O}\left(\beta(\alpha_s)\right) + \mathcal{O}\left(\alpha_s^4\right), \tag{5.37a}$$

where each dot marks the derivative over the moment N. Solving (5.37a) iteratively gives

$$\gamma_\sigma = \mathcal{P} + \sigma\mathcal{P}\dot{\mathcal{P}} + \left[\mathcal{P}\dot{\mathcal{P}}^2 + \tfrac{1}{2}\mathcal{P}^2\ddot{\mathcal{P}}\right] + \cdots . \tag{5.37b}$$

Now, restricting to the first loop, $\mathcal{P}^{(1)} = \mathcal{O}\,(as)$,

$$\gamma_\sigma = \mathcal{P}^{(1)} + \sigma\mathcal{P}^{(1)}\dot{\mathcal{P}}^{(1)} + \cdots , \tag{5.38}$$

the second term on the r.h.s. of (5.38) generates the GLR breaking observed in the second loop for non-singlet quark→quark evolution [Curci, Furmanski and Petronzio (1980)]. This shows that the GLR violation is not a dynamical higher order effect but is *inherited* from the previous loop(s) via a non-linear relation. In three loops the GLR breaking predicted by (5.35) was verified in [Mitov, Moch and Vogt (2006)].

The same structure of the GLR violation holds for the gluon→gluon evolution as well. Strictly speaking, this is true only for two colour structures that emerge in the second loop anomalous dimensions $\mathcal{P}_{gg}^{(2)}$ namely, C_A^2 and $C_A C_F$. The third one, $2n_f T_R C_F$, corresponds to the $g \to q(\bar{q}) \to g$ two-step transition that mixes gluon and quark states. This colour factor is also present in the *singlet* quark evolution, $q \to g \to q$, described by the two-loop anomalous dimensions $\mathcal{P}_{qq,s}^{(2)}$. As a result, the structure of the GLR breaking in this specific colour structure turned out to be more involved though natural:

$$\tfrac{1}{2}[\mathcal{P}_{qq,s}^{(2,T)} - \mathcal{P}_{gg}^{(2,S)}] \Longrightarrow \mathcal{P}_{gq}^{(1)}\dot{\mathcal{P}}_{qg}^{(1)}, \quad \tfrac{1}{2}[\mathcal{P}_{gg}^{(2,T)} - \mathcal{P}_{qq,s}^{(2,S)}] \Longrightarrow \mathcal{P}_{qg}^{(1)}\dot{\mathcal{P}}_{gq}^{(1)}.$$

The analysis of non-diagonal parton transitions is more difficult since here the scheme dependence is more pronounced [Stratmann and Vogelsang (1997)].

Deep universal nature of the reciprocity respecting evolution equation (RREE) was elucidated by Basso and Korchemsky who have derived (5.36) from the *conformal invariance*. They have verified validity of the RREE for various anomalous dimensions that had been calculated to higher orders in different quantum field theories [Basso and Korchemsky (2007)].

For extension of the GLR to higher orders and link to string theory see [Beccaria, Forini and Macorini (2010); Velizhanin (2011)].

5.7.1 *Applications*

An important aspect of the RREE is the *dual nature* of the expansion (5.37): it is obviously a series in the coupling ($\mathcal{P} \propto \alpha_s$); at the same time, it is an expansion in N^{-1}. Indeed, the leading behaviour of the evolution kernel is logarithmic, $\mathcal{P}(N) \propto \alpha_s \ln N$, so that each successive term in (5.37b) acquires a $1/N$ suppression factor: $\dot{\mathcal{P}} \sim N^{-1}$, $\ddot{\mathcal{P}} \sim \dot{\mathcal{P}}^2 \sim N^{-2}$, etc. In the x space, N^{-1} translates into $(1-x)$ — another expansion parameter that becomes small in the quasi-elastic kinematics when the invariant mass of the final state hadron system is much smaller than the hardness scale of the process: $W^2 \simeq Q^2(1-x)/x \ll Q^2$.

In the $x \to 1$ limit (large moments N) the diagonal anomalous dimensions ($q \to q$, $g \to g$) have a general structure

$$\gamma(x) = \frac{Ax}{(1-x)_+} + B\delta(1-x) + C\ln(1-x) + D + \mathcal{O}\left((1-x)\log^p(1-x)\right), \tag{5.39}$$

where A, B, C, D, are series in α known nowadays to three loops. In particular,

$$C_F{}^{-1} \cdot \frac{A_q}{4\pi} = C_A{}^{-1} \cdot \frac{A_g}{4\pi} \equiv \alpha_{\text{phys}} \tag{5.40}$$

is the *physical coupling* (cusp anomalous dimension [Korchemsky (1989)]) that determines the radiation intensity of relatively soft gluons [Catani, Marchesini and Webber (1991); Dokshitzer, Khoze and Troian (1996)]. It universally appears in higher order corrections to observables that are sensitive to soft gluon emission: anomalous dimensions and coefficient functions in the quasi-elastic limit, Sudakov quark and gluon form factors (threshold resummation, the Drell–Yan K-factor, distributions of jet event shapes in the near-to-two-jet kinematics, heavy quark fragmentation functions, etc.), in quark and gluon Regge trajectories, even in power suppressed effects in jet shapes and elsewhere.

Soft radiation has a *classical nature* and is governed by the celebrated Low (LBK) theorem [Low (1958)]. LBK taught us that both the singular, $(1-x)^{-1}$, and the constant terms, $(1-x)^0$, in the photon/gluon emission are universal and contained by the first (A) structure in (5.39), while the quantum contributions vanish in the $x \to 1$ limit as $(1-x)$:

$$\frac{dw}{dx} = C_p \frac{\alpha_{\text{phys}}}{\pi}\left[\frac{x}{1-x} + (1-x)g_p(x;\alpha_{\text{s}})\right]. \tag{5.41}$$

The first singular term has classical nature and is universal. Meanwhile, quantum effects are encoded in g_p and depend on the nature of the radiating particle p ($g_{1/2} = \frac{1}{2} + \mathcal{O}(\alpha_{\text{s}})$ for quark/electron, $g_1 = x + x^{-1} + \mathcal{O}(\alpha_{\text{s}})$ for gluon, etc.).

The terms C and D in (5.39) fall into the gap between the classical and quantum physics. In the leading order, $C = D = 0$; beyond the leading order, these structures must be therefore *deducible* rather than genuine higher order corrections. Indeed, given the physical coupling $A(\alpha_{\text{s}})$ and the virtual correction term $B(\alpha_{\text{s}})$, from RREE follow [Dokshitzer, Marchesini and Salam (2006); Basso and Korchemsky (2007)] the *all-order* relations

$$C = -\sigma A^2, \tag{5.42a}$$

$$D = -\sigma A\, B + \mathcal{O}(\beta). \tag{5.42b}$$

These relations were found to hold in three loops by Moch, Vermasseren and Vogt [Moch, Vermaseren and Vogt (2004)].

Another interesting RREE offering is a possibility to link together two puzzling perturbative results that were never thought to be of a common origin. These are:

the absence of the α_s^2 and α_s^3 terms in the BFKL anomalous dimension in the DIS problem, and, on the other hand, the phenomenon of the *exact angular ordering* that was found to miraculously apply down to the next-to-next-to-leading order in e^+e^- parton cascades (for discussion see [Dokshitzer, Marchesini and Salam (2006)]).

Supersymmetric QFTs, and maximally super-symmetric $\mathcal{N} = 4$ Yang–Mills model in particular, occupy a special place in this list. Here, turning from $\gamma_\pm$ to the RREE kernel $\mathcal{P}$ reduces complexity of high order contributions by an order of magnitude[5] [Dokshitzer and Marchesini (2007)]. GLR has been checked in various multi-loop calculations of weakly coupled $\mathcal{N} = 4$ SUSY (for review see [Beccaria, Forini and Macorini (2010)]).

5.7.2 $\mathcal{N} = 4$ *SUSY*

Soft gluon radiation is responsible for the major part of high order contributions to the QCD anomalous dimensions. The fact that soft radiation is classical by nature (in the LBK sense) may feed an ambitious but still viable hope to be able to grasp its structure in all orders in α_s. $\mathcal{N} = 4$ SUSY provides a perfect ground for testing this idea.

In this model quantum effects due to boson and fermion fields compensate each other at one loop, resulting in $g_p = 0$ in (5.41). This leaves us with pure "classical physics" at the Born level. Quantum effects cancel also in the vacuum polarisation, resulting in $\beta(\alpha) \equiv 0$. Is this enough to suspect that the $\mathcal{N} = 4$ SUSY is actually devoid of quantum physics, being purely classical?

Another argument in favour of such suspicion goes as follows. Kotikov and Lipatov have observed that the two-loop anomalous dimension γ_{SYM} is built of the functions characterised by the same level of *transcendentality* [Kotikov and Lipatov (2000)]. The observation gave rise to the hypothesis that the kth order γ_{SYM} contains exclusively functions of transcendentality $\tau = 2k-1$ [Kotikov, Lipatov, Onishchenko and Velizhanin (2004)]. This allowed extraction of the third loop contribution to the $\mathcal{N} = 4$ SYM anomalous dimension by simply picking up the "most complicated terms" from the corresponding QCD result [Moch, Vermaseren and Vogt (2004)].

In fact, such functions (combinations of Euler–Zagier harmonic sums) are nothing but *finite leftovers* of real–virtual cancellation of infrared gluon divergences.

Classicality and Integrability (QCD). There are examples where significant simplifications occur as soon as LBK-classical gluons enter the stage. In the first place, it is the famous Parke–Taylor QCD amplitudes [Parke and Taylor (1986)]. They describe radiation of arbitrary number of gluons with "classical" helicities and are known *exactly* in *all orders*. A number of specific QCD problems happen to be

[5] When measured by amount of ink necessary to write down the answer.

fully solvable as they reduce to *integrable* dynamical models. The list includes (but is not exhausted by)

- high energy Regge asymptotics of scattering amplitudes in planar limit ($N_c \to \infty$) [Lipatov (1993); Faddeev and Korchemsky (1995)],
- spin-3/2 baryon distribution function in $z \to 1$ kinematics [Braun, Derkachov and Manashov (1998)],
- maximal helicity multi-parton operators [Braun, Derkachov, Korchemsky and Manashov (1999)],
- evolution equations for maximal helicity three–parton states [Belitsky (1999)], etc.

In all these cases the problem is such that it is classical gluons only that contribute to the answer in the chosen approximation.

Integrability and Classicality ($\mathcal{N} = 4$ SUSY). Turning to $\mathcal{N} = 4$ SUSY, it is tempting to revert the argument. Indeed, there is a strong evidence for this theory to be *integrable* via string–field duality, the AdS/CFT correspondence (see [Beisert (2004)] for discussion and references).

Integrability may mean that the theory is essentially classical, devoid of truly quantum effects. It is imaginable that one day a clever formulation of the theory will be found, reducing $\mathcal{N} = 4$ SUSY to a free-field dynamics, allowing to set the Planck constant to zero.

Should such a programme work, this would explain many a mysterious result, unprecedented for QFTs, such as calculability of high order multi-leg amplitudes, existence of an all-order expansion for the physical coupling [Beisert, Eden and Staudacher (2007)], all the way up to amazingly simple relations between all–order expressions for *event shapes* [Belitsky, Hohenegger, Korchemsky, Sokatchev and Zhiboedov (2014)].

On a practical side, full understanding of self-interacting classical gluon fields via $\mathcal{N} = 4$ SUSY should help QCD practitioners to control, in *all orders* of perturbative theory, the major part of the quark–gluon parton dynamics.

5.8 Non-scientific conclusion

In 1976 I found myself in Lev's shoes, about to be kicked out from the LNPI theory group. Two thirds of the time allocated for my PhD thesis were over, and I still could not find an interesting project to work on. Alexei Anselm took a pity on me.

— *No, it's too easy. Virtually nothing to do.*
— *No, too difficult. All those matrices...*

These were two independent expert advises that he received from Lev Lipatov and Vladimir Gribov, respectively. The question was: is it worth proposing to a failing student to carry out the Gribov–Lipatov programme in non-Abelian field theory? Anselm wisely took an average; this is how I got involved with QCD for life.

Bibliography

Abreu, P. *et al.*, DELPHI Collab. (2000). Measurement of the gluon fragmentation function and a comparison of the scaling violation in gluon and quark jets, *Eur. Phys. J. C* **13**, pp. 573–589.

Altarelli, G. (2009). DIS 2009 Concluding talk: Outlook and perspective, *Prog. High Energy Phys.* **3**, p. 238.

Altarelli, G. and Parisi, G. (1977). Asymptotic freedom in parton language, *Nucl. Phys. B* **126**, pp. 298–318.

Basso, B. and Korchemsky, G. P. (2007). Anomalous dimensions of high-spin operators beyond the leading order, *Nucl. Phys. B* **775**, pp. 1–30.

Beccaria, M., Forini, V., and Macorini, G. (2010). Generalized Gribov–Lipatov reciprocity and AdS/CFT, *Adv. High Energy Phys.*, p. 753248.

Beisert, N. (2004). The dilatation operator of $\mathcal{N} = 4$ super Yang–Mills theory and integrability, *Phys. Rept.* **405**, pp. 1–202.

Beisert, N., Eden, B., and Staudacher, M. (2007) *J. Stat. Mech.* **0701**.

Belitsky, A. V. (1999). Fine structure of spectrum of twist–three operators in QCD Phys. Lett. **B453**, pp. 59–72; Integrability and WKB solution of twist–three evolution equations, *Nucl. Phys. B* **558**, pp. 259–284.

Belitsky, A. V. Hohenegger, S., Korchemsky, G. P. Sokatchev, E., and Zhiboedov, A. (2014). Event shapes in $\mathcal{N} = 4$ Super–Yang–Mills theory, *Nucl. Phys. B* **884**, pp. 206–256.

Blümlein, J., Ravindran, V., and van Neerven, W. L. (2000). On the Drell–Levy–Yan relation to $\mathcal{O}\left(\alpha_s^2\right)$, *Nucl. Phys. B* **586**, pp. 349–381.

Braun, V. M., Derkachov, S. E., Korchemsky, G. P., and Manashov, A. N. (1999). Baryon distribution amplitudes in QCD, *Nucl. Phys. B* **553**, pp. 355–426.

Braun, V. M., Derkachov, S. E., and Manashov, A. N. (1998). Integrability of three particle evolution equations in QCD, *Phys. Rev. Lett. B* **81**, pp. 2020–2023.

Braun, V. M., Korchemsky, G. P., and Mueller, D. (2003). The uses of conformal symmetry in QCD, *Prog. Part. Nucl. Phys.* **51**, pp. 311–398.

Bukhvostov, A. P., Lipatov, L. N., and Popov, N. P. (1974). Parton distribution functions in perturbation theory, *Yad. Fiz.* **20**, pp. 532–548.

Catani, S., Marchesini, G., and Webber, B. R. (1991). QCD coherent branching and semiinclusive processes at large x, *Nucl. Phys. B* **349**, pp. 635–654.

Christ, N. H., Hasslacher, B., and Mueller, A. H. (1972). Light cone behavior of perturbation theory, *Phys. Rev. D* **6**, p. 3543.

Curci, G., Furmanski, W., and Petronzio, R. (1980). Evolution of parton densities beyond leading order: The nonsinglet case, *Nucl. Phys. B* **175**, pp. 27–92.

Dokshitzer, Yu. L. (1977). Calculation of the structure functions for deep inelastic scattering and e^+e^- annihilation by perturbation theory in quantum chromodynamics, *Sov. Phys. JETP* **46**, pp. 641–653.

Dokshitzer, Yu. L., Diakonov, D. I., and Troian, S. I. (1980). Hard processes in quantum chromodynamics, *Phys. Rept.* **58**, pp. 269–395.

Dokshitzer, Yu. L., Khoze, V. A., Mueller, A. H., and Troian, S. I. (1988). QCD coherence in high-energy reactions, *Rev. Mod. Phys.* **60**, p. 373.

Dokshitzer, Yu. L., Khoze, V. A., and Troian, S. I. (1996). Specific features of heavy quark production. LPHD approach to heavy particle spectra, *Phys. Rev. D* **53**, pp. 89–119.

Dokshitzer, Yu. L. and Marchesini, G. (2007) N=4 SUSY Yang-Mills: three loops made simple(r), *Phys. Lett. B* **646**, pp. 189–201.

Dokshitzer, Yu. L., Marchesini, G., and Salam, G. P. (2006). Revisiting parton evolution and the large-x limit *Phys. Lett. B* **634**, pp. 504–507.

Drell, S. D., Levy, D. J., and Yan, T.-M. (1970). A Theory Of Deep inelastic lepton nucleon scattering and lepton pair annihilation processes. 2. Deep inelastic electron scattering, *Phys. Rev. D* **1**, pp. 1035–1068; A Theory of Deep Inelastic Lepton–Nucleon Scattering and Lepton Pair Annihilation Processes. 3. Deep Inelastic Electron–Positron Annihilation, *ibid*: pp. 1617–1639.

Ermolaev, B. I. and Fadin, V. S. (1981). Log-Log asymptotic form of exclusive cross-sections in quantum chromodynamics, *JETP Lett.* **33**, pp. 269–272.

Etim, E. (1973). Gribov-Lipatov reciprocity from conservation constraints and $O(4)$ symmetry, *Phys. Lett. B* **45**, pp. 478–480.

Faddeev, L. D. and Korchemsky, G. P. (1995). High-Energy QCD as a completely integrable model, *Phys. Lett. B* **342**, pp. 311–322.

Ferrara, S., Gatto, R., and Parisi, G. (1973). A group theoretical derivation of the Gribov–Lipatov reciprocity relation, *Phys. Lett. B* **44**, pp. 381–382.

Furmanski, W. and Petronzio, R. (1980). Singlet parton densities beyond leading order, *Phys. Lett. B* **97**, pp. 437–442.

Goulianos, D. (1997). Jet fragmentation properties at the tevatron, in: *QCD and High Energy Hadronic Interactions.* Proceedings, 32nd Rencontres de Moriond, Les Arcs, France, March 22–29, 1997.

Gribov, V. N. and Lipatov, L. N. (1972). Deep inelastic *ep* scattering in perturbation theory, *Sov. J. Nucl. Phys.* **15**, pp. 438–450; e^+e^- Pair Annihilation and Deep Inelastic *ep* Scattering in Perturbation Theory, *ibid.*: pp. 675–684.

Gross, D. J. and Wilczek, F. (1973). Asymptotically free gauge theories – I, *Phys. Rev. D* **8**, pp. 3633–3652.

Korchemsky, G. P. (1989). Asymptotics of the Altarelli–Parisi–Lipatov evolution kernels of parton distributions, *Mod. Phys. Lett. A* **4**, pp. 1257–1276; Korchemsky, G. P. and Marchesini, G. (1993). Structure function for large x and renormalization of Wilson Loop, *Nucl. Phys. B* **406**, pp. 225–258.

Korytov, A. (1997). Inclusive momentum distributions of charged particles in jets at CDF, *Nucl. Phys. Proc. Suppl. A* **54**, pp. 67–70.

Kotikov, A. V. and Lipatov, L. N. (2000). NLO corrections to the BFKL equation in QCD and in supersymmetric gauge theories, *Nucl. Phys. B* 582, pp. 19 –43.

Kotikov, A. V., Lipatov, L. N., Onishchenko, A. I., and Velizhanin, V. N. (2004). Three loop universal anomalous dimension of the Wilson operators in $N = 4$ SUSY Yang-Mills model, *Phys. Lett. B* **595**, pp. 521–529; Erratum-*ibid.* **632** (2006) 754–756.

Lipatov, L. N. (1974). The parton model and perturbation theory, *Yad. Fiz.* **20**, pp. 181–198. [*Sov. J. Nucl. Phys.* **20**, (11975), pp. 94–102].

Lipatov, L. N. (1993). High-Energy asymptotics of multicolor QCD and two-dimensional conformal field theories, *Phys. Lett. B* **309**, pp. 394–396; Asymptotic behavior of multicolor QCD at high energies in connection with exactly solvable spin models, *JETP Lett.* **59** (1994), pp. 596–599.

Low, F. E. (1958). Bremsstrahlung of very low-energy quanta in elementary particle collisions, *Phys. Rev.* **110**, p. 974; Burnett, T. H. and Kroll, N. M. (1968). Extension of the Low Soft Photon Theorem. *Phys. Rev. Lett.* **20**, p. 86.

Mitov, A., Moch, S., and Vogt, A. (2006). Next-to-Next-to-Leading order evolution of non-singlet fragmentation functions, *Phys. Lett. B* **638**, pp. 61–67.

Moch, S., Vermaseren, J. A. M. and Vogt, A. (2004). The three loop splitting functions in QCD: The nonsinglet case, *Nucl. Phys. B* **688**, pp. 101–134.

Mueller, A. H. (1981). On the multiplicity of hadrons in QCD Jets, *Phys. Lett. B* **104**, pp. 161–164.

Parke, S. J. and Taylor, T. R. (1986). An amplitude for n Gluon scattering, *Phys. Rev. Lett.* **56**, pp. 2459.

Safonov, A. (2000). Fragmentation of CDF jets: Perturbative or non-perturbative?, *Nucl. Phys. Proc. Suppl. A* **86**, pp. 55–64.

Stratmann, M. and Vogelsang, W. (1997). Next-to-Leading order evolution of polarized and unpolarized fragmentation functions, *Nucl. Phys. B* **496**, pp. 41–65.

Velizhanin, V. N. (2011). Double-Logs, Gribov–Lipatov reciprocity and wrapping, *JHEP* **08**, p. 092.

Vogt, A., Moch, S., and Vermaseren, J. A. M. (2004). The three loop splitting functions in QCD: The singlet case, *Nucl. Phys. B* **691**, pp. 129–181.

https://doi.org/10.1142/9789811231124_0006

Chapter 6

Calculation of Green's Functions in $\mathcal{PT}$-Symmetric Quantum Field Theory

Carl M. Bender

Physics Department, Washington University,
St. Louis, MO 63130, USA

Abstract

$\mathcal{PT}$-symmetric quantum theory began with a study of the non-Hermitian Hamiltonian $H = p^2 + x^2(ix)^\varepsilon$. This Hamiltonian is $\mathcal{PT}$ invariant and the eigenvalues of H are *discrete, real, and positive* when $\varepsilon \geq 0$. Here we describe the properties of the corresponding quantum-field-theoretic Hamiltonian $H = \frac{1}{2}(\nabla\phi)^2 + \frac{1}{2}\phi^2(i\phi)^\varepsilon$ in D-dimensional Euclidean space-time, where ϕ is a pseudoscalar field. We show how to calculate the Green's functions as series in powers of ε directly from the partition function. We derive exact finite expressions for the vacuum energy density, the renormalized mass, and the connected n-point Green's functions to first order in ε for all n and for $0 \leq D < 2$. For $D \geq 2$ the one-point Green's function and the renormalized mass become infinite, but perturbative renormalization can be performed. The surprising spectral properties of $\mathcal{PT}$-symmetric quantum mechanics appear to persist in $\mathcal{PT}$-symmetric quantum field theory.

6.1 Introduction

I first met Lev Lipatov in the very early 1990s when he invited me to visit him at the Leningrad Nuclear Physics Institute. Lipatov then made an extended visit to me at

Washington University in St. Louis. Lipatov was a joy to work with because he had such a profound understanding of perturbation theory, complex variables, functional integration, and mathematical analysis. We had very strong common interests; we had both worked on the large-order behaviour of perturbation theory. Our collaboration led to two papers in which we proposed an unusual new perturbative approach in quantum mechanics and quantum field theory; namely, perturbation expansions in powers of D, the dimension of space-time [Bender *et al.* (1992a,b); Bender (2010)]. A surprising quantum-field-theoretic result was that we were able to calculate the coefficients of D^0 and D^1 in the dimensional expansion *exactly and in closed form.* This paper describes another unusual perturbation expansion in which the Green's functions of a $\mathcal{PT}$-symmetric quantum field theory are expressed in powers of ε, where ε is a measure of the *nonlinearity* of the theory. Once again, we can calculate the first two terms in this expansion exactly and in closed form for all of the Green's functions of the theory.

To explain the origin of $\mathcal{PT}$-symmetric quantum theory we consider the simple problem of finding the positive root of the polynomial equation

$$x^5 + x - 1 = 0. \tag{6.1}$$

We use a quintic polynomial for this illustration because there is no general formula for the roots of such a polynomial. Thus, (6.1) is a "difficult" problem and we use perturbative methods to find the root rather than trying to find an exact expression. We first try a *strong-coupling* expansion; that is, we insert a small parameter ε in front of the *lower* power of x,

$$x^5 + \varepsilon x - 1 = 0,$$

and then to seek a perturbative expansion of the form

$$x(\varepsilon) = \sum_{n=0}^{\infty} a_n \varepsilon^n, \quad a_0 = 1.$$

The perturbation coefficients a_n are easy to find because they satisfy a sequence of *linear* equations:

$$\begin{aligned} 5a_1 + 1 &= 0, \\ 5a_2 + 10a_1^2 + a_1 &= 0, \\ 5a_3 + 20a_1a_2 + a_2 + 10a_1^3 &= 0, \\ 5a_4 + 20a_1a_3 + a_3 + 10a_2^2 + 30a_1^2a_2 + 5a_1^4 &= 0, \end{aligned}$$

and so on. We solve these equations sequentially for the coefficients $\{a_n\}$ and give the perturbation series to sixth order in ε (the coefficient a_4 is zero):

$$x(\varepsilon) = 1 - \frac{1}{5}\varepsilon - \frac{1}{25}\varepsilon^2 - \frac{1}{125}\varepsilon^3 + \frac{21}{15625}\varepsilon^5 + \frac{78}{78125}\varepsilon^6 + \cdots. \tag{6.2}$$

We obtain the positive root of (6.1) by setting $\varepsilon = 1$ in (6.2) and summing the perturbation series at $\varepsilon = 1$. This procedure works well because the radius of convergence of the perturbation series (6.2) is $5/4^{4/5} = 1.64938\cdots$, which is greater than 1. Thus, to sixth order in powers of ε the sum of this series is $x(1) = 0.75434$ and the exact (numerical) answer is $x = 0.75488\cdots$. To sixth order in perturbation theory we get a relative error of 0.07%, which is very good.

On the other hand, a high-energy physicist might use a *weak-coupling* expansion in which the small parameter ε is inserted as the coefficient of the *highest* power of x in (6.1):

$$\varepsilon x^5 + x - 1 = 0.$$

The resulting perturbation series now has the form

$$x(\varepsilon) = 1 - \varepsilon + 5\varepsilon^2 - 35\varepsilon^3 + 285\varepsilon^4 - 2530\varepsilon^5 + 23751\varepsilon^6 - \cdots . \tag{6.3}$$

The coefficients in this series grow rapidly and unfortunately the radius of convergence of the series is $4^4/5^5 = 0.08192$, which is much less than 1. Thus, if we sum this sixth-order perturbation series directly at $\varepsilon = 1$ we get the ridiculous answer $x(1) = 21476$. However, a (3,3)-Padé evaluated at $\varepsilon = 1$ has a relative error of 0.7%, which is not bad.

An unusual way to introduce a small parameter ε into (6.1) is to insert it in the *exponent*:

$$x^{1+\varepsilon} + x - 1 = 0.$$

Now the perturbation parameter ε plays a new role; rather than acting as a coupling constant that measures the strength of an interaction, ε measures the *degree of nonlinearity* of the problem, that is, how much it departs from linearity. The first few terms in the perturbation series for $x(\varepsilon)$ are

$$x(\varepsilon) = \frac{1}{2} + \frac{\log 2}{4}\varepsilon - \frac{\log 2}{8}\varepsilon^2 + \left[-\frac{(\log 2)^3}{48} + \frac{(\log 2)^2}{32} + \frac{\log 2}{16}\right]\varepsilon^3 + \cdots .$$

The radius of convergence of this perturbation series is 1, but if we again construct a (3,3)-Padé and evaluate it at $\varepsilon = 4$, the accuracy is 0.5%, which is better than the weak-coupling result.

This last procedure, namely, introducing a perturbation parameter into the exponent so that it measures the nonlinearity of the problem rather than the strength of the coupling is a powerful new way to solve difficult nonlinear problems. It can be applied to both classical and quantum-mechanical problems.

For a classical illustration we consider the Thomas-Fermi equation,

$$y''(x) = [y(x)]^{3/2}/\sqrt{x}, \quad y(0) = 1, \quad y(\infty) = 0,$$

which is an approximate description of the charge density in a nucleus. The objective in solving this difficult and delicate boundary-value problem is to determine the initial condition $y'(0) = -1.5880710\cdots$. No exact analytical solution to this equation has been found but a novel way to attack this problem is to introduce the small parameter ε into the exponent

$$y''(x) = y(x)[y(x)/x]^{\varepsilon}$$

and to seek a perturbative solution in powers of ε:

$$y(x) = y_0(x) + y_1(x)\varepsilon + y_2(x)\varepsilon^2 + y_3(x)\varepsilon^3 + \cdots . \tag{6.4}$$

Note that ε measures the departure of this problem from the trivial linear problem $y''(x) = y(x)$, so the zeroth term in the perturbation expansion is $y_0(x) = e^{-x}$. It is easy to calculate the higher coefficients in this expansion and a (2,1)-Padé approximation evaluated at $\varepsilon = 1/2$ gives an accuracy of 1.1% [Bender *et al.* (1989)], which is quite impressive.

Another classical example is the Blasius equation, which describes the boundary layer in fluid flow across a flat surface:

$$y'''(x) = y(x)y''(x), \quad y(0) = y'(0) = 0, \quad y'(\infty) = 1.$$

While this is a difficult boundary value problem, it is easy to find the first three coefficients in the perturbation expansion (6.4) for the problem

$$y'''(x) = y''(x)[y(x)]^{\varepsilon}.$$

Once again, the parameter ε measures the departure from linearity. The (1,1)-Padé gives a numerical accuracy of 8.7%, which is impressive for this formidable problem [Bender *et al.* (1989)].

Other classical equations for which this perturbation technique gives accurate results are the Lane-Emden equation (stellar structure) $y''(x) + 2y'(x)/x + [y(x)]^n = 0$, which we solve by introducing ε as [Bender *et al.* (1989)]

$$y''(x) + 2y'(x)/x + [y(x)]^{1+\varepsilon} = 0,$$

and the Korteweg-de Vries equation (nonlinear waves) $u_t + uu_x + u_{xxx} = 0$, which we solve perturbatively by introducing ε as [Bender *et al.* (1989)]

$$u_t + u^{\varepsilon}u_x + u_{xxx} = 0.$$

6.2 Quantum mechanics

We emphasize that introducing a small parameter ε in the exponent so that ε measures the departure of an equation from linearity is a powerful technique for finding accurate solutions to nonlinear problems. In this section we show how this technique can be applied in quantum mechanics and in the next section we show how these ideas can be extended to quantum field theory.

Consider the anharmonic-oscillator Hamiltonian $H = p^2 + x^4$. This is a tough problem and no analytical expression for the eigenvalues of H has been found.

However, accurate numerical results can be obtained by studying the quantum-mechanical Hamiltonian $H = p^2 + x^2\left(x^2\right)^{\varepsilon}$. Here ε measures the departure of this Hamiltonian from the quantum-harmonic-oscillator Hamiltonian. Similarly, we can study a quantum field theory with a quartic selfinteraction term ϕ^4 by replacing ϕ^4 with $\phi^2\left(\phi^2\right)^{\varepsilon}$ [Bender *et al.* (1987, 1988)].

Note that we have inserted the small parameter ε so that x^2 and ϕ^2 and not x and ϕ that are raised to the power ε. This is because x and ϕ can be negative, and a negative number raised to a noninteger power is complex. It is crucial to avoid introducing complex numbers as an artefact of the perturbation procedure.

However, a extraordinary discovery was made in 1998. It was shown that while x^{ε} does indeed introduce spurious complex numbers as a consequence of the perturbation technique, one can avoid the appearance of complex numbers if one raises ix and not x to the power ε [Bender and Boettcher (1998)]. Indeed, as Fig. 6.1 shows, the spectrum of the non-Hermitian Hamiltonian

$$H = p^2 + x^2(ix)^{\varepsilon} \qquad (\varepsilon \text{ real}) \tag{6.5}$$

is entirely real, positive, and discrete for all $\varepsilon \geq 0$.

Note that when $\varepsilon \geq 0$, the eigenvalues are all real and positive, and they increase with increasing ε. When ε decreases below 0, the eigenvalues disappear into the complex plane as complex conjugate pairs. Eventually, only one real eigenvalue remains when ε is less than about -0.57, and as ε approaches -1 from above, this eigenvalue becomes infinite. The region $\varepsilon \geq 0$ in which the eigenvalues are all real is called the *region of unbroken $\mathcal{PT}$ symmetry* and the region $\varepsilon < 0$ in which some

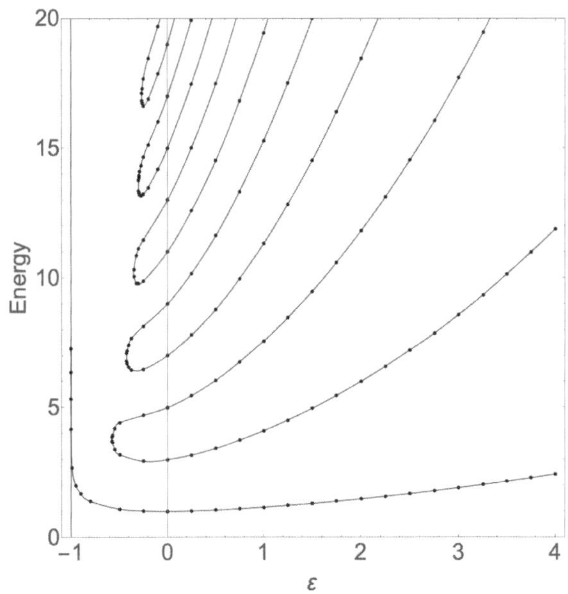

Fig. 6.1 Energy levels of the parametric family of Hamiltonians $H = p^2 + x^2(ix)^{\varepsilon}$ in (6.5).

of the eigenvalues are complex is called the *region of broken $\mathcal{PT}$ symmetry*. The transition point between these two regions is called the *$\mathcal{PT}$ phase transition*, and this transition has been observed in many published laboratory experiments.

In Ref. [Bender and Boettcher (1998)] it is argued that the reality of the spectrum is a consequence of the $\mathcal{PT}$ invariance ($\mathcal{PT}$ symmetry) of the Hamiltonian. (Under parity $\mathcal{P}$, $x \to -x$, and under time reversal $\mathcal{T}$, $i \to -i$, so the combination ix is $\mathcal{PT}$ invariant.) A proof that the eigenvalues are real was first formulated in Refs. [Dorey *et al.* (2001, 2007)]. The positivity of the spectrum is particularly remarkable for the case $\varepsilon = 2$ because in this case we have an "upside-down" $-x^4$ potential, and the conventional view is that such a potential is repulsive and could not have discrete positive bound-state energies!

In the region of unbroken $\mathcal{PT}$ symmetry ($\varepsilon \geq 0$), where the spectrum is entirely real, the Hamiltonian in (6.5) defines a fully consistent unitary quantum-mechanical theory [Bender (2007)]. One may think of a $\mathcal{PT}$-symmetric quantum theory as a complex deformation of a conventional quantum theory. Specifically, for the case of H in (6.5), ε plays the role of a deformation parameter and the theory that is being deformed is the quantum harmonic oscillator.[1]

There have been dozens of experimental studies of $\mathcal{PT}$-symmetric systems published in *Nature*, *Science*, and *Physical Review Letters* in diverse research areas such as optics [Guo *et al.* (2009); Longhi (2009); Rueter *et al.* (2010); Longhi (2010); Lin *et al.* (2011); Feng *et al.* (2011); Ramezani *et al.* (2012); Regensberger *et al.* (2012); Razzari and Morandotti (2012); Liang and Chong (2013); Luo *et al.* (2013); Regensberger *et al.* (2013); Hang *et al.* (2013)], NMR [Zheng *et al.* (2017)], superconductivity [Rubinstein *et al.* (2007); Chtchelkatchev *et al.* (2012)], microwave cavities [Dietz *et al.* (2011); Bittner *et al.* (2012)], lasers [Chong *et al.* (2011); Liertzer *et al.* (2012)], atomic diffusion [Zhao *et al.* (2010)], electronic circuits [Schindler *et al.* (2011); Bender *et al.* (2013b)], chaos and noise [Schomerus (2010); West *et al.* (2010)], mechanical systems [Bender *et al.* (2013a)], and graphene [Fagotti *et al.* (2011); Szameit *et al.* (2012)]. Some particularly interesting recent experimental work involves the study of $\mathcal{PT}$-symmetric whispering-gallery microcavities [Bender and Gianfreda (2013); Peng *et al.* (2014a,b)]. New experiments involving Bose-Einstein condensates are in the planning stage [Cartarius and Wunner (2012); Kreibich *et al.* (2013); Kartashov *et al.* (2013)] and new $\mathcal{PT}$-symmetric metamaterials are being developed [Castaldi *et al.* (2013); Lazarides and Tsironis (2013); Yin and Zhang (2013); Feng *et al.* (2013)]. There are also beautiful experiments on $\mathcal{PT}$-symmetric wireless power transfer [Assawawarrait *et al.* (2017)]. For a brief nontechnical description of $\mathcal{PT}$ symmetry see Ref. [Bender (2016)] and for detailed studies see [Bender *et al.* (2019); Christodoulides and Yang (2018)].

[1] A simpler way to perform a complex deformation of the harmonic oscillator is to use an *additive* rather than a *multiplicative* deformation: $H = p^2 + x^2 + ix\varepsilon$. The eigenvalues of this Hamiltonian are real and positive for all real ε: $E_n = 2n + 1 + \varepsilon^2/4$.

6.3 $\mathcal{PT}$-Symmetric quantum field theory

The main point of this paper is to show how to solve a $\mathcal{PT}$-symmetric quantum field theory described by the Euclidean Lagrangian density

$$\mathcal{L} = \tfrac{1}{2}(\nabla\phi)^2 + \tfrac{1}{2}\phi^2(i\phi)^\varepsilon \quad (\varepsilon \geq 0) \tag{6.6}$$

as a perturbation series in powers of ε [Bender *et al.* (2018)]. We treat the parameter ε as small and demonstrate how to calculate the vacuum energy density E_0, the connected n-point Green's functions G_n, and the renormalized mass $M_{\rm R}$ as series in powers of ε. We assume here that $0 \leq D < 2$ to avoid the appearance of renormalization infinities. Later we comment briefly on the perturbative renormalization procedure for the case $D \geq 2$.

Because we treat ε as small ($\varepsilon << 1$), the unusual Lagrangian density $\mathcal{L}$ in (6.6) develops a *logarithmic* self-interaction term:

$$\mathcal{L} = \tfrac{1}{2}(\nabla\phi)^2 + \tfrac{1}{2}\phi^2 + \tfrac{1}{2}\varepsilon\phi^2\log(i\phi) + \mathrm{O}(\varepsilon^2). \tag{6.7}$$

For a quantum field theory having a complex logarithmic interaction term it is not obvious how to find Feynman rules for performing perturbative diagrammatic calculations. We show here how to construct such Feynman rules. To begin, we replace the complex logarithm with a real logarithm in such a way as to preserve $\mathcal{PT}$ invariance. We define $\log(i\phi) \equiv \frac{1}{2}i\pi + \log(\phi)$ if $\phi > 0$ and we define $\log(i\phi) \equiv -\frac{1}{2}i\pi + \log(-\phi)$ if $\phi < 0$. From these two equations, we get

$$\log(i\phi) = \tfrac{1}{2}i\pi|\phi|/\phi + \tfrac{1}{2}\log(\phi^2). \tag{6.8}$$

In (6.8) the imaginary part is *odd* in ϕ and the real part is *even* in ϕ. Equation (6.8) enforces the $\mathcal{PT}$ symmetry because the pseudoscalar field ϕ changes sign under parity $\mathcal{P}$ and i changes sign under time reversal $\mathcal{T}$.

To derive (6.8) we assume that ϕ is *real*. This is justified because the functional integration

$$Z(\varepsilon) = \int \mathcal{D}\phi\, e^{-\int d^Dx\mathcal{L}} \tag{6.9}$$

is performed along the *real-ϕ* axis and not in the complex-ϕ domain. We are treating ε as small so we need not be concerned with complex functional integration paths that terminate in complex Stokes sectors. This is because the functional integral (6.9) converges term-by-term in powers of ε. In $\mathcal{PT}$-symmetric quantum mechanics the boundary conditions on the Schrödinger equation associated with the Hamiltonian $H = p^2 + x^2\left(x^2\right)^\varepsilon$ are imposed in complex Stokes sectors [Bender and Boettcher (1998)]. However, in the context of quantum field theory it would be difficult to consider *functional* Stokes sectors. This is why we treat ε as small.

We are concerned here with calculating the coefficients in the ε series and we defer the mathematical issues involved with the summation of such a series for large values of ε.

Graphical techniques were developed in Ref. [Bender *et al.* (1989)] to handle real logarithmic interaction terms. These techniques are generalizations of the replica trick [Mezard *et al.* (1987)], which has been used in the study of spin glasses. The idea of the replica trick is that a logarithmic term $\log A$ can be reformulated as the limit $\log A = \lim_{N\to 0} \frac{1}{N}\left(A^N - 1\right)$, or equivalently and slightly more simply as the limit

$$\log A = \lim_{N\to 0} \tfrac{d}{dN} A^N. \tag{6.10}$$

One then regards N as an integer and identifies A^N as an N-point vertex in a graphical expansion. This procedure is not rigorous because it requires taking the *continuous* limit $N \to 0$. The validity of this approach has not been established rigorously, but when we compare with exactly known results in low-dimensional theories, the replica trick gives the correct answer. We verify our field-theoretic results by comparing them with exact results for $D = 0$ (where the functional integral becomes an ordinary integral) and for $D = 1$ (quantum mechanics).

Our graphical calculations are done in coordinate space. Once the vertices are identified, we only need the free propagator in D-dimensional Euclidean space $\Delta(x-y)$, which satisfies the differential equation

$$\left(-\nabla_x^2 + 1\right)\Delta(x-y) = \delta^{(D)}(x-y). \tag{6.11}$$

The Fourier transform of this equation gives the amplitude for the free propagator of a particle of mass 1 in momentum space: $\tilde{\Delta}(p) = 1/\left(p^2+1\right)$. The D-dimensional inverse Fourier transform of this expression then gives the D-dimensional coordinate-space propagator in terms of an associated Bessel function:

$$\Delta(x_1 - x_2) = (2\pi)^{-D/2}|x_1 - x_2|^{1-D/2}\mathrm{K}_{1-D/2}(|x_1 - x_2|). \tag{6.12}$$

Letting $x_1 \to x_2$, we obtain $\Delta(0)$ for a *self loop* (the amplitude for a line to originate from and return to the same point in coordinate space):

$$\Delta(0) = (4\pi)^{-D/2}\Gamma(1 - D/2). \tag{6.13}$$

Note that $\Delta(0)$ is finite and nonsingular for $0 \le D < 2$.

Ground-state energy density: The partition function is the exponential of the ground-state energy density E_0 multiplied by the volume of space-time

V: $Z = e^{-E_0 V}$. The *shift* in the ground-state energy density ΔE to order ε is given by

$$\Delta E = \frac{\varepsilon}{4Z(0)V} \int \mathcal{D}\phi \, e^{-\int d^D x \mathcal{L}_0} \int d^D y \phi^2(y) \log\left[\phi^2(y)\right].$$

Hence, from (6.10) we obtain

$$\Delta E = \lim_{N\to 1} \frac{\varepsilon}{4Z(0)V} \frac{d}{dN} \int \mathcal{D}\phi \, e^{-\int d^D x \mathcal{L}_0} \int d^D y \phi^{2N}(y).$$

This expression has a graphical interpretation as the product of N self loops from the space-time point y back to y. There are exactly $(2N-1)!!$ ways to construct these self loops, so the expression for ΔE simplifies to

$$\Delta E = \lim_{N\to 1} \frac{\varepsilon}{4V} \frac{d}{dN} \int d^D y [\Delta(0)]^N (2N-1)!!.$$

The integral $\int d^D y$ is the volume of space-time V, so this simplifies to:

$$\Delta E = \lim_{N\to 1} \frac{\varepsilon}{4} \frac{d}{dN} [\Delta(0)]^N (2N-1)!!.$$

Finally, we use the identity $(2N-1)!! = 2^N \Gamma\left(N + \frac{1}{2}\right)/\sqrt{\pi}$ and then take the derivative with respect to N to get

$$\Delta E = \tfrac{1}{4}\varepsilon\Delta(0) \left\{\log[2\Delta(0)] + \Gamma'\left(\tfrac{3}{2}\right)/\Gamma\left(\tfrac{3}{2}\right)\right\}, \tag{6.14}$$

where $\Gamma'\left(\frac{3}{2}\right)/\Gamma\left(\frac{3}{2}\right) = 2 - \gamma - 2\log 2$. The result in (6.14) may be verified exactly for the special cases of $D = 0$ and $D = 1$.

One-point Green's function: Let us now perform a more sophisticated calculation. The one-point Green's function G_1 in $\mathcal{PT}$-symmetric quantum theory is *nonperturbative* in character but it can be calculated by following the approach used above to calculate ΔE. We keep terms that do not vanish under $\phi \to -\phi$, and evaluate directly the functional-integral representation

$$G_1(a) = -\frac{\varepsilon}{4Z(0)} \int \mathcal{D}\phi\phi(a) e^{-\int d^D x \mathcal{L}_0} \int d^D y i\pi\phi(y)|\phi(y)|, \tag{6.15}$$

where $\mathcal{L}_0 = \frac{1}{2}(\nabla\phi)^2 + \frac{1}{2}\phi^2$ is the free Euclidean Lagrangian.

We introduce the identity

$$\phi|\phi| = \frac{2}{\pi}\phi^2 \int_0^\infty \frac{dt}{t}\sin(t\phi) \tag{6.16}$$

to replace $\phi(y)|\phi(y)|$ in the functional integral (6.15) and then replace $\sin(t\phi)$ by its Taylor series:

$$\sin(t\phi) = \sum_{n=0}^{\infty} \frac{(-1)^n t^{2n+1}}{(2n+1)!}\phi^{2n+1}.$$

This converts (6.15) to the product of a one-dimensional integral in t, an infinite sum over n, a D-dimensional integral in y, and a functional integral in ϕ:

$$\begin{aligned} G_1(a) = &-\frac{i\varepsilon}{2}\int_{t=0}^{\infty} dt \sum_{n=0}^{\infty} \frac{(-1)^n t^{2n}}{(2n+1)!}\int d^D y \\ &\times \frac{1}{Z(0)}\int \mathcal{D}\phi\, e^{-\int d^D x \mathcal{L}_0}\phi(a)[\phi(y)]^{2n+3}. \end{aligned} \tag{6.17}$$

Like the calculation of the ground-state energy density, the functional integral in the second line of (6.17) has a graphical interpretation; it is the product of the free propagator $\Delta(a-y)$ representing a line from a to y multiplied by $n+1$ self loops from y to y, and this product is accompanied by the combinatorial factor $(2n+3)!!$. Thus, the second line in (6.17) reduces to $(2n+3)!!\Delta(y-a)\Delta^{n+1}(0)$.

This result simplifies further because the D-dimensional integral is trivial: $\int d^D y \Delta(y-a) = 1$. This establishes the translation invariance of G_1. The rest is straightforward:

$$\begin{aligned} G_1 &= -\frac{i\varepsilon}{2}\int_{t=0}^{\infty} dt \sum_{n=0}^{\infty} \frac{(-1)^n t^{2n}\Delta^{n+1}(0)(2n+3)!!}{(2n+1)!} \\ &= -\frac{i\varepsilon}{2}\int_{t=0}^{\infty} dt\, \Delta(0)\left[3-\Delta(0)t^2\right]e^{-\frac{1}{2}\Delta(0)t^2} \\ &= -i\varepsilon\sqrt{\pi\Delta(0)/2}. \end{aligned} \tag{6.18}$$

This expression for G_1 is *exact* to order ε.

Note that G_1 is a *negative imaginary* number. This is precisely what we would expect based on previous studies of classical $\mathcal{PT}$-symmetric systems. At a heuristic level, we can calculate the *classical* trajectories in complex coordinate space of a particle described by the Hamiltonian $H = p^2 + x^2(ix)^\varepsilon$. These trajectories are left-right symmetric but they lie mostly in the lower-half plane [Bender (2007)]. Thus, the average value of the classical orbits is a negative-imaginary number.

Two-point Green's function: We wish to obtain the *connected* two-point Green's function $G_2(a,b)$, and to do so we subtract G_1^2 from the vacuum expectation value of $\phi(a)\phi(b)$. Since G_1 is of order ε, we see that to first order in ε we need only evaluate $Z(\varepsilon)$ in (6.9) with $\phi(a)\phi(b)$ inserted after $\mathcal{D}\phi$ and then divide this integral by $Z(\varepsilon)$. We neglect the imaginary terms in these integrals because they are odd under $\phi \to -\phi$. Thus, to obtain $G_2(a,b)$ we must calculate

$$\frac{\int \mathcal{D}\phi\phi(a)\phi(b)e^{-\int d^Dx\mathcal{L}_0}\left\{1-\frac{\varepsilon}{4}\int d^Dy\phi^2(y)\log\left[\phi^2(y)\right]\right\}}{\int \mathcal{D}\phi e^{-\int d^Dx\mathcal{L}_0}\left\{1-\frac{\varepsilon}{4}\int d^Dy\phi^2(y)\log\left[\phi^2(y)\right]\right\}}.$$

We expand this expression to first order in ε as a sum of three functional integrals: $G_2(a,b) = A + B + C$, where

$$\begin{aligned}
A &= \frac{1}{Z(0)}\int \mathcal{D}\phi\phi(a)\phi(b)e^{-\int d^Dx\mathcal{L}_0},\\
B &= -\frac{\varepsilon}{4Z(0)}\int \mathcal{D}\phi\phi(a)\phi(b)e^{-\int d^Dx\mathcal{L}_0}\int d^Dy\phi^2(y)\log\left[\phi^2(y)\right],\\
C &= \frac{\varepsilon}{4Z(0)}\int \mathcal{D}\phi\phi(a)\phi(b)e^{-\int d^Dx\mathcal{L}_0}\\
&\quad\times\frac{1}{Z(0)}\int \mathcal{D}\phi e^{-\int d^Dx\mathcal{L}_0}\int d^Dy\phi^2(y)\log\left[\phi^2(y)\right].
\end{aligned}\tag{6.19}$$

We then evaluate the three contributions A, B, and C. The functional integral A in (6.19) is just the free propagator $\Delta(a-b)$ and this verifies that if we set $\varepsilon = 0$ in the Lagrangian (6.6), we obtain a free field theory; the two-point Green's function for such a field theory is $\Delta(a-b)$.

The double functional integral C is slightly more complicated. The first line of C is proportional to A and evaluates to $\frac{1}{4}\varepsilon\Delta(a-b)$. However, we saw in our calculation of the ground-state energy that the next two lines of C evaluate to $\frac{4}{\varepsilon}V\Delta E$, where ΔE is the first-order shift in the ground-state energy density (6.14) and V is the volume of Euclidean space-time Thus, $C = \Delta(a-b)V\Delta E$. This quantity is *divergent* because V is infinite.

To resolve this divergence problem we calculate B. We use the identity in (6.10) to express the B integral as a limit:

$$B = -\frac{\varepsilon}{4Z(0)}\lim_{N\to 1}\frac{d}{dN}\int \mathcal{D}\phi\, e^{-\int d^Dx\mathcal{L}_0}\int d^Dy\phi(a)\phi(b)\phi^{2N}(y).\tag{6.20}$$

The interpretation of this functional integral is that we must connect the set of $2N+2$ points consisting of a, b, and the $2N$ points y pairwise with the free propagator Δ in (6.12). We must consider two cases: In the first case a is connected to b and the remaining $2N$ points at y are connected in pairs. This reproduces the result above for C except with the opposite sign. Thus, the volume divergences exactly cancel!

In the second case a is not connected to b. Instead, a connects to a point y (there are $2N$ ways to do this) and b connects to one of the remaining $2N-1$ points y (there are $2N-1$ ways to do this). The rest of the $2N-2$ points y are joined in pairs [there are $(2N-3)!!$ ways to do this]. Thus, the amplitude for this case is

$$-\frac{\varepsilon}{2}\int d^D y \Delta(a-y)\Delta(y-b) \lim_{N\to 1}\frac{d}{dN}N(2N-1)!!\Delta^{N-1}(0),$$

which simplifies to $-\varepsilon K\int d^D y\Delta(a-y)\Delta(y-b)$, where

$$K=\tfrac{1}{2}+\tfrac{1}{2}\Gamma'\left(\tfrac{3}{2}\right)/\Gamma\left(\tfrac{3}{2}\right)+\tfrac{1}{2}\log[2\Delta(0)]=\tfrac{3}{2}-\tfrac{1}{2}\gamma+\tfrac{1}{2}\log\left[\tfrac{1}{2}\Delta(0)\right]. \quad (6.21)$$

Hence, our result for the coordinate-space two-point Green's function to order ε is

$$G_2(a-b)=\Delta(a-b)-\varepsilon K\int d^D y\Delta(a-y)\Delta(y-b), \quad (6.22)$$

and in momentum space this becomes

$$\tilde{G}_2(p)=\frac{1}{p^2+1}-\varepsilon\frac{K}{(p^2+1)^2}+\mathrm{O}(\varepsilon^2). \quad (6.23)$$

From (6.23) we can construct the $(0,1)$ Padé approximant, which is merely the geometric sum of a chain of bubbles:

$$\tilde{G}_2(p)=\frac{1}{p^2+1+\varepsilon K+\mathrm{O}(\varepsilon^2)}. \quad (6.24)$$

We then read off the *square of the renormalized mass* to first order in ε:

$$M_{\mathrm{R}}^2=1+K\varepsilon+\mathrm{O}(\varepsilon^2). \quad (6.25)$$

Higher Green's functions: The *connected* Green's functions are obtained by evaluating cumulants. For example, the connected three-point Green's function is given by the cumulant

$$\begin{aligned}G_3(x,y,z)=&\frac{\langle\phi(x)\phi(y)\phi(z)\rangle}{Z}-\frac{\langle\phi(x)\phi(y)\rangle\langle\phi(z)\rangle}{Z^2}\\&-\frac{\langle\phi(x)\phi(z)\rangle\langle\phi(y)\rangle}{Z^2}-\frac{\langle\phi(y)\phi(z)\rangle\langle\phi(x)\rangle}{Z^2}+2\frac{\langle\phi(x)\rangle\langle\phi(y)\rangle\langle\phi(z)\rangle}{Z^3},\end{aligned}$$

and the connected four-point function is defined by the cumulant

$$\begin{aligned}G_4(x,y,z,w)=&-\frac{\langle\phi(x)\phi(y)\phi(z)\phi(w)\rangle}{Z}\\&-\frac{\langle\phi(x)\phi(y)\phi(z)\rangle\langle\phi(w)\rangle}{Z^2}-\text{(three permutations)}\\&-\frac{\langle\phi(x)\phi(y)\rangle\langle\phi(z)\phi(w)\rangle}{Z^2}-\text{(two permutations)}\\&+2\frac{\langle\phi(x)\phi(y)\rangle\langle\phi(z)\rangle\langle\phi(w)\rangle}{Z^3}+\text{(five permutations)}\\&-6\frac{\langle\phi(x)\rangle\langle\phi(y)\rangle\langle\phi(z)\rangle\langle\phi(w)\rangle}{Z^4}.\end{aligned}$$

Evaluating the integrals is tedious but not difficult, and we get an exact general formula for the nth Green's function that is valid for all $n \neq 2$:

$$G_n(x_1, x_2, \ldots x_n) = -\tfrac{1}{2}\varepsilon(-i)^n \Gamma\left(\tfrac{1}{2}n - 1\right)\left[\tfrac{1}{2}\Delta(0)\right]^{1-n/2} \times \int d^D u \prod_{k=1}^{n} \Delta\left(x_k - u\right). \tag{6.26}$$

All of the connected Green's functions are of order ε except for G_2 in (6.22), which is of order 1. Note that (6.26) reduces to (6.18) for $n = 1$.

The formula (6.26) holds for both odd and even n. Our calculation of the odd-n Green's functions uses the integral representation in (6.16). By contrast, our calculation of the even-n Green's functions uses the identity in (6.10). It is quite amazing that these two extremely different techniques lead to the single universal formula in (6.26) for G_n.

6.4 Discussion and future work

In this paper we have presented the machinery necessary to calculate the Green's functions of the $\mathcal{PT}$-symmetric quantum field theory in (6.6) to any order in powers of ε. Thus, we have opened a vast area for future study and investigation; one can calculate the masses of the theory (poles of the Green's functions), scattering amplitudes, critical indices, and so on. The Green's-function calculations presented here are exact to first order in ε. However, the procedures discussed generalize to all orders in powers of ε.

This is useful because even in low orders the perturbation series in powers of ε is highly accurate and it continues to be accurate for large ε. To demonstrate this we calculate the one-point Green's function G_1 in $D = 0$ to *second* order in ε. In $D = 0$ this Green's function is a ratio of two *ordinary integrals* (not functional integrals):

$$G_1 = \frac{\int_{-\infty}^{\infty} dx\, x \exp\left[-\frac{1}{2}x^2\left(1 + \varepsilon L + \frac{1}{2}\varepsilon^2 L^2\right)\right]}{\int_{-\infty}^{\infty} dx\, \exp\left[-\frac{1}{2}x^2\left(1 + \varepsilon L\right)\right]}, \tag{6.27}$$

where $L = \log(ix) = \frac{1}{2}i\pi|x|/x + \frac{1}{2}\log\left(x^2\right)$. Evaluating these integrals is easy and we get

$$G_1 = -i\varepsilon\sqrt{\tfrac{\pi}{2}}\left[1 + \tfrac{1}{4}\varepsilon\left(\gamma - 2 - 3\log 2\right) + \mathrm{O}\left(\varepsilon^2\right)\right]$$
$$= -i\varepsilon\sqrt{\tfrac{\pi}{2}}\left[1 - 0.8756\varepsilon + \mathrm{O}\left(\varepsilon^2\right)\right]. \tag{6.28}$$

To check the accuracy of (6.28) we calculate the one-point Green's function for a cubic theory ($\varepsilon = 1$). We convert the expansion in (6.28) to a $[0, 1]$-Padé approximant,

$$G_1 = -i\varepsilon\sqrt{\tfrac{\pi}{2}}\frac{1}{1+0.8756\varepsilon},$$

and then set $\varepsilon = 1$ to obtain the approximate result $G_1 = -0.6682i$.

The exact value of G_1 for the zero-dimensional cubic theory $\varepsilon = 1$ is the ratio of integrals

$$G_1 = \frac{\int_{-\infty}^{\infty} dx\, x \exp\left(-\frac{1}{2}ix^3\right)}{\int_{-\infty}^{\infty} dx \exp\left(-\frac{1}{2}ix^3\right)} = -i2^{\frac{1}{3}}\frac{\Gamma(2/3)}{\Gamma(1/3)} = -0.6369i.$$

Thus, the *two*-term ε expansion (6.28) has an accuracy of 5%, which is impressive for such a large value of ε. This good result is consistent with the results found in previous studies of the accuracy of the ε expansion for various classical equations (see Ref. [Bender *et al.* (1989)]).

The second-order version of (6.28) is

$$\begin{aligned} G_1 &= -i\varepsilon\sqrt{\tfrac{\pi}{2}}\left[1+\tfrac{1}{4}\varepsilon(\gamma-2-3\log 2)+\tfrac{1}{192}\varepsilon^2\left(54\log^2 2+144\log 2\right.\right.\\ &\qquad \left.\left.-36\gamma\log 2-\pi^2+6\gamma^2-48\gamma+48\right)+\mathrm{O}\left(\varepsilon^3\right)\right]\\ &= -i\varepsilon\sqrt{\tfrac{\pi}{2}}\left[1-0.8756\varepsilon+0.6447\varepsilon^2+\mathrm{O}\left(\varepsilon^3\right)\right]. \end{aligned}$$

Converting this expansion in square brackets to a $[1,1]$ Padé approximant and setting $\varepsilon = 1$, we obtain the result that $G_1 = -0.6213i$, which now differs from the exact result by only -2%. These nice numerical results motivate us to extend our studies of the ε expansion of $\mathcal{PT}$-symmetric quantum field theories to higher order in ε.

A second issue that should be examined in depth is that of renormalization. Because $\Delta(0)$ becomes singular when the dimension D of Euclidean space-time reaches 2, the one-point Green's function G_1 and the renormalized mass M_{R} become singular. (To first order in ε the higher Green's functions do not become infinite when $D = 2$.) Thus, for $D \geq 2$ we must undertake a perturbative renormalization procedure.

In this presentation we have worked entirely with dimensionless quantities. However, to carry out a perturbative renormalization one must use the Lagrangian

$$\mathcal{L} = \tfrac{1}{2}(\nabla\phi)^2 + \tfrac{1}{2}\mu^2\phi^2 + \tfrac{1}{2}g\mu_0^2\phi^2\left(i\mu_0^{1-D/2}\phi\right)^{\varepsilon}$$

for which the dimensional parameters are explicit: μ is the unrenormalized mass μ_0 is a fixed parameter having dimensions of mass, and g is a dimensionless unrenormalized coupling constant. The mass renormalization procedure consists of expressing the renormalized mass M_{R} in terms of these Lagrangian parameters and absorbing the divergence that arises when $D \geq 2$ into parameter μ. Coupling-constant renormalization is similar; we define the renormalized coupling constant G_{R} as the value of the three-point or four-point Green's functions at particular values of the external momentum and again absorb the divergence that arises into the Lagrangian parameter g. One must then verify that all higher Green's functions are finite when expressed in terms of M_{R} and G_{R}. This program will be studied in detail in the future.

CMB is supported in part by a research award from the Alexander von Humboldt Foundation.

Bibliography

Assawawarrait, S., Yu, X., and Fan, S. (2017). Robust wireless power transfer using a nonlinear parity-time-symmetric circuit, *Nature* **546**, p. 387.

Bender, C. M. (2007). Making sense of non-Hermitian Hamiltonians, *Rept. Prog. Phys.* **70**, p. 947, doi:10.1088/0034-4885/70/6/R03, arXiv:hep-th/0703096.

Bender, C. M. (2010). Almost zero-dimensional PT-symmetric quantum field theories, pp. 7–16, arXiv:1003.3881 [hep-th].

Bender, C. M. (2016). $\mathcal{PT}$ symmetry in quantum physics, *Europhysics News* **472**, p. 17.

Bender, C. M., Berntson, B., Parker, D., and Samuel, E. (2013a). Observation of PT phase transition in a simple mechanical system, *Am. J. Phys.* **81**, pp. 173–179, doi: 10.1119/1.4789549, arXiv:1206.4972 [math-ph].

Bender, C. M. and Boettcher, S. (1998). Real spectra in nonHermitian Hamiltonians having PT symmetry, *Phys. Rev. Lett.* **80**, pp. 5243–5246, doi:10.1103/PhysRevLett.80. 5243, arXiv:physics/9712001.

Bender, C. M., Boettcher, S., and Lipatov, L. (1992a). Almost zero-dimensional quantum field theories, *Phys. Rev. D* **46**, pp. 5557–5573, doi:10.1103/PhysRevD.46.5557.

Bender, C. M., Boettcher, S., and Lipatov, L. (1992b). Dimensional expansions, *Phys. Rev. Lett.* **68**, pp. 3674–3677, doi:10.1103/PhysRevLett.68.3674.

Bender, C. M., Dorey, P. E., Dunning, C., Fring, A., Hook, D. W., Jones, H. F., Kuzhel, S., Lvai, G., and Tateo, R. (2019). *PT Symmetry* (WSP), doi:10.1142/q0178.

Bender, C. M. and Gianfreda, M. (2013). Twofold transition in PT-symmetric coupled oscillators, *Phys. Rev. A* **88**, 6, p. 062111, doi:10.1103/PhysRevA.88.062111, arXiv:1305.7107 [hep-th].

Bender, C. M., Hassanpour, N., Klevansky, S., and Sarkar, S. (2018). *PT*-symmetric quantum field theory in *D* dimensions, *Phys. Rev. D* **98**, 12, p. 125003, doi: 10.1103/PhysRevD.98.125003, arXiv:1810.12479 [hep-th].

Bender, C. M., Milton, K., Moshe, M., Pinsky, S., and Simmons, L. (1987). Logarithmic approximations to polynomial lagrangians, *Phys. Rev. Lett.* **58**, pp. 2615–2618, doi: 10.1103/PhysRevLett.58.2615.

Bender, C. M., Milton, K. A., Moshe, M., Pinsky, S. S., and Simmons, J., L.M. (1988). Novel perturbative scheme in quantum field theory, *Phys. Rev. D* **37**, p. 1472, doi: 10.1103/PhysRevD.37.1472.

Bender, C. M., Milton, K. A., Pinsky, S. S., and Simmons, J., L.M. (1989). A new perturbative approach to nonlinear problems, *J. Math. Phys.* **30**, p. 1447, doi: 10.1063/1.528326.

Bender, N., Factor, S., Bodyfelt, J. D., Ramezani, H., Christodoulides, D. N., Ellis, F. M., and Kottos, T. (2013b). Observation of asymmetric transport in structures with active nonlinearities, *Phys. Rev. Lett.* **110**, p. 2341012.

Bittner, S., Dietz, B., Guenther, U., Harney, H. L., Miski-Oglu, M., Richter, A., and Schaefer, F. (2012). $\mathcal{PT}$ symmetry and spontaneous symmetry breaking in microwave billiards, *Phys. Rev. Lett.* **108**, p. 024101.

Cartarius, H. and Wunner, G. (2012). Model of a PT-symmetric Bose-Einstein condensate in a delta-function double-well potential, *Phys. Rev. A* **86**, p. 013612, doi:10.1103/PhysRevA.86.013612, arXiv:1203.1885 [quant-ph].

Castaldi, G., Savoia, S., Galdi, V., Alu, A., and Engheta, N. (2013). $\mathcal{PT}$ metamaterials via complex coordinate transformation optics, *Phys. Rev.Lett.* **110**, p. 173901.

Chong, Y. D., Ge, L., and Stone, A. D. (2011). $\mathcal{PT}$-symmetry breaking and laser-absorber modes in optical scattering systems, *Phys. Rev. Lett.* **106**, p. 093902.

Christodoulides, D. N. and Yang, J. (2018). *Parity-Time Symmetry and its Applications*, Vol. 280 (Springer Tracts in Modern Physics, Heidelberg).

Chtchelkatchev, N., Golubov, A., Baturina, T., and Vinokur, V. (2012). Stimulation of the fluctuation superconductivity by $\mathcal{PT}$ symmetry, *Phys.Rev. Lett.* **109**, p. 150305.

Dietz, B., Harney, H., Kirillov, O., Miski-Oglu, M., Richter, A., and Schafer, F. (2011). Exceptional points in a microwave billiard with time-reversal invariance violation, *Phys. Rev. Lett.* **106**, p. 150403, doi:10.1103/PhysRevLett.106.150403, arXiv:1008.2623 [nlin.CD].

Dorey, P., Dunning, C., and Tateo, R. (2001). Spectral equivalences, Bethe Ansatz equations, and reality properties in PT-symmetric quantum mechanics, *J. Phys. A* **34**, pp. 5679–5704, doi:10.1088/0305-4470/34/28/305, arXiv:hep-th/0103051.

Dorey, P., Dunning, C., and Tateo, R. (2007). The ODE/IM Correspondence, *J. Phys. A* **40**, p. R205, doi:10.1088/1751-8113/40/32/R01, arXiv:hep-th/0703066.

Fagotti, M., Bonatti, C., Logoteta, D., Marconcini, P., and Macucci, M. (2011). Armchair graphene nanoribbons: $\mathcal{PT}$-symmetry breaking and exceptional points without dissipation, *Phys.Rev. B* **83**, p. 241406(R).

Feng, L., Ayache, M., Huang, J., Xu, Y.-L., Lu, M.-H., Chen, Y.-F., Fainman, Y., and Scherer, A. (2011). Nonreciprocal light propagation in a silicon photonic circuit, *Science* **333**, p. 729.

Feng, L., Xu, Y.-L., Fegadolli, W. S., Lu, M.-H., Oliveira, J. E. B., Almeida, V. R., Chen, Y.-F., and Scherer, A. (2013). Experimental demonstration of a unidirectional reflectionless parity-time metamaterial at optical frequencies, *Nature* **12**, p. 108.

Guo, A., Salamo, G., Duchesne, D., Morandotti, R., Volatier-Ravat, M., Aimez, V., Siviloglou, G., and Christodoulides, D. (2009). Observation of PT-Symmetry Breaking in Complex Optical Potentials, *Phys. Rev. Lett.* **103**, 9, p. 093902, doi: 10.1103/PhysRevLett.103.093902.

Hang, C., Huang, G., and Konotop, V. V. (2013). $\mathcal{PT}$ symmetry with a system of three-level atoms, *Phys. Rev. Lett.* **110**, p. 083604.

Kartashov, Y. V., Konotop, V. V., and Abdullaev, F. K. (2013). Gap solitons in a spin-orbit-coupled Bose-Einstein condensate, *Phys. Rev. Lett.* **111**, p. 060402.

Kreibich, J., M. ad Mnain, Cartarius, H., and Wunner, G. (2013). Hermitian four-well potential as a realization of a $\mathcal{PT}$-symmetric system, *Phys. Rev. A* **87**, p. 051601.

Lazarides, N. and Tsironis, G. p. (2013). Gain-driven discrete breathers in $\mathcal{PT}$-symmetric nonlinear metamaterials, *Phys. Rev. Lett.* **110**, p. 053901.

Liang, G. and Chong, Y. (2013). Optical resonator analog of a two-dimensional topological insulator, *Phys. Rev. Lett.* **110**, 20, p. 203904, doi:10.1103/PhysRevLett.110.203904, arXiv:1212.5034 [physics.optics].

Liertzer, M., Ge, L., Cerjan, A., Stone, A. D., Tureci, H., and Rotter, S. (2012). Pump induced exceptional points in lasers, *Phys. Rev. Lett.* **108**, p. 173901.

Lin, Z., Ramezani, H., Eichelkraut, T., Kottos, T., Cao, H., and Christodoulides, D. N. (2011). Unidirectional invisibility induced by $\mathcal{PT}$-symmetric periodic structures, *Phys. Rev. Lett.* **106**, p. 213901.

Longhi, S. (2009). Bloch oscillations in complex crystals with $\mathcal{PT}$ symmetry, *Phys. Rev. Lett.* **103**, p. 123601.

Longhi, S. (2010). Optical realization of relativistic non-Hermitian quantum mechanics, *Phys. Rev. Lett.* **105**, p. 013903.

Luo, X., Huang, J., Zhong, H., Qin, X., Xie, Q., Kivshar, Y. S., and Lee, C. (2013). Pseudo-parity-time symmetry in optical systems, *Phys. Rev. Lett.* **110**, p. 243902.

Mezard, M., Parisi, G., and Virasoro, M. (1987). *Spin Glass Theory and Beyond* (World Scientific, Singapore).

Peng, B., Oezdemir, S. K., Lei, F., Monifi, F., Gianfreda, M., Long, G. L., Fan, S., Nori, F., Bender, C. M., and Yang, L. (2014a). Parity-time-symmetric whispering-gallery microcavities, *Nature Physics* **10**, p. 394.

Peng, B., Oezdemir, S. K., Rotter, S., Yilmaz, H., Liertzer, M., Monifi, F., Bender, C. M., Nori, F., and Yang, L. (2014b). Loss-induced suppression and revival of lasing, *Science* **346**, p. 328.

Ramezani, H., Christodoulides, D. N., Kovanis, V., Vitebskiy, I., and Kottos, T. (2012). $\mathcal{PT}$-symmetric Talbot effects. *Phys. Rev. Lett.* **109**, p. 033902.

Razzari, L. and Morandotti, R. (2012). Gain and loss mixed in the same cauldron, *Nature* **488**, p. 163.

Regensberger, A., Bersch, C., Miri, M. A., Onishchukov, G., and Christodoulides, D. N. (2012). Parity-time synthetic photonic lattices, *Nature* **488**, p. 167.

Regensberger, A., Miri, M. A., Bersch, C., Nager, J., Onishchukov, G., Christodoulides, D. N., and Peschel, U. (2013). Observation of defect states in $\mathcal{PT}$-symmetric optical lattices, *Phys. Rev. Lett.* **110**, p. 223902.

Rubinstein, J., Sternberg, P., and Ma, Q. (2007). Bifurcation diagram and pattern formation of phase slip centers in superconducting wires driven by electric currents, *Phys. Rev. Lett* **99**, p. 167003.

Rueter, C. E., Makris, K. G., El-Ganainy, R., Christodoulides, D. N., Segev, M., and Kip, D. (2010). Observation of $\mathcal{PT}$ symmetry in optics,, *Nat.Phys.* **6**, p. 192.

Schindler, J., Li, A., Zheng, M. C., Ellis, F., and Kottos, T. (2011). Experimental Study of Active LRC Circuits with PT-Symmetries, *Phys. Rev. A* **84**, 4, p. 040101, doi: 10.1103/PhysRevA.84.040101, arXiv:1109.2913 [cond-mat.other].

Schomerus, H. . (2010). Quantum noise and self-sustained radiation of $\mathcal{PT}$-symmetric systems, *Phys. Rev. Lett.* **104**, p. 2336012.

Szameit, A., Rechtsman, M. C., Bahat-Treidel, O., and Segev, M. (2012). $\mathcal{PT}$-symmetry in honeycomb photonic lattices, *Phys. Rev. A* **84**, p. 021806.

West, C. T., Kottos, T., and Prosen, T. (2010). $\mathcal{PT}$-symmetric wave chaos, *Phys. Rev. Lett.* **104**, p. 0541022.

Yin, X. and Zhang, X. (2013). Unidirectional light propagation at exceptional points, *Nature Materials* **12**, p. 175.

Zhao, K. F., Schaden, M., and Wu, Z. (2010). Enhanced magnetic resonance signal of spin-polarized Rb atoms near surfaces of coated cells, *Phys.Rev. A* **81**, p. 042903.

Zheng, C., Hao, L., and Long, G. L. (2017). Observation of a fast evolution in a parity-time-symmetric system, *Phil. Trans. R. Soc* **371**, p. 20120053.

https://doi.org/10.1142/9789811231124_0007

Chapter 7

Rapidity Evolution and Small-x Physics

Ian Balitsky
JLab, 12000 Jefferson Ave, Newport News, VA 23606, USA
and
Phys. Department, Old Dominion Univ., Norfolk, VA 243529, USA

7.1 Lev Lipatov and the birth of low-x physics in LNPI

I came to LNPI Theory Group as a graduate student in 1976 after completion of MS thesis. My MS thesis advisor A. Anselm told me that there is a new rising star in LNPI theory group — Lev Nikolaevich Lipatov, and suggested to try to write Ph.D. under his supervision. That was very exciting time for high-energy physics in LNPI since QCD just came into being as a theory of strong interactions. Due to asymptotic freedom, perturbative QCD was much alike QED, and in LNPI theory group there was a lot of expertise in resummation of QED diagrams, sometimes easily (and sometimes not easily) translated to QCD. One of the leading experts in perturbative calculations was Lipatov, who, after a series of papers on high-energy behavior in QED, applied that expertise to get high-energy behavior of Yang-Mills amplitudes in a theory with massive gluons. In collaboration with E. A. Kuraev and V.S. Fadin he published a series of pioneering papers on the subject. When I came to him and asked what a young student can do, he said that there is a supposedly manageable problem: after their papers the high-energy asymptotics in the leading logarithmic approximation is known in a models with massive gluons, but what will happen if gluons are massless? Fortunately for me, he did not have sufficient interest to take limit $m_g \to 0$ himself since at the time he was very much involved in analysis of asymptotics of perturbation theory at high orders determined by classical solutions. I decided to try that problem and in about a year learned from Lipatov the LNPI know-how tricks for Feynman diagrams in the Regge limit. After that, I calculated the first nontrivial QCD diagram and realized that IR singularities cancel. When I told that to Lipatov, he immediately wrote what is known nowadays as LO BFKL kernel in QCD and the solution at zero transverse momentum. Unfortunately, at that time we were unable to solve the equation at $t \neq 0$ — it was done by Lipatov several years later.

During my years at LNPI I tried to the best of my ability to imitate Lipatov's style: to find an interesting problem, try some bright ideas, and solve the problem without fear of calculations if necessary. Well, Lipatov definitely taught me to be not afraid of long calculations. As to bright ideas, it is difficult to teach a person to have ones, but even later, when I met Lipatov on conferences, I always tried to talk to him and listened very carefully since sometimes even a couple of words by Lipatov can give you an idea. Looking back, I think I was extremely fortunate to work with and learn from a scientist of such high caliber as Lev Nikolaevich Lipatov.

One of the greatest achievements of Lev Nikolaevich Lipatov was the development of perturbative methods at high energy from the pioneering paper on the subject [Lipatov (1976)] to last applications in QCD, $\mathcal{N} = 4$ SYM and quantum gravity. In this article I discuss an approach to small-x physics based on rapidity factorization and evolution. Although L.N. Lipatov never used this approach himself, he encouraged his Ph.D. student (the author) to learn the effective action language from the papers of Shifman, Vainshtein, and Zakharov. It turned out that using Wilson lines as an effective degrees of freedom at high energy it became possible to reformulate resummation of Lipatov logarithms in terms of rapidity evolution of Wilson lines. Later this approach proved out to be crucial in the description of non-linear phenomena in high-energy scattering from dense QCD systems.

7.2 Introduction

The standard way to analyze the high-energy behavior of pQCD amplitudes is the direct summation of Feynman diagrams. In the Regge limit $s \gg t, m^2$ (where m is a characteristic mass or virtuality of colliding particles) there is an extra parameter in addition to α_s, namely $\alpha_s \ln \frac{s}{m^2}$. At pre-asymptotic energies we may have a window where the leading logarithmic approximation (LLA)

$$\alpha_s \ll 1, \quad \alpha_s \ln \frac{s}{m^2} \sim 1 \tag{7.1}$$

is valid. Following the pioneeing work by Lipatov [Lipatov (1976)], the amplitude in this region of energies was shown to be determined by the BFKL pomeron [Kuraev *et al.* (1976, 1977); Balitsky and Lipatov (1978)] which leads to the behavior of QCD cross sections of the type

$$\sigma(s) = \int d\nu \Big(\frac{s}{m^2}\Big)^{\omega(\nu)} F(\nu) \sim \Big(\frac{s}{m^2}\Big)^{4\frac{\alpha_s}{\pi}N_c \ln 2} \tag{7.2}$$

where N_c is a number of colors ($N_c = 3$ for QCD),

$$\omega(\nu) = \frac{\alpha_s}{\pi} N_c \left[2\psi(1) - \psi\left(\frac{1}{2} + i\nu\right) - \psi\left(\frac{1}{2} - i\nu\right)\right] + O(\alpha_s^2) \tag{7.3}$$

is the BFKL intercept, and $F(\nu)$ is the "pomeron residue" which depends on the process.

The power behavior of BFKL cross section (7.2) violates the Froissart bound and, therefore, the BFKL pomeron describes only the pre-asymptotic behavior at intermediate energies when the cross sections are small in comparison to the geometric cross section $2\pi R^2$. In order to find the true high-energy asymptotics by analysis of Feynman diagrams we should sum up not only the leading logarithms $(\alpha_s \ln s)^n$ but also the sub-leading ones $\alpha_s(\alpha_s \ln s)^n$, then the sub-sub-leading terms $\alpha_s^2(\alpha_s \ln s)^n$, etc.

This program is, at best, very challenging. In QCD, up to now only next-to-leading (NLO) pomeron intercept was found [Fadin and Lipatov (1998); Ciafaloni and Camici (1998)]. More progress was made in $\mathcal{N} = 4$ SYM where NNLO BFKL kernel was obtained by comparison with non-global logs [Caron-Huot and Herranen (2018)]. Also, NNLO pomeron intercept was obtained from integrability-based calculation [Gromov *et al.* (2014)] of anomalous dimensions of twist-2 operators in the BFKL limit [Gromov *et al.* (2015)].

Probably, a more realistic approach is to unitarize the BFKL pomeron, i.e. to sum up the subset of sub-leading logarithms which restores the unitarity in s channel. Still, it is a difficult problem which has been in a need of a solution for more than 40 years. One of the most popular ideas on solving this problem is reducing QCD at high energies to some sort of low-dimensional effective theory which will be simpler than original QCD, maybe even to the extent of exact solvability. The first step on this road is to identify proper degrees of freedom for this effective theory. One of the possible choices is to formulate high-energy scattering in terms of "reggeized gluons." [Kirschner *et al.* (1994); Lipatov (1995, 1997); Bondarenko *et al.* (2017)] An alternative and related approach [Balitsky (1996)] is based on so-called Wilson lines — infinite gauge links corresponding to fast gluons moving along the straight-line classical trajectories.

A general feature of high-energy scattering is that a fast particle moves along its straight-line classical trajectory and the only quantum effect is the eikonal phase factor acquired along this propagation path. In QCD, for the fast quark or gluon scattering off some target, this eikonal phase factor is a Wilson line — the infinite gauge link ordered along the straight line collinear to particle's velocity n^μ:

$$U^Y(x_\perp) = \mathrm{Pexp}\left\{ig\int_{-\infty}^{\infty} du\ n_\mu A^\mu(un + x_\perp)\right\}. \tag{7.4}$$

Here A_μ is the gluon field of the target, $x_\perp$ is the transverse position of the particle which remains unchanged throughout the collision, and the index Y labels the rapidity of the particle. Repeating the above argument for the target (moving fast in the spectator's frame) we see that particles with very different rapidities perceive each other as Wilson lines and therefore these Wilson-line operators form the convenient effective degrees of freedom in high-energy QCD (for a review, see ref. [Balitsky (2001)]). The corresponding effective action should be 2+1 dimensional — two dimensions for transverse coordinate and one for rapidity. There were few attempts to construct such effective action in terms of Wilson lines [Balitsky (1999, 2005); Hatta *et al.* (2005, 2006); Kovner and Lublinsky (2005); Altinoluk *et al.* (2009)], but unfortunately the full solution of the problem still eludes us.

However, the description of high-energy scattering in terms of Wilson lines turn out to be relevant for the description of so-called dilute-dense scattering, the typical example being small-x deep inelastic scattering from large nuclei. This scattering can be studied by two different but equivalent approaches. The first one gets the small-x behavior of DIS amplitude from the the evolution of "color dipole" made from two Wilson-line operators

$$\hat{\mathcal{U}}^Y(x_\perp, y_\perp) = 1 - \frac{1}{N_c}\mathrm{Tr}\{\hat{U}^Y(x_\perp)\hat{U}^{\dagger Y}(y_\perp)\} \tag{7.5}$$

corresponding to two fast quarks produced by virtual photon ($N_c = 3$ for QCD). At relatively high energies and for sufficiently small dipoles we can use the leading logarithmic approximation (7.1) ($\alpha_s \ll 1$, $\alpha_s \ln x_B \sim 1$) and get the non-linear BK evolution equation for the color dipoles [Balitsky (1996, 1999); Kovchegov (1999, 2000)]:

$$\begin{aligned}\frac{d}{dY}\hat{\mathcal{U}}^Y(z_1,z_2) = \frac{\alpha_s N_c}{2\pi^2}\int d^2z_3 \frac{z_{12}^2}{z_{13}^2 z_{23}^2} \times [&\hat{\mathcal{U}}^Y(z_1,z_3) + \hat{\mathcal{U}}^Y(z_3,z_2) - \hat{\mathcal{U}}^Y(z_1,z_3)\\ &-\hat{\mathcal{U}}^Y(z_1,z_3)\hat{\mathcal{U}}^Y(z_3,z_2)]\end{aligned} \tag{7.6}$$

where $Y = \ln\frac{1}{x_B}$ and $z_{12} \equiv z_1 - z_2$ etc. The first three terms in square brackets correspond to the linear BFKL evolution [Balitsky and Lipatov (1978)] and describe the parton emission while the last term is responsible for the parton annihilation. For sufficiently low x_B the parton emission balances the parton annihilation so the partons reach the state of saturation [Gribov *et al.* (1983); Mueller and Qiu (1986); Mueller (1990)] with the characteristic transverse momentum Q_s growing with energy $1/x_B$ (for reviews, see Refs. [Iancu and Venugopalan (2003); Weigert (2005); Kovchegov and Levin (2012)]). It should be mentioned that on the operator level the equation (7.6) is not closed: the evolution of four-Wilson-lne operators in the r.h.s. gives six-Wilson-line operators etc. leading to a hierarchy of Wilson-lines evolution [Balitsky (1996, 1999)]. However, in the case of large nucleus one can rewrite Eq. (7.6) for matrix elements leading to closed non-linear equation for structure functions of large nuclei [Kovchegov (1999, 2000)].

It should be mentioned that alternative (and equivalent) approach is to consider the small-x evolution from the target side. The obtained JIMWLK evolution [Jalilian-Marian *et al.* (1997, 1998, 1999); Weigert (2002); Iancu *et al.* (2001); Ferreiro *et al.* (2002)] of target's wave function with respect to rapidity cutoff of gluons populating the wave function turns out to be equivalent to the hierarchy of Wilson-lines evolution.

7.3 Operator expansion for high-energy scattering

The simple idea of rapidity factorization using Wilson lines leads to scheme of calculations based on methods familiar in operator product expansions. Let me first remind the logic of usual operator expansion near the light cone.

7.3.1 *The logic of OPE*

A typical example of using OPE to find the behavior of amplitudes is the calculation of the structure functions of deep inelastic scattering (DIS) at moderate $x_B = Q^2/(2p \cdot q)$ determined by the T-product of two electromagnetic currents $T\{j_\mu(x)j_\nu(y)\}$ sandwitched between the target states. The first step is to identify the relevant operators. To this end, we formally put Q^2 to infinity, then $(x-y)^2 = 0$ and we see that the relevant operators are the light-ray ones of the type of $\bar{\psi}(x)\gamma_\mu[x,y]\psi(y)$ where

$$[x,y] \equiv \mathrm{Pexp}\left\{-ig\int_0^1 du\ (x-y)^\mu A_\mu(ux-uy+y)\right\} \tag{7.7}$$

is a standard notation for a straight-line gauge link connecting points x and y.

Now, to get the Q^2 behavior of structure functions one needs to perform the following four steps:

- Separate relevant Feynman loop integrals over $k_\perp$ in two parts — the coefficient functions (with transverse momenta $k_\perp^2$ greater than the factorization scale μ^2) and parton densities — matrix elements of light-ray operators with $k_\perp^2 < \mu^2$ (see Fig. 7.1). Technically, the cutoff of the transverse momenta in these matrix elements is done by adding counterterms with normalization point μ^2.
- Find the (DGLAP) evolution equations of light-ray operators with respect to μ^2.
- Solve the corresponding evolution equations.
- Assemble the result for structure functions: take some initial conditions at low normalization point (usually around $\mu^2 = 1\,\mathrm{GeV}^2$), evolve parton densities to high Q^2 and multiply the result by the coefficient function.

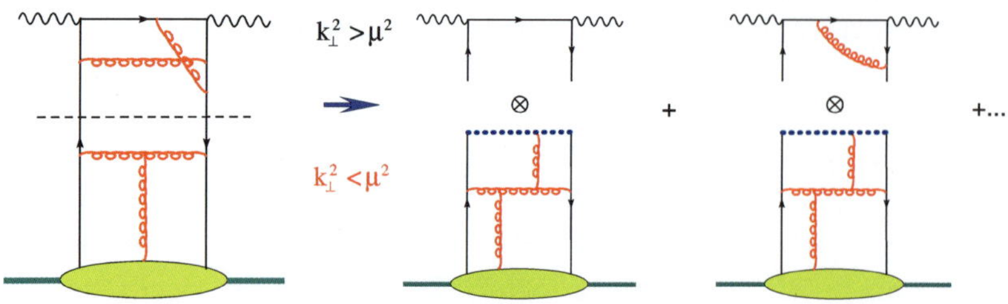

Fig. 7.1 Light-ray operator expansion in for DIS. Gauge link is denoted by a dotted line.

One can calculate the coefficient functions by direct computation of Feynman diagrams with quark and gluon tails or by using the Feynman diagrams in external quark and gluon fields (see e.g. [Balitsky (1983); Balitsky and Braun (1989a)]). A typical NLO contribution coming from the upper part of last diagram in Fig. 7.1 has the form

$$T\{j_\mu(x)j_\mu(0)\} = \frac{-i}{\pi^2 x^4}\int_0^1 du\left[\frac{1}{u}\right]_+ [\bar\psi(x)[x,ux]\,\not x\psi(ux)]^{\text{l.t.}} \times\left(1+\frac{\alpha_s}{4\pi}[\ln x^2\mu^2+\ln u]\right)+\cdots \tag{7.8}$$

where $\left[\frac{1}{u}\right]_+$ is a standard "plus" prescription

$$\int_0^1 du\,\left[\frac{1}{u}\right]_+ f(u) \equiv \int_0^1 du\frac{f(u)-f(0)}{u} \tag{7.9}$$

and the leading-twist light-ray operator $[\bar\psi(x)[x,ux]\,\not x\psi(ux)]^{\text{l.t.}}$ is taken at $x^2=0$. Such light-ray operator $[\bar\psi(x)[x,ux]\,\not x\psi(ux)]^{\text{l.t.}}$ has specific light-cone UV divergencies (in addition to usual UV contributions assembled to the running coupling constant) which are regularized by adding light-ray counterterms. The resulting composite operator (original operator minus counterterms) depends on the renormalization point μ and the renorm-group equation for these operators leads to the DGLAP evolution equations for parton densities. These linear equations are easily solved by Mellin transformation and multiplying the solution by the coefficient function one obtains the DIS structure functions with great accuracy.

7.3.2 *High-energy OPE in Wilson lines*

Now we want to extend the four-step program from the previous section to describe the high-energy amplitudes. A typical process is deep inelastic scattering at small values of Bjorken x. Since we are interested now in x_B evolution rather than Q^2 evolution, as step one we factorize in rapidity instead of transverse momenta. As a preliminary step we identify the relevant operators. The virtual photon splits into quark-antiquark pair, and if we set formally the energy of incoming photon to infinity we see that these quark and antiquark travel along the light-like classical

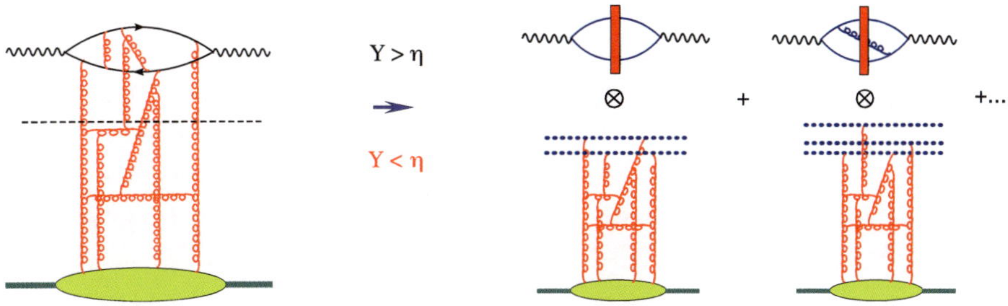

Fig. 7.2 High-energy operator expansion in Wilson lines

trajectories. This is the well-known general result from quantum mechanics — the fast particle moves along its straight-line classical trajectory and the only quantum effect is the eikonal phase factor acquired along this propagation path. In QCD, for fast quark or gluon scattering off some target, this eikonal phase factor is a Wilson line (7.4) — an infinite gauge link ordered along the straight line collinear to particle's velocity n^μ:

Having identified the operators, we can repeat the 4 steps which give us the high-energy (small-x for DIS) evolution of the amplitude.

- Separate the relevant Feynman loop integrals over the longitudinal momentum α in two parts — the coefficient functions ("impact factors") with α greater than the rapidity factorization scale $\sigma = e^\eta$ and matrix elements of Wilson-line operators with $\alpha < \sigma$ (see Fig. 7.2). We were not able to find the analog of dimensional regularization for longitudinal divergence so we cut the integration over $\alpha < \sigma$ "by hand".
- Find the evolution equations of color dipoles with respect to the cutoff in rapidity Y.
- Solve of the corresponding evolution equation(s).
- Assemble the result for structure functions: take the initial conditions at low energy, evolve color dipoles to high rapidity $Y \sim Y_A$ and multiply the result by the corresponding impact factor.

The first two steps are performed by factorization in rapidity (=log of longitudinal momentum) rather than factorization in transverse momenta as in the light-ray OPE case. We introduce the rapidity divide η which separates the "fast" gluons from the "slow" ones. As we discussed, slow and fast gluons perceive each other as Wilson lines which greatly simplifies the analysis of high-energy behavior of small-x DIS amplitude. Let me describe use of rapidity factorization for calculation of impact factors and evolution of Wilson lines.

7.3.2.1 *First step: Impact factors*

To calculate the impact factor(s), we integrate over gluons with rapidities $Y > \eta$ and leave the integration over $Y < \eta$ for the later time, see Fig. 7.2. It is convenient to use the background field formalism: we integrate over gluons with $\alpha > \sigma = e^\eta$ and leave gluons with $\alpha < \sigma$ as a background field, to be integrated over later. Since the rapidities of the background gluons are very different from the rapidities of gluons in our Feynman diagrams, the background field can be taken in the form of a shock wave due to the Lorentz contraction. To derive the expression of a quark (or gluon) propagator in this shock-wave background we represent the propagator as a path integral over various trajectories, each of them weighed with the gauge factor $\mathrm{Pexp}(ig \int dx_\mu A^\mu)$ ordered along the propagation path. Now, since the shock wave is very thin, quarks (or gluons) do not have time to deviate in transverse direction so their trajectory inside the shock wave can be approximated by a segment of the straight line. Moreover, since there is no external field outside the shock wave, the integral over the segment of straight line can be formally extended to $\pm\infty$ limits yielding the Wilson-line gauge factor

$$\begin{aligned} U_x^\eta &= \mathrm{Pexp}\Big[ig \int_{-\infty}^{\infty} du\; p_1^\mu A_\mu^\sigma(up_1 + x_\perp)\Big], \\ A_\mu^\eta(x) &= \int d^4k\; \theta(e^\eta - |\alpha_k|) e^{ik\cdot x} A_\mu(k) \end{aligned} \tag{7.10}$$

where the Sudakov variable α_k is defined as usual, $k = \alpha_k p_1 + \beta_k p_2 + k_\perp$. We define the light-like vectors p_1 and p_2 such that $q = p_1 - x_B p_2$ and $p = p_2 + \frac{m_N^2}{s} p_1$ where q is the virtual photon momentum, p is the momentum of the target particle, and $x_B = Q^2/s \ll 1$ is the Bjorken variable (at large energies $s \simeq 2p\cdot q$). Our metric is $g^{\mu\nu} = (1, -1, -1, -1)$ so that $p\cdot q = (\alpha_p\beta_q + \alpha_q\beta_p)\frac{s}{2} - (p, q)_\perp$ where $(p, q)_\perp \equiv -p_i q^i$. Also, we use the notations $x_\bullet \equiv x_\mu p_1^\mu$ and $x_* \equiv x_\mu p_2^\mu$ for the dimensionless light-cone coordinates ($x_* = \sqrt{\frac{s}{2}} x_+$ and $x_\bullet = \sqrt{\frac{s}{2}} x_-$). Throughout the paper, the sum over the Latin indices i, j... runs over the two transverse components while the sum over Greek indices runs over the four components as usual.

The structure of the propagator in a shock-wave background looks as follows (see Fig. 7.3):

[Free propagation from initial point x to the point of intersection with the shock wave z]
$\times$ [Interaction with the shock wave described by the Wilson-line operator U_z]
$\times$ [Free propagation from point of interaction z to the final point y].

Fig. 7.3 Propagator in a shock-wave background

The explicit form of quark and gluon propagators in a shock wave located at $z_* = 0$ is

$$\langle \mathrm{T}\{\hat{\psi}(x)\hat{\bar{\psi}}(y)\}\rangle_A \overset{x_*>0>y_*}{=} \frac{i}{4\pi^{d-1}}\int d^{d-2}z_\perp U_z \times \frac{\left(\frac{2}{s}x_*\not{p}_1 + \not{X}_\perp\right)\not{p}_2\left(\frac{2}{s}|y_*|\not{p}_1 - \not{Y}_\perp\right)(x_*|y_*|)^{\frac{d}{2}-1}\Gamma\left(\frac{d}{2}\right)}{\left[-\frac{4}{s}(x-y)_\bullet x_*|y_*| + X^2|y_*| + Y^2 x_* + i\epsilon\right]^{\frac{d}{2}}} \tag{7.11}$$

for quark propagator at $x_* > 0 > y_*$ and

$$\langle \mathrm{T}\{\hat{A}^a_\mu(x)\hat{A}^b_\nu(y)\}\rangle_A \overset{x_*>0>y_*}{=} -\frac{1}{4\pi^3}\int d^{d-2}z_\perp U^{ab}_z \frac{(x_*|y_*|)^{\frac{d}{2}-2}\Gamma\left(\frac{d}{2}-1\right)}{[-\frac{4}{s}(x-y)_\bullet x_*|y_*| + X^2|y_*| + Y^2x_* + i\epsilon]^{\frac{d}{2}-1}} \times \left[g^\perp_{\mu\nu}x_*|y_*| - p_{2\mu}X_\nu|y_*| + Y_\mu p_{2\nu}x_* + p_{2\mu}p_{2\nu}(X,Y)_\perp\right] \tag{7.12}$$

for the gluon propagator in the light-like $p_2^\mu A_\mu = 0$ gauge.

The result of the integration over gluons with rapidities $Y > \eta$ gives the impact factor — the amplitude of the transition of virtual photon in two-Wilson-lines operators (sometimes called "color dipole"). The LO impact factor is a product of two quark propagators in a shock-wave background shown in Fig. 7.2 while the NLO impact factor is represented by a product of two quark and one gluon propagator in the same Fig. 7.2. It is convenient to introduce the conformal vectors [Cornalba *et al.* (2008)]

$$\kappa = \kappa_x - \kappa_y, \quad \kappa_x = \frac{\sqrt{s}}{2x_*}\left(\frac{p_1}{s} - x^2 p_2 + x_\perp\right)$$
$$\zeta_i = \left(\frac{p_1}{s} + z^2_{i\perp}p_2 + z_{i\perp}\right), \tag{7.13}$$

and the notation $\mathcal{R} \equiv \frac{\kappa^2(\zeta_1\cdot\zeta_2)}{2(\kappa\cdot\zeta_1)(\kappa\cdot\zeta_2)}$. The result for the impact factor is

$$(x-y)^4 T\{\hat{\bar{\psi}}(x)\gamma^\mu\hat{\psi}(x)\hat{\bar{\psi}}(y)\gamma^\nu\hat{\psi}(y)\} = \int\frac{d^2z_1d^2z_2}{z_{12}^4}\Big\{I^{\rm LO}_{\mu\nu}(z_1,z_2)\Big[1+\frac{3\alpha_s}{4\pi}\Big]\mathrm{tr}\{\hat{U}^\eta_{z_1}\hat{U}^{\dagger\eta}_{z_2}\} + \int d^2z_3\, I^{\rm NLO}_{\mu\nu}(z_1,z_2,z_3;\eta)[\mathrm{tr}\{\hat{U}^\eta_{z_1}\hat{U}^{\dagger\eta}_{z_3}\}\mathrm{tr}\{\hat{U}^\eta_{z_3}\hat{U}^{\dagger\eta}_{z_2}\} - N_c\mathrm{tr}\{\hat{U}^\eta_{z_1}\hat{U}^{\dagger\eta}_{z_2}\}]\Big\} \tag{7.14}$$

where

$$I^{\rm LO}_{\mu\nu}(z_1,z_2) = \frac{\mathcal{R}^2}{\pi^6(\kappa\cdot\zeta_1)(\kappa\cdot\zeta_2)}\frac{\partial^2}{\partial x^\mu\partial y^\nu}\Big[(\kappa\cdot\zeta_1)(\kappa\cdot\zeta_2) - \frac{1}{2}\kappa^2(\zeta_1\cdot\zeta_2)\Big], \tag{7.15}$$

which corresponds to the well-known expression for the LO impact factor in the momentum space. Note, however, that the NLO impact factor depends on the "rapidity divide" η in Fig. 7.2 and on the way in which this rapidity divide is introduced. This corresponds to the choice of "energy scale" for NLO BFKL [Fadin and Lipatov (1998)]. To clarify this subtle point, we performed similar calculation in $\mathcal{N} = 4$ SYM where we have additional restriction due to conformal invariance. It turns out that the rapidity cutoff (7.10) violates conformal invariance of a color dipole $\mathrm{tr}\{\hat{U}^\eta_{z_1}\hat{U}^{\dagger\eta}_{z_2}\}$ so to restore this invariance at the NLO level we considered the "composite conformal dipole" [Balitsky and Chirilli (2009)]

$$\begin{aligned}\left[\mathrm{tr}\{\hat{U}_{z_1}\hat{U}^\dagger_{z_2}\}\right]_a = &\ \mathrm{tr}\{\hat{U}^\sigma_{z_1}\hat{U}^{\dagger\sigma}_{z_2}\} + \frac{\alpha_s}{4\pi^2}\int d^2z_3\,\frac{z_{12}^2}{z_{13}^2z_{23}^2}\\ &\times\left[\mathrm{tr}\{\hat{U}^\sigma_{z_1}\hat{U}^{\dagger\sigma}_{z_3}\}\mathrm{tr}\{\hat{U}^\sigma_{z_3}\hat{U}^{\dagger\sigma}_{z_2}\} - N_c\mathrm{tr}\{\hat{U}^\sigma_{z_1}\hat{U}^{\dagger\sigma}_{z_2}\}\right]\\ &\times\ln\frac{4az_{12}^2}{s\sigma^2z_{13}^2z_{23}^2} + O(\alpha_s^2)\end{aligned}\tag{7.16}$$

where the parameter a serves as a rapidity cutoff in place of $\eta = \ln\sigma$. It turns out that the new rapidity divide "a" between the impact factor and the composite dipole (7.16) is uniquely fixed by two requirements: (i) the impact factor should not scale with energy, all energy dependence should come from matrix element of the dipole, and (ii) impact factor should be conformally invariant with respect to the $SL(2,C)$ (Möbius) group. Incidentally, this conformal choice leads to symmetric energy scale in Lipatov-Fadin NLO BFKL kernel [Fadin and Lipatov (1998)]. The explicit form of such defined NLO impact factor is presented in Refs. [Balitsky and Chirilli (2011, 2013a)].

7.3.2.2 *Second step: evolution of Wilson lines*

Next step is to obtain the evolution equation for color dipoles in the leading order in α_s. To find the evolution of the color dipole (7.5) with respect to rapidity of the Wilson lines in the leading log approximation we consider the matrix element of the color dipole between (arbitrary) target states and integrate over the gluons with rapidities $Y_1 > Y > Y_2 = Y_1 - \Delta Y$ leaving the gluons with $Y < Y_2$ as a background field (to be integrated over later). In the frame of gluons with $Y \sim Y_1$ the fields with $Y < Y_2$ shrink to a pancake and we obtain the four diagrams shown in Fig. 7.4. Technically, to find the kernel in the leading-order approximation we write down the general form of the operator equation for the evolution of the color dipole

$$\frac{d}{dY}\mathrm{Tr}\{\hat{U}^Y_{z_1}\hat{U}^{\dagger Y}_{z_2}\} = K_{\mathrm{LO}}\mathrm{Tr}\{\hat{U}^Y_{z_1}\hat{U}^{\dagger Y}_{z_2}\} + \cdots\tag{7.17}$$

(where dots stand for the higher orders of the expansion) and calculate the l.h.s. of Eq. (7.17) in the shock-wave background

$$\frac{d}{dY}\langle\mathrm{Tr}\{\hat{U}^Y_{z_1}\hat{U}^{\dagger Y}_{z_2}\}\rangle_{\mathrm{shockwave}} = \langle K_{\mathrm{LO}}\mathrm{Tr}\{\hat{U}^Y_{z_1}\hat{U}^{\dagger Y}_{z_2}\}\rangle_{\mathrm{shockwave}}.\tag{7.18}$$

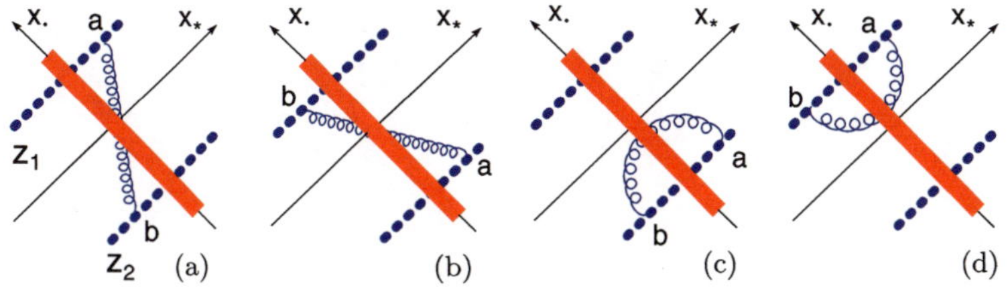

Fig. 7.4 Leading-order diagrams for the small-x evolution of color dipole. Gauge links are denoted by dotted lines.

Just by looking at Fig. 7.4 one sees that the evolution equation for color dipoles should be non-linear since the first step of evolution of two Wilson lines yields two Wilson lines in quark representation plus Wilson line in gluon representation equivalent to two quark lines. After using gluon propagators in external shock-wave background (7.12) one obtains BK equation (7.6). It should be noted that the equation (7.6) is not a closed one: the evolution of the two-dipole operator in the r.h.s. will give three-dipole operators and so on. So instead of one non-linear equation we have a hierarchy of Wilson-lines evolution, fortunately with pair-wise kernels (at the LO level) shown in Fig. 7.4 [Balitsky (1996)]. However, as demonstrated in Refs. [Kovchegov (1999, 2000)], for the large nucleus and and large N_c matrix element of two-dipole operator in the r.h.s. of Eq. (7.6) factorizes leading to closed non-linear equation for structure functions of large nuclei at low x.

In the NLO there is about a hundred diagrams with different types shown in Fig. 7.5. The result of the evolution of the color dipole with rapidity cutoff (7.10) is [Balitsky and Chirilli (2008)]:

$$
\begin{aligned}
&\frac{d}{dY}\mathrm{tr}\{\hat{U}_{z_1}\hat{U}^\dagger_{z_2}\}\\
&\quad=\frac{\alpha_s}{2\pi^2}\int d^2z\frac{z_{12}^2}{z_{13}^2z_{23}^2}\Big\{1+\frac{\alpha_s}{4\pi}\Big[b\ln z_{12}^2\mu^2-b\frac{z_{13}^2-z_{23}^2}{z_{12}^2}\ln\frac{z_{13}^2}{z_{23}^2}\\
&\qquad+\Big(\frac{67}{9}-\frac{\pi^2}{3}\Big)N_c-\frac{10}{9}n_f\\
&\qquad-2N_c\ln\frac{z_{13}^2}{z_{12}^2}\ln\frac{z_{23}^2}{z_{12}^2}\Big]\Big\}[\mathrm{tr}\{\hat{U}_{z_1}\hat{U}^\dagger_{z_3}\}\mathrm{tr}\{\hat{U}_{z_3}\hat{U}^\dagger_{z_2}\}-N_c\mathrm{tr}\{\hat{U}_{z_1}\hat{U}^\dagger_{z_2}\}]\\
&\qquad+\frac{\alpha_s^2}{16\pi^4}\int d^2z_3d^2z_4\Bigg[\Big(-\frac{4}{z_{34}^4}+\Big\{2\frac{z_{13}^2z_{24}^2+z_{14}^2z_{23}^2-4z_{12}^2z_{34}^2}{z_{34}^4[z_{13}^2z_{24}^2-z_{14}^2z_{23}^2]}\\
&\qquad+\frac{z_{12}^4}{z_{13}^2z_{24}^2-z_{14}^2z_{23}^2}\Big[\frac{1}{z_{13}^2z_{24}^2}+\frac{1}{z_{23}^2z_{14}^2}\Big]\\
&\qquad+\frac{z_{12}^2}{z_{34}^2}\Big[\frac{1}{z_{13}^2z_{24}^2}-\frac{1}{z_{14}^2z_{23}^2}\Big]\Big\}\ln\frac{z_{13}^2z_{24}^2}{z_{14}^2z_{23}^2}\Big)[\mathrm{tr}\{\hat{U}_{z_1}\hat{U}^\dagger_{z_3}\}\mathrm{tr}\{\hat{U}_{z_3}\hat{U}^\dagger_{z_4}\}\mathrm{tr}\{\hat{U}_{z_4}\hat{U}^\dagger_{z_2}\}
\end{aligned}
$$

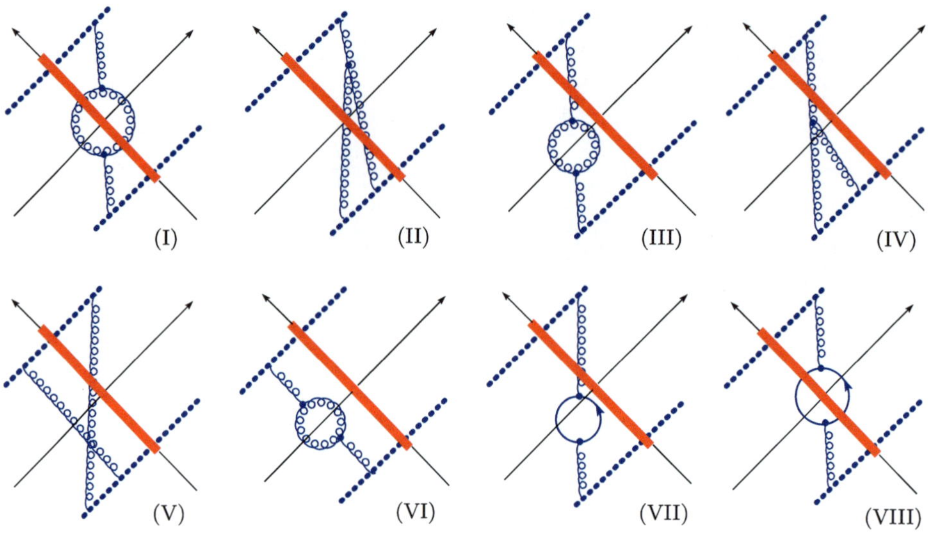

Fig. 7.5 Different types of diagrams for the NLO evolution of color dipole.

$$
\begin{aligned}
&- \mathrm{tr}\{\hat{U}_{z_1}\hat{U}^\dagger_{z_3}\hat{U}_{z_4}U^\dagger_{z_2}\hat{U}_{z_3}\hat{U}^\dagger_{z_4}\} - (z_4 \to z_3)] \\
&+ \left\{\frac{z_{12}^2}{z_{34}^2}\left[\frac{1}{z_{13}^2 z_{24}^2} + \frac{1}{z_{23}^2 z_{14}^2}\right] - \frac{z_{12}^4}{z_{13}^2 z_{24}^2 z_{14}^2 z_{23}^2}\right\} \ln\frac{z_{13}^2 z_{24}^2}{z_{14}^2 z_{23}^2} \\
&\times \mathrm{tr}\{\hat{U}_{z_1}\hat{U}^\dagger_{z_3}\}\mathrm{tr}\{\hat{U}_{z_3}\hat{U}^\dagger_{z_4}\}\mathrm{tr}\{\hat{U}_{z_4}\hat{U}^\dagger_{z_2}\} \\
&+ 4n_f\left\{\frac{4}{z_{34}^4} - 2\frac{z_{14}^2 z_{23}^2 + z_{24}^2 z_{13}^2 - z_{12}^2 z_{34}^2}{z_{34}^4(z_{13}^2 z_{24}^2 - z_{14}^2 z_{23}^2)} \ln\frac{z_{13}^2 z_{24}^2}{z_{14}^2 z_{23}^2}\right\} \\
&\times \mathrm{tr}\{t^a\hat{U}_{z_1}t^b\hat{U}^\dagger_{z_2}\}[\mathrm{tr}\{t^a\hat{U}_{z_3}t^b\hat{U}^\dagger_{z_4}\} - (z_4 \to z_3)]\Bigg]
\end{aligned}
\tag{7.19}
$$

where we use the $\overline{MS}$ scheme ($b = \frac{11}{3}N_c - \frac{2}{3}n_f$ as usual). This NLO kernel is a sum of the running-coupling part (proportional to b), the non-conformal double-log term $\sim \ln\frac{z_{12}^2}{z_{13}^2}\ln\frac{z_{12}^2}{z_{13}^2}$, and conformal terms proportional to log of conformal ratio $\ln\frac{z_{13}^2 z_{24}^2}{z_{14}^2 z_{23}^2}$. This is the QCD result, in $\mathcal{N} = 4$ the kernel is without terms proportional to b and n_f but still has the non-conformal double-log term $\sim \ln\frac{z_{12}^2}{z_{13}^2}\ln\frac{z_{12}^2}{z_{13}^2}$. This is not surprising since, as we discussed above, the cutoff (7.10) violates conformal invariance of the dipole. It turns out, however, that the evolution of composite dipole (7.16) is conformally invariant, the double log is replaced by log of the second conformal ratio $\ln\frac{z_{12}^2 z_{34}^2}{z_{14}^2 z_{23}^2}$. Because of approximate conformal symmetry of pQCD at the tree level, a natural guess for QCD result would be some sort of tree-level conformal structure which "gets dressed" by the running coupling constant. At the NLO level, this would correspond to the sum of the conformal part and the running-coupling part proportional to b. As we see, the Eq. (7.19) does not quite look like such sum due to the presence of the double-log term. It turns out, however, that if one rewrites the evolution equation (7.19) in terms of composite operators (7.16), the NLO kernel separates into the sum of conformal and running-coupling parts [Balitsky and Chirilli (2009)]:

$$
\begin{aligned}
&2a\frac{d}{da}\big[\mathrm{tr}\{\hat U_{z_1}\hat U^\dagger_{z_2}\}\big]_a\\
&=\frac{\alpha_s}{2\pi^2}\int d^2z_3\frac{z_{12}^2}{z_{13}^2z_{23}^2}\big[\mathrm{tr}\{\hat U_{z_1}\hat U^\dagger_{z_3}\}\mathrm{tr}\{\hat U_{z_3}\hat U^\dagger_{z_2}\}-N_c\mathrm{tr}\{\hat U_{z_1}\hat U^\dagger_{z_2}\}\big]_a\\
&\times\Big\{1+\frac{\alpha_s}{4\pi}\Big[b(\ln\frac{z_{12}^2\mu^2}{4}+2C)-b\frac{z_{13}^2-z_{23}^2}{z_{12}^2}\ln\frac{z_{13}^2}{z_{23}^2}+\Big(\frac{67}{9}-\frac{\pi^2}{3}\Big)N_c-\frac{10}{9}n_f\Big]\Big\}\\
&+\frac{\alpha_s^2}{16\pi^4}\int\frac{d^2z_3d^2z_4}{z_{34}^4}\Bigg[\Big\{\Big(-2+2\frac{z_{12}^2z_{34}^2}{z_{13}^2z_{24}^2}\ln\frac{z_{12}^2z_{34}^2}{z_{14}^2z_{23}^2}\\
&+\Big[\frac{z_{12}^2z_{34}^2}{z_{13}^2z_{24}^2}\Big(1+\frac{z_{12}^2z_{34}^2}{z_{13}^2z_{24}^2-z_{14}^2z_{23}^2}\Big)\\
&+\frac{2z_{13}^2z_{24}^2-4z_{12}^2z_{34}^2}{z_{13}^2z_{24}^2-z_{14}^2z_{23}^2}\Big]\ln\frac{z_{13}^2z_{24}^2}{z_{14}^2z_{23}^2}\Big)+\ (z_3\leftrightarrow z_4)\Big\}\\
&\times\big[\big(\mathrm{tr}\{\hat U_{z_1}\hat U^\dagger_{z_3}\}\mathrm{tr}\{\hat U_{z_3}\hat U^\dagger_{z_4}\}\mathrm{tr}\{\hat U_{z_4}\hat U^\dagger_{z_2}\}\\
&-\mathrm{tr}\{\hat U_{z_1}\hat U^\dagger_{z_3}\hat U_{z_4}\hat U^\dagger_{z_2}\hat U_{z_3}\hat U^\dagger_{z_4}\}\big)-(z_4\to z_3)\big]_a\\
&+\frac{z_{12}^2z_{34}^2}{z_{13}^2z_{24}^2}\Big\{2\ln\frac{z_{12}^2z_{34}^2}{z_{14}^2z_{23}^2}+\Big[1+\frac{z_{12}^2z_{34}^2}{z_{13}^2z_{24}^2-z_{14}^2z_{23}^2}\Big]\ln\frac{z_{13}^2z_{24}^2}{z_{14}^2z_{23}^2}\Big\}\\
&\times\big[\mathrm{tr}\{\hat U_{z_1}\hat U^\dagger_{z_3}\}\mathrm{tr}\{\hat U_{z_3}\hat U^\dagger_{z_4}\}\mathrm{tr}\{\hat U_{z_4}\hat U^\dagger_{z_2}\}-z_3\leftrightarrow z_4\big]_a\Bigg]\\
&+\frac{\alpha_s^2n_f}{2\pi^4}\int\frac{d^2z_3d^2z_4}{z_{34}^4}\Big\{2-\frac{z_{13}^2z_{24}^2+z_{23}^2z_{14}^2-z_{12}^2z_{34}^2}{z_{13}^2z_{24}^2-z_{14}^2z_{23}^2}\ln\frac{z_{13}^2z_{24}^2}{z_{14}^2z_{23}^2}\Big\}\\
&\times[\mathrm{tr}\{t^a\hat U_{z_1}t^b\hat U^\dagger_{z_2}\}\mathrm{tr}\{t^a\hat U_{z_3}t^b(\hat U^\dagger_{z_4}-\hat U_{z_3})\}]_a. \qquad (7.20)
\end{aligned}
$$

In QCD, this composite operator is no longer conformal but still more convenient than the original dipole since the evolution equation for this operator splits into the sum of the conformal part and the running-coupling part proportional to $b=\frac{11}{3}N_c-\frac{2}{3}n_f$. The linearized version of this evolution equation reproduces NLO BFKL result (with symmetric energy scale) [Fadin and Lipatov (1998)].

It should be noted, however, that there is an approach to improve the non-conformal NLO BK equation (7.19) by collinear resummation of terms leading to double-log contribution in all orders in perturbation theory (see Refs. [Avsar *et al.* (2011); Beuf (2014); Ducloue *et al.* (2019a)]). This "collinearly improved" NLO BK equation seems to agree with HERA data on low-x DIS structure functions [Ducloue *et al.* (2019b)]. Still, if one considers $\mathcal{N}=4$ SYM, one should start with

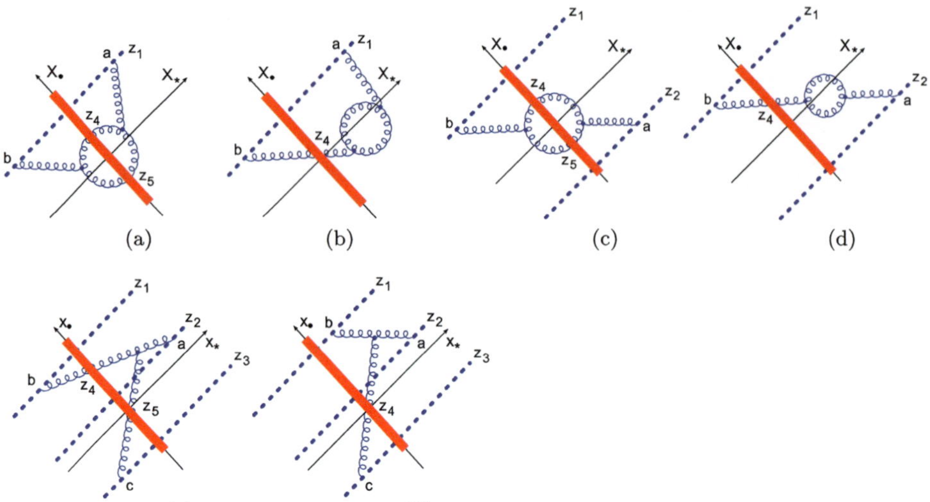

Fig. 7.6 Typical diagrams for NLO hierarchy: self-interation (a,b), pairwise (c,d), and triple interaction (e,f).

conformally invariant kernel: so, in my opinion, the equation (7.20) is a better zero-order approximation for collinear improvement in QCD as well.

Similarly to the leading order, the NLO equation (7.20) is not closed, and the hierarchy of NLO Wilson-lines evolution includes triple interaction in addition to pairwise one, see Fig. 7.6. The hierarchy of NLO Wilson-lines evolution is presented in Ref. [Balitsky and Chirilli (2013b)].

It should be mentioned that NLO hierarchy was conveniently expressed in terms of NLO JIMWLK Hamiltonian in the papers [Kovner *et al.* (2014b,a)]. Also, the NLO JIMWLK was calculated "from scratch" in the paper [Lublinsky and Mulian (2017)].

7.3.3 *Further steps of HE OPE program*

Unfortunately, for small-x DIS the execution of the third and fourth part of OPE program — to solve evolution equation and convolute with initial conditions — is very complicated due to non-linearity of the equation (7.19). However, it is possible to do this in the case when one can linearize the NLO BK equation and get NLO BFKL in the dipole representation. This was performed for the correlation function of four scalar operators in $\mathcal{N} = 4$ SYM [Balitsky and Chirilli (2010)] and later for the physical case of $\gamma^*\gamma^*$ scattering in Refs. [Chirilli and Kovchegov (2013, 2014)].

7.4 Rapidity factorization and evolution for particle production in hadron collisions

In the previous section I demonstrated how to get high-energy behavior of amplitudes like low-x DIS by rapidity evolution. In the next sections I will discuss how to use rapidity evolution to get the amplitude of particle production in hadron-hadron collisions (e.g. inclusive Higgs production at LHC). The goal will be to develop a formalism which bridges between small-x amplitudes described by k_t-factorization and moderate-x amplitudes usually studied in terms of TMDs based on CSS approach.

Particle production in hadron-hadron scattering with transverse momentum of produced particle much smaller than its invariant mass is described in the framework of TMD factorization [Collins (2013); Collins and Soper (1982); Collins *et al.* (1985); Ji *et al.* (2005); Echevarria *et al.* (2012)]. The typical example is the Higgs production at LHC through gluon-gluon fusion. Factorization formula for particle production in hadron-hadron scattering looks like [Collins (2013); Collins and Rogers (2015)]

$$\frac{d\sigma}{d\eta d^2q_\perp} = \sum_f \int d^2b_\perp e^{i(q,b)_\perp}\mathcal{D}_{f/A}(x_A, b_\perp, \eta)\mathcal{D}_{f/B}(x_B, b_\perp, \eta)\sigma(ff \to H)$$
$$+\text{power corrections} + \text{Y} - \text{terms} \tag{7.21}$$

where η is the rapidity, $\mathcal{D}_{f/A}(x, z_\perp, \eta)$ is the TMD density of a parton f in hadron A, and $\sigma(ff \to H)$ is the cross section of production of particle H of invariant mass $m_H^2 = Q^2$ in the scattering of two partons. (For simplicity, we consider the scattering of unpolarized hadrons.)

We will develop a rapidity factorization for particle production which is applicable at $Q^2 \gg q_\perp^2$ and an arbitrary relation between s and Q^2. In a sense, this is the (first) application of rapidity factorization outside of small-x regime.

To obtain formula (7.21) with first corrections we use factorization in rapidity [Balitsky (1999)]. We denote quarks and gluons with rapidity close to the rapidity of the projectile and target protons as A-fields and B-fields, respectively. We call the remaining fields in the central region of rapidity by the name C-fields and integrate over them in the corresponding functional integral. At this step, we get the effective action depending on A and B fields. The subsequent integration over A fields gives matrix elements of some TMD-like operators switched between projectile proton states, and integration over B fields will give matrix elements between target states.[1]

7.4.1 *TMD factorization from functional integral*

We consider production of an (imaginary) scalar particle Φ in proton-proton scattering. This particle is connected to gluons by the vertex

$$\mathcal{L}_\Phi = g_\Phi \int d^4x\ \Phi(x)g^2F^2(x), \quad F^2(x) \equiv F^a_{\mu\nu}(x)F^{a\mu\nu}(x). \tag{7.22}$$

---

[1]It should be noted that due to the kinematics $Q^2 \gg Q_\perp^2, m^2$ we will not need the explicit form of the high-energy effective action which is much sought after in the small-x physics but not known up to now except a couple of first perturbative terms [Balitsky (1999, 2005); Hatta *et al.* (2006, 2005); Kovner and Lublinsky (2005); Altinoluk *et al.* (2009)].

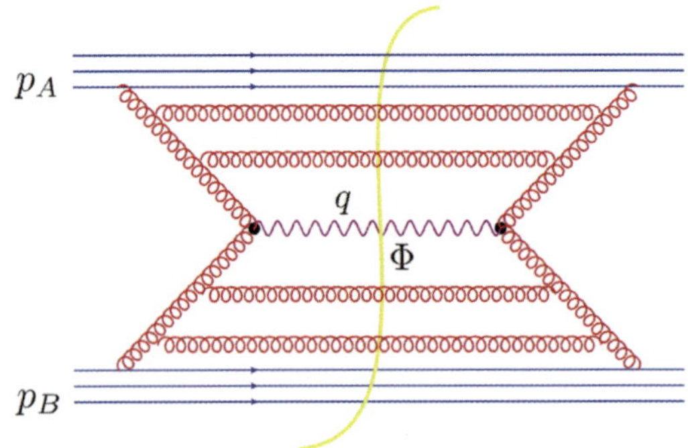

Fig. 7.7 Particle production by gluon-gluon fusion

(This is a $\frac{m_H}{m_t} \ll 1$ approximation for Higgs production via gluon fusion at LHC). The differential cross section of Φ production has the form

$$d\sigma = \frac{d^3q}{2E_q(2\pi)^3}\frac{g_\Phi^2}{2s}W(p_A, p_B, q) \tag{7.23}$$

where we defined the "hadronic tensor" $W(p_A, p_B, q)$ as

$$\begin{aligned} W(p_A, p_B, q) &\stackrel{\text{def}}{=} \sum_X \int d^4x\, e^{-iqx}\langle p_A, p_B|g^2F^2(x)|X\rangle\langle X|g^2F^2(0)|p_A, p_B\rangle \\ &= \int d^4x\ e^{-iqx}\langle p_A, p_B|g^4F^2(x)F^2(0)|p_A, p_B\rangle. \end{aligned} \tag{7.24}$$

As usual, $\sum_X$ denotes the sum over full set of "out" states. It can be represented by double functional integral

$$\begin{aligned} &W(p_A, p_B, q) \\ &= \sum_X \int d^4x\, e^{-iqx}\langle p_A, p_B|g^2F^2(x)|X\rangle\langle X|g^2F^2(0)|p_A, p_B\rangle \\ &= \lim_{\substack{t_i\to-\infty}}^{t_f\to\infty} g^4 \int d^4x\, e^{-iqx} \int^{\tilde{A}(t_f)=A(t_f)} D\tilde{A}_\mu DA_\mu \int^{\tilde{\psi}(t_f)=\psi(t_f)} D\tilde{\bar{\psi}}D\tilde{\psi}D\bar{\psi}D\psi \\ &\quad\times e^{-iS_{\text{QCD}}(\tilde{A},\tilde{\psi})}e^{iS_{\text{QCD}}(A,\psi)}\Psi^*_{p_A}(\vec{\tilde{A}}(t_i), \tilde{\psi}(t_i))\Psi^*_{p_B}(\vec{\tilde{A}}(t_i), \tilde{\psi}(t_i)) \\ &\quad\times \tilde{F}^2(x)F^2(0)\Psi_{p_A}(\vec{A}(t_i), \psi(t_i))\Psi_{p_B}(\vec{A}(t_i), \psi(t_i)). \end{aligned} \tag{7.25}$$

Here the fields A, ψ correspond to the amplitude $\langle X|F^2(0)|p_A, p_B\rangle$, fields $\tilde{A}, \tilde{\psi}$ correspond to complex conjugate amplitude $\langle p_A, p_B|F^2(x)|X\rangle$ and $\Psi_p(\vec{A}(t_i), \psi(t_i))$ denote the proton wave function at the initial time t_i. The boundary conditions $\tilde{A}(t_f) = A(t_f)$ and $\tilde{\psi}(t_f) = \psi(t_f)$ reflect the sum over all states X, cf. Refs. [Balitsky and Braun (1989b)], [Balitsky and Braun (1991)], [Balitsky and Braun (1992)].

To derive the factorization formula, we separate the (quark and gluon) fields in the functional integral (7.25) into three sectors: "projectile" fields A_μ, ψ_a with

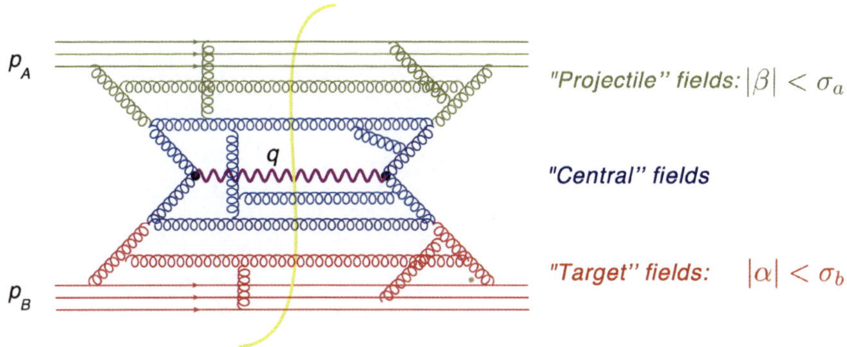

Fig. 7.8 Rapidity factorization for particle production

$|\beta| < \sigma_a$, "target" fields with $|\alpha| < \sigma_b$ and "central rapidity" fields C_μ, ψ with $|\alpha| > \sigma_b$ and $|\beta| > \sigma_a$.[2]

Our goal is to integrate over central fields and get the amplitude in the factorized form, as a (sum of) products of functional integrals over A fields representing projectile matrix elements (TMDs) and functional integrals over B fields representing target matrix elements. In the spirit of background-field method, we "freeze" projectile and target fields (and denote them by $\bar{A}$, $\bar{\xi}_a$, ξ_a and $\bar{B}$, $\bar{\xi}_b$, ξ_b, respectively) and get a sum of diagrams in these external fields. Since $|\beta| < \sigma_a$ in the projectile fields and $|\alpha| < \sigma_b$ in the target fields, at the tree-level one can set with power accuracy $\beta = 0$ for the projectile fields and $\alpha = 0$ for the target fields — the corrections will be $O\big(\frac{m^2}{\sigma_a s}\big)$ and $O\big(\frac{m^2}{\sigma_b s}\big)$. Beyond the tree level, one should expect that the integration over C fields will produce the logarithms of the cutoffs σ_a and σ_b which will cancel with the corresponding logs in gluon TMDs of the projectile and the target.

It should be emphasized that the rapidity factorization here is slightly different from the double high-energy OPE in Sect. 7.3. In the calculation of 4-point correlator we integrated over $\alpha > a$ for the projectile fields and $\beta > b$ and later multiplied the obtained impact factors by dipole-dipole amplitude with longitudinal momenta $|\alpha| < a$ and $|\beta| < b$. If we were interested in pure perturbative processes, say Higgs production in $\gamma^*\gamma^*$ scattering, we could proceed in the same way and multiply two photon impact factors by the amplitude of Higgs production in dipole-dipole scattering. However, in the case of hadron-hadron scattering the impact factors are non-perturbative, so it is convenient to integrate first over perturbative central fields and leave the integration over projectile and target field for later time (actually,

[2]The standard factorization scheme for particle production in hadron-hadron scattering is splitting the diagrams in collinear to projectile part, collinear to target part, hard factor, and soft factor [Collins (2013)]. Here we factorize only in rapidity. For our purpose of calculation of power corrections in the tree approximation it is sufficient; however, we hope to treat possible logs of transverse scales in loop corrections in the same way as it was done in our rapidity evolution equations for gluon TMDs in the next Section.

parametrize them as TMDs of the hadrons). A presumably more important reason is that the method presented in this section is not restricted to small x. It is applicable in principle to any x provided that $s, Q^2 \gg q_\perp^2$, although, as we will see below, the explicit formulas are much simpler in $s \gg Q^2$ case.

As usual, diagrams disconnected from the vertices $F^2(x)$ and $F^2(0)$ ("vacuum bubbles" in external fields) exponentiate so the result has the schematic form

$$\int DC_\mu \int^{\tilde{C}(t_f)=C(t_f)} D\tilde{C}_\mu \int D\bar{\psi}_C D\psi_C \int^{\tilde{\psi}_c(t_f)=\psi_c(t_f)} D\tilde{\bar{\psi}}_C D\tilde{\psi}_C \; g^4 \tilde{F}_C^2(x) F_C^2(0)$$
$$\times e^{-i\tilde{S}_C + iS_C} = e^{S_{\rm eff}(U,V,\tilde{U},\tilde{V})} \mathcal{O}(q, x; A, \tilde{A}, \psi_a \tilde{\psi}_a; B, \tilde{B}, \psi_b, \tilde{\psi}_b) \tag{7.26}$$

where $S_C = S_{\rm QCD}(C+A+B) - S_{\rm QCD}(A) - S_{\rm QCD}(B)$ and $\tilde{S}_C = S_{\rm QCD}(\tilde{C}+A+B) - S_{\rm QCD}(A) - S_{\rm QCD}(B)$. In the r.h.s. of Eq. (7.26) $\mathcal{O}^{\mu\nu}(q, x; A, \psi_A; B, \psi_B)$ is a sum of diagrams connected to $\tilde{F}^2(x)F^2(0)$. Since rapidities of central fields and A, B fields are very different, one should expect the result of integration over C-fields to be represented in terms of Wilson-line operators constructed form A and B fields.

After integration over C fields the amplitude (7.25) can be rewritten as

$$W(p_A, p_B, q)$$
$$= \int d^4x e^{-iqx} \int^{\tilde{A}(t_f)=A(t_f)} D\tilde{A}_\mu DA_\mu \int^{\tilde{\psi}_a(t_f)=\psi_a(t_f)} D\bar{\psi}_a D\psi_a$$
$$\times D\tilde{\bar{\psi}}_a D\tilde{\psi}_a e^{-iS_{\rm QCD}(\tilde{A},\tilde{\psi}_a)} e^{iS_{\rm QCD}(A,\psi_a)} \Psi^*_{p_A}(\vec{\tilde{A}}(t_i), \tilde{\psi}_a(t_i)) \Psi_{p_A}(\vec{A}(t_i), \psi(t_i))$$
$$\times \int^{\tilde{B}(t_f)=B(t_f)} D\tilde{B}_\mu DB_\mu \int^{\tilde{\psi}_b(t_f)=\psi_b(t_f)} D\bar{\psi}_b D\psi_b D\tilde{\bar{\psi}}_b D\tilde{\psi}_b$$
$$\times e^{-iS_{\rm QCD}(\tilde{B},\tilde{\psi}_b)} e^{iS_{\rm QCD}(B,\psi_b)} \Psi^*_{p_B}(\vec{\tilde{B}}(t_i), \tilde{\psi}_b(t_i)) \Psi_{p_B}(\vec{B}(t_i), \psi_b(t_i))$$
$$\times e^{S_{\rm eff}(U,V,\tilde{U},\tilde{V})} \mathcal{O}(q, x; A, \psi_a, \tilde{A}, \tilde{\psi}_a; B, \psi_b, \tilde{B}, \tilde{\psi}_b) \tag{7.27}$$

where $\Psi_{p_A}(\vec{A}(t_{i,f}), \psi(t_{i,f}))$ and $\Psi_{p_B}(\vec{B}(t_{i,f}), \psi_b(t_{i,f}))$ are proton wave functionals. Note that due to boundary conditions at t_f in the above integral, the functional integral over C fields in Eq. (7.26) should be done in the background of the A and B fields satisfying

$$\tilde{A}(t_f) = A(t_f), \quad \tilde{\psi}_a(t_f) = \psi_a(t_f) \quad \text{and}$$
$$\tilde{B}(t_f) = B(t_f), \quad \tilde{\psi}_b(t_f) = \psi_b(t_f). \tag{7.28}$$

Our approximation at the tree level is that $\beta = 0$ for $A, \tilde{A}$ fields and $\alpha = 0$ for $B, \tilde{B}$ fields which corresponds to $A = A(x_\bullet, x_\perp)$, $\tilde{A} = \tilde{A}(x_\bullet, x_\perp)$ and $B = B(x_*, x_\perp)$, $\tilde{B} = \tilde{B}(x_*, x_\perp)$.

Now comes the important point: because of boundary conditions (7.28), for the purpose of calculating the integral (7.26) over central fields one can set

$$A(x_\bullet, x_\perp) = \tilde{A}(x_\bullet, x_\perp), \quad \psi_a(x_\bullet, x_\perp) = \tilde{\psi}_a(x_\bullet, x_\perp),$$
$$B(x_*, x_\perp) = \tilde{B}(x_*, x_\perp), \quad \psi_b(x_*, x_\perp) = \tilde{\psi}_b(x_*, x_\perp). \tag{7.29}$$

Indeed, because A, ψ and $\tilde{A}, \tilde{\psi}$ do not depend on x_*, if they coincide at $x_* = \infty$ they should coincide everywhere. Similarly, if B, ψ_b and $\tilde{B}, \tilde{\psi}_b$ do not depend on $x_\bullet$, if they coincide at $x_\bullet = \infty$ they should be equal.

Summarizing, we see that at the tree level in our approximation

$$\int DC_\mu \int^{\tilde{C}(t_f)=C(t_f)} D\tilde{C}_\mu \int D\bar{\psi}_C D\psi_C \int^{\tilde{\psi}_c(t_f)=\psi_c(t_f)} D\bar{\tilde{\psi}}_C D\tilde{\psi}_C$$
$$\times\, g^4 \tilde{F}_C^2(x) F_C^2(0)\, e^{-i\tilde{S}_C + iS_C} = \mathcal{O}(q, x; A, \psi_a; B, \psi_b). \tag{7.30}$$

It is known that in the tree approximation the double functional integral (7.30) is given by a set of retarded Green functions in the background fields [Gelis and Venugopalan (2004, 2006); Gelis *et al.* (2007)]. Since the double functional integral (7.30) is given by a set of retarded Green functions (in the background field $A + B$), the calculation of tree-level contributions to, say, $F^2(x)$ in the r.h.s. of Eq. (7.30) is equivalent to solving YM equation for $A_\mu(x)$ (and $\psi(x)$) with boundary conditions that the solution has the same asymptotics at $t \to -\infty$ as the superposition of incoming projectile and target background fields.

The hadronic tensor (7.27) can now be represented as

$$W(p_A, p_B, q) = \int d^4x e^{-iqx} \langle p_A|\langle p_B|\hat{\mathcal{O}}(q, x; \hat{A}, \hat{\psi}_a; \hat{B}, \hat{\psi}_b)|p_A\rangle|p_B\rangle \tag{7.31}$$

where $\hat{\mathcal{O}}(q, x; \hat{A}, \hat{\psi}_a; \hat{B}, \hat{\psi}_b)$ should be expanded in a series in $\hat{A}, \hat{\psi}_a; \hat{B}, \hat{\psi}_b$ operators and evaluated between the corresponding (projectile or target) states. As we will demonstrate below, the relevant operators are quark and gluon fields with Wilson-line type gauge links collinear to either p_2 for A fields or p_1 for B fields.

7.4.2 *Power corrections and solution of classical YM equations*

7.4.2.1 *Approximate solution of classical equations*

As we discussed in Sect 7.4.1, the calculation of the functional integral (7.30) over C-fields in the tree approximation reduces to finding fields C_μ and ψ_c as solutions of Yang-Mills equations for the action $S_C = S_{\rm QCD}(C+A+B) - S_{\rm QCD}(A) - S_{\rm QCD}(B)$

$$D^\nu F^a_{\mu\nu}(\bar{A} + \bar{B} + C) = g \textstyle\sum_f (\bar{\psi}^f_a + \bar{\psi}^f_b + \bar{\psi}^f_c)\gamma_\mu t^a (\psi^f_a + \psi^f_b + \psi^f_c)$$
$$(i\,\not{\partial} + g\,\not{\bar{A}} + g\,\not{\bar{B}} + g\,\not{C})(\psi^f_a + \psi^f_b + \psi^f_c) = m(\psi^f_a + \psi^f_b + \psi^f_c). \tag{7.32}$$

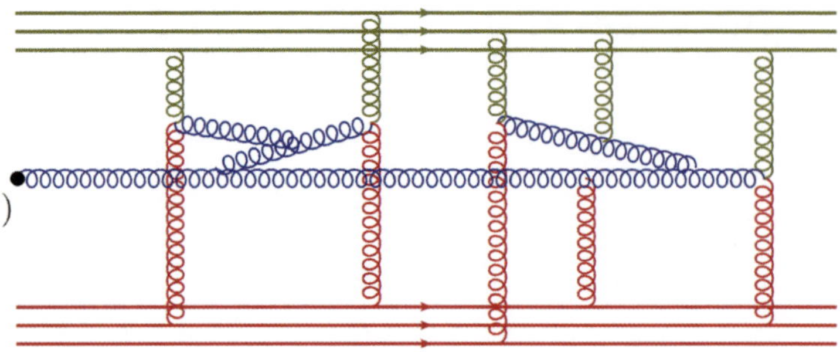

Fig. 7.9 Typical diagram for the classical field with projectile/target sources. The Green functions of the central fields are given by retarded propagators.

As we discussed above the solution of Eq. (7.32) which we need corresponds to the sum of set of diagrams in background field $\bar{A}+\bar{B}$ with *retarded* Green functions (see Fig. 7.9). The retarded Green functions (in the background-Feynman gauge) are defined as

$$\begin{aligned}&\left(x|\frac{1}{\bar{P}^2g^{\mu\nu}+2ig\bar{F}^{\mu\nu}+i\epsilon p_0}|y\right)\\&\quad\equiv\left(x|\frac{1}{p^2+i\epsilon p_0}|y\right)-g\left(x|\frac{1}{p^2+i\epsilon p_0}\mathcal{O}_{\mu\nu}\frac{1}{p^2+i\epsilon p_0}|y\right)\\&\qquad+g^2\left(x|\frac{1}{p^2+i\epsilon p_0}\mathcal{O}_{\mu\xi}\frac{1}{p^2+i\epsilon p_0}\mathcal{O}^{\xi}_{\nu}\frac{1}{p^2+i\epsilon p_0}|y\right)+\cdots \qquad (7.33)\end{aligned}$$

where

$$\begin{aligned}\bar{P}_\mu &\equiv i\partial_\mu+g\bar{A}_\mu+g\bar{B}_\mu,\\ \bar{F}_{\mu\nu}&=\partial_\mu(\bar{A}+\bar{B})_\nu-\mu\leftrightarrow\nu-ig[\bar{A}_\mu+\bar{B}_\mu,\bar{A}_\nu+\bar{B}_\nu] \qquad (7.34)\\ \mathcal{O}_{\mu\nu}&\equiv(\{p^\xi,\bar{A}_\xi+\bar{B}_\xi\}+g(\bar{A}+\bar{B})^2)g_{\mu\nu}+2i\bar{F}_{\mu\nu}\end{aligned}$$

and similarly for quarks.

The solutions of Eqs. (7.32) in terms of retarded Green functions give fields C_μ and ψ_c that vanish at $t\to-\infty$. Thus, we are solving the usual classical YM equations

$$D^\nu F^a_{\mu\nu}=\sum_f g\bar{\psi}^f t^a\gamma_\mu\psi^f,\quad (\not{P}-m_f)\psi^f=0 \qquad (7.35)$$

with boundary conditions

$$\begin{aligned}A_\mu(x)&\overset{x_*\to-\infty}{=}\bar{A}_\mu(x_\bullet,x_\perp),\quad \psi(x)\overset{x_*\to-\infty}{=}\psi_a(x_\bullet,x_\perp)\\ A_\mu(x)&\overset{x_\bullet\to-\infty}{=}\bar{B}_\mu(x_*,x_\perp),\quad \psi(x)\overset{x_\bullet\to-\infty}{=}\psi_b(x_*,x_\perp)\end{aligned} \qquad (7.36)$$

following from $C_\mu, \psi_c \stackrel{t\to-\infty}{\to} 0$. These boundary conditions reflect the fact that at $t \to -\infty$ we have only incoming hadrons with "A" and "B" fields.

The solution of YM equations (7.35) in general case is yet unsolved problem, especially important for scattering of two heavy nuclei in semiclassical approximation. Fortunately, for our case of particle production with $\frac{q_\perp}{Q} \ll 1$ we can construct the approximate solution of (7.35) as a series in this small parameter. Also, it is convenient to use the gauge where the projectile fields satisfy $A_*(x_\bullet, x_\perp) = 0$ and target fields satisfy $B_\bullet(x_*, x_\perp) = 0$. (The existence of such gauge was proved in Ref. [Balitsky and Tarasov (2015)] by explicit construction). In this gauge the initial conditions for YM equations (7.32) are

$$\begin{aligned} gA_\mu(x) &\stackrel{x_*\to-\infty}{=} U_\mu(x_\bullet, x_\perp), \quad \psi(x) \stackrel{x_*\to-\infty}{=} \Sigma_a(x_\bullet, x_\perp) \\ gA_\mu(x) &\stackrel{x_\bullet\to-\infty}{=} V_\mu(x_*, x_\perp), \quad \psi(x) \stackrel{x_\bullet\to-\infty}{=} \Sigma_b(x_*, x_\perp) \end{aligned} \tag{7.37}$$

where

$$\begin{aligned} U_\mu(x_\bullet, x_\perp) &\equiv \frac{2}{s} p_{2\mu} U_\bullet(x_\bullet, x_\perp) + U_{\mu_\perp}(x_\bullet, x_\perp) \\ V_\mu(x_*, x_\perp) &\equiv \frac{2}{s} p_{1\mu} V_*(x_*, x_\perp) + V_{\mu_\perp}(x_*, x_\perp) \\ U_i(x_\bullet, x_\perp) &\equiv \frac{2}{s} \int_{-\infty}^{x_\bullet} dx'_\bullet \; U_{*i}(x'_\bullet, x_\perp), \\ V_i(x_*, x_\perp) &\equiv \frac{2}{s} \int_{-\infty}^{x_*} dx'_* \; V_{\bullet i}(x'_*, x_\perp) \\ U_\bullet(x_\bullet, x_\perp) &\equiv \frac{2}{s} \int_{-\infty}^{x_\bullet} dx'_\bullet \; U_{*\bullet}(x'_\bullet, x_\perp), \\ V_*(x_*, x_\perp) &\equiv -\frac{2}{s} \int_{-\infty}^{x_*} dx'_* \; V_{*\bullet}(x'_*, x_\perp) \end{aligned} \tag{7.38}$$

and Σ_a, Σ_b are defined as

$$\begin{aligned} \Sigma_a(z_\bullet, z_\perp) &\equiv [-\infty_\bullet, z_\bullet]_z \psi_a(z_\bullet, z_\perp), \\ \Sigma_b(z_*, z_\perp) &\equiv [-\infty_*, z_*]_z \psi_b(z_*, z_\perp). \end{aligned} \tag{7.39}$$

The relative strength of Lorentz components of projectile and target fields can be read from the comparison of leading-twist TMDs with typical power corrections. One obtains [Balitsky and Tarasov (2015)]

$$U_\bullet \sim V_* \sim m_\perp^2, \quad U_i \sim V_i \sim m_\perp. \tag{7.40}$$

Finally, the classical equations for projectile and target fields in this gauge read[3]:

$$
\begin{aligned}
D_U^\nu U^a_{\mu\nu} &= g^2 \sum_f \bar\Sigma^f_a \gamma_\mu t^a \Sigma^f_a, \quad i\not{D}_U \Sigma_a = 0 \\
D_V^\nu V^a_{\mu\nu} &= g^2 \sum_f \bar\Sigma^f_b \gamma_\mu t^a \Sigma^f_b, \quad i\not{D}_V \Sigma_b = 0
\end{aligned}
\tag{7.41}
$$

where $U_{\mu\nu} \equiv \partial_\mu U_\nu - \partial_\nu U_\mu - i[U_\mu, U_\nu]$, $D_U^\mu \equiv (\partial^\mu - i[U^\mu,)$ and similarly for V fields.

We solve Eqs. (7.35) iteratively, order by order in perturbation theory, starting from the zero-order approximation in the form of the sum of projectile and target fields

$$
\begin{aligned}
g\mathcal{A}^{[0]}_\mu(x) &= U_\mu(x_\bullet, x_\perp) + V_\mu(x_*, x_\perp) \\
\Psi^{[0]}(x) &= \Sigma_a(x_\bullet, x_\perp) + \Sigma_b(x_*, x_\perp)
\end{aligned}
\tag{7.42}
$$

and improving it by calculation of Feynman diagrams with retarded propagators in the background fields (7.42).

The first step is the calculation of the linear term for the trial configuration (7.42). It is convenient to sort out linear terms according to their order in small parameter $\frac{m_\perp^2}{s}$:

$$
\begin{aligned}
L^a_\bullet &\equiv \mathcal{D}^\mu \mathcal{F}^{[0]a}_{\mu\bullet} + g\bar\Psi^{[0]}\gamma_\bullet t^a \Psi^{[0]} = L^{(-1)a}_\bullet + L^{(0)a}_\bullet + L^{(1)a}_\bullet, \\
L^a_* &\equiv \mathcal{D}^\mu \mathcal{F}^{[0]a}_{\mu *} + g\bar\Psi^{[0]}\gamma_* t^a \Psi^{[0]} = L^{(-1)a}_* + L^{(0)a}_* + L^{(1)a}_*, \\
L^a_i &\equiv \mathcal{D}^\mu \mathcal{F}^{[0]a}_{\mu i} + g\bar\Psi^{[0]}\gamma_i t^a \Psi^{[0]} = L^{(0)a}_i + L^{(1)a}_i.
\end{aligned}
\tag{7.43}
$$

The contribution to the leading power correction is coming from

$$
L^{(-1)a}_\bullet = \frac{i}{g} U^{jab} V^b_{\bullet j}, \quad L^{(-1)a}_* = \frac{i}{g} V^{jab} U^b_{*j}.
\tag{7.44}
$$

The explicit form of other terms in Eq. (7.43) is presented in Ref. [Balitsky and Tarasov (2017)]. Hereafter the order of perturbation theory is labeled by $(...)^{[n]}$ and the order of expansion in the parameter $\frac{m_\perp^2}{s}$ by $(...)^{(n)}$.

With the linear term (7.43), a couple of first terms in perturbative series are

$$
\begin{aligned}
A^{[1]a}_\mu(x) &= \int d^4z (x|\frac{1}{\mathcal{P}^2 g^{\mu\nu} + 2ig\mathcal{F}^{[0]\mu\nu}}|z)^{ab} L^{b\nu}(z) \\
A^{[2]a}_\mu(x) &= g\int d^4z \Big[-i(x|\frac{1}{\mathcal{P}^2 g^{\mu\eta} + 2ig\mathcal{F}^{[0]\mu\eta}}\mathcal{P}^\xi|z)^{aa'} f^{a'bc} A^{[1]b}_\xi A^{[1]c\eta} \\
&\quad + (x|\frac{1}{\mathcal{P}^2 g^{\mu\eta} + 2ig\mathcal{F}^{[0]\mu\eta}}|z)^{aa'} f^{a'bc} A^{[1]b\xi} (\mathcal{D}_\xi A^{[1]c\eta} - \mathcal{D}^\eta A^{[1]c}_\xi) \Big]
\end{aligned}
\tag{7.45}
$$

---

[3]Here we consider only u, d, and s quarks which can be regarded as massless.

for gluon fields (in the background-Feynman gauge) and

$$\begin{aligned}\Psi_f^{[1]}(x) &= -\int d^4z (x|\frac{1}{\not{\mathcal{P}}}|z) L_\psi(z),\\ \Psi_f^{[2]}(x) &= -g\int d^4z (x|\frac{1}{\not{\mathcal{P}}}|z) \not{A}^{[1]}(z)\Psi_f^{[0]}(z)\end{aligned} \tag{7.46}$$

for quarks where

$$\mathcal{P}_\bullet = i\partial_\bullet + U_\bullet, \quad \mathcal{P}_* = i\partial_* + V_*, \quad \mathcal{P}_i = i\partial_i + U_i + V_i \tag{7.47}$$

are operators in external zero-order fields (7.42). Hereafter we use Schwinger's notations for propagators in external fields normalized according to $(x|F(p)|y) \equiv \int d\!\!^-{}^4 p\ e^{-ip(x-y)} F(p)$. Moreover, when it will not lead to a confusion, we will use short-hand notation $\frac{1}{\mathcal{O}}\mathcal{O}'(x) \equiv \int d^4z\ (x|\frac{1}{\mathcal{O}}|z)\mathcal{O}'(z)$. Next iterations will give us a set of tree-level Feynman diagrams in the background field $U_\mu + V_\mu$ and $\Sigma_a + \Sigma_b$.

Let us consider the fields in the first order in perturbation theory:

$$\begin{aligned}A_\mu^{[1]} &= \frac{1}{\mathcal{P}^2 g^{\mu\nu} + 2ig\mathcal{F}^{[0]\mu\nu}} L^\nu\\ &= \frac{1}{[\{\alpha + \frac{2}{s}V_*, \beta + \frac{2}{s}U_\bullet\}\frac{s}{2} - (P+U+V)_\perp^2]g^{\mu\nu} + 2ig\mathcal{F}^{[0]\mu\nu} + i\epsilon p_0} L^\nu\\ \Psi_f^{[1]}(x) &= -\frac{1}{\not{\mathcal{P}}} L_\psi = -\frac{(\alpha + \frac{2}{s}V_*)\not{p}_1 + (\beta + \frac{2}{s}U_\bullet)\not{p}_2 + \not{\mathcal{P}}_\perp}{\{\alpha + \frac{2}{s}V_*, \beta + \frac{2}{s}U_\bullet\}\frac{s}{2} - (p+U+V)_\perp^2 + i\epsilon p_0} L_\psi.\end{aligned} \tag{7.48}$$

Here α, β, and $p_\perp$ are understood as differential operators $\alpha = i\frac{\partial}{\partial x_\bullet}$, $\beta = i\frac{\partial}{\partial x_*}$ and $p_i = i\frac{\partial}{\partial x^i}$.

Now comes the central point of our approach. Let us expand quark and gluon propagators in powers of background fields, then we get a set of diagrams shown in Fig. 7.9. The typical bare gluon propagator in Fig. 7.9 is

$$\frac{1}{p^2 + i\epsilon p_0} = \frac{1}{\alpha\beta s - p_\perp^2 + i\epsilon(\alpha+\beta)}. \tag{7.49}$$

Since we do not consider loops of C-fields, the transverse momenta in tree diagrams are determined by further integration over projectile ("A") and target ("B") fields in Eq. (7.27) which converge on either $q_\perp$ or m. On the other hand, the integrals over α converge on either α_q or $\alpha \sim 1$ and similarly the characteristic β's are either β_q or ~ 1. Since $\alpha_q\beta_q s = Q_\parallel^2 \gg Q_\perp^2$, one can expand gluon and quark propagators in powers of $\frac{p_\perp^2}{\alpha\beta s}$

$$\begin{aligned}\frac{1}{p^2 + i\epsilon p_0} &= \frac{1}{s(\alpha+i\epsilon)(\beta+i\epsilon)}\Big(1 + \frac{p_\perp^2/s}{(\alpha+i\epsilon)(\beta+i\epsilon)} + \cdots\Big)\\ \frac{\not{p}}{p^2 + i\epsilon p_0} &= \frac{1}{s}\Big(\frac{\not{p}_1}{\beta+i\epsilon} + \frac{\not{p}_2}{\alpha+i\epsilon} + \frac{\not{p}_\perp}{(\alpha+i\epsilon)(\beta+i\epsilon)}\Big)\Big(1 + \frac{p_\perp^2/s}{(\alpha+i\epsilon)(\beta+i\epsilon)} + \cdots\Big).\end{aligned} \tag{7.50}$$

After the expansion (7.50), the dynamics in the transverse space effectively becomes trivial: all background fields stand either at x or at 0.

One may wonder why we do not cut the integrals over longitudinal variables to $|\alpha| > \sigma_b$ and $|\beta| > \sigma_a$ according to the definition of C fields in Sect. 7.4.1.[4] The reason is that in the diagrams like Fig. 7.9 with retarded propagators (7.50) one can shift the contour of integration over α and/or β to the complex plane away to avoid the region of small α or β.[5]

Note that the background fields are also smaller than typical $p_\parallel^2 \sim s$. Indeed, from Eq. (7.40) we see that $p_\bullet = \frac{s}{2}\beta \gg U_\bullet \sim m^2$ (because $\alpha \geq \alpha_q \gg \frac{m^2}{s}$) and similarly $p_* \gg V_*$. Also $(p_i+U_i+V_i)^2 \sim q_\perp^2 \ll p_\parallel^2$. The only exception are the fields $V_{\bullet i}$ or U_{*i} which are of order of $sm_\perp$ but we will see that effectively the expansion in powers of these fields is cut at the second term with our accuracy.

The corresponding expansion of field strengths reads

$$
\begin{aligned}
&gF_{\bullet i}^{(-1)a}(x) = V_{\bullet i}^a(x), \quad gF_{*i}^{(-1)a}(x) = U_{*i}^a(x)\\
&gF_{\bullet i}^{(0)a}(x) = U_{\bullet i}^a(x) - iU_\bullet^{ab}(x)V_i^b(x) - \frac{ig}{2\alpha}\tilde{L}_i^{(0)a}(x) + \mathcal{D}_i^{ab}V_j^{bc}(x)\frac{1}{2\alpha}U^{cj}(x)\\
&gF_{*i}^{(0)a}(x) = V_{*i}^a(x) - iV_*^{ab}(x)U_i^b(x) - \frac{ig}{2\beta}\tilde{L}_i^{(0)a}(x) + \mathcal{D}_i^{ab}U_j^{bc}(x)\frac{1}{2\beta}V^{cj}(x).\\
&gF_{*\bullet}^{(-1)a}(x) = U_{*\bullet}^a(x) + V_{*\bullet}^a(x) - \frac{is}{2}U_j^{ab}(x)V^{bj}(x)\\
&gF_{ik}^{(0)a}(x) = U_{ik}^a(x) + V_{ik}^a(x) - i\big(U_i^{ab}(x)V_k^b(x) - i \leftrightarrow k\big).
\end{aligned}
\tag{7.51}
$$

Power corrections to hadronic tensor are proportional to

$$
\begin{aligned}
F^2(x) &\equiv F_{\mu\nu}^a(x)F^{a\mu\nu}(x)\\
&= \frac{8}{s}F_{\bullet i}^a(x)F_*^{ai}(x) + F_{ik}^a(x)F^{aik}(x) - \frac{8}{s^2}F_{*\bullet}^a(x)F_{*\bullet}^a(x)
\end{aligned}
\tag{7.52}
$$

so

$$
\begin{aligned}
(F^2(x))^{(-1)} &= \frac{8}{sg^2}U_{*i}^aV_\bullet^{ai}\\
(F^2(x))^{(0)} &= F_{ik}^{(0)a}(x)F^{(0)aik} - \frac{8}{s^2}F_{*\bullet}^{(-1)a}(x)F_{*\bullet}^{(-1)a}(x)\\
&\quad + \frac{8}{sg}V_\bullet^{ai}(x)F_{*i}^{(0)a}(x) + \frac{8}{sg}U_*^{ai}(x)F_{\bullet i}^{(0)a}(x)
\end{aligned}
\tag{7.53}
$$

---

[4]Such cutoffs for integrals over C fields are introduced explicitly in the framework of soft-collinear effective theory, see the review [Rothstein and Stewart (2016)].

[5]This may be wrong if there is pinching of poles in the integrals over α or β but we will see that in our integrals for the tree-level power corrections the pinching of poles never occurs. In the higher orders in perturbation theory the pinching may occur and in such case one needs to formulate a subtraction program to avoid double counting.

and the leading higher-twist correction is proportional to

$$
\begin{aligned}
&(F^2(x))^{(0)}(F^2(0))^{(-1)} + (x \leftrightarrow 0) \\
&= \Big[F_{ik}^{(0)a}(x)F^{(0)aik}(x) - \frac{8}{s^2}F_{*\bullet}^{(-1)a}(x)F_{*\bullet}^{(-1)a}(x) \\
&\quad + \frac{8}{sg}V_\bullet^{ai}(x)F_{*i}^{(0)a}(x) + \frac{8}{sg}U_*^{ai}(x)F_{\bullet i}^{(0)a}(x)\Big] \\
&\quad \times \frac{8}{sg^2}U_{*i}^a(0)V_\bullet^{ai}(0) \; + \; (x \leftrightarrow 0).
\end{aligned}
\tag{7.54}
$$

7.4.3 *Leading higher-twist correction at* $s \gg Q^2 \gg Q_\perp^2 \gg m^2$

As we mentioned in the beginning of Sect. 7.4, our method is relevant for calculation of higher-twist corrections at any $s, Q^2 \gg Q_\perp^2, m^2$. However, the expressions become manageable in the physically interesting case $s \gg Q^2 \gg Q_\perp^2 \gg m^2$ which we consider in this Section.[6] As demonstrated in Ref. [Balitsky and Tarasov (2015)], the leading correction in this region comes from the following part of Eq. (7.52)

$$
g^2F^2(x) = \frac{8}{s}U_*^{ai}(x)V_{\bullet i}^a(x) + 2f^{mac}f^{mbd}\Delta^{ij,kl}U_i^a(x)U_j^b(x)V_k^c(x)V_l^d(x) + \cdots \tag{7.55}
$$

where $\Delta^{ij,kl} \equiv g^{ij}g^{kl} - g^{ik}g^{jl} - g^{il}g^{jk}$. The higher-twist correction coming from the second term in the r.h.s. will be $\sim \frac{Q_\perp^2}{Q^2}$ whereas other terms in the r.h.s. of Eq. (7.52) yield contributions $\sim \frac{Q_\perp^2}{s}$, $\sim \frac{Q_\perp^2}{\alpha_q s}$, or $\sim \frac{Q_\perp^2}{\beta_q s}$ all of which are small (see the footnote 6). In this approximation we get

$$
\begin{aligned}
&g^4F^2(x)F^2(0) \\
&= \frac{64}{s^2}U_*^{mi}(x)V_{\bullet i}^m(x)U_*^{nj}(0)V_{\bullet j}^n(0) \\
&\quad + \frac{16}{s}f^{mac}f^{mbd}\Delta^{ij,kl}\big[U_i^a(x)U_j^b(x)V_k^c(x)V_l^d(x)U_*^{nr}(0)V_{\bullet r}^n(0) \\
&\quad + U_*^{nr}(x)V_{\bullet r}^n(x)U_i^a(0)U_j^b(0)V_k^c(0)V_l^d(0)\big]
\end{aligned}
\tag{7.56}
$$

where the first term is the leading order and the second is the higher-twist correction.

Substituting our approximation (7.55) to Eq. (7.24) and promoting background fields to operators we get (note that $\alpha_q\beta_q s = Q_\parallel^2 \simeq Q^2$):

$$
\begin{aligned}
&W(p_A, p_B, q) \\
&= \frac{64/s^2}{N_c^2 - 1}\int d^2x_\perp \frac{2}{s}\int dx_\bullet dx_* \cos\big(\alpha_q x_\bullet + \beta_q x_* - (q,x)_\perp\big)
\end{aligned}
$$

[6] We also assume that the scalar particle is emitted in the central region of rapidity, so $\alpha_q s \sim \beta_q s \gg Q^2$.

$$\times\Big\{\langle p_A|\hat U_*^{mi}(x_\bullet,x_\perp)\hat U_*^{mj}(0)|p_A\rangle\langle p_B|\hat V_{\bullet i}^{n}(x_*,x_\perp)\hat V_{\bullet j}^{n}(0)|p_B\rangle$$
$$-\frac{4N_c^2}{N_c^2-4}\frac{\Delta^{ij,kl}}{Q^2}\frac{2}{s}\int_{-\infty}^{x_\bullet}dx'_\bullet d^{abc}\langle p_A|\hat U_{*i}^{a}(x_\bullet,x_\perp)\hat U_{*j}^{b}(x'_\bullet,x_\perp)\hat U_{*r}^{c}(0)|p_A\rangle$$
$$\times\frac{2}{s}\int_{-\infty}^{x_*}dx'_* d^{mpq}\langle p_B|\hat V_{\bullet k}^{m}(x_*,x_\perp)\hat V_{\bullet l}^{p}(x'_*,x_\perp)\hat V_\bullet^{qr}(0)|p_B\rangle\Big\}. \tag{7.57}$$

Unfortunately, not much is known about three-gluon matrix elements so it is hard to estimate at what $q_\perp^2$ power corrections become essential. The situation is better for quark TMDs as we will see in the next section where we present the result for power corrections for Z-boson production.

7.4.4 *Power corrections for Z-boson production*

The analysis performed in the previous section can be repeated for Z-boson production.

The "hadronic tensor" $W(p_A,p_B,q)$ for Z-boson production is defined as

$$W(p_A,p_B,q)\stackrel{\text{def}}{=}\frac{1}{(2\pi)^4}\sum_X\int d^4x\; e^{-iqx}\langle p_A,p_B|J_\mu(x)|X\rangle\langle X|J^\mu(0)|p_A,p_B\rangle$$
$$=\frac{1}{(2\pi)^4}\int d^4x\, e^{-iqx}\langle p_A,p_B|J_\mu(x)J^\mu(0)|p_A,p_B\rangle \tag{7.58}$$

where

$$J_\mu=-\frac{e}{2s_Wc_W}\sum_f\bar\psi_f\gamma_\mu(g_f^V-g_f^A\gamma_5)\psi_f \tag{7.59}$$

and $g_f^A=(t_3^L)_f$, $g_f^V=(t_3^L)_f-2q_fs_W^2$ as usual.

Power corrections to the hadronic tensor (7.58) are obtained by approximate solution of YM equations, now with emphasis on the quark fields. It turns out that power corrections can be sorted out by powers of N_c and terms leading in N_c can be expressed in terms of leading-twist TMDs due to QCD equations of motion. The sum of leading-twist contribution and main power correction reads [Balitsky and Tarasov (2018)]

$$W(\alpha_z,\beta_z,q_\perp)=-\frac{e^2}{8s_W^2c_W^2N_c}\int d^2k_\perp\bigg[\Big\{(1+a_u^2)\Big[1-2\frac{(k,q-k)_\perp}{Q^2}\Big]$$
$$\times f_{1u}(\alpha_z,k_\perp)\bar f_{1u}(\beta_z,q_\perp-k_\perp)$$
$$+(\alpha_z\leftrightarrow\beta_z)\Big\}+\Big\{u\leftrightarrow c\Big\}+\Big\{u\leftrightarrow d\Big\}+\Big\{u\leftrightarrow s\Big\}\bigg]$$
$$+O\Big(\frac{m_\perp^8}{s}\Big)+O\Big(\frac{1}{N_c^2}\Big) \tag{7.60}$$

where $a_{u,c} \equiv (1 - \frac{8}{3}s_W^2)$, $a_{d,s} \equiv (1 - \frac{4}{3}s_W^2)$ and $f, \bar{f}$ are standard parametrizations of leading-twist TMDs

$$\frac{1}{16\pi^3}\int dx_\bullet d^2x_\perp\, e^{-i\alpha x_\bullet + i(k,x)_\perp}\langle A|\hat{\bar{\psi}}_f(x_\bullet, x_\perp)\gamma^\mu\hat{\psi}_f(0)|A\rangle$$

$$= p_1^\mu f_1^f(\alpha, k_\perp^2) + k_\perp^\mu f_\perp^f(\alpha, k_\perp^2) + p_2^\mu \frac{2m^2}{s} f_3^f(\alpha, k_\perp^2),$$

$$\frac{1}{16\pi^3}\int dx_\bullet d^2x_\perp\, e^{i\alpha x_\bullet - i(k,x)_\perp}\langle A|\hat{\bar{\psi}}_f(x_\bullet, x_\perp)\gamma^\mu\hat{\psi}_f(0)|A\rangle$$

$$= -p_1^\mu \bar{f}_1^f(\alpha, k_\perp^2) - k_\perp^\mu \bar{f}_\perp^f(\alpha, k_\perp^2) - p_2^\mu \frac{2m^2}{s} \bar{f}_3^f(\alpha, k_\perp^2). \tag{7.61}$$

Next, power corrections become sizable at $q_\perp^2 \gg m^2$ where we probe the perturbative tails of TMD's $f_1 \sim \frac{1}{k_\perp^2}$ and $h_1^\perp \sim \frac{1}{k_\perp^4}$ [Zhou *et al.* (2008)]. So, as long as $k_\perp^2 \leq \alpha_z\sigma_a s = Q^2$ we can approximate

$$f_1(\alpha_z, k_\perp^2) \simeq \frac{f(\alpha_z)}{k_\perp^2}, \quad \bar{f}_1 \simeq \frac{\bar{f}(\alpha_z)}{k_\perp^2} \tag{7.62}$$

(up to logarithmic corrections). Similarly, for the target we can use the estimate

$$f_1(\beta_z, k_\perp^2) \simeq \frac{f(\beta_z)}{k_\perp^2}, \quad \bar{f}_1 \simeq \frac{\bar{f}(\beta_z)}{k_\perp^2} \tag{7.63}$$

as long as $k_\perp^2 \leq \beta_z\sigma_b s = Q^2$.

Substituting this to Eq. (7.60) we get the following estimate of the strength of power corrections for Z-boson production

$$W(\alpha_z, \beta_z, q_\perp) \simeq -\frac{e^2}{8s_W^2 c_W^2 N_c}\int d^2k_\perp \frac{1}{k_\perp^2(q-k)_\perp^2}\Big[1 - 2\frac{(k, q-k)_\perp}{Q^2}\Big]$$
$$\times\Big[\big\{(1 + a_u^2)[f_u(\alpha_z)\bar{f}_u(\beta_z) + \bar{f}_u(\alpha_z)f_u(\beta_z)]\big\}$$
$$+\{u \leftrightarrow c\} + \{u \leftrightarrow d\} + \{u \leftrightarrow s\}\Big].$$

Thus, the relative weight of the leading twist and power correction is determined by the factor $1 - 2\frac{(k,q-k)_\perp}{Q^2}$. The integrals over $k_\perp$ are logarithmic and should be cut from below by m^2 and from above by Q^2 so we get an estimate

$$W(\alpha_z, \beta_z, q_\perp) = -\frac{\pi e^2}{4s_W^2 c_W^2 N_c}\Big[\frac{1}{q_\perp^2}\ln\frac{q_\perp^2}{m^2} + \frac{1}{Q^2}\ln\frac{Q^2}{q_\perp^2}\Big]$$
$$\times\Big[\big\{(1 + a_u^2)[f_u(\alpha_z)\bar{f}_u(\beta_z) + \bar{f}_u(\alpha_z)f_u(\beta_z)]\big\}$$
$$+\{u \leftrightarrow c\} + \{u \leftrightarrow d\} + \{u \leftrightarrow s\}\Big] \tag{7.64}$$

where we assumed that the first term is determined by the logarithmical region $q_\perp^2 \gg k_\perp^2 \gg m^2$ and the second by $Q^2 \gg k_\perp^2 \gg q_\perp^2$. By this estimate, the power correction reaches the level of few percent at $q_\perp \geq 20$ GeV. Of course, when $q_\perp^2$ increases, the correction becomes bigger, but the validity of the approximation $\frac{q_\perp^2}{Q^2} \ll 1$ worsens. Moreover, we have ignored all logarithmic (and double-log) evolutions which can significantly change the relative strength of power corrections.

7.5 Rapidity evolution of TMDs

As we discussed in section 7.4.1, we factorize the integrals over gluon fields in "projectile" fields A_μ, ψ_a with $|\beta| < \sigma_a$, "target" fields with $|\alpha| < \sigma_b$ and "central rapidity" fields C_μ, ψ with $|\alpha| > \sigma_b$ and $|\beta| > \sigma_a$. For the purpose of calculating power corrections, it is sufficient to assume that projectile fields have zero components along the target direction, and *vice versa*, see Eq. (7.29). However, to use factorization formula like (7.57) we need to find out the evolution of TMD operators with respect to cutoffs σ_a and σ_b. This problem was solved in Ref. [Balitsky and Tarasov (2015, 2016)] and here we briefly remind main ideas and results.

7.5.1 *Rapidity factorization and evolution of TMDs in the leading order*

We will study the rapidity evolution of the operator

$$\tilde{\mathcal{F}}_i^{a\eta}(\beta_B, x_\perp)[x_\perp, y_\perp]_{-\infty}^{ab} \mathcal{F}_j^{b\eta}(\beta_B, y_\perp). \tag{7.65}$$

Note, however, that forward matrix element of this operator has an unbounded integration over $x_* - y_*$. It is convenient to introduce the notation $\langle\langle p|O|p\rangle\rangle$ for the forward matrix element of the operator O stripped of this integration

$$\begin{aligned}&\langle p|\tilde{\mathcal{F}}_i^{a\eta}(\beta_B, z_\perp)\mathcal{F}^{ai\eta}(\beta_B, 0_\perp)|p + \xi p_2\rangle \\ &\quad = 2\pi\delta(\xi)\langle\langle p|\tilde{\mathcal{F}}_i^{a\eta}(\beta_B, z_\perp)\mathcal{F}^{ai\eta}(\beta_B, 0_\perp)|p\rangle\rangle.\end{aligned} \tag{7.66}$$

With this notation the unintegrated gluon TMD can be parametrized as [Mulders and Rodrigues (2001)]

$$\begin{aligned}&\int d^2z_\perp\, e^{i(k,z)_\perp} \langle\langle p|\tilde{\mathcal{F}}_i^{a\eta}(\beta_B, z_\perp)\mathcal{F}_j^{a\eta}(\beta_B, 0_\perp)|p\rangle\rangle^\eta \\ &\quad = \pi g^2 \mathcal{R}_{ij}(\beta_B, k_\perp; \eta) \\ &\mathcal{R}_{ij}(\beta_B, k_\perp; \eta) \\ &\quad = -g_{ij}\beta_B \mathcal{D}(\beta_B, k_\perp, \eta) + \Big(\frac{2k_ik_j}{m^2} + g_{ij}\frac{k_\perp^2}{m^2}\Big)\beta_B \mathcal{H}(\beta_B, k_\perp, \eta)\end{aligned} \tag{7.67}$$

where m is the mass of the target hadron (typically proton).

The reason we study the evolution of the operator (7.65) with non-convoluted indices i and j is that, as we shall see below, the rapidity evolution mixes functions $\mathcal{D}$ and $\mathcal{H}$. It should be also noted that our final equation for the evolution of the operator (7.65) is applicable for polarized targets as well.

In the spirit of rapidity factorization, in order to find the evolution of TMD

$$\langle\!\langle p|\mathcal{F}_i^a(x_*,x_\perp)[x_\perp,y_\perp]^{ab}_{-\infty}\mathcal{F}_j^b(y_*,y_\perp)|p\rangle\!\rangle^\eta \tag{7.68}$$

with respect to rapidity cutoff η (see Eq. (7.10)) one should integrate in the matrix element (7.68) over gluons and quarks with rapidities $\eta > Y > \eta'$ and temporarily "freeze" fields with $Y < \eta'$ to be integrated over later. (For a review, see Refs. [Balitsky (2001, 2010)].) In this case, we obtain functional integral over fields with $\eta > Y > \eta'$ in the "external" fields with $Y < \eta'$. In terms of Sudakov variables we integrate over gluons with α between $\sigma = e^\eta$ and $\sigma' = e^{\eta'}$ and, in the leading order, only the diagrams with gluon emissions are relevant — the quark diagrams will enter as loops at the next-to-leading (NLO) level.

To calculate diagrams, one needs to return to a double functional integral representation of gluon TMD (7.68):

$$\begin{aligned}&\langle p|\mathcal{F}_i^a(x_*,x_\perp)[x_\perp,y_\perp]^{ab}_{-\infty}\mathcal{F}_j^b(y_*,y_\perp)|p'\rangle^\eta\\&=\int^{\tilde{A}(\infty)=A(\infty)} D\tilde{A}D\bar{\tilde{\psi}}D\tilde{\psi}DAD\bar{\psi}D\psi\Psi_p^*(\tilde{A},\tilde{\psi})|_{t_i=-\infty}\, e^{-iS_{\mathrm{QCD}}(\tilde{A},\tilde{\psi})}\\&\tilde{\mathcal{F}}_i^a(x_*,x_\perp)[x_\perp,y_\perp]^{ab}_{-\infty}e^{iS_{\mathrm{QCD}}(A,\psi)}\mathcal{F}_j^b(y_*,y_\perp)\Psi_{p'}(A,\psi)|_{t_i=-\infty}.\end{aligned} \tag{7.69}$$

Now, in accordance with general background-field formalism we separate the gluon field into the "classical" background part with $Y < \eta'$ and "quantum" part with $\eta > Y > \eta'$ and integrate over quantum fields. In the leading order there are two types of diagrams: with and without gluon production, see Fig. 7.10 (we assume that there are no gluons with $\eta > Y > \eta'$ in the proton wave function). However, the application of small-x methods discussed in Sect. 7.3 to evolution of TMD (7.68) is not straightforward as discussed in the next section.

7.5.1.1 *Shock-wave formalism and transverse momenta*

To make connections with parton model we will have in mind the frame where target's velocity is large and call the small-α "quantum" fields by the name "fast fields" and large-α "classical" fields by the name "slow" fields. Of course, "fast" *vs* "slow" depends on frame but we will stick to naming fields as they appear in the projectile's frame. (Note that in Ref. [Balitsky (1996)] the terminology is

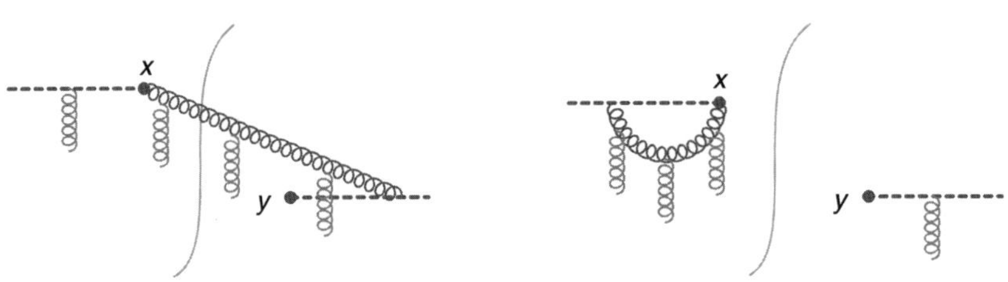

Fig. 7.10 Typical diagrams for production (a) and virtual (b) contributions to the evolution kernel. The dashed lines denote gauge links.

opposite, as appears in the target's frame). As discussed in Ref. [Balitsky (1996)], the interaction of "slow" gluons of large α with "fast" fields of small α is described by eikonal gauge factors and the integration over slow fields results in Feynman diagrams in the background of fast fields which form a thin shock wave due to Lorentz contraction. However, in Ref. [Balitsky (1996)] (as well as in all small-x literature) it was assumed that the characteristic transverse momenta of fast and slow fields are of the same order of magnitude. For our present purposes we need to relax this condition and consider cases where the transverse momenta of fast and slow fields do differ. In this case, we need to rethink the shock-wave approach.

Let us figure out how the relative longitudinal size of fast and slow fields depends on their transverse momenta. The typical longitudinal size of fast fields is $\sigma_* \sim \frac{\sigma' s}{l_\perp^2}$ where $l_\perp$ is the characteristic scale of transverse momenta of fast fields. The typical distances traveled by slow gluons are $\sim \frac{\sigma s}{k_\perp^2}$ where $k_\perp$ is the characteristic scale of transverse momenta of slow fields. Effectively, the large-α gluons propagate in the external field of the small-α shock wave as shown in Fig. 7.11, except the case $l_\perp^2 \ll k_\perp^2$ which should be treated separately since the "shock wave" is not necessarily thin in this case, see Fig. 7.12.

Fortunately, when $l_\perp^2 \ll k_\perp^2$ one can use the light-cone expansion of slow fields and leave at the leading order only the light-ray operators of the leading twist. As demonstrated in Ref. [Balitsky and Tarasov (2015)], the combination of shock-wave and light-cone expansions enables one to write the interpolating formulas which describe the leading-order contributions in both cases. The detailed calculation of the evolution of operator (7.68) is performed in Ref. [Balitsky and Tarasov (2016)] while here I will present the final result for evolution equation.

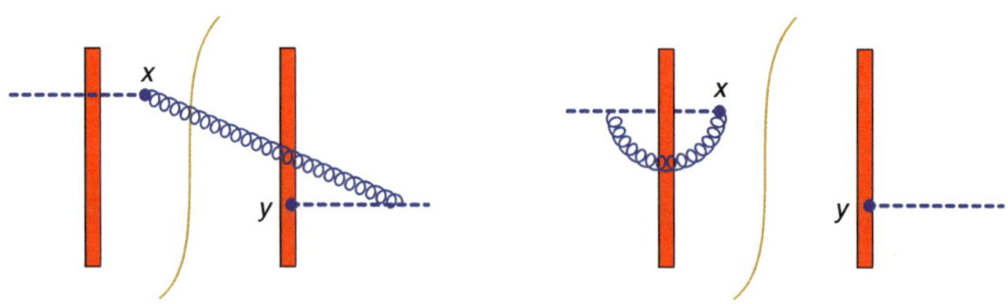

Fig. 7.11 Typical diagrams in a shock-wave background at $k_\perp \sim l_\perp$

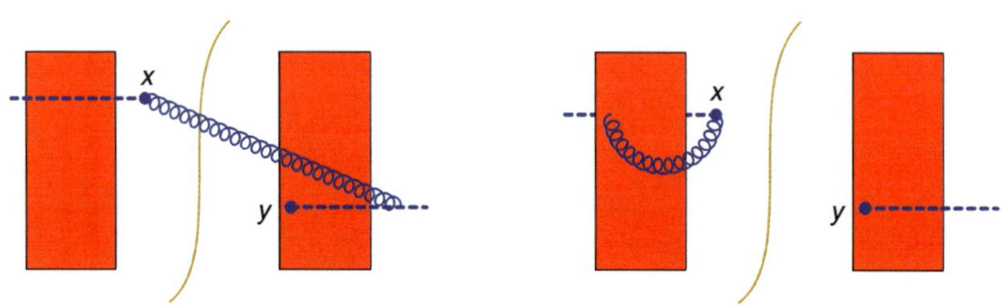

Fig. 7.12 Typical diagrams at $k_\perp \gg l_\perp$

7.5.2 *Evolution equation for gluon TMD*

The rapidity evolution of gluon TMD (7.66) has the form [Balitsky and Tarasov (2016)]

$$
\begin{aligned}
&\frac{d}{d\eta}\langle\!\langle p|\tilde{\mathcal{F}}_i^a(\beta_B,x_\perp)\mathcal{F}_j^a(\beta_B,y_\perp)|p\rangle\!\rangle^\eta\\
&\quad= -\alpha_s\langle\!\langle p|\mathrm{Tr}\Big\{\int d^2k_\perp\theta\big(1-\beta_B-\frac{k_\perp^2}{\sigma s}\big)\Big[(x_\perp|\Big(U^\dagger\frac{1}{\sigma\beta_Bs+p_\perp^2}(Uk_k+p_kU)\\
&\quad\times\frac{\sigma\beta_Bsg_{\mu i}-2k_\mu^\perp k_i}{\sigma\beta_Bs+k_\perp^2}-2k_\mu^\perp g_{ik}U^\dagger\frac{1}{\sigma\beta_Bs+p_\perp^2}U\\
&\quad-2g_{\mu k}U^\dagger\frac{p_i}{\sigma\beta_Bs+p_\perp^2}U\Big)\tilde{\mathcal{F}}^k\big(\beta_B+\frac{k_\perp^2}{\sigma s}\big)|k_\perp)\\
&\quad\times(k_\perp|\mathcal{F}^l\big(\beta_B+\frac{k_\perp^2}{\sigma s}\big)\Big(\frac{\sigma\beta_Bs\delta_j^\mu-2k_\perp^\mu k_j}{\sigma\beta_Bs+k_\perp^2}(k_lU^\dagger+U^\dagger p_l)\frac{1}{\sigma\beta_Bs+p_\perp^2}U\\
&\quad-2k_\perp^\mu g_{jl}U^\dagger\frac{1}{\sigma\beta_Bs+p_\perp^2}U-\ 2\delta_l^\mu U^\dagger\frac{p_j}{\sigma\beta_Bs+p_\perp^2}U\Big)|y_\perp)\\
&\quad+2(x_\perp|\tilde{\mathcal{F}}_i\big(\beta_B+\frac{k_\perp^2}{\sigma s}\big)|k_\perp)(k_\perp|\mathcal{F}^l\Big(\beta_B+\frac{k_\perp^2}{\sigma s}\Big)\\
&\quad\times\Big(\frac{k_j}{k_\perp^2}\frac{\sigma\beta_Bs+2k_\perp^2}{\sigma\beta_Bs+k_\perp^2}(k_lU^\dagger+U^\dagger p_l)\frac{1}{\sigma\beta_Bs+p_\perp^2}U\\
&\quad+2U^\dagger\frac{g_{jl}}{\sigma\beta_Bs+p_\perp^2}U-2\frac{k_l}{k_\perp^2}U^\dagger\frac{p_j}{\sigma\beta_Bs+p_\perp^2}U\Big)|y_\perp)\\
&\quad+2(x_\perp|\Big(U^\dagger\frac{1}{\sigma\beta_Bs+p_\perp^2}(Uk_k+p_kU)\frac{k_i}{k_\perp^2}\frac{\sigma\beta_Bs+2k_\perp^2}{\sigma\beta_Bs+k_\perp^2}+2U^\dagger\frac{g_{ik}}{\sigma\beta_Bs+p_\perp^2}U\\
&\quad-2U^\dagger\frac{p_i}{\sigma\beta_Bs+p_\perp^2}U\frac{k_k}{k_\perp^2}\Big)\tilde{\mathcal{F}}^k\big(\beta_B+\frac{k_\perp^2}{\sigma s}\big)|k_\perp)(k_\perp|\mathcal{F}_j\big(\beta_B+\frac{k_\perp^2}{\sigma s}\big)|y_\perp)\Big]\\
&\quad+2\tilde{\mathcal{F}}_i(\beta_B,x_\perp)(y_\perp|\frac{p^m}{p_\perp^2}\mathcal{F}_k(\beta_B)(i\overleftarrow{\partial}_l+U_l)(2\delta_m^k\delta_j^l-g_{jm}g^{kl})\\
&\quad\times U^\dagger\frac{1}{\sigma\beta_Bs-p_\perp^2+i\epsilon}U|y_\perp)\\
&\quad-2(x_\perp|U^\dagger\frac{1}{\sigma\beta_Bs-p_\perp^2-i\epsilon}U(2\delta_i^k\delta_m^l-g_{im}g^{kl})(i\partial_k-U_k)\\
&\quad\times\tilde{\mathcal{F}}_l(\beta_B)\frac{p^m}{p_\perp^2}|x_\perp)\mathcal{F}_j(\beta_B,y_\perp)\\
&\quad-4\int\frac{d^2k_\perp}{k_\perp^2}\Big[\theta\big(1-\beta_B-\frac{k_\perp^2}{\sigma s}\big)\tilde{\mathcal{F}}_i\big(\beta_B+\frac{k_\perp^2}{\sigma s},x_\perp\big)\mathcal{F}_j\big(\beta_B+\frac{k_\perp^2}{\sigma s},y_\perp\big)e^{i(k,x-y)_\perp}\\
&\quad-\mathrm{V.p.}\frac{\sigma\beta_Bs}{\sigma\beta_Bs-k_\perp^2}\tilde{\mathcal{F}}_i(\beta_B,x_\perp)\mathcal{F}_j(\beta_B,y_\perp)\Big]\Big\}|p\rangle\!\rangle^\eta\ +\ O(\alpha_s^2).
\end{aligned}
\tag{7.70}
$$

This equation describes the rapidity evolution of gluon TMD (7.68) with rapidity cutoff (7.10) in the whole range of $\beta_B = x_B$ and $k_\perp$ ($\sim |x-y|_\perp^{-1}$). In the next section we will consider some specific cases.

7.5.3 *BK, DGLAP, and Sudakov limits of TMD evolution equation*

7.5.3.1 *Small-x case: BK evolution of the Weizsacker-Williams distribution*

First, let us consider the evolution of the Weizsäcker-Williams (WW) unintegrated gluon distribution

$$\alpha_s x_B \mathcal{D}(x_B, z_\perp)|_{x_B\to 0}$$
$$= -\frac{1}{8\pi^2(p\cdot n)}\int du \sum_X \langle p|\tilde{\mathcal{F}}_\xi^a(z_\perp + un)|X\rangle\langle X|\mathcal{F}^{a\xi}(0)|p\rangle \tag{7.71}$$

which can be obtained from Eq. (7.70) by setting $\beta_B = 0$. Moreover, in the small-x regime it is assumed that the energy is much higher than anything else, so the characteristic transverse momenta $p_\perp^2 \sim (x-y)_\perp^{-2} \ll s$ and in the whole range of evolution ($1 \gg \sigma \gg \frac{(x-y)_\perp^{-2}}{s}$) we have $\frac{p_\perp^2}{\sigma s} \ll 1$, hence the kinematical constraint $\theta\big(1-\beta_B-\frac{k_\perp^2}{\sigma s}\big)$ in Eq. (7.70) can be omitted. Under these assumptions, all $\mathcal{F}_i\big(\beta_B + \frac{p_\perp^2}{\sigma s}\big)$ and $\mathcal{F}_i(\beta_B)$ can be replaced by $U^\dagger i\partial_i U$ (and similarly for $\tilde{\mathcal{F}}_i$). After some algebra one obtains (cf. Ref. [Balitsky (1999)])

$$\frac{d}{d\eta}U_i^a(z_1)U_j^a(z_2)$$
$$= -\frac{g^2}{8\pi^3}\mathrm{Tr}\Big\{(i\partial_i^{z_1} + U_i^{z_1})\Big[\int d^2z_3(U_{z_1}^\dagger U_{z_3} - 1)\frac{z_{12}^2}{z_{13}^2 z_{23}^2}(U_{z_3}^\dagger U_{z_2} - 1)\Big](-i\overleftarrow{\partial}_j^{z_2} + U_j^{z_2})\Big\} \tag{7.72}$$

where all indices are 2-dimensional and Tr stands for the trace in the adjoint representation. Note that the expression in the square brackets is actually the BK kernel [Balitsky (1996); Kovchegov (1999, 2000)]. One should also mention that Eq. (7.72) coincides with Eq. (12) from Ref. [Dominguez *et al.* (2011)] after some algebra.

It should be noted that the Eq. (7.72) holds true also at small β_B up to $\beta_B \sim \frac{(x-y)_\perp^{-2}}{s}$ since in the whole range of evolution $1 \gg \sigma \gg \frac{(x-y)_\perp^{-2}}{s}$ one can neglect $\sigma\beta_B s$ in comparison to $p_\perp^2$ in Eq. (7.70). This effectively reduces β_B to 0, so one reproduces Eq. (7.72).

7.5.3.2 *Large transverse momenta and the light-cone limit*

Now let us discuss the case when $\beta_B = x_B \sim 1$ and $(x-y)_\perp^{-2} \sim s$. At the start of the evolution (at $\sigma \sim 1$) the cutoff in $p_\perp^2$ in the integrals Eq. (7.70) is $\sim (x-y)_\perp^{-2}$. However, as the evolution in rapidity ($\sim \ln\sigma$) progresses the characteristic $p_\perp^2$ become smaller due to the kinematical constraint $p_\perp^2 < \sigma(1-\beta_B)s$. Due to this kinematical constraint evolution in σ is correlated with the evolution in $p_\perp^2$: if $\sigma \gg \sigma'$ the corresponding transverse momenta of background fields ${p'_\perp}^2$ are much smaller than $p_\perp^2$ in quantum loops. This means that during the evolution we are always in the light-cone case and the evolution equation for $\beta_B = x_B \sim 1$ and $(x-y)_\perp^{-2} \sim s$ has the form [Balitsky and Tarasov (2016)]

$$\begin{aligned}
&\frac{d}{d\ln\sigma}\langle\!\langle p|\tilde{\mathcal{F}}_i^a(\beta_B,x_\perp)\mathcal{F}_j^a(\beta_B,y_\perp)|p\rangle\!\rangle \\
&= \frac{g^2N_c}{\pi}\int d^2k_\perp\Big\{e^{i(k,x-y)_\perp}\langle\!\langle p|\tilde{\mathcal{F}}_k^a\big(\beta_B+\frac{k_\perp^2}{\sigma s},x_\perp\big) \\
&\quad\times\mathcal{F}_l^a\Big(\beta_B+\frac{k_\perp^2}{\sigma s},y_\perp\Big)|p\rangle\!\rangle\Big[\frac{\delta_i^k\delta_j^l}{k_\perp^2}-\frac{2\delta_i^k\delta_j^l}{\sigma\beta_B s+k_\perp^2} \\
&\quad+\frac{k_\perp^2\delta_i^k\delta_j^l+\delta_j^k k_ik^l+\delta_i^l k_jk^k-\delta_j^l k_ik^k-\delta_i^k k_jk^l-g^{kl}k_ik_j-g_{ij}k^kk^l}{(\sigma\beta_B s+k_\perp^2)^2} \\
&\quad+k_\perp^2\frac{2g_{ij}k^kk^l+\delta_i^k k_jk^l+\delta_j^l k_ik^k-\delta_j^k k_ik^l-\delta_i^l k_jk^k}{(\sigma\beta_B s+k_\perp^2)^3}-\frac{k_\perp^4 g_{ij}k^kk^l}{(\sigma\beta_B s+k_\perp^2)^4}\Big] \\
&\quad\times\theta\Big(1-\beta_B-\frac{k_\perp^2}{\sigma s}\Big)-\text{V.p.}\frac{\sigma\beta_B s}{k_\perp^2(\sigma\beta_B s-k_\perp^2)} \\
&\quad\langle\!\langle p|\tilde{\mathcal{F}}_i^a(\beta_B,x_\perp)\mathcal{F}_j^a(\beta_B,y_\perp)|p\rangle\!\rangle\Big\}
\end{aligned} \tag{7.73}$$

which reduces to the system of evolution equations for gluon TMDs $\mathcal{D}(\beta_B,|z_\perp|,\ln\sigma)$ and $\mathcal{H}(\beta_B,|z_\perp|,\ln\sigma)$ in the case of unpolarized hadron. The evolution equation (7.73) can be rewritten as a system of evolution equations for $\mathcal{D}$ and $\mathcal{H}''$ functions ($z' \equiv \frac{\sigma s\beta_B}{k_\perp^2+\sigma s\beta_B}$):

$$\begin{aligned}
&\frac{d}{d\eta}\alpha_s\mathcal{D}(\beta_B,z_\perp,\eta) \\
&= \frac{\alpha_sN_c}{\pi}\int_{\beta_B}^1\frac{dz'}{z'}\Big\{J_0\Big(|z_\perp|\sqrt{\sigma s\beta_B\frac{1-z'}{z'}}\Big)\Big[\Big(\frac{1}{1-z'}\Big)_+ \\
&\quad+\frac{1}{z'}-2+z'(1-z')\Big]\alpha_s\mathcal{D}\Big(\frac{\beta_B}{z'},z_\perp,\eta\Big) \\
&\quad+\frac{4}{m^2}(1-z')z'z_\perp^2 J_2\Big(|z_\perp|\sqrt{\sigma s\beta_B\frac{1-z'}{z'}}\Big)\alpha_s\mathcal{H}''\Big(\frac{\beta_B}{z'},z_\perp,\eta\Big)\Big\}, \\
&\frac{d}{d\eta}\alpha_s\mathcal{H}''(\beta_B,z_\perp,\eta) \\
&= \frac{\alpha_sN_c}{\pi}\int_{\beta_B}^1\frac{dz'}{z'}\Big\{J_0\Big(|z_\perp|\sqrt{\sigma s\beta_B\frac{1-z'}{z'}}\Big)\Big[\Big(\frac{1}{1-z'}\Big)_+-1\Big]\alpha_s\mathcal{H}''\Big(\frac{\beta_B}{z'},z_\perp,\eta\Big) \\
&\quad+\frac{m^2}{4z_\perp^2}\frac{1-z'}{z'}J_2\Big(|z_\perp|\sqrt{\sigma s\beta_B\frac{1-z'}{z'}}\Big)\alpha_s\mathcal{D}\Big(\frac{\beta_B}{z'},z_\perp,\eta\Big)\Big\}
\end{aligned} \tag{7.74}$$

where $\int_x^1 dz f(z)g(z)_+ = \int_x^1 dz f(z)g(z) - \int_0^1 dz f(1)g(z)$. The above equation is our final result for the rapidity evolution of gluon TMDs (7.68) in the near-light-cone case.

If we take the light-cone limit $x_\perp = y_\perp$ ($\Leftrightarrow z_\perp = 0$) we get the (one-loop) DGLAP equation:

$$\frac{d}{d\eta}\alpha_s\mathcal{D}(\beta_B, 0_\perp, \eta) = \frac{\alpha_s}{\pi}N_c\int_{\beta_B}^1 \frac{dz'}{z'}\Big[\Big(\frac{1}{1-z'}\Big)_+ + \frac{1}{z'} - 2 + z'(1-z')\Big]$$
$$\times\,\alpha_s\mathcal{D}\Big(\frac{\beta_B}{z'}, 0_\perp, \eta\Big). \tag{7.75}$$

One immediately recognizes the expression in the square brackets as gluon-gluon DGLAP kernel (the term $\frac{11}{12}\delta(1-z')$ is absent since we consider the gluon light-ray operator multiplied by an extra α_s).

7.5.3.3 *Sudakov logarithms*

Finally, let us consider the evolution of $\mathcal{D}(x_B, k_\perp, \eta = \ln\sigma)$ in the region where $x_B \equiv \beta_B \sim 1$ and $k_\perp^2 \sim (x-y)_\perp^{-2} \sim$ few GeV^2. In this case the integrals over $p_\perp^2$ in the production part of the kernel (7.70) are $\sim (x-y)_\perp^{-2} \sim k_\perp^2$ so that $p_\perp^2 \ll \sigma\beta_B s$ for the whole range of evolution $1 > \sigma > \frac{k_\perp^2}{s}$. With these approximations we get the following rapidity evolution equation in the Sudakov region [Balitsky and Tarasov (2016)]:

$$\frac{d}{d\ln\sigma}\langle\!\langle p|\tilde{\mathcal{F}}_i^a(\beta_B, x_\perp)\mathcal{F}_j^a(\beta_B, y_\perp)|p\rangle\!\rangle$$
$$= 4\alpha_s N_c\int\frac{d^2p_\perp}{p_\perp^2}\Big[e^{i(p,x-y)_\perp}\langle\!\langle p|\tilde{\mathcal{F}}_i^a\Big(\beta_B + \frac{p_\perp^2}{\sigma s}, x_\perp\Big)\mathcal{F}_j^a\Big(\beta_B + \frac{p_\perp^2}{\sigma s}, y_\perp\Big)|p\rangle\!\rangle$$
$$-\,\text{V.p.}\frac{\sigma\beta_B s}{\sigma\beta_B s - p_\perp^2}\langle\!\langle p|\tilde{\mathcal{F}}_i^a(\beta_B, x_\perp)\mathcal{F}_j^a(\beta_B, y_\perp)|p\rangle\!\rangle\Big]. \tag{7.76}$$

In the double-log region where $1 \gg \sigma \gg \frac{(x-y)_\perp^{-2}}{s}$ and $\sigma\beta_B s \gg p_\perp^2 \gg (x-y)_\perp^{-2}$ only the second term in the r.h.s. of Eq. (7.76) survives and the evolution equation reduces to

$$\frac{d}{d\ln\sigma}\langle\!\langle p|\tilde{\mathcal{F}}_i^a(\beta_B, x_\perp)\mathcal{F}_j^a(\beta_B, y_\perp)|p\rangle\!\rangle^{\eta=\ln\sigma}$$
$$= -\frac{g^2N_c}{\pi}\int\frac{d^2p_\perp}{p_\perp^2}\big[1 - e^{i(p,x-y)_\perp}\big]\langle\!\langle p|\tilde{\mathcal{F}}_i^a(\beta_B, x_\perp)\mathcal{F}_j^a(\beta_B, y_\perp)|p\rangle\!\rangle^{\eta} \tag{7.77}$$

which can be rewritten for the TMD (7.66) as

$$\frac{d}{d\ln\sigma}\mathcal{D}(x_B, z_\perp, \ln\sigma) = -\frac{\alpha_s N_c}{\pi^2}\mathcal{D}(x_B, z_\perp, \ln\sigma)\int\frac{d^2p_\perp}{p_\perp^2}\big[1 - e^{i(p,z)_\perp}\big] \tag{7.78}$$

leading to the usual Sudakov double-log result

$$\mathcal{D}(x_B, k_\perp, \ln\sigma) \sim \exp\Big\{-\frac{\alpha_s N_c}{2\pi}\ln^2\frac{\sigma s}{k_\perp^2}\Big\}\mathcal{D}\Big(x_B, k_\perp, \ln\frac{k_\perp^2}{s}\Big). \tag{7.79}$$

It is worth noting that the coefficient in front of $\ln^2\frac{\sigma s}{k_\perp^2}$ is determined by the cusp anomalous dimension of two light-like Wilson lines going from point y to ∞p_1 and ∞p_2 directions (with our cutoff $\alpha < \sigma$), see the discussion in Ref. [Balitsky and Tarasov (2015)].

With the single-log accuracy the rapidity evolution of gluon TMD in the Sudakov region is given by [Balitsky and Chirilli (2019)]

$$\begin{aligned}
&\mathcal{O}^{\sigma_2}(z_{1*}, z_{1*})\\
&\quad= e^{-2\bar\alpha_s\ln\frac{\sigma_2}{\sigma_1}[\ln\sigma_1\sigma_2\frac{s}{4}z_{12\perp}^2+4\gamma_E]}\frac{1}{4\pi^2}\int dz'_{1*}dz'_{2*}\mathcal{O}^{\sigma_1}(z'_{1*}, z'_{1*})\\
&\quad\times\left[\frac{i\Gamma\big(1-2\bar\alpha_s\ln\frac{\sigma_2}{\sigma_1}\big)}{(z_{1*}-z'_{1*}+i\epsilon)^{1-2\bar\alpha_s\ln\frac{\sigma_2}{\sigma_1}}}+c.c.\right]\left[\frac{i\Gamma\big(1-2\bar\alpha_s\ln\frac{\sigma_2}{\sigma_1}\big)}{(z_{2*}-z'_{2*}+i\epsilon)^{1-2\bar\alpha_s\ln\frac{\sigma_2}{\sigma_1}}}+c.c.\right]
\end{aligned} \tag{7.80}$$

where

$$\mathcal{O}_{ij}(z_{1*}, z_{1\perp}; z_{2*}, z_{2\perp}) = \mathcal{F}_i^a(z_1)[z_1-\infty p_1, z_2-\infty p_1]^{ab}\mathcal{F}_j^b(z_2)\Big|_{z_{1\bullet}=z_{2\bullet}=0}. \tag{7.81}$$

As the result of evolution (7.80) the transverse separation between gluon operators $\mathcal{F}_i$ and $\mathcal{F}_j$ remains intact while the longitudinal separation increases.

It should be mentioned that to get the conformal invariance of the leading-order evolution equation (7.80) we should impose the rapidity cutoff which depends on the transverse size of TMD [Balitsky and Chirilli (2019)]:

$$\begin{aligned}
\mathcal{F}_i^a(z_\perp, z_*)^\sigma &\equiv gF_{\bullet i}^m(z)\big[\mathrm{P}e^{ig\frac{2}{s}\int_{-\infty}^{z_*}dz'_*A_\bullet(z_*, z_\perp)}\big]^{ma},\\
A_\mu^\sigma(x) &= \int\frac{d^4k}{16\pi^4}\theta\Big(\frac{2\sigma}{z_{12\perp}\sqrt{s}}-|\alpha|\Big)e^{-ik\cdot x}A_\mu(k).
\end{aligned} \tag{7.82}$$

Note that the situation here is quite different from the dipole evolution where LO BK is conformally invariant with the cutoff $\alpha < \sigma$ and only the NLO BK depends on the details of the cutoff. Probably, at the NLO level one should expect some correction of the operator (7.81) similarly to the "composite dipole" (7.16).

For usual momentum-space TMD one obtains

$$\frac{D^{\sigma_2}(x, q_\perp)}{D^{\sigma_1}(x, q_\perp)} = e^{-2\bar\alpha_s\ln\frac{Q^2}{q_\perp^2}\left[\ln\frac{Q^2}{q_\perp^2}+4\gamma_E-\ln 2\right]} \tag{7.83}$$

which coincides with usual one-loop evolution of TMDs [Aybat and Rogers (2011)] up to replacement $4\gamma_E - 2\ln 2 \to 4\gamma_E - 4\ln 2$. As we discussed, such constant depends on the way of cutting α-integration which should be coordinated with the cutoffs in the "coefficient function" $\sigma(ff \to H)$ in Eq. (7.21). Thus, the discrepancy is just like using two different schemes for usual renormalization.

7.5.4 *Evolution beyond Sudakov region*

As discussed in Refs. [Balitsky and Tarasov (2015, 2016)], the Sudakov approximation can be trusted until the upper cutoff in α integrals is greater than $\frac{q_\perp^2}{x_B s}$. If $x_B \sim 1$ and $q_\perp \sim m_N$, the relative energy between Wilson-line operators $\mathcal{F}$ and target nucleon at the final point of evolution is $\sim m_N^2$ so one should use phenomenological models of TMDs with this low rapidity cutoff as a starting point of the evolution (7.80). If, however, $x_B \ll 1$, this relative energy is $\frac{q_\perp^2}{x_B} \gg m_N^2$ so one can continue the rapidity evolution in the region $\frac{q_\perp^2}{x_B s} > \sigma > \frac{m_N^2}{s}$ beyond the Sudakov region into the small-x region. The evolution in a "proper" small-x region is given by Eq. (7.72) — the TMD operator, known also as Weizsäcker-Williams distribution, will produce a hierarchy of color dipoles as a result of the non-linear evolution. However, the transition between Sudakov region and "proper" small-x region is described by the rather complicated interpolation formula (7.70). In the coordinate space this transition region means $z_\parallel^2 \sim z_\perp^2$ and I think that conformal considerations can help us to obtain the TMD evolution in that region. Indeed, if one considers the 6-point correlation function corresponding to "particle production" in $\mathcal{N} = 4$ SYM, the restrictions imposed by conformal invariance may help to match the results of the Sudakov evolution (7.80) to the evolution of the Weizsäcker-Williams distribution (7.72). The study is in progress.

7.6 Conclusions

The rapidity evolution is the most powerful method to study small-x behavior of amplitudes in QCD. [7] Although the NLO BFKL can be obtained by direct summation of Feynman diagrams [Fadin and Lipatov (1998); Ciafaloni and Camici (1998)], the non-linear NLO BK evolution was never obtained by direct summation of Feynman diagrams (although the LO BK can be restored from three-pomeron vertex [Bartels and Wusthoff (1995)]). However, I believe that rapidity factorization and rapidity evolution can be used not only at low x, but in the intermediate region between low and moderate x. The "golden standard" at moderate x is collinear factorization and TMD factorization based on CSS [Collins *et al.* (1985)] or SCET [Rothstein and Stewart (2016)] formalism. The analysis of TMD evolution at $x \sim 1$ is performed in that formalism with a combination of UV and rapidity cutoffs which gives two evolution equations, in μ^2 and ζ (related to rapidity). The results of these two evolutions are known at two- [Echevarria *et al.* (2016a); Li *et al.* (2016); Luebbert *et al.* (2016)] and three-loop [Echevarria *et al.* (2016b)] levels.

[7] As was mentioned in the Introduction, in $\mathcal{N} = 4$ SYM there are even more powerful methods which give us NNLO accuracy [Gromov *et al.* (2015); Caron-Huot and Herranen (2018)] but they do not generalize to QCD.

It should be mentioned, however, that their relation to conformal properties of TMD operators is not obvious. It would be interesting to check if our cutoff corresponds to some "conformal" evolution path in two-dimensional (μ^2, ζ) plane [Scimemi and Vladimirov (2018)].

Unfortunately, the CSS formalism is based on summation of logarithms of DGLAP type and cannot be generalized to the small-x region where one must sum up BFKL-type logarithms (7.1). I believe that methods based on rapidity factorizations discussed here have a better chance to serve as a bridge between small-x and moderate-x regions. This transition is especially interesting in view of a future EIC collider which will will probe that range of energies.

Acknowledgements

The author is grateful to his collaborators G.A. Chirilli and A. Tarasov for developing methods of rapidity factorization and evolution outlined in this article. The work was supported by DOE contract DE-AC05-06OR23177 and by the grant DE-FG02-97ER41028.

Bibliography

Altinoluk, T., Kovner, A., Lublinsky, M., and Peressutti, J. (2009). QCD Reggeon Field Theory for every day: Pomeron loops included, *JHEP* **03**, p. 109, doi:10.1088/1126-6708/2009/03/109, arXiv:0901.2559 [hep-ph].

Avsar, E., Stasto, A. M., Triantafyllopoulos, D. N., and Zaslavsky, D. (2011). Next-to-leading and resummed BFKL evolution with saturation boundary, *JHEP* **10**, p. 138, doi:10.1007/JHEP10(2011)138, arXiv:1107.1252 [hep-ph].

Aybat, S. M. and Rogers, T. C. (2011). TMD Parton Distribution and Fragmentation Functions with QCD Evolution, *Phys. Rev.* **D83**, p. 114042, doi:10.1103/PhysRevD.83.114042, arXiv:1101.5057 [hep-ph].

Balitsky, I. (1996). Operator expansion for high-energy scattering, *Nucl. Phys.* **B463**, pp. 99–160, doi:10.1016/0550-3213(95)00638-9, arXiv:hep-ph/9509348 [hep-ph].

Balitsky, I. (1999). Factorization and high-energy effective action, *Phys. Rev.* **D60**, p. 014020, doi:10.1103/PhysRevD.60.014020, arXiv:hep-ph/9812311 [hep-ph].

Balitsky, I. (2001). High-energy QCD and Wilson lines, doi:10.1142/9789812810458_0030, arXiv:hep-ph/0101042 [hep-ph].

Balitsky, I. (2005). High-energy effective action from scattering of QCD shock waves, *Phys. Rev.* **D72**, p. 074027, doi:10.1103/PhysRevD.72.074027, arXiv:hep-ph/0507237 [hep-ph].

Balitsky, I. (2010). High-energy amplitudes in the next-to-leading order, in *Subtleties in quantum field theory: Lev Lipatov Festschrift*, pp. 104–149, arXiv:1004.0057 [hep-ph], http://www1.jlab.org/Ul/publications/view_pub.cfm?pub_id=9545.

Balitsky, I. and Chirilli, G. A. (2008). Next-to-leading order evolution of color dipoles, *Phys. Rev.* **D77**, p. 014019, doi:10.1103/PhysRevD.77.014019, arXiv:0710.4330 [hep-ph].

Balitsky, I. and Chirilli, G. A. (2009). NLO evolution of color dipoles in N=4 SYM, *Nucl. Phys.* **B822**, pp. 45–87, doi:10.1016/j.nuclphysb.2009.07.003, arXiv:0903.5326 [hep-ph].

Balitsky, I. and Chirilli, G. A. (2010). High-energy amplitudes in N=4 SYM in the next-to-leading order, *Phys. Lett.* **B687**, pp. 204–213, doi:10.1016/j.physletb.2010.02.084, arXiv:0911.5192 [hep-ph].

Balitsky, I. and Chirilli, G. A. (2011). Photon impact factor in the next-to-leading order, *Phys. Rev.* **D83**, p. 031502, doi:10.1103/PhysRevD.83.031502, arXiv:1009.4729 [hep-ph].

Balitsky, I. and Chirilli, G. A. (2013a). Photon impact factor and k_T-factorization for DIS in the next-to-leading order, *Phys. Rev.* **D87**, 1, p. 014013, doi:10.1103/PhysRevD.87.014013, arXiv:1207.3844 [hep-ph].

Balitsky, I. and Chirilli, G. A. (2013b). Rapidity evolution of Wilson lines at the next-to-leading order, *Phys. Rev.* **D88**, p. 111501, doi:10.1103/PhysRevD.88.111501, arXiv:1309.7644 [hep-ph].

Balitsky, I. and Chirilli, G. A. (2019). Conformal invariance of transverse-momentum dependent parton distributions rapidity evolution, *Phys. Rev.* **D100**, 5, p. 051504, doi:10.1103/PhysRevD.100.051504, arXiv:1905.09144 [hep-ph].

Balitsky, I. and Tarasov, A. (2015). Rapidity evolution of gluon TMD from low to moderate x, *JHEP* **10**, p. 017, doi:10.1007/JHEP10(2015)017, arXiv:1505.02151 [hep-ph].

Balitsky, I. and Tarasov, A. (2016). Gluon TMD in particle production from low to moderate x, *JHEP* **06**, p. 164, doi:10.1007/JHEP06(2016)164, arXiv:1603.06548 [hep-ph].

Balitsky, I. and Tarasov, A. (2017). Higher-twist corrections to gluon TMD factorization, *JHEP* **07**, p. 095, doi:10.1007/JHEP07(2017)095, arXiv:1706.01415 [hep-ph].

Balitsky, I. and Tarasov, A. (2018). Power corrections to TMD factorization for Z-boson production, *JHEP* **05**, p. 150, doi:10.1007/JHEP05(2018)150, arXiv:1712.09389 [hep-ph].

Balitsky, I. I. (1983). String operator expansion of the T product of two currents near the light cone, *Phys. Lett.* **124B**, pp. 230–236, doi:10.1016/0370-2693(83)91442-9.

Balitsky, I. I. and Braun, V. M. (1989a). Evolution Equations for QCD String Operators, *Nucl. Phys.* **B311**, pp. 541–584, doi:10.1016/0550-3213(89)90168-5.

Balitsky, I. I. and Braun, V. M. (1989b). Nonlocal Operator Expansion for Structure Functions of e^+e^- Annihilation, *Phys. Lett.* **B222**, pp. 123–131, doi:10.1016/0370-2693(89)90733-8.

Balitsky, I. I. and Braun, V. M. (1991). The Nonlocal operator expansion for inclusive particle production in e+ e- annihilation, *Nucl. Phys.* **B361**, pp. 93–140, doi:10.1016/0550-3213(91)90618-8.

Balitsky, I. I. and Braun, V. M. (1992). Valleys in Minkowski space and instanton induced cross-sections, *Nucl. Phys.* **B380**, pp. 51–82, doi:10.1016/0550-3213(92)90515-D.

Balitsky, I. I. and Lipatov, L. N. (1978). The Pomeranchuk Singularity in Quantum Chromodynamics, *Sov. J. Nucl. Phys.* **28**, pp. 822–829, [Yad. Fiz.28,1597(1978)].

Bartels, J. and Wusthoff, M. (1995). The Triple Regge limit of diffractive dissociation in deep inelastic scattering, *Z. Phys.* **C66**, pp. 157–180, doi:10.1007/BF01496591.

Beuf, G. (2014). Improving the kinematics for low-x QCD evolution equations in coordinate space, *Phys. Rev.* **D89**, 7, p. 074039, doi:10.1103/PhysRevD.89.074039, arXiv:1401.0313 [hep-ph].

Bondarenko, S., Lipatov, L., and Prygarin, A. (2017). Effective action for reggeized gluons, classical gluon field of relativistic color charge and color glass condensate approach, *Eur. Phys. J.* **C77**, 8, p. 527, doi:10.1140/epjc/s10052-017-5101-5, arXiv:1706.00278 [hep-ph].

Caron-Huot, S. and Herranen, M. (2018). High-energy evolution to three loops, *JHEP* **02**, p. 058, doi:10.1007/JHEP02(2018)058, arXiv:1604.07417 [hep-ph].

Chirilli, G. A. and Kovchegov, Y. V. (2013). Solution of the NLO BFKL Equation and a Strategy for Solving the All-Order BFKL Equation, *JHEP* **06**, p. 055, doi:10.1007/JHEP06(2013)055, arXiv:1305.1924 [hep-ph].

Chirilli, G. A. and Kovchegov, Y. V. (2014). $\gamma^*\gamma^*$ Cross Section at NLO and Properties of the BFKL Evolution at Higher Orders, *JHEP* **05**, p. 099, doi:10.1007/JHEP05(2014)099,10.1007/JHEP08(2015)075, arXiv:1403.3384 [hep-ph], [Erratum: JHEP08,075(2015)].

Ciafaloni, M. and Camici, G. (1998). Energy scale(s) and next-to-leading BFKL equation, *Phys. Lett.* **B430**, pp. 349–354, doi:10.1016/S0370-2693(98)00551-6, arXiv:hep-ph/9803389 [hep-ph].

Collins, J. (2013). *Foundations of perturbative QCD* (Cambridge University Press), ISBN 9781107645257, 9781107645257, 9780521855334, 9781139097826, http://www.cambridge.org/de/knowledge/isbn/item5756723.

Collins, J. and Rogers, T. (2015). Understanding the large-distance behavior of transverse-momentum-dependent parton densities and the Collins-Soper evolution kernel, *Phys. Rev.* **D91**, 7, p. 074020, doi:10.1103/PhysRevD.91.074020, arXiv:1412.3820 [hep-ph].

Collins, J. C. and Soper, D. E. (1982). Parton Distribution and Decay Functions, *Nucl. Phys.* **B194**, pp. 445–492, doi:10.1016/0550-3213(82)90021-9.

Collins, J. C., Soper, D. E., and Sterman, G. F. (1985). Transverse Momentum Distribution in Drell-Yan Pair and W and Z Boson Production, *Nucl. Phys.* **B250**, pp. 199–224, doi:10.1016/0550-3213(85)90479-1.

Cornalba, L., Costa, M. S., and Penedones, J. (2008). Eikonal Methods in AdS/CFT: BFKL Pomeron at Weak Coupling, *JHEP* **06**, p. 048, doi:10.1088/1126-6708/2008/06/048, arXiv:0801.3002 [hep-th].

Dominguez, F., Mueller, A. H., Munier, S., and Xiao, B.-W. (2011). On the small-x evolution of the color quadrupole and the Weizscker?Williams gluon distribution, *Phys. Lett.* **B705**, pp. 106–111, doi:10.1016/j.physletb.2011.09.104, arXiv:1108.1752 [hep-ph].

Ducloue, B., Iancu, E., Mueller, A. H., Soyez, G., and Triantafyllopoulos, D. N. (2019a). Non-linear evolution in QCD at high-energy beyond leading order, *JHEP* **04**, p. 081, doi:10.1007/JHEP04(2019)081, arXiv:1902.06637 [hep-ph].

Ducloue, B., Iancu, E., Soyez, G., and Triantafyllopoulos, D. N. (2019b). HERA data and collinearly-improved BK dynamics, arXiv:1912.09196 [hep-ph].

Echevarria, M. G., Idilbi, A., and Scimemi, I. (2012). Factorization Theorem For Drell-Yan At Low q_T And Transverse Momentum Distributions On-The-Light-Cone, *JHEP* **07**, p. 002, doi:10.1007/JHEP07(2012)002, arXiv:1111.4996 [hep-ph].

Echevarria, M. G., Scimemi, I., and Vladimirov, A. (2016a). Universal transverse momentum dependent soft function at NNLO, *Phys. Rev.* **D93**, 5, p. 054004, doi:10.1103/PhysRevD.93.054004, arXiv:1511.05590 [hep-ph].

Echevarria, M. G., Scimemi, I., and Vladimirov, A. (2016b). Unpolarized Transverse Momentum Dependent Parton Distribution and Fragmentation Functions at next-to-next-to-leading order, *JHEP* **09**, p. 004, doi:10.1007/JHEP09(2016)004, arXiv:1604.07869 [hep-ph].

Fadin, V. S. and Lipatov, L. N. (1998). BFKL pomeron in the next-to-leading approximation, *Phys. Lett.* **B429**, pp. 127–134, doi:10.1016/S0370-2693(98)00473-0, arXiv:hep-ph/9802290 [hep-ph].

Ferreiro, E., Iancu, E., Leonidov, A., and McLerran, L. (2002). Nonlinear gluon evolution in the color glass condensate. 2. *Nucl. Phys.* **A703**, pp. 489–538, doi:10.1016/S0375-9474(01)01329-X, arXiv:hep-ph/0109115 [hep-ph].

Gelis, F., Lappi, T., and Venugopalan, R. (2007). High energy scattering in Quantum Chromodynamics, *Int. J. Mod. Phys.* **E16**, pp. 2595–2637, doi:10.1142/S0218301307008331, arXiv:0708.0047 [hep-ph].

Gelis, F. and Venugopalan, R. (2004). Large mass q anti-q production from the color glass condensate, *Phys. Rev.* **D69**, p. 014019, doi:10.1103/PhysRevD.69.014019, arXiv:hep-ph/0310090 [hep-ph].

Gelis, F. and Venugopalan, R. (2006). Particle production in field theories coupled to strong external sources, *Nucl. Phys.* **A776**, pp. 135–171, doi:10.1016/j.nuclphysa.2006.07.020, arXiv:hep-ph/0601209 [hep-ph].

Gribov, L. V., Levin, E. M., and Ryskin, M. G. (1983). Semihard Processes in QCD, *Phys. Rept.* **100**, pp. 1–150, doi:10.1016/0370-1573(83)90022-4.

Gromov, N., Kazakov, V., Leurent, S., and Volin, D. (2014). Quantum Spectral Curve for Planar $\mathcal{N} = 4$ Super-Yang-Mills Theory, *Phys. Rev. Lett.* **112**, 1, p. 011602, doi:10.1103/PhysRevLett.112.011602, arXiv:1305.1939 [hep-th].

Gromov, N., Levkovich-Maslyuk, F., and Sizov, G. (2015). Pomeron Eigenvalue at Three Loops in $\mathcal{N} = 4$ Supersymmetric Yang-Mills Theory, *Phys. Rev. Lett.* **115**, 25, p. 251601, doi:10.1103/PhysRevLett.115.251601, arXiv:1507.04010 [hep-th].

Hatta, Y., Iancu, E., McLerran, L., and Stasto, A. (2005). Color dipoles from Bremsstrahlung in QCD evolution at high energy, *Nucl. Phys.* **A762**, pp. 272–297, doi:10.1016/j.nuclphysa.2005.08.008, arXiv:hep-ph/0505235 [hep-ph].

Hatta, Y., Iancu, E., McLerran, L., Stasto, A., and Triantafyllopoulos, D. N. (2006). Effective Hamiltonian for QCD evolution at high energy, *Nucl. Phys.* **A764**, pp. 423–459, doi:10.1016/j.nuclphysa.2005.09.006, arXiv:hep-ph/0504182 [hep-ph].

Iancu, E., Leonidov, A., and McLerran, L. D. (2001). Nonlinear gluon evolution in the color glass condensate. 1. *Nucl. Phys.* **A692**, pp. 583–645, doi:10.1016/S0375-9474(01)00642-X, arXiv:hep-ph/0011241 [hep-ph].

Iancu, E. and Venugopalan, R. (2003). The Color glass condensate and high-energy scattering in QCD, in R. C. Hwa and X.-N. Wang (eds.), *Quark-gluon plasma 4*, pp. 249–3363, doi:10.1142/9789812795533_0005, arXiv:hep-ph/0303204 [hep-ph].

Jalilian-Marian, J., Kovner, A., Leonidov, A., and Weigert, H. (1997). The BFKL equation from the Wilson renormalization group, *Nucl. Phys.* **B504**, pp. 415–431, doi:10.1016/S0550-3213(97)00440-9, arXiv:hep-ph/9701284 [hep-ph].

Jalilian-Marian, J., Kovner, A., Leonidov, A., and Weigert, H. (1998). The Wilson renormalization group for low x physics: Towards the high density regime, *Phys. Rev.* **D59**, p. 014014, doi:10.1103/PhysRevD.59.014014, arXiv:hep-ph/9706377 [hep-ph].

Jalilian-Marian, J., Kovner, A., Leonidov, A., and Weigert, H. (1999). Unitarization of gluon distribution in the doubly logarithmic regime at high density, *Phys. Rev.* **D59**, p. 034007, doi:10.1103/PhysRevD.59.034007,10.1103/PhysRevD.59.099903, arXiv:hep-ph/9807462 [hep-ph], [Erratum: Phys. Rev.D59,099903(1999)].

Ji, X.-d., Ma, J.-p., and Yuan, F. (2005). QCD factorization for semi-inclusive deep-inelastic scattering at low transverse momentum, *Phys. Rev.* **D71**, p. 034005, doi:10.1103/PhysRevD.71.034005, arXiv:hep-ph/0404183 [hep-ph].

Kirschner, R., Lipatov, L. N., and Szymanowski, L. (1994). Effective action for multi-Regge processes in QCD, *Nucl. Phys.* **B425**, pp. 579–594, doi:10.1016/0550-3213(94)90288-7, arXiv:hep-th/9402010 [hep-th].

Kovchegov, Y. V. (1999). Small x F(2) structure function of a nucleus including multiple pomeron exchanges, *Phys. Rev.* **D60**, p. 034008, doi:10.1103/PhysRevD.60.034008, arXiv:hep-ph/9901281 [hep-ph].

Kovchegov, Y. V. (2000). Unitarization of the BFKL pomeron on a nucleus, *Phys. Rev.* **D61**, p. 074018, doi:10.1103/PhysRevD.61.074018, arXiv:hep-ph/9905214 [hep-ph].

Kovchegov, Y. V. and Levin, E. (2012). Quantum chromodynamics at high energy, *Camb. Monogr. Part. Phys. Nucl. Phys. Cosmol.* **33**, pp. 1–350, doi:10.1017/CBO9781139022187.

Kovner, A. and Lublinsky, M. (2005). From target to projectile and back again: Selfduality of high energy evolution, *Phys. Rev. Lett.* **94**, p. 181603, doi:10.1103/PhysRevLett.94.181603, arXiv:hep-ph/0502119 [hep-ph].

Kovner, A., Lublinsky, M., and Mulian, Y. (2014a). Conformal symmetry of JIMWLK Evolution at NLO, *JHEP* **04**, p. 030, doi:10.1007/JHEP04(2014)030, arXiv:1401.0374 [hep-th].

Kovner, A., Lublinsky, M., and Mulian, Y. (2014b). Jalilian-Marian, Iancu, McLerran, Weigert, Leonidov, Kovner evolution at next to leading order, *Phys. Rev.* **D89**, 6, p. 061704, doi:10.1103/PhysRevD.89.061704, arXiv:1310.0378 [hep-ph].

Kuraev, E. A., Lipatov, L. N., and Fadin, V. S. (1976). Multi-Reggeon processes in the Yang-Mills theory, *Sov. Phys. JETP* **44**, pp. 443–450, [Zh. Eksp. Teor. Fiz.71,840(1976)].

Kuraev, E. A., Lipatov, L. N., and Fadin, V. S. (1977). The Pomeranchuk singularity in nonabelian gauge theories, *Sov. Phys. JETP* **45**, pp. 199–204, [Zh. Eksp. Teor. Fiz.72,377(1977)].

Li, Y., Neill, D., and Zhu, H. X. (2016). An exponential regulator for rapidity divergences, *Submitted to: Phys. Rev. D* arXiv:1604.00392 [hep-ph].

Lipatov, L. N. (1976). Reggeization of the vector meson and the vacuum singularity in nonabelian gauge theories, *Sov. J. Nucl. Phys.* **23**, pp. 338–345, [Yad. Fiz.23,642(1976)].

Lipatov, L. N. (1995). Gauge invariant effective action for high-energy processes in QCD, *Nucl. Phys.* **B452**, pp. 369–400, doi:10.1016/0550-3213(95)00390-E, arXiv:hep-ph/9502308 [hep-ph].

Lipatov, L. N. (1997). Small x physics in perturbative QCD, *Phys. Rept.* **286**, pp. 131–198, doi:10.1016/S0370-1573(96)00045-2, arXiv:hep-ph/9610276 [hep-ph].

Lublinsky, M. and Mulian, Y. (2017). High energy QCD at NLO: from light-cone wave function to JIMWLK evolution, *JHEP* **05**, p. 097, doi:10.1007/JHEP05(2017)097, arXiv:1610.03453 [hep-ph].

Luebbert, T., Oredsson, J., and Stahlhofen, M. (2016). Rapidity renormalized TMD soft and beam functions at two loops, *JHEP* **03**, p. 168, doi:10.1007/JHEP03(2016)168, arXiv:1602.01829 [hep-ph].

Mueller, A. H. (1990). Small x Behavior and Parton Saturation: A QCD Model, *Nucl. Phys.* **B335**, pp. 115–137, doi:10.1016/0550-3213(90)90173-B.

Mueller, A. H. and Qiu, J.-w. (1986). Gluon Recombination and Shadowing at Small Values of x, *Nucl. Phys.* **B268**, pp. 427–452, doi:10.1016/0550-3213(86)90164-1.

Mulders, P. J. and Rodrigues, J. (2001). Transverse momentum dependence in gluon distribution and fragmentation functions, *Phys. Rev.* **D63**, p. 094021, doi:10.1103/PhysRevD.63.094021, arXiv:hep-ph/0009343 [hep-ph].

Rothstein, I. Z. and Stewart, I. W. (2016). An Effective Field Theory for Forward Scattering and Factorization Violation, *JHEP* **08**, p. 025, doi:10.1007/JHEP08(2016)025, arXiv:1601.04695 [hep-ph].

Scimemi, I. and Vladimirov, A. (2018). Systematic analysis of double-scale evolution, *JHEP* **08**, p. 003, doi:10.1007/JHEP08(2018)003, arXiv:1803.11089 [hep-ph].

Weigert, H. (2002). Unitarity at small Bjorken x, *Nucl. Phys.* **A703**, pp. 823–860, doi: 10.1016/S0375-9474(01)01668-2, arXiv:hep-ph/0004044 [hep-ph].

Weigert, H. (2005). Evolution at small x(bj): The Color glass condensate, *Prog. Part. Nucl. Phys.* **55**, pp. 461–565, doi:10.1016/j.ppnp.2005.01.029, arXiv:hep-ph/0501087 [hep-ph].

Zhou, J., Yuan, F., and Liang, Z.-T. (2008). Hyperon Polarization in Unpolarized Scattering Processes, *Phys. Rev.* **D78**, p. 114008, doi:10.1103/PhysRevD.78.114008, arXiv:0808.3629 [hep-ph].

https://doi.org/10.1142/9789811231124_0008

Chapter 8

Resummation at Small x

Anna Maria Stasto

Department of Physics, Penn State University, University Park, PA 16802, USA

Abstract

In this contribution we briefly review the concept of resummation at small x. The problems with large NLL corrections are first discussed, together with their kinematical origin. Different forms of kinematical constraints are studied, and a resummation scheme reviewed which takes into account these constraints as well as other corrections due to the collinear dynamics. Finally, implications of resummation for phenomenology are reviewed, in the context of the inclusive DIS data and the high energy neutrino interaction and production cross sections.

I first had a chance to meet Professor Lev Lipatov in January 1996 at the Epiphany Conference on *Proton Structure* which took place in Kraków, Poland. As an undergraduate student of physics at that time, I remember Lev's talk *High Energy Limit in QCD* very vividly. Before Lev's talk, my advisor Professor Jan Kwieciński told me: 'now we shall hear the real theoretical physics talk'.

I did not quite understand what it meant at that time but soon I found out. Lev put out on the projector his first transparency, which was densely packed with complicated mathematical formulae written in black. He moved from formula to formula, equation to equation, from transparency to transparency smoothly and with elegance. He was completely immersed in the subject, one could almost see how he lived within all the mathematical structures during the talk. His talk filled me with awe, though I did not understand much of it. This was my first encounter with Lev Lipatov, brilliant theoretical physicist with extraordinary mathematical talent.

8.1 The issue of large NLL corrections to BFKL

The calculation of NLL corrections to BFKL by Fadin and Lipatov [Fadin and Lipatov (1998)] and in parallel by Camici and Ciafaloni [Ciafaloni and Camici (1998)] revealed that the NLL corrections are very large as compared to the LL approximation. Therefore it became evident quickly that they will need to be resummed. Although it was known that the LL intercept leads to too strong of a rise as compared with the experimental data, see for example [Bojak and Ernst (1997)], and one would expect the NLL corrections to moderate it, the size of these corrections was too large and lead to some instabilities in the cross sections. It thus became necessary to understand the origins of these corrections and to resum them. Therefore it was soon realized that the resummation is necessary to stabilize the result.

The resummation program was developed over several years, starting from an important observation that kinematical constraints yield numerically large corrections when imposed onto the BFKL cascade [Andersson *et al.* (1996); Kwiecinski *et al.* (1996, 1997)], even though they are formally of subleading order. Full resummation, which included matching to NLL BFKL, and appropriate subtractions to guarantee the momentum sum rule and avoid double counting, was developed in a series of works [Salam (1998, 1999); Ciafaloni *et al.* (1999a,b, 2003a,b, 2004, 2007)], [Altarelli *et al.* (2000, 2001, 2002, 2003, 2008)], [Thorne (2001); White and Thorne (2007)] and [Sabio Vera (2005)]. Below we shall briefly review the properties of the NLL evolution and some issues related to it.

We shall briefly review the basic notation for the case of the BFKL evolution in the forward case $t = 0$. The BFKL [Lipatov (1986)] equation can be formulated as an evolution equation which is differential in rapidity, defined as $y = \ln 1/x$. The evolution equation for the gluon Green's function can be cast into the following generic form

$$\frac{\partial}{\partial y} G(y; \mathbf{k}, \mathbf{k}_0) = \delta^2(\mathbf{k} - \mathbf{k}_0) + \int \frac{d^2\mathbf{k}'}{\pi} K(\mathbf{k}, \mathbf{k}') G(y; \mathbf{k}', \mathbf{k}_0), \tag{8.1}$$

where $G(y; \mathbf{k}', \mathbf{k}_0)$ is the forward BFKL gluon Green's function and $K(\mathbf{k}, \mathbf{k}')$ is the BFKL kernel which depends on the transverse momenta of the (reggeized) gluons. It can be computed in perturbation theory and has the following perturbative expansion

$$K(\mathbf{k},\mathbf{k}') = \bar{\alpha}_s(\mu^2)K_0(\mathbf{k},\mathbf{k}') + \bar{\alpha}_s^2(\mu^2)K_1(\mathbf{k},\mathbf{k}') + \cdots . \tag{8.2}$$

In the above expansion the first term, K_0, is the leading logarithmic LL and the second term, K_1, is the next-to-leading logarithmic NLL BFKL kernel, respectively. The variables $\mathbf{k}, \mathbf{k}', \mathbf{k}_0$ are transverse momenta of the reggeized gluons. In the following, we shall also use the following notation: $|\mathbf{k}^2| = k^2$. It is customary to use the rescaled strong coupling constant, $\bar{\alpha}_s(\mu^2) = \frac{N_c}{\pi}\alpha_s(\mu^2)$, with N_c the number of colors. In order to analyze the structure of the branching kernels and the differences between LL and NLL order it is useful to go to the Mellin space. The Mellin transform of the kernels K_0, K_1 is defined by

$$f(\gamma) = \int_0^\infty dk'^2 K(k,k')\left(\frac{k'^2}{k^2}\right)^\gamma . \tag{8.3}$$

In the following, it will be assumed that the kernel has been averaged over the angles. The Mellin space expansion of the kernel reads

$$\chi(\gamma) = \bar{\alpha}_s\chi_0(\gamma) + \bar{\alpha}_s^2\chi_1(\gamma) + \mathcal{O}(\bar{\alpha}_s^3). \tag{8.4}$$

In QCD the kernels are known up to NLL order [Fadin and Lipatov (1998); Ciafaloni and Camici (1998)], and the corresponding terms at LL and NLL level have the following form

$$\chi_0(\gamma) = 2\psi(1) - \psi(\gamma) - \psi(1-\gamma) \tag{8.5}$$

and

$$\begin{aligned}\chi_1(\gamma =) &- \frac{b}{2}[\chi_0^2(\gamma) + \chi_0'(\gamma)] - \frac{1}{4}\chi_0''(\gamma) - \frac{1}{4}\\ &\times\left(\frac{\pi}{\sin\pi\gamma}\right)^2 \frac{\cos\pi\gamma}{3(1-2\gamma)}\left(11 + \frac{\gamma(1-\gamma)}{(1+2\gamma)(3-2\gamma)}\right)\\ &+\left(\frac{67}{36} - \frac{\pi^2}{12}\right)\chi_0(\gamma) + \frac{3}{2}\zeta(3) + \frac{\pi^3}{4\sin\pi\gamma}\\ &-\sum_{n=0}^{\infty}(-1)^n\left[\frac{\psi(n+1+\gamma)-\psi(1)}{(n+\gamma)^2} + \frac{\psi(n+2-\gamma)-\psi(1)}{(n+1-\gamma)^2}\right],\end{aligned} \tag{8.6}$$

where $b = (33-2N_f)/(12\pi)$ with N_f the number of (active) flavors, and $\psi(\gamma) = \frac{1}{\Gamma(\gamma)}\frac{d\Gamma(\gamma)}{d\gamma}$ is the polygamma function.

In the supersymmetric N = 4 Yang-Mills theory (N = 4 sYM), see next section, the BFKL kernel χ is known up to NNLL order [Velizhanin (2015); Gromov *et al.* (2015); Caron-Huot and Herranen (2018)]. In the LL order, the functions $(k'^2)^\gamma$ are the eigenfunctions of the equation, and $\chi_0(\gamma)$ is the eigenvalue at this order. This is not the case at higher orders, since in QCD the coupling starts to run within the BFKL framework. One needs to construct different eigenfunctions, which include

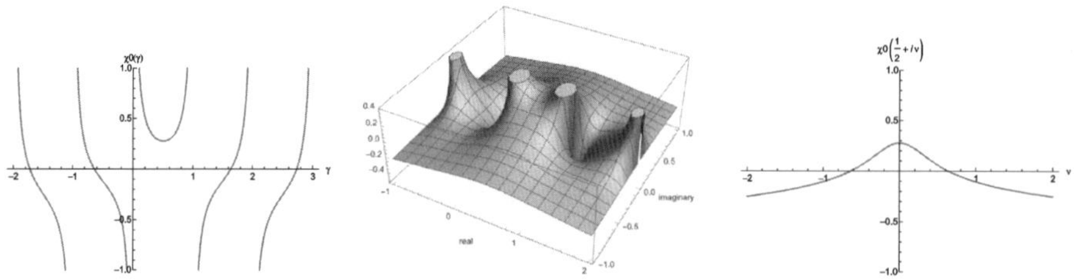

Fig. 8.1 The LLx BFKL kernel eigenvalue $\chi_0(\gamma)$ as a function of γ on the real axis (left plot), along the imaginary axis $\gamma = 1/2 + i\nu$ (right plot), and in the complex plane (center plot).

the running of the QCD coupling, see [Chirilli and Kovchegov (2013)]. Still the Mellin space, as defined by transformation (8.3), is very useful, since it allows to analyze the analytic structure of the NLL kernel (8.6) more easily and it is easier to construct the resummation.

The Mellin space is particularly useful for analyzing the behavior of the kernels (8.5) and (8.6) near the poles which govern the behavior in the collinear and anti-collinear limits. The collinear logarithms, which are logarithms of ratios of two transverse scales, i.e. $\ln k^2/k'^2$, manifest themselves as poles in the γ plane. The collinear poles, which correspond to the strong ordering along the gluon ladder when $k^2 \gg k'^2$, appear as $1/\gamma$ and the anti-collinear poles, which correspond to inverse strong ordering $k^2 \ll k'^2$, as $1/(1-\gamma)$.

In Fig. 8.1 we show the $\chi_0(\gamma)$ eigenvalue at LL order, see Eq. 8.5, on the real axis (left plot), along the imaginary axis at $\gamma = 1/2 + i\nu$ (right plot), and as a function of complex variable (center plot). The leading collinear poles near $\gamma = 0$ and $\gamma = 1$ are clearly visible, as well as the saddle point structure around $\gamma = 1/2$. This allows, for example, to find the approximate solution using the saddle point approximation. There are also infinitely many other single poles $\gamma = 2, 3, \ldots$ and $\gamma = -1, -2, \ldots$ which correspond to higher twists, i.e. roughly speaking correspond to the expansion in higher powers of k^2/k'^2 or k'^2/k^2.

The structure of the kernel at NLL is much more complicated since in addition to single poles, there are double and triple collinear poles appearing. The double poles arise due to the running coupling and the non-singular (in $1/z$) part of the DGLAP splitting function which appear at NLL order in BFKL. To be precise, keeping the most singular $\gamma \to 0$ and $\gamma \to 1$ contributions gives the following approximation for the corresponding terms

$$-\frac{b}{2}[\chi_0^2(\gamma) + \chi_0'(\gamma)] \;\longrightarrow\; -b\frac{1}{(1-\gamma)^2} \tag{8.7}$$

for the running coupling, and

$$\begin{aligned}
&-\frac{1}{4}\left(\frac{\pi}{\sin\pi\gamma}\right)^2 \frac{\cos\pi\gamma}{3(1-2\gamma)}\left(11 + \frac{\gamma(1-\gamma)}{(1+2\gamma)(3-2\gamma)}\right)\\
&\to -\frac{11}{12}\frac{1}{\gamma^2} - \frac{11}{12}\frac{1}{(1-\gamma)^2}
\end{aligned} \tag{8.8}$$

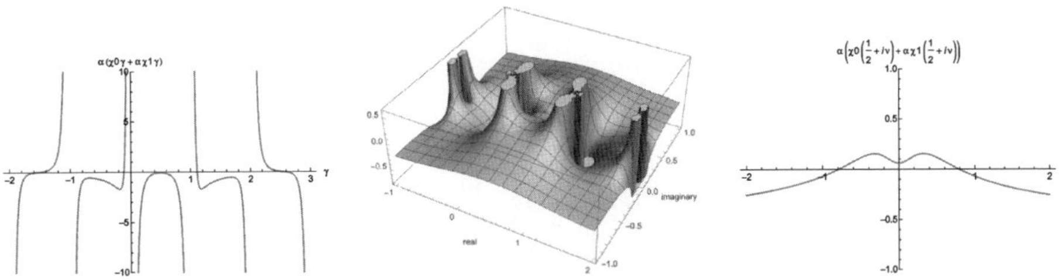

Fig. 8.2 The NLL BFKL kernel eigenvalue $\bar{\alpha}_s(\chi_0(\gamma)+\bar{\alpha}_s\chi_1(\gamma))$ as a function of γ on the real axis (left plot), along the imaginary axis $\gamma = 1/2 + i\nu$ (right plot), and in the complex plane (center plot). Strong coupling is fixed.

for the DGLAP contribution. There are also triple collinear poles which appear due to the kinematical constraints [Ciafaloni *et al.* (2003a,b)]. Such constraints were discussed in the BFKL context as originating from the improved kinematics, and more precisely by the requirement that the exchanged momenta are dominated by the transverse components [Andersson *et al.* (1996); Kwiecinski *et al.* (1996)], for more recent work on different forms of kinematical constraints see [Deak *et al.* (2019)]. These contributions generate the double transverse logarithms, and in the Mellin space they exhibit the most singular behavior, the triple collinear poles

$$-\frac{1}{4}\chi_0''(\gamma) \longrightarrow -\frac{1}{2}\frac{1}{\gamma^3} - \frac{1}{2}\frac{1}{(1-\gamma)^3}. \tag{8.9}$$

As it has been demonstrated in [Salam (1998)] the collinear approximation, which is the sum of Eqs. (8.7), (8.8), and (8.9)

$$\chi(\gamma)_{\rm coll} = -b\frac{1}{(1-\gamma)^2} - \frac{11}{12}\frac{1}{\gamma^2} - \frac{11}{12}\frac{1}{(1-\gamma)^2} - \frac{1}{2}\frac{1}{\gamma^3} - \frac{1}{2}\frac{1}{(1-\gamma)^3}, \tag{8.10}$$

is good approximation to the NLL kernel. The LL+NLL kernel is shown in Fig. 8.2, where again on the left plot we show the slice of the function on the real axis, in the right plot we show the dependence on the imaginary axis, and on the middle plot we show the full dependence in the complex plane. One observes a much more complicated structure of poles of the kernel. In particular, one sees that the NLL correction is huge, for small values of the strong coupling $\bar{\alpha}_s = 0.1$. On top of it, the saddle point is not on a real axis, but in fact there are two complex saddle points. That feature can lead to the oscillating cross sections.

8.2 Resumming large corrections at small x

The NLL corrections at low x are thus large and can lead to instabilities in the evolution, and eventually to the negative cross sections. It thus soon became clear that they need to be resummed. The origin of the large corrections was subsequently understood as the running coupling, kinematical constraints and the contribution from the DGLAP evolution.

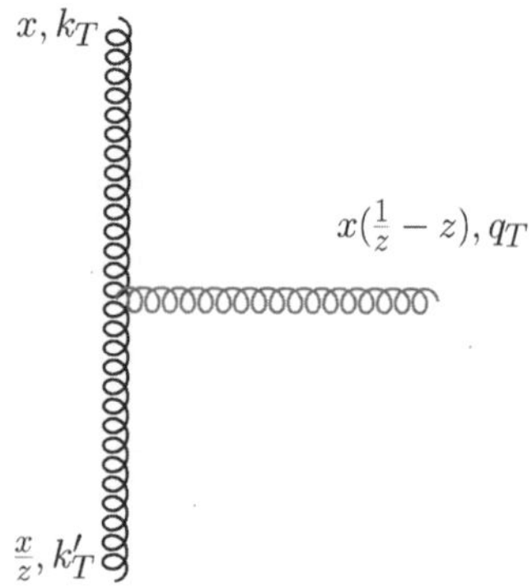

Fig. 8.3 Schematic diagram representing gluon emission in the BFKL chain. x and x/z are the longitudinal momentum fractions of the target's momentum carried by the exchanged gluon. $\mathbf{k}, \mathbf{k}'$ and $\mathbf{q}$ denote the two-dimensional transverse momenta of the exchanged gluons and the emitted gluon, respectively.

8.2.1 *Kinematical constraint*

The kinematical constraint was probably first considered in [Ciafaloni (1988)], and later-on elaborated in [Andersson *et al.* (1996); Kwiecinski *et al.* (1996)]. The derivation sketched below follows [Kwiecinski *et al.* (1996)]. Let us consider one rung of the evolution, schematically shown in Fig. 8.3 which shows a real emission of one gluon. In Fig. 8.3, $\mathbf{k}$ and $\mathbf{k}' = \mathbf{k} + \mathbf{q}$ are the transverse momenta of the exchanged gluons and $\mathbf{q}$ is the transverse momentum of the gluon emitted. The longitudinal momentum fractions of the exchanged gluons are x and $\frac{x}{z}$, respectively.

In the leading logarithmic approximation the integration over $\mathbf{k}'$ in the Eq. (8.1), is not constrained by an upper limit and thus violates the energy-momentum conservation. Thus one expects kinematical constraints, that should be imposed onto the integrations over the transverse momenta. In the case of the DIS process there should be a limit on the integration over the transverse momentum

$$Q^2/x \sim W^2, \tag{8.11}$$

where Q^2 is the hard scale of the DIS process, x is the Bjorken variable and W^2 is the c.m.s energy squared of the photon-proton system in DIS. This constraint stems from the general bound on the total available energy for the gluon cascade in the DIS collision. In addition to that global bound, there is a stronger constraint arising from the requirement that in the low x formalism the exchanged gluons have off-shellness dominated by the transverse components, i.e. one keeps terms that obey $|k^2| \simeq \mathbf{k}^2$. The argument is as follows, see [Kwiecinski *et al.* (1996)] and also [Deak *et al.* (2019)].

The four momentum of a particle can be decomposed into light cone and transverse components

$$k = (k^+, k^-, \mathbf{k}), \tag{8.12}$$

where $k^{\pm} = k^0 \pm k^3$. The exchanged gluon virtuality in these variables is equal to

$$k^2 = k^+ k^- - \mathbf{k}^2. \tag{8.13}$$

The condition $|k^2| \simeq \mathbf{k}^2$ leads to

$$\mathbf{k}^2 > k^+ k^-. \tag{8.14}$$

The emitted gluon is on-shell, thus we have $q^2 = q^+ q^- - \mathbf{q}^2 = 0$, which gives

$$q^- = \mathbf{q}^2 / q^+. \tag{8.15}$$

In the multi-Regge kinematics there is a strong ordering of longitudinal momenta along the ladder, so one can use the approximation

$$k^- = k'^- - q^- \simeq q^-. \tag{8.16}$$

Combining (8.15) and (8.14) together with (8.16) results in the following constraint

$$\mathbf{k}^2 > k^+ \frac{\mathbf{q}^2}{q^+} = \frac{z}{1-z} \mathbf{q}^2, \tag{8.17}$$

or as a limit of the $\mathbf{q}^2$ integration

$$\mathbf{q}^2 < \frac{1-z}{z} \mathbf{k}^2. \tag{8.18}$$

In the literature, there are several versions of this condition, which utilize different approximations. In the small x limit, the constraint (8.18) can be approximated to

$$\mathbf{q}^2 < \frac{\mathbf{k}^2}{z}. \tag{8.19}$$

Further approximations allow to rewrite the kinematical constraint as a condition on the transverse momentum of the exchanged gluon $\mathbf{k}'$,

$$\mathbf{k}'^2 < \frac{\mathbf{k}^2}{z}. \tag{8.20}$$

The latter constraint is obtained since for a given value of $\mathbf{k}$ a high value of $\mathbf{k}'$ means also high value of $\mathbf{q}$.

This form of the constraint was used for example in [Andersson *et al.* (1996)] and in [Kwiecinski *et al.* (1997); Ciafaloni *et al.* (2003a,b); Golec-Biernat *et al.* (2002); Kutak and Kwiecinski (2003)]. The nice feature of (8.20) is the fact that the kernel with kinematical constraint has a Mellin representation which is identical to the LL kernel eigenvalue but includes a simple shift of γ poles in the Mellin space.

8.2.2 ω shift of collinear poles in Mellin space

Imposing the kinematical constraint on the BFKL equation (8.1) leads to the strong reduction of the intercept which governs the growth of the gluon Green's function and hence it will slow down the growth of the corresponding cross sections which are computed within this small-x framework. In the Mellin space the condition for the intercept resulting from the BFKL equation is given by the famous equation

$$\omega = \bar{\alpha}_s \chi_0(\gamma) + \frac{\bar{\alpha}_s^2}{4} \chi_1(\gamma), \tag{8.21}$$

where ω is the Mellin variable conjugate to the longtudinal momentum fraction. The kinematical constraint, like Eq. (8.20), when imposed onto the transverse integrals in the BFKL equation, turns this equation into the integral equation over longitudinal momentum fraction (instead of differential in $\ln 1/x$). The kernel becomes z dependent and in the Mellin space this results in the ω dependence of the kernel $\chi(\gamma) \to \chi(\gamma, \omega)$. The condition for the intercept is now a transcendental equation of the form

$$\omega = \bar{\alpha}_s \chi(\gamma, \omega), \tag{8.22}$$

which in practice can be solved exactly only numerically. In the case of the constraint Eq. (8.20), the kernel in Mellin space is modified by a shift of the poles in γ by ω, [Andersson *et al.* (1996); Kwiecinski *et al.* (1996)]

$$\chi(\gamma, \omega) = 2\psi(1) - \psi(\gamma) - \psi(1 - \gamma + \omega). \tag{8.23}$$

When $\omega = 0$ the above result coincides with the LL BFKL eigenvalue $\chi_0(\gamma)$ in QCD and in N = 4 sYM, Eq. (8.5).

The ω shift of the poles generates an entire tower of subleading corrections to the BFKL kernel. At NLL and NNLL it generates the highest poles in $\gamma = 0$ and $\gamma = 1$, both in QCD and N = 4 sYM. It is worth to note that at NLL these poles coincide in QCD and in supersymmetric theory. In order to compare with the explicit results for the fixed order eigenvalues at NLL and NNLL order it is easier to rewrite the Eq. (8.23) into a symmetric form by performing scale changing transformation. The symmetric modified kernel eigenvalue reads thus

$$\chi(\gamma, \omega) = 2\psi(1) - \psi\left(\gamma + \frac{\omega}{2}\right) - \psi\left(1 - \gamma + \frac{\omega}{2}\right). \tag{8.24}$$

We shall show that by taking Eq. (8.24) one can retrieve the leading and the vanishing subleading poles of the N = 4 sYM case by doing an ω expansion of the shifted eigenvalue. Since it is a symmetric case the coefficients of the poles around $\gamma = 0$ and $\gamma = 1$ are identical, and we focus here on the poles around $\gamma = 0$.

Expanding in small ω gives

$$\chi(\gamma,\omega) = 2\psi(1) - \psi\left(\gamma + \frac{\omega}{2}\right) - \psi\left(1 - \gamma + \frac{\omega}{2}\right)$$
$$= \chi_0 + \chi^{(1)}\frac{\omega}{2} + \frac{1}{2!}\chi^{(2)}\left(\frac{\omega}{2}\right)^2 + \cdots, \tag{8.25}$$

where the $\chi^{(i)}$ is the i-th derivative of χ^{ω} with respect to ω. The term lowest in order $\bar{\alpha}_s$ in the expansion is simply χ_0 which is the LL eigenvalue given by Eq. (8.5). In order to retrieve the term contributing to the NLL order one can use the solution to the equation for the intercept in the lowest order of the coupling, i.e. $\omega_0 = \bar{\alpha}_s\chi_0$, and substitute it into (8.25). At this order one needs to keep terms up to first power in $\bar{\alpha}_s$

$$\chi(\gamma) = \chi_0 + \frac{1}{2}\bar{\alpha}_s\chi^{(1)}\chi_0. \tag{8.26}$$

This gives the contribution to the NLL order

$$\chi_1(\gamma) = \frac{1}{2}\chi^{(1)}\chi_0 = \frac{1}{2}\left[\psi^{(1)}(\gamma) + \psi^{(1)}(1-\gamma)\right]\left[2\psi(1) - \psi(\gamma) - \psi(1-\gamma)\right]. \tag{8.27}$$

Extracting the singular behavior around $\gamma = 0$ yields

$$\chi_1(\gamma) = -\frac{1}{2\gamma^3} - \frac{\zeta(2)}{\gamma} + O(1). \tag{8.28}$$

Similarly, one can derive the term contributing at NNLL level from the ω expansion to the second order by taking $\omega_1 = \bar{\alpha}_s\chi_0 + \bar{\alpha}_s^2\chi_1$, substituting again into (8.25) and extracting terms proportional to $\bar{\alpha}_s^2$ which results in

$$\chi_2 = \frac{1}{4}\left(\chi^{(1)}\right)^2\chi_0 + \frac{1}{8}\chi^{(2)}\left(\chi_0\right)^2. \tag{8.29}$$

In this case the pole structure is

$$\chi_2 = \frac{1}{2\gamma^5} + \frac{\zeta(2)}{\gamma^3} + \frac{2\zeta(3)}{\gamma^2} + O\left(\frac{1}{\gamma}\right). \tag{8.30}$$

The poles in Eqs. (8.28) and (8.30) should be compared to the exact results at NLL and NNLL order. Since in QCD there are other contributions, we shall compare these results to the N=4 sYM theory. The poles around $\gamma = 0$ of NLL [Kotikov and Lipatov (2003)] and NNLL kernel in N = 4 sYM are given by [Gromov *et al.* (2015); Velizhanin (2015); Caron-Huot and Herranen (2018)]

$$\chi_1^{sYM} = -\frac{1}{2\gamma^3} - 1.79 + \mathcal{O}(\gamma), \tag{8.31}$$

$$\chi_2^{sYM} = \frac{1}{2\gamma^5} - \frac{\zeta(2)}{\gamma^3} - \frac{9\zeta(3)}{4\gamma^2} - \frac{29\zeta(4)}{8\gamma} + \mathcal{O}(1). \tag{8.32}$$

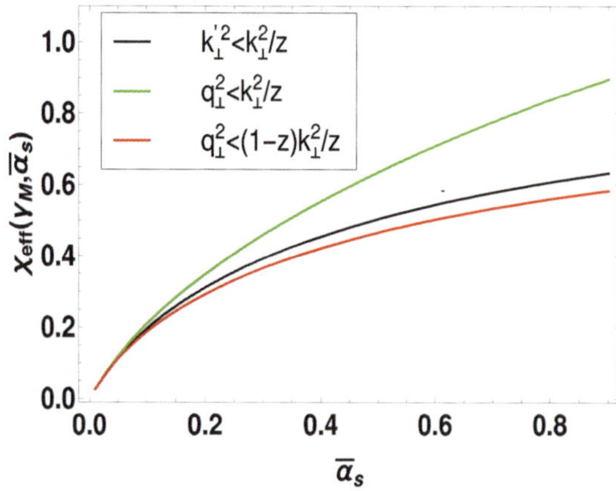

Fig. 8.4 The values of the intercept $\omega_{I\!P} = \chi_{\rm eff}(\gamma_M, \bar\alpha_s)$, Eq. (8.34), as a function of fixed $\bar\alpha_s$ for different choices of the kinematical constraint. Figure from Ref. [Deak *et al.* (2019)].

We observe that the leading poles $-1/2\gamma^3$ (8.28) coincide with the exact result at NLL in $N = 4$ sYM and in QCD. The poles $1/2\gamma^5$ (8.30) coincide with the result at NNLL order in $N = 4$ sYM (8.31). What is also interesting is the absence of subleading poles, i.e. $1/\gamma^2$ and $1/\gamma^4$ at NLL and NNLL respectively in the ω expansion of kinematical constraint which again coincides with the exact results in $N = 4$ sYM (8.31). It was demonstrated in [Deak *et al.* (2019)] that this is a general expectation from the kinematical constraint.

In Fig. 8.4 we show the solution of the intercept in the case of the LL BFKL equation with kinematical constraint, where different constraints have been compared, i.e. defined by Eqs. (8.18), (8.19), (8.20). They all have been evaluated by solving the equation (8.22) for the effective kernel eigenvalue

$$\chi_{\rm eff}(\gamma, \bar\alpha_s), \tag{8.33}$$

and then evaluating this kernel at the minimum to find the value of the intercept

$$\omega_{I\!P} = \chi_{\rm eff}(\gamma_{\rm min}, \bar\alpha_s). \tag{8.34}$$

It has been shown [Deak *et al.* (2019)] that all the constraints give similar intercepts for values of $\bar\alpha_s$ up to about 0.2. The constraint (8.19) gives somewhat larger values, when $\bar\alpha_s$ is further increased, while the two constraints (8.20) and (8.18) give comparable values for the intercepts even for very large values of the strong coupling.

8.3 CCSS resummation scheme

As shown in the previous section, the kinematical constraint is responsible for a part of the NLL corrections to the kernel. By taking into account this constraint these corrections can be resummed to all orders. There are also other corrections that need to be resummed. As demonstrated in [Ciafaloni *et al.* (2003b)], part of the double poles in the NLL kernel stem from the non-singular part of the DGLAP splitting function. These corrections can be included as well, by taking instead the non-singular part of the splitting function and subtracting the corresponding double poles. This was done in earlier work [Kwiecinski *et al.* (1997)] in the momentum space, where the unified DGLAP and BFKL equation was proposed. In addition to the DGLAP splitting function, the running coupling was incorporated into the evolution. In the BFKL formalism the strong coupling is fixed at LL order, and then it only starts to run at NLL order. This is manifest in the appearance of the term which depends on the beta function in the NLL kernel. It is another source of the double poles in the Mellin space in γ. This term can be effectively resummed by including the scale of the running coupling in front of the leading logarithmic kernel. It was argued in [Ciafaloni *et al.* (2003b)] that the choice of scale of q^2, i.e. the transverse momentum squared of the emitted gluon, is the correct choice. This means that when the coupling in front of the LL part of the kernel is evaluated at this scale, then there is no β_0 dependent term in the NLL part of the kernel. Summarizing, the CCSS framework for the resummation proposed in [Ciafaloni *et al.* (2003b)] consisted of several improvements, as listed below.

(1) The kinematical constraint on the real emission part of the BFKL kernel.
(2) The non singular part of the DGLAP splitting function.
(3) The running of the coupling.
(4) Subtraction of the triple and double poles corresponding to the above terms in order to avoid the double counting.
(5) Inclusion of the rest of the NLL kernel.
(6) Additional subtractions to remove the remaining single poles, which also ensure that the momentum sum rule was satisfied for the resummed anomalous dimension.

The effective intercept of the resummed equation within the CCSS resummation (for the fixed coupling case) is shown in Fig. 8.5, as compared to the LL and NLL result.

We see that the effective resummed intercept is reduced to about 0.3 for values of $\bar{\alpha}_s \simeq 0.2$, from the LL value which is 0.55. When comparing Fig. 8.5 with the calculations which only include the kinematical constraint, Fig. 8.4, we see that the latter is indeed leading to the suppression which is very close in magnitude to the full resummed result. In addition, we see that the NLL value of the intercept has a much smaller value than the LL and resummed cases. The resummed small x equation was formulated and solved in momentum space [Kwiecinski *et al.* (1996); Ciafaloni *et al.* (2003b)]. An example of a solution for the gluon Green's function is shown in Fig. 8.6. The left plot is for the case of the fixed coupling, $\bar{\alpha}_s = 0.2$, and the right plot is for the case of the running coupling. The resummed calculation in the fixed coupling case is compared to the LL approximation. We see that the resummed calculation is much lower than the LL case. There are two effects evident: one is

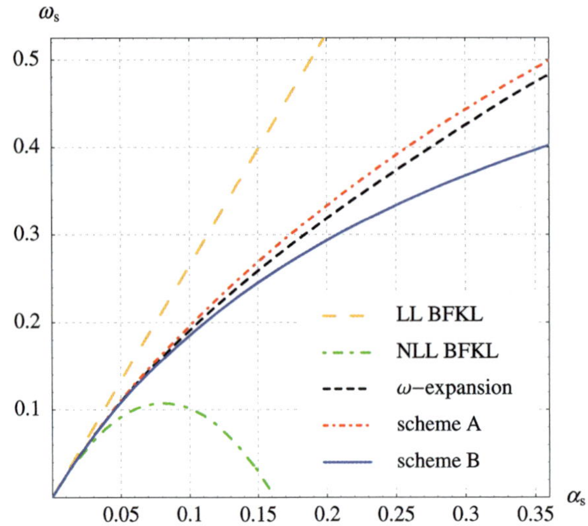

Fig. 8.5 The values of the intercept $\omega_{I\!P}$ for the LL, NLL, and the resummed case with different subtractions. Figure taken from Ref. [Ciafaloni *et al.* (2003a)].

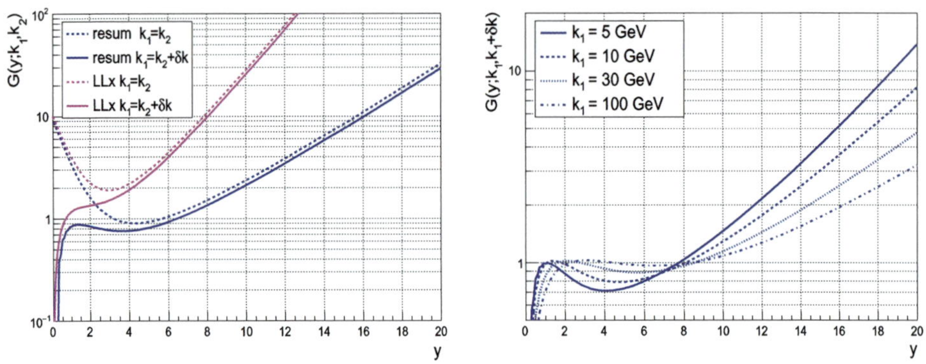

Fig. 8.6 Left: gluon Green's function $G(y;k_1,k_2)$ as a function of rapidity y for fixed transverse momentum $k_1 = 5\,\text{GeV}$ and fixed coupling constant $\bar{\alpha}_s = 0.2$. LL (magenta) and resummed (blue) calculations are shown for comparison. Dashed lines correspond to $k_1 = k_2$, and solid lines $k_1 = k_2 + \delta k$ with δk an offset which is given by $\delta k = k_1(\exp(\delta \ln k) - 1)$ where $\delta \ln k$ is the spacing in the logarithmic grid used in the numerical solution. Right: gluon Green's function $G(y;k_1,k_2)$ as a function of rapidity y for fixed transverse momenta $k_1 = 5, 10, 30, 100$ GeV and running strong coupling constant. Figures taken from [Deak *et al.* (2020)].

the lowering of the intercept, which leads to the slow down of the evolution. The other effect is the delay in the onset of the power-like growth. It is evident that the resummed gluon Green's function develops a plateau or a dip, for several units of rapidity. This is also illustrated in Fig. 8.6 (right plot) where the calculation based on the resummed evolution is shown for the running coupling case. In this plot, the gluon Green's function is shown for different external scales k_1. We see that the onset of the growth is delayed to higher energies, for higher values of the external scales. We thus see that the preasymptotic effects in the low x evolution are very prominent and vital for phenomenology, see further discussion in [Deak *et al.* (2020)].

8.4 Phenomenological applications

8.4.1 *Small x dynamics in inclusive HERA data*

The calculations based on the LL BFKL equation failed to describe the HERA data, due to the much too strong increase of the gluon density with decreasing x, see for example [Bojak and Ernst (1997)]. Since the 90's, several fits were performed to the HERA data on inclusive structure function, which incoporated the partial resummation of the small x effects. These included both BFKL as well as CCFM equations with kinematical constraint. In the early works, for example in [Kwiecinski *et al.* (1996)], it was already demonstrated that the kinematical constraint plays a big role in taming the growth of the structure function with small x and gives results more aligned with phenomenology. The unified DGLAP/BFKL fit was developed in [Kwiecinski *et al.* (1997)], where it was demonstrated that the kinematical constraint is crucial for improving the quality of the fit. Also, the presence of the DGLAP terms allowed to satisfy the momentum sum rule. The unified DGLAP/BFKL framework was successfully applied to the more recent, precise HERA data [Kutak and Sapeta (2012)], using also terms which contain the small x saturation. A very interesting analysis was performed more recently in [Ball *et al.* (2018)] where the ABF resummation [Altarelli *et al.* (2000, 2001, 2002)] was used in the fits to the structure function data, and included the DGLAP up to NNLO [Bonvini *et al.* (2016)]. In that analysis, which uses the machinery of the NNPDF [Forte *et al.* (2002)], several different fits were performed. The standard fixed order fits, using DGLAP evolution up to NLO or NNLO order, were compared with the fits which used small x resummation and DGLAP up to the same order. It turned out that the quality of fits was best for the case of the resummed fit matched to the NNLO DGLAP evolution. The improvement actually came from the small x and small Q^2 region of the data. To test this, series of different fits were performed to restricted sets of data, where more of the low x and Q^2 data were systematically included. It turned out that the χ^2 of the fits started to increase for the case of the fixed NNLO DGLAP case when more data from this region were included in the fit. This clearly indicated the tension in the fit in this scenario. On the other hand, the χ^2 remained unchanged, for the fit with the resummation included, when the same analysis was performed. That meant that the improvement came from the region which is crucial for the small x physics. It is expected that such an effect would be much more pronounced in the future high energy electron -proton machines like LHeC [Abelleira Fernandez *et al.* (2012)] or FCC-eh [Abada *et al.* (2019b,a)], which can potentially reach much lower values of x.

8.4.2 *Ultra-high energy neutrino physics*

Understanding the small-x behavior of the gluon density is crucial in the context of ultra-high energy neutrino physics and astroparticle physics. There is a vigorous theoretical and experimental effort to develop ultra-high energy neutrino astronomy. Very highly energetic neutrinos provide a unique window into the Universe, and objects within it. This is mainly due to the fact that neutrinos interact only weakly (and gravitationally) with matter, for an excellent review see [Gandhi *et al.* (1998)]. Thus weakly interacting neutrinos can travel long distances from distant astrophysical sources, undeflected by the magnetic fields inside and in between galaxies. The ultra-high energy neutrinos, measured for example in the IceCube observatory on the Antarctica [Aartsen *et al.* (2017)], can therefore provide complementary information to cosmic rays, gamma rays and gravitational wave signals. The detailed knowledge about low-x dynamics is indispensable in two contexts of ultra-high energy astronomy: neutrino interaction with matter and neutrino production, either on Earth or in other objects in the Universe.

Ultra-high energy neutrino cross section When neutrinos travel through the Earth, they can either interact with electrons or with nuclei. At energies beyond TeV the dominant part of the cross section is due to the neutrino DIS Charged Current and Neutral Current interaction with the nuclei [Gandhi *et al.* (1998)].

In Fig. 8.7, the charged current neutrino cross section as a function of the neutrino energy for an isoscalar target is shown. The neutrino energy is given in the laboratory frame where the target nucleus is at rest. The particular calculation [Kwiecinski *et al.* (1999)] shown in Fig. 8.7 was based on the unified BFKL/DGLAP evolution with kinematical constraint developed in [Kwiecinski *et al.* (1997)]. The charged current cross section, has a monotonic energy dependence with two distinct regions. At energies below $\sim 50\,\text{TeV}$ the cross section grows linearly with energy. This linear dependence on energy is purely kinematical. In this region, the cross section is dominated by the contributions from the large-x region, where valence partons dominate. This is illustrated in Fig. 8.7 by the solid red line, which becomes independent of energy. This flat behavior is due to the lack of growth of the valence distribution, i.e. the number of valence partons is fixed. Beyond that energy, of about $\sim 50\,\text{TeV}$, the energy dependence becomes milder, and the cross section grows roughly as a power $\sim E_\nu^\lambda$ with $\lambda \simeq 0.3$. This high energy behavior of the cross section is completely controlled by the small-x dynamics of the parton distributions. The dominance at high energies of the sea (up-down, charm-strange, top-bottom) contributions to the cross section is clearly seen in Fig. 8.7.

It is interesting to investigate the contributing regions in x and Q^2. This is done in Fig. 8.8, where the differential cross section for the CC interaction $xQ^2 d\sigma^{CC}/dxdQ^2$ is shown, for a neutrino energy $E_\nu = 10^{11}\,\text{GeV}$. The diagonal line visible both in the surface and in the contour plot is due to the kinematic limit.

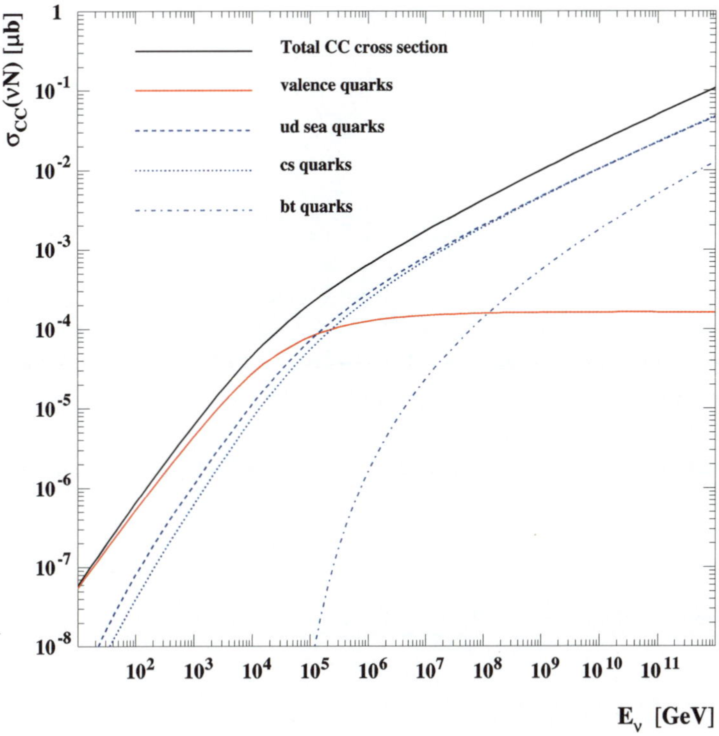

Fig. 8.7 Charged current (CC) cross section for the neutrino — nucleon interaction on a isoscalar target as a function of neutrino energy in the laboratory frame where the target is at rest. The total CC cross section is broken down into several contributions due to valence, up-down, strange-charm and bottom-top quarks. The calculation was based on a unified DGLAP/BFKL approach developed in Ref. [Kwiecinski *et al.* (1997)]. Figure from Ref. [Kwiecinski *et al.* (1999)].

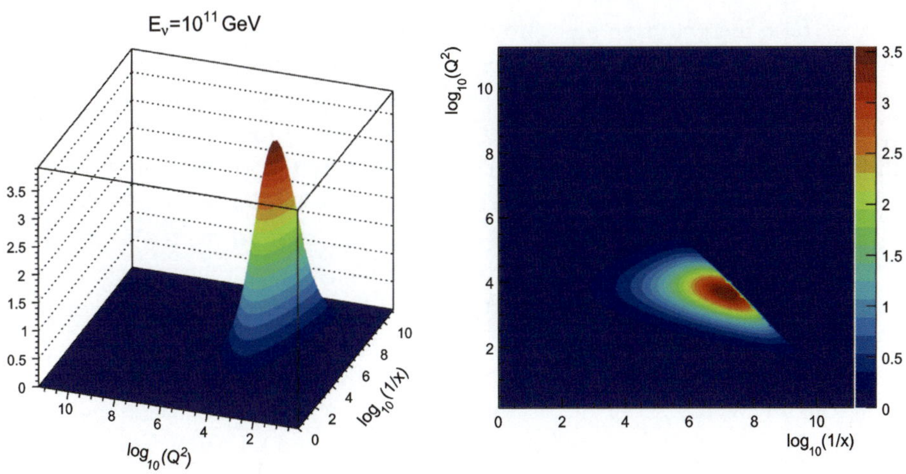

Fig. 8.8 Differential charged current neutrino cross section $10^5 \cdot xQ^2 d\sigma^{CC}/dxdQ^2$ [nb] as a function of Q^2 and x for fixed neutrino energy $E_\nu = 10^{11}$ GeV. Left: surface plot; right: contour plot.

We see a clear peak of the cross section at roughly a value of $Q^2 = M_W^2$ and an x value

$$x \simeq \frac{M_W^2}{2ME_\nu}, \tag{8.35}$$

which for this energy is 3×10^{-8}.

We thus see that at such high energies the neutrino cross section is dominated by the values of x which have never been measured in accelerators. The standard fits based on DGLAP dynamics, yield large error bands due to the uncertainties in the parton distribution functions. Thus extrapolations based on robust theory which include the small x resummation become essential for the reliable prediction of the neutrino-nucleon cross section at these ultra-high energies.

Prompt atmospheric neutrino flux There is another instance where gluon dynamics at small x is perhaps even more essential. This is the process of (forward) neutrino production in ultra-high energy hadronic interactions. The cosmic neutrinos at high energies are produced in interactions which involve hadrons, either in γp or in pp collisions. As a result of these collisions of pions, kaons, and at higher energies charmed and beauty mesons are produced with high energies [Gaisser (1990)]. The neutrinos emerge as final decay products of these hadrons, and also as further decays of muons. As one example, in the Earth's atmosphere neutrinos are produced in the interactions of the highly energetic cosmic rays with nitrogen and oxygen nuclei. Traditionally, the atmospheric neutrino flux is described as a superposition of two components which have somewhat distinct energy behavior. Up to about 100 TeV the atmospheric neutrino spectrum originates from the decay of pions and kaons. This part of the atmospheric flux is called *conventional*. Because of the life-time $\tau \sim 10^{-8}\,s$ of pions and kaons, at these energies these mesons have high probability of interacting before decay. This will lead to the degradation of their energy and thus resulting in the steeply falling flux of the produced neutrinos $\sim E^{-3.7}$. The second part of the neutrino flux, which dominates at energies above 100 TeV is called *prompt* neutrino flux. The neutrinos in this part of the spectrum originate as decay products of shorter-lived $\tau \sim 10^{-12}\,s$ charmed (and bottom) hadrons. The cross section for the production of heavy mesons (or baryons) is obviously smaller than that for pions and kaons, but the charmed mesons $D^\pm, D^0, D_s$ and baryon Λ_c live shorter than pions and kaons, and thus they will decay prior to the interaction. This means that they will not loose energy, and thus through their decay they will produce highly energetic neutrinos. Because of that mechanics at energies about 100 TeV the prompt neutrino flux, which roughly behaves as $\sim E_\nu^{-2.7}$ starts to dominate over the conventional part of the flux.

Therefore, the knowledge of this part of the spectrum is essential as it provides an important background for the sought-after astrophysical neutrinos [Aartsen *et al.* (2014)]. The charmed quarks are produced in hadron-hadron collisions through gluon-gluon fusion. In this scenario, where the highly energetic proton hits the proton at rest, the main interest is in the forward production where the charm quark(anti-quark) carries most of the momentum. Therefore in the gluon-gluon fusion mechanism in this kinematics, one gluon has longitudinal momentum fraction which is rather large, and the other gluon very small x. Since the scales are very small, of the order of the charm masses, the values of the longitudinal momentum fractions involved are also extremely small at these energies. Therefore the knowledge of the parton distributions in the small x region is essential for the predictions of the prompt neutrino flux [Gelmini *et al.* (2000)]. One sees that this process probes the gluon distribution function precisely in the region where BFKL dynamics is very important. In Fig. 8.9 the differential cross section for $d\sigma/dx_F$ is shown, where $x_F = E_c/E_p$ is the fraction of the energy of the proton carried by the charmed quark. The calculation [Bhattacharya *et al.* (2016)] is based on k_T factorization, together with the BFKL/DGLAP evolution with (k_T nonlinear) and without (k_T linear) the additional effects (implemented in the framework of [Kutak and Kwiecinski (2003); Kutak and Sapeta (2012)]). This calculation is compared to the collinear calculation based on DGLAP evolution. We see that as the energies become higher the small x effects become more prominent. The calculation denoted by k_T linear starts to dominate the collinear calculation already at $10^6\, GeV$. We also observe that, if present, the saturation effects at small x will become very important at these energies. They would lead to the substantial decrease of the cross section and thus the resulting neutrino flux.

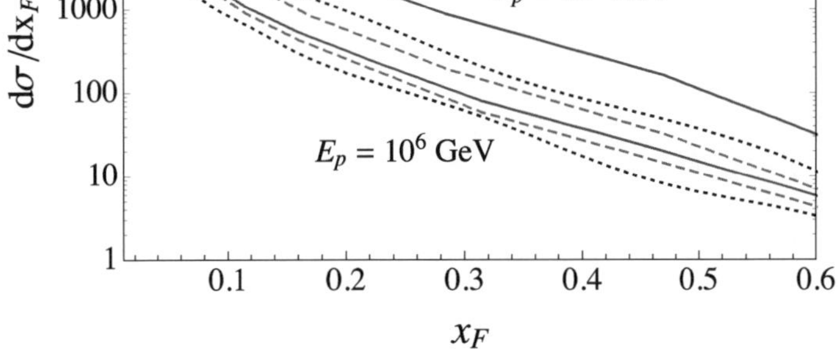

Fig. 8.9 The differential cross section $d\sigma/dx_F$ as a function of x_F for two energies $E = 10^6$ GeV and $E = 10^9$ GeV from k_T factorization, with linear evolution (solid upper blue), and non-linear evolution (lower dashed magenta). Shown for comparison is the perturbative differential cross section based on the standard collinear formalism from Ref. [Bhattacharya *et al.* (2015)]. Figure from Ref.[Bhattacharya *et al.* (2016)].

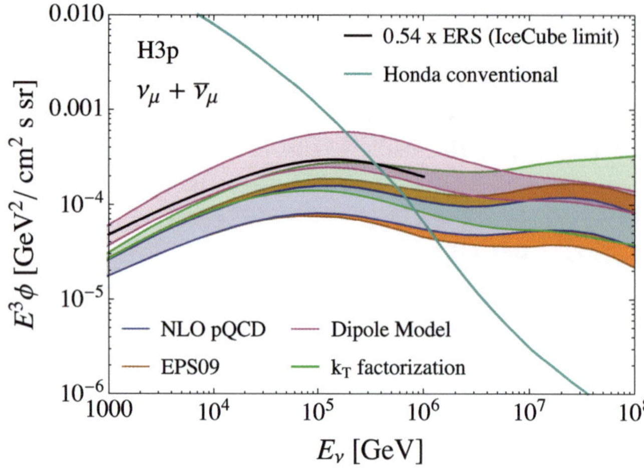

Fig. 8.10 A comparison of the neutrino fluxes from different QCD models with the IceCube upper limit. Also shown in the conventional vertical neutrino flux. Figure from Ref.[Bhattacharya *et al.* (2016)].

In Fig. (8.10) we show the prompt neutrino fluxes obtained from various calculations which incorporate small x dynamics. They are compared with the conventional neutrino flux (steeply dropping curve). We observe that the prompt neutrino fluxes dominate over the conventional flux for neutrino energies above $10^5 - 10^6$ GeV.

Bibliography

Aartsen, M. G. *et al.* (2014). Observation of high-energy astrophysical neutrinos in three years of IceCube data, *Phys. Rev. Lett.* **113**, p. 101101, doi:10.1103/PhysRevLett. 113.101101, arXiv:1405.5303 [astro-ph.HE].

Aartsen, M. G. *et al.* (2017). The IceCube neutrino observatory: Instrumentation and online systems, *JINST* **12**, 03, p. P03012, doi:10.1088/1748-0221/12/03/P03012, arXiv:1612.05093 [astro-ph.IM].

Abada, A. *et al.* (2019a). FCC-hh: The hadron collider, *Eur. Phys. J. ST* **228**, 4, pp. 755–1107, doi:10.1140/epjst/e2019-900087-0.

Abada, A. *et al.* (2019b). FCC physics opportunities, *Eur. Phys. J.* **C79**, 6, p. 474, doi: 10.1140/epjc/s10052-019-6904-3.

Abelleira Fernandez, J. L. *et al.* (2012). A large hadron electron collider at CERN: Report on the physics and design concepts for machine and detector, *J. Phys.* **G39**, p. 075001, doi:10.1088/0954-3899/39/7/075001, arXiv:1206.2913 [physics.acc-ph].

Altarelli, G., Ball, R. D., and Forte, S. (2000). Resummation of singlet parton evolution at small x, *Nucl. Phys.* **B575**, pp. 313–329, doi:10.1016/S0550-3213(00)00032-8, arXiv:hep-ph/9911273 [hep-ph].

Altarelli, G., Ball, R. D., and Forte, S. (2001). Small x resummation and HERA structure function data, *Nucl. Phys.* **B599**, pp. 383–423, doi:10.1016/S0550-3213(01)00023-2, arXiv:hep-ph/0011270 [hep-ph].

Altarelli, G., Ball, R. D., and Forte, S. (2002). Factorization and resummation of small x scaling violations with running coupling, *Nucl. Phys.* **B621**, pp. 359–387, doi: 10.1016/S0550-3213(01)00563-6, arXiv:hep-ph/0109178 [hep-ph].

Altarelli, G., Ball, R. D., and Forte, S. (2003). An anomalous dimension for small x evolution, *Nucl. Phys.* **B674**, pp. 459–483, doi:10.1016/j.nuclphysb.2003.09.040, arXiv:hep-ph/0306156 [hep-ph].

Altarelli, G., Ball, R. D., and Forte, S. (2008). Small x resummation with quarks: Deep-inelastic scattering, *Nucl. Phys.* **B799**, pp. 199–240, doi:10.1016/j.nuclphysb.2008.03.003, arXiv:0802.0032 [hep-ph].

Andersson, B., Gustafson, G., and Samuelsson, J. (1996). The linked dipole chain model for DIS, *Nucl. Phys.* **B467**, pp. 443–478, doi:10.1016/0550-3213(96)00114-9.

Ball, R. D., Bertone, V., Bonvini, M., Marzani, S., Rojo, J., and Rottoli, L. (2018). Parton distributions with small-x resummation: evidence for BFKL dynamics in HERA data, *Eur. Phys. J.* **C78**, 4, p. 321, doi:10.1140/epjc/s10052-018-5774-4, arXiv:1710.05935 [hep-ph].

Bhattacharya, A., Enberg, R., Jeong, Y. S., Kim, C. S., Reno, M. H., Sarcevic, I., and Stasto, A. (2016). Prompt atmospheric neutrino fluxes: perturbative QCD models and nuclear effects, *JHEP* **11**, p. 167, doi:10.1007/JHEP11(2016)167, arXiv:1607.00193 [hep-ph].

Bhattacharya, A., Enberg, R., Reno, M. H., Sarcevic, I., and Stasto, A. (2015). Perturbative charm production and the prompt atmospheric neutrino flux in light of RHIC and LHC, *JHEP* **06**, p. 110, doi:10.1007/JHEP06(2015)110, arXiv:1502.01076 [hep-ph].

Bojak, I. and Ernst, M. (1997). Limitations of small x resummation methods from F2 data, *Nucl. Phys.* **B508**, pp. 731–752, doi:10.1016/S0550-3213(97)00656-1, arXiv:hep-ph/9702282 [hep-ph].

Bonvini, M., Marzani, S., and Peraro, T. (2016). Small-x resummation from HELL, *Eur. Phys. J.* **C76**, 11, p. 597, doi:10.1140/epjc/s10052-016-4445-6, arXiv:1607.02153 [hep-ph].

Caron-Huot, S. and Herranen, M. (2018). High-energy evolution to three loops, *Journal of High Energy Physics* **2018**, 2, p. 58.

Chirilli, G. A. and Kovchegov, Y. V. (2013). Solution of the NLO BFKL equation and a strategy for solving the all-order BFKL equation, *JHEP* **06**, p. 055, doi:10.1007/JHEP06(2013)055, arXiv:1305.1924 [hep-ph].

Ciafaloni, M. (1988). Coherence effects in initial jets at Small q**2 / s, *Nucl. Phys.* **B296**, pp. 49–74, doi:10.1016/0550-3213(88)90380-X.

Ciafaloni, M. and Camici, G. (1998). Energy scale(s) and next-to-leading BFKL equation, *Phys. Lett.* **B430**, pp. 349–354, doi:10.1016/S0370-2693(98)00551-6, arXiv:hep-ph/9803389 [hep-ph].

Ciafaloni, M., Colferai, D., Colferai, D., Salam, G. P., and Stasto, A. M. (2003a). Extending QCD perturbation theory to higher energies, *Phys. Lett.* **B576**, pp. 143–151, doi: 10.1016/j.physletb.2003.09.078, arXiv:hep-ph/0305254 [hep-ph].

Ciafaloni, M., Colferai, D., and Salam, G. P. (1999a). A collinear model for small x physics, *JHEP* **10**, p. 017, doi:10.1088/1126-6708/1999/10/017, arXiv:hep-ph/9907409 [hep-ph].

Ciafaloni, M., Colferai, D., and Salam, G. P. (1999b). Renormalization group improved small x equation, *Phys. Rev.* **D60**, p. 114036, doi:10.1103/PhysRevD.60.114036, arXiv:hep-ph/9905566 [hep-ph].

Ciafaloni, M., Colferai, D., Salam, G. P., and Stasto, A. M. (2003b). Renormalization group improved small x Green's function, *Phys. Rev.* **D68**, p. 114003, doi:10.1103/PhysRevD.68.114003, arXiv:hep-ph/0307188 [hep-ph].

Ciafaloni, M., Colferai, D., Salam, G. P., and Stasto, A. M. (2004). The Gluon splitting function at moderately small x, *Phys. Lett.* **B587**, pp. 87–94, doi:10.1016/j.physletb.2004.02.054, arXiv:hep-ph/0311325 [hep-ph].

Ciafaloni, M., Colferai, D., Salam, G. P., and Stasto, A. M. (2007). A matrix formulation for small-x singlet evolution, *JHEP* **08**, p. 046, doi:10.1088/1126-6708/2007/08/046, arXiv:0707.1453 [hep-ph].

Deak, M., Frankfurt, L., Strikman, M., and Stasto, A. M. (2020). Taming the small x growth with resummation, *in preparation* .

Deak, M., Kutak, K., Li, W., and Stato, A. M. (2019). On the different forms of the kinematical constraint in BFKL, *Eur. Phys. J.* **C79**, 8, p. 647, doi:10.1140/epjc/s10052-019-7171-z, arXiv:1906.09062 [hep-ph].

Fadin, V. S. and Lipatov, L. N. (1998). BFKL pomeron in the next-to-leading approximation, *Phys. Lett.* **B429**, pp. 127–134, doi:10.1016/S0370-2693(98)00473-0, arXiv:hep-ph/9802290 [hep-ph].

Forte, S., Garrido, L., Latorre, J. I., and Piccione, A. (2002). Neural network parametrization of deep inelastic structure functions, *JHEP* **05**, p. 062, doi:10.1088/1126-6708/2002/05/062, arXiv:hep-ph/0204232 [hep-ph].

Gaisser, T. K. (1990). *Cosmic rays and particle physics*, ISBN 9780521339315, http://www.cambridge.org/uk/catalogue/catalogue.asp?isbn=0521326672.

Gandhi, R., Quigg, C., Reno, M. H., and Sarcevic, I. (1998). Neutrino interactions at ultrahigh-energies, *Phys. Rev.* **D58**, p. 093009, doi:10.1103/PhysRevD.58.093009, arXiv:hep-ph/9807264 [hep-ph].

Gelmini, G., Gondolo, P., and Varieschi, G. (2000). Prompt atmospheric neutrinos and muons: Dependence on the gluon distribution function, *Phys. Rev.* **D61**, p. 056011, doi:10.1103/PhysRevD.61.056011, arXiv:hep-ph/9905377 [hep-ph].

Golec-Biernat, K. J., Motyka, L., and Stasto, A. M. (2002). Diffusion into infrared and unitarization of the BFKL pomeron, *Phys. Rev.* **D65**, p. 074037, doi:10.1103/PhysRevD.65.074037, arXiv:hep-ph/0110325 [hep-ph].

Gromov, N., Levkovich-Maslyuk, F., and Sizov, G. (2015). Pomeron eigenvalue at three loops in $\mathcal{N} = 4$ supersymmetric Yang-Mills theory, *Phys. Rev. Lett.* **115**, 25, p. 251601, doi:10.1103/PhysRevLett.115.251601, arXiv:1507.04010 [hep-th].

Kotikov, A. V. and Lipatov, L. N. (2003). DGLAP and BFKL evolution equations in the N=4 supersymmetric gauge theory, *Nuclear Physics B* **661**, 1-2, pp. 19–61, doi:10.1016/S0550-3213(03)00264-5, http://arxiv.org/abs/hep-ph/0208220, arXiv:hep-ph/0208220.

Kutak, K. and Kwiecinski, J. (2003). Screening effects in the ultrahigh-energy neutrino interactions, *Eur. Phys. J.* **C29**, p. 521, doi:10.1140/epjc/s2003-01236-y, arXiv:hep-ph/0303209 [hep-ph].

Kutak, K. and Sapeta, S. (2012). Gluon saturation in dijet production in p-Pb collisions at Large Hadron Collider, *Phys. Rev.* **D86**, p. 094043, doi:10.1103/PhysRevD.86.094043, arXiv:1205.5035 [hep-ph].

Kwiecinski, J., Martin, A. D., and Stasto, A. M. (1997). A Unified BFKL and GLAP description of F2 data, *Phys. Rev.* **D56**, pp. 3991–4006, doi:10.1103/PhysRevD.56.3991, arXiv:hep-ph/9703445 [hep-ph].

Kwiecinski, J., Martin, A. D., and Stasto, A. M. (1999). Penetration of the earth by ultrahigh-energy neutrinos predicted by low x QCD, *Phys. Rev.* **D59**, p. 093002, doi:10.1103/PhysRevD.59.093002, arXiv:astro-ph/9812262 [astro-ph].

Kwiecinski, J., Martin, A. D., and Sutton, P. J. (1996). Constraints on gluon evolution at small x, *Z. Phys.* **C71**, pp. 585–594, doi:10.1007/BF02907019,10.1007/s002880050206, arXiv:hep-ph/9602320 [hep-ph].

Lipatov, L. N. (1986). The bare Pomeron in quantum chromodynamics, *Sov. Phys. JETP* **63**, pp. 904–912, [Zh. Eksp. Teor. Fiz.90,1536(1986)].

Sabio Vera, A. (2005). An 'All-poles' approximation to collinear resummations in the Regge limit of perturbative QCD, *Nucl. Phys.* **B722**, pp. 65–80, doi:10.1016/j.nuclphysb.2005.06.003, arXiv:hep-ph/0505128 [hep-ph].

Salam, G. P. (1998). A Resummation of large subleading corrections at small x, *JHEP* **07**, p. 019, doi:10.1088/1126-6708/1998/07/019, arXiv:hep-ph/9806482 [hep-ph].

Salam, G. P. (1999). An introduction to leading and next-to-leading BFKL, *Acta Phys. Polon.* **B30**, pp. 3679–3705, arXiv:hep-ph/9910492 [hep-ph].

Thorne, R. S. (2001). The Running coupling BFKL anomalous dimensions and splitting functions, *Phys. Rev.* **D64**, p. 074005, doi:10.1103/PhysRevD.64.074005, arXiv:hep-ph/0103210 [hep-ph].

Velizhanin, V. N. (2015). BFKL pomeron in the next-to-next-to-leading approximation in the planar N=4 SYM theory, arXiv:1508.02857 [hep-th].

White, C. D. and Thorne, R. S. (2007). A global fit to scattering data with NLL BFKL resummations, *Phys. Rev.* **D75**, p. 034005, doi:10.1103/PhysRevD.75.034005, arXiv:hep-ph/0611204 [hep-ph].

https://doi.org/10.1142/9789811231124_0009

Chapter 9

From Parton Saturation to Proton Spin: The Impact of BFKL Equation and Reggeon Evolution

Yuri V. Kovchegov
Department of Physics, The Ohio State University, Columbus, OH 43210, USA

Lev Lipatov was a giant in the field of strong interactions and a dominant force in high-energy QCD for many decades. His work deeply influenced both how we think about QCD and how we perform calculations in the theory. Below we describe the work in two related research directions: the physics of parton saturation and proton spin at small-x. Both developments would have been impossible without Lipatov's groundbreaking work. Saturation physics would not have happened without the Balitsky–Fadin–Kuraev–Lipatov (BFKL) equation [Kuraev *et al.* (1977); Balitsky and Lipatov (1978)]. The recent progress in our theoretical understanding of the proton spin contribution coming from small-x partons started with the seminal paper by Kirschner and Lipatov [Kirschner and Lipatov (1983)] resumming double logarithms of energy in the Reggeon evolution (see also [Gorshkov *et al.* (1968)]).

9.1 Saturation and Unitarity

With the advent of QCD as the theory of strong interactions, an important question arose: what is the correct high-energy asymptotics of QCD? Lev Lipatov and collaborators tackled this question head on by deriving the BFKL equation [Kuraev *et al.* (1977); Balitsky and Lipatov (1978)], an equation describing the high-energy behavior of cross sections and parton distribution functions in QCD. The BFKL equation was a breakthrough in our understanding of QCD at high energy. The solution of BFKL equation [Kuraev *et al.* (1977); Balitsky and Lipatov (1978)] leads to a cross section which grows as a small positive power of the center-of-mass energy squared s,

$$\sigma \sim s^{\text{const}\,\alpha_s}, \tag{9.1}$$

where the constant in the exponent is positive (const > 0) and α_s is the strong coupling. This behavior violates the unitarity bound [Heisenberg (1952); Froissart (1961); Martin (1969)]

$$\sigma \leq \ln^2 s. \tag{9.2}$$

Therefore, the high energy QCD community understood that BFKL equation needed to be unitarized. That is, at very high energies, some corrections to BFKL evolution must become important, restoring unitarity.

At the same time, it is known that the gluon and quark parton distribution functions (PDFs) given by the solution of BFKL evolution, or extracted from experimental data, grow as positive powers of $1/x$. While there is no unitarity bound for the PDFs, very high numbers of quarks and gluons resulting from the power-of-$1/x$ growth at small x must lead to density of quarks and gluons in the proton becoming large. The dynamics of the resulting ultra-dense parton system has to differ from that of the dilute system at larger x. Indeed, as was observed by Gribov, Levin and Ryskin [Gribov *et al.* (1983)] shortly after the BFKL equation was derived, at very high parton density the parton mergers have to eventually compensate for splittings, leading to a slowdown of the PDFs' growth with decreasing x. This phenomenon is known as the parton saturation.

Below we describe how the saturation of gluons leads to unitarization of QCD cross sections, thus allowing us to understand the high-energy asymptotics of strong interactions. The reader interested in a more detailed exposition of the material presented in this Section is referred to [Kovchegov and Levin (2012)] and references therein.

9.1.1 *Deep Inelastic Scattering in the quasi-classical approximation*

Consider the Deep Inelastic Scattering (DIS) process, where a virtual photon (projectile) scatters on a proton or a nucleus (target). Our usual way of thinking about DIS is in terms of the "handbag" diagram pictured in the left panel of Fig. 9.1, where the oval at the bottom represents the target proton or nucleus, the wavy line is the virtual photon (γ^*), and the straight lines are quark propagators. The diagrams in Fig. 9.1 contribute to the forward $\gamma^* + p$ scattering amplitude, with the center of mass energy squared $s = (p+q)^2$ of the virtual photon scattering on the target. The leading-order "handbag" diagram results in the DIS structure functions F_1 and F_2 expressed in terms of the quark distribution in the target.

However, if one is interested in DIS at small values of the Bjorken x variable (corresponding to large center-of-mass energy squared s for fixed photon virtuality $Q^2 = -q^2$ with $x \approx Q^2/s$), then a straightforward calculation of the Born-level quark and gluon scattering cross sections demonstrates that t-channel exchanges of gluons dominate over the t-channel exchanges of quarks. Indeed, if one compares the right panel in Fig. 9.1 to the "handbag" diagram on the left of this figure, one sees that, being a one-loop correction, the right diagrams is suppressed by a power of the strong coupling constant α_s, while simultaneously being enhanced by a power of $1/x$. That is, for the two diagrams in Fig. 9.1 we have

$$\frac{\text{quark loop diagram}}{\text{“handbag” diagram}} \propto \frac{\alpha_s}{x}. \tag{9.3}$$

Clearly, for $x < 0.01$, the $1/x$ enhancement overcomes the $\sim \alpha_s$ suppression factor, and the quark loop diagram in Fig. 9.1 dominates. The physics behind the quark loop diagram in Fig. 9.1, evaluated at small x, is as follows: the virtual photon (produced by the scattering lepton in DIS) fluctuates into a quark–anti-quark pair long before the interaction with the target proton or nucleus. The resulting quark–anti-quark *dipole* goes on to interact with the target. This is the dipole picture of DIS [Gribov (1970); Bjorken and Kogut (1973); Bertsch *et al.* (1981); Frankfurt and Strikman (1988); Kopeliovich *et al.* (1981); Mueller (1990); Nikolaev and Zakharov (1991)]. This interaction is limited to a two-gluon exchange in Fig. 9.1, but can be more complicated in general. Since we are interested in the forward scattering amplitude, long after the interaction the quark–anti-quark pair recombines back into the virtual photon. This interpretation is valid in the rest frame of the target or in the frame where both the virtual photon and the target are relativistic, moving along two different light-cones.

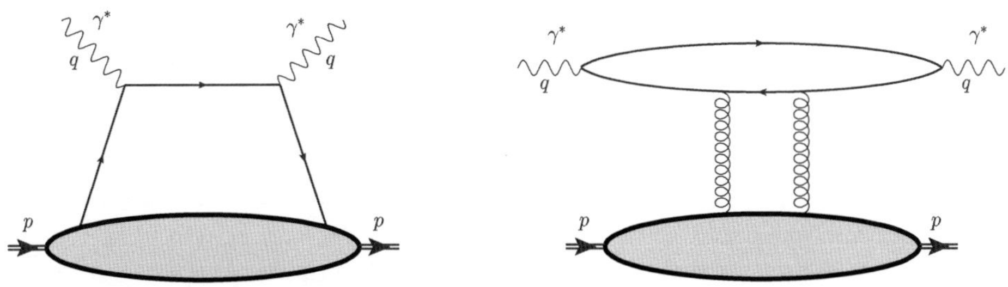

Fig. 9.1 DIS process in the standard "handbag" diagram approximation (left panel) and due to the exchange of gluons in the t-channel (right panel).

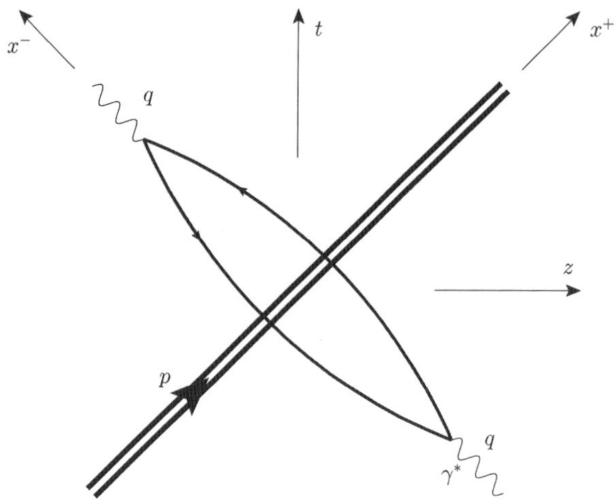

Fig. 9.2 DIS process at small x in the space-time representation.

The space-time picture of the process from the right panel of Fig. 9.1 is shown in Fig. 9.2. We are working in the frame where the proton and virtual photon momenta are

$$p^\mu \approx (p^+, 0, 0_\perp), \quad q^\mu = \left(-\frac{Q^2}{2\,q^-}, q^-, 0_\perp\right) \tag{9.4}$$

with p^+ and q^- large. Our light-cone components are defined by $x^\pm = (x^0 \pm x^3)/\sqrt{2}$ while the transverse vectors are denoted by $\vec{x}_\perp = (x^1, x^2)$ with $x_\perp = |\vec{x}_\perp|$. Figure 9.2 illustrates that the lifetime (or, more precisely, x^--extent) of the quark loop is much longer than the width of the Lorentz-contracted proton in the x^- direction. The two-gluon interaction between the quark loop and the proton from Fig. 9.1 happens instantaneously in x^- (near $x^- = 0$), and does not violate the separation of time scales into the $\gamma^* \to q\bar{q}$ fluctuation, followed by the interaction and the $q\bar{q} \to \gamma^*$ recombination. Further, one can argue that the high energy interaction of the quark–anti-quark dipole and the proton does not change the transverse positions of the quark and the anti-quark. That is, the interaction is diagonal in the transverse plane [Kopeliovich *et al.* (1981); Levin and Ryskin (1987); Mueller (1990); Nikolaev and Zakharov (1991)]. We therefore write down the following formula for the DIS cross section at small x:

$$\sigma_{T,\,L}^{\gamma^* A}(x, Q^2) = \int \frac{d^2 x_\perp}{4\pi} \int_0^1 \frac{dz}{z\,(1-z)}\, |\Psi_{T,\,L}^{\gamma^* \to q\bar{q}}(\vec{x}_\perp, z)|^2\, \sigma_{tot}^{q\bar{q}A}(\vec{x}_\perp, Y), \tag{9.5}$$

where subscripts T, L denote the transverse and longitudinal polarizations of the virtual photon. While the target is labeled by the superscript A, which, in general, indicates a nucleus, it is also understood that Eq. (9.5) applies to DIS on a proton as well. The functions $\Psi_{T,\,L}^{\gamma^* \to q\bar{q}}(\vec{x}_\perp, z)$ are the light-cone wave functions of the virtual photon fluctuating into the $q\bar{q}$ pair. They can be calculated using the rules of the light-cone (or light-front) perturbation theory [Lepage and Brodsky (1980); Brodsky *et al.* (1998)]. The wave functions depend on the fraction z of the photon's light-cone momentum q^- carried by the quark in the pair and on the transverse separation $\vec{x}_\perp$ between the quark and the anti-quark. At the leading order one has [Bjorken *et al.* (1971); Nikolaev and Zakharov (1991)]

$$|\Psi_T^{\gamma^*\to q\bar{q}}(\vec{x}_\perp, z)|^2 = 2N_c \sum_f \frac{\alpha_{EM} Z_f^2}{\pi} z\,(1-z) \times \left\{ a_f^2 \left[K_1(x_\perp\, a_f)\right]^2 \left[z^2 + (1-z)^2\right] + m_f^2 \left[K_0(x_\perp\, a_f)\right]^2 \right\}, \tag{9.6}$$

and

$$|\Psi_L^{\gamma^*\to q\bar{q}}(\vec{x}_\perp, z)|^2 = 2N_c \sum_f \frac{\alpha_{EM}\, Z_f^2}{\pi} 4\, Q^2\, z^3\, (1-z)^3 \left[K_0(x_\perp\, a_f)\right]^2, \tag{9.7}$$

where

$$a_f^2 = Q^2 z(1-z) + m_f^2. \tag{9.8}$$

Here N_c is the number of quark colors, m_f is the mass of quarks with flavor f, Z_f is the fractional charge of the quark (in the units of the absolute value of the electron charge), and α_{EM} is the fine-structure constant.

Employing Eq. (9.5) one can construct the DIS structure function via

$$F_2(x, Q^2) = \frac{Q^2}{4\pi^2 \alpha_{EM}} \sigma_{tot}^{\gamma^* A} = \frac{Q^2}{4\pi^2 \alpha_{EM}} \left[\sigma_T^{\gamma^* A} + \sigma_L^{\gamma^* A}\right], \tag{9.9a}$$

$$2x\, F_1(x, Q^2) = \frac{Q^2}{4\pi^2 \alpha_{EM}} \; \sigma_T^{\gamma^* A}. \tag{9.9b}$$

The interaction between the dipole and the proton or nucleus is described by the cross section $\sigma_{tot}^{q\bar{q}A}(\vec{x}_\perp, Y)$ in Eq. (9.5), which depends on $\vec{x}_\perp$ and rapidity $Y = \ln(1/x) \approx \ln(s/Q^2)$. All the strong interaction dynamics for the DIS process at small x is contained in this cross section. For further transparency, let us first rewrite this cross section as an integral over the impact parameter $\vec{b}_\perp$,

$$\sigma_{tot}^{q\bar{q}A}(\vec{x}_\perp, Y) = 2 \int d^2 b_\perp \left[1 - \mathrm{Re}\, S(\vec{x}_\perp, \vec{b}_\perp, Y)\right]. \tag{9.10}$$

Here $S(\vec{x}_\perp, \vec{b}_\perp, Y)$ is the forward (in the transverse position space) matrix element of the S-matrix of the dipole-target scattering, defined by

$$S(\vec{x}_\perp, \vec{b}_\perp, Y) = 1 + i\, A(\vec{x}_\perp, \vec{b}_\perp, Y), \tag{9.11}$$

with

$$A = \frac{M}{2\, s}, \tag{9.12}$$

where M is the dipole-target forward scattering amplitude.

At the leading order in high-energy scattering A is always imaginary, and, hence, S is real. Therefore, we will drop the Re sign in Eq. (9.10). Defining the imaginary part of A by $N = \mathrm{Im}A$ such that

$$N(\vec{x}_\perp, \vec{b}_\perp, Y) = 1 - S(\vec{x}_\perp, \vec{b}_\perp, Y) \tag{9.13}$$

we rewrite Eq. (9.10) as

$$\sigma_{tot}^{q\bar{q}A}(\vec{x}_\perp, Y) = 2 \int d^2b \, N(\vec{x}_\perp, \vec{b}_\perp, Y). \tag{9.14}$$

At this point let us put our calculation momentarily on hold, and spend a little time discussing unitarity constraints. Unitarity requires that $|S| \leq 1$. This leads to $0 \leq N \leq 2$ and

$$\sigma_{tot}^{q\bar{q}A}(\vec{x}_\perp, Y) \leq 4 \int d^2b = 4\pi R^2, \tag{9.15}$$

where R is the radius of the (circular) region in the transverse plane where the interaction is strong enough for $S \approx -1$ and $N \approx 2$. This region of strong scattering is known as the *black disk*. In the case of high energy scattering one can argue that, due to the dominance of inelastic processes over the elastic once, $N \leq 1$ and

$$\sigma_{tot}^{q\bar{q}A}(\vec{x}_\perp, Y) \leq 2 \int d^2b = 2\pi R^2, \tag{9.16}$$

where now $S \approx 0$ and $N \approx 1$ in the black disk region. Eq. (9.16) is known as the black disk limit for the total cross section. Observing that at high energy the radius of the black disk in QCD grows at most logarithmically with energy, $R = R_0 + a \ln(s/s_0)$, one arrives at the Froissart-Martin bound on the total cross section [Heisenberg (1952); Froissart (1961); Martin (1969)]

$$\sigma_{tot}^{q\bar{q}A}(\vec{x}_\perp, Y) \leq 2\pi a^2 \ln^2 \frac{s}{s_0}. \tag{9.17}$$

Here $a \sim 1/m_\pi$ with m_π the pion mass.

Returning to Eq. (9.14), one can show that the two-gluon exchange from the right panel of Fig. 9.1 gives

$$N_{LO}(\vec{x}_\perp, \vec{b}_\perp, Y) = \frac{\pi \, \alpha_s^2 \, C_F}{N_c} \, T(\vec{b}_\perp) \, x_\perp^2 \, \ln \frac{1}{x_\perp \Lambda}, \tag{9.18}$$

where $C_F = (N_c^2 - 1)/(2N_c)$ is the fundamental Casimir of SU(N_c) and Λ is some infrared (IR) cutoff. In arriving at Eq. (9.18) we had to sum over all possible connections of the gluons to the quark and anti-quark line in the dipole. Equation (9.18) is valid for $x_\perp < 1/\Lambda$, that is, the dipole should be perturbatively small. We assumed the target to be a single quark in the proton. The latter could be inside a nucleus. In preparation for the scattering on a nucleus, we have also employed the *nuclear profile function* $T(\vec{b}_\perp)$ defined by

$$T(\vec{b}_\perp) \equiv \int_{-\infty}^{\infty} db_3 \, \rho_A(\vec{b}_\perp, b_3) \tag{9.19}$$

with $\rho_A(\vec{b}_\perp, b_3)$ the number density of the nucleons in the nucleus normalized as $\int d^3b \, \rho_A = A$, where A is the atomic number of the nucleus (the number of protons and neutrons in the nucleus). For the proton target $A = 1$. For a large nucleus $T(\vec{b}_\perp) \sim A^{1/3}$ and the nuclear profile function may become numerically large.

It appears that we have an immediate problem: as we increase $x_\perp$, the forward amplitude N_{LO} from Eq. (9.18) grows, potentially exceeding the black-disk limit ($N \leq 1$) for some dipole sizes. One may counter this worry by suggesting that dipole sizes should be non-perturbatively large, $x_\perp \gtrsim 1/\Lambda$, for this violation of the black-disk limit to happen. However, for large nuclei with $A \gg 1$ this is not so. In fact, no matter how small the dipole size $x_\perp$ is, we can always increase A until $N_{LO} > 1$ in Eq. (9.18). (Indeed this may require unrealistically large nuclei, but this is a theoretical exercise.) Conversely, for a fixed large A, as we increase $x_\perp$ we may reach the region where $N_{LO} > 1$ while still being in the perturbative QCD regime of relatively small dipole sizes, $x_\perp < 1/\Lambda$.

In the case of a very large nucleus, our problem above can be resolved in the following way. Large nucleus has many nucleons. When N_{LO} approaches unity, interaction with each nucleon happens with the probability close to one. This means, interactions with multiple nucleons are also becoming likely. Hence, as N_{LO} increases, we need to consider multiple rescatterings of the $q\bar{q}$ dipole in the nucleus, since those are becoming order-one corrections to the single scattering contained in N_{LO} from Eq. (9.18). Diagrammatically we need to sum up the graphs of the type shown in Fig. 9.3, where each oval at the bottom represents a nucleon in the nucleus, and disconnected gluon lines at the top imply summation over all possible connections to the quark and anti-quark lines.

Interaction with each nucleon is local in x^-, such that the gluon exchanges with different nucleons do not get entangled and we also do not have any gluon-gluon

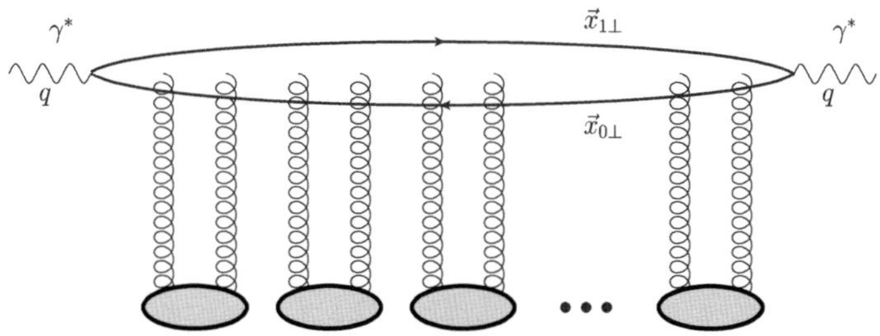

Fig. 9.3 DIS on a nucleus with multiple rescattering on the nucleons (ovals at the bottom). Disconnected gluon lines at the top denote a sum over all possible connections of the gluons to the quark and anti-quark lines.

interactions. (We are working in the Feynman gauge, $\partial_\mu A^\mu = 0$ or in the light-cone gauge of the projectile, $A^- = 0$.) This means that we exchange exactly two gluons with each nucleon, as shown in Fig. 9.3. This allows us to define the resummation parameter: two gluons bring in α_s^2, while there are $\sim A^{1/3}$ nucleons in the nucleus at a given impact parameter. Hence, the resummation parameter of the multiple rescattering calculation is $\alpha_s^2 \, A^{1/3}$ [Kovchegov (1997)]. When $\alpha_s^2 \, A^{1/3} \gtrsim 1$, multiple rescatterings are important and have to be re-summed to all orders.

The result of resumming diagrams in Fig. 9.3 is a simple exponentiation of Eq. (9.18). It reads [Mueller (1990)]

$$N(\vec{x}_\perp, \vec{b}_\perp, Y = 0) \; = \; 1 - \exp\left\{ - \frac{\alpha_s^2 \, C_F \, \pi}{N_c} \, T(\vec{b}_\perp) \, x_\perp^2 \, \ln \left(\frac{1}{x_\perp \, \Lambda} \right) \right\}. \tag{9.20}$$

Defining *the saturation scale* by

$$Q_s^2(\vec{b}_\perp) \; \equiv \; \frac{4 \, \pi \, \alpha_s^2 \, C_F}{N_c} \, T(\vec{b}_\perp) \tag{9.21}$$

we rewrite Eq. (9.20) as

$$N(\vec{x}_\perp, \vec{b}_\perp, Y = 0) \; = \; 1 - \exp\left\{ - \frac{1}{4} \, x_\perp^2 \, Q_s^2(\vec{b}_\perp) \, \ln \frac{1}{x_\perp \, \Lambda} \right\}. \tag{9.22}$$

This is the Gribov–Glauber–Mueller (GGM) [Mueller (1990)] formula for the dipole amplitude N.

The GGM formula is sketched in Fig. 9.4 as a function of $x_\perp$. We see that, just like for N_{LO} from Eq. (9.18), the amplitude N from Eq. (9.22) approaches zero as $x_\perp \to 0$. Indeed, this makes perfect physical sense: when the quark and the anti-quark are close to each other, their colors cancel and their interaction with the target tends to zero. This phenomenon is known as color transparency [Kopeliovich *et al.* (1981); Nikolaev and Zakharov (1991); Heiselberg *et al.* (1991); Frankfurt *et al.* (1993)].

Remarkably, when $x_\perp$ increases, the amplitude in Eq. (9.22) and Fig. 9.4 never exceeds one, that is, one always has $N < 1$ and the black disk limit is not violated (in the perturbative region, $x_\perp < 1/\Lambda$). This behavior is due to the exponentiation in Eq. (9.22), that is, due to multiple rescatterings. We see that inclusion of multiple

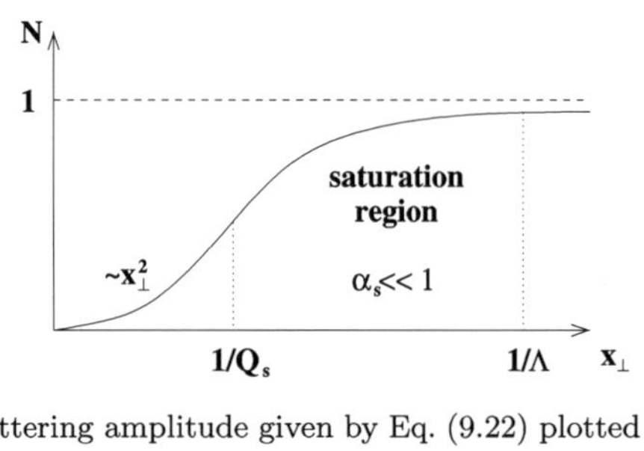

Fig. 9.4 The forward scattering amplitude given by Eq. (9.22) plotted as a function of the dipole transverse size $x_\perp$.

rescatterings indeed resolved the problem of the black-disk-limit violation and of unitarity violation by the leading-order expression (9.18).

Since $T(\vec{b}_\perp) \sim A^{1/3}$ we have the following scaling:

$$Q_s^2 \sim A^{1/3}. \tag{9.23}$$

Therefore, for large enough nucleus, one has $Q_s^2 \gg \Lambda^2$, and the saturation scale is perturbatively large. Since the transition from the region where N is rising with increasing $x_\perp$ to the region where the growth is gradually tamed happens at $x_\perp \approx 1/Q_s \ll 1/\Lambda$, the unitarization takes place in the perturbative region where $\alpha_s \ll 1$ still. The region $1/Q_s \lesssim x_\perp \ll 1/\Lambda$ is called *the saturation region.* There the strong coupling constant α_s is small, but, due to the large size of the nucleus, the interactions are strong and the resulting dipole–nucleus cross section approaches the black disk limit of $N = 1$ (without exceeding it). We conclude that the saturation region appears to be weakly-coupled, but strongly-interacting.

The GGM approach resumming the parameter $\alpha_s^2 \, A^{1/3}$ is equivalent to the quasi-classical description of the small-x nuclear wave function in the framework of the McLerran–Venugopalan (MV) model [McLerran and Venugopalan (1994a,b,c)]. Gluon fields in the MV model are classical at the leading order in the resummation of powers of $\alpha_s^2 \, A^{1/3}$, and have to be found by solving the classical Yang–Mills equations. Further details on this very important model for saturation physics can be found in the reviews [Iancu and Venugopalan (2003); Weigert (2005); Jalilian-Marian and Kovchegov (2006); Gelis *et al.* (2010); Albacete and Marquet (2014)] and in the book [Kovchegov and Levin (2012)].

Last but not least, let us formulate our results in terms of operators. The propagator for a high-energy quark (or anti-quark) can be written as (proportional to) an eikonal Wilson line along the corresponding light-cone trajectory. Since the quark and anti-quark are generated in our small-x scattering process long before the interaction with the target, and are absorbed back into the virtual photon long after that interaction, we can think of their (close to) light-cone trajectories as being infinite in x^-. Define a light-cone Wilson line for quarks by

$$V_{\vec{x}_\perp}[b^-, a^-] = \mathrm{P} \exp \left\{ ig \int\limits_{a^-}^{b^-} dx^- \, A^+(x^+ = 0, x^-, \vec{x}_\perp) \right\} \tag{9.24}$$

with $A^\mu = t^a \, A^{a\mu}$ and t^a the generators of SU(N_c) in the fundamental representation. Infinite Wilson line is denoted by

$$V_{\vec{x}_\perp} \equiv V_{\vec{x}_\perp}[+\infty, -\infty]. \tag{9.25}$$

With the help of these operators, we can write down the expectation value for the S-matrix of the dipole–target scattering as

$$S(\vec{x}_{1\perp}, \vec{x}_{0\perp}, Y) = \left\langle \frac{1}{N_c} \, \mathrm{tr} \left[V_{\vec{x}_{1\perp}} \, V^\dagger_{\vec{x}_{0\perp}} \right] \right\rangle. \tag{9.26}$$

We have switched notation in Eq. (9.26): instead of using the dipole separation $\vec{x}_\perp$ and the impact parameter $\vec{b}_\perp$, we now are using the transverse positions of the quark and the anti-quark,

$$\vec{x}_{1\perp} = \vec{b}_\perp + \tfrac{1}{2}\,\vec{x}_\perp, \quad \vec{x}_{0\perp} = \vec{b}_\perp - \tfrac{1}{2}\,\vec{x}_\perp, \tag{9.27}$$

in the argument of S, as shown in Fig. 9.3.

The angle brackets in Eq. (9.26) indicate averaging in the target proton or nucleus state (see e.g. [Kovchegov (1996)]). They can be related to the standard expectation value in the proton (nucleus) state by

$$\left\langle \hat{\mathcal{O}}(b,r) \right\rangle = \frac{1}{2p^+} \int \frac{d^2\Delta_\perp\, d\Delta^+}{(2\pi)^3}\, e^{ib\cdot\Delta} \left\langle p + \frac{\Delta}{2} \right| \hat{\mathcal{O}}(0,r) \left| p - \frac{\Delta}{2} \right\rangle \tag{9.28}$$

for an arbitrary operator $\hat{\mathcal{O}}(b,r)$ [Kovchegov (2019)]. Here $b\cdot\Delta = b^-\Delta^+ - \vec{b}_\perp \cdot \vec{\Delta}_\perp$.

With the help of Eq. (9.26) we re-write the dipole scattering amplitude as

$$N(\vec{x}_{1\perp}, \vec{x}_{0\perp}, Y) = 1 - \left\langle \frac{1}{N_c}\, \mathrm{tr}\left[V_{\vec{x}_{1\perp}}\, V^\dagger_{\vec{x}_{0\perp}} \right] \right\rangle. \tag{9.29}$$

Equation (9.22) evaluates the correlation function in Eq. (9.29) in the quasi-classical GGM/MV approximation.

In the operator language, formula (9.5) can be interpreted as a statement of factorization between the light-cone wave functions squared (the so-called impact factor) and the correlation function of two infinite light-cone Wilson lines. The latter contains the bulk of the strong interactions dynamics. It is interesting to note that this factorization, while naturally appearing in this quasi-classical calculation due to the separation between the $\gamma^* \to q\bar{q}$ splitting time scale and the interaction time scale, will remain valid once the small-x evolution corrections are included.

The operator definitions (9.26) and (9.29) are strictly-speaking not gauge-invariant: to make them gauge-invariant one needs to insert into the trace a pair of transverse-pointing gauge links connecting the two light-cone Wilson lines at the infinities $x^- = \pm\infty$. However, in the $A^- = 0$ and $\partial_\mu A^\mu = 0$ gauges that we employ here such gauge links do not contribute: thus, we will follow the standard notation and will not show these links explicitly. (The gauge links at infinity become very important in $A^+ = 0$ gauge.)

The operator formalism we have just introduced may not seem particularly useful in the quasi-classical limit considered in this Section. It is, nevertheless, a very powerful method, whose benefits we will demonstrate later in this Chapter.

9.1.2 *Nonlinear small-x BK evolution equation at large-N_c and Mueller's dipole model*

The GGM formula (9.22) derived above describes how saturation effects prevent the cross section from violating the black disk limit. However it suffers from one significant shortcoming: the amplitude N given by the GGM formula is independent of the center-of-mass energy, or, equivalently, of rapidity Y. Indeed the saturation scale (9.21) is energy-independent, and so is the amplitude N in Eq. (9.22). This is why we put $Y = 0$ in the argument of N in Eq. (9.22): we will shortly see that this implies that Eq. (9.22) will serve as the initial condition for the evolution equation we are about to derive. This energy independence of the GGM formula (9.22) contradicts experimental data: for instance, the measured F_2 structure function is indeed a function of x, which grows rapidly with decreasing x. Theoretically, it would be important to find radiative corrections to Eq. (9.22): above they were neglected as being suppressed by an additional power of α_s, and not enhanced by $A^{1/3}$. However, at small x there exists another large parameter, $\ln(1/x)$, and some of the radiative corrections to the GGM formula come in with a factor of $\alpha_s \ln(1/x)$. That is, the suppression introduced by $\alpha_s \ll 1$ can be compensated by the large logarithm $\ln(1/x) \gg 1$, such that the resulting parameter is order one, $\alpha_s \ln(1/x) \sim 1$, and the powers of this parameter need to be summed up to all orders.

One can show that radiative corrections bringing in powers of $\alpha_s \ln(1/x)$ in the $A^- = 0$ gauge are generated by the long lived gluons emitted (and absorbed) by the projectile dipole (moving in the x^- direction, as per Fig. 9.2). This is illustrated in Fig. 9.5. The x^--lifetime of those gluons should be much longer than the interaction time between the projectile and the target via the GGM gluons. Therefore, we can combine the interaction with the GGM gluons into an almost instantaneous interaction with the "shock wave", as depicted in by the red rectangle in Fig. 9.5. The meaning of the red rectangle is clarified in Fig. 9.6: at this point it represents all GGM-like interaction of the projectile with the target. If the projectile is not just a $q\bar{q}$ pair, and if it includes long-lived gluons as in Fig. 9.5, those gluons can also interact with the target by exchanging instantaneous (GGM) gluons.

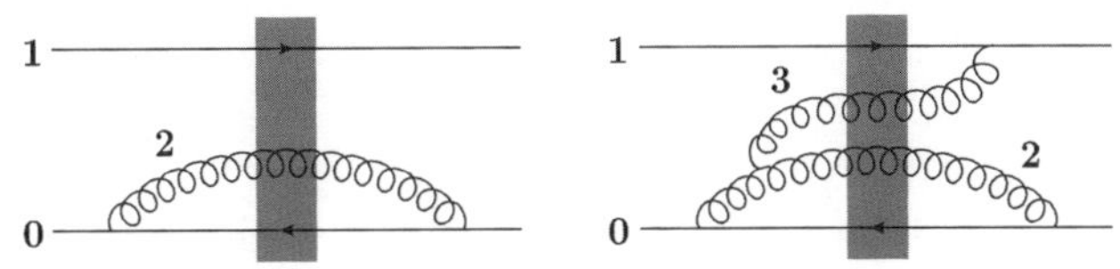

Fig. 9.5 One and two-gluon emissions by the $q\bar{q}$ dipole scattering on a target shock wave.

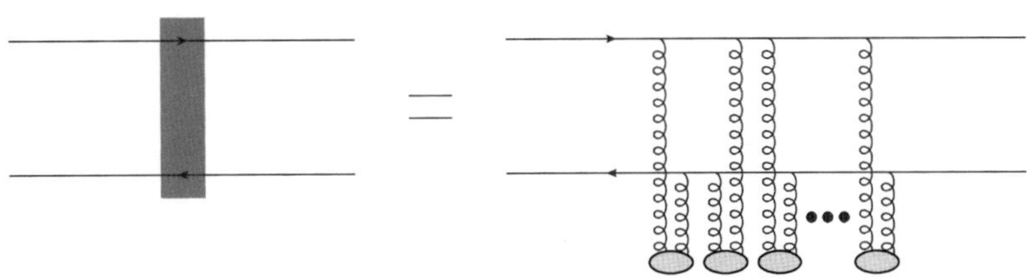

Fig. 9.6 The abbreviated notation for a shock wave, with the red rectangle representing all interactions with the target due to GGM gluons exchanges.

Each long-lived gluon in Fig. 9.5 brings in a power of $\alpha_s \ln(1/x)$, with the logarithm of $1/x$ arising from the phase space integral as we will see shortly. One can show that the gluons emitted or absorbed when the dipole is inside the shock wave do not generate $\ln(1/x)$: hence we need to consider only the gluons emitted and absorbed when the dipole is outside the shock wave. From Fig. 9.5 we conclude that to resum all powers of $\alpha_s \ln(1/x)$ we need to sum up an infinite cascade of these long-lived gluons scattering on the shock wave target.

Resummation of a gluon cascade is not easy in general. To simplify it, we will employ the 't Hooft's large-N_c limit ['t Hooft (1974)] following the prescription of the Mueller's dipole model [Mueller (1994); Mueller and Patel (1994); Mueller (1995)] (see also [Nikolaev *et al.* (1994)]). In the large-N_c limit only the planar diagrams survive. This means that at large N_c the diagrams in Fig. 9.5 can be redrawn as shown in Fig. 9.7, where the double lines represent the gluons. (At large N_c each gluon can be represented as a quark and anti-quark pair in the $N_c^2 - 1$ adjoint representation of SU(N_c).)

Let us start with the left panel in Fig. 9.7. There, the emission of gluon 2 in the large-N_c limit splits the original color dipole (made out of the quark 1 and the anti-quark 0) into two color dipoles, one made out of the quark 1 and the anti-quark part of the gluon line 2, and the other one made out of the quark part of the gluon line 2 and the anti-quark 0. It is important to point out that, in order to generate logarithms of x, the emitted gluon 2 has the minus component of its momentum much smaller than the minus momenta of the quark and anti-quark in the original dipole, $k_2^- \ll k_1^-, k_0^-$. An essential consequence of this ordering is that the transverse position of the anti-quark 0 (in the left panel of Fig. 9.7) does not change after emitting and absorbing the gluon 2. The same applies to the quark 1 and the gluon 2 in the right panel of Fig. 9.7: their transverse positions do not change after the gluon 3 emission and absorption due to the $k_3^- \ll k_2^- \ll k_1^-, k_0^-$ ordering. Similar to the left panel, in the right panel of Fig. 9.7 we observe that the original color dipole 10 is split into three color dipoles 13, 32, and 20, by the time the projectile system crosses the shock wave. We see that, at large N_c, our gluon cascade turns into a cascade of color dipoles. This is the essence of Mueller's dipole model. Moreover, note that the dipoles do not interact with each other: such interactions would be non-planar, and hence N_c-suppressed. We conclude that we need to sum up a cascade of dipoles, in which each dipole may produce other ("daughter") dipoles, but the dipoles do not interact with each other otherwise. In the end, each dipole

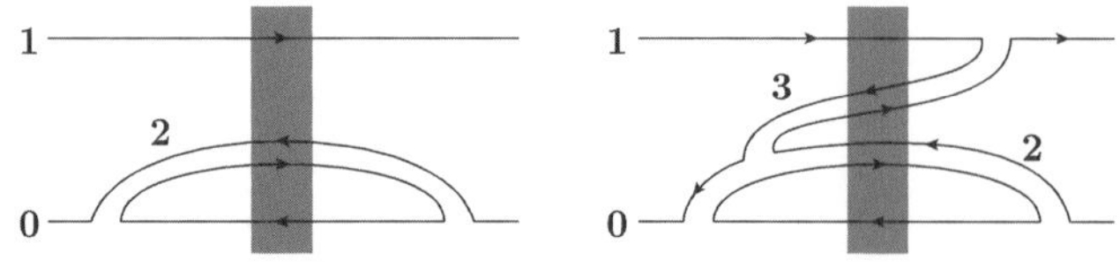

Fig. 9.7 One and two-gluon emissions by the $q\bar{q}$ dipole scattering on a target shock wave in the large-N_c limit.

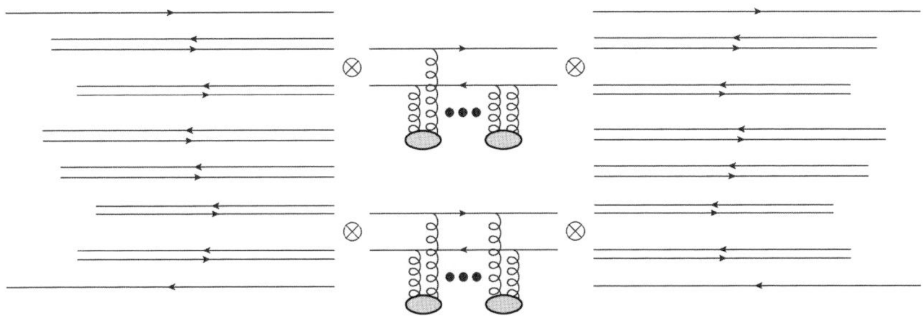

Fig. 9.8 The color dipole cascade, with each dipole interacting independently with the target nucleus.

interacts with the shock wave independently via the GGM multiple rescatterings. This physical picture is illustrated in Fig. 9.8, where the disconnected gluon (double straight) lines represent the sum over all possible (planar) ways each gluon can be emitted by the gluons which were emitted earlier in the cascade. The $\otimes$ signs in Fig. 9.8 denote the convolution between the GGM formula (9.22) and the dipole cascade over the transverse positions of the quark and anti-quark in the dipoles. (In addition to the gluons crossing the shock wave, as shown in Fig. 9.7, there may also be virtual gluons, which do not cross the shock wave: those are both emitted and absorbed either to the left or to the right of the shock wave. While these gluons need to be included as well, and will be included shortly, we have not explicitly discussed them here, since they do not modify the qualitative picture of the dipole cascade we have described.)

We are now ready to resum the cascade. Employing the above observations that the dipoles do not interact with each other and that they also interact with the shock wave target independently, we arrive at the evolution equation demonstrated diagrammatically in Fig. 9.9. There we describe one step of small-x evolution for a color dipole. Disconnected gluon (double) lines again denote all possible ways of emitting or absorbing the gluon by the quark and the anti-quark lines in the dipole. The last two diagrams on the right of Fig. 9.9 denote the virtual gluon correction we have just mentioned above. The meaning of the shock wave has changed as compared to Fig. 9.6: now it includes not only the GGM gluon exchanges, but also the subsequent gluon emissions in the cascade. Indeed, as we pointed out before, the minus momenta of the gluons in the cascade are ordered,

$$k_0^-, k_1^- \gg k_2^- \gg k_3^- \gg \ldots, \tag{9.30}$$

while the transverse momenta are comparable to each other,

$$k_{0\perp}, k_{1\perp} \sim k_{2\perp} \sim k_{3\perp} \sim \ldots. \tag{9.31}$$

These conditions lead to the ordering of the x^--lifetimes of the cascade gluons,

$$\frac{2k_2^-}{k_{2\perp}^2} \gg \frac{2k_3^-}{k_{3\perp}^2} \gg \ldots. \tag{9.32}$$

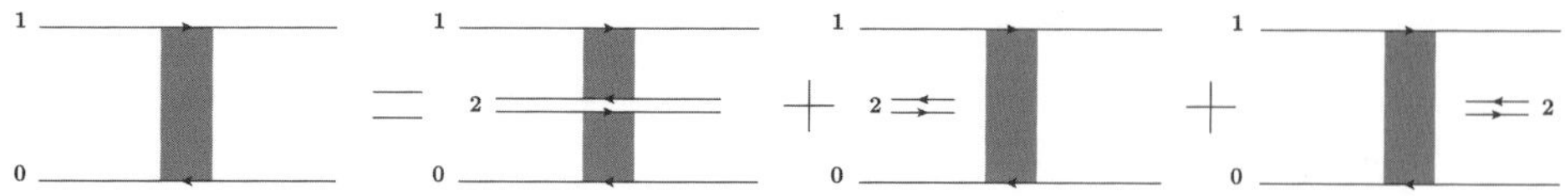

Fig. 9.9 Diagrammatic illustration of the dipole evolution at small x in the large-N_c limit.

Since the softer (smaller-k^-) gluons have much shorter lifetimes, they can be considered as a part of the shock wave from the standpoint of the harder, longer-living gluons.

In Fig. 9.9 we have also modified the shock wave notation to underline the fact that each dipole interacts independently with the shock wave by drawing the shock wave only inside each dipole in the first diagram on the right-hand side. The physical picture of the evolution in Fig. 9.9 is then rather transparent: in one step of small-x evolution at leading order in α_s only one gluon can be emitted. The gluon can be "real", splitting the original 10 dipole into two dipoles, 12 and 20 by the time the system crosses the shock wave, as shown in the first diagram on the right of Fig. 9.9. Alternatively, the gluon can be virtual, and be both emitted and absorbed either to the left or to the right of the shock wave, as shown in the last two diagrams in Fig. 9.9 respectively. Virtual gluons leave the original dipole 10 to interact with the target shock wave.

An explicit calculation of the diagrams pictured in Fig. 9.9 leads to the following evolution equation for the dipole S-matrix:

$$\partial_Y S(\vec{x}_{1\perp}, \vec{x}_{0\perp}, Y) = \frac{\alpha_s N_c}{2\pi^2} \int d^2 x_2 \, \frac{x_{10}^2}{x_{20}^2 \, x_{21}^2} \times [S(\vec{x}_{1\perp}, \vec{x}_{2\perp}, Y) S(\vec{x}_{2\perp}, \vec{x}_{0\perp}, Y) - S(\vec{x}_{1\perp}, \vec{x}_{0\perp}, Y)], \tag{9.33}$$

where $\partial_Y \equiv \partial/\partial Y$ and $x_{ij} = |\vec{x}_{i\perp} - \vec{x}_{j\perp}|$ are the transverse sizes of the dipoles. Equation (9.33) is the Balitsky–Kovchegov (BK) equation [Balitsky (1996, 1999); Kovchegov (1999, 2000)] for the dipole S-matrix. We are using the notation where the arguments of S are the positions of the quark and anti-quark in the dipole, along with $Y = \ln(1/x)$.

Employing Eq. (9.13), the BK equation (9.33) can be re-written in a somewhat more conventional form as

$$\partial_Y N(\vec{x}_{1\perp}, \vec{x}_{0\perp}, Y) = \frac{\alpha_s N_c}{2\pi^2} \int d^2 x_2 \, \frac{x_{10}^2}{x_{20}^2 \, x_{21}^2} \Big[N(\vec{x}_{1\perp}, \vec{x}_{2\perp}, Y) + N(\vec{x}_{2\perp}, \vec{x}_{0\perp}, Y) - N(\vec{x}_{1\perp}, \vec{x}_{0\perp}, Y) - N(\vec{x}_{1\perp}, \vec{x}_{2\perp}, Y) \, N(\vec{x}_{2\perp}, \vec{x}_{0\perp}, Y) \Big]. \tag{9.34}$$

As we mentioned above, the initial condition for the BK evolution (9.34) at $Y = 0$ can be given by the GGM formula (9.22). In this case, the solution of the BK equation would resum the powers of $\alpha_s N_c Y$ due to the evolution (9.34) along with the powers of $\alpha_s^2 \, A^{1/3}$ due to the GGM initial condition (9.22).

We now have the following formalism to calculate DIS structure functions of proton and nuclei at small x. One first has to solve the BK equation (9.34) using some initial conditions at $Y = 0$, preferably the GGM formula (9.22). Then, using the resulting dipole amplitude N in Eqs. (9.14), (9.5), and (9.9), while also employing Eqs. (9.6) and (9.7) one would arrive at predictions for the DIS structure functions F_2 and F_1. Non-perturbative QCD physics may enter through the initial conditions for the BK evolution. (Though they are suppressed when the saturation scale is sufficiently large.) In addition, the contribution of large impact parameters b in the Eq. (9.14) may suffer from large non-perturbative corrections [Kovner and Wiedemann (2002)]. However, for a sufficiently large nuclear target such peripheral contributions are suppressed by a power of $A^{1/3}$ and can be neglected.

If we neglect the quadratic in N term in Eq. (9.34) retaining only the linear terms on its right-hand side (for instance, in the $N \ll 1$ approximation), we would obtain the BFKL equation [Kuraev *et al.* (1977); Balitsky and Lipatov (1978)] in the dipole form. However, for larger values of the dipole scattering amplitude N, the quadratic term in Eq. (9.34) becomes important. Hence, the BK evolution generalizes the BFKL equation to the case of the scattering amplitude approaching the black-disk limit $N \lesssim 1$. The BK equation has the same structure (BFKL equation minus a quadratic term) as the Gribov-Levin-Ryskin (GLR) equation [Gribov *et al.* (1983)], which laid the foundation for the saturation physics. The double logarithmic limit of GLR equation arising at large values of Q^2 was independently studied in [Mueller and Qiu (1986)], leading to the GLR-MQ equation.The GLR and GLR-MQ equations were derived by summing the so-called BFKL pomeron "fan diagrams", with the quadratic term considered to be only the first saturation correction: the BK equation above shows that the quadratic term accounts for all saturation effects at large N_c and in the leading logarithmic approximation (LLA), in which we resum powers of $\alpha_s \ln(1/x)$.

An important conclusion we derive from the above derivation of the BK equation is that color dipoles are a useful degree of freedom in high energy QCD at large N_c. Instead of gluon degrees of freedom, one seems to be better off using dipoles. Below we discuss how this conclusion can be generalized to light-cone Wilson lines.

While Lev Lipatov was not actively involved in the saturation research, he, together with collaborators, re-derived the BK equation in [Bartels *et al.* (2005)], establishing an important connection between the s-channel evolution involving the shock wave that we presented here and the traditional t-channel formalism used in the original derivation of the BFKL equation.

9.1.3 *Light cone Wilson line operators as new degrees of freedom; JIMWLK equation*

Let us revisit the above gluon/dipole cascade resummation and try to avoid taking the large N_c limit. Physically very little would change in the above picture. Indeed, instead of a dipole cascade we would have a gluon cascade. However, just like at

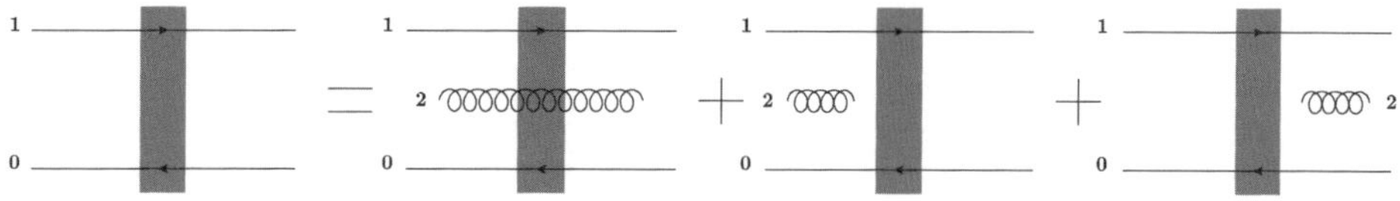

Fig. 9.10 Diagrammatic illustration of one step of the dipole evolution at small x beyond the large-N_c limit.

large N_c, the harder gluons would be longer-lived than the softer gluons. Hence the latter can be absorbed in the shock wave. While dipoles are no longer the right degrees of freedom, one can take a color dipole as a starting point and imagine one step of evolution for the dipole in the shock wave background without taking the large-N_c limit. This is depicted in Fig. 9.10 which looks somewhat similar to Fig. 9.9, except the gluon line now is not replaced by the quark an anti-quark lines. Again, disconnected gluon lines imply all possible connections of the gluon 2 to the quark 1 and the anti-quark 0.

Another similarity between the large-N_c and any-N_c evolution is that the emission of softer gluons does not change the transverse positions of the harder gluons or quarks. Hence, the interaction of quarks and gluons with the shock wave can be described by Wilson lines. Define the adjoint light-cone Wilson line by

$$U_{\vec{x}_\perp}[b^-, a^-] = \mathrm{P}\exp\left\{ ig \int_{a^-}^{b^-} dx^- \, \mathcal{A}^+(x^+ = 0, x^-, \vec{x}_\perp) \right\} \tag{9.35}$$

with $\mathcal{A}^\mu = T^a \, A^{a\mu}$ and T^a the generators of SU(N_c) in the adjoint representation. Similar to the case of the fundamental Wilson lines (9.25), the infinite adjoint light-cone Wilson line is denoted by

$$U_{\vec{x}_\perp} \equiv U_{\vec{x}_\perp}[+\infty, -\infty]. \tag{9.36}$$

In terms of Wilson lines, one step of a dipole evolution depicted in Fig. 9.10 leads to the following equation:

$$\partial_Y \left\langle \frac{1}{N_c} \mathrm{tr}\left[V_{\vec{x}_{1\perp}} V^\dagger_{\vec{x}_{0\perp}} \right] \right\rangle_Y = \frac{\alpha_s}{\pi^2} \int d^2x_2 \, \frac{x_{10}^2}{x_{20}^2 \, x_{21}^2} \left[\left\langle \frac{1}{N_c} \mathrm{tr}\left[t^b \, V_{\vec{x}_{1\perp}} \, t^a \, V^\dagger_{\vec{x}_{0\perp}} \right] U^{ba}_{\vec{x}_{2\perp}} \right\rangle_Y - \left\langle \frac{C_F}{N_c} \mathrm{tr}\left[V_{\vec{x}_{1\perp}} V^\dagger_{\vec{x}_{0\perp}} \right] \right\rangle_Y \right]. \tag{9.37}$$

Here we have explicitly shown the indices a, b of the adjoint Wilson line accounting for the gluon 2 interaction with the shock wave. The rapidity dependence of the Wilson line correlators can be introduced by tilting the Wilson lines contours away from the light cone. It is denoted by the subscript Y on the angle brackets.

Equation (9.37) captures the difficulty one has when trying to generalize the BK evolution to all N_c: the first correlator on its right-hand side is not the same as the dipole from its left-hand side (cf. Eq. (9.26)). It is a correlator describing the interaction of a quark, anti-quark and a gluon with the shock wave, as depicted in the first diagram on the right of Fig. 9.10. Hence, Eq. (9.37) is not a closed equation that one can solve: instead, it relates different correlators of light-cone Wilson lines to each other. If one tries to write an evolution equation for the new correlator,

$$\left\langle \frac{1}{N_c} \operatorname{tr}\left[t^b\, V_{\vec{x}_{1\perp}}\, t^a\, V^\dagger_{\vec{x}_{0\perp}}\right] U^{ba}_{\vec{x}_{2\perp}} \right\rangle_Y , \tag{9.38}$$

one would generate correlators with more gluon Wilson lines on that equation's right-hand side, again obtaining an equation which does not close, even if combined with Eq. (9.37). This way one generates an infinite hierarchy of equations for correlators with progressively higher numbers of Wilson lines. This hierarchy is known as the Balitsky hierarchy [Balitsky (1996)] and Eq. (9.37) is the first equation in this hierarchy.

Note that one can use

$$U^{ba}_{\vec{x}_\perp} = 2\operatorname{tr}\left[t^a\, V^\dagger_{\vec{x}_\perp}\, t^b V_{\vec{x}_\perp}\right] \tag{9.39}$$

along with the Fierz identity to rewrite Eq. (9.37) as

$$\begin{aligned} \partial_Y \left\langle \frac{1}{N_c} \operatorname{tr}\left[V_{\vec{x}_{1\perp}}\, V^\dagger_{\vec{x}_{0\perp}}\right]\right\rangle_Y &= \frac{\alpha_s\, N_c}{2\,\pi^2} \int d^2x_2\, \frac{x_{10}^2}{x_{20}^2\, x_{21}^2} \\ &\quad \times \left[\left\langle \frac{1}{N_c} \operatorname{tr}\left[V_{\vec{x}_{1\perp}}\, V^\dagger_{\vec{x}_{2\perp}}\right] \frac{1}{N_c} \operatorname{tr}\left[V_{\vec{x}_{2\perp}}\, V^\dagger_{\vec{x}_{0\perp}}\right]\right\rangle_Y \right. \\ &\quad \left. - \left\langle \frac{1}{N_c} \operatorname{tr}\left[V_{\vec{x}_{1\perp}}\, V^\dagger_{\vec{x}_{0\perp}}\right]\right\rangle_Y\right]. \end{aligned} \tag{9.40}$$

Even in this form, this is not a closed equation. At large N_c and for large nuclear target one writes

$$\begin{aligned} &\left\langle \operatorname{tr}\left[V_{\vec{x}_{1\perp}}\, V^\dagger_{\vec{x}_{2\perp}}\right] \operatorname{tr}\left[V_{\vec{x}_{2\perp}}\, V^\dagger_{\vec{x}_{0\perp}}\right]\right\rangle_Y \\ &= \left\langle \operatorname{tr}\left[V_{\vec{x}_{1\perp}}\, V^\dagger_{\vec{x}_{2\perp}}\right]\right\rangle_Y \left\langle \operatorname{tr}\left[V_{\vec{x}_{2\perp}}\, V^\dagger_{\vec{x}_{0\perp}}\right]\right\rangle_Y + \mathcal{O}\left(\frac{1}{N_c^2}, \frac{1}{A^{1/3}}\right) \end{aligned} \tag{9.41}$$

and the BK equation (9.33) is recovered. Outside the large-N_c limit the equation (9.40) is still hard to simplify.

It appears that to generalize the BK equation to the case of all N_c one needs to solve an infinite hierarchy of coupled integro-differential equations. The degrees of freedom in this hierarchy are the fundamental and adjoint infinite light-cone Wilson lines $V_{\vec{x}_\perp}$ and $U_{\vec{x}_\perp}$. They replace the color dipoles as the right degrees of freedom for an arbitrary N_c.

An alternative way to generate any-N_c small-x evolution is to think of energy dependence of the Wilson line correlators as coming from some rapidity-dependent weight functional $W_Y[\alpha]$ needed to obtain expectation values of operators by using the functional averaging

$$\langle \hat{O}_\alpha \rangle_Y = \int \mathcal{D}\alpha \, \hat{O}_\alpha \, W_Y[\alpha] \tag{9.42}$$

with $\hat{O}_\alpha$ an arbitrary operator. Here

$$\alpha(x^-, \vec{x}_\perp) \equiv A^+(x^+ = 0, x^-, \vec{x}_\perp) \tag{9.43}$$

and the functional is normalized to one,

$$\int \mathcal{D}\alpha \, W_Y[\alpha] = 1. \tag{9.44}$$

The rapidity dependence of the functional $W_Y[\alpha]$ needs to be obtained from an evolution equation, which one has to construct for it. Following [Mueller (2001)] one can take some test operator $\hat{O}_\alpha$ made out of infinite light-cone Wilson lines (say, $\hat{O}_{\vec{x}_{1\perp}, \vec{x}_{0\perp}} = V_{\vec{x}_{1\perp}} \otimes V^\dagger_{\vec{x}_{0\perp}}$, a product of two Wilson lines without a color trace) and construct its evolution, obtaining an equation of the type

$$\partial_Y \, \langle \hat{O}_\alpha \rangle_Y = \langle \mathcal{K}_\alpha \otimes \hat{O}_\alpha \rangle_Y = \int \mathcal{D}\alpha \, \left[\mathcal{K}_\alpha \otimes \hat{O}_\alpha \right] \, W_Y[\alpha], \tag{9.45}$$

with some kernel $\mathcal{K}_\alpha$, which is a differential in α operator whose action on the operator $\hat{O}_\alpha$ is denoted by $\otimes$. At the same time, differentiating Eq. (9.42) with respect to Y we get

$$\partial_Y \, \langle \hat{O}_\alpha \rangle_Y = \int \mathcal{D}\alpha \, \hat{O}_\alpha \, \partial_Y \, W_Y[\alpha]. \tag{9.46}$$

Integrating Eq. (9.45) by parts on its right-hand side we arrive at the evolution equation for $W_Y[\alpha]$,

$$\partial_Y \, W_Y[\alpha] = \mathcal{K}_\alpha \otimes W_Y[\alpha]. \tag{9.47}$$

(We assume that the form of the kernel $\mathcal{K}_\alpha$ does not change in the integration by parts, which is indeed the case.)

An explicit calculation of the kernel yields (for details see e.g. [Mueller (2001); Kovchegov and Levin (2012)])

$$\mathcal{K}_\alpha = \frac{\alpha_s}{\pi^2} \int d^2x_\perp \, d^2y_\perp \, d^2w_\perp \, \frac{(\vec{x}_\perp - \vec{w}_\perp) \cdot (\vec{y}_\perp - \vec{w}_\perp)}{|\vec{x}_\perp - \vec{w}_\perp|^2 \, |\vec{y}_\perp - \vec{w}_\perp|^2} \left(U_{\vec{w}_\perp} - \frac{U_{\vec{x}_\perp} + U_{\vec{y}_\perp}}{2} \right)^{ba}$$
$$\times \frac{(ig)^{-2} \, \delta^2}{\delta \, \alpha^a(x^- < 0, \vec{x}_\perp) \, \delta \, \alpha^b(y^- > 0, \vec{y}_\perp)}. \tag{9.48}$$

(This kernel is given in the form recently written in [Cougoulic and Kovchegov (2019)].) Equation (9.47) with the kernel from Eq. (9.48) is the Jalilian-Marian–Iancu–McLerran–Weigert–Leonidov–Kovner (JIMWLK) [Jalilian-Marian *et al.* (1998b,a); Weigert (2002); Iancu *et al.* (2001b,a); Ferreiro *et al.* (2002)] functional evolution equation. The initial condition for JIMWLK evolution is given by the Gaussian functional of the MV model. Since JIMWLK is a functional differential equation, its solution is hard to find analytically, but is possible to construct numerically [Weigert (2002); Rummukainen and Weigert (2004); Schlichting and Schenke (2014)].

9.1.4 *Unitarization of the BFKL evolution and saturation*

No analytic solution of the BK evolution equation (9.34) for all dipole sizes and rapidities exists at this time. Instead, there exist approximate analytic solutions which apply in various subset of the $(x_{10}, b_\perp, Y)$ phase space for $N(\vec{x}_{10}, \vec{b}_\perp, Y)$. There are also numerical solutions. A pedagogical presentation of the analytical attempts to solve the BK equation, along with a subset of numerical solutions, can be found in [Kovchegov and Levin (2012)].

Here we want to convey the essential physical properties of the BK solution. To do so, let us consider the following toy version of the BK equation (9.34),

$$\partial_Y N = \alpha_s N - \alpha_s N^2 \tag{9.49}$$

with $N(Y = 0) = N_0 \ll 1$ as the initial condition. In arriving at Eq. (9.49) we have essentially discarded the integral kernel of the full BK equation (9.34), suppressing the transverse coordinate dependence in N. The solution of Eq. (9.49) is straightforward to find. It reads

$$N(Y) = \frac{N_0\, e^{\alpha_s Y}}{1 + N_0\, (e^{\alpha_s Y} - 1)}. \tag{9.50}$$

We see that at moderately large rapidity, when $\alpha_s Y \gtrsim 1$, the equation (9.50) can be linearized in N_0, yielding

$$N(Y = \ln(1/x)) \approx N_0\, e^{\alpha_s Y} \propto e^{\alpha_s Y} \propto \left(\frac{1}{x}\right)^{\alpha_s}. \tag{9.51}$$

Similar power-law growth of the amplitudes and cross sections follows from the BFKL equation. Indeed, linearizing Eq. (9.34) by discarding the quadratic in N term on the right, and solving the resulting linear (dipole BFKL) equation, one obtains

$$N(\vec{x}_{10}, \vec{b}_\perp, Y = \ln(1/x)) \propto \left(\frac{1}{x}\right)^{\alpha_P - 1} \tag{9.52}$$

with

$$\alpha_P - 1 = \frac{4\alpha_s N_c}{\pi} \ln 2. \tag{9.53}$$

This is the standard solution of the leading-order BFKL equation, with Eq. (9.53) giving the so-called BFKL pomeron intercept. Employing Eq. (9.14) we see that

Eq. (9.52) leads to

$$\sigma_{tot}^{q\bar{q}A}(\vec{x}_\perp, Y = \ln(1/x)) \propto \left(\frac{1}{x}\right)^{\alpha_P - 1} \propto s^{\alpha_P - 1}. \tag{9.54}$$

Clearly, such a power-law growth of the cross section with the center-of-mass energy squared would violate both the black disk limit (9.16) and the Froissart-Martin bound (9.17) at a sufficiently high s. This violation of unitarity has been a problem for BFKL evolution: while the BFKL equation was a great breakthrough in our understanding of high-energy QCD, it was also clear that the equation violated unitarity, and, therefore, something else had to come in at higher energies and modify BFKL equation to make it preserve unitarity.

Let us return to the full solution (9.50) of the toy BK equation (9.49). We see that $N(Y)$ given by this solution preserves unitarity. That is, $N(Y) \leq 1$ for all values of Y. Moreover, $N(Y)$ asymptotically approaches 1 at very large rapidities,

$$N(Y) \to 1 \quad \text{as} \quad Y \to \infty. \tag{9.55}$$

We see that the toy BK evolution preserves the black-disk limit. Moreover, the dipole scattering amplitude approaches the black disk limit for very large Y, corresponding to very high energies. Thus, our toy BK equation (9.49) does not have the unitarity violation problem that BFKL equation had.

The same conclusion can be applied to the full BK equation (9.34). In Fig. 9.11 we plot a numerical solution of the leading-order BK equation versus the dipole size $x_\perp$. This plot and the underlying solution were constructed by Javier Albacete for [Kovchegov and Levin (2012)] (see also [Albacete *et al.* (2005); Albacete and Kovchegov (2007); Albacete *et al.* (2009, 2011)]). The b-dependence has been suppressed in the solution, such that $N(x_\perp, b_\perp, Y) \approx N(x_\perp, Y)$, which is valid for a large nucleus target. The initial condition for the evolution corresponding to $Y = 0$ is given by a GGM-type formula and is plotted by the dashed line in Fig. 9.11 (cf. Fig. 9.4 above). The solid curves are given by the BK solution at four different values of $\alpha_s Y$ listed in the legend of Fig. 9.11. There are two ways of thinking about this plot: first of all, one can fix $x_\perp$, and see that with increasing Y the amplitude N raises as well, but never exceeds 1, in qualitative agreement with our toy model (9.50). Alternatively, one can think of evolution in Fig. 9.11 moving the initial condition (dashed) curve to the left, in the process somewhat modifying its shape. Again, one always has $N \leq 1$ and the black disk limit is not violated. We see that the BK equation leads to unitarization of the BFKL evolution, resolving the problem of unitarity violation by the latter.

Let us define the saturation scale $Q_s(Y)$ like in Fig. 9.4 as a transition point between the region where $N(x_\perp, Y)$ grows with $x_\perp$ and the region where this growth is tamed, that is, by requiring that $N(x_\perp = 1/Q_s(Y), Y) = \text{const}$, with some order-one constant. From Fig. 9.11 one can clearly see that $1/Q_s(Y)$ decreases with $\alpha_s Y$. This is why the curves in Fig. 9.11 look like they are moving to the left as Y increases. Therefore, $Q_s(Y)$ increases with $\alpha_s Y$. A more detailed analytic calculation leads

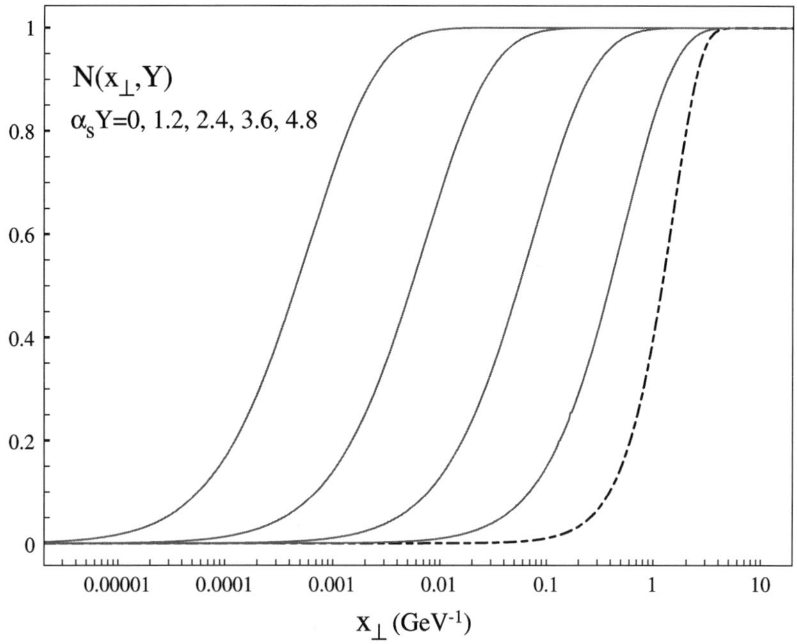

Fig. 9.11 Solution of the BK equation plotted as a function of the dipole size $x_\perp$ for several values of $\alpha_s Y$ specified in the legend. The GGM-type initial condition is given by the dashed line.

to [Gribov *et al.* (1983); Iancu *et al.* (2002); Mueller and Triantafyllopoulos (2002)]

$$Q_s(Y) \approx Q_s(0)\, e^{2.44 \frac{\alpha_s N_c}{\pi} Y}, \tag{9.56}$$

where $Q_s(0)$ is the GGM saturation scale (9.21). The saturation scale extracted by Albacete from the numerical solution of the BK equation is shown in Fig. 9.12, and also exhibits growth with rapidity Y.

We see that the non-linear small-x evolution (9.34) makes the saturation scale larger, driving it deeper into the perturbative region. Combining Eqs. (9.21) and (9.56) we see that

$$Q_s^2(x) \propto A^{1/3} \left(\frac{1}{x}\right)^{4.88 \frac{\alpha_s N_c}{\pi}}. \tag{9.57}$$

We conclude that there are two ways of obtaining a large saturation scale, justifying perturbative approach to small-x physics: one can either use a large nucleus as a target, thus increasing the atomic number A. Alternatively (or simultaneously) one can increase the center-of-mass energy, decreasing x and driving Q_s up. The saturation region, defined by $1/Q_s(Y) < x_\perp < 1/\Lambda$, also increases and becomes more prominent with increasing A and decreasing x.

We summarize the above discussion by the plot in Fig. 9.13, depicting various regimes of high-energy QCD in the $(\ln Q^2, \ln(1/x))$ plane. First of all

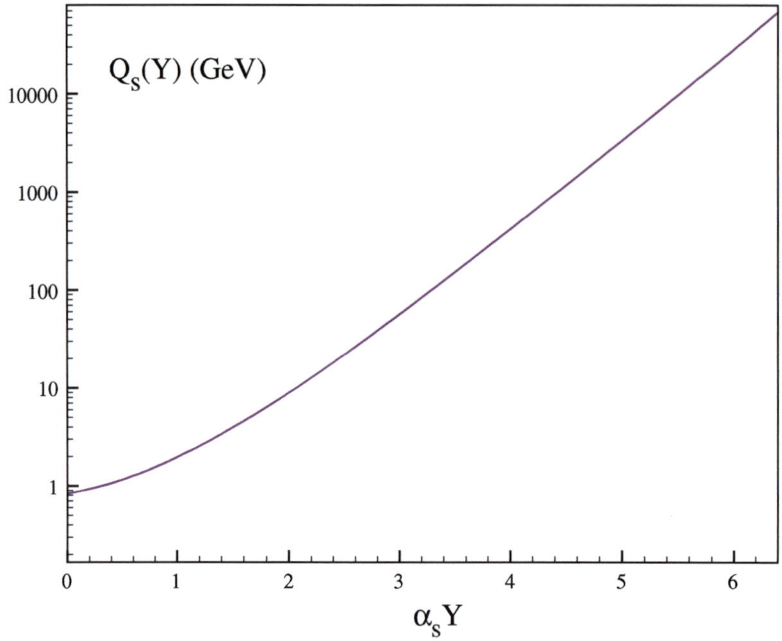

Fig. 9.12 Saturation scale as a function of $\alpha_s Y$ given by the numerical solution of the BK equation from Fig. 9.11.

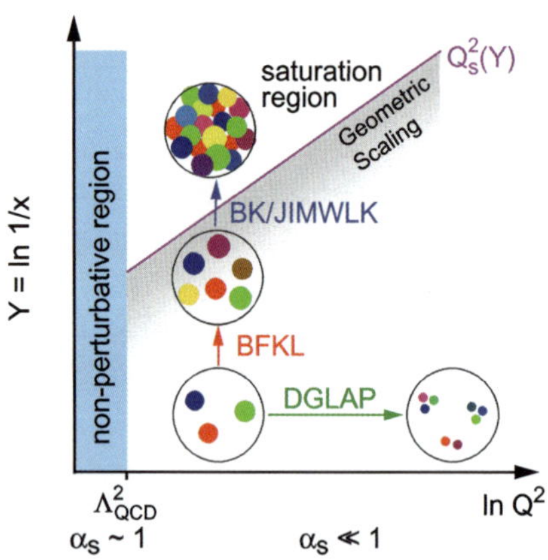

Fig. 9.13 High energy QCD summarized in the $(\ln Q^2, \ln(1/x))$ plane.

we exclude the non-perturbative region with $Q^2 \lesssim \Lambda_{QCD}^2$ as the region where our perturbative methods do not apply. We concentrate on the $Q^2 > \Lambda_{QCD}^2$ perturbative QCD region where $\alpha_s \ll 1$. If one is interested in evolution in Q^2, it is accomplished by the famous Dokshitzer-Gribov-Lipatov-Altarelli-Parisi (DGLAP) evolution equation [Gribov and Lipatov (1972); Altarelli and Parisi (1977); Dokshitzer (1977)]. Description of this equation is outside our scope here but, for completeness, it is shown in Fig. 9.13; this is another important equation which bears Lev Lipatov's name.

To study the high energy asymptotics of QCD one needs small-x evolution equations. The BFKL equation allows one to evolve the dipole amplitude or gluon distribution function toward small-x. However, at very small x, near the saturation region, it has to be replaced by the non-linear BK and JIMWLK evolution equations, as shown in Fig. 9.13. These latter equations are valid both inside and outside of the saturation region, describing non-linear QCD interactions at high energy.

We refer the readers interested in the phenomenology of saturation physics to the reviews [Gribov *et al.* (1983); Iancu and Venugopalan (2003); Weigert (2005); Jalilian-Marian and Kovchegov (2006); Gelis *et al.* (2010); Albacete and Marquet (2014)] and the book [Kovchegov and Levin (2012)]. Let us note that a significant part of the physics program of the Electron-Ion Collider (EIC) [Accardi *et al.* (2016)] proposed in the US will be dedicated to the search for and study of the saturation physics.

9.2 Spin at small x

Now let us apply the above formalism to the proton spin problem. For several decades it has been known that the measured spin of quarks and gluons inside the proton does not add up to 1/2 required by helicity sum rules [Jaffe and Manohar (1990); Ji (1997); Ji *et al.* (2012)]. This is known as the proton spin puzzle or the proton spin crisis (see [Accardi *et al.* (2016); Aschenauer *et al.* (2016)] and references therein for the current status of the spin puzzle).

One of the two commonly used helicity sum rules is the Jaffe-Manohar sum rule [Jaffe and Manohar (1990)]

$$S_q + L_q + S_G + L_G = \frac{1}{2}. \tag{9.58}$$

It contains the spin contributions carried by quarks and gluons in the proton

$$S_q(Q^2) = \frac{1}{2}\int_0^1 dx\, \Delta\Sigma(x, Q^2), \quad S_G(Q^2) = \int_0^1 dx\, \Delta G(x, Q^2), \tag{9.59}$$

along with the quark and gluon orbital angular momenta L_q and L_G. Here we have employed helicity parton distribution functions (hPDFs)

$$\Delta f(x, Q^2) \equiv f^+(x, Q^2) - f^-(x, Q^2), \tag{9.60}$$

where f^+ (f^-) denote the number density of partons with the same (opposite) helicity as the proton, and $f = u, \bar{u}, d, \bar{d}, \ldots, G$. In addition,

$$\Delta\Sigma(x, Q^2) = \left[\Delta u + \Delta\bar{u} + \Delta d + \Delta\bar{d} + \cdots\right](x, Q^2). \tag{9.61}$$

Since no experiment can measure helicity PDFs down to $x = 0$, any given experiment can measure the integrands in Eq. (9.59) only down to some x_{min} determined by the experiment's acceptance. Theory input appears to be needed to assess the amount of spin carried by quarks and gluons with $x \in [0, x_{min}]$. Thus, the small-x asymptotics of helicity PDFs needs to be under theoretical control in order to resolve the proton spin puzzle.

Below we will construct the small-x evolution equations which determine the small-x asymptotics of $\Delta\Sigma$ and ΔG. They are different from the BFKL, BK and JIMWLK evolution equations in several aspects. First of all, at small x the helicity distributions are sub-eikonal: they are suppressed by a power of x as compared to the unpolarized observables we explored above. Another difference, partially compensating for this suppression, is that the resummation parameter of the small-x evolution for helicity is double-logarithmic: the parameter is $\alpha_s \ln^2(1/x)$, instead of $\alpha_s \ln(1/x)$ resummed by the BFKL, BK and JIMWLK equations.

Lev Lipatov was the first to explore resummation of such double-logarithmic parameter, $\alpha_s \ln^2(1/x)$, in QCD. In [Kirschner and Lipatov (1983)], Kirschner and Lipatov were the first to resum double logarithms of energy for the flavor non-singlet structure function dominated by the QCD Reggeon exchange (see also [Gorshkov *et al.* (1968)] for similar resummation in QED). The resummation was accomplished using the so-called infrared evolution equations (IREE), which are a technique different from the one used in the original derivation of the BFKL equation and from the shock-wave formalism we described above. While Lipatov himself never applied the IREE technique to the evolution of helicity distributions, it was done by Bartels, Ermolaev and Ryskin in [Bartels *et al.* (1996b,a)]. Unfortunately, the small-x asymptotics of $\Delta\Sigma$ and ΔG we construct below using the shock wave approach differs from that found in [Bartels *et al.* (1996a)] by using IREE. Understanding the origin of this discrepancy is an open problem in the field at the time of writing.

The details of the calculations leading to the results we present below can be found in [Kovchegov *et al.* (2016, 2017a,c,d,b); Kovchegov and Sievert (2019); Cougoulic and Kovchegov (2019)].

9.2.1 *Polarized "Wilson lines" and dipoles*

Our analysis here will be parallel to the one above for the unpolarized BK and JIMWLK evolution. As one can show, at small x the quark helicity PDF can be written as [Kovchegov *et al.* (2016); Kovchegov and Sievert (2019)]

$$\Delta\Sigma(x, Q^2) = \frac{N_c N_f}{2\pi^3} \int_{\Lambda^2/s}^{1} \frac{dz}{z} \int_{\frac{1}{zs}}^{\frac{1}{zQ^2}} \frac{dx_{10}^2}{x_{10}^2} \, d^2 b_\perp \, G_{10}(zs), \tag{9.62}$$

where N_f is the number of quark flavors, Λ is an infrared (IR) cutoff, and the center of mass energy squared is $s \approx Q^2/x$. As before, the impact parameter is defined by $\vec{b}_\perp = (\vec{x}_{1\perp} + \vec{x}_{0\perp})/2$.

Equation (9.62) is the helicity analogue of Eqs. (9.9) and (9.14) above. It relates a physical observable of interest, the quark helicity distribution, to the amplitude of a quark–anti-quark dipole scattering on a target. However, in the case of Eq. (9.62), the dipole-target scattering has to depend on helicity. More precisely, we are interested in the term in the dipole-target forward scattering amplitude proportional to the product of the projectile and target helicities. The eikonal scattering considered above is independent of helicity. Hence, to obtain a non-zero contribution to Eq. (9.62) we have to introduce sub-eikonal interactions between the dipole and the longitudinally polarized target. This is formally accomplished by defining the so-called polarized dipole scattering amplitude $G_{10}(zs)$ as

$$\begin{aligned} &G_{10}(zs = \min\{z_1, z_0\}s) \\ &\quad \equiv \frac{1}{2N_c} \left\langle\!\!\left\langle \mathrm{T}\,\mathrm{tr}\left[V^{pol}_{\vec{x}_{1\perp}} V^{\dagger}_{\vec{x}_{0\perp}}\right] + \mathrm{T}\,\mathrm{tr}\left[V_{\vec{x}_{0\perp}} V^{pol\,\dagger}_{\vec{x}_{1\perp}}\right] \right\rangle\!\!\right\rangle (\min\{z_1, z_0\}s) \\ &\quad \equiv \frac{z_1 s}{2N_c} \left\langle \mathrm{T}\,\mathrm{tr}\left[V^{pol}_{\vec{x}_{1\perp}} V^{\dagger}_{\vec{x}_{0\perp}}\right] + \mathrm{T}\,\mathrm{tr}\left[V_{\vec{x}_{0\perp}} V^{pol\,\dagger}_{\vec{x}_{1\perp}}\right] \right\rangle, \end{aligned} \tag{9.63}$$

where the double-angle brackets are defined to scale out the center-of-mass energy $z_1 s$ between the polarized (anti-)quark and the target. Here the quark/anti-quark 1 carries momentum fraction z_1 of some projectile's light-cone (minus) momentum, while the anti-quark/quark 0 carries momentum fraction z_0. As usual, T denotes time-ordering. The polarized dipole amplitude is illustrated in Fig. 9.14. In the polarization-dependent scattering, one (anti-)quark line in the dipole carries polarization information. This is line 1 in our notation. This line interacts with the target in the polarization-dependent sub-eikonal way, which we describe by the so-called "polarized Wilson line" operator $V^{pol}_{\vec{x}_{1\perp}}$ in the left panel of Fig. 9.14 and by its conjugate in the right panel. This sub-eikonal helicity-dependent interaction is denoted by the box in Fig. 9.14; its structure will be clarified shortly. The unpolarized line 0 interacts with the shock wave target in the standard eikonal way described by the infinite light-cone Wilson line in Eq. (9.25).

To determine the form of the "polarized Wilson line" operator $V^{pol}_{\vec{x}_{1\perp}}$ we again use the analogue to the unpolarized case and consider scattering of a quark on a GGM type of a target, say a large nucleus with many nucleons. The difference now is that one of the nucleons has to be longitudinally polarized, while the interaction with this nucleon should be helicity-dependent, and, hence, sub-eikonal. The scattering amplitude giving the helicity-dependent contribution to the quark scattering on a shock wave target is shown in Fig. 9.15, where the black circles in the left panel denote sub-eikonal helicity-dependent parts of the quark-gluon interactions.

A detailed calculation of the "polarized Wilson line" operator in the fundamental representation can be found in [Kovchegov *et al.* (2017b); Kovchegov and Sievert

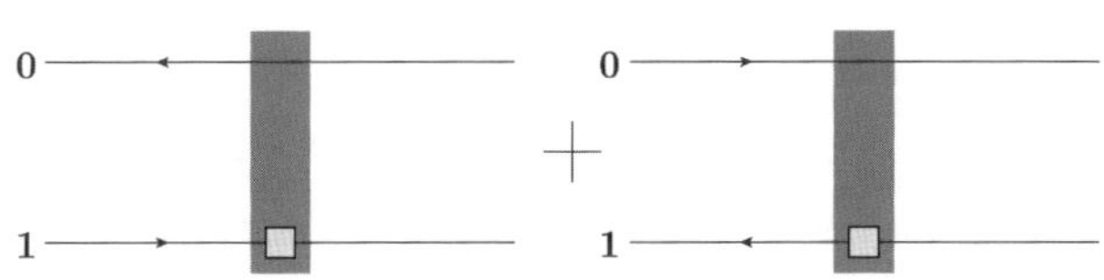

Fig. 9.14 Diagrammatic illustration of the polarized dipole amplitude defined in Eq. (9.63).

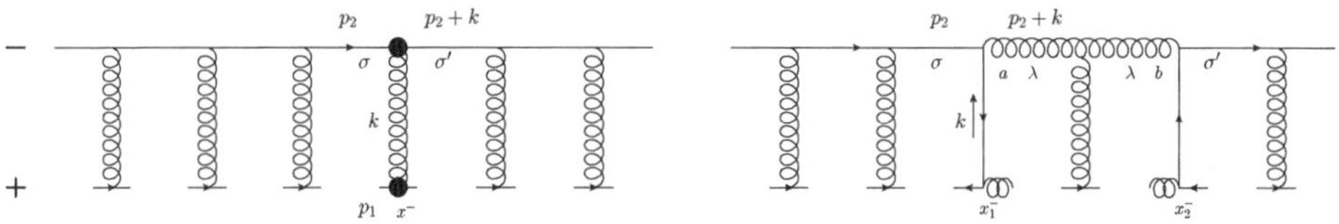

Fig. 9.15 Two contributions to the polarized fundamental Wilson line in a background field. Filled circles at the quark-gluon vertices denote the spin-dependent sub-eikonal scattering.

(2019)] (see also [Chirilli (2019)]) and yields

$$V_{\vec{x}_\perp}^{pol} = \frac{igp_1^+}{s} \int_{-\infty}^{\infty} dx^- \, V_{\vec{x}_\perp}[+\infty, x^-] \, F^{12}(x^-, \vec{x}_\perp) \, V_{\vec{x}_\perp}[x^-, -\infty]$$
$$- \frac{g^2 p_1^+}{s} \int_{-\infty}^{\infty} dx_1^- \int_{x_1^-}^{\infty} dx_2^- V_{\vec{x}_\perp}[+\infty, x_2^-] \, t^b \, \psi_\beta(x_2^-, \vec{x}_\perp)$$
$$\times U_{\vec{x}_\perp}^{ba}[x_2^-, x_1^-] \left[\frac{1}{2}\gamma^+ \gamma^5\right]_{\alpha\beta} \bar{\psi}_\alpha(x_1^-, \vec{x}_\perp) \, t^a \, V_{\vec{x}_\perp}[x_1^-, -\infty]. \tag{9.64}$$

While the target still generates the eikonal gluon field A^+ contributing to the Wilson lines in Eq. (9.64), at the sub-eikonal level it also generates a (helicity-dependent part of the) transverse gluon field $\vec{A}_\perp$ which gives the 12 component of the gluon field strength tensor F^{12} in the first term of Eq. (9.64). In addition, the target generates the (sub-eikonal) quark and anti-quark fields ψ_β and $\bar{\psi}_\alpha$ with α, β the Dirac spinor indices: these quark fields lead to the second term in Eq. (9.64). Here p_1^+ is the large momentum component of the target nucleons (see Fig. 9.15).

A similar calculation for a longitudinally polarized gluon scattering on a polarized target leads to the adjoint "polarized Wilson line" operator [Kovchegov and Sievert (2019)]

$$(U_{\vec{x}_\perp}^{pol})^{ab} = \frac{2i\, g\, p_1^+}{s} \int_{-\infty}^{+\infty} dx^- \, \left(U_{\vec{x}_\perp}[+\infty, x^-] \, \mathcal{F}^{12}(x^+ = 0, x^-, \vec{x}_\perp) \, U_{\vec{x}_\perp}[x^-, -\infty]\right)^{ab}$$
$$- \frac{g^2\, p_1^+}{s} \int_{-\infty}^{\infty} dx_1^- \int_{x_1^-}^{\infty} dx_2^- \, U_{\vec{x}_\perp}^{aa'}[+\infty, x_2^-] \, \bar{\psi}(x_2^-, \vec{x}_\perp) t^{a'}$$
$$\times V_{\vec{x}_\perp}[x_2^-, x_1^-] \, \frac{1}{2}\gamma^+ \gamma_5 \, t^{b'} \, \psi(x_1^-, \vec{x}_\perp) \, U_{\vec{x}_\perp}^{b'b}[x_1^-, -\infty] - \text{c.c.}. \tag{9.65}$$

Now $\mathcal{F}^{12}$ is the gluon field strength tensor in the adjoint representation.

Apart from some formal complexity, the overall picture of helicity-dependent scattering is very similar to the unpolarized case considered above. We need to construct and solve the evolution equation for the polarized dipole amplitude $G_{10}(zs)$. Equation (9.62) would then give us the flavor-singlet quark helicity distribution. Similar calculation, with a slightly different polarized dipole amplitude, would give us the gluon helicity distribution ΔG at small x [Kovchegov *et al.* (2017b)]. For the quark flavor non-singlet helicity PDF, one needs a different polarized dipole definition from Eq. (9.63), which, in turn, satisfies different evolution equations from what we construct below, see [Kovchegov *et al.* (2017a); Kovchegov and Sievert (2019)] along with [Bartels *et al.* (1996b)]: however, the flavor non-singlet distribution, while an interesting quantity in its own right, is not a part of the helicity sum rules such as the one in Eq. (9.58) and will not be discussed here.

9.2.2 *Small-x evolution for polarized dipoles*

Our next step is to construct an evolution equation for the polarized dipole amplitude $G_{10}(zs)$ in Eq. (9.63). In the operator language for the background field method, like that employed in [Balitsky (1996)] for the derivation of the unpolarized small-x evolution equations, one separates the gluon field into the classical A^μ_{cl} and quantum a^μ parts by writing $A^\mu = A^\mu_{cl} + a^\mu$, and integrates out the quantum field keeping only the terms up to the required order in the coupling constant α_s (ditto for the quark fields). Equivalently, one can think of this evolution in terms of Feynman diagrams, similar to the unpolarized case discussed above. Because the expression for the evolution of both terms in the polarized dipole amplitude in Fig. 9.14 is somewhat lengthy, we show the (flavor-singlet) evolution only for one of those terms in Fig. 9.16. In the first row of Fig. 9.16, shaded circles denote the sub-eikonal helicity-dependent gluon emissions, just like in Fig. 9.15. These vertices originate in the F^{12} contribution in Eq. (9.64). The second row of Fig. 9.16 contains emission of the soft quark, originating in the second term of Eq. (9.64). The remaining rows three to five in Fig. 9.16 correspond to the standard eikonal unpolarized BK/JIMWLK evolution described above.

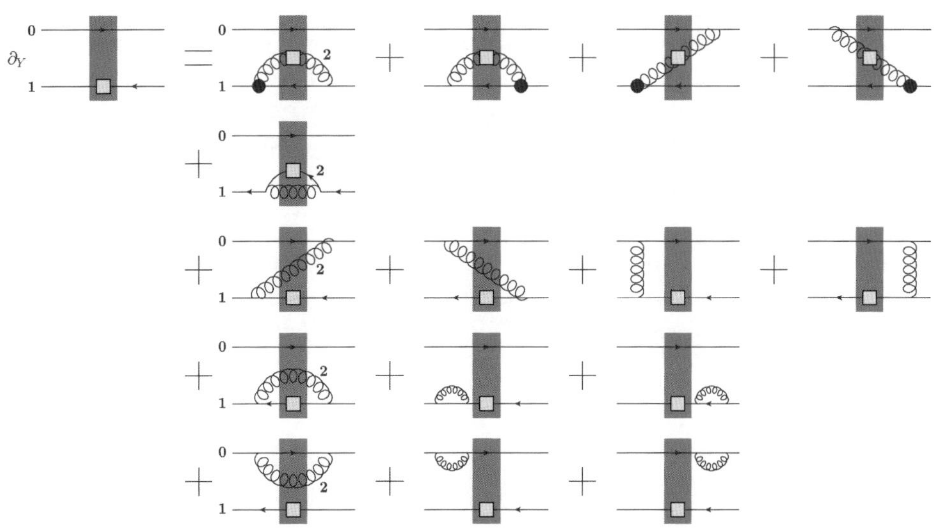

Fig. 9.16 One step of the polarized fundamental dipole small-x evolution. Black circles represent spin-dependent (sub-eikonal) soft gluon emission vertices.

The diagrams in Fig. 9.16 lead to the following evolution equation [Kovchegov *et al.* (2016)]:

$$
\begin{aligned}
&\frac{1}{N_c} \left\langle\!\!\left\langle \mathrm{tr}\left[V_{\vec{x}_{0\perp}} V_{\vec{x}_{1\perp}}^{pol\,\dagger}\right]\right\rangle\!\!\right\rangle (zs) \\
&= \frac{1}{N_c} \left\langle\!\!\left\langle \mathrm{tr}\left[V_{\vec{x}_{0\perp}} V_{\vec{x}_{1\perp}}^{pol\,\dagger}\right]\right\rangle\!\!\right\rangle_0 (zs) + \frac{\alpha_s}{2\pi} \int_{\Lambda^2/s}^{z} \frac{dz_2}{z_2} \int_{1/(z_2 s)} \frac{dx_{21}^2}{x_{21}^2} \\
&\times \Bigg\{ \theta(x_{10} - x_{21}) \frac{2}{N_c} \left\langle\!\!\left\langle \mathrm{tr}\left[t^b V_{\vec{x}_{0\perp}} t^a V_{\vec{x}_{1\perp}}^{\dagger}\right] U_{\vec{x}_{2\perp}}^{pol\,ba}\right\rangle\!\!\right\rangle (z_2) \\
&+ \theta(x_{10}^2 z - x_{21}^2 z_2) \frac{1}{N_c} \left\langle\!\!\left\langle \mathrm{tr}\left[t^b V_{\vec{x}_{0\perp}} t^a V_{\vec{x}_{2\perp}}^{pol\,\dagger}\right] U_{\vec{x}_{1\perp}}^{ba}\right\rangle\!\!\right\rangle (z_2) \\
&+ \theta(x_{10} - x_{21}) \frac{1}{N_c} \left[\left\langle\!\!\left\langle \mathrm{tr}\left[V_{\vec{x}_{0\perp}} V_{\vec{x}_{2\perp}}^{\dagger}\right] \mathrm{tr}\left[V_{\vec{x}_{2\perp}} V_{\vec{x}_{1\perp}}^{pol\,\dagger}\right]\right\rangle\!\!\right\rangle (z_2) \right. \\
&\left. - N_c \left\langle\!\!\left\langle \mathrm{tr}\left[V_{\vec{x}_{0\perp}} V_{\vec{x}_{1\perp}}^{pol\,\dagger}\right]\right\rangle\!\!\right\rangle (z_2) \right] \Bigg\}.
\end{aligned}
\tag{9.66}
$$

Lines 2 and 3 of Eq. (9.66) correspond to rows 1 and 2 in Fig. 9.16 respectively, while lines 4 and 5 in Eq. (9.66) are given by the unpolarized small-x evolution in rows 3-5 of Fig. 9.16 (cf. Eq. (9.40)). (Here z_2 is the longitudinal momentum fraction carried by the soft parton 2, which can be a gluon or a quark.) The evolution equation (9.66) is written in the integral form, with the $\langle\!\langle \ldots \rangle\!\rangle_0$ inhomogeneous term given by the initial conditions for the evolution. (The BK equation can also be written in an integral form [Kovchegov (1999)].)

As we mentioned above, there is an important difference between the helicity evolution (9.66) and the BK/JIMWLK equations. Let us recall the lifetime ordering (9.32) of the soft gluon emissions in the unpolarized evolution. The same lifetime ordering applies here to spin evolution. In the unpolarized evolution case, the lifetime ordering was included by simply imposing the longitudinal momentum ordering (9.30), or, equivalently, the longitudinal momentum fraction ordering

$$z_0, z_1 \gg z_2 \gg z_3 \gg \ldots . \tag{9.67}$$

Equations (9.32) and (9.67) are equivalent only if the partons' transverse momenta are comparable. (Note that $z_i = k_i^- / p_2^-$, where p_2^- is the large light-cone momentum of the projectile.) In general, the condition (9.32) translates into

$$\min\{z_1, z_0\}\, x_{10}^2 \gg z_2\, x_{21}^2 \gg z_3\, x_{32}^2 \gg \ldots \tag{9.68}$$

in the transverse coordinate space. (At this point we have randomly picked the distance x_{21} over x_{20} and picked x_{32} over other transverse distances at that step of evolution.)

The unpolarized evolution equation (9.40) is both the IR and ultra-violet (UV) finite: the expression in the square brackets on the right goes to zero when

$\vec{x}_{2\perp} \to \vec{x}_{1\perp}$ and $\vec{x}_{2\perp} \to \vec{x}_{0\perp}$ regulating the divergence in the integral kernel in those limits. The same integral kernel goes to zero reasonably fast when $\vec{x}_{2\perp} \to \infty$, ensuring the IR convergence of the integral. Since the x_2 integral in Eq. (9.40) is thus convergent, the typical "daughter" dipole sizes are not much larger or smaller than the "parent" dipole size, that is, $x_{21} \approx x_{20} \approx x_{10}$. This way, Eq. (9.67) automatically leads to Eq. (9.68). Now, let us consider the evolution in Eq. (9.66). The x_2 integral in it is no longer UV convergent. For instance, analyzing the terms in the last two lines of Eq. (9.66) we see that (unlike the unpolarized case) they do not cancel in the $\vec{x}_{2\perp} \to \vec{x}_{1\perp}$ limit. This UV divergence is regularized by the highest energy scale in the problem, the effective center-of-mass energy squared $z_2 s$ between the soft parton and the target: that is why we have the $1/(z_2 s)$ lower limit on the x_{21} integral in Eq. (9.66). Moreover, the divergence in the $x_{21}^2 \to 1/(z_2 s)$ limit is logarithmic: hence, in the case of helicity evolution the transverse integral generates another logarithm of energy. We have simplified the integral kernel in arriving at Eq. (9.66) to extract this logarithm while discarding non-logarithmic transverse integral contributions. That is why the x_{21} integral is logarithmic in Eq. (9.66). We conclude that $x_{21} \ll x_{20} \approx x_{10}$ in the last two terms on the right of Eq. (9.66): this condition is enforced by $\theta(x_{10} - x_{21})$ multiplying those terms. In addition, consider the second term in the curly brackets of Eq. (9.66). The x_{21} integral in its kernel is UV and IR divergent. The IR divergence means that one may have $x_{21} \approx x_{20} \gg x_{10}$, such that the condition (9.67) no longer automatically leads to the condition (9.68). Hence, the condition (9.68) has to be imposed explicitly in Eq. (9.66). This is what gives rise to $\theta(x_{10}^2 z - x_{21}^2 z_2)$ in Eq. (9.66). This theta-function in fact multiplies all terms on the right of Eq. (9.66): however, in some terms it has been replaced by a stronger condition, $\theta(x_{10} - x_{21})$, resulting from the simplification of the integral kernel to extract the second logarithm of energy, as we have just described.

Let us pose here to describe the resummation parameter. At each step of small-x helicity evolution we emit a soft parton, generating a power of the coupling α_s. Similar to the unpolarized evolution, the integral over the longitudinal momentum fraction z_2 of this parton gives a logarithm of energy, or, equivalently, of Bjorken x. In addition, as we have just discussed, and unlike the unpolarized evolution, the transverse momentum/position integral gives another $\ln(1/x)$ for helicity evolution. The resulting resummation parameter for helicity evolution is $\alpha_s \ln^2(1/x)$: the powers of this parameter are resummed by the equation (9.66). This parameter does not exist for the unpolarized evolution, which resums powers of $\alpha_s \ln(1/x)$ at leading order, which is a smaller parameter at small x. Hence, the resummation parameter for helicity evolution is larger than the one for BFKL, BK and JIMWLK equations. At the same time, helicity evolution is sub-eikonal, and helicity distributions are suppressed by an extra power of x as compared to the unpolarized parton distributions at small x. As we mentioned above, Lev Lipatov, together with Kirschner, were the first to discover this double-logarithmic $\alpha_s \ln^2(1/x)$ resummation parameter in [Kirschner and Lipatov (1983)].

Returning to Eq. (9.66) we notice that it is not closed, just like the first equation in Balitsky hierarchy (9.40). Once again, the equation closes if we take the large-N_c limit, and, in addition, in the large-N_c&N_f limit (with N_f the number of quark flavors) [Kovchegov *et al.* (2016)]. In the large-N_c limit we obtain (adding the contribution of the other term in Fig. 9.14) [Kovchegov *et al.* (2016, 2017a)]

$$G_{10}(zs) = G_{10}^{(0)}(zs) + \frac{\alpha_s N_c}{2\pi} \int_{\frac{1}{s x_{10}^2}}^{z} \frac{dz'}{z'} \int_{\frac{1}{z's}}^{x_{10}^2} \frac{dx_{21}^2}{x_{21}^2} \left[\Gamma_{10,21}(z's) + 3\, G_{21}(z's)\right], \tag{9.69a}$$

$$\Gamma_{10,21}(z's) = \Gamma_{10,21}^{(0)}(z's) + \frac{\alpha_s N_c}{2\pi} \int_{\min\{\Lambda^2, \frac{1}{x_{10}^2}\}/s}^{z'} \frac{dz''}{z''} \int_{\frac{1}{z''s}}^{\min\{x_{10}^2, x_{21}^2 z'/z''\}} \frac{dx_{32}^2}{x_{32}^2} \times \left[\Gamma_{10,32}(z''s) + 3\, G_{32}(z''s)\right]. \tag{9.69b}$$

Here we had to introduce an auxiliary function Γ, termed the "neighbor dipole amplitude", in which further evolution is constrained by the lifetime of an adjacent dipole: the evolution in dipole 10 may depend in the size of the neighbor dipole 21. This happens when the lifetime ordering condition (9.68) for the evolution in dipole 10 is dominated by the lifetime of this neighbor dipole, that is, if the lifetime of dipole 21, $z'x_{21}^2$, is much shorter than the lifetime of dipole 10, $z'x_{10}^2$. The resulting neighbor dipole amplitude depends on $\vec{x}_{10}$, x_{21} and z' and is denoted by $\Gamma_{10,21}(z's)$ (for more details, see [Kovchegov *et al.* (2016)]). The inhomogeneous terms $G_{10}^{(0)}(zs) = \Gamma_{10,21}^{(0)}(z's)$ in Eqs. (9.69) are determined by the initial conditions for the helicity evolution.

Note that after taking the large-N_c limit of Eq. (9.66) we have also put all the unpolarized S-matrices (9.26) to 1, which corresponds to neglecting saturation effects in the resulting equations (9.69). This is done in order to keep only the leading double-logarithmic evolution (powers of $\alpha_s \ln^2(1/x)$); saturation corrections appear at the leading-$\ln(1/x)$ level, that is, at order-$\alpha_s \ln(1/x)$, which is a higher-order correction for helicity evolution.

9.2.3 *Small-x asymptotics for quark and gluon helicity distributions*

To determine the small-x asymptotics of the quark helicity distribution one has to solve Eqs. (9.69). These equations can be solved numerically [Kovchegov *et al.* (2017c)] and analytically [Kovchegov *et al.* (2017d)] giving

$$G(x_{10}^2, zs) \propto (zs\, x_{10}^2)^{\alpha_h^q} \tag{9.70}$$

with

$$\alpha_h^q = \frac{4}{\sqrt{3}} \sqrt{\frac{\alpha_s N_c}{2\pi}} \approx 2.31 \sqrt{\frac{\alpha_s N_c}{2\pi}}. \tag{9.71}$$

Employing Eq. (9.62) this result gives us the small-x asymptotics for the quark helicity distribution

$$\Delta\Sigma(x, Q^2) \sim \left(\frac{1}{x}\right)^{\alpha_h^q}. \tag{9.72}$$

Note that this result was derived in the large-N_c limit. At the time of writing, the solution of the large-N_c&N_f helicity evolution equations from [Kovchegov *et al.* (2016); Kovchegov and Sievert (2019)] has not been constructed, and the resulting corrections to Eq. (9.72) due to bringing back the quarks into the evolution are not yet known.

Also note that Eq. (9.72) gives us the behavior of $\Delta\Sigma$ at small x, but still outside of the saturation region. While detailed inclusion of saturation effects into equations (9.69) can only be accomplished when the leading-logarithmic corrections (powers of $\alpha_s \ln(1/x)$) are summed up consistently, the preliminary conclusion one can draw from the version of equations (9.69) with the saturation corrections included, as derived in [Kovchegov *et al.* (2016)], is that $\Delta\Sigma \lesssim$ const inside the saturation region and, therefore, this region contributes very little to the net quark spin in Eq. (9.59).

An analysis similar to above can be applied to the gluon helicity distribution [Kovchegov *et al.* (2017b)]. Again, in the large-N_c limit and outside the saturation region, one obtains

$$\Delta G(x, Q^2) \sim \left(\frac{1}{x}\right)^{\alpha_h^G} \quad \text{with} \quad \alpha_h^G = \frac{13}{4\sqrt{3}}\sqrt{\frac{\alpha_s N_c}{2\pi}} \approx 1.88\sqrt{\frac{\alpha_s N_c}{2\pi}}. \tag{9.73}$$

The small-x asymptotics of ΔG in the large-N_c&N_f limit is also not yet known. Similar to the quark helicity distribution, we expect $\Delta G \lesssim$ const in the saturation region.

Let us note again that the results shown in Eqs. (9.72) and (9.73) disagree with the earlier conclusions regarding the small-x asymptotics of helicity distributions reached in [Bartels *et al.* (1996a)]. This discrepancy is an open problem in the field. A possible resolution of this problem was proposed in [Kovchegov *et al.* (2017a)].

Equations (9.72) and (9.73) are the leading perturbative results for the small-x asymptotics of $\Delta\Sigma$ and ΔG at large N_c and outside the saturation region. These results can be used to estimate the amount of proton spin carried by the small-x quarks and gluons in the proton. Initial phenomenological analysis has been performed in [Kovchegov *et al.* (2017c,b)], indicating a potential numerical importance of the small-x region for the proton spin balance. Higher-order corrections to the helicity evolution equations derived above, along with a more detailed phenomenology implementing those corrections, may reduce the theoretical error bars, allowing one to make precise predictions for the helicity PDFs at small x to be measured at EIC. If such predictions are confirmed by the EIC data, one may then gain enough confidence to extrapolate the helicity PDFs down to $x \to 0$, thus providing a robust theoretical assessment of the amount of proton spin at small x.

9.3 Conclusions

In this Chapter we have presented two developments initiated by the works of Lev Lipatov. The work on saturation and unitarity started shortly after the derivation of the BFKL equation by Lipatov and collaborators. It would have been impossible without BFKL. These days the field of parton saturation at small-x is a developed field, with a large number of active researchers and an integral component of most QCD-themed conferences and workshops. The search for and discovery of saturation physics is a large part of the physics case for the proposed Electron-Ion Collider in the US [Accardi *et al.* (2016)].

While Lev Lipatov himself has not worked on the proton spin puzzle, he was the first to study the the double-logarithmic resummation parameter $\alpha_s \ln^2(1/x)$ in QCD. Resummation of this parameter is important for the determination of small-x asymptotics of helicity PDFs and may prove to be a crucial tool for resolving the proton spin puzzle. The work on helicity PDFs at small x is only beginning, but has a promising future, especially when the EIC produces new data on longitudinal spin at small x.

Acknowledgments

I would like to thank Al Mueller, Genya Levin, and Larry McLerran for their collaboration and for teaching me a lot about high energy QCD when I was a junior researcher. I also thank Dan Pitonyak and Matt Sievert for their collaboration on the papers which led to the second part of this chapter.

This material is based upon work supported by the U.S. Department of Energy, Office of Science, Office of Nuclear Physics under Award Number DE-SC0004286.

Bibliography

Accardi, A. *et al.* (2016). Electron ion collider: The next QCD frontier, *Eur. Phys. J. A* **52**, 9, p. 268, doi:10.1140/epja/i2016-16268-9, arXiv:1212.1701 [nucl-ex].

Albacete, J. L., Armesto, N., Milhano, J. G., Quiroga-Arias, P., and Salgado, C. A. (2011). AAMQS: A non-linear QCD analysis of new HERA data at small-x including heavy quarks, *Eur. Phys. J. C* **71**, p. 1705, doi:10.1140/epjc/s10052-011-1705-3, arXiv:1012.4408 [hep-ph].

Albacete, J. L., Armesto, N., Milhano, J. G., and Salgado, C. A. (2009). Non-linear QCD meets data: A global analysis of lepton- proton scattering with running coupling BK evolution, *Phys. Rev. D* **80**, p. 034031, doi:10.1103/PhysRevD.80.034031, arXiv:0902.1112 [hep-ph].

Albacete, J. L., Armesto, N., Milhano, J. G., Salgado, C. A., and Wiedemann, U. A. (2005). Numerical analysis of the Balitsky-Kovchegov equation with running coupling: Dependence of the saturation scale on nuclear size and rapidity, *Phys. Rev. D* **71**, p. 014003, hep-ph/0408216.

Albacete, J. L. and Kovchegov, Y. V. (2007). Solving high energy evolution equation including running coupling corrections, *Phys. Rev. D* **75**, p. 125021, arXiv:0704.0612 [hep-ph].

Albacete, J. L. and Marquet, C. (2014). Gluon saturation and initial conditions for relativistic heavy ion collisions, *Prog.Part.Nucl.Phys.* **76**, pp. 1–42, doi:10.1016/j.ppnp.2014.01.004, arXiv:1401.4866 [hep-ph].

Altarelli, G. and Parisi, G. (1977). Asymptotic freedom in Parton language, *Nucl. Phys. B* **126**, p. 298, doi:10.1016/0550-3213(77)90384-4.

Aschenauer, E.-C. *et al.* (2016). The RHIC Cold QCD Plan for 2017 to 2023: A portal to the EIC, arXiv:1602.03922 [nucl-ex].

Balitsky, I. (1996). Operator expansion for high-energy scattering, *Nucl. Phys. B* **463**, pp. 99–160, doi:10.1016/0550-3213(95)00638-9, arXiv:hep-ph/9509348 [hep-ph].

Balitsky, I. (1999). Factorization and high-energy effective action, *Phys. Rev. D* **60**, p. 014020, hep-ph/9812311.

Balitsky, I. and Lipatov, L. (1978). The Pomeranchuk singularity in quantum chromodynamics, *Sov.J.Nucl.Phys.* **28**, pp. 822–829.

Bartels, J., Ermolaev, B., and Ryskin, M. (1996a). Flavor singlet contribution to the structure function G(1) at small x, *Z.Phys. C* **72**, pp. 627–635, doi:10.1007/s002880050285, arXiv:hep-ph/9603204 [hep-ph].

Bartels, J., Ermolaev, B., and Ryskin, M. (1996b). Nonsinglet contributions to the structure function g1 at small x, *Z.Phys. C* **70**, pp. 273–280, arXiv:hep-ph/9507271 [hep-ph].

Bartels, J., Lipatov, L. N., and Vacca, G. P. (2005). Interactions of reggeized gluons in the Mobius representation, *Nucl. Phys. B* **706**, pp. 391–410, doi:10.1016/j.nuclphysb.2004.10.008, arXiv:hep-ph/0404110 [hep-ph].

Bertsch, G., Brodsky, S. J., Goldhaber, A. S., and Gunion, J. F. (1981). Diffractive excitation in QCD, *Phys. Rev. Lett.* **47**, p. 297, doi:10.1103/PhysRevLett.47.297.

Bjorken, J. D. and Kogut, J. B. (1973). Correspondence arguments for high-energy collisions, *Phys. Rev. D* **8**, p. 1341, doi:10.1103/PhysRevD.8.1341.

Bjorken, J. D., Kogut, J. B., and Soper, D. E. (1971). Quantum electrodynamics at infinite momentum: scattering from an external field, *Phys. Rev. D* **3**, p. 1382, doi:10.1103/PhysRevD.3.1382.

Brodsky, S. J., Pauli, H.-C., and Pinsky, S. S. (1998). Quantum chromodynamics and other field theories on the light cone, *Phys.Rept.* **301**, pp. 299–486, doi:10.1016/S0370-1573(97)00089-6, arXiv:hep-ph/9705477 [hep-ph].

Chirilli, G. A. (2019). Sub-eikonal corrections to scattering amplitudes at high energy, *JHEP* **01**, p. 118, doi:10.1007/JHEP01(2019)118, arXiv:1807.11435 [hep-ph].

Cougoulic, F. and Kovchegov, Y. V. (2019). Helicity-dependent generalization of the JIMWLK evolution, arXiv:1910.04268 [hep-ph].

Dokshitzer, Y. L. (1977). Calculation of the structure functions for deep inelastic scattering and e^+e^- Annihilation by perturbation theory in quantum chromodynamics, *Sov. Phys. JETP* **46**, pp. 641–653.

Ferreiro, E., Iancu, E., Leonidov, A., and McLerran, L. (2002). Nonlinear gluon evolution in the color glass condensate. II, *Nucl. Phys. A* **703**, pp. 489–538, hep-ph/0109115.

Frankfurt, L., Miller, G. A., and Strikman, M. (1993). Coherent nuclear diffractive production of mini - jets: Illuminating color transparency, *Phys. Lett. B* **304**, pp. 1–7, doi:10.1016/0370-2693(93)91390-9, arXiv:hep-ph/9305228 [hep-ph].

Frankfurt, L. and Strikman, M. (1988). Hard nuclear processes and microscopic nuclear structure, *Phys.Rept.* **160**, pp. 235–427, doi:10.1016/0370-1573(88)90179-2.

Froissart, M. (1961). Asymptotic behavior and subtractions in the mandelstam representation, *Phys. Rev.* **123**, pp. 1053–1057.

Gelis, F., Iancu, E., Jalilian-Marian, J., and Venugopalan, R. (2010). The color glass condensate, *Ann.Rev.Nucl.Part.Sci.* **60**, pp. 463–489, doi:10.1146/annurev.nucl.010909.083629, arXiv:1002.0333 [hep-ph].

Gorshkov, V. G., Gribov, V. N., Lipatov, L. N., and Frolov, G. V. (1968). Doubly logarithmic asymptotic behavior in quantum electrodynamics, *Sov. J. Nucl. Phys.* **6**, p. 95, [Yad. Fiz.6,129(1967)].

Gribov, L. V., Levin, E. M., and Ryskin, M. G. (1983). Semihard processes in QCD, *Phys. Rept.* **100**, pp. 1–150.

Gribov, V. N. (1970). Interaction of gamma quanta and electrons with nuclei at high-energies, *Sov. Phys. JETP* **30**, pp. 709–717, [Zh. Eksp. Teor. Fiz.57,1306(1969)].

Gribov, V. N. and Lipatov, L. N. (1972). Deep inelastic e p scattering in perturbation theory, *Sov. J. Nucl. Phys.* **15**, pp. 438–450.

Heiselberg, H., Baym, G., Blaettel, B., Frankfurt, L. L., and Strikman, M. (1991). Color transparency, color opacity, and fluctuations in nuclear collisions, *Phys. Rev. Lett.* **67**, pp. 2946–2949, doi:10.1103/PhysRevLett.67.2946.

Heisenberg, W. (1952). Mesonenerzeugung als Stosswellenproblem, *Z. Phys.* **133**, p. 65, doi:10.1007/BF01948683.

Iancu, E., Itakura, K., and McLerran, L. (2002). Geometric scaling above the saturation scale, *Nucl. Phys. A* **708**, pp. 327–352, hep-ph/0203137.

Iancu, E., Leonidov, A., and McLerran, L. D. (2001a). Nonlinear gluon evolution in the color glass condensate. I, *Nucl. Phys. A* **692**, pp. 583–645, hep-ph/0011241.

Iancu, E., Leonidov, A., and McLerran, L. D. (2001b). The renormalization group equation for the color glass condensate, *Phys. Lett. B* **510**, pp. 133–144, doi: 10.1016/S0370-2693(01)00524-X.

Iancu, E. and Venugopalan, R. (2003). The color glass condensate and high energy scattering in QCD, hep-ph/0303204.

Jaffe, R. L. and Manohar, A. (1990). The G(1) Problem: Fact and Fantasy on the Spin of the Proton, *Nucl. Phys. B* **337**, pp. 509–546, doi:10.1016/0550-3213(90)90506-9.

Jalilian-Marian, J. and Kovchegov, Y. V. (2006). Saturation physics and deuteron gold collisions at RHIC, *Prog. Part. Nucl. Phys.* **56**, pp. 104–231, hep-ph/0505052.

Jalilian-Marian, J., Kovner, A., Leonidov, A., and Weigert, H. (1998a). The Wilson renormalization group for low x physics: Towards the high density regime, *Phys. Rev. D* **59**, p. 014014, hep-ph/9706377.

Jalilian-Marian, J., Kovner, A., and Weigert, H. (1998b). The Wilson renormalization group for low x physics: Gluon evolution at finite parton density, *Phys. Rev. D* **59**, p. 014015, hep-ph/9709432.

Ji, X., Xiong, X., and Yuan, F. (2012). Proton spin structure from measurable parton distributions, *Phys. Rev. Lett.* **109**, p. 152005, doi:10.1103/PhysRevLett.109.152005, arXiv:1202.2843 [hep-ph].

Ji, X.-D. (1997). Gauge-invariant decomposition of nucleon spin, *Phys. Rev. Lett.* **78**, pp. 610–613, doi:10.1103/PhysRevLett.78.610, arXiv:hep-ph/9603249 [hep-ph].

Kirschner, R. and Lipatov, L. (1983). Double logarithmic asymptotics and regge singularities of quark amplitudes with flavor exchange, *Nucl.Phys. B* **213**, pp. 122–148, doi:10.1016/0550-3213(83)90178-5.

Kopeliovich, B. Z., Lapidus, L. I., and Zamolodchikov, A. B. (1981). Dynamics of color in Hadron diffraction on nuclei, *JETP Lett.* **33**, pp. 595–597, [Pisma Zh. Eksp. Teor. Fiz.33,612(1981)].

Kovchegov, Y. V. (1996). Non-abelian Weizsäcker-Williams field and a two- dimensional effective color charge density for a very large nucleus, *Phys. Rev. D* **54**, pp. 5463–5469, hep-ph/9605446.

Kovchegov, Y. V. (1997). Quantum structure of the non-Abelian Weizsäcker-Williams field for a very large nucleus, *Phys. Rev. D* **55**, pp. 5445–5455, hep-ph/9701229.

Kovchegov, Y. V. (1999). Small-x F_2 structure function of a nucleus including multiple pomeron exchanges, *Phys. Rev. D* **60**, p. 034008, hep-ph/9901281.

Kovchegov, Y. V. (2000). Unitarization of the BFKL pomeron on a nucleus, *Phys. Rev. D* **61**, p. 074018, hep-ph/9905214.

Kovchegov, Y. V. (2019). Orbital angular momentum at small x, *JHEP* **03**, p. 174, doi: 10.1007/JHEP03(2019)174, arXiv:1901.07453 [hep-ph].

Kovchegov, Y. V. and Levin, E. (2012). *Quantum Chromodynamics at High Energy*, Vol. 33 (Cambridge University Press), ISBN 9780521112574, 9780521112574, 9781139557689, http://www.cambridge.org/de/knowledge/isbn/item6803159.

Kovchegov, Y. V., Pitonyak, D., and Sievert, M. D. (2016). Helicity evolution at small-x, *JHEP* **01**, p. 072, doi:10.1007/JHEP01(2016)072, arXiv:1511.06737 [hep-ph].

Kovchegov, Y. V., Pitonyak, D., and Sievert, M. D. (2017a). Helicity evolution at small x: Flavor singlet and non-singlet observables, *Phys. Rev. D* **95**, 1, p. 014033, doi: 10.1103/PhysRevD.95.014033, arXiv:1610.06197 [hep-ph].

Kovchegov, Y. V., Pitonyak, D., and Sievert, M. D. (2017b). Small-x asymptotics of the gluon helicity distribution, *JHEP* **10**, p. 198, doi:10.1007/JHEP10(2017)198, arXiv:1706.04236 [nucl-th].

Kovchegov, Y. V., Pitonyak, D., and Sievert, M. D. (2017c). Small-x asymptotics of the quark helicity distribution, *Phys. Rev. Lett.* **118**, 5, p. 052001, doi:10.1103/PhysRevLett.118.052001, arXiv:1610.06188 [hep-ph].

Kovchegov, Y. V., Pitonyak, D., and Sievert, M. D. (2017d). Small-x asymptotics of the quark helicity distribution: Analytic results, *Phys. Lett. B* **772**, pp. 136–140, doi: 10.1016/j.physletb.2017.06.032, arXiv:1703.05809 [hep-ph].

Kovchegov, Y. V. and Sievert, M. D. (2019). Small-x Helicity Evolution: An operator treatment, *Phys. Rev. D* **99**, 5, p. 054032, doi:10.1103/PhysRevD.99.054032, arXiv:1808.09010 [hep-ph].

Kovner, A. and Wiedemann, U. A. (2002). Nonlinear qcd evolution: Saturation without unitarization, *Phys. Rev. D* **66**, p. 051502, hep-ph/0112140.

Kuraev, E. A., Lipatov, L. N., and Fadin, V. S. (1977). The Pomeranchuk singlularity in non-Abelian gauge theories, *Sov. Phys. JETP* **45**, pp. 199–204.

Lepage, G. P. and Brodsky, S. J. (1980). Exclusive processes in perturbative quantum chromodynamics, *Phys. Rev. D* **22**, p. 2157.

Levin, E. M. and Ryskin, M. G. (1987). Diffraction dissociation on nuclei in the QCD leading logarithmic approximation. (In Russian), *Yad. Fiz.* **45**, pp. 234–244, [Sov. J. Nucl. Phys.45,150(1987)].

Martin, A. (1969). Scattering Theory: Unitarity, analyticity and crossing, *Lect. Notes Phys.* **3**, pp. 1–117, doi:10.1007/BFb0101043.

McLerran, L. D. and Venugopalan, R. (1994a). Computing quark and gluon distribution functions for very large nuclei, *Phys. Rev. D* **49**, pp. 2233–2241, hep-ph/9309289.

McLerran, L. D. and Venugopalan, R. (1994b). Gluon distribution functions for very large nuclei at small transverse momentum, *Phys. Rev. D* **49**, pp. 3352–3355, hep-ph/9311205.

McLerran, L. D. and Venugopalan, R. (1994c). Green's functions in the color field of a large nucleus, *Phys. Rev. D* **50**, pp. 2225–2233, hep-ph/9402335.

Mueller, A. H. (1990). Small x behavior and parton saturation: A QCD model, *Nucl. Phys. B* **335**, p. 115.

Mueller, A. H. (1994). Soft gluons in the infinite momentum wave function and the BFKL pomeron, *Nucl. Phys. B* **415**, pp. 373–385, doi:10.1016/0550-3213(94)90116-3.

Mueller, A. H. (1995). Unitarity and the BFKL pomeron, *Nucl. Phys. B* **437**, pp. 107–126, doi:10.1016/0550-3213(94)00480-3, arXiv:hep-ph/9408245 [hep-ph].

Mueller, A. H. (2001). A Simple derivation of the JIMWLK equation, *Phys.Lett. B* **523**, pp. 243–248, doi:10.1016/S0370-2693(01)01343-0, arXiv:hep-ph/0110169 [hep-ph].

Mueller, A. H. and Patel, B. (1994). Single and double BFKL pomeron exchange and a dipole picture of high-energy hard processes, *Nucl. Phys. B* **425**, pp. 471–488, hep-ph/9403256.

Mueller, A. H. and Qiu, J.-W. (1986). Gluon recombination and shadowing at small values of x, *Nucl. Phys. B* **268**, pp. 427–452, doi:10.1016/0550-3213(86)90164-1.

Mueller, A. H. and Triantafyllopoulos, D. N. (2002). The energy dependence of the saturation momentum, *Nucl. Phys. B* **640**, pp. 331–350, hep-ph/0205167.

Nikolaev, N. N. and Zakharov, B. G. (1991). Colour transparency and scaling properties of nuclear shadowing in deep inelastic scattering, *Z. Phys. C* **49**, pp. 607–618, doi: 10.1007/BF01483577.

Nikolaev, N. N., Zakharov, B. G., and Zoller, V. R. (1994). The s channel approach to Lipatov's pomeron and hadronic cross-sections, *JETP Lett.* **59**, pp. 6–12, arXiv:hep-ph/9312268 [hep-ph].

Rummukainen, K. and Weigert, H. (2004). Universal features of JIMWLK and BK evolution at small x, *Nucl. Phys. A* **739**, pp. 183–226, hep-ph/0309306.

Schlichting, S. and Schenke, B. (2014). The shape of the proton at high energies, *Phys. Lett. B* **739**, pp. 313–319, doi:10.1016/j.physletb.2014.10.068, arXiv:1407.8458 [hep-ph].

't Hooft, G. (1974). A two-dimensional model for mesons, *Nucl. Phys. B* **75**, pp. 461–470, doi:10.1016/0550-3213(74)90088-1.

Weigert, H. (2002). Unitarity at small Bjorken x, *Nucl. Phys. A* **703**, pp. 823–860, hep-ph/0004044.

Weigert, H. (2005). Evolution at small x_{bj}: The color glass condensate, *Prog. Part. Nucl. Phys.* **55**, pp. 461–565, hep-ph/0501087.

https://doi.org/10.1142/9789811231124_0010

Chapter 10

The Odderon and BKP States in Quantum Chromodynamics

M. A. Braun* and G. P. Vacca†

**Dep. of High Energy physics, Saint-Petersburg State University, 198504 S. Petersburg, Russia*

†INFN Sezione di Bologna, Via Irnerio 46, 40126, Bologna, Italy

Personal remarks of the authors:
I have known Lev Lipatov since a long time ago. He was one of my first students at the Department of Theoretical Physics of the then State Leningrad University. In fact I was the supervisor of his diploma (master thesis) which he defended on his graduation in 1962. He was certainly a brilliant student.

At that time, I was too young (I was only 27 years old) to fully recognize his powerful, creative and hard-working mind. One has to know that Soviet universities on the whole were very conservative institutions in those years. The fear of dissident students caused the Communist Party and the government to impose very stringent restrictions on all sorts of university activities. It was stressed that scientific research was a secondary task for them, relegated to the Academy of Science which was a much more liberal and independent place. Nevertheless, the theoretical department at Leningrad University was a good beginning for Lev, mainly due to the influence of the mighty Academician V. A. Fock, and was one of the centers of active studies in problems within quantum field theory, particularly in strong interactions. This made it an excellent place for nurturing talented scientists in this domain, exemplified by graduates like Vladimir Gribov and Ludwig Faddeev who became renowned physicists. So I believe, in spite of a certain modesty in his first scientific work through his diploma thesis (for which I certainly bear most of the responsibility), the experience at Leningrad University had fully prepared Lev for Big Science, which he dived into when he was adopted as a Ph.D. student by the V. N. Gribov group in the Academy of Science at Gatchina.

In the following years I had little contact with Lev, limited to our meetings at various seminars. Much later in the 90's of the past century, Lev was invited to teach Weak Interactions at our department of High-Energy Physics, which was then headed by myself. Lev used to be a very diligent and devoted teacher, although his lack of experience made his lectures very difficult for the students (and I guess also for him). He remained rather detached from people at the university, while focusing his interests on those at Gatchina and his close collaborators abroad.

Still, I was in touch with Lev, mostly during our common stay in Hamburg with Jochen Bartels where on one occasion I had the pleasure to work with Lev on a certain problem within his effective theory. It was then that I could clearly recognize his deepness and power as he easily resolved our current problems which were for some time unclear to me and my collaborators.

Over time, our relations grew warm but not quite intimate, for which I blame the both of us with a certain sadness. Last time I met Lev was two days before his death on the eve of his journey to the Dubna conference. He was very weak (and ill in my opinion). I tried to convince him to give up attending this conference but he refused (in fact he liked conferences very much). We know the tragic end.

Lev was a towering figure in our field. We shall certainly miss him very much.

M. A. Braun

I have had the privilege to know Lev since 20 years, after having spent my first months in Hamburg as a postdoc. Together with Jochen Bartels we unexpectedly discovered some leading QCD Odderon states and we started a correspondence with Lev, who found a very different approach to see such solutions. We met a few months later and started a long term collaboration, meeting each other in Hamburg, St. Petersburg, Bologna... I always remember as a real pleasure the long scientific discussions at the blackboard with him, from which there was always something to learn, including his gentle personality. We all miss Lev, as a great scientist and man.

Gian Paolo Vacca

Abstract

We discuss the Odderon in QCD and analyze the effects of the running coupling introduced in a particular way to preserve within the leading logarithmic description the reggeization of the gluon. This idea is also applied to a family of BKP states with an arbitrary number of gluons in the planar limit. The numerical analysis shows that, contrary to the Pomeron case where the leading states are discrete the Odderon states still remain a continuous family starting at intercept one. The resulting rapidity dependence of the amplitude is studied. For the Pomeron-Odderon system the relation between the description in the reggeized gluon (BFKL) framework and the color dipole picture/CGC is investigated.

10.1 Introduction

The Odderon has a long history in the high energy physics. It started in 1973 and its first 30 years of existence are exposed in the comprehensive review [Ewerz (2003)]. So here we only mention some principal points and developments after 2003. The Odderon was born as a result of an abstract idea that "strong interactions are as strong as they can be" in the paper by L. Lukaszuk and B. Nicolescu [Lukaszuk and Nicolescu (1973)] and it was given a name "'maximal Odderon".

The famous Froissart theorem establishes a limit on the behavior of the scattering amplitudes governed by strong interactions as

$$|A(s,t)| < C s \log^2 s, \quad C < \pi/m_\pi^2 \simeq 62\ mbn. \tag{10.1}$$

In fact, the amplitude can be split into $C = +1$ and $C = -1$ parts (or signatures $\xi = \pm$):

$$A^{(\pm)}(s) = \frac{1}{2}[A(s) \pm A(-s)], \tag{10.2}$$

and the long-standing Pomeranchuk theorem asserted

$$\frac{A^-(s)}{A^+(s)}\bigg|_{s\to\infty} \to 0, \tag{10.3}$$

where the leading part of the $A^{(+)}$ and $A^{(-)}$ in the high energy limit are associated to the Pomeron and to the Odderon exchange, respectively. The results following from Regge theory and based on the leading ω, ρ-Regge trajectory in the $C = -1$ sector fully confirm this rule. The ω, ρ intercept is $\alpha_{\rho,\omega}(0) \simeq 1/2$ so that the corresponding reggeon interchange leads to the amplitude $A^{(-)} \sim \sqrt{s}$, which is fully confirmed by experiment.

The idea of the maximal Odderon proclaimed that both amplitudes $A^{(\pm)}$ take their maximal values at $s \to \infty$, with the only difference that $A^{(+)}(s)$ is positive imaginary and $A^{(-)}(s)$ real:

$$A^{(+)}(s) \sim is(\log^2 s - i\pi \log s), \quad A^{(-)} \sim s(\ln^2 s - i\pi \log s), \quad s \gg m^2. \tag{10.4}$$

Let us briefly recall some simple and evident properties of the maximal Odderon amplitude (10.4) (at $t = 0$). As mentioned it is mostly real. Its imaginary part is proportional to $s \log s$, which corresponds to the difference between the particle-particle and particle-antiparticle cross-sections growing like $\log s$:

$$\sigma_{a\bar{b}}(s) - \sigma_{ab} \sim \log s, \quad s \gg m^2. \tag{10.5}$$

Since the total cross-section for each of the two reactions grows like $\log^2 s$, the Pomeranchuk theorem is fulfilled in the sense

$$\frac{\sigma_{a\bar{b}}(s)}{\sigma_{ab}(s)} \to 1, \quad s \to \infty. \tag{10.6}$$

The most striking feature of the asymptotic Odderon is a nonvanishing ratio of the real and imaginary parts of the amplitudes:

$$\frac{\mathrm{Re}A(s)}{\mathrm{Im}A(s)} \sim const, \quad s \gg m^2. \tag{10.7}$$

In the course of time, and depending on the experimental situation, these properties served as a dominant motive for and against the introduction of the Odderon. We shall turn to the experimental evidence for the Odderon at the end of this section.

In an attempt to realize this picture in the complex angular j-plane the authors were obliged to introduce rather exotic singularities. Addressing the interested author to the review [Ewerz (2003)] to see the details, we only mention that typically higher order poles at $j = 1$ were introduced to generate the desired $\log(s)$ dependence. The authors had also to struggle against the singularities which might appear in the physical amplitude, since for $A^{(-)}$ the point $j = 1$ is physical.

With the advent of Quantum Chromodynamics the Odderon acquired a new life. In the lowest order there one finds both $C = \pm 1$ amplitudes on equal footing, the $C = +1$ one coming from the exchange of two gluons and the $C = -1$ from the exchange of three gluons. Both amplitudes are linear in s, so that their behavior at $s \to \infty$ is similar and corresponds to a pole in the j plane at $j = 1$. Ever since the theoretical efforts were directed to find corrections to this simple results in higher orders. Remarkable success has been achieved in the so-called leading-log approximation, in which terms of the order $(\alpha_s \ln(s))^n$ are summed at high s and at small α_s. The well-known BFKL equation which describes the evolution of the Pomeron was set up and solved as early as 1975 [Kuraev *et al.* (1977); Balitsky and Lipatov (1978)], and its generalization to the next-to leading order in was derived in 1995 [Fadin and Lipatov (1998); Ciafaloni and Gianni (1998)]. The vertex for splitting of the Pomeron into two Pomerons was constructed in 1995 in two approaches, the dispersion one in [Bartels and Wusthoff (1995)], following the derivation of the BFKL equation, and in the approach based on the color dipole picture in [Mueller and Patel (1994)]. On the basis of this vertex equations for the scattering of a small object on the nucleus were proposed by I. I. Balitsky and Yu. Kovchegov (the BK equation) [Balitsky (1996); Kovchegov (1999)], later on generalized to the CGC-JINWLK equation (see e.g. [Ferreiro *et al.* (2002)]). Analogous developments were realized in the BFKL-Bartels approach [Braun (2000); Bartels *et al.* (2005b)]. Parallel to this the Odderon was investigated both in the BFKL-Bartels and in the dipole approach. The equation for the Odderon in the leading-log approximation was set up as one of the so-called BKP equations proposed in [Bartels (1980); Jaroszewicz (1980); Kwiecinski and Praszalowicz (1980)]. L. N. Lipatov analyzed the conformal properties of the Odderon and related its equation to that of the solvable chain of conformal spins [Lipatov (1994)]. This allowed to R. A. Janik and J. Wosiek to obtain an explicit equation for the Odderon wave function (the JW Odderon) and find its maximal intercept [Wosiek and Janik (1997)]

$$\alpha_O = 1 - 0.24717\frac{\alpha_s N_c}{\pi},$$

where $\alpha_s = g_s^2/(4\pi)$. Somewhat later a new solution of the Odderon equation was obtained by J. Bartels, L. N. Lipatov and G. P. Vacca (the BLV Odderon) with $\alpha_O = 1$ in which two of the three reggeons forming the odderon are fused into one [Bartels *et al.* (2000)]. This simple odderon was later also found in the dipole approach in [Kovchegov *et al.* (2004)]. On the basis of the latter the evolution equation was set up for the combined Pomeron-Odderon system in [Hatta *et al.* (2005)] and attempts of solving it were made in [Hatta *et al.* (2005); Motyka (2006)], although in a very simplified (one-dimensional) approach. As to the more complicated JW odderon, the necessary Pomeron-two Odderon vertex was found in [Bartels and Ewerz (1999)], although its application has not been attempted so far, given that these Odderon states where shown to decouple from some perturbative impact factors, while the BLV Odderon states are maximally coupled to them [Bartels *et al.* (2001)]. Later, using the effective action for reggeized gluons constructed by L. N. Lipatov, the BKP kernel for three reggeized gluon systems in the NLL approximation was constructed [Bartels *et al.* (2013)]. This can be used to analyze the Odderon both in QCD, but in the large N_c limit, or in general supersymmetric extensions, in particular $N = 4$ SYM for any N_c. Then, starting from generalized bootstrap equations involving inelastic production amplitudes which were conjectured in [Bartels and Vacca (2013)] but have still to be proven, a construction of a family of Odderon solutions at NLL was given. They are extensions of the BLV solution to the NLL approximation and have also intercept starting strictly at one [Bartels and Vacca (2013)].

We stress that in all developments of the Odderon theory seminal contributions were made by L. N. Lipatov: beginning with the study of the BFKL Pomeron and including the understanding of the conformal properties of the Pomeron and Odderon, then the equivalence of the Odderon equation to the solvable chain of conformal spins, and also the construction of the BLV odderon solution.

On the experimental side the history of the Odderon is just as long but marked by much less clarity, novelty and success. Manifestations of the Odderon can be present either in the processes realized by the purely Odderon exchange or in the interference of the Pomeron and Odderon exchanges. The typical example of the first case is the production of a pseudoscalar (PS) meson (C = +1) in the interactions starting with the photon (C $= -1$) either real or virtual. Discovery of this process would be a straightforward detection of the Odderon. There were various theoretical estimates of the rate of this transition, e.g. for the process $\gamma \to \eta_c$ [Bartels *et al.* (2001, 2004)] (see also the review [Ewerz (2003)] for earlier works). Unfortunately they all gave estimations for the cross-sections far below the present experimental possibilities. More informative were attempts to see the interference of the Pomeron and Odderon exchanges primarily in the comparison of pp and $p\bar{p}$ elastic scattering, In fact, this comparison was the basic motivation for the introduction of the Odderon in the first papers in which this notion was introduced. Already at that time (in the seventies) a significant difference between these two cross-sections was discovered. In particular the dip present in the pp cross-section as a function of t was not seen in the $p\bar{p}$ cross-section. At comparatively low energies this difference could be satisfactorily explained by the exchange of the well-known $C = -1$ reggeons $\omega - \rho$ with an intercept much below unity. However the difference did not disappear at higher energies. The situation at present date can be seen from Fig. 10.1 in which the latest data on the elastic pp scattering data from the TOTEM collaboration at 2.76 TeV [Antchev *et al.* (2020)] are compared to the $p\bar{p}$ cross-sections from D0 at 1.96 Tev [Abazov *et al.* (2012)]. Remarkably, the picture is nearly identical to the similar one for the experiments at 53 GeV presented in the review [Ewerz (2003)] in 2003. So the difference between pp and $p\bar{p}$, if indeed it exists, does not depend on energy once it is high enough. These latest experiments renewed the extensive discussion of the Odderon contribution at high energies, however, not yet with conclusive arguments, especially in view of considerable experimental

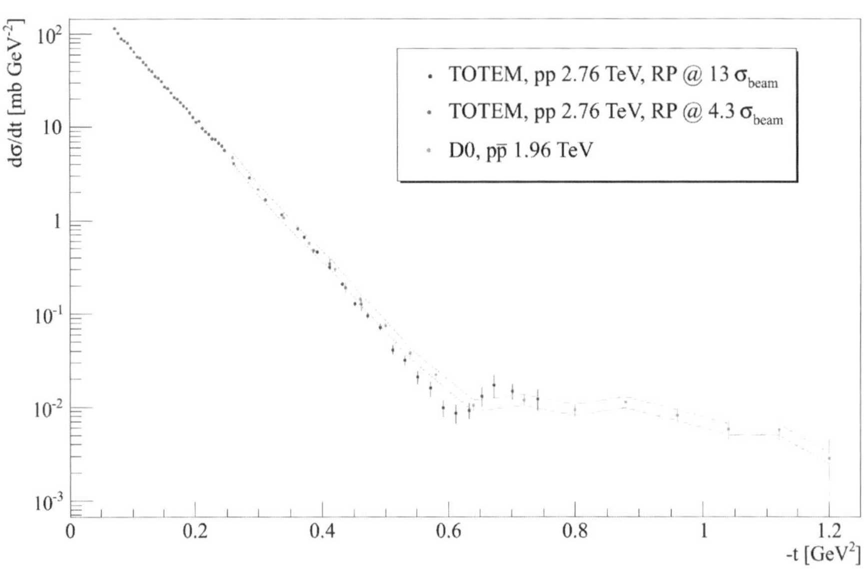

Fig. 10.1 The differential cross sections $d\sigma/dt$ at $\sqrt{s} = 2.76$ TeV measured by the TOTEM experiment and the elastic $p\bar{p}$ measurement of the D0 experiment at 1.96 TeV [Abazov *et al.* (2012)]. The green dashed line indicates the normalization uncertainty of the D0 measurement.

uncertainties [Khoze *et al.* (2018, 2014); Martynov and Nicolescu (2018)], not to speak about the still open theoretical issues.

In the present paper we try to cover some theoretical questions left untouched in the previous studies, which may throw additional light on the Odderon as the QCD object. In the first place we consider interacting Pomeron-Odderon systems in QCD in the generalized leading logarithmic approximation, in order to discuss the link between the reggeized gluon BFKL approach and the findings in the dipole/CGC picture. In particular we try to construct the equations for the combined evolution of the Pomeron and of the BLV Odderon in the dense (nuclear) target, in the BFKL-Bartels approach for comparison with similar equations in the dipole/CGC picture [Hatta *et al.* (2005)]. We shall be able to derive only the first of these equations: the evolution of the Pomeron in the presence of the Odderon field. The second equation for the Odderon itself requires knowledge of the vertex for the transition of three reggeons to five reggeons: this task has not been realized so far and is evidently beyond the scope of this study. We hope to return to this problem in the future. We also discuss what, as one may think, could be an effective useful description in the non perturbative regime and in the large transverse distance limit, the so called Reggeon Field Theory developed by V. Gribov for which we review some details of the universal critical behavior. Then we study the influence of the running coupling on the Odderon and on higher BKP states, which has not received much attention in the past. We also present in this framework a numerical study of both the Pomeron and the Odderon spectrum.

10.2 Interacting Pomeron-Odderon systems

We start discussing the perturbative QCD behavior of the Pomeron-Odderon system, interactions which go into the direction of "unitarizing" the theory in the sense of removing the violation of the Froissart bound. The full evolution of effective interacting Pomeron-Odderon systems is too hard to be investigated in small x QCD, since it should involve also loops. Nevertheless, an effective tree level description has been used to study some features of the large rapidity evolution and propagation in a dense nuclear medium. A simplified model, the so called Reggeon Field Theory, is subsequently presented. In this framework computation with loops are at reach and we give some recent perturbative results for the critical properties.

10.2.1 *Fan diagrams for the Pomeron-Odderon system*

The distribution of gluons (and quarks) in hadronic scattering processes depends on the rapidity under consideration and can be associated to phases characterized by high densities of strongly interacting quanta. Collisions with the center of mass energy dominating over other momentum scales require summation of contributions proportional to powers of $\log s$. This task was realized by L. N. Lipatov and collaborators in the so called BFKL approach. The leading log (LL) and next-to-leading-log (NLL) approximations resum $(\alpha_s \log s)^2$ and $\alpha_s(\alpha_s \log s)^2$ contributions, respectively. In the lowest order of perturbation theory it is sufficient to take into account only two reggeized gluons exchanged in the colorless t-channel. Then it was discovered that the cross sections grow as a power of energy in violation of the unitarity restrictions. This clearly showed the necessity to include more complicated structures into the t-channel, larger numbers of reggeized gluons and transitions changing these numbers. Of course, the final answer to the problem requires fully solving Quantum Chromodynamics in the relevant kinematical conditions (Regge kinematics), which does not seem to be realistic. So at present one has to rely on certain approximations starting from the simplest BFKL approach. Their improvememnt can be realized by different simplified models based on diagrams constructed from reggeized gluons in planar approximation or from Pomeron diagrams without loops.

Considerable progress has been achieved in the study of the scattering of compact objects, quark-antiquark loops or "onia". One starts from the scattering amplitudes for the collision of two such objects. In the BFKL approach it can be factorized into impact factors (local in rapidity) and a BFKL Pomeron Green function where the rapidity dependence is encoded, both objects depending on the transverse momenta $\boldsymbol{q}$ and $\boldsymbol{k}$

$$\mathcal{A}(Y,\boldsymbol{q}) = i\int d^2\boldsymbol{k}\Phi_1(\boldsymbol{q},\boldsymbol{k})\,G_Y(\boldsymbol{q},\boldsymbol{k})\,\Phi_1(\boldsymbol{q},\boldsymbol{k}). \tag{10.8}$$

The Green function G_Y is associated to the evolution kernel described later in Section 10.3.2, restricted to a fixed coupling constant. Including more t channel gluons in the analysis and taking into account the change of their numbers has been a long lasting program started by J. Bartels. Essentially the analysis was pushed up to six t-channel reggeized gluons. It was based on the study of multi-cut amplitudes D_n for the transition of the quark-antiquark loop into n reggeized gluons (actually their multiple energy discontinuities). They can be related to the total cross sections by the AGK rules [Abramovsky *et al.* (1973); Bartels *et al.* (2005c)]. A set of coupled equations has been constructed for D_n with the following ingredients: impact factors $D_n^{(0)}$ for n reggeized gluons directly attached to the quark-antiquark loop, the reggeized gluon trajectories, the $K_{n\to m}$ elementary transition vertices from n to m reggeized gluons and the so-called bootstrap relations.

As the outcome, the amplitude D_n can be presented as a sum of two contributions, the so-called reggeized one D_N^R for which evolution in rapidity is governed by a single Pomeron Green function for two reggeized gluons G_2 (BFKL Green function), and the other, irreducible one, D_n^I which evolves in rapidity via Green functions G_n for $n > 2$ reggeized gluons. In particular for the 4-gluon amplitude one finds

$$D_4 = D_4^R + D_4^I, \quad D_4^I = \int G_4(Y-Y')V_{2\to4}D_2(Y') = G_4 \otimes V_{2\to4}D_2, \tag{10.9}$$

where $D_2(y') = G_2(y')D_2(0)$, the vertex $V_{2\to4}$ describes the transition from 2 to 4 reggeized gluons, and G_4 is the Green function for the general evolution of

4 reggeized gluons. Similarly, for the 6-gluon amplitude one finds [Bartels and Ewerz (1999)]

$$D_6 = D_6^R + D_6^I + D_6^E, \tag{10.10}$$

where

$$D_6^I = \int G_6(Y - Y')V_{2\to 6}D_2(Y') = G_6 \otimes V_{2\to 6}D_2. \tag{10.11}$$

It is remarkable that in the vertex $V_{2\to 6}$ a term appears with a colour structure containing a product of two coefficients d_{abc}. It describes the transition of a Pomeron into two Odderons. The last term

$$D_6^E = \sum G_6 \otimes V_{2\to 4}G_4 \otimes V_{2\to 4}D_2 + D^{\text{extra}} \tag{10.12}$$

contains contributions related to a double $2 \to 4$ splitting including the BKP evolution (fan structure) and, finally, another term with a peculiar tensor color structure which was not completely analyzed [Bartels and Ewerz (1999)].

In an alternative, dipole or CGC approach, started by A. H. Mueller with collaborators, the scattering of an "onium" on a dense nuclear target was described (in the large N_C limit) by the Balitski-Kovchegov evolution equation [Balitsky (1996); Kovchegov (1999)]. From the viewpoint of the BFKL-Bartels approach this equation is equivalent to summation of fan diagrams made of BFKL Green functions and triple Pomeron vertices $V_{2\to 4}$ [Bartels *et al.* (2005b)].

From the start in the BFKL theory apart from the Pomeron also the Odderon was introduced as a compound state of three reggeized gluons with a color factor d_{abc}. In its general form its wave function depends on the three gluon momenta (or coordinates). The equation for this wave function was studied by L. N. Lipatov [Lipatov (1994)] and, as mentioned in the Introduction, later solved by Janik and Woiscek (JW-Odderon) who showed that its ground state has an intercept smaller than unity. So the JW-Odderon generates cross-sections vanishing at high energies.

However, later a different Odderon state was found, which essentially depends on only two coordinates with a fusing pair of the three gluons, the BLV-Odderon [Bartels *et al.* (2000)]. Its wave function is essentially the antisymmetric Pomeron wave function whose ground state has the intercept exactly equal to one. The contribution from the BLV-Odderon remains constant at high energies and so dominates over the JW-Odderon. Moreover, some leading order impact factors are known to decouple from JW-Odderon while couple maximally with the BLV-Odderon [Bartels *et al.* (2001)]. Explicitly the BLV-Odderon eigenstate Ψ is written in terms of the odd Pomeron eigenstate ψ as

$$\Psi = \mathcal{N}\hat{S}\psi, \tag{10.13}$$

where $\hat{S}$ is an operator acting on 2-gluon states resulting in 3-gluon states, which performs an antisymmetrization of the 2 incoming gluons, splits the first of them in two, and sums over the cyclic permutations of the outgoing gluons:

$$\hat{S}(1,2,3|1',2')\phi(1',2') = \frac{1}{2}\sum_{(123)}[\phi(12,3) - \phi(3,12)]. \tag{10.14}$$

$\mathcal{N}$ is a normalization factor. One finds

$$\langle\Psi|\Psi\rangle = c_o\langle\psi|\mathcal{N}^2 H_{12}|\psi\rangle, \tag{10.15}$$

where $c_o = 3g_s^2(N_c^2-4)/N_c^2$. If one wants to have the same normalized scalar product for Ψ as for ψ one has to choose $\mathcal{N} = (c_o H_{12})^{-1/2}$.

Somewhat later this Odderon was also found in the dipole approach [Kovchegov *et al.* (2004)], also associated with the antisymmetric Pomeron state. Subsequently this dipole Odderon was included into the fan diagrams giving rise to the evolution equations for the combined Pomeron plus Odderon evolution. Denoting the Pomeron and Odderon density fields as N and O respectively, the equations are [Hatta *et al.* (2005)]

$$\begin{aligned}\frac{d}{dY}N_{\boldsymbol{x},\boldsymbol{y}} = \bar{\alpha}_s \int \frac{d^2z}{2\pi}\frac{|x-y|^2}{|x-z|^2|y-z|^2}\\ \times\left(N_{\boldsymbol{x},\boldsymbol{z}} + N_{\boldsymbol{y},\boldsymbol{z}} - N_{\boldsymbol{x},\boldsymbol{y}} - N_{\boldsymbol{x},\boldsymbol{z}}N_{\boldsymbol{y},\boldsymbol{z}} + O_{\boldsymbol{x},\boldsymbol{z}}O_{\boldsymbol{y},\boldsymbol{z}}\right)\end{aligned} \tag{10.16}$$

and

$$\begin{aligned}\frac{d}{dY}O_{\boldsymbol{x},\boldsymbol{y}} = \bar{\alpha}_s \int \frac{d^2z}{2\pi}\frac{|x-y|^2}{|x-z|^2|y-z|^2}\\ \times\left(O_{\boldsymbol{x},\boldsymbol{z}} + O_{\boldsymbol{y},\boldsymbol{z}} - O_{\boldsymbol{x},\boldsymbol{y}} - O_{\boldsymbol{x},\boldsymbol{z}}N_{\boldsymbol{y},\boldsymbol{z}} - N_{\boldsymbol{x},\boldsymbol{z}}O_{\boldsymbol{y},\boldsymbol{z}}\right).\end{aligned} \tag{10.17}$$

Here we show that at least the first of these equations can be derived in the BFKL-Bartels framework on the basis of the known transition vertex $V_{2\to 6}$ constructed in [Bartels and Ewerz (1999)].

First we recall how the Pomeron part of Eq. (10.16) was derived in [Bartels *et al.* (2005b)]. The linear part on the right hand side of both equations (10.16) and (10.17) is just the BFKL evolution of the Pomeron and Odderon fields. To see the link between the reggeized gluon BFKL description and the dipole picture consider the evolution of the BFKL Pomeron wave function in the form of the quasi-Schroedinger equation

$$\frac{d}{dY}\psi(\rho_1,\rho_2) = -H_{12}\psi(\rho_1,\rho_2) \tag{10.18}$$

with the kernel $H_{12} = \frac{\bar{\alpha}_s}{2}h_{12}$, where $\bar{\alpha}_s = \alpha_s N_c/\pi$ and

$$h_{12} = \ln|p_1|^2 + \ln|p_2|^2 + \frac{1}{p_1p_2^*}\ln|\rho_{12}|^2\,p_1p_2^* + \frac{1}{p_1^*p_2}\ln|\rho_{12}|^2\,p_1^*p_2 - 4\Psi(1). \tag{10.19}$$

Here $\Psi(x) = d\ln\Gamma(x)/dx$, gluon holomorphic momenta are used and the first two log terms belong to the virtual corrections. Passing to the coordinate representation and taking into account that the coordinate wave function of the Pomeron vanishes when the two reggeized gluon are located at the same point (termed "the Möbius representation in [Bartels *et al.* (2005b,a)]") the action of this kernel on a state can be rewritten as

$$H_{12}\,\psi(\boldsymbol{\rho}_1,\boldsymbol{\rho}_2) = \bar{\alpha}_s \int \frac{d^2\rho_3}{2\pi} \frac{|\rho_{12}|^2}{|\rho_{13}|^2\,|\rho_{23}|^2} \left(\psi(\boldsymbol{\rho}_1,\boldsymbol{\rho}_2) - \psi(\boldsymbol{\rho}_1,\boldsymbol{\rho}_3) - \psi(\boldsymbol{\rho}_2,\boldsymbol{\rho}_3)\right). \tag{10.20}$$

This has the same form as the linear part of Eqs. (10.16) and (10.17).

Now consider the first non-linear term on the right-hand side of Eq. (10.16). The irreducible term D_4^I of Eq. (10.9) contains the transition from 2 to 4 reggeized gluon states. The latter evolve in rapidity with the 4-gluon Green function, which in the planar (large N_c) limit factorizes into the product of two independent BFKL Green functions: $G_4 = G_2 \times G_2$. Then the second term of Eq. (10.9) acquires the structure of the triple Pomeron vertex. In the planar limit $V_{2\to4}D_2$ becomes

$$\begin{aligned} V(1234)D_2 = \frac{1}{2}g^2\Big[&G(1,2+3,4) + G(2,1+3,4) + G(1,2+4,3) \\ &+ G(2,1+4,3) - G(1+2,3,4 - G(1+2,4,3) - G(1,2,3+4) \\ &- G(2,1,3+4) + G(1+2,0,3+4)\Big], \end{aligned} \tag{10.21}$$

where the numbers i denote the momenta q_i. The general function $G(1,2,3)$ was defined in [Braun and Vacca (1999a)] and consists of two pieces

$$G(\boldsymbol{k}_1,\boldsymbol{k}_2,\boldsymbol{k}_3) = G_1(\boldsymbol{k}_1,\boldsymbol{k}_2,\boldsymbol{k}_3) + G_2(\boldsymbol{k}_1,\boldsymbol{k}_2,\boldsymbol{k}_3). \tag{10.22}$$

The first term is constructed from real diagrams

$$G_1(\boldsymbol{k}_1,\boldsymbol{k}_2,\boldsymbol{k}_3) = g^2 N_c \int \frac{d^2\boldsymbol{q}_1 d^2\boldsymbol{q}_2}{(2\pi)^3}\delta^2(\boldsymbol{q}_1+\boldsymbol{q}_2-\boldsymbol{q}) W(\boldsymbol{k}_1,\boldsymbol{k}_2,\boldsymbol{k}_3|\boldsymbol{q}_1,\boldsymbol{q}_2) D_2(\boldsymbol{q}_1,\boldsymbol{q}_2), \tag{10.23}$$

where $\boldsymbol{q} = \boldsymbol{k}_1 + \boldsymbol{k}_2 + \boldsymbol{k}_3$ and

$$W(\boldsymbol{k}_1,\boldsymbol{k}_2,\boldsymbol{k}_3|\boldsymbol{q}_1,\boldsymbol{q}_2) = \left(\frac{(\boldsymbol{k}_2+\boldsymbol{k}_3)^2}{(\boldsymbol{q}_1-\boldsymbol{k}_1)^2\boldsymbol{q}_2^2} + \frac{(\boldsymbol{k}_1+\boldsymbol{k}_2)^2}{\boldsymbol{q}_1^2(\boldsymbol{q}_2-\boldsymbol{k}_3)^2} - \frac{\boldsymbol{k}_2^2}{(\boldsymbol{q}_1-\boldsymbol{k}_1)^2(\boldsymbol{q}_2-\boldsymbol{k}_3)^2} - \frac{\boldsymbol{q}^2}{\boldsymbol{q}_1^2\boldsymbol{q}_2^2}\right). \tag{10.24}$$

The second term is related to the virtual correction present in the reggeized gluon trajectory:

$$\begin{aligned} G_2(\boldsymbol{k}_1,\boldsymbol{k}_2,\boldsymbol{k}_3) = &-[\omega(\boldsymbol{k}_2) - \omega(\boldsymbol{k}_2+\boldsymbol{k}_3)]\,D_2(\boldsymbol{k}_1,\boldsymbol{k}_2+\boldsymbol{k}_3) \\ &-[\omega(\boldsymbol{k}_2) - \omega(\boldsymbol{k}_1+\boldsymbol{k}_2)]\,D_2(\boldsymbol{k}_1+\boldsymbol{k}_2,\boldsymbol{k}_3). \end{aligned}$$

Fourier transforming the vertex (10.21), D_2, and the two Pomeron subsystems $G_2(12) \times G_2(34)$ (in fact passing to the Möbius representation for the states) one finds

$$T_4 \equiv \langle \psi(12)|(V(1234)D_2)|\psi(34)\rangle = 2g_s^2 \langle \psi(12)|(G(1,2+3,4)|\psi(34)\rangle$$
$$= \psi(\rho_{13})\psi(\rho_{23}) \left[-c \frac{\rho_{12}^2}{\rho_{13}^2 \rho_{23}^2} \right] D_2(\rho_{12}), \tag{10.25}$$

where $c = g_s^4 N_c/(4\pi^3)$ and $\rho_{12} = \rho_1 - \rho_2$. This has exactly the same form of the interaction term present in Eq. (10.16), with a suitable rescaling (by $\frac{\bar{\alpha}_s}{2\pi c}$) of the two dipole (Pomeron) fields $N(\rho_1, \rho_3)N(\rho_2, \rho_3)$.

Now we perform a similar derivation for the last term on the right-hand side of Eq. (10.17), made of two Odderon fields. It comes from the $V_{2\to 6}$ vertex present in the D_6^I of Eq. (10.11) discussed before. Consider the six gluons $(1-6)$ grouped into two Odderon fields as (123) and (456). In the planar limit the Odderon term of the vertex can be written in a compact notation [Bartels *et al.* (2004)] as

$$W_6^{odd}(1,2,3|4,5,6) = -\frac{1}{8} g_s^4 (\hat{S}_1 - \hat{P}_1) f_{12} (\hat{S}_2^\dagger - \hat{P}_2^\dagger). \tag{10.26}$$

Here the indices 1 and 2 refer to the triplets of gluons (123) and (456), the function f_{12} depends on four gluon momenta

$$f(1,2|3,4) = G(1,2+3,4) - G(2,1+3,4) - G(1,2+4,3) + G(2,1+4,3). \tag{10.27}$$

The operator $\hat{S}$ was defined in (10.14) and $\hat{P}$ is an operator which, when acting on a function of two gluon momenta, antisymmetrizes them and splits the first momentum into three outgoing momenta, while putting to zero the second one:

$$\hat{P}(1,2,3|1',2')\phi(1',2') = \frac{1}{2}[\phi(123,0) - \phi(0,123)]. \tag{10.28}$$

(The presence of the subtraction term ($\hat{P}$) in the definition of W_6^{odd} is due to the fact that the function f_{12} is not zero in the two 2-gluons subsectors when one the momenta is zero). Calling φ_- a generic two gluon antisymmetric function one has [Bartels *et al.* (2000)]

$$\langle \Psi|(\hat{S} - \hat{P})\varphi_-\rangle = c_o \langle \psi|\mathcal{N} H_{12}|\varphi_-\rangle \,. \tag{10.29}$$

Note that in the Odderon channel the eigenvalues of H_{12} are non negative. The BLV states start with the intercept exactly equal to unity.

One then finds

$$\langle \Psi_{n,\nu}(1,2,3)|(\hat{S} - \hat{P}) = c_o^{1/2} \langle \psi_{n,\nu}(1,2)|H_{12}^{1/2} \,, \tag{10.30}$$

where n, ν are the conformal quantum numbers of the Odderon state Ψ and the related Pomeron state ψ. Using (10.30) the contraction of the vertex W_6^{odd} with two

Odderon states $\Psi_1(1,2,3)$ and $\Psi_2(4,5,6)$, normalized as described above, becomes

$$\begin{aligned} T_6 &\equiv -\frac{g_s^2}{8}\langle\Psi_1|(\hat{S}_1-\hat{P}_1)f_{12}(\hat{S}_2^\dagger-\hat{P}_2^\dagger)|\Psi_2\rangle \\ &= -\frac{g_s^2}{8}c_O\langle\psi_{1\nu_1,n_1}|H_{12}^{1/2}\, f_{12}\, H_{12}^{1/2}|\psi_{2\nu_2,n_2}\rangle \\ &= -\frac{g_s^2}{2}c_O\psi_{\nu_1,n_1}(1,2)H_{12}^{1/2}\; G(1,2+3,4)H_{34}^{1/2}\psi_{\nu_2,n_2}(3,4), \end{aligned} \tag{10.31}$$

where for the last equality the antisymmetry of the 2 gluon state ψ associated to the Odderon state is used. Then performing a Fourier transform one finally finds, similarly to Eq. (10.25),

$$T_6 = \psi(\rho_{13})H_{13}^{1/2}\psi(\rho_{23})H_{23}^{1/2}\left[c'\frac{\rho_{12}^2}{\rho_{13}^2\rho_{23}^2}\right]D_2(\rho_{12}), \tag{10.32}$$

where $c' = \frac{g_s^4 N_c}{16\pi^3}c_O$, and in the square brackets one can identify the Pomeron → two Odderon vertex in the coordinate representation, which in Eq. (10.16) is multiplied by the two Odderon fields $O_{\boldsymbol{\rho}_1,\boldsymbol{\rho}_3}O_{\boldsymbol{\rho}_2,\boldsymbol{\rho}_3}$. In order to match the dipole picture representation in the large N_c limit, one needs as before to renormalize the BLV Odderon field, rescaling it by the extra operator factor $H_{ij}^{1/2}$ as well as by the constant factor $\frac{\bar{\alpha}_s}{2\pi\sqrt{cc'}}$.

It is remarkable and important that the Pomeron-two-Odderons vertex turns out to be of the opposite sign compared to the triple Pomeron vertex. In this framework this fact comes automatically from the expressions of the two vertices $V_{2\to4}$ and $V_{2\to6}$. This property is related to the different signatures of the Pomeron and Odderon fields and, in particular, to the fact that in a Pomeron amplitude the two Pomeron cut is negative while the two Odderon cut is positive. In the next subsection we show how these properties are realized in an effective local description called Reggeon Field Theory.

In Eq. (10.17) the interaction terms ON and NO could be computed in the reggeized gluon approach starting from the transition vertex $V_{3\to5}$, which unfortunately has not been computed yet in the reggeized gluon approach.

10.2.2 *A simplified picture: The Reggeon field theory*

The equations above are used to study Pomeron-Odderon fields in the regime where saturation effects set in taming their growth with rapidity. This behavior is induced by the fan structure which is resummed by the differential equations. Nevertheless, as always stressed by L. N. Lipatov, the real picture is more complicated since one should include also loops involving Pomeron and Odderon fields (and more in general reggeized gluons). Moreover at high energy (large rapidities) but long transverse distances non perturbative QCD physics, where hadrons are involved, cannot be ignored. This is a problem whose solution at present is well out of reach.

Before the QCD era strong interactions were studied in the context of S-matrix theory and the scattering in the Regge limit was described by simpler systems derived from the analytic structure of the partial wave amplitudes, in particular focusing on the so called reggeons associated to their poles. As the next natural step, reggeon interactions were considered by V. N. Gribov and encoded in a $2+1$ dimensional Reggeon Field Theory, with one time (rapidity) and two transverse space dimensions. This is a framework were Pomeron and Odderon interactions can be studied, in particular using renormalization group (RG) techniques. Recently an investigation using a functional RG approach was carried out [Bartels *et al.* (2016, 2017)], including the case of a system of one Pomeron and one Odderon local fields, denoted as $\psi, \psi^\dagger$ and $\chi, \chi^\dagger$, respectively. The RFT action in its simpler version, with just one local cubic interactions, has the form:

$$S[\psi^\dagger, \psi, \chi^\dagger, \chi] = \int \mathrm{d}^D x\, \mathrm{d}\tau \left(Z_P \left(\frac{1}{2}\psi^\dagger \overleftrightarrow{\partial}_\tau \psi - \alpha'_P \psi^\dagger \nabla^2 \psi \right) \right.$$
$$\left. + Z_O \left(\frac{1}{2}\chi^\dagger \overleftrightarrow{\partial}_\tau \chi - \alpha'_O \chi^\dagger \nabla^2 \chi \right) + V[\psi, \psi^\dagger, \chi, \chi^\dagger] \right) \quad (10.33)$$

with the potential $V = V_3$ given by

$$V_3 = -\mu_P \psi^\dagger \psi + i\lambda \psi^\dagger (\psi + \psi^\dagger)\psi - \mu_O \chi^\dagger \chi + i\lambda_2 \chi^\dagger (\psi + \psi^\dagger)\chi$$
$$+ \lambda_3 (\psi^\dagger \chi^2 + {\chi^\dagger}^2 \psi). \quad (10.34)$$

Here D is the number of spatial dimensions ($D = 2$ is the physical case), which can be conveniently considered as a continuous parameter. The potential (allowed interactions) is constrained by signature conservation (even for the Pomeron, and odd for the Odderon) and by the overall sign of the multi-Reggeon discontinuity amplitudes: $-i\prod_j (i\xi_j)$, where ξ_j are the signature factors $\xi = (\tau - e^{-i\pi\omega})/\sin \pi\omega$ with $\omega = \alpha(0) - 1$. Note that for the Pomeron ξ_P is almost imaginary while for the Odderon ξ_O is almost real. This implies that t-channel states with odd and even number of Odderons never mix. Another constraint on the potential is that the transitions $P \to PP$ are imaginary (two Pomeron cut is negative), $P \to OO$ are real (two Odderon cut is positive) and $O \to OP$ is imaginary (Odderon-Pomeron cut is negative). These considerations, implemented in Eq. (10.34), are easily generalized to higher order (subleading) interactions [Bartels *et al.* (2017)]. As a consequence, we can write the potential in terms of different contributions with operators which relate states differing by an integer number of Odderon pairs. In perturbative QCD the $P \to OO$ vertex has been computed in the generalized leading-log approximation. We shall see that the Reggeon interactions nevertheless are dominated by a fixed point in which operators changing the number of Odderon pairs are not present. They can be present only in deformations of the critical theory.

Since $D = 4$ is the scaling dimension (critical dimension) of reggeon field theory, critical behaviour can be studied in perturbation theory with the ϵ expansion in $D = 4 - \epsilon$ dimensions. Moreover functional renormalization group techniques based on the use of the effective average action can be used in $D = 2$ in a truly non perturbative context. We present here the results of a one-loop renormalization group analysis in $D = 4 - \epsilon$, i.e. around the critical dimension 4 of the transverse space. We do not report the beta functions here but just the results for the fixed points and the linear behavior around them. In [Bartels *et al.* (2017)] also the case $D = 2$ was studied using non perturbative methods based on functional renormalization group techniques, which shows that the perturbative results are qualitatively correct but quantitatively different. Using the cubic truncation one finds, besides a fixed point solution related to the pure Pomeron theory, a second non trivial fixed point also in the Odderon sector, such that $\lambda^2, \lambda_2^2, \lambda_3^2, \mu_P, \mu_O = O(\epsilon)$:

$$\mu_P = \frac{\epsilon}{12}, \quad \mu_O = \frac{95 + 17\sqrt{33}}{2304}\epsilon, \quad \lambda^2 = \frac{8\pi^2}{3}\epsilon, \quad \eta_P = -\frac{\epsilon}{6}, \quad \eta_O = -\frac{7 + \sqrt{33}}{72}\epsilon,$$

$$\lambda_2^2 = \frac{23\sqrt{6} + 11\sqrt{22}}{48}\epsilon, \quad \lambda_3 = 0, \quad \zeta_P = \zeta_O = \frac{\epsilon}{12}, \quad r = \frac{3}{16}(\sqrt{33}-1). \tag{10.35}$$

Here $\eta_{P,O} = -\mu\frac{d}{d\mu}\log Z_{P,O}$ are the corresponding field anomalous dimensions, $\zeta_{P,O} = -\mu\frac{d}{d\mu}\log \alpha'_{P,O}$ are the corresponding anisotropic anomalous dimensions, and $r = \alpha'_O/\alpha'_P$ is the ratio between the slopes. One immediately notes that $\lambda_3 = 0$ so that the critical theory has no operator changing the number of Odderon pairs. Apart from the anomalous dimensions, the universal critical exponents are extracted with a spectral analysis of the stability matrix, obtained by perturbing the beta functions around the fixed point. We find two negative eigenvalues, associated to two relevant directions, which give the standard corresponding critical exponents ν related to the behavior of the two point functions for the Pomeron and Odderon:

$$\lambda^{(1)} = -2 + \frac{\epsilon}{4} \quad \to \nu_P = \frac{1}{2} + \frac{\epsilon}{16}, \quad \lambda^{(2)} = -2 + \frac{\epsilon}{12} \to \nu_O = \frac{1}{2} + \frac{\epsilon}{48}. \tag{10.36}$$

This fixed point (as well as its non perturbative counterpart for $D = 2$) is important in order to understand the renormalization group flow properties. Indeed we expect that from the QCD description at short distances moving to larger and larger transverse distances one gets closer to an effective RFT formulation. As a last remark we can add that this system is in the same universality class of an out of equilibrium directed percolating system, as it was shown by J. Cardy for a purely single Pomeron RFT.

10.3 The running coupling constant

10.3.1 *Lipatov's proposal*

The perturbative QCD Pomeron and Odderon in the LLA come from a summation of terms of the order $(\alpha_s \ln s)^n$ but neglect terms of the order $(\alpha_s \ln k^2)^n$ responsible for the running of the coupling. So in this theory α_s is assumed to be a fixed (small) parameter. As a result, the BFKL theory neglects one of the basic properties of QCD — the running of the coupling and so the asymptotic freedom. It also automatically neglects its other basic properties — the confinement, which manifests itself in the growth of the coupling at small k^2. It is then natural that nearly immediately after the introduction of the BFKL Pomeron attempts to introduce the running of the coupling in the theory appeared .

The first of these attempts was made by L. N. Lipatov as early as in 1986 [Lipatov (1986)]. It was rather straightforward. Let the BFKL Hamiltonian H be written with the separated factor α_s, namely $H = \alpha_s h$. Consider the forward case, so that the BFKL equation is given by

$$\alpha_s(h\psi)(k) = E\psi(k), \quad \text{or} \quad (h\psi)(k) = \chi\,\psi(k) \tag{10.37}$$

with the set of eigenvalues $\chi_{\nu,n}$ and eigenfunctions $\psi_{\nu,n}$

$$\chi_{\nu,n} = 2\hat{\psi}(1) - \hat{\psi}\left(\frac{1}{2} + i\nu\right) - \hat{\psi}\left(\frac{1}{2} - i\nu\right), \quad \psi_{\nu,n}(k) = ck^{2i\nu-1/2}e^{in\phi}, \tag{10.38}$$

where $\hat{\psi}$ is the logarithmic derivative of the Γ-function, and $-\infty < \nu < +\infty$, $n = 0, \pm1, \pm2, \ldots$. The idea of L. N. Lipatov was to change the fixed α_s in (10.37) into the running one as a function of k

$$\alpha_s \to \frac{1}{\beta_0 \ln(k^2/\Lambda_{QCD}^2)}, \quad \beta_0 = \frac{11N_c - 2N_f}{12\pi} = \frac{b}{\pi}. \tag{10.39}$$

Rescaling $k \to k/\Lambda_{QCD}$ then turns Eq. (10.37) into

$$(h\psi)(k) = \omega\,\beta_0 \ln(k^2)\psi(k). \tag{10.40}$$

Now one passes to the (ν, n) representation

$$\psi(k) = \int d\nu \sum_n \psi_{\nu,n} a_{\nu,n} \tag{10.41}$$

and obtains an equation for $a_{n,nu}$

$$i\omega\beta_0 \frac{da_{\nu,n}}{d\nu} = \chi(\nu, n)a_{\nu,n}, \tag{10.42}$$

which evidently admits an explicit solution.

The next step is to impose the appropriate boundary conditions. Addressing the reader to the original paper for details, we only mention that they are two conditions, following from the behavior of the resulting $\psi(k)$ at large and small k. At large k it is required that the solution passes to the known form for the DGLAP equation in the double log approximation. At small k it is required that the phase of ψ should be fixed from the non-perturbative region. The two boundary conditions make the spectrum a discrete one, which for the n-th state with $n \gg 1$ is

$$\omega_n = -\frac{0.4085}{n - \frac{1}{4} + \frac{\phi_0 - \nu t_0}{\pi}}, \tag{10.43}$$

where ϕ_0 is the non-perturbative phase and the small $t_0 = k_0^2/\Lambda_{QCD}^2$ is the point at which the low boundary condition is imposed. As a result one finds an infinite sequence of poles in the j plane converging to the point $j = 1$.

This way of introducting the running coupling raises questions. The operator h is in fact an integral operator with a kernel $h(k, k')$. The change (10.39) implies that one argument is used for the running. Such a choice, apart from being arbitrary, violates the intrinsic symmetry between k and k' which is essential for the interchange projectile $\leftrightarrow$ target. The form of the boundary at small k does not seem to be unique. Still the conclusion of the splitting of the cut into a series of poles as a result of the running of the coupling was later confirmed in other versions of the running coupling, and so it was an important prediction in this line of approach. Also the Lipatov running coupling allows to obtain many explicit results, in contrast to more elaborate versions. Later it was extensively used to obtain physically important results (the latest are [Kowalski *et al.* (2014, 2016, 2010)]). See also [Bartels *et al.* (2019)].

10.3.2 *The running coupling constant from the bootstrap*

In our search of a more fundamental basis for the introduction of the running coupling constant many years ago we turned to the bootstrap property of the BFKL dynamics, which is one of the pillars on which the idea of gluon reggeization rests [Braun (1995)]. As it is well known, the bootstrap discovered in the original papers [Kuraev *et al.* (1977); Balitsky and Lipatov (1978)] is the unitarity condition for the t-channel with the gluon color number. It guarantees that the unitarity in this section gives rise to the solution in the form of the reggeized gluon. It was natural to expect that this condition should be preserved also with the running coupling. So preservation of the bootstrap could be the guiding principle for the introduction of the running coupling constant. Later the idea of a strong bootstrap condition in perturbation theory up to next-to-leading order was proposed in [Braun and Vacca (1999b, 2000)], where non trivial relations for the NLO BFKL in the octet channel, the reggeized gluon trajectory and any impact factor were conjectured. They were later on veryfied by V. Fadin and collaborators [Fadin and Papa (2002)].

To realize this idea one can write the inhomogeneous equation for two reggeons as

$$(H - E)\psi = F \tag{10.44}$$

with the Hamiltonian for the color group $SU(N_c)$

$$H = -\omega(q_1) - \omega(q_2) + (T_1T_2)V. \tag{10.45}$$

The form of $\omega(q)$ and of the interaction V are in fact interrelated. To make it explicit we express both via a single function $\eta(q)$.[1] Namely

$$\omega(q) = \frac{1}{2} N_c \int \frac{d^2 q_1}{(2\pi)^2} \frac{\eta(q)}{\eta(q_1)\eta(q - q_1)}. \tag{10.46}$$

The interaction term, apart from the product of the gluon color vectors T_i^a, $i = 1, 2$, $a = 1, \ldots N_c$, involves the interaction kernel V which we also express via the same function $\eta(q)$

$$V(q_1, q_2|q_1' q_2') = \frac{\eta(q_1 + q_2)}{\eta(q_1')\eta(q_2')} - \left(\frac{\eta(q_1)}{\eta(q_1')} + \frac{\eta(q_2)}{\eta(q_2')} \right) \frac{1}{\eta(q_1 - q_1')}. \tag{10.47}$$

It conserves the momentum: $q_1 + q_2 = q_1' + q_2'$. Combining (10.46) with (10.47) for the gluon channel with $(T_1 T_2) = -\frac{1}{2} N_c$ one finds the bootstrap relation for any $\eta(q)$:

$$\frac{1}{2} N_c \int \frac{d^2 q_1}{(2\pi)^2} V(q_1, q_2|q_1', q_2') = \omega(q_1) + \omega(q_2) - \omega(q_1 + q_2)\,. \tag{10.48}$$

Then inserting $F = F(q_1 + q_2)$ into the inhomogeneous equation one finds a solution to Eq. (10.44):

$$\psi(q_1, q_2) = \psi(q_1 + q_2) = \frac{F(q_1 + q_2)}{\omega(q_1 + q_2) - E}. \tag{10.49}$$

This means that the two gluons 1 and 2 have fused into a single one with the momentum $q_1 + q_2$ and the moving pole at $j = 1 - E$. This phenomenon is true for arbitrary $\eta(q)$.

For the original BFKL equation with the fixed coupling we have a particular $\eta(q)$:

$$\eta(q) = \frac{2\pi}{g^2} q^2, \quad \eta(0) = 0\,. \tag{10.50}$$

Then one finds the standard LL BFKL expressions for the trajectory $\omega(q)$ and interaction $V(q_1, q_2|q_1', q_2')$. Also the trajectory passes through zero at $q = 0$ to correspond to the real gluon with a vanishing mass.

Using the fact that the bootstrap is fulfilled for arbitrary $\eta(q)$ we can choose $\eta(q)$ to satisfy the high-momentum behavior of the gluon distribution with a running coupling [Braun (1995)]

$$\eta(q) = \frac{b}{2\pi} q^2 \ln \frac{q^2}{\Lambda_{QCD}^2}, \quad q^2 \gg \Lambda^2, \tag{10.51}$$

where Λ is the standard QCD parameter, and $b = (11Nc - 2Nf)/12$ is defined in (10.39).

[1] We stress that this is just one possible way, but not the unique one, to satisfy the bootstrap relation.

The asymptotic behavior (10.51) and the condition $\eta(0) = 0$ are the only properties of $\eta(q)$ which follow from the theoretical reasoning. A concrete form of $\eta(q)$ interpolating between $q^2 = 0$ and (10.51) may be chosen differently. One hopes that physical results will not too strongly depend on the choice.

With $\eta(q)$ satisfying (10.51) we can build the trajectory $\omega(q)$ and the interaction V between two reggeons with the running coupling constant. In this way we obtain equations for both Pomeron, Odderon and higher BKP states with the running coupling constant. Note that, as with the fixed coupling, the resulting equations remain infrared stable if $\eta(0) = 0$ as dictated by the zero gluon mass. However, we can improve the infrared behavior to mimic confinement and introduce a finite gluon mass m dropping the latter condition. In our practical applications we use

$$\eta(q) = \frac{b}{2\pi} f(q), \quad f(q) = (q^2 + m^2) \ln \frac{q^2 + m^2}{\Lambda^2}. \tag{10.52}$$

With this choice $\eta(0) \neq 0$, and this has to be taken into account in the interaction.

It is interesting to note that this very simple ansatz given in Eqs. (10.46) and (10.47) in order to satisfy the bootstrap condition, when feeded with the choice of Eq. (10.51), can be studied to compare to the results of the NLL BFKL kernel and the external particle impact factors. This can be conveniently done using dimensional regularization [Braun and Vacca (2000)]: there it was shown that fermionic contributions (the N_f dependent part) were correctly reproduced, as well as the gluon sector related to the running coupling. Only a gluon contribution of the NLL approximation is actually missing. This is therefore an interesting approximation containing an important part of the NLL corrections and can be considered as a good starting point to study the running coupling effects.

10.3.3 *The BLV Odderon with the running coupling constant*

For the Odderon the color factor of the wave function is $d_{a_1 a_2 a_3}$ where a_i is the color of the i-th reggeon. Separating it, the equation for the Odderon momentum wave function is

$$(H_{od} - E)\psi_{od} = 0 \tag{10.53}$$

with a Hamiltonian

$$H_{od} = -\sum_{i=1}^{3} \omega(q_i) - \frac{1}{2} N_c \sum_{i<k}^{3} V_{ik}. \tag{10.54}$$

Consider the Odderon equation with a specific inhomogeneous term

$$(H_{od} - E)\psi_{12} = F_{12}, \tag{10.55}$$

where

$$F_{12}(1,2,3) = \int \frac{d^2 q_3'}{(2\pi)^2} \hat{W}(1,2,3|1',3')\psi(1',3') + F_2(1,2,3), \tag{10.56}$$

where for brevity we denote the momenta q_1, q_2, q_3 just by their numbers 1, 2, 3 and so on. Here $q_1' + q_3' = q_1 + q_2 + q_3$. In Eq. (10.56) $\psi(1', 3')$ is some function of two momenta q_1' and q_3'. The kernel $\hat{W}$ is a sum of two terms

$$\hat{W}(1,2,3|1',3') = -\frac{1}{2}N_c\Big(W(2,1,3|1',3') + W(1,2,3|1',3')\Big), \tag{10.57}$$

and W is the difference between two kernels (10.47), expressed via the function $\eta(q)$. In full notation

$$W(q_2, q_1, q_3|q_1', q_3') = V(q_1, q_3|q_1' - q_2, q_3') - V(q_1 + q_2, q_3|q_1', q_i'). \tag{10.58}$$

It possesses an important symmetry property:

$$W(1,2,3|1',3') = W(3,2,1|3',1'). \tag{10.59}$$

Note that in the fixed coupling case this kernel reduces to the Bartels kernel $K_{2\to 3}$ which describes the transition from two reggeized gluons tlo three [Bartels and Ewerz (1999)]. Therefore one could be tempted to introduce in a similar way the running coupling also in the effective vertices which change the number of (t-channel) reggeized gluons, and in particular the Pomeron $\to$ 2 Pomeron and the Pomeron $\to$ 2 Odderon vertices whose expressions are known at fixed coupling [Bartels and Wusthoff (1995); Bartels and Ewerz (1999)].

It can be shown [Braun (2008)] that the inhomogeneous equation (10.55) can be solved by

$$\psi_{12}(q_1, q_2, q_3) = \psi(q_1 + q_2, q_3), \tag{10.60}$$

provided the function ψ is a Pomeron wave function which satisfies the homogeneous Schroedinger equation (10.44) with a Hamiltonian (10.45) in which $T_1T_2 = -N_c$ (the BFKL equation in the case of a fixed coupling).

Using this, we cyclically permute the gluons 1, 2, 3 to obtain two more relations

$$(H_{od} - E)\psi_{23} = F_{23} \tag{10.61}$$

and

$$(H_{od} - E)\psi_{31} = F_{31}, \tag{10.62}$$

where for instance

$$\psi_{23}(q_1, q_2, q_3) = \psi(q_2 + q_3, q_1) \tag{10.63}$$

and

$$F_{23}(1,2,3) = -\frac{1}{2}N_c \int \frac{d^2q_3'}{(2\pi)^2}\Big(W(3,2,1|1',3') + W(2,3,1|1',3')\Big)\psi(1',3') \tag{10.64}$$

(the integration momenta q_3' and $q_1' = q_1 + q_2 + q_3 - q_3'$ do not change under permutations of the external momenta). We add all the three equations obtained in this manner together to obtain

$$(H_{od} - E)\Big(\psi_{12} + \psi_{23} + \psi_{31}\Big) = F_{12} + F_{23} + F_{31} \equiv F_{tot}. \tag{10.65}$$

The total inhomogeneous term is

$$F_{tot}(1,2,3) = -\frac{N_c}{2}\int \frac{d^2q_3'}{(2\pi)^2} U(1,2,3|1',3')\psi(1',3'), \tag{10.66}$$

where

$$\begin{aligned} U(1,2,3|1',3') &= W(2,1,3|1',3') + W(1,2,3|1',3') + W(3,2,1|1',3') \\ &\quad + W(2,3,1|1',3') + W(1,3,2|1',3') + W(3,1,2|1',3'). \end{aligned} \tag{10.67}$$

Using the property (10.59) one easily finds that U is symmetric in the last pair of arguments

$$U(1,2,3|1',3') = U(1,2,3|3',1'). \tag{10.68}$$

It follows that if the function $\psi(q_1, q_2)$ is antisymmetric in its variables, the total inhomogeneous term (10.66) in Eq. (10.65) vanishes and the sum

$$\psi_{od} = \psi_{12} + \psi_{23} + \psi_{31} \tag{10.69}$$

is a solution of the Odderon equation

$$(H_{od} - E)\psi_{od} = 0. \tag{10.70}$$

Thus from any antisymmetric solution of the Pomeron equation one gets a corresponding solution of the Odderon equation.

For a fixed coupling this result was found in [Bartels *et al.* (2000)]. Our derivation generalizes it to the case of a running coupling, provided it is introduced in the manner which preserves the bootstrap.

10.3.4 *A general family of BKP states in the planar limit with the running coupling constant*

We consider here an n gluon composite state in color singlet which has the high energy behavior in the Regge limit described by the following kernel

$$H_n = -\sum_{i=1}^{n} \omega(q_i) + \sum_{i<j} T_i T_j V_{ij}, \tag{10.71}$$

which in the planar limit reduces to

$$H_n^{\infty} = \frac{1}{2}\left[H_{12}^{(1)} + H_{23}^{(1)} + \cdots + H_{n1}^{(1)}\right], \tag{10.72}$$

and where H_{ij} is a two gluon Hamiltonian acting on the gluons i and j. As for the case of the previous section we shall proceed with the replacement of the fixed coupling case of Eq. (10.50) by the running coupling one of Eq. (10.51), this time following an approach similar to [Vacca (2000)]. Starting from the ansatz

$$\psi_n(k_1,k_2,\ldots,k_n) = \sum_{i=0}^{n-1} (R_n)^i c_i\, \psi_{n-1}(k_1+k_2,k_3,\ldots,k_n), \tag{10.73}$$

where R_n is the generator of a cyclic shift along the cylinder, i.e. $(R_n)^n = 1$, and c_i are constant to be determined, and rearranging the Hamiltonian in the form

$$\bar{H}_n^{\infty}(1,2,\ldots,n) = \frac{1}{2}\left(\bar{H}_{12}^{(1)} + \bar{H}_{1n}^{(1)} - \bar{H}_{2n}^{(1)}\right) + \bar{H}_{n-1}^{\infty}(2,3,\ldots,n), \tag{10.74}$$

let us consider the action of the Hamiltonian on the function $g = \psi_{n-1}(k_1 + k_2, k_3, \ldots, k_n)$. One finds

$$\begin{aligned}
&\bar{H}_n^{\infty}(1,2,\ldots,n) \otimes g \\
&= \int \{d^2k_i'\} \bar{H}_{n-1}^{\infty}(k_1+k_2,k_3,\ldots,k_n|k_2',k_3',\ldots,k_n')\, \psi_{n-1}(k_2',k_3',\ldots,k_n') \\
&\quad + \frac{1}{2}\int \{d^2k_i'\}\, W(k_1,k_2,k_3|k_1',k_3')\psi_{n-1}(k_1',k_3',\ldots,k_n') \\
&\quad + \frac{1}{2}\int \{d^2k_i'\}\, W(k_2,k_1,k_n|k_2',k_n')\, \psi_{n-1}(k_2',k_3',\ldots,k_n'),
\end{aligned} \tag{10.75}$$

where we denote with $\{d^2k_i'\}$ the necessary integral measure. Imposing for ψ_{n-1} the following properties: (1) it is an eigenstate, i.e. $\bar{K}_{n-1}^{\infty}\, \psi_{n-1} = E\, \psi_{n-1}$ and (2) has definite symmetry such that $R_{n-1}\, \psi_{n-1} = r_{n-1}\, \psi_{n-1}$, one can write the previous relation as

$$\begin{aligned}
&\bar{H}_n^{\infty}(1,2,\ldots,n) \otimes \psi_{n-1}(k_1+k_2,k_3,\ldots,k_n) \\
&= E\, \psi_{n-1}(k_1+k_2,k_3,\ldots,k_n) + \frac{1+r_{n-1}(R_n)^{-1}}{2} \\
&\quad \times \int \{d^2k_i'\}\, W(k_1,k_2,k_3|k_1',k_3')\psi_{n-1}(k_1',k_3',\ldots,k_n'),
\end{aligned} \tag{10.76}$$

so that, using the definition of Eq. (10.73) we can have the relation

$$\bar{H}_n^{\infty}\psi_n = E\, \psi_n, \tag{10.77}$$

provided the unknown coefficients c_i satisfy the secular equation

$$\sum_{i=0}^{n-1} (R_n)^i \left(1 + r_{n-1}(R_n)^{-1}\right) c_i = 0, \tag{10.78}$$

where r_{n-1} plays the role of an eigenvalue and c_i are the components of the corresponding eigenvector. Using the constraint $(r_{n-1})^{n-1} = 1$, one finds $r_{n-1} = (-1)^n$. Therefore for n even one has $r_{n-1} = +1$ and $c_i = (-1)^i$, while for n odd we get $r_{n-1} = +1$ and $c_i = +1$. The case with n even is unphysical because the state ψ_n would not be Bose symmetric. Instead the case with n odd is physical, with the solution constructed from another solution for $n-1$ gluons which is odd under cyclic permutations. In particular the Odderon corresponds to the special case $n = 3$. In summary, given for an even number of gluons a solution of the kernel, improved to include running coupling effects using the bootstrap approach, with some intercept eigenvalue, then one can automatically construct another solution with one more gluon with the same intercept.

10.3.5 *Equations for the Pomeron and the BLV Odderon with the running coupling*

The BFKL equation admits explicit solutions both in the forward and non-forward directions based on its conformal invariance. With the running coupling this invariance is lost and one has to rely on numerical studies. The first calculations of the Pomeron ground states intercept and slope were done nearly immediately after the introduction of the running coupling from the bootstrap [Braun *et al.* (1996)]. Here we expand these calculations and extend them to the BLV Odderon. We borrow from [Braun *et al.* (1996)] the starting equations in the form suitable for numerical calculations.

We put $N_c = 3$, present $\eta(q)$ according to (10.52) and remove all numerical factors including b from the Hamiltonian redefining the eigenvalue. We also symmetrize the integral kernel in V by introducing a new wave function ϕ by

$$\psi(q_1) = \phi(q_1)\sqrt{f(q_1)f(q_2)}, \quad q_1 + q_2 = q, \tag{10.79}$$

where q is the total momentum. Then the equation for ϕ takes the form

$$\int d^2q_1' h_q(q_1, q_1')\phi(q_1') = A_q(q_1)\phi(q_1) + \int d^2q_1' L_q(q_1, q_1')\phi(q_1') = \epsilon(q)\phi(q_1). \tag{10.80}$$

Here the "kinetic energy" is

$$A_q(q_1) = (1/2)\int \frac{d^2q_1' f(q_1)}{f(q_1')f(q_1 - q_1')} + (1/2)\int \frac{d^2q_2' f(q_2)}{f(q_2')f(q_2 - q_2')}. \tag{10.81}$$

The interaction kernel consists of two parts, a quasilocal and a separable one:

$$L = L^{(ql)} + L^{(sep)}. \tag{10.82}$$

They are given by

$$L_q^{(ql)}(q_1, q_1') = -\sqrt{\frac{f(q_1)}{f(q_2)}}\frac{1}{f(q_1 - q_1')}\sqrt{\frac{f(q_2')}{f(q_1')}} - \sqrt{\frac{f(q_2)}{f(q_1)}}\frac{1}{f(q_2 - q_2')}\sqrt{\frac{f(q_1')}{f(q_2')}} \tag{10.83}$$

and

$$L_q^{(sep)}(q_1, q_1') = \frac{f(q)}{\sqrt{f(q_1)f(q_2)f(q_1')f(q_2')}}. \tag{10.84}$$

Both parts are evidently symmetric in q_1 and q_1'. The scaled energy ϵ is related to the initial one by

$$E = \frac{3}{2\pi b}\epsilon = \frac{6}{\pi(11-(2/3)N_F)}\epsilon. \tag{10.85}$$

Symmetric solutions of Eq. (10.80) give the Pomeron energies and wave functions. Its solutions being antisymmetric in $q_1 \leftrightarrow q_2$ give energies and wave functions of the BLV Odderon.

The nonforward equation (10.80) with $q \neq 0$ is too complicated for the numerical study due to three independent variables. So we concentrate on the forward case $q = 0$.

At $q = 0$ the equation retains its form (10.80) with

$$A_0(q_1) = \int \frac{d^2q_1' f(q_1)}{f(q_1')f(q_1-q_1')} \tag{10.86}$$

and the interaction given by (10.82) where now

$$L_0^{(ql)}(q_1,q_1') = -\frac{2}{f(q_1-q_1')} \tag{10.87}$$

has really become local and

$$L_0^{(sep)}(q_1,q_1') = \frac{f(0)}{f(q_1)f(q_1')}. \tag{10.88}$$

In the following we omit the subindex 0 implying $q = 0$.

To keep under control the behavior of the kernel at extremely large values of momenta we transform the trajectory term $A(q)$ similar to what is standardly done in the BFKL equation. Using the identity

$$\int d^2q_1 \frac{f(q)}{f(q_1)f(q-q_1)} = 2\int d^2q_1 \frac{f(q)}{f(q-q_1)\Big(f(q_1)+f(q-q_1)\Big)},$$

we find

$$A(q) = \int dq_1^2 \int_0^{2\pi} d\chi_1 \frac{f(q)}{f(q-q_1)\Big(f(q_1)+f(q-q_1)\Big)}, \tag{10.89}$$

where χ_1 is the azimuthal angle of q_1.

To make the equation for ϕ one-dimensional, the angular momentum n of the gluons is introduced

$$\phi(q) = \phi_n(q^2)e^{in\chi}, \tag{10.90}$$

where χ is the azimuthal angle. Integrating over it in the equation, one obtains an one-dimensional integral equation for the radial function $\phi_n(q^2)$:

$$A(q)\phi_n(q^2) + \int dq_1^2 L_n(q^2, q_1^2)\phi_n(q_1^2) = \epsilon\phi_n(q^2) \tag{10.91}$$

with the kernel L_n now given by

$$L_n(q^2, q_1^2) = -B_n(q^2, q_1^2) + \delta_{n0}\pi \frac{f(0)}{f(q)f(q_1)}, \tag{10.92}$$

where

$$B_n(q^2, q_1^2) = \int_0^{2\pi} d\chi \frac{\cos n\chi}{f(q^2 + q_1^2 - 2qq_1 \cos\chi)}. \tag{10.93}$$

The kinetic term $A(q)$ after the angular integration in Eq. (10.89) can be written as

$$A(q) = \int dq_1^2 \frac{f(q)}{f(q_1)} \Big(B_0(q, q_1) - C_0(q, q_1)\Big), \tag{10.94}$$

where

$$C_0(q, q_1) = \int_0^{2\pi} d\chi \frac{1}{f(q_1) + f(q - q_1)}. \tag{10.95}$$

These are the equations we are solving numerically. We shall see that the ground state of the Pomeron with negative energies indeed splits into an infinite series of discrete states, with energies accumulating at $E = 0$. In contrast, the Odderon states remain continuous with energies along a cut in the j plane starting from $j = 1$. As a result we expect that in the non-forward direction the discrete Pomeron states will move with the total momentum q with the intercept $\alpha(q)$. As to the continuous Odderon states we expect them to stay on the cut above $E = 0$. Should they move, it may introduce moving unphysical cuts in the t-plane and violate the unitarity.

As mentioned the equations become very difficult to solve at $q \neq 0$ even numerically. So, rather than to attempt to solve them for all q, we limit ourselves to small values of q and determine not the whole trajectory $\alpha(q)$ but only the intercept α' defined by

$$\alpha(q) = 1 - E(q) = 1 + \Delta - \alpha'(0)q^2. \tag{10.96}$$

This can be done in a much simpler manner using a perturbative approach. We present "the Hamiltonian" in (10.80)

$$H_q = A_q + L_q = H_0 + W(q) \tag{10.97}$$

and calculate analytically $W(q)$ up to terms of the second order in q. Then for small q the value of the energy $\epsilon(q)$ will be given by the standard perturbation formula

$$\epsilon(q) = \epsilon(0) + \langle W(q) \rangle, \tag{10.98}$$

where $\langle\rangle$ here means taking the average with the wave function at $q = 0$, determined from the numerical solution of the equation discussed above. Note that for the continuous and non-normalizable states $\langle W(q)\rangle = 0$ is expected unless the interaction is too bad. Our numerical studies confirm this result.

Technically it is natural to write the momenta as

$$q_1 = \frac{1}{2}q + l, \quad q_2 = \frac{1}{2}q - l, \quad q_1' = \frac{1}{2}q + l', \quad q_2' = \frac{1}{2}q - l' \tag{10.99}$$

and expand in power series all expressions in (10.80) up to the second order in q. The corresponding calculations are straightforward but tedious, and so together with the results they are briefly presented in the Appendix.

10.3.6 *Change of variable and discretization*

We pass to the logarithmic variable $t = \ln q^2$ (the unit in which q^2 is measured is inferred from the numerical values of the dimensional parameters m and Λ_{QCD}). To retain the symmetry of the kernel we express $\psi(q)$ as

$$\phi_n(q^2) = \tilde{\phi}_n(q^2)/q. \tag{10.100}$$

The equation becomes

$$A(q)\tilde{\phi}_n(q^2) + \int dt_1 q q_1 L_n(q^2, q_1^2)\tilde{\phi}_n(q_1^2) = \epsilon\tilde{\phi}_n(q^2), \tag{10.101}$$

where $q^2 = e^t$ and $q_1^2 = e_1^t$. Here

$$A(q) = \int dt_1 q_1^2 \frac{f(q)}{f(q_1)}\Big(B_0(q^2, q_1^2) - C_0(q^2, q_1^2)\Big). \tag{10.102}$$

The discretization is realized using a uniform grid in t

$$t_i = t_{min} + id, \quad i = 0, 1, \ldots, N, \quad d = \frac{t_{max} - t_{min}}{N}. \tag{10.103}$$

We approximate the integrals over t by finite sums

$$\int_{-\infty}^{\infty} dt\, F(t) \simeq \sum_{i=0}^{N} w_i F(t_i) \tag{10.104}$$

with points t_i and weights w_i depending on the chosen approximation scheme.

Then Eq. (10.101) becomes a finite linear system of equations, where the kernel has become a matrix. To symmetrize it we finally introduce

$$\tilde{\phi}_i = \frac{v_i}{\sqrt{w_i}}, \quad \text{or} \quad \phi(q_i^2) = \frac{1}{\sqrt{q_i^2 w_i}} v_i, \tag{10.105}$$

and the final equation is

$$A_i v_i + \sum_{j=0}^{N} \sqrt{w_i w_j} q_i q_j L_n(q_i^2, q_j^2) v_j = \epsilon v_i \tag{10.106}$$

with

$$A_i = \sum_{j=0}^{N} w_j q_j^2 \frac{f(q_i)}{f(q_j)} \Big(B_0(q_i^2, q_j^2) - C_0(q_i^2, q_j^2) \Big). \tag{10.107}$$

We are going to find the eigenvalues and eigenvectors of the matrix associated to the linear operator of the equation. The matrix is

$$M_{ij} = A_i \delta_{ij} + \sqrt{w_i w_j} q_i q_j \left(-B_n(q_i^2, q_j^2) + \delta_{n0} \pi \frac{f(0)}{f(q_i) f(q_j)} \right). \tag{10.108}$$

Its diagonal term is

$$M_{ii} = A_i + w_i q_i^2 \left(-B_n(q_i^2, q_i^2) + \delta_{n0} \pi \frac{f(0)}{f^2(q_i)} \right). \tag{10.109}$$

In order to make explicit the behavior at large momentum, i.e. at large t, we separate in A_i from the sum over j the term $j = i$ and write

$$A_i = w_i q_i^2 \Big(B_0(q_i^2, q_i^2) - C_0(q_i^2, q_i^2) \Big) + S, \tag{10.110}$$

where

$$S = \sum_{j=0, j \neq i} w_j q_j^2 \frac{f(q_i)}{f(q_j)} \Big(B_0(q_i^2, q_j^2) - C_0(q_i^2, q_j^2) \Big), \tag{10.111}$$

and combine the separated term with the second term in (10.109) to obtain

$$M_{ii} = w_i q_i^2 \left(B_0(q_i^2, q_i^2) - B_n(q_i^2, q_i^2) - C_0(q_i^2, q_i^2) + \delta_{n0} \pi \frac{f(0)}{f(q_i)^2} \right) + S. \tag{10.112}$$

Here the extra large term at large q_i is cancelled between $B_0(q_i^2, q_i^2)$ and $B_n(q_i^2, q_i^2)$. So denoting

$$B_{0n}(q_i^2) = B_0(q_i^2, q_i^2) - B_n(q_i^2, q_i^2) \tag{10.113}$$

we find finally

$$M_{ii} = w_i q_i^2 \left(B_{0n}(q_i^2) - C_0(q_i^2, q_i^2) + \delta_{n0} \pi \frac{f(0)}{f(q_i)^2} \right) + S. \tag{10.114}$$

In this form the calculation of eigenvalues of M can be implemented numerically with not much efforts.[2]

[2]E.g. it can be computed with a FORTRAN program using only double precision.

Note that for the Pomeron with $n = 0$ the first term in (10.114) is zero and we get

$$M_{ii}^{n=0} = w_i q_i^2 \Big(- C_0(q_i^2, q_i^2) + \pi \frac{f(0)}{f(q_i)^2} \Big) + S. \tag{10.115}$$

For the Odderon ($n = 1$) the last term in (10.114) is zero and one has

$$M_{ii}^{n=1} = w_i q_i^2 \Big(B_{01}(q_i^2) - C_0(q_i^2, q_i^2) \Big) + S. \tag{10.116}$$

In both cases S is given by (10.111).

In this form the matrix M is suitable for the determination of eigenvalues and eigenvectors with standard numerical techniques. We shall use a grid up to $t_{max} = 270$.

10.4 Numerical results

Our numerical results were obtained by the standard FORTRAN programs for finding all eigenvalues and eigenvectors of a real symmetric matrix. In the calculations made up to now we have taken

$$m = 0.82 GeV, \quad \Lambda = 0.2 GeV.$$

The lower limit in t was taken in all cases as

$$t_{min} = -90,$$

while the upper limit was taken as

$$t_{max} = 90, \ 180, \ 270.$$

Moreover the number of divisions N was taken as

$$N = 1000, \text{ and, } 2000.$$

It turned out that the results practically do not change with the rise of N from 1000 to 2000, so that $N = 1000$ proved to be fair enough for a reasonable accuracy. On the other hand, some results do depend on the value of t_{max}. It is worth mentioning that within the used FORTRAN program we were not able to raise it above 300.

We considered four cases. First we repeated our old calculations for the Pomeron with $n = 0$ and the running coupling. Second, we considered the Odderon ($n = 1$) with the running coupling. Finally we studied both the Pomeron and Odderon with a fixed coupling regulated by the "gluon mass" m in the infrared.

10.4.1 *Pomeron*

For the Pomeron we have found that, in agreement with the L. N. Lipatov picture, the introduction of the running coupling splits the negative part of the cut in energy into a sequence of poles converging to the start of the cut at $E = 0$. The number of negative poles rises from 19 at $t_{max} = 90$ to 55 at $t_{max} = 270$. The negative part of the spectrum at $t_{max} = 270$ is shown in Fig. 10.2. Energies and slopes of the first ten states at $t_{max} = 90$ and $t_{max} = 270$ are presented in Table 10.1. One observes that their values practically do not depend on t_{max} Note that the first two energies are in agreement with our old calculations. However their slopes are not, because of a past error in the code which we have corrected. We can study the energy ω_n and the slope α'_n of the state ϕ_n, as a function of n. From a fit on the first 10 leading states we find for the "energies"

$$\omega_n \simeq -\frac{0.4141}{n + 0.1383}, \tag{10.117}$$

which should be compared to the behavior reported in Eq. (10.43) and which is qualitatively similar also to what has been found in [Bartels *et al.* (2019)]. Also the slope can be fitted. It goes roughly as $\alpha'_n = a/(n^2 + b)$ for some a and b, even if the fit is a bit less accurate compared to what happens for the energies.

For comparison we calculated energies (in units $|E_{BFKL}|$) and slopes for the Pomeron with a fixed coupling. Here we observe a cut in energy starting at exactly

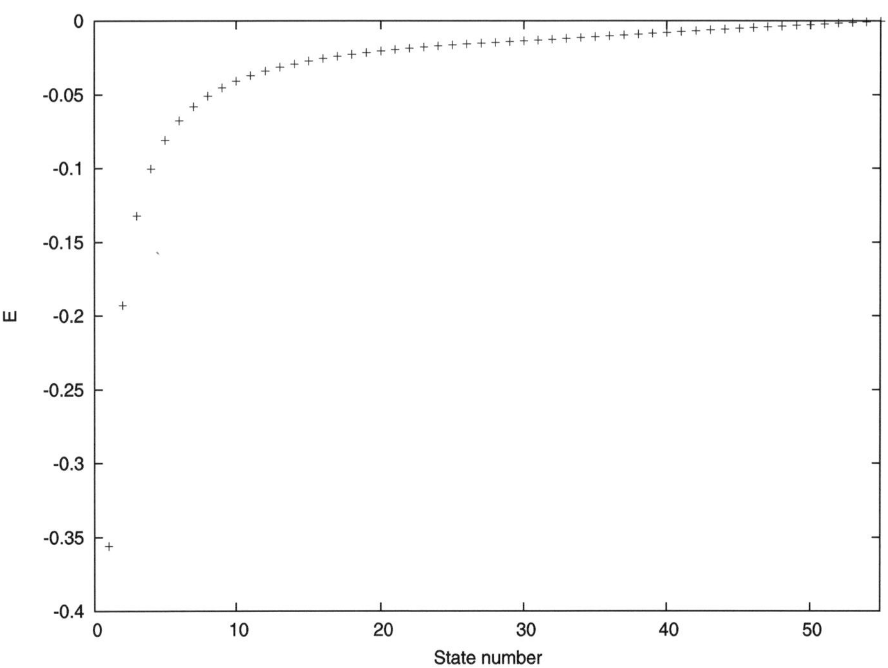

Fig. 10.2 Energies of the first 55 states of the Pomeron with the running coupling, $t_{max} = 270$.

Table 10.1 Energies and slopes of the first 10 states of the Pomeron (running coupling, $t_{max} = 270$).

$t_{max} = 90$		$t_{max} = 270$	
−0.35582E+00	0.61753E−01	−0.35584E+00	0.61731E−01
−0.19280E+00	0.20731E−01	−0.19296E+00	0.20732E−01
−0.13205E+00	0.10719E−01	−0.13215E+00	0.10713E−01
−0.10025E+00	0.65802E−02	−0.10066E+00	0.65780E−02
−0.80711E−01	0.44542E−02	−0.80777E−01	0.44549E−02
−0.67512E−01	0.32153E−02	−0.67568E−01	0.32168E−02
−0.58005E−01	0.24300E−02	−0.58053E−01	0.24317E−02
−0.50834E−01	0.19017E−02	−0.50878E−01	0.19025E−02
−0.45214E−01	0.15478E−02	−0.45275E−01	0.15290E−02
−0.40496E−01	0.14027E−02	−0.40780E−01	0.12555E−02

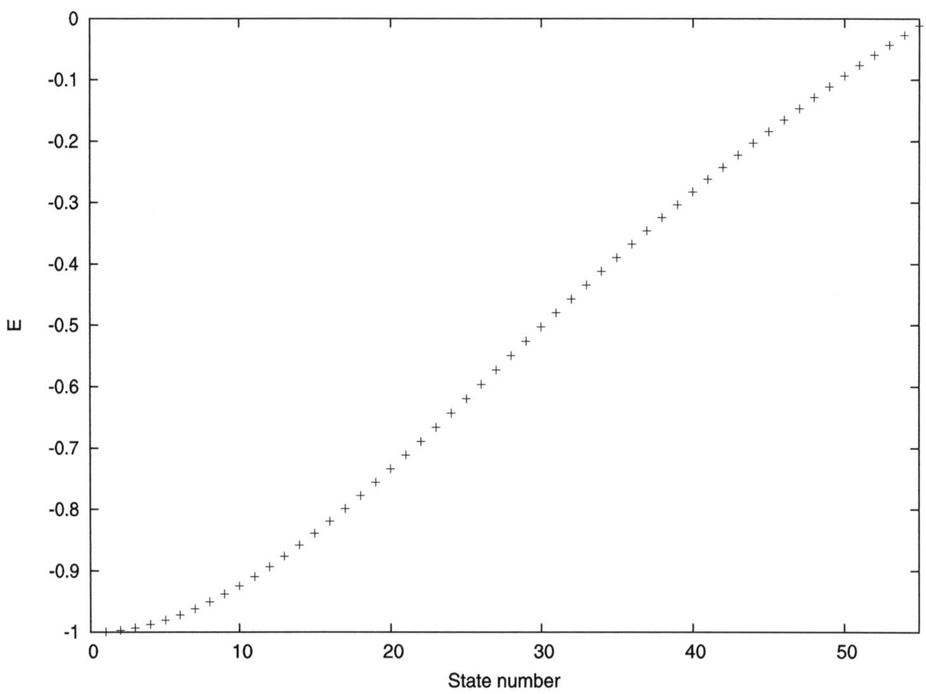

Fig. 10.3 Energies of the first 55 states of the Pomeron with the fixed coupling, $t_{max} = 270$.

the BFKL endpoint and going upwards, as predicted by previous calculation. Remarkably the number of negative states is the same as for the running coupling: 19 with $t_{max} = 90$ and 55 with $t_{max} = 270$. However the location of energies is quite different indicating a cut, as illustrated in Fig. 10.3. These intercepts belonging to the cut correspond to non-normalizable states. They are not zero but strongly diminish with the growth of t_{max}, which can be seen from Table 10.2. Such behavior $\propto 1/t_{max}$ is of course expected on general grounds.

Table 10.2 Energies and slopes of the first 10 states of the Pomeron (fixed coupling).

$t_{max} = 90$		$t_{max} = 270$	
−0.99335E+00	0.25232E−03	−0.99953E+00	0.10250E−04
−0.97347E+00	0.98302E−03	−0.99714E+00	0.40871E−04
−0.94143E+00	0.21184E−02	−0.99318E+00	0.91482E−04
−0.89868E+00	0.35509E−02	−0.98767E+00	0.16146E−03
−0.84704E+00	0.51579E−02	−0.98064E+00	0.24995E−03
−0.78851E+00	0.68197E−02	−0.97213E+00	0.35588E−03
−0.72505E+00	0.84346E−02	−0.96219E+00	0.47801E−03
−0.65850E+00	0.99274E−02	−0.95088E+00	0.61492E−03
−0.59048E+00	0.11252E−01	−0.93826E+00	0.76509E−03
−0.52233E+00	0.12388E−01	−0.92439E+00	0.92686E−03

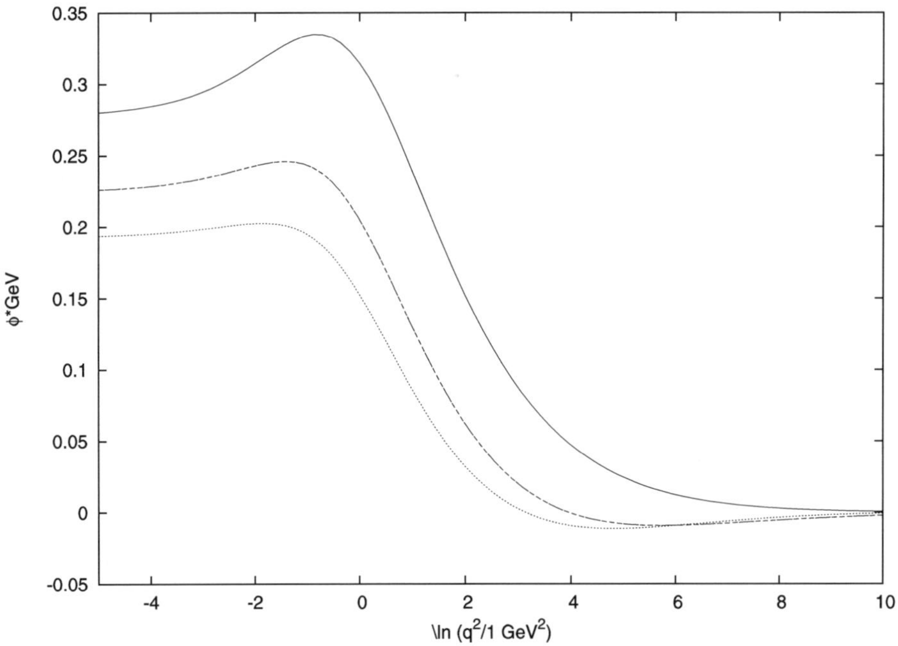

Fig. 10.4 Wave functions $\phi_n(q^2)$ as functions of q^2 for the first three states with the lowest energy. From top to bottom: curves correspond to the number of state $n = 1, 2, 3$.

We also studied the behavior of the Pomeron wave functions $\phi(q^2)$ as functions of q^2 for the three states with the lowest energy. It is shown in Fig. 10.4. Remarkably all three wave functions are very similar in their q^2 dependence in the low momentum region.

In contrast, at larger momenta different states have a different number of nodes (zeros). In order to see this let us analyze in detail how the support in momentum space of the discrete states differ. This is best understood analyzing the rescaled eigenfunctions $\tilde{\phi}_j(q^2)$. The square of the first, second and sixth states

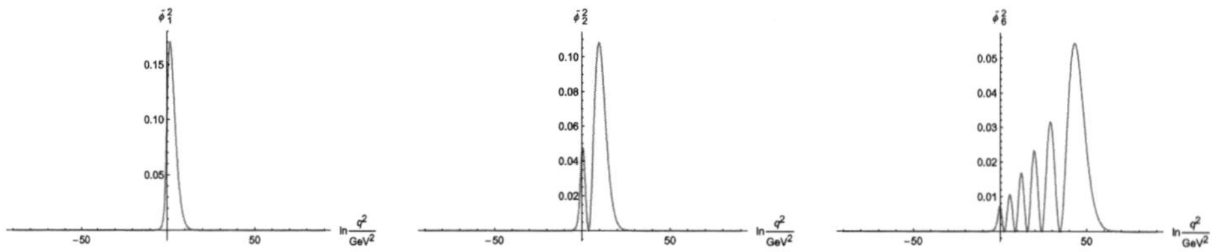

Fig. 10.5 Square of the Pomeron eigenfunctions $\tilde{\phi}_i(q^2)$ for the first, second and sixth leading energy states. The region of highest momentum support is pushed rapidly towards the UV with the increasing order of the eigenstate.

with lowest energies are shown in Fig 10.5. This behaviour is similar to the one found in [Kowalski *et al.* (2016)] where excited states show a clear shift towards higher values of q^2. One can see the dependence on the state number $\tilde{\phi}_j(q^2)$ of the position $q_{UV}(j)$ of the center of the most UV region of support. The dependence is very well approximated with a simple linear fit. For example taking the first ten states one gets

$$\ln q_{UV}^2(j)/\text{GeV}^2 \simeq -7.278 + 8.456j. \tag{10.118}$$

Finally let us comment on the dependence on the mass parameter m which also appears inside the logarithmic dependence of the running coupling (keeping the same value of Λ_{QCD}), and whose value fixes the order of the scale where α_s freezes. Lowering m from 0.82 GeV to 0.52 Gev, for the leading state the energy ($= 1-$ intercept) increases (in absolute values) by about 20% while the slope changes much more, becoming almost 4 times larger, which is close to the well known value $0.25GeV^{-2}$.

10.4.2 *Odderon*

Passing to the Odderon let us first discuss the intercept. We have found that both with the running and fixed coupling constants energies lie on the cut starting at $E = 0$. The density of states rises with t_{max}. In particular with the running coupling the number of states with $E < 0.02$ rises from 18 at $t_{max} = 90$ to 135 at $t_{max} = 270$. With the fixed coupling constant the density of states is considerably smaller and rises not so fast. At $t_{max} = 90$ we found only 5 states with $E < 0.02$ and at $t_{max} = 270$ their number rises to 13. The location of energies smaller than 0.02 for low-lying states for the running and fixed coupling constants is illustrated in Figs. 10.6 and 10.7, respectively.

For the Odderon we studied the character of the singularity of its spectrum at the branch-point $E = 0$. It is revealed in the behavior of the amplitude governed by the Odderon exchange at high s, Namely if the spectral density $\tau(E)$ has a singularity E^β at $E = 0$ the amplitude will diminish as $y^{-(1+\beta)}$ at large $y = \ln s$. In our calculations the continuous spectrum is approximated by a set of poles at E_i, $i = 1, 2, 3, \ldots$ with very small distances from one another. We introduced the coupling a_i of the i-th Odderon state to the projectile or target taking for each of

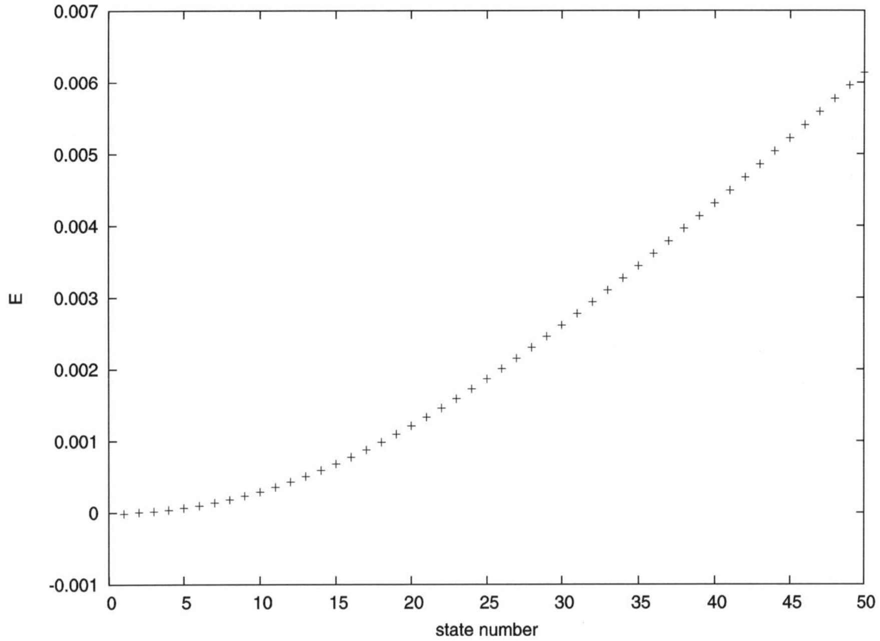

Fig. 10.6 Energies of the first 50 states of the Odderon with the running coupling. $t_{max} = 270$.

them a certain distribution of colour $\rho(q^2) = c\exp(-r_0 q^2)$ with r_0 of the order of the proton radius. Then

$$a_i = \sum_j \sqrt{w_j q_j}\rho(q_j^2)v_j^{(i)},$$

where $v^{(i)}$ is the eigenvector for the i-th Odderon (actually its wave function on the grid). The discrete spectral density is accordingly taken as $\tau_i = a_i^2$. With this we studied the y-dependence of the amplitude coming from the exchange of all Odderon states

$$\mathcal{A}(y) = \sum_i a_i^2 e^{-E_i y}\,. \tag{10.119}$$

The ratios $r(y) = \mathcal{A}(y)/\mathcal{A}(0)$ are plotted in Fig. 10.8 for the odderon with running and fixed coupling constants. It is clearly seen that running of the coupling makes the amplitude fall somewhat weaker at high y. This shows how important is the inclusion of running coupling effects for any quantitative estimate of an Odderon exchange cross section. Fits to the curves give the behavior $\sim y^{-0.38}$ and $\sim y^{-0.86}$ for the running and fixed coupling respectively. This implies that the singularity at $E = 0$ of the spectral density is stronger with the running coupling $E^{-0.62}$ than with the fixed one $E^{-0.14}$.

These results are qualitatively similar to the ones recently found using another prescription to introduce the running coupling [Bartels *et al.* (2020)].

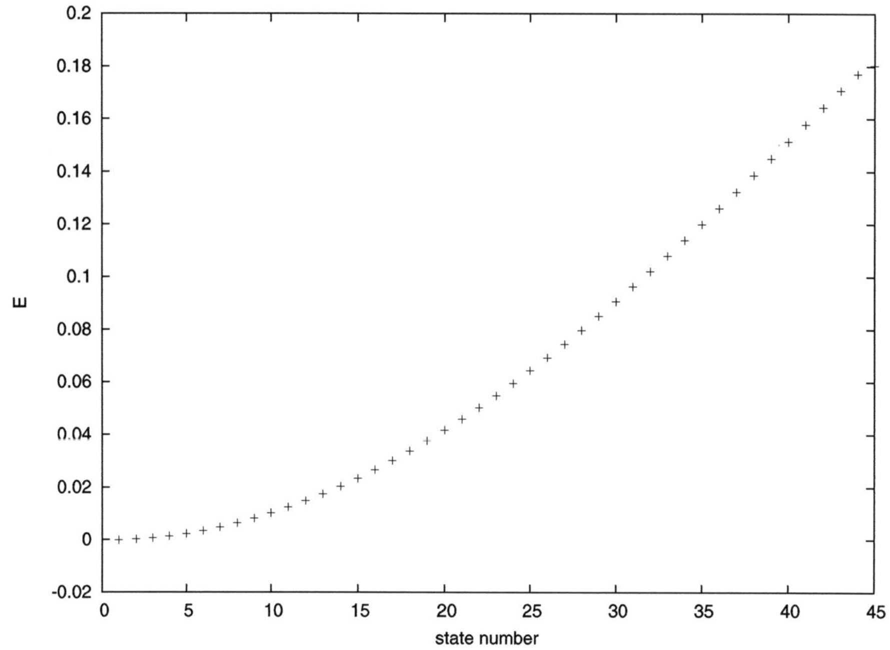

Fig. 10.7 Energies of the first 55 states of the Odderon with the fixed coupling, $t_{max} = 270$.

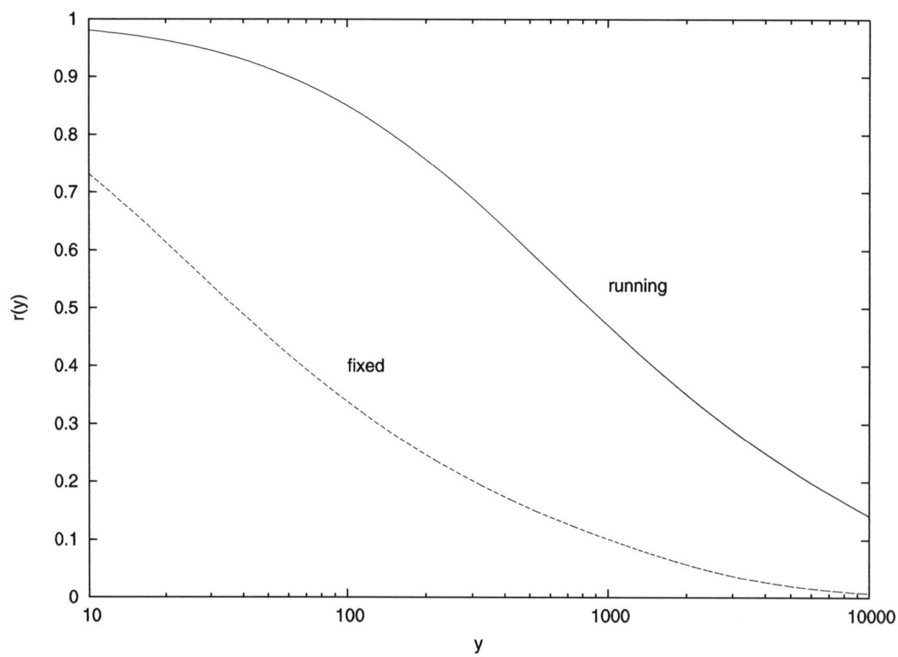

Fig. 10.8 The behavior of the amplitude with the Odderon exchange $A(y)$ as a function of rapidity y. The upper and lower curves show the ratios $A(y)/A(0)$ with and without running of the coupling.

10.5 Conclusions

We have focused on some aspects of QCD strong interactions involving the so called Odderon exchange, which may be relevant in the scattering process in the Regge limit and also in presence of dense nuclear media, and are still posing challenging problems, both theoretically and experimentally. It is still to be understood and investigated more at experimental level the difference at high energies in the scattering of pp and $p\bar{p}$, even if new data started to arrive from the TOTEM experiment. Non perturbative physics and the uncertainties in non linear effects of parton distributions are the main theoretical obstacles.

We have discussed, in the context of perturbative QCD and BFKL physics, a path to improve the description of a family of known and leading Odderon solutions (BLV) as well as more general BKP states in the LL approximation and large N_c limit to account for running coupling effects. This is obtained requiring a match with standard one loop running coupling properties, which leaves the freedom to model the IR non perturbative region and at the same time respects s-channel unitarity for the reggeization of the gluon as the basis of the BFKL approach. This so called bootstrap approach was shown at the perturbative level to reproduce running coupling properties of the NLL BFKL. For the Pomeron channel we explicitly show that this procedures leads to a series of discrete leading poles with intercept greater than one which accumulate at one.

For the Odderon the inclusion of the running coupling increases the density of states close to intercept one, where the cut starts. Therefore the rapidity dependence of the scattering amplitudes is affected by making the cross section larger, so that running coupling effects should not be neglected.

Pomeron and Odderon fields can be elements of an effective description to study rapidity dependence and saturation effects in dense nuclear media or in extreme high energy scattering processes. Descriptions alternative to the reggeized gluon approach are the color dipole picture (large N_c) and the CGC approach. We have shown how to reproduce in the BFKL description the non linear evolution of the BFKL Pomeron induced by the Odderon splitting which was known from previous CGC studies. To derive in the BFKL framework also a non linear evolution for the Odderon [Hatta *et al.* (2005)] the $V_{3\to 5}$ reggeized gluon vertex should be computed. One of the main theoretical problems is that these effective theories do not take into account Pomeron or Odderon loops since only tree level effective equations are available. The loops were formally considered in the old days, before the advent of QCD, in the effective reggeon field theory where local fields for Pomeron and Odderon were considered. This description may be valid only for very large transverse distances. We have recalled how one can compute the critical properties of such a theory and extract the universal behavior (critical exponents) even if this regime is far from being probed by any experimental setup. It is interesting that there is a universal link with statistical mechanical systems, in particular with the out of equilibrium dynamics in generalized directed percolating systems.

Finally we note that in the LL generalized BFKL framework all the known gauge invariant effective vertices are build using the bootstrap condition, and involve always reggeized gluon trajectories and the real production part of the two gluon BFKL kernel. Therefore one can envisage a way to introduce also there the running coupling by the same procedure to take into account this part of the effects which would otherwise appear only at the level of the NLL approximation.

To conclude we want to stress the most important and seminal contributions made by Lev Lipatov in this field and discussed here. Apart from being one of the father-founders of the whole reggeized gluons approach, he specifically actively dealt with Odderon and BKP states during his career with outstanding results. He studied the conformal invariance of the Odderon wave function and formulated equations which later were used to find the JW-odderon. He discovered that in the planar (multicolor) limit the BKP equations become fully integrable and showed their relation to the non-compact Heisenberg spin chain, which formed the basis of the subsequent study of their properties. This was the first demonstration of integrability in gauge theories. Lev Lipatov was the first to discuss the influence of the running of the coupling and to conclude that it splits the negative cut in the energy spectrum of the Pomeron into a sequence of poles, the results being fully confirmed later, irrespective of the way of how to introduce this running. Lev Lipatov was one of the authors who showed the equivalence of the reggeized gluons and dipole approaches for the fan diagram BK evolution equation, and in relation to this he formulated the correct transition to the coordinate representation as the transition to "the Möbius representation". He was also one of the authors to investigate the next-to-leading order for the Odderon structure, the result which will certainly play its role in future studies.

10.6 Appendix: The slope

The slope can be obtained starting from Eq. (10.80), expanding up to second order in q the relation

$$\langle (h_q - h_0) \rangle_0 = \epsilon(q) - \epsilon(0), \tag{10.120}$$

where the average is done on the eigenstates in the forward direction ($q = 0$), and according to the previous notation as in Eq. (10.82) we consider separate contributions and write

$$h_q - h_0 = W^{kin}(q) + W^{ql}(q) + W^{sep}(q). \tag{10.121}$$

First we consider the function $f(p)$ of Eq. (10.52) which is a kind of building block appearing in different instances. According to the decomposition of Eq. (10.99), in the first two orders of q we get

$$f(q_1) = f(l) \left[1 + a_1 \Big((ql) + \frac{1}{4} q^2 \Big) + \frac{1}{2} a_2 (ql)^2 \right], \tag{10.122}$$

where we denote by $ql = q \cdot l$ the scalar product of the two dimensional vectors, and

$$a_1 = \frac{1 + \ln[(l^2 + m^2)/\Lambda^2]}{f(l)}, \quad a_2 = \frac{1}{(l^2 + m^2) f(l)}. \tag{10.123}$$

From this result one directly finds also $f(q_2)$, $f(q_1')$ and $f(q_2')$, by simply changing $l \to -l$, $l \to l'$ and $l \to -l'$, respectively. Coefficients $a_1(l)$ and $a_2(l)$ do not change under $l \to -l$. As in [Braun *et al.* (1996)] we denote them simply by a_1 and a_2. The coefficients $a_1(l')$ and $a_2(l')$ will be denoted by a_1' and a_2'. Also

$$f(q) = f(0)(1 + a_3 q^2), \quad a_3 = \frac{1 + \ln m^2/\Lambda^2}{f(0)}. \tag{10.124}$$

Note also that $q_1 - q_1' = q_2' - q_2 = l - l'$ do not depend on q.

10.6.1 W^{kin}

The kinetic term (10.81) contains two terms corresponding to trajectories ω_1 and ω_2. For the first part of W^{kin} coming from ω_1 we find

$$\begin{aligned} W_1^{kin} &= \frac{1}{2} \int d^2 l' \frac{f(l)}{f(l') f(l - l')} \left[a_1 \left((ql) + \frac{1}{4} q^2 \right) + \frac{1}{2} a_2 (ql)^2 \right. \\ &\left. - a_1' \left((ql') + \frac{1}{4} q^2 \right) + \left(a_1'^2 - \frac{1}{2} a_2' \right) (ql')^2 - a_1 a_1' (ql)(ql') \right]. \end{aligned} \tag{10.125}$$

The second part of W^{kin} is found to be

$$\begin{aligned} W_2^{kin} &= \frac{1}{2} \int d^2 l' \frac{f(l)}{f(l') f(l - l')} \left[a_1 \left(-(ql) + \frac{1}{4} q^2 \right) + \frac{1}{2} a_2 (ql)^2 \right. \\ &\left. - a_1' \left(-(ql') + \frac{1}{4} q^2 \right) + \left(a_1'^2 - \frac{1}{2} a_2' \right) (ql')^2 - a_1 a_1' (ql)(ql') \right]. \end{aligned} \tag{10.126}$$

In the total W^{kin} terms linear in q cancel and quadratic ones are doubled. Thus at second order in q we find

$$\begin{aligned} W^{kin} &= \int d^2 l' \frac{f(l)}{f(l') f(l - l')} \left[\frac{1}{4} q^2 (a_1 - a_1') + \frac{1}{2} a_2 (ql)^2 \right. \\ &\left. + \left(a_1'^2 - \frac{1}{2} a_2' \right) (ql')^2 - a_1 a_1' (ql)(ql') \right]. \end{aligned} \tag{10.127}$$

10.6.2 W^{ql}

At finite q this part is given by Eq. (10.83). One finds at second order in q

$$\sqrt{\frac{f(q_1)}{f(q_2)}} = 1 + a_1(ql) + \frac{1}{2} a_1^2 (ql)^2, \quad \sqrt{\frac{f(q_2')}{f(q_1')}} = 1 - a_1'(ql') + \frac{1}{2} a_1'^2 (ql')^2,$$

and for the product

$$\sqrt{\frac{f(q_1)f(q_2')}{f(q_2)f(q_1')}} = 1 + a_1(ql) - a_1'(ql') + \frac{1}{2}\Big(a_1(ql) - a_1'(ql')\Big)^2.$$

In the second term of L in Eq. (10.83) we have to change $l \to -l$ and $l' \to -l'$. So the total q-dependent factor becomes

$$2 + \Big(a_1(ql) - a_1'(ql')\Big)^2,$$

from which one finally reads at second order in q

$$W^{ql}(l,l') = -\frac{\Big(a_1(ql) - a_1'(ql')\Big)^2}{f(l-l')}. \tag{10.128}$$

10.6.3 W^{sep}

The separable part of the interaction is given by (10.84). Here one obtains in a straightforward manner

$$\frac{f(q)}{\sqrt{f(q1)f(q_2)f(q'1)f(q_2')}} = \frac{f(0)}{f(l)f(l')}\left[1 + q^2\left(a_3 - \frac{1}{4}(a_1 + a_1')\right) - \frac{1}{2}(a_2 - a_1^2)(ql)^2 \right.$$
$$\left. - \frac{1}{2}(a_2' - a_1'^2)(ql')^2\right].$$

From this one gets at second order in q

$$W^{sep}(l,l') = \frac{f(0)}{f(l)f(l')}\left[q^2\left(a_3 - \frac{1}{4}(a_1 + a_1')\right) - \frac{1}{2}(a_2 - a_1^2)(ql)^2 \right.$$
$$\left. - \frac{1}{2}(a_2' - a_1'^2)(ql')^2\right]. \tag{10.129}$$

10.6.4 *Averaging in the state with* $n = 0$

The first task is to fix the normalization. Let

$$\mathcal{N} = \int d^2q\, \phi^2(q^2) = \pi \int dq^2 \phi^2(q^2) = \pi \int dt q^2 \phi^2(q^2). \tag{10.130}$$

On the grid we get

$$\mathcal{N} = \pi \sum_{i=0}^{N} w_i q_i^2 \phi^2(q_i^2) = \pi \sum_{i=0}^{N} v_i^2, \tag{10.131}$$

where we used (10.105). The normalization can be fixed by requiring

$$\sum_{i=0}^{N} v_i^2 = 1, \tag{10.132}$$

in accordance with the standard programs for the search of eigenvalues and eigenvectors. This implies $\mathcal{N} = \pi$ and means that after the integration of our W with the functions $\phi(q^2)$ one has to divide the result by π.

We proceed in several steps:

1. $\langle W^{kin}\rangle$: In this way we find

$$\langle W^{kin}\rangle = \frac{1}{4}q^2 \int dl^2\phi^2(l) \int dl'^2 \frac{f(l)}{f(l')} \left[\left(\frac{1}{2}(a_1 - a_1') + \frac{1}{2}a_2 l^2 + (a_1'^2 - \frac{1}{2}a_2')l'^2\right) B_0(l^2, l'^2) - a_1 a_1' ll' B_1(l^2, l'^2)\right]. \quad (10.133)$$

Passing first to the integration over t and t' and to summation on the grid we get

$$\langle W^{kin}\rangle = \frac{1}{4}q^2 \sum_{i,j} \phi^2(l_i) l_i^2 l_j^2 w_i w_j \frac{f(l_i)}{f(l_j)} \left[\left((a_{li} - a_{lj}) + \frac{1}{2}a_{2j} l_i^2 + (a_{1j}^2 - \frac{1}{2}a_{2j})l_j^2\right) B_0(l_i^2, l_j^2) - a_{1i}a_{1j} l_i l_j B_1(l_i^2, l_j^2)\right]. \quad (10.134)$$

The final expression in terms of v is obtained after using (10.105)

$$\langle W^{kin}\rangle = \frac{1}{4}q^2 \sum_{i,j} v_i^2 l_j^2 w_j \frac{f(l_i)}{f(l_j)} \left[\left(\frac{1}{2}(a_{1i} - a_{1j}) + \frac{1}{2}a_{2i} l_i^2 + (a_{1j}^2 - \frac{1}{2}a_{2j})l_j^2\right) \times B_0(l_i^2, l_j^2) - a_{1i}a_{1j} l_i l_j B_1(l_i^2, l_j^2)\right]. \quad (10.135)$$

2. $\langle W^{ql}\rangle$: Doing the angular integrations and dividing by π we obtain

$$\langle W^{ql}\rangle = -\frac{1}{4}q^2 \int dl^2 dl'^2 \phi(l)\phi(l') \left[\left(a_1^2 l^2 + a_1'^2 l'^2\right) B_0(l^2, l'^2) - 2a_1 a_1' ll' B_1(l^2, l'^2)\right].$$

In going to variables t and t' and then to the summation on the grid, using (10.105) we find

$$\langle W^{ql}\rangle = -\frac{1}{4}q^2 \sum_{i,j} \sqrt{w_i w_j} l_i l_j v_i v_j \left[\left(a_{1i}^2 l_i^2 + a_{1j}^2 l_j^2\right) B_0(l_i^2, l_j^2) - 2a_{1i}a_{1j} l_i l_j B_1(l_i^2, l_j^2)\right]. \quad (10.136)$$

3. $\langle W^{sep}\rangle$: Starting from the expression (10.129) we have

$$\langle W^{sep}\rangle = \frac{1}{4\pi} \int dl^2 dl'^2 \phi(l)\phi(l') d\chi d\chi' \frac{f(0)}{f(l)f(l')} \times \left[q^2\left(a_3 - \frac{1}{4}(a_1 + a_1')\right) - \frac{1}{2}(a_2 - a_1^2)(ql)^2 - \frac{1}{2}(a_2' - a_1'^2)(ql')^2\right].$$

Here all the angular dependence is trivial. After angular integration terms not depending on the angles give $4\pi^2$ and those containing $\cos^2\chi$ or $\cos^2\chi'$ give $2\pi^2$.

So we find

$$\langle W^{sep}\rangle = \pi q^2 \int dl^2 dl'^2 \frac{\psi(l)\psi(l')f(0)}{f(l)f(l')}\Big[a_3 - \frac{1}{4}(a_1 + a_1') - \frac{l^2}{4}(a_2 - a_1^2) - \frac{l'^2}{4}(a_2' - a_1'^2)\Big].$$

Passing first to integration over t and t' and then on the grid we finally find

$$\langle W^{sep}\rangle = \pi q^2 \sum_{i,j} \sqrt{w_i w_j} \frac{v_i v_j l_i l_j f(0)}{f(l_i)f(l_j)} \Bigg[a_3 - \frac{1}{4}(a_{1i} + a_{1j}) - \frac{l_i^2}{4}(a_{2i} - a_{1i}^2) - \frac{l_j^2}{4}(a_{2j} - a_{1j}^2)\Bigg]. \quad (10.137)$$

The final rescaled slope is given by the sum

$$\tilde{\alpha}' = \frac{1}{q^2}\langle W^{kin} + W^{ql} + W^{sep}\rangle. \quad (10.138)$$

10.6.5 *Fixed coupling*

For comparison it is instructive to consider the simple case of the fixed coupling with the infrared regularization:

$$f(q) = q^2 + m^2. \quad (10.139)$$

The energy E is then related to ϵ as

$$E = \frac{3\alpha_s}{2\pi^2}\epsilon, \quad (10.140)$$

where α_s is the fixed coupling constant. To eliminate the dependence on α_s it is convenient to measure its relation to the energy of the BFKL ground state $E_{BFKL} = 12\alpha_s \ln 2/\pi$. Then

$$E = \epsilon \frac{\epsilon}{8\pi \ln 2} E_{BFKL}. \quad (10.141)$$

With (10.139) the functions B_n and C_0 become known analytically. In particular,

$$B_0(q_1^2, q_2^2) = \frac{2\pi}{r}, \quad r^2 = (q_1^2 - q_2^2)^2 + 2m^2(q_1^2 + q_2^2) + m^4, \quad (10.142)$$

$$B_1(q_1^2, q_2^2) = \frac{2\pi b}{r(a+r)}, \quad a = q_1^2 + q_2^2 + m^2, \quad b = 2q_1 q_2, \quad (10.143)$$

$$B_{01}(q_1^2, q_1^2) = \frac{2\pi}{b}\left(1 - \frac{r}{a+b}\right), \quad (10.144)$$

and

$$C_0(q2, q_1^2) = \frac{2\pi}{r_1}, \quad r_1^2 = (q^2 + 2q_1^2 + 2m^2)^2 - 4q^2 q_1^2. \quad (10.145)$$

In the calculation of slopes we find

$$a_1(q) = \frac{1}{f(q)}, \quad a_2 = 0, \quad a_3 = \frac{1}{f(0)}. \tag{10.146}$$

Bibliography

Abazov, V. M. *et al.* (2012). Measurement of the differential cross section $d\sigma/dt$ in elastic $p\bar{p}$ scattering at $\sqrt{s} = 1.96$ TeV, *Phys. Rev. D* **86**, p. 012009, doi:10.1103/PhysRevD.86.012009, arXiv:1206.0687 [hep-ex].

Abramovsky, V., Gribov, V., and Kancheli, O. (1973). Character of inclusive spectra and fluctuations produced in inelastic processes by multi — Pomeron exchange, *Yad. Fiz.* **18**, pp. 595–616.

Antchev, G. *et al.* (2020). Elastic differential cross-section dσ/dt at $\sqrt{s} = 2.76$ TeV and implications on the existence of a colourless C-odd three-gluon compound state, *Eur. Phys. J. C* **80**, 2, p. 91, doi:10.1140/epjc/s10052-020-7654-y, arXiv:1812.08610 [hep-ex].

Balitsky, I. (1996). Operator expansion for high-energy scattering, *Nucl. Phys. B* **463**, pp. 99–160, doi:10.1016/0550-3213(95)00638-9, arXiv:hep-ph/9509348.

Balitsky, I. and Lipatov, L. (1978). The Pomeranchuk singularity in quantum chromodynamics, *Sov. J. Nucl. Phys.* **28**, pp. 822–829.

Bartels, J. (1980). High-energy behavior in a Nonabelian Gauge Theory (II): First Corrections to $T_{n\to m}$ beyond the leading $\ln s$ approximation, *Nucl. Phys. B* **175**, pp. 365–401, doi:10.1016/0550-3213(80)90019-X.

Bartels, J., Braun, M., Colferai, D., and Vacca, G. (2001). Diffractive eta(c) photoproduction and electroproduction with the perturbative QCD odderon, *Eur. Phys. J. C* **20**, pp. 323–331, doi:10.1007/s100520100676, arXiv:hep-ph/0102221.

Bartels, J., Braun, M., and Vacca, G. (2004). The Process gamma(*) + p —¿ eta(c) + X: A Test for the perturbative QCD odderon, *Eur. Phys. J. C* **33**, pp. 511–521, doi:10.1140/epjc/s2004-01589-7, arXiv:hep-ph/0304160.

Bartels, J., Contreras, C., and Vacca, G. (2017). Pomeron — Odderon interactions in a reggeon field theory, *Phys. Rev. D* **95**, 1, p. 014013, doi:10.1103/PhysRevD.95.014013, arXiv:1608.08836 [hep-th].

Bartels, J., Contreras, C., and Vacca, G. P. (2016). Could reggeon field theory be an effective theory for QCD in the Regge limit? *JHEP* **03**, p. 201, doi:10.1007/JHEP03(2016)201, arXiv:1512.07182 [hep-th].

Bartels, J., Contreras, C., and Vacca, G. P. (2019). A functional RG approach for the BFKL Pomeron, *JHEP* **01**, p. 004, doi:10.1007/JHEP01(2019)004, arXiv:1808.07517 [hep-ph].

Bartels, J., Contreras, C., and Vacca, G. P. (2020). The Odderon in QCD with running coupling, *JHEP* **04**, p. 183, doi:10.1007/JHEP04(2020)183, arXiv:1910.04588 [hep-th].

Bartels, J. and Ewerz, C. (1999). Unitarity corrections in high-energy QCD, *JHEP* **09**, p. 026, doi:10.1088/1126-6708/1999/09/026, arXiv:hep-ph/9908454.

Bartels, J., Fadin, V., Lipatov, L., and Vacca, G. (2013). NLO Corrections to the kernel of the BKP-equations, *Nucl. Phys. B* **867**, pp. 827–854, doi:10.1016/j.nuclphysb.2012.10.024, arXiv:1210.0797 [hep-ph].

Bartels, J., Lipatov, L., Salvadore, M., and Vacca, G. (2005a). Deformed spectral representation of the BFKL kernel and the bootstrap for gluon reggeization, *Nucl. Phys. B* **726**, pp. 53–74, doi:10.1016/j.nuclphysb.2005.07.033, arXiv:hep-ph/0506235.

Bartels, J., Lipatov, L., and Vacca, G. (2000). A New odderon solution in perturbative QCD, *Phys. Lett. B* **477**, pp. 178–186, doi:10.1016/S0370-2693(00)00221-5, arXiv:hep-ph/9912423.

Bartels, J., Lipatov, L., and Vacca, G. (2005b). Interactions of reggeized gluons in the Mobius representation, *Nucl. Phys. B* **706**, pp. 391–410, doi:10.1016/j.nuclphysb.2004.10.008, arXiv:hep-ph/0404110.

Bartels, J., Salvadore, M., and Vacca, G. (2005c). AGK cutting rules and multiple scattering in hadronic collisions, *Eur. Phys. J. C* **42**, pp. 53–71, doi:10.1140/epjc/s2005-02258-1, arXiv:hep-ph/0503049.

Bartels, J. and Vacca, G. (2013). Generalized Bootstrap Equations and possible implications for the NLO Odderon, *Eur. Phys. J. C* **73**, p. 2602, doi:10.1140/epjc/s10052-013-2602-8, arXiv:1307.3985 [hep-th].

Bartels, J. and Wusthoff, M. (1995). The Triple Regge limit of diffractive dissociation in deep inelastic scattering, *Z. Phys. C* **66**, pp. 157–180, doi:10.1007/BF01496591.

Braun, M. (1995). Pomeron with a running coupling constant: Intercept and slope, *Phys. Lett. B* **345**, pp. 155–160, doi:10.1016/0370-2693(94)01545-N, arXiv:hep-ph/9410279.

Braun, M. (2000). Structure function of the nucleus in the perturbative QCD with N(c) —¿ infinity (BFKL Pomeron fan diagrams), *Eur. Phys. J. C* **16**, pp. 337–347, doi:10.1007/s100520050026, arXiv:hep-ph/0001268.

Braun, M. (2008). Odderon with a running coupling constant, *Eur. Phys. J. C* **53**, pp. 59–63, doi:10.1140/epjc/s10052-007-0428-y, arXiv:0707.2314 [hep-ph].

Braun, M. and Vacca, G. (1999a). Triple Pomeron vertex in the limit N(c) —¿ infinity, *Eur. Phys. J. C* **6**, pp. 147–157, doi:10.1007/s100520050328, arXiv:hep-ph/9711486.

Braun, M. and Vacca, G. P. (1999b). The 2nd order corrections to the interaction of two Reggeized gluons from the bootstrap, *Phys. Lett. B* **454**, pp. 319–327, doi:10.1016/S0370-2693(99)00306-8, arXiv:hep-ph/9810454.

Braun, M. and Vacca, G. P. (2000). The Bootstrap for impact factors and the gluon wave function, *Phys. Lett. B* **477**, pp. 156–162, doi:10.1016/S0370-2693(00)00196-9, arXiv:hep-ph/9910432.

Braun, M., Vacca, G. P., and Venturi, G. (1996). Properties of the hard Pomeron with a running coupling constant and the high-energy scattering, *Phys. Lett. B* **388**, pp. 823–831, doi:10.1016/S0370-2693(96)01214-2, arXiv:hep-ph/9605304.

Ciafaloni, M. and Camici, G. (1998). Energy scale(s) and next-to-leading BFKL equation, *Phys. Lett. B* **430**, pp. 349–354, doi:10.1016/S0370-2693(98)00551-6, arXiv:hep-ph/9803389.

Ewerz, C. (2003). The Odderon in quantum chromodynamics, arXiv:hep-ph/0306137.

Fadin, V. and Papa, A. (2002). A Proof of fulfillment of the strong bootstrap condition, *Nucl. Phys. B* **640**, pp. 309–330, doi:10.1016/S0550-3213(02)00579-5, arXiv:hep-ph/0206079.

Fadin, V. S. and Lipatov, L. (1998). BFKL Pomeron in the next-to-leading approximation, *Phys. Lett. B* **429**, pp. 127–134, doi:10.1016/S0370-2693(98)00473-0, arXiv:hep-ph/9802290.

Ferreiro, E., Iancu, E., Leonidov, A., and McLerran, L. (2002). Nonlinear gluon evolution in the color glass condensate. 2. *Nucl. Phys. A* **703**, pp. 489–538, doi:10.1016/S0375-9474(01)01329-X, arXiv:hep-ph/0109115.

Hatta, Y., Iancu, E., Itakura, K., and McLerran, L. (2005). Odderon in the color glass condensate, *Nucl. Phys. A* **760**, pp. 172–207, doi:10.1016/j.nuclphysa.2005.05.163, arXiv:hep-ph/0501171.

Jaroszewicz, T. (1980). Infrared divergences and Regge behavior in QCD, *Acta Phys. Polon. B* **11**, p. 965.

Khoze, V., Martin, A., and Ryskin, M. (2014). High energy elastic and diffractive cross sections, *Eur. Phys. J. C* **74**, 2, p. 2756, doi:10.1140/epjc/s10052-014-2756-z, arXiv:1312.3851 [hep-ph].

Khoze, V., Martin, A., and Ryskin, M. (2018). Elastic proton-proton scattering at 13 TeV, *Phys. Rev. D* **97**, 3, p. 034019, doi:10.1103/PhysRevD.97.034019, arXiv:1712.00325 [hep-ph].

Kovchegov, Y. V. (1999). Small x F(2) structure function of a nucleus including multiple Pomeron exchanges, *Phys. Rev. D* **60**, p. 034008, doi:10.1103/PhysRevD.60.034008, arXiv:hep-ph/9901281.

Kovchegov, Y. V., Szymanowski, L., and Wallon, S. (2004). Perturbative odderon in the dipole model, *Phys. Lett. B* **586**, pp. 267–281, doi:10.1016/j.physletb.2004.02.036, arXiv:hep-ph/0309281.

Kowalski, H., Lipatov, L., and Ross, D. (2014). The green function for the BFKL Pomeron and the transition to DGLAP evolution, *Eur. Phys. J. C* **74**, 6, p. 2919, doi:10.1140/epjc/s10052-014-2919-y, arXiv:1401.6298 [hep-ph].

Kowalski, H., Lipatov, L., and Ross, D. (2016). The behaviour of the green function for the BFKL Pomeron with running coupling, *Eur. Phys. J. C* **76**, 1, p. 23, doi:10.1140/epjc/s10052-015-3865-z, arXiv:1508.05744 [hep-ph].

Kowalski, H., Lipatov, L., Ross, D., and Watt, G. (2010). Using HERA data to determine the infrared behaviour of the BFKL amplitude, *Eur. Phys. J. C* **70**, pp. 983–998, doi:10.1140/epjc/s10052-010-1500-6, arXiv:1005.0355 [hep-ph].

Kuraev, E., Lipatov, L., and Fadin, V. S. (1977). The Pomeranchuk singularity in nonabelian gauge theories, *Sov. Phys. JETP* **45**, pp. 199–204.

Kwiecinski, J. and Praszalowicz, M. (1980). Three gluon integral equation and odd c singlet Regge singularities in QCD, *Phys. Lett. B* **94**, pp. 413–416, doi:10.1016/0370-2693(80)90909-0.

Lipatov, L. (1986). The bare Pomeron in quantum chromodynamics, *Sov. Phys. JETP* **63**, pp. 904–912.

Lipatov, L. (1994). Asymptotic behavior of multicolor QCD at high energies in connection with exactly solvable spin models, *JETP Lett.* **59**, pp. 596–599, arXiv:hep-th/9311037.

Lukaszuk, L. and Nicolescu, B. (1973). A Possible interpretation of p p rising total cross-sections, *Lett. Nuovo Cim.* **8**, pp. 405–413, doi:10.1007/BF02824484.

Martynov, E. and Nicolescu, B. (2018). Did TOTEM experiment discover the Odderon? *Phys. Lett. B* **778**, pp. 414–418, doi:10.1016/j.physletb.2018.01.054, arXiv:1711.03288 [hep-ph].

Motyka, L. (2006). Nonlinear evolution of Pomeron and Odderon in momentum space, *Phys. Lett. B* **637**, pp. 185–191, doi:10.1016/j.physletb.2006.04.039, arXiv:hep-ph/0509270.

Mueller, A. H. and Patel, B. (1994). Single and double BFKL Pomeron exchange and a dipole picture of high-energy hard processes, *Nucl. Phys. B* **425**, pp. 471–488, doi:10.1016/0550-3213(94)90284-4, arXiv:hep-ph/9403256.

Vacca, G. (2000). Properties of a family of n reggeized gluon states in multicolor QCD, *Phys. Lett. B* **489**, pp. 337–344, doi:10.1016/S0370-2693(00)00947-3, arXiv:hep-ph/0007067.

Wosiek, J. and Janik, R. (1997). Solution of the odderon problem for arbitrary conformal weights, *Phys. Rev. Lett.* **79**, pp. 2935–2938, doi:10.1103/PhysRevLett.79.2935, arXiv:hep-th/9610208.

https://doi.org/10.1142/9789811231124_0011

Chapter 11

Lipatov's QCD High Energy Effective Action: Past and Future

Martin Hentschinski

Departamento de Actuaria,
Fsica y Matemticas Universidad de las Amricas Puebla,
Ex-Hacienda Santa Catarina Martir S/N,
San Andrs Cholula 72820 Puebla, Mexico

When I met Lev Lipatov in 2006, I had just started working on my doctoral thesis at the II. Institute for Theoretical Physics at Hamburg University. While working with him was at first a challenge, I soon realized what a great teacher he was and how many insights he was generously sharing with me. I must admit that I could not always follow all of his explanations and that sometimes it took me years until I was able to finally understand the full impact of one or the other comment he made to me during these days. While the high energy effective action might be not among the achievements Lipatov is most famous for, I am deeply convinced that it constitutes an important part of his legacy. It allows us to benefit from his deep insights into the Regge limit of QCD even after he went away from us.

Abstract

In this contribution we briefly review some aspects of Lipatov's gauge invariant QCD high energy effective action. The high energy effective action provides a field theory framework to systematically factorize QCD scattering amplitudes and related theories in the limit of high center of mass energies. After a short review of the underlying concepts, we address the question how the high energy effective action can be used for actual calculations. As an explicit example we review the derivation of the gluon Regge trajectory up to 2 loops. We then review two topics of current interest: automatic amplitude generation of high energy effective action amplitudes, and a discussion of the relation between the so-called Color-Glass-Condensate framework and the high energy effective action.

11.1 Introduction

Effective field theories are highly successful tools to study dedicated kinematic limits of quantum field theories. This is in particular true for the study of Quantum Chromodynamics (QCD), the fundamental theory of strong interactions. In 1995, Lev Lipatov formulated a gauge invariant high energy effective action [Lipatov (1995, 1997)], which provides the effective field theory description of QCD in the high energy or Regge limit. This effective action is an effective field theory which describes the interaction between the conventional QCD degrees of freedom, *i.e.* quarks, gluons and ghost, and a newly introduced degree of freedom, the reggeized gluon. The formulation of the high energy effective action by Lev Lipatov is not the only effective field theory description of the QCD high energy limit: most notably one should mention first attempts by Verlinde and Verlinde [Verlinde and Verlinde (1993)], see also [Arefeva (1994)] as well as an earlier version of an high energy effective action by Lev Lipatov [Lipatov (1991)] and collaborators [Kirschner *et al.* (1994, 1995)]. Another important framework, which has been developed around the same time as Lipatov's high energy effective action, is provided by the Color Glass Condensate effective theory and its ensuing calculational framework. We will comment on recent results concerning the relation of both frameworks further down in the text.

The benefits of such an effective field theory framework are manifold: first and most important they provide a field theory definition of the underlying degrees of freedom in the t-channel of high energy factorized QCD amplitudes: the reggeized gluon. While the bare reggeized gluon field is introduced as a new auxiliary degree in the formulation of the high energy effective action, reggeization of the gluon, *i.e.* the nonzero Regge trajectory arises within the high energy effective action as a direct consequence of the calculation of perturbative higher order corrections to the propagator of this new "auxiliary" degree of freedom. Introducing this degree of freedom is therefore crucial for the formulation of the high energy effective action. By definition this new degree of freedom is invariant under local gauge transformations which in turn is essential for achieving gauge invariant factorization of scattering amplitudes in the high energy limit.

The outline of this short review is as follows: in section Sec. 11.2 we briefly review the formulation of the high energy effective action. In Sec. 11.3 we review essential ingredients for the use of the high energy effective action in actual calculations. In Sec. 11.4 we give a short overview of the determination of the 2-loop gluon trajectory. Section 11.5 provides a short introduction into the construction of scattering amplitudes for the high energy effective action, while in Sec. 11.6 we address the high energy effective action in the presence of a dense reggeized gluon field and its relation to the framework of the Color Glass Condensate effective theory. In Sec. 11.7 we summarize our results and most importantly highlight those results which we could not cover in this short review. In particular we would like to note that the topics covered in this review are highly biased by the own work of the author. This is mainly due to the limitations of the author and implies by no means that work not mentioned would be of less relevance.

11.2 The high energy effective action: derivation

The high energy effective action aims at a description of scattering amplitudes which exhibit regions where plus and minus components are strongly ordered. It is instructive to consider the example of an explicit scattering process $p_A + p_B \to X$ in the high energy limit: ('any mass scale')$^2/s \to 0$ with $s = (p_A + p_B)^2$ denoting the center of mass energy squared. In this limit, the momenta of scattering particles $p_{A,B}$ are close to the opposite sides of the light cone. It is therefore natural to define dimensionless light-cone vectors associated with these momenta. If $p_{A,B}$ are already light-like, these are simply obtained through the following re-scaling

$$n_\pm = \frac{2}{\sqrt{s}} p_{B,A}, \qquad n_+ \cdot n_- = 2. \tag{11.1}$$

Plus- and minus components of momenta (and general four vectors) are defined through

$$k^\pm = n^\pm \cdot k = n_\mp \cdot k = k_\mp. \tag{11.2}$$

This implies the following Sudakov decomposition of a four vector,

$$k = k^+ \frac{n^-}{2} + k^- \frac{n^+}{2} + \mathbf{k} = k_- \frac{n_+}{2} + k_+ \frac{n_-}{2} + \mathbf{k}, \tag{11.3}$$

where $\mathbf{k}$ denotes the momentum perpendicular to the 2 light-cone directions with $\mathbf{k}^2 > 0$. For derivatives the following relations hold:

$$\partial_\pm x^\pm = 2 \quad \partial_\mp x^\pm = 0. \tag{11.4}$$

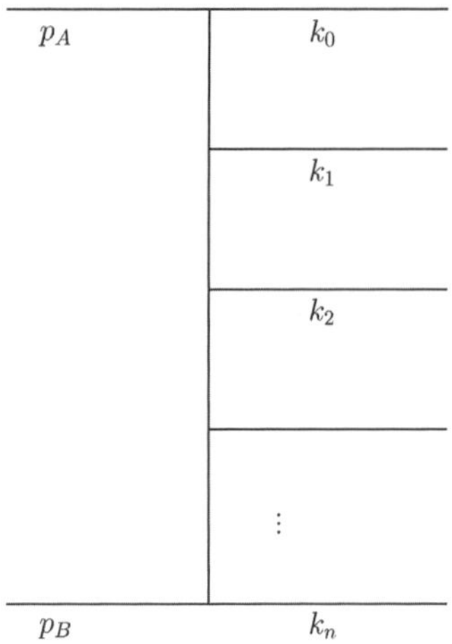

Fig. 11.1 The multi-Regge-kinematics.

The kinematic limit of interest is then defined as follows: a genuine scattering process is organized into several sub-sectors of particles which are close, *i.e.* local, in rapidity $\eta = \frac{1}{2}\ln\frac{k^+}{k^-}$. These sectors are strongly ordered with respect to each other in both plus and minus momenta (and therefore in rapidity). The situation is most easily illustrated using the tree diagram depicted in Fig. 11.1: the momenta of each of the sectors k_i with $i = 0, \ldots, n$ obey the conditions $k_0^+ \gg k_1^+ \gg \ldots k_n^+$ and $k_0^- \ll k_1^- \ll \ldots k_n^-$. As a side remark we note that for real particle production, this condition corresponds to the well known Multi-Regge-Kinematics, *i.e.* strong ordering in rapidity, while all transverse momenta are of the same order of magnitude.

11.2.1 *The high energy effective action for gluons*

Following closely [Lipatov (1995, 1997)], the essential idea is to decompose the complete gluonic field V_μ field into a sum of fields $v_\mu^{(s)}$ and $A_\mu^{(s)}$, where the upper index 's' denotes a certain sector s which is local in rapidity, *i.e.* a sector which has no regions with strong ordering in plus and minus momenta. The index 's' indicates that the fields $v_\mu^{(s)}$ are restricted to such a sector, and for each such sector there exists an individual field $v_\mu^{(s)}$. The fields A_μ^s differ from these fields: while the index 's' indicates that these fields also couple to a certain rapidity sector, these fields never occur *inside* a certain sector. They rather transmit the interaction between different sectors. They are "non-local in rapidity". They generally appear in the t-channels of scattering processes and do not appear as asymptotic states. Since the fields A_μ^s connect to sectors which are strongly ordered in plus and minus momenta with respect to typical momenta in the sector s, their coordinate dependence is strongly boosted with respect to the regarding light-cone direction. As a consequence, only certain components of the the fields A^s_μ are of relevance. In particular,

$$A^{\mu(s)}(x) = A_+^{(s)}(x)\frac{n_-^\mu}{2} + A_-^{(s)}(x)\frac{n_+^\mu}{2}, \tag{11.5}$$

while the dependence of the fields on light-cone coordinates is frozen:

$$\partial_+ A_-^{(s)}(x) = 0 = \partial_- A_+^{(s)}. \tag{11.6}$$

In the momentum representations, to which we will turn soon, this encodes strong ordering of different sectors (which the $A_\pm$ fields connect) with respect to their plus and minus momenta. As we will see below, the field A_μ directly yields the so-called reggeized gluon, which has been identified by Lipatov and collaborators as the fundamental degree of freedom in the t-channel of high energy factorized scattering amplitudes. Although the fields A_μ initially are not reggeized, *i.e.* they have no non-zero Regge trajectory, reggeization of this fields will result from the resummation of radiative corrections. Following [Lipatov (1995, 1997)], we therefore will generally refer to the fields A_μ as "reggeized gluon" fields.

The essential goal is now to construct the effective action which describes the local interaction, *i.e.* inside the regarding sectors, between $v^{(s)}$ and $A_\pm^{(s)}$ fields and $v^{(s)}$ fields themselves. The interactions among the $v_\mu^{(s)}$ fields is described by the conventional QCD Lagrangian, with the implicit constraint that the resulting QCD dynamics must be local in rapidity. In addition to the QCD Lagrangian, one further requires a term which describes the interaction among $v_\mu^{(s)}$ and $A_\mu^{(s)}$ fields. The resulting Lagrangian of the effective theory is therefore obtained as the sum of the conventional QCD Lagrangian and a so-called *induced* Lagrangian $\mathcal{L}_{\text{ind.}}$, which couples the gluon field to the A_μ field:

$$\mathcal{L}_{\text{eff}} = \mathcal{L}_{\text{QCD}}[v_\mu^{(s)}, \psi^{(s)}, \bar{\psi}^{(s)}] + \mathcal{L}_{\text{ind.}}[v_\mu^{(s)}, A_+^{(s)}, A_+^{(s)}]. \tag{11.7}$$

Here the QCD Lagrangian is as usually given by

$$\mathcal{L}_{\text{QCD}}[v_\mu^{(s)}, \psi^{(s)}, \bar{\psi}^{(s)}] = -\frac{1}{2}\text{tr}\,[G_{\mu\nu}G^{\mu\nu}] + \sum_{f=1}^{n_f} \bar{\psi}_f(i\not{D} - m_f)\psi_f. \tag{11.8}$$

In the effective Lagrangian, vectors fields are generally understood as anti-hermitian matrices in the fundamental representation of $\text{SU}(N_c)$, *i.e.* $A_\pm(x) = -it^a A_\pm^a(x)$ and $v_\mu = -it^a v_\mu(x)$ with t^a the $\text{SU}(N_c)$ generators in the fundamental representation $\text{tr}(t^a t^b) = \delta^{ab}/2$; the trace is therefore in the fundamental representation. Furthermore

$$D_\mu = \partial_\mu + g v_\mu, \quad G_{\mu\nu} = \frac{1}{g}[D_\mu, D_\nu]. \tag{11.9}$$

From now on we will drop the index 's' and implicitly assume that the fields of the effective action are limited to a certain sector. As a first naïve attempt to construct the induced Lagrangian, one might start with the term

$$\mathrm{tr}[2v^\mu \partial^2 A_\mu] = \mathrm{tr}[v_+ \partial^2 A_- + v_- \partial^2 A_+], \tag{11.10}$$

which provides the simplest coupling between the two fields. The differential operator '∂^2' has been introduced to ensure that the field A_μ cannot appear as an asymptotic state: whenever the momentum of the A_μ is the on the mass-shell, $\partial^2 = 0$, the field will decouple from QCD dynamics. To obtain the correct normalization for the propagator of the A_μ-field, one further introduces a corrective term such that the coupling between both fields reads

$$\mathrm{tr}\left[(v_+ - A_+)\partial^2 A_- + (v_- - A_-)\partial^2 A_+\right]. \tag{11.11}$$

The above term is sufficient to construct the Lagrangian of the effective theory, as long as we ignore questions related to gauge invariance: the term induces a transition between the conventional gluon field v_μ inside a certain sector and the field A_μ, which connects sectors strongly ordered in rapidity with respect to each other. The strong ordering condition is imposed on the field A^μ by both polarization and the kinematic constraint Eq. (11.6). For a gauge theory, the above proposal comes with an essential deficit: factorization in the high energy limit of a gauge independent quantity, *i.e.* a scattering amplitude, must happen in an gauge independent way. To achieve this, Lipatov proposed to define the reggeized gluon fields A^μ to be invariant under local gauge transformations

$$\delta_L A_\pm = [A_\pm, \chi_L] = 0, \tag{11.12}$$

with χ_L denoting the parameter of local gauge transformation which decreases for $x \to \infty$. QCD fields, on the other hand, are as usual subject to such local gauge transformations,

$$\delta_L v_\pm = [D_\pm, \chi], \quad \delta_L \psi = -\chi, \quad D_\mu = \partial_\mu + g v_\mu. \tag{11.13}$$

As a consequence, the proposal Eq. (11.11) for the coupling between the v^μ and the A^μ fields violates local gauge invariance. The gauge invariant high energy effective action proposed by Lipatov solves this problems by adding the minimal necessary terms to re-establish gauge invariance, leading to the following induced Lagrangian:

$$\mathcal{L}_{\mathrm{ind}}[v^\mu, A_+, A_-] = \Big\{ \mathrm{tr}\left[\left(T_-[v(x)] - A_-(x)\right) \partial_\perp^2 A_+(x)\right] + \mathrm{tr}\left[\left(T_+[v(x)] - A_+(x)\right) \partial_\perp^2 A_-(x)\right] \Big\}. \tag{11.14}$$

The functionals $T_\pm[v]$ can be obtained from the following operator definition

$$T_\pm[v] = -\frac{1}{g}\partial_\pm \frac{1}{1 + \frac{g}{\partial_\pm} v_\pm} = v_\pm - g v_\pm \frac{1}{\partial_\pm} v_\pm + g^2 v_\pm \frac{1}{\partial_\pm} v_\pm \frac{1}{\partial_\pm} v_\pm - \ldots, \tag{11.15}$$

where the integral operator $1/\partial_\pm$ is implied to act on a unit constant matrix from the left. Boundary conditions of the $1/\partial_\pm$ are fixed through

$$\frac{1}{1+\frac{g}{\partial_\pm}v_\pm} = U[v_\pm(x)] = \mathcal{P}\exp\left(-\frac{g}{2}\int_{-\infty}^{x^\pm} dx'^\pm v_\pm(x')\right)$$
$$= 1 - \frac{g}{2}\int_{-\infty}^{x^\pm} dx'^\pm v_\pm(x') + \frac{g^2}{4}\int_{-\infty}^{x^\pm} dx'^\pm \int_{-\infty}^{x'^\pm} dx''^\pm v_\pm(x')v_\pm(x'') + \ldots. \tag{11.16}$$

The functional $T[v_\pm]$ therefore generalizes the naïve coupling between A_μ and v_μ fields. In particular

$$\delta_L(\mathrm{tr}[T[v_\pm]\partial^2 A_\mp]) = 0, \tag{11.17}$$

and the induced Lagrangian Eq. (11.14) is therefore invariant under local gauge transformations.

11.2.2 *The high energy effective action for quarks*

Following the same ideas, one can write down the high energy effective action for quarks, meant to describe the high energy limit of processes with flavor exchange [Lipatov and Vyazovsky (2001)]. The degree of freedom, $Q_\pm$, which is used to factorize amplitudes in the high energy limit, is in this case not bosonic, but fermionic. In correspondence to Eq. (11.5) this field satisfies:

$$\not{n}_\pm Q_\mp = 0, \quad \bar{Q}_\mp \not{n}_\pm = 0, \tag{11.18}$$

while the dependence on light-cone coordinates is frozen,

$$\partial_\pm Q_\mp(x) = 0 = \partial_\pm \bar{Q}_\mp(x) = 0, \tag{11.19}$$

in accordance with Eq. (11.6). The complete effective Lagrangian is given by

$$\mathcal{L}_{\text{eff}} = \mathcal{L}_{\text{QCD}}[v_\mu, \psi, \bar{\psi}] + \sum_{f=1}^{n_f} \mathcal{L}_{\text{ind.}}^{(Q_f)}[\psi, \bar{\psi}, v_\mu, Q_\pm, \bar{Q}_\pm], \tag{11.20}$$

where the induced Lagrangian $\mathcal{L}_{\text{ind.}}^{(Q)}$ again describes the coupling of the $Q_\pm$ and $\bar{Q}_\pm$ fields to QCD fields:

$$\mathcal{L}_{\text{ind.}}^{(Q_f)}[\psi, \bar{\psi}, v_\mu, Q_\pm, \bar{Q}_\pm] = \bar{Q}_-^f(x)\,(i\not{\partial} - m_f)\left[Q_+^f(x) - U^\dagger[v_+(x)]\psi_f(x)\right]$$
$$+ \left[\bar{Q}_+^f(x) - \bar{\psi}(x)U[v_+(x)]\right](i\not{\partial} - m_f)\,Q_-^f(x)$$
$$+ \{\text{“}+\text{”} \leftrightarrow \text{“}-\text{”}\}. \tag{11.21}$$

For the definition of the Wilson line operator $U[v_\pm(x)]$ we refer to Eq. (11.16). At leading order in perturbation theory the Wilson line operator $U[v_\pm(x)]$ is equal to unity, and one finds a direct coupling between conventional quark fields ψ and reggeized quark fields $Q_\pm$. The operator $i\partial\!\!\!/ - m_f$ ensures again the vanishing of the coupling for asymptotic states. Considering the all order Wilson line operator, the induced Lagrangian turns invariant under local gauge transformations, Eq. (11.13).

11.3 Working with the high energy effective action

The determination of scattering amplitudes requires at first the determination of Feynman rules, which has been pioneered in [Antonov *et al.* (2005)]. The Feynman rules of the effective action Eq. (11.7) comprise, in addition to the usual QCD Feynman rules, the propagator of the reggeized gluon and an infinite number of so-called induced vertices. The leading order induced vertices and propagators are summarized in Fig. 11.2. In general it is possible to avoid the presence of a direct transition vertex between gluon and reggeized gluon field. This is most easily achieved through a shift $v_\pm \to v_\pm + A_\pm$ in the effective action which removes this transition vertex and gives a direct coupling of the reggeized gluon field as one of the external fields of conventional QCD vertices. We will comment on this option further below. Working with the direct transition vertex, the reggeized gluon propagator receives also a correction from such a direct transition, which is already included in the above reggeized gluon propagator. A similar set of Feynman rules can be derived for the quark high energy effective action. In the following we will focus on the high energy effective action for reggeized gluon exchange, while the bulk of observations equally applies to the high energy effective action for reggeized quarks.

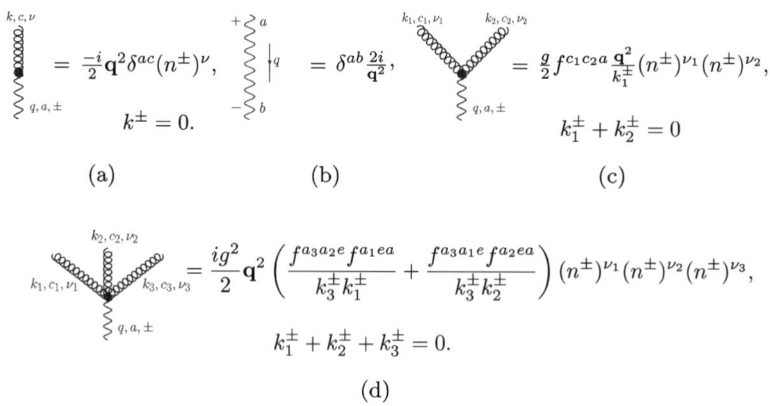

Fig. 11.2 Feynman rules for the lowest-order effective vertices of the effective action. Wavy lines denote reggeized fields and curly lines gluons. Note that in comparison with the Feynman rules used in [Hentschinski and Sabio Vera (2012); Chachamis *et al.* (2013b, 2012, 2013a)] we absorbed a factor 1/2 into the vertices which is compensated by changing the residue of the reggeized gluon propagator from 1/2 to 2.

For a very nice recent treatment of the high energy effective action for quarks see [Nefedov and Saleev (2017); Nefedov (2019)].

To start we note that the set of Feynman rules arising from the high energy effective actions Eqs. (11.7) and (11.20) cannot be used in a straightforward manner. This is immediately clear from the particular form of the effective action: They contain both the complete QCD Lagrangian and on top of it the coupling the auxiliary fields $A_\pm$. The solution to this problem is naturally resolved through the crucial observation that all the fields inside the high energy effective action are to be understood as to be restricted to a certain region local in rapidity: No strong ordering in light-cone momenta occurs within such a sector; only the entire set of light-cone momenta of a certain sector is strongly ordered with respect to other sectors. Interaction between those sectors takes place through reggeized gluon exchange. A closely related problem arises due to the manifestation of so-called rapidity divergences in certain correlators, which are not regularized by standard methods such as, *e.g.*, dimensional regularization. Last but not least it is necessary to provide a suitable pole-description to the light-cone denominators, which appear in the induced vertices Fig. 11.2 to define properly integrations over loop momenta. We will shortly review these problems in the following and discuss the commonly used solutions to such problems employed in the literature.

11.3.1 *Locality in rapidity and subtraction*

The problem is most easily illustrated using a particular example: We consider the additional emission of a gluon in the interaction of a quark with a reggeized gluon. For full details and explicit expressions we refer to [Hentschinski and Sabio Vera (2012); Chachamis *et al.* (2014)]. The scattering amplitude can be obtained from the high energy effective action through evaluating the sets of diagrams, shown in Fig. 11.3. The resulting scattering amplitude does not vanish in the strongly ordered region $p^+ \gg q^+$; it is therefore non-local in rapidity. Apart from the above matrix element in Fig. 11.3, the effective action allows also for the construction of a scattering amplitude for the same process, which combines two different sectors, connected through an internal reggeized gluon line: the two sectors are in this case given by the coupling of a quark to a reggeized gluon and the emission of a gluon from a reggeized gluon, the famous gluon production or Lipatov vertex. For a derivation of this vertex from the effective action we refer to [Antonov *et al.* (2005)]. Analyzing the explicit expressions corresponding to Fig. 11.3, it is evident that Fig. 11.3 turns into Fig. 11.5 in the strongly ordered region $p^+ \gg q^+$, where gluon and quark are widely separated in rapidity. In this particular example, it would be straightforward to include an explicit cut-off on the final state that separates Fig. 11.3 from Fig. 11.4. With increasing complexity of the final states and/or the case of higher-loop corrections, this becomes rather cumbersome. It is therefore more suitable, to simple *subtract* diagrams with an *internal* reggeized gluon line from the regarding correlator, removing in this way the contribution which causes

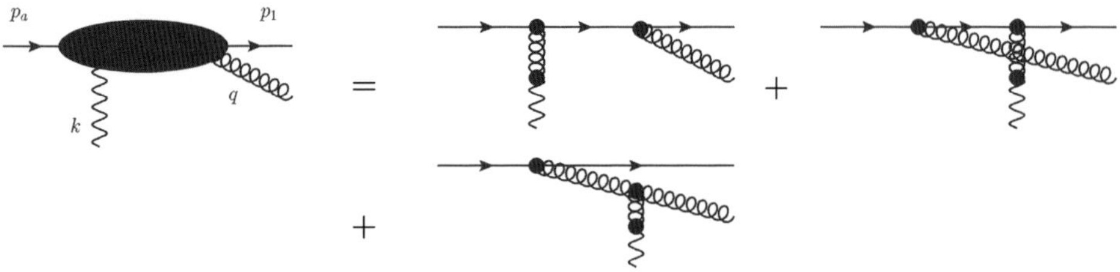

Fig. 11.3 Effective action diagrams for the process quark + reggeized gluon → quark + gluon.

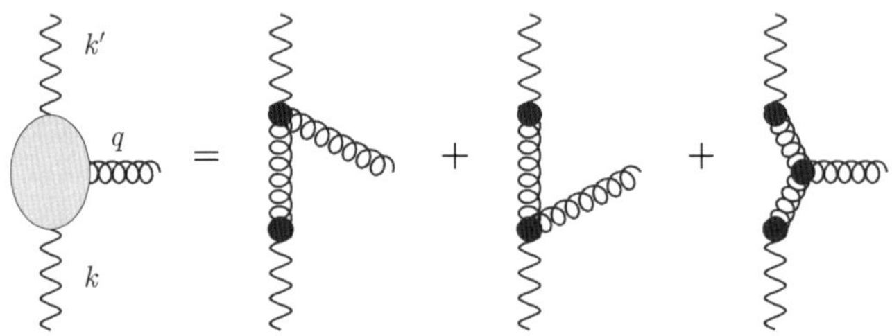

Fig. 11.4 The central gluon production vertex, *i.e.* the Lipatov vertex, in terms of effective action diagrams.

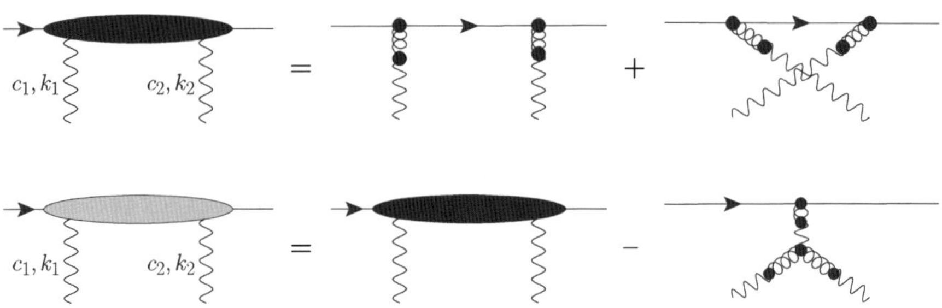

Fig. 11.5 First line: Amplitude for two reggeized gluons coupling to a quark. Second line: Amplitude with the factorized contribution subtracted.

the non-locality in rapidity. For the above example one finds that the combination

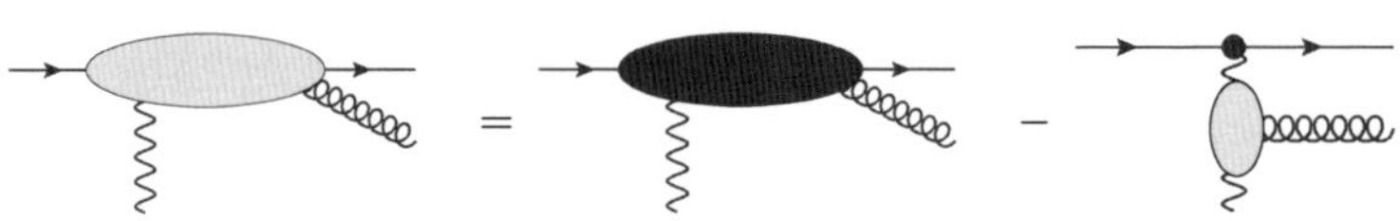 (11.22)

is local in rapidity, in the sense that the resulting expression is not growing in the strongly ordered region $p^+ \gg q^+$ but instead assumes a constant value.

In general, the following set of rules can be formulated for the determination of a (sub-) amplitude, local in rapidity within the effective action, using perturbation theory:

a) determine the correlator of interest from the high energy effective action, to the desired order in perturbation theory, with the reggeized gluon field treated as a background field
b) subtract disconnected contributions which contain internal reggeized gluon lines (which are already themselves subtracted) to the desired order in perturbation theory.

A similar situation arises for diagrams with multiple reggeized gluon exchange, such as Fig. 11.5, first line. In the limit where the squared center-of-mass energy $s = (p_a + k_1)^2$ tends to infinity, the contribution diverges logarithmically $1/k_1^-$. Similar to the previous example, the high energy limit of this correlator is captured by the corresponding diagram with an internal reggeized gluon exchange, such that the subtracted contribution, Fig. 11.5, second line, is local in rapidity in the sense that the expression vanishes up to terms $\mathcal{O}(1/(k_1^-))^2$ in the high energy limit [Hentschinski (2009)]. See also the discussion of a related observation in [Braun and Vyazovsky (2019)].

11.3.2 *Pole prescription for induced vertices*

At first sight there is no need to chose a particular pole prescription: the poles in $1/\partial_\pm$ arise from propagators of particles with significantly different rapidity, which emit gluons into the rapidity cluster under consideration. The corresponding particle virtualities $k^\pm$ are therefore large. Based on this argument, the pole prescription of these denominators can be ignored and one arrives directly at the induced vertices in Fig. 11.2. Most notably, the color structure of these vertices is given by structure constants f^{abc} only, which arise from the commutators of $\mathrm{SU}(N_c)$ generators of the fields $v(x) = -it^a v^a(x)$ in the functional $T_\pm(v_\pm)$. Since the color structure of induced vertices arises from commutators of $\mathrm{SU}(N_c)$ generators, the color structure is independent of the representation of the fields $v(x)$. The resulting induced vertices are therefore universal. They do not depend on the color representation of the specific particle from which the gluons have been emitted.

Even though the overall result must be independent of the pole prescription due to the above arguments, it is nevertheless necessary to chose a certain prescription in actual calculations. To have consistency for the evaluation of different elements (which eventually one needs to combine) it is highly recommendable to do so at the level of the effective Lagrangian or, equivalently, with the resulting Feynman rules. At first sight the problem appears to be well defined: Requiring convergence of the path ordered exponential, Eq. (11.16) at $-\infty$ naturally defines a pole prescription of the light-cone denominator: it induces the replacement $\partial_\pm \to \partial_\pm + \epsilon$. There are various reasons why this is not the generally accepted treatment (see in particular the discussion in [Lipatov (2016)]): At the very first, using such a pole prescription, the effective Lagrangian is no longer hermitian and unitarity of the theory is formally lost (which itself is an essential requirement for any effective field theory description to study QCD high energy limit). In practice this problem can be overcome by replacing the path ordered exponential Eq. (11.16) by the combination

$$U[v_\pm(x)] \to \frac{1}{2}\left(U[v_\pm(x)] + U^\dagger[v_\pm(x)]\right). \tag{11.23}$$

A similar prescription has been adapted in the original formulation of the quark effective action [Lipatov and Vyazovsky (2001)]; for a related discussion in the case of the effective action for gluons see [Nefedov and Saleev (2017)]. For the induced vertex Fig. 11.2(c), such a prescription leads to a principal value prescription, *i.e.*

$$\frac{1}{k_1^\pm} \to \frac{1}{2}\left(\frac{1}{k_1^\pm + i\epsilon} + \frac{1}{k_1^\pm - i\epsilon}\right) = \frac{k_1^\pm}{(k_1^\pm)^2 + \epsilon^2}. \tag{11.24}$$

Such a prescription is not only attractive from the point of view of hermicity of the high energy effective action, it further provides vertices which are odd under $k_1^\pm \to -k_1^\pm$ — a property in accordance with negative signature of reggeized gluon (which eventually arises from the reggeized gluon fields). While this pole prescription provides a simple and straightforward definition of the pole prescription and appears suitable for the use in calculations, it gives rise to certain symmetric color structures which cannot be expressed as commutators (and therefore enhances the color structure of the induced vertices in Fig. 11.2). Since symmetric color tensors depend in general on the representation of the SU(N_c) generators, universality of these vertices is therefore lost.[1] To avoid this problem, in [Hentschinski (2012)] a perturbative prescription has been developed which projects out, order by order in perturbation theory, the symmetric color structures. The resulting pole prescription respects then Bose symmetry and universality of the induced vertices. For the $\mathcal{O}(g)$ vertex (see Fig. 11.2(c)), this corresponds to replacing the denominator $1/k_1^\pm$ by a Cauchy principal value. For the $\mathcal{O}(g^2)$ vertex (see Fig. 11.2(d)), the combinations of denominators $1/(k_3^\pm k_1^\pm)$ and $1/(k_3^\pm k_2^\pm)$ are replaced by functions $g_2^\pm(3,2,1)$ and $g_2^\pm(3,1,2)$, respectively, with

$$g_2^\pm(i,j,m) = \left[\frac{1}{[k_i^\pm][k_m^\pm]} + \frac{\pi^2}{3}\delta(k_i^\pm)\delta(k_m^\pm)\right], \quad \frac{1}{[k]} \equiv \frac{1}{2}\left(\frac{1}{k+i0} + \frac{1}{k-i0}\right). \tag{11.25}$$

Similar functions can be constructed for the higher order induced vertices. One of the main advantages over simply defining all denominators as principal values is that the function g_2 (and corresponding higher order versions) satisfy eikonal identities in the sense of the theory of distributions *i.e.*

$$g_2^\pm(1,2,3) + g_2^\pm(2,3,1) + g_2^\pm(3,1,2) = 0. \tag{11.26}$$

[1] Applying a consistent subtraction mechanism, the dependence on such color tensors must drop out in the final result, *i.e.* on the level of the observable. It should therefore be possible to carry out actual higher order calculations within this prescription.

Finally we would like to highlight that Lipatov himself proposed a certain pole-prescription in [Lipatov (2016)]

$$T_\pm[v_\pm] = \frac{1}{2}\mathcal{P}\frac{e^{-\frac{g}{4}\int_{-\infty}^{x^\pm} dx^{\pm\prime} v_\pm}}{e^{-\frac{g}{4}\int_{x^\pm}^{\infty} dx^{\pm\prime} v_\pm}}\left[\mathcal{P}e^{\frac{g}{4}\int_{-\infty}^{x^\pm} dx^{\pm\prime} v_\pm} + \bar{\mathcal{P}}e^{-\frac{g}{4}\int_{-\infty}^{x^\pm} dx^{\pm\prime} v_\pm}\right], \tag{11.27}$$

where $\mathcal{P}$ ($\bar{\mathcal{P}}$) denotes (anti-)path ordering. To the best our knowledge the vertices resulting from this particular pole prescription have not been explored so far.

11.3.3 *Regularization of high energy singularities*

Last but not least it is necessary to address the problem that higher order corrections lead to so-called rapidity divergences which are not regularized by conventional methods, such as dimensional regularization. A very nice review of different regulators which can be used can be found in [Nefedov (2019)]. A very convenient choice is given by tilting the light-cone direction of the effective action slightly against the light cone through introducing a regulator ρ, evaluated in the limit $\rho \to \infty$

$$\begin{aligned} n^- &\to n_a = e^{-\rho}n^+ + n^-, \\ n^+ &\to n_b = n^+ + e^{-\rho}n^-. \end{aligned} \tag{11.28}$$

This regulator has an important advantage over other regulators (such as regulation by a hard cut-off or analytic regulations) since it respects gauge invariance properties of the high energy effective action.

11.4 The gluon Regge trajectory and the effective action

As an example of an explicit higher order calculation which combines all of the above aspects, we review in the following the determination of the gluon Regge trajectory from the high energy effective action, see [Chachamis *et al.* (2012, 2013a)] for full details. This calculation requires

- the determination of the propagator of the reggeized gluon to the desired order in perturbation theory;
- a transition function for its rapidity divergences.

As a start it is necessary to to determine the self-energies of the reggeized gluon. Within the subtraction procedure Sec. 11.3.1, this requires

(a) the determination of the self-energy of the reggeized gluon from the effective action, with the reggeized gluon being treated as a background field;
(b) the subtraction of all disconnected contributions which contain internal reggeized gluon lines.

Using the pole prescription of [Hentschinski (2012)], at 1-loop, all diagrams with internal reggeized gluon lines vanish, and no subtraction occurs; the contributing diagrams are

$$\Sigma^{(1)}\left(\rho;\epsilon,\frac{\mathbf{q}^2}{\mu^2}\right) = \quad (11.29)$$

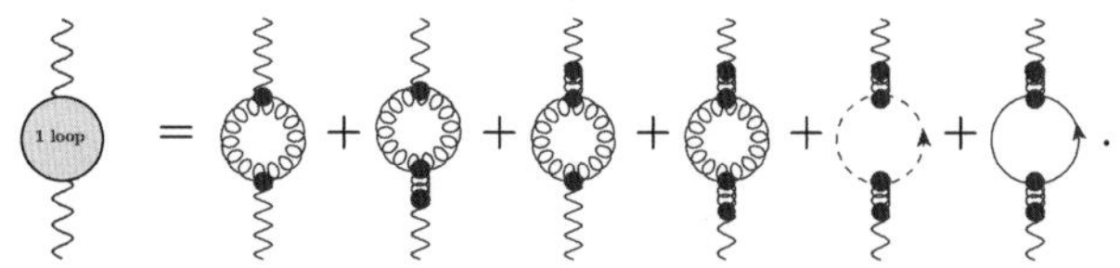

The first of the above diagrams is divergent and requires regularization. To achieve a consistent treatment, we tilt the light-cone direction for all diagrams and take the limit $\rho \to \infty$ after the evaluation of integrals. Keeping only finite and divergent terms, one finds in $d = 4 + 2\epsilon$ dimensions[2]:

$$\frac{\Sigma^{(1)}\left(\rho;\epsilon,\frac{\mathbf{q}^2}{\mu^2}\right)}{(-2i\mathbf{q}^2)} = \bar{g}^2 c_\Gamma \left(\frac{\mathbf{q}^2}{\mu^2}\right)^{\epsilon}\left\{\frac{i\pi - 2\rho}{\epsilon} - \frac{5+3\epsilon-\frac{n_f}{N_c}(2+2\epsilon)}{(1+2\epsilon)(3+2\epsilon)\epsilon}\right\}, \quad (11.30)$$

and

$$\bar{g}^2 = \frac{g^2 N_c \Gamma(1-\epsilon)}{(4\pi)^{2+\epsilon}}, \qquad c_\Gamma = \frac{\Gamma^2(1+\epsilon)}{\Gamma(1+2\epsilon)}. \quad (11.31)$$

To determine the 2-loop self energy it is now needed to subtract disconnected diagrams, whereas diagrams with multiple internal reggeized gluons can be shown to yield zero result. Schematically one has

$$\Sigma^{(2)}\left(\rho;\epsilon,\frac{\mathbf{q}^2}{\mu^2}\right) = \quad (11.32)$$

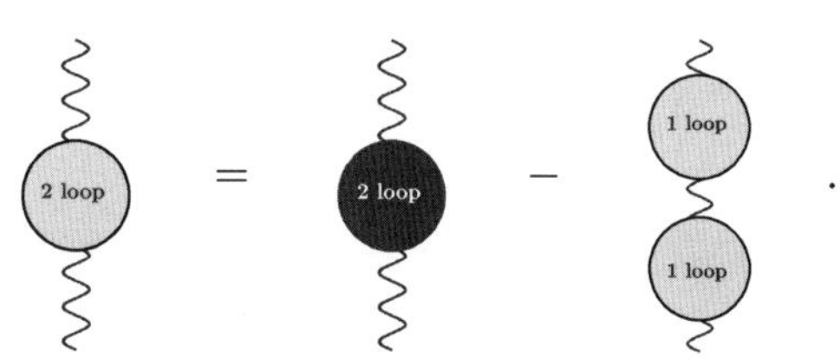

The black blob denotes the unsubtracted 2-loop reggeized gluon self-energy. It is obtained through the direct application of the Feynman rules of the effective action (with the reggeized gluon itself being treated as a background field). The resulting 2-loop integrals have been determined in [Chachamis *et al.* (2012, 2013a)].

[2] In the original result present in [Hentschinski and Sabio Vera (2012)] and reproduced in [Chachamis *et al.* (2014)], a finite result for the second and third diagram has been erroneously included, see also [Chachamis *et al.* (2013b, 2012, 2013a)]

The result for n_f flavor reads

$$\Sigma^{(2)}\left(\rho, \frac{\mathbf{q^2}}{\mu^2}\right) = (-2i\mathbf{q}^2)\frac{g^4 N_c^2}{(4\pi)^4}\Bigg\{ - \left[\frac{2}{\epsilon^2} + \frac{4(1-\tau)}{\epsilon} + 4(1-\tau)^2 - \frac{\pi^2}{3}\right]\rho^2$$
$$+ \left[\frac{1}{3\epsilon^2} + \frac{1}{9\epsilon} + \frac{\pi^2}{3\epsilon} - \frac{2\tau}{3\epsilon} + \frac{\pi^2(11-12\tau)}{18} + \frac{16}{27} - \frac{2}{9}\tau\right.$$
$$\left. + \frac{2}{3}\tau^2 - 2\zeta(3)\right]\rho\Bigg\} + \frac{n_f}{N_c}\left(\frac{2}{3\epsilon} + \frac{n_f(6-36\tau)}{27\epsilon}\right.$$
$$\left. + \frac{32 - 3\pi^2 - 12\tau + 36\tau^2}{27}\right) + \mathcal{O}(\epsilon) + \mathcal{O}(\rho^0) \qquad (11.33)$$

with $\tau = 1 - \ln \frac{\mathbf{q}^2 e^{\gamma_E}}{4\pi\mu^2}$. To obtain the gluon trajectory, we need to construct next the (bare) two-loop reggeized gluon propagators

$$G\left(\rho; \epsilon, \mathbf{q}^2, \mu^2\right) = \frac{i/2}{\mathbf{q}^2}\left\{1 + \frac{i/2}{\mathbf{q}^2}\Sigma\left(\rho; \epsilon, \frac{\mathbf{q}^2}{\mu^2}\right) + \left[\frac{i/2}{\mathbf{q}^2}\Sigma\left(\rho; \epsilon, \frac{\mathbf{q}^2}{\mu^2}\right)\right]^2 + \dots\right\} \qquad (11.34)$$

with the reggeized gluon self energy

$$\Sigma\left(\rho; \epsilon, \frac{\mathbf{q}^2}{\mu^2}\right) = \Sigma^{(1)}\left(\rho; \epsilon, \frac{\mathbf{q}^2}{\mu^2}\right) + \Sigma^{(2)}\left(\rho; \epsilon, \frac{\mathbf{q}^2}{\mu^2}\right) + \dots . \qquad (11.35)$$

Apparently Eq. (11.29) is divergent in the limit $\rho \to \infty$. In [Hentschinski and Sabio Vera (2012); Chachamis *et al.* (2013b)] it has been demonstrated by explicit calculations that these divergences cancel at one-loop level against divergences in the couplings of the reggeized gluon to external particles. The entire one-loop amplitude is then found to be free of any high energy singularity in ρ. Consistency of high energy factorization as formulated within the high energy effective action requires that such a cancellation holds also beyond one loop. To make this cancellation explicit, we introduce transition functions $Z^\pm$. For explicit examples we refer the reader to [Hentschinski and Sabio Vera (2012); Chachamis *et al.* (2013b)]. In particular we define the renormalized reggeized gluon propagator as

$$G^{\mathrm{R}}(\eta; \epsilon, \mathbf{q^2}, \mu^2) = \frac{G(\rho; \epsilon, \mathbf{q^2}, \mu^2)}{Z^+\left(\eta, \rho; \epsilon, \frac{\mathbf{q^2}}{\mu^2}\right) Z^-\left(\eta, \rho; \epsilon, \frac{\mathbf{q^2}}{\mu^2}\right)}. \qquad (11.36)$$

In their most general form these transition functions are parametrized as

$$Z^\pm\left(\eta, \rho; \epsilon, \frac{\mathbf{q^2}}{\mu^2}\right) = \exp\left[\frac{\rho - \eta}{2}\omega\left(\epsilon, \frac{\mathbf{q^2}}{\mu^2}\right) + f^\pm\left(\epsilon, \frac{\mathbf{q^2}}{\mu^2}\right)\right]. \qquad (11.37)$$

The coefficient of the ρ-divergent term defines the gluon Regge trajectory $\omega(\epsilon, \mathbf{q}^2)$,

$$\omega\left(\epsilon, \frac{\mathbf{q}^2}{\mu^2}\right) = \omega^{(1)}\left(\epsilon, \frac{\mathbf{q}^2}{\mu^2}\right) + \omega^{(2)}\left(\epsilon, \frac{\mathbf{q}^2}{\mu^2}\right) + \dots, \qquad (11.38)$$

which is determined by the requirement that the renormalized reggeized gluon propagator must be free of high energy divergences, *i.e.* ρ independent. At one loop one obtains

$$\omega^{(1)}\left(\epsilon, \frac{\mathbf{q}^2}{\mu^2}\right) = -\frac{2\bar{g}^2\Gamma^2(1+\epsilon)}{\Gamma(1+2\epsilon)\epsilon}\left(\frac{\mathbf{q}^2}{\mu^2}\right)^{\epsilon}. \tag{11.39}$$

The function $f^{\pm}(\epsilon, \mathbf{q}^2)$ parametrizes finite contributions and is, in principle, arbitrary. Symmetry of the scattering amplitude requires $f^+ = f^- = f$. Regge theory suggests fixing it in such a way that terms which are not enhanced in ρ are entirely transferred from the reggeized gluon propagators to the vertices, to which the reggeized gluon couples. With

$$f\left(\epsilon, \frac{\mathbf{q}^2}{\mu^2}\right) = f^{(1)}\left(\epsilon, \frac{\mathbf{q}^2}{\mu^2}\right) + f^{(2)}\left(\epsilon, \frac{\mathbf{q}^2}{\mu^2}\right)\ldots \tag{11.40}$$

we obtain from Eq. (11.29)

$$f^{(1)}\left(\epsilon, \frac{\mathbf{q}^2}{\mu^2}\right) = \frac{\bar{g}^2\Gamma^2(1+\epsilon)}{\Gamma(1+2\epsilon)}\left(\frac{\mathbf{q}^2}{\mu^2}\right)^{\epsilon}\frac{(-1)}{(1+2\epsilon)2\epsilon}\left[\frac{5+3\epsilon}{3+2\epsilon} - \frac{n_f}{N_c}\left(\frac{2+2\epsilon}{3+2\epsilon}\right)\right], \tag{11.41}$$

and consequently

$$\omega^{(2)}\left(\epsilon, \frac{\mathbf{q}^2}{\mu^2}\right) = \lim_{\rho\to\infty}\frac{1}{\rho}\left[\frac{\Sigma^{(2)}}{(-2i\mathbf{q}^2)} + \frac{\rho^2}{2}\left(\omega^{(1)}\right)^2 + 2\rho f^{(1)}\omega^{(1)}\right], \tag{11.42}$$

where on the right hand side we omitted the dependencies on ϵ and $\mathbf{q}^2/\mu^2$ and expanded $\Sigma^{(1)}$ in terms of the functions $\omega^{(1)}$ and $f^{(1)}$. We obtain

$$\omega^{(2)}(\mathbf{q}^2) = \frac{(\omega^{(1)}(\mathbf{q}^2))^2}{4}\left[\frac{11}{3} - \frac{2n_f}{3N_c} + \left(\frac{\pi^2}{3} - \frac{67}{9}\right)\epsilon + \left(\frac{404}{27} - 2\zeta(3)\right)\epsilon^2\right], \tag{11.43}$$

in agreement with the results in the literature [Fadin *et al.* (1996)]. The parameters η in the transition functions are arbitrary; their role is analogous to the renormalization scale in UV renormalization or to the factorization scale in collinear factorization. It gives rise to a dependence of the reggeized gluon propagator on the factorization parameter η from which a renormalization group equation (RGE) results:

$$\frac{d}{d\eta}G^R\left(\eta; \epsilon, \frac{\mathbf{q}^2}{\mu^2}\right) = \omega\left(\epsilon, \frac{\mathbf{q}^2}{\mu^2}\right)G^R\left(\eta; \epsilon, \frac{\mathbf{q}^2}{\mu^2}\right). \tag{11.44}$$

With the above choice for the function $f^{\pm}$, one has

$$G^R\left(0; \epsilon, \frac{\mathbf{q}^2}{\mu^2}\right) = \frac{i/2}{\mathbf{q}^2} \tag{11.45}$$

and

$$G^R\left(\eta; \epsilon, \frac{\mathbf{q}^2}{\mu^2}\right) = \frac{i/2}{\mathbf{q}^2}\exp\left[\eta\cdot\omega\left(\epsilon, \frac{\mathbf{q}^2}{\mu^2}\right)\right]. \tag{11.46}$$

The value of the parameter η is to be fixed by the impact factors which describe the coupling of the reggeized gluon to external particles. In the case of partonic scattering the natural choice is given by $\eta = \ln s/|t|$, with $t = -\mathbf{q}^2$ and s the center-of-mass energy squared. In general the value of η is however arbitrary and must be constrained through the evaluation of higher order corrections to impact factors. It goes without saying that Eq. (11.46) gives nothing but reggeization of the gluon.

11.5 The generation of scattering amplitudes for the high energy effective action

For a successful phenomenology of the QCD high energy limit, an efficient determination of the high energy factorized QCD scattering amplitudes is of high importance. An important step in this direction has been achieved through the parton-level event generator KaTie [van Hameren (2018)]. In a nut-shell, one uses the fact that conventional collinear tree-level scattering amplitudes can be very effectively determined via numerical Dyson-Schwinger recursion. The state of a reggeized gluon is then generated through an auxiliary quark-antiquark pair, which allows to obtain the reggeized gluon state through a conventional colllinear scattering amplitude. In more detail: Due to the kinematic constraint Eq. (11.6), the momenta of reggeized gluons satisfy $k \cdot n^{\pm} = 0$. Relating light-cone directions of reggeized gluons to actual momenta in a scattering process we have $n^{\pm} = \sqrt{2}\frac{p_{1,2}}{\sqrt{p_1 \cdot p_2}}$ and a reggeized gluon momentum can be parametrized as

$$k^{\mu} = x p_1^{\mu} + \mathbf{k}^{\mu}, \quad k \cdot n^{+} = 0. \tag{11.47}$$

To generate the scattering amplitude of an off-shell reggeized gluon with quarks and gluons, the reggeized gluon is then embedded into a purely partonic scattering amplitude with the help of an auxiliary quark-antiquark pair with momenta p_A^{μ} and $p_{A'}^{\mu}$. Both auxiliary momenta are light-like $p_A^2 = 0 = p_{A'}^2$. In more detail, one defines

$$p_A^{\mu} = \Lambda p_1^{\mu} + \alpha p_2^{\mu} + \beta \mathbf{k}. \tag{11.48}$$

The requirement $p_A^2 = 0$ fixes then $\alpha = \frac{\beta^2 \mathbf{k}^2}{\Lambda (p+q)^2}$ while $p_{A'}^2 = 0 = (p_A - k)^2$ yields $\beta = \frac{1}{1+\sqrt{1-x/\Lambda}}$. In this way the momenta p_A and $p_{A'}$ fulfill the important requirement

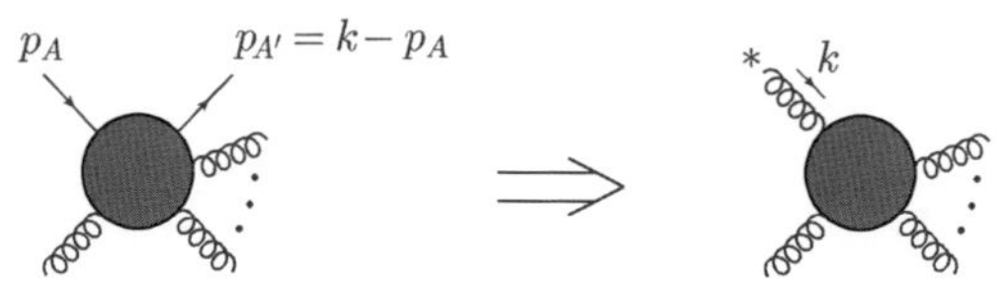

Fig. 11.6 Left: A QCD scattering amplitude of arbitrary many external QCD states. A certain quark-antiquark pair is identified. Following the prescription given in the text, it is reduced to a reggeized gluon which then yields a corresponding scattering amplitude of a reggeized gluon with arbitrary many external partons [figure taken from [van Hameren (2017)]].

of light-likeness for all values of Λ, as well as $p_A + p_{A'} = k$. The desired scattering amplitude of a single reggeized gluon with arbitrary many partons is then obtained from the limit[3]

$$\mathcal{M}\left(0 \to r^*(xp_1 + \mathbf{k}) + X\right) = \lim_{\Lambda\to\infty} \frac{|\mathbf{k}|}{\Lambda} \cdot \mathcal{M}\left(0 \to q(p_A(\Lambda)), \bar{q}(p_{A'}(\Lambda)) + X\right). \tag{11.49}$$

Within the resulting scattering amplitude, it is then straightforward to distinguish gluons which are emitted from the auxiliary quark line from those which are not. The latter play the rôle of the gluons which are generated by the QCD Lagrangian within the Lagrangian of the high energy effective action Eq. (11.7). The former yield the contributions stemming from the induced Lagrangian. To make this relation more explicit, it is useful to consider a typical propagator of the auxiliary quark line. With a generic momentum in the auxiliary quark line given by

$$p^\mu = (\Lambda + x_p)p_1^\mu + y_p p_2^\mu + \mathbf{p} \tag{11.50}$$

one then defines

$$\begin{aligned} p' &= p - p_{A'} = p - k + p_A \\ &= (2\Lambda - x)p_1 + (y_q + \alpha)p_2 + \mathbf{p} - \mathbf{k}(1-\beta) \end{aligned} \tag{11.51}$$

as the momentum flowing in the (auxiliary) quark propagator, if the external quark p_A would give the momentum k and the anti-quark momentum 0, *i.e.* the total momentum of the gluons emitted by the quark sum up to the reggeized gluon momentum. p' is therefore the total momentum of gluons "so far" emitted from the auxiliary quark line. In particular

$$\lim_{\Lambda\to\infty} p_1 \cdot p' = y_p 2 p_1 \cdot q. \tag{11.52}$$

One then finds for the quark propagator with momentum p in the limit $\Lambda \to \infty$:

$$\lim_{\Lambda\to\infty} \frac{\not{p}}{p^2} = \lim_{\Lambda\to\infty} \frac{(\Lambda + x_p)\not{p}_1 + y_p \not{p}_2 + \not{\mathbf{p}}}{2(\Lambda + x_p)y_p p_1 \cdot p_2 + \mathbf{p}^2} = \frac{\not{p}_1}{2p_1 \cdot p'} = \frac{\not{n}^+}{2p^{+'}}. \tag{11.53}$$

Making now use of well-known identities such as

$$\not{n}^+ \gamma_\mu \not{n}^+ = 2n^\mu \cdot \not{n}^+, \tag{11.54}$$

one re-obtains from these contributions the induced vertices[4] of the high energy effective action, Fig. 11.2. The whole procedure can be generalized in a

[3] We use the symbol r^* to denote the external state of an (off-shell) reggeized gluon

[4] In more detail: one first obtains the term corresponding to one particular ordering of the operator Eq. (11.15); summing over all possible orderings, the full color structure of Fig. 11.2 is obtained.

straightforward manner to the case of two reggeized gluons with opposite, *i.e.* plus- and minus-polarization. In this way it is then possible to generate all tree-level amplitudes of the high-energy effective action. This is of particular use, since there exist very powerful algorithms which allow to generate the underlying scattering amplitudes numerically. It is therefore correct to claim that the problem of efficiently computing gauge invariant tree level scattering amplitudes within high energy factorization in the region of phase space where one reggeized gluon exchange provides a suitable approximation, has been completely solved in recent years, both analytically and numerically. For details we refer the interested reader to the original papers [van Hameren *et al.* (2012, 2013b,a); van Hameren (2017, 2014); Bury and van Hameren (2015); van Hameren and Serino (2015); Kutak *et al.* (2017)]. The method has been used in the recent past for a large number of phenomenological studies, but also for theoretical exploration, see for instance the calculation of [Hentschinski *et al.* (2018)].

11.6 The high energy effective action and high parton densities

An area of high phenomenological interest is given by the physics of high parton densities: occurrence of high parton densities is closely related to the physics of gluon saturation which is assumed to be the relevant mechanism to ensure unitarity of QCD scattering amplitudes in the high energy limit. The bulk of theoretical efforts in recent years to describe this region of phase space within high energy factorization, can be summarized under the term "Color Glass Condensate effective field theory" (see [Gelis *et al.* (2010)] for a review). In the following we summarize the essential results of [Hentschinski (2018)] which demonstrate these efforts with the high energy effective action.

11.6.1 *A special parametrization of the gluonic field*

The starting point for this exploration has already been laid in the original publication [Lipatov (1995, 1997)] by Lipatov himself: there it has been noted that while a shift in the gauge field $v^\mu \to v^\mu + n_-^\mu A_+/2 + n_+^\mu A_-/2$ provides a possibility to avoid the appearance of a direct transition vertex between reggeized gluon field and gluon field, Fig. 11.2(a), it mixes two fields with different gauge transformation properties: v_μ transforms as a gauge field, while $A_\pm$ is by definition invariant under local gauge transformations. The solution put forward in [Lipatov (1995, 1997)] suggests the following parametrization of the gluonic field:

$$V^\mu(x) = v^\mu(x) + \frac{n_+^\mu}{2} B_-(x) + \frac{n_-^\mu}{2} B_+(x), \tag{11.55}$$

where

$$B_\pm[v_\mp] = U[v_\mp] A_\pm U^{-1}[v_\mp]. \tag{11.56}$$

and (inverse) Wilson line operators are defined as

$$U[v_\pm] = \frac{1}{1 + \frac{g}{\partial_\pm} v_\pm}, \qquad U^{-1}[v_\pm] = 1 + \frac{g}{\partial_\pm} v_\pm. \tag{11.57}$$

For the above composite field $B_\pm[v_\mp]$, one finds the following gauge transformation properties:

$$\delta_L B_\pm = \delta_L U[v_\mp] A_\pm U^{-1}[v_\mp] + U[v_\mp] A_\pm \delta_L U^{-1}[v_\mp] = [g B_\pm, \chi_L], \tag{11.58}$$

and as a consequence the shifted gluonic field Eq. (11.55) transforms like a gluonic field:

$$\delta V_\pm = [D_\pm, \chi] + [g B_\pm, \chi] = [D_\pm + g B_\pm, \chi]. \tag{11.59}$$

In the following we will use the above parametrization of the gluonic field to expand the high energy effective action for the quasi-elastic case around the reggeized gluon field A_+ which we treat as a strong classical background field $gA_+ \sim 1$. We therefore consider the high energy effective action for the quasi-elastic case

$$S_{\text{eff}}^{\text{q.e.}} = S_{\text{QCD}} + S_{\text{ind.}}^{\text{q.e.}} \tag{11.60}$$

with

$$S_{\text{ind.}}^{\text{q.e.}} = \int d^4x \,\text{tr}\left[\{T_-[v] - A_-(x)\} \partial^2 A_+(x)\right]. \tag{11.61}$$

Keeping fields A_+ to all orders and expanding in quantum fluctuations v_μ and ψ, $\bar\psi$ to quadratic order, we obtain

$$S_{\text{eff}}^{\text{q.e.}} = \int d^4x \left[\mathcal{L}_0 + \mathcal{L}_1 - \text{tr}\left(A_- \partial^2 A_+\right)\right] + \mathcal{O}(v_\mu^3), \tag{11.62}$$

with the kinetic term of the gluonic and quark field

$$\mathcal{L}_0 = \text{tr}\left(-v^\mu [g_{\mu\nu}\partial^2 - \partial_\mu \partial_\nu] v^\nu\right) + \bar\psi i \partial\!\!\!/ \psi, \tag{11.63}$$

and the quadratic terms which describe the interaction with the reggeized gluon field:

$$\mathcal{L}_1 = g \cdot \left\{ \frac{i}{2} \bar\psi n\!\!\!/_- A_+ \psi + \text{tr}\left[\partial_- v_\mu [A_+, v^\mu] + 2\partial_\mu v_- [v^\mu, A_+] + \right.\right.$$
$$\left.\left. + \partial^2 v_- \left[\left(\frac{1}{\partial_-} v_-\right), A_+\right] - v_- \left(\frac{1}{\partial_-} v_-\right) \partial^2 A_+ \right]\right\}. \tag{11.64}$$

11.6.2 *Resummation of a dense reggeized gluon field*

The quasi-elastic effective action allows then for the extraction of quark-quark-reggeized gluon (QQR) and gluon-gluon-reggeized gluon vertices,

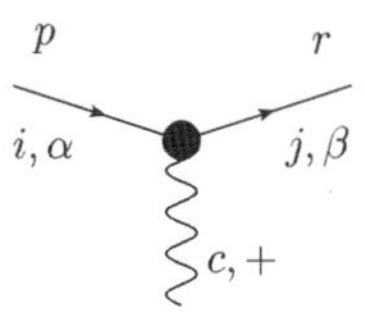

$$= -igt^c_{ji}\Gamma_{\beta\alpha}(r,p)\int d^4z\, e^{-iz\cdot(p-r)}A^c_+(z),$$

$$\Gamma_{\beta\alpha}(r,p) = -\frac{1}{2}\not{n}^+_{\alpha\beta}, \tag{11.65}$$

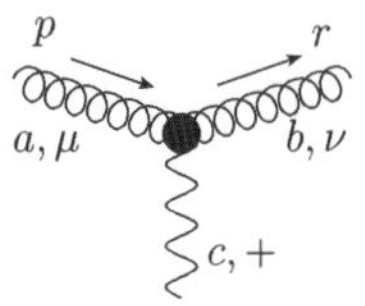

$$= -igT^c_{ba}\Gamma^{\nu\mu}(r,p)\int d^4z\, e^{-iz\cdot(p-r)}A^c_+(z),$$

$$\Gamma^{\nu\mu}_+(r,p) = p^+g^{\mu\nu} - (n^+)^\mu p^\nu - (n^+)^\nu r^\mu + \frac{r\cdot p}{p^+}(n^+)^\mu(n^+)^\nu, \tag{11.66}$$

with $T^c_{ab} = -if^{abc}$. The QQR and GGR vertices obey various properties:

$$\begin{aligned}\Gamma_{\beta\gamma'}(r,p)\not{n}_{\gamma'\gamma} &= 0 = \not{n}_{\beta\beta'}\Gamma_{\beta'\gamma}(r,p),\\ \Gamma_{\beta\gamma}(r,k)\not{k}_{\gamma\gamma'}\Gamma_{\gamma'\alpha}(k,p) &= -p^+\Gamma_{\beta\alpha}(r,p),\end{aligned} \tag{11.67}$$

and

$$\begin{aligned} n^+_\nu\cdot\Gamma^{\nu\mu}_+(r,p) &= 0 = \Gamma^{\nu\mu}_+(r,p)\cdot n^+_\mu,\\ \Gamma^{\nu\alpha}_+(r,k)\cdot(-g_{\alpha\alpha'})\cdot\Gamma^{\alpha'\mu}_+(k,p) &= -p^+\Gamma^{\nu\mu}_+(r,p),\\ r_\nu\cdot\Gamma^{\nu\mu}_+(r,p) &= 0 = \Gamma^{\nu\mu}_+(r,p)\cdot p_\mu.\end{aligned} \tag{11.68}$$

Note that the above GGR-vertex was already obtained in [Lipatov (1995)]. A potential disadvantage of this vertex, pointed out in [Lipatov (1995)], is that it can yield individual Feynman diagrams which contain singularities in overlapping channels and therefore violate the Steinmann-relations [Steinmann (1960)]. Since the vertex is obtained from an effective action which explicitly respects the Steinmann relations, such effects must cancel at the level of the observable. The properties Eq. (11.67) are of great use for the resummation of a strong reggeized gluon field $gA_+ \sim 1$ to all orders. Such an exercise has been carried out in [Hentschinski (2018)]. As a result one obtains the following effective vertices which sum up the interaction of a quark and a gluon with an arbitrary number of reggeized gluon fields:

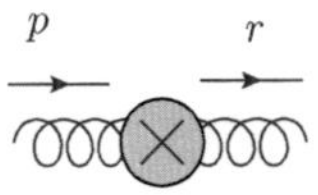

$$= \tau^{ab}_{G,\nu\mu}(p,-r) = -4\pi\delta(p^+ - r^+)\Gamma_{\nu\mu}(r,p)e^{-ix_0^+(p^- - r^-)}$$

$$\cdot \int d^2\mathbf{z} e^{i\mathbf{z}\cdot(\mathbf{p}-\mathbf{r})}\left[\theta(p^+)\left[U^{ba}(\mathbf{z}) - \delta^{ab}\right] - \theta(-p^+)\left[[U^{ba}(\mathbf{z})]^\dagger - \delta^{ab}\right]\right], \quad (11.69)$$

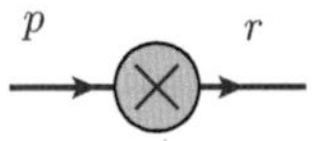

$$= \tau_F(q,-r) = 2\pi\delta(p^+ - r^+)\not{n}^+ e^{-ix_0^+(p^- - r^-)}$$

$$\cdot \int d^2\mathbf{z} e^{i\mathbf{z}\cdot(\mathbf{p}-\mathbf{r})}\left[\theta(p^+)\left[W(\mathbf{z}) - 1\right] - \theta(-p^+)\left[[W(\mathbf{z})]^\dagger - 1\right]\right]. \quad (11.70)$$

To write down the above expressions, we have introduced Wilson lines in the adjoint representation

$$U^{ab}(\mathbf{z}) = \mathrm{P}\exp\left(-\frac{g}{2}\int_{-\infty}^{\infty} dz^+ \tilde{A}_+\right), \quad \tilde{A}_+ = -iT^c_{ab}A^c_+, \quad (11.71)$$

and the fundamental representation

$$W(\mathbf{z}) = \mathrm{P}\exp\left(-\frac{g}{2}\int_{-\infty}^{\infty} dz^+ A_+\right), \quad A_+ = -it^c_{ij}A^c_+. \quad (11.72)$$

The above vertices Eq. (11.69) allow to construct partonic propagators in the presence of a (reggeized gluon) background field and can be compared with corresponding propagators derived for the resummation of a background field in the light-cone gauge [McLerran and Venugopalan (1994); Baltz *et al.* (2001); Gelis and Peshier (2002); Balitsky and Belitsky (2002); Ayala *et al.* (2016, 2017)]. It turns out that the regarding propagators agree with each other, if the gluonic field is taken in the same gauge, *i.e.* light-cone gauge. The interpretation is however slightly different for the Wilson lines which resum the gluonic background field in the light-cone gauge and the reggeized gluon field, respectively. For a more detailed discussion we refer the interested reader to [Hentschinski (2018)]. As a first application it was possible to use the vertices Eq. (11.69) to re-derive the leading order Balitsky-JIMWLK equation from the high energy effective action: Calculating quantum fluctuations of an ensemble of Wilson lines of the reggeized gluon fields

$$U^{ab}(\mathbf{z}) = 2\mathrm{tr}\left[t^a W(\mathbf{z}) t^b W^\dagger(\mathbf{z})\right], \quad (11.73)$$

one finds for the evolution with respect to the regulator used for the high energy singularities, the Balitsky-JIMWLK equation

$$-\Lambda_a \frac{d}{d\Lambda_a}\left[W(\mathbf{x}_1)\otimes\ldots\otimes W(\mathbf{x}_n)\right] = \sum_{i,j=1} H_{ij}\left[W(\mathbf{x}_1)\otimes\ldots\otimes W(\mathbf{x}_n)\right], \quad (11.74)$$

with the Balitsky-JIMWLK Hamiltonian

$$H_{ij} = \frac{\alpha_s \Gamma^2(1+\epsilon)}{2\pi^2 \Gamma(1-\epsilon)} \left(\frac{4}{\pi\mu^2}\right)^{\epsilon} \int d^{2+2\epsilon}\mathbf{z} \frac{(\mathbf{x}_i - \mathbf{z}) \cdot (\mathbf{x}_j - \mathbf{z})}{[(\mathbf{x}_i - \mathbf{z})^2]^{1+\epsilon} [(\mathbf{x}_j - \mathbf{z})^2]^{1+\epsilon}}$$
$$\times \left[T^a_{i,L} T^a_{j,L} + T^a_{i,R} T^a_{j,R} - U^{ab}(\mathbf{z}) \left(T^a_{i,L} T^b_{j,R} + T^a_{j,L} T^b_{i,R} \right) \right], \tag{11.75}$$

where $T^a_{L,i}$ and $T^a_{R,j}$ are the group generators acting to the left (L) or to the right (R) on the Wilson line $W(\mathbf{x}_i)$,

$$T^a_{L,i}[W(\mathbf{z}_i)] \equiv t^a W(\mathbf{z}_i), \quad T^a_{R,i}[W(\mathbf{z}_i)] \equiv W(\mathbf{z}_i) t^a. \tag{11.76}$$

11.7 Conclusions

The topics covered in this short review provide apparently only a small subset of results achieved within the high energy effective action (with the choice heavily biased by the author's own work). Among the results which we did not review here, are the effective action for Regge processes in gravity [Lipatov (2013)], earlier versions of an high energy effective action [Lipatov (1991); Kirschner *et al.* (1994, 1995)] as well as the determination of the $3 \to 3$ reggeized gluon transition kernel, needed for the NLO Bartels-Kwiecinski-Praszalowicz equation in [Bartels *et al.* (2013)] and the NLO corrections of the Mueller-Tang jet impact factor [Hentschinski *et al.* (2014b,c,a)]. Another line of research has been developed in a series of papers [Braun and Vyazovsky (2007); Braun *et al.* (2010, 2011, 2012b,a, 2013, 2014, 2016); Braun and Vyazovsky (2019)] dedicated to the study of production processes at central rapidities in the presence of multiple reggeized gluon exchange. The relation of the high energy effective action to the Color Glass Condensate approach as well as the further development of the underlying formulation has been explored in [Bondarenko *et al.* (2017a,b)] and continued in [Bondarenko and Zubkov (2018); Bondarenko and Pozdnyakov (2019b, 2018b,a, 2019a)]. In parallel to the high energy effective action for gluons, the corresponding effective action for quarks has been explored in series of papers, at first mainly focusing on phenomenological applications in combination with transverse momentum dependent parton distribution functions [Saleev (2008, 2009); Kniehl *et al.* (2010, 2011); Saleev *et al.* (2012); Saleev and Shipilova (2012); Nefedov *et al.* (2013b,c); Saleev and Shipilova (2013); Nefedov *et al.* (2013a,d); Kniehl *et al.* (2014); Nefedov and Saleev (2015); Karpishkov *et al.* (2015a,b, 2016, 2017); Nefedov and Saleev (2019)], while recent efforts address the determination of higher order perturbative corrections [Nefedov and Saleev (2017); Nefedov (2019)].

Bibliography

Antonov, E. N., Lipatov, L. N., Kuraev, E. A., and Cherednikov, I. O. (2005). Feynman rules for effective Regge action, *Nucl. Phys. B* **721**, pp. 111–135, doi:10.1016/j.nuclphysb.2005.05.013,10.1016/j.nuclphysb.2005.013, arXiv:hep-ph/0411185 [hep-ph].

Arefeva, I. Ya. (1994). Large N QCD at high-energies as two-dimensional field theory, *Phys. Lett. B* **328**, pp. 411–419, doi:10.1016/0370-2693(94)91499-0, arXiv:hep-th/9306014 [hep-th].

Ayala, A., Hentschinski, M., Jalilian-Marian, J., and Tejeda-Yeomans, M. E. (2016). Polarized 3 parton production in inclusive DIS at small x, *Phys. Lett. B* **761**, pp. 229–233, doi:10.1016/j.physletb.2016.08.035, arXiv:1604.08526 [hep-ph].

Ayala, A., Hentschinski, M., Jalilian-Marian, J., and Tejeda-Yeomans, M. E. (2017). Spinor helicity methods in high-energy factorization: Efficient momentum-space calculations in the color glass condensate formalism, *Nucl. Phys. B* **920**, pp. 232–255, doi:10.1016/j.nuclphysb.2017.03.028, arXiv:1701.07143 [hep-ph].

Balitsky, I. I. and Belitsky, A. V. (2002). Nonlinear evolution in high density QCD, *Nucl. Phys. B* **629**, pp. 290–322, doi:10.1016/S0550-3213(02)00149-9, arXiv:hep-ph/0110158 [hep-ph].

Baltz, A. J., Gelis, F., McLerran, L. D., and Peshier, A. (2001). Coulomb corrections to e+ e− production in ultrarelativistic nuclear collisions, *Nucl. Phys. A* **695**, pp. 395–429, doi:10.1016/S0375-9474(01)01109-5, arXiv:nucl-th/0101024 [nucl-th].

Bartels, J., Fadin, V. S., Lipatov, L. N., and Vacca, G. P. (2013). NLO Corrections to the kernel of the BKP-equations, *Nucl. Phys. B* **867**, pp. 827–854, doi:10.1016/j.nuclphysb.2012.10.024, arXiv:1210.0797 [hep-ph].

Bondarenko, S., Lipatov, L., Pozdnyakov, S., and Prygarin, A. (2017a). One loop light-cone QCD, effective action for reggeized gluons and QCD RFT calculus, *Eur. Phys. J. C* **77**, 9, p. 630, doi:10.1140/epjc/s10052-017-5208-8, arXiv:1708.05183 [hep-th].

Bondarenko, S., Lipatov, L., and Prygarin, A. (2017b). Effective action for reggeized gluons, classical gluon field of relativistic color charge and color glass condensate approach, *Eur. Phys. J. C* **77**, 8, p. 527, doi:10.1140/epjc/s10052-017-5101-5, arXiv:1706.00278 [hep-ph].

Bondarenko, S. and Pozdnyakov, S. (2018a). On correlators of Reggeon fields and operators of Wilson lines in high energy QCD, *Int. J. Mod. Phys. A* **33**, 35, p. 1850204, doi:10.1142/S0217751X18502044, arXiv:1806.02563 [hep-th].

Bondarenko, S. and Pozdnyakov, S. (2018b). S-matrix and productions amplitudes in high energy QCD, *Phys. Lett. B* **783**, pp. 207–211, doi:10.1016/j.physletb.2018.06.054, arXiv:1803.04131 [hep-th].

Bondarenko, S. and Pozdnyakov, S. (2019a). Effective action and classical solutions, *Phys. Part. Nucl. Lett.* **16**, 5, pp. 433–435, doi:10.1134/S1547477119050066.

Bondarenko, S. and Pozdnyakov, S. S. (2019b). NNLO classical solution for Lipatovs effective action for reggeized gluons, *Int. J. Mod. Phys. A* **34**, 20, p. 1950111, doi:10.1142/S0217751X19501112, arXiv:1802.05508 [hep-ph].

Bondarenko, S. and Zubkov, M. A. (2018). The dimensionally reduced description of the high energy scattering and the effective action for the reggeized gluons, *Eur. Phys. J. C* **78**, 8, p. 617, doi:10.1140/epjc/s10052-018-6089-1, arXiv:1801.08066 [hep-ph].

Braun, M. A., Lipatov, L. N., Salykin, M. Yu., and Vyazovsky, M. I. (2011). Gluon production on two centers and the effective action approach, *Eur. Phys. J. C* **71**, p. 1639, doi:10.1140/epjc/s10052-011-1639-9, arXiv:1103.3618 [hep-ph].

Braun, M. A., Pozdnyakov, S. S., Salykin, M. Yu., and Vyazovsky, M. I. (2013). Gluon production in the Lipatov effective action formalism, *Eur. Phys. J. C* **73**, 9, p. 2572, doi:10.1140/epjc/s10052-013-2572-x, arXiv:1306.3583 [hep-ph].

Braun, M. A., Pozdnyakov, S. S., Salykin, M. Yu., and Vyazovsky, M. I. (2014). On the application of the effective action approach to amplitudes with reggeon splitting, *Eur. Phys. J. C* **74**, 8, p. 2989, doi:10.1140/epjc/s10052-014-2989-x, arXiv:1402.4786 [hep-ph].

Braun, M. A., Salykin, M. Yu., Pozdnyakov, S. S., and Vyazovsky, M. I. (2012a). Production of a gluon with the exchange of three reggeized gluons in the Lipatov effective action approach, *Eur. Phys. J. C* **72**, p. 2223, doi:10.1140/epjc/s10052-012-2223-7, arXiv:1209.2490 [hep-ph].

Braun, M. A., Salykin, M. Yu., and Vyazovsky, M. I. (2010). Production of two gluons in the Lipatov effective action formalism, *Eur. Phys. J. C* **65**, pp. 385–394, doi: 10.1140/epjc/s10052-009-1184-y, arXiv:0906.3555 [hep-ph].

Braun, M. A., Salykin, M. Yu., and Vyazovsky, M. I. (2012b). On the inclusive gluon production in the Lipatov effective action formalism, *Eur. Phys. J. C* **72**, p. 1864, doi:10.1140/epjc/s10052-012-1864-x, arXiv:1109.1340 [hep-ph].

Braun, M. A., Vyazovsky, M. B., Pozdnyakov, S. S., and Salykin, M. Yu. (2016). An effective two reggeon two reggeon + particle vertex in Lipatovs effective action and regge kinematics, *Bull. Russ. Acad. Sci. Phys.* **80**, 8, pp. 959–965, doi: 10.3103/S1062873816080062, [Izv. Ross. Akad. Nauk Ser. Fiz.80,no.8,1047(2016)].

Braun, M. A. and Vyazovsky, M. I. (2007). The Reggeon —> 2 Reggeons + Particle vertex in the Lipatov effective action formalism, *Eur. Phys. J. C* **51**, pp. 103–111, doi:10.1140/epjc/s10052-007-0270-2, arXiv:hep-ph/0612323 [hep-ph].

Braun, M. A. and Vyazovsky, M. I. (2019). Gluon emission at small longitudinal momenta in the QCD effective action approach, *Eur. Phys. J. C* **79**, 5, p. 392, doi:10.1140/epjc/s10052-019-6897-y, arXiv:1901.08844 [hep-ph].

Bury, M. and van Hameren, A. (2015). Numerical evaluation of multi-gluon amplitudes for high energy factorization, *Comput. Phys. Commun.* **196**, pp. 592–598, doi:10.1016/j.cpc.2015.06.023, arXiv:1503.08612 [hep-ph].

Chachamis, G., Hentschinski, M., Madrigal Martinez, J. D., and Sabio Vera, A. (2012). Quark contribution to the gluon Regge trajectory at NLO from the high energy effective action, *Nucl. Phys. B* **861**, pp. 133–144, doi:10.1016/j.nuclphysb.2012.03.015, arXiv:1202.0649 [hep-ph].

Chachamis, G., Hentschinski, M., Madrigal Martinez, J. D., and Sabio Vera, A. (2013a). Gluon Regge trajectory at two loops from Lipatov's high energy effective action, *Nucl. Phys. B* **876**, pp. 453–472, doi:10.1016/j.nuclphysb.2013.08.013, arXiv:1307.2591 [hep-ph].

Chachamis, G., Hentschinski, M., Madrigal Martnez, J. D., and Sabio Vera, A. (2013b). Next-to-leading order corrections to the gluon-induced forward jet vertex from the high energy effective action, *Phys. Rev. D* **87**, 7, p. 076009, doi:10.1103/PhysRevD.87.076009, arXiv:1212.4992 [hep-ph].

Chachamis, G., Hentschinski, M., Madrigal Martnez, J. D., and Sabio Vera, A. (2014). Forward jet production & quantum corrections to the gluon Regge trajectory from Lipatov's high energy effective action, *Phys. Part. Nucl.* **45**, 4, pp. 788–799, doi: 10.1134/S1063779614040030, arXiv:1211.2050 [hep-ph].

Fadin, V. S., Fiore, R., and Kotsky, M. I. (1996). Gluon Regge trajectory in the two loop approximation, *Phys. Lett. B* **387**, pp. 593–602, doi:10.1016/0370-2693(96)01054-4, arXiv:hep-ph/9605357 [hep-ph].

Gelis, F., Iancu, E., Jalilian-Marian, J., and Venugopalan, R. (2010). The color glass condensate, *Ann. Rev. Nucl. Part. Sci.* **60**, pp. 463–489, doi:10.1146/annurev.nucl.010909.083629, arXiv:1002.0333 [hep-ph].

Gelis, F. and Peshier, A. (2002). Probing colored glass via q anti-q photoproduction, *Nucl. Phys. A* **697**, pp. 879–901, doi:10.1016/S0375-9474(01)01264-7, arXiv:hep-ph/0107142 [hep-ph].

Hentschinski, M. (2009). *The high energy behavior of QCD: The effective action and the triple-Pomeron-vertex*, Ph.D. thesis, Hamburg U., arXiv:0908.2576 [hep-ph], http://www-library.desy.de/cgi-bin/showprep.pl?thesis09-025.

Hentschinski, M. (2012). Pole prescription of higher order induced vertices in Lipatov's QCD effective action, *Nucl. Phys. B* **859**, pp. 129–142, doi:10.1016/j.nuclphysb.2012.02.001, arXiv:1112.4509 [hep-ph].

Hentschinski, M. (2018). Color glass condensate formalism, Balitsky-JIMWLK evolution, and Lipatov's high energy effective action, *Phys. Rev. D* **97**, 11, p. 114027, doi:10.1103/PhysRevD.97.114027, arXiv:1802.06755 [hep-ph].

Hentschinski, M., Kusina, A., Kutak, K., and Serino, M. (2018). TMD splitting functions in k_T factorization: The real contribution to the gluon-to-gluon splitting, *Eur. Phys. J. C* **78**, 3, p. 174, doi:10.1140/epjc/s10052-018-5634-2, arXiv:1711.04587 [hep-ph].

Hentschinski, M., Madrigal Martinez, J. D., Murdaca, B., and Sabio Vera, A. (2014a). The gluon-induced MuellerTang jet impact factor at next-to-leading order, *Nucl. Phys. B* **889**, pp. 549–579, doi:10.1016/j.nuclphysb.2014.10.026, arXiv:1409.6704 [hep-ph].

Hentschinski, M., Madrigal Martnez, J. D., Murdaca, B., and Sabio Vera, A. (2014b). The next-to-leading order vertex for a forward jet plus a rapidity gap at high energies, *Phys. Lett. B* **735**, pp. 168–172, doi:10.1016/j.physletb.2014.06.022, arXiv:1404.2937 [hep-ph].

Hentschinski, M., Madrigal Martnez, J. D., Murdaca, B., and Sabio Vera, A. (2014c). The quark induced MuellerTang jet impact factor at next-to-leading order, *Nucl. Phys. B* **887**, pp. 309–337, doi:10.1016/j.nuclphysb.2014.08.010, arXiv:1406.5625 [hep-ph].

Hentschinski, M. and Sabio Vera, A. (2012). NLO jet vertex from Lipatov's QCD effective action, *Phys. Rev. D* **85**, p. 056006, doi:10.1103/PhysRevD.85.056006, arXiv:1110.6741 [hep-ph].

Karpishkov, A., Saleev, V., and Shipilova, A. (2016). Large-p_T production of D mesons at the LHCb in the parton Reggeization approach, *Phys. Rev. D* **94**, 11, p. 114012, doi:10.1103/PhysRevD.94.114012, arXiv:1610.04975 [hep-ph].

Karpishkov, A. V., Nefedov, M. A., and Saleev, V. A. (2017). $B\bar{B}$ angular correlations at the LHC in parton Reggeization approach merged with higher-order matrix elements, *Phys. Rev. D* **96**, 9, p. 096019, doi:10.1103/PhysRevD.96.096019, arXiv:1707.04068 [hep-ph].

Karpishkov, A. V., Nefedov, M. A., Saleev, V. A., and Shipilova, A. V. (2015a). B-meson production in the parton Reggeization approach at Tevatron and the LHC, *Int. J. Mod. Phys. A* **30**, 04n05, p. 1550023, doi:10.1142/S0217751X15500232, arXiv:1411.7672 [hep-ph].

Karpishkov, A. V., Nefedov, M. A., Saleev, V. A., and Shipilova, A. V. (2015b). Open charm production in the parton Reggeization approach: Tevatron and the LHC, *Phys. Rev. D* **91**, 5, p. 054009, doi:10.1103/PhysRevD.91.054009, arXiv:1410.7139 [hep-ph].

Kirschner, R., Lipatov, L. N., and Szymanowski, L. (1994). Effective action for multi-Regge processes in QCD, *Nucl. Phys. B* **425**, pp. 579–594, doi:10.1016/0550-3213(94)90288-7, arXiv:hep-th/9402010 [hep-th].

Kirschner, R., Lipatov, L. N., and Szymanowski, L. (1995). Symmetry properties of the effective action for high-energy scattering in QCD, *Phys. Rev. D* **51**, pp. 838–855, doi:10.1103/PhysRevD.51.838, arXiv:hep-th/9403082 [hep-th].

Kniehl, B. A., Nefedov, M. A., and Saleev, V. A. (2014). Prompt-photon plus jet associated photoproduction at HERA in the parton Reggeization approach, *Phys. Rev. D* **89**, 11, p. 114016, doi:10.1103/PhysRevD.89.114016, arXiv:1404.3513 [hep-ph].

Kniehl, B. A., Saleev, V. A., and Shipilova, A. V. (2010). Inclusive b and b anti-b production with quasi-multi-Regge kinematics at the Tevatron, *Phys. Rev. D* **81**, p. 094010, doi:10.1103/PhysRevD.81.094010, arXiv:1003.0346 [hep-ph].

Kniehl, B. A., Saleev, V. A., Shipilova, A. V., and Yatsenko, E. V. (2011). Single jet and prompt-photon inclusive production with multi-Regge kinematics: From Tevatron to LHC, *Phys. Rev. D* **84**, p. 074017, doi:10.1103/PhysRevD.84.074017, arXiv:1107.1462 [hep-ph].

Kutak, K., Hameren, A., and Serino, M. (2017). QCD amplitudes with 2 initial spacelike legs via generalised BCFW recursion, *JHEP* **02**, p. 009, doi:10.1007/JHEP02(2017) 009, arXiv:1611.04380 [hep-ph].

Lipatov, L. N. (1991). High-energy scattering in QCD and in quantum gravity and two-dimensional field theories, *Nucl. Phys. B* **365**, pp. 614–632, doi:10.1016/ 0550-3213(91)90512-V.

Lipatov, L. N. (1995). Gauge invariant effective action for high-energy processes in QCD, *Nucl. Phys. B* **452**, pp. 369–400, doi:10.1016/0550-3213(95)00390-E, arXiv:hep-ph/9502308 [hep-ph].

Lipatov, L. N. (1997). Small x physics in perturbative QCD, *Phys. Rept.* **286**, pp. 131–198, doi:10.1016/S0370-1573(96)00045-2, arXiv:hep-ph/9610276 [hep-ph].

Lipatov, L. N. (2013). Effective action for the Regge processes in gravity, *Phys. Part. Nucl.* **44**, pp. 391–413, doi:10.1134/S1063779613020214, arXiv:1105.3127 [hep-th].

Lipatov, L. N. (2016). Euler-Lagrange equations for high energy actions in QCD and in gravity, *EPJ Web Conf.* **125**, p. 01010, doi:10.1051/epjconf/201612501010.

Lipatov, L. N. and Vyazovsky, M. I. (2001). QuasimultiRegge processes with a quark exchange in the t channel, *Nucl. Phys. B* **597**, pp. 399–409, doi:10.1016/ S0550-3213(00)00709-4, arXiv:hep-ph/0009340 [hep-ph].

McLerran, L. D. and Venugopalan, R. (1994). Green's functions in the color field of a large nucleus, *Phys. Rev. D* **50**, pp. 2225–2233, doi:10.1103/PhysRevD.50.2225, arXiv:hep-ph/9402335 [hep-ph].

Nefedov, M. and Saleev, V. (2015). Diphoton production at the Tevatron and the LHC in the NLO approximation of the parton Reggeization approach, *Phys. Rev. D* **92**, 9, p. 094033, doi:10.1103/PhysRevD.92.094033, arXiv:1505.01718 [hep-ph].

Nefedov, M. and Saleev, V. (2017). On the one-loop calculations with Reggeized quarks, *Mod. Phys. Lett. A* **32**, 40, p. 1750207, doi:10.1142/S0217732317502078, arXiv:1709.06246 [hep-th].

Nefedov, M. and Saleev, V. (2019). Off-shell initial state effects, gauge invariance and angular distributions in the DrellYan process, *Phys. Lett. B* **790**, pp. 551–556, doi: 10.1016/j.physletb.2018.12.071, arXiv:1810.04061 [hep-ph].

Nefedov, M., Saleev, V., and Shipilova, A. (2013a). Prompt *Upsilon*(nS) production at the LHC in the Regge limit of QCD, *Phys. Rev. D* **88**, 1, p. 014003, doi:10.1103/ PhysRevD.88.014003, arXiv:1305.7310 [hep-ph].

Nefedov, M. A. (2019). Computing one-loop corrections to effective vertices with two scales in the EFT for Multi-Regge processes in QCD, *Nucl. Phys. B* **946**, p. 114715, doi: 10.1016/j.nuclphysb.2019.114715, arXiv:1902.11030 [hep-ph].

Nefedov, M. A., Nikolaev, N. N., and Saleev, V. A. (2013b). Drell-Yan lepton pair production at high energies in the parton Reggeization approach, *Phys. Rev. D* **87**, 1, p. 014022, doi:10.1103/PhysRevD.87.014022, arXiv:1211.5539 [hep-ph].

Nefedov, M. A., Saleev, V. A., and Shipilova, A. V. (2013c). Charmonium production at the Tevatron and Large Hadron Collider in the Regge limit of QCD, *Phys. Atom. Nucl.* **76**, pp. 1546–1553, doi:10.1134/S1063778813110124, [Yad. Fiz.76,615(2013)].

Nefedov, M. A., Saleev, V. A., and Shipilova, A. V. (2013d). Dijet azimuthal decorrelations at the LHC in the parton Reggeization approach, *Phys. Rev. D* **87**, 9, p. 094030, doi:10.1103/PhysRevD.87.094030, arXiv:1304.3549 [hep-ph].

Saleev, V. and Shipilova, A. (2012). Inclusive b-jet and $b\bar{b}$-dijet production at the LHC via Reggeized gluons, *Phys. Rev. D* **86**, p. 034032, doi:10.1103/PhysRevD.86.034032, arXiv:1201.4640 [hep-ph].

Saleev, V. A. (2008). Prompt photon photoproduction at HERA within the framework of the quark Reggeization hypothesis, *Phys. Rev. D* **78**, p. 114031, doi:10.1103/PhysRevD.78.114031, arXiv:0812.0946 [hep-ph].

Saleev, V. A. (2009). Diphoton production at Tevatron in the quasi-multi-Regge-kinematics approach, *Phys. Rev. D* **80**, p. 114016, doi:10.1103/PhysRevD.80.114016, arXiv:0911.5517 [hep-ph].

Saleev, V. A., Nefedov, M. A., and Shipilova, A. V. (2012). Prompt J/psi production in the Regge limit of QCD: From Tevatron to LHC, *Phys. Rev. D* **85**, p. 074013, doi:10.1103/PhysRevD.85.074013, arXiv:1201.3464 [hep-ph].

Saleev, V. A. and Shipilova, A. V. (2013). Production of b-quark jets at the large Hadron Collider in the parton-reggeization approach, *Phys. Atom. Nucl.* **76**, pp. 1391–1397, doi:10.1134/S1063778813100177, [Yad. Fiz.76,1472(2013)].

Steinmann, O. (1960). *Ueber den Zusammenhang zwischen den Wightmanfunktionen und den retardierten Kommutatoren*, Ph.D. thesis, ETH Zurich, Zrich, doi:10.3929/ethz-a-000107369, sA aus: Helvetica physica acta, vol.33, fasc.4, pp. 257–298. Diss. Naturwiss. ETH Zrich, Nr. 2910, 0000. Ref.: Jost, R. ; Korref.: Specker, E..

van Hameren, A. (2014). BCFW recursion for off-shell gluons, *JHEP* **07**, p. 138, doi:10.1007/JHEP07(2014)138, arXiv:1404.7818 [hep-ph].

van Hameren, A. (2017). Calculating off-shell one-loop amplitudes for k_T-dependent factorization: A proof of concept, arXiv:1710.07609 [hep-ph].

van Hameren, A. (2018). KaTie: For parton-level event generation with k_T-dependent initial states, *Comput. Phys. Commun.* **224**, pp. 371–380, doi:10.1016/j.cpc.2017.11.005, arXiv:1611.00680 [hep-ph].

van Hameren, A., Kotko, P., and Kutak, K. (2012). Multi-gluon helicity amplitudes with one off-shell leg within high energy factorization, *JHEP* **12**, p. 029, doi:10.1007/JHEP12(2012)029, arXiv:1207.3332 [hep-ph].

van Hameren, A., Kotko, P., and Kutak, K. (2013a). Helicity amplitudes for high-energy scattering, *JHEP* **01**, p. 078, doi:10.1007/JHEP01(2013)078, arXiv:1211.0961 [hep-ph].

van Hameren, A., Kutak, K., and Salwa, T. (2013b). Scattering amplitudes with off-shell quarks, *Phys. Lett. B* **727**, pp. 226–233, doi:10.1016/j.physletb.2013.10.039, arXiv:1308.2861 [hep-ph].

van Hameren, A. and Serino, M. (2015). BCFW recursion for TMD parton scattering, *JHEP* **07**, p. 010, doi:10.1007/JHEP07(2015)010, arXiv:1504.00315 [hep-ph].

Verlinde, H. L. and Verlinde, E. P. (1993). QCD at high-energies and two-dimensional field theory, arXiv:hep-th/9302104 [hep-th].

https://doi.org/10.1142/9789811231124_0012

Chapter 12

High–Energy Scattering Amplitudes in QED, QCD and Supergravity

Agustín Sabio Vera
CERN, Theoretical Physics Department, Geneva, Switzerland
Instituto de Física Teórica UAM/CSIC
Universidad Autónoma de Madrid, Madrid, Spain

Abstract

This is a personal recollection of several results related to the study of the high energy limit of scattering amplitudes in gravitational theories. They would not have been possible without the encouragement and constant support from Lev Nikolaevich Lipatov.

12.1 Lipatov's high energy effective action for gravity

Phenomenology at the LHC and the AdS/CFT correspondence [Maldacena (1999); Gubser *et al.* (1998); Witten (1998)] have motivated a plethora of new results related to the calculation of scattering amplitudes. This is, in particular, the case for $\mathcal{N} = 4$ supersymmetric Yang–Mills theory [Bern *et al.* (2010)] and $\mathcal{N} = 8$ supergravity [Bern *et al.* (1998, 2009)]. An important finding in this context was the double–copy structure connecting both theories [Bern (2002)].

An interesting limit in graviton scattering at high center–of–mass energy, s, is that of multi–Regge kinematics, where the amplitudes factorize, with Reggeized gravitons exchanged in the t–channel [Grisaru *et al.* (1975); Grisaru and Schnitzer (1981)]. There are also contributions related to eikonal and double–logarithmic in s (DL) terms [Lipatov (1982a,b, 1991)] which can be calculated using the high energy effective action proposed by Lipatov [Lipatov (2013)]. Following Gribov's proposal for the construction of an effective quantum field theory for Pomeron interactions, Lipatov worked on building it for gravity. The basic idea underlying this approach is that gravitons are not elementary particles when radiative corrections are taken into account, they lie on their Regge trajectories.

Lipatov's effective action for gravity shares common ground with the Quantum Chromodynamics (QCD) case. It is based on the concept that the leading in energy contributions to the cross section, besides eikonal contributions, are those with gravitons produced at bunches well separated in rapidity from each other. Each gap in rapidity between subsequent bunches is generated by the exchange of a Reggeized graviton. There exist generally covariant fields which create, A^{++}, or annihilate, A^{--}, them in the t–channel and are subject to kinematic constraints $\partial_+ A^{++} = \partial_- A^{--} = 0$ to create the rapidity gaps. Within a bunch there appear local and non–local interactions described by the action [Lipatov (2013)]

$$S = -\frac{1}{2\kappa}\int d^4x\left(\sqrt{-g}R + \partial_\sigma A^{++}\partial_\sigma A^{--} + \frac{1}{2}\left(\partial_+ j^- \partial_\mu^2 A^{++} + [+ \leftrightarrow -]\right)\right), \tag{12.1}$$

where $\kappa^2 = 8\pi G = 8\pi^2\alpha$, and G is Newton's constant. It corresponds to the Einstein–Hilbert action plus a kinetic term for the Reggeon fields and induced contributions. In the latter, the currents are related to the usual graviton field $h_{\mu\nu}$ in the form

$$\partial_\pm j^\mp = h_{\pm\pm} - \left(h_{\rho\pm} - \frac{1}{2}\frac{\partial_\rho}{\partial_\pm}h_{\pm\pm}\right)^2 + \cdots . \tag{12.2}$$

This accounts for the interactions among usual gravitons and Reggeized ones when inserted back in the action.

From this approach it is possible to obtain the graviton Regge trajectory and effective vertices needed to calculate elastic and inelastic amplitudes at any order in the theory. It is a compact formalism which allows for the resummation of the leading contributions from an infinite number of Feynman diagrams. In particular, the non–local $1/\partial_\pm$ operators generate graviton emissions from hard lines far in rapidity from the local clusters. In order to preserve general covariance, Lipatov's high energy effective action also contains eikonal contributions.

In relation with the underlying double–copy structure of gravity, it is worth noting that the graviton emission vertex in multi–Regge kinematics is directly related to the equivalent [Lipatov (1976); Fadin *et al.* (1975); Kuraev *et al.*

(1976, 1977); Balitsky and Lipatov (1978)] gluon emission vertex in QCD [Lipatov (1982a,b, 1991); Sabio Vera *et al.* (2012)] as it was originally found by Lipatov.

12.2 Double logarithms in supergravity

There exists a subtle interplay between the Reggeization of the graviton and the DL contributions to the scattering amplitudes. In [Bartels *et al.* (2014)] these DLs in Einstein–Hilbert gravity and different supergravities were calculated to all orders. The results there presented are in agreement up to two loops with those obtained for $\mathcal{N} = 4, 5, 6$ supergravities using a double–copy procedure [Boucher-Veronneau and Dixon (2011)]. They also agree with a recent three loop calculation for $\mathcal{N} = 8$ supergravity [Henn and Mistlberger (2019)]. A remarkable feature of the all–orders resummation in the DL sector of the amplitudes is that, when compared to their Born counterparts, they grow with s whenever $\mathcal{N} < 4$ but rapidly decrease with s if $\mathcal{N} > 4$.

Let us take the normalizations (L=loop order, N=number of gravitinos) for the 4–graviton scattering amplitude with helicities (++;++),

$$\mathcal{A}_{4,(N)} = \mathcal{A}_4^{\rm Born}\mathcal{M}_{4,(N)} = \kappa^2\frac{s^3}{tu}\left(1+\sum_{L=1}^{\infty}\mathcal{M}_{4,(N)}^{(L)}\right), \tag{12.3}$$

where $s = (p_1+p_2)^2$, $t = (p_1-p_3)^2$ and $u = (p_1-p_4)^2$. Lipatov found that the Regge limit, $s \gg -t = \vec{q}^{\,2}$, of the one loop amplitude contains the graviton Regge trajectory [Lipatov (1982a,b, 1991)],

$$\begin{aligned}\omega(q) = {}& \frac{\alpha\,\vec{q}^{\,2}}{\pi}\int\frac{d^2\vec{k}}{\vec{k}^2(\vec{q}-\vec{k})^2}\\ &\times\left((\vec{k}\cdot(\vec{q}-\vec{k}))^2\left(\frac{1}{\vec{k}^2}+\frac{1}{(\vec{q}-\vec{k})^2}\right)+\frac{N}{2}\left(\vec{k}\cdot(\vec{q}-\vec{k})\right)-\vec{q}^{\,2}\right).\end{aligned} \tag{12.4}$$

Its infrared divergence is regularized by a cut–off, λ, while the ultraviolet one, which is due to the use of an effective action at high energies, can be regularized by s, taken as an upper cut–off. One then gets

$$\omega(q) = \alpha\,t\left(\ln\left(\frac{-t}{\lambda^2}\right)+\frac{N-4}{2}\ln\left(\frac{s}{-t}\right)\right). \tag{12.5}$$

A similar structure remains to all orders in perturbation theory. With DL accuracy, it is useful to use the representation

$$\mathcal{A}_{4,(N)} = \mathcal{A}_4^{\rm Born}\left(\frac{s}{-t}\right)^{\alpha t\ln\left(\frac{-t}{\lambda^2}\right)}\int_{\delta-i\infty}^{\delta+i\infty}\frac{d\omega}{2\pi i}\left(\frac{s}{-t}\right)^{\omega}\frac{f_\omega^{(N)}}{\omega},\ \delta>0, \tag{12.6}$$

where the t–channel partial wave $f_\omega^{(N)}$ has the expansion

$$f_\omega^{(N)} = \sum_{n=0}^{\infty} \mathcal{C}_n^{(N)} \left(\frac{\alpha t}{\omega^2}\right)^n . \tag{12.7}$$

In [Lipatov (1982a)] Lipatov calculated the DL contributions to the 4–graviton amplitude stemming from ladder diagrams. In [Bartels *et al.* (2014)] it was shown that non–ladder contributions should also be taken into account. These extra terms already appear in Quantum Electrodynamics (QED) where, *e.g.*, in e^+e^- forward scattering the scattering amplitude factorizes at DL accuracy: $\mathcal{A}_{\rm DL}(s) = \mathcal{A}_{\rm Born}\mathcal{R}(s)$. An infrared cut–off μ can be introduced for the transverse momenta of the virtual contributions, $\vec{p}_T^{\,2} > \mu^2$. It is then possible to write $\mathcal{R}(s) \to \mathcal{R}(s,\mu^2)$ and the following equation holds in Sudakov variables,

$$\mathcal{R}(s,\mu^2) = 1 + e^2 \int_{-\infty}^{\infty} \frac{d(s\alpha)}{s\alpha} \int_{-\infty}^{\infty} \frac{d(s\beta)}{s\beta} \int \frac{d^2\vec{k}}{i(2\pi)^4} \frac{\vec{k}^2\,\theta(\vec{k}^2-\mu^2)}{(s\alpha\beta-\vec{k}^2)^2}$$
$$\times\, \mathcal{R}(s\alpha,\vec{k}^2)\mathcal{R}(s\beta,\vec{k}^2). \tag{12.8}$$

Its solution can be expressed in terms of a Bessel function. For e^+e^- backward scattering the contributions from Sudakov photon lines attached to the external fermions kick in, and the corresponding equation has a solution written in terms of a parabolic cylinder function which will be described below.

These are classical results which were obtained by Gorshkov, Gribov, Lipatov and Frolov from 1966 to 1970 [Gorshkov *et al.* (1968b, 1966, 1968a)]. In QCD, in deep inelastic electron–hadron scattering at large photon virtualities, one can modify the usual DGLAP [Gribov and Lipatov (1972); Dokshitzer (1977); Altarelli and Parisi (1977)] evolution equation for parton distributions introducing a dependence mixing longitudinal and transverse degrees of freedom to generate the DL contributions [Kirschner and Lipatov (1982a,b, 1983)]. Similar techniques have been applied to the electroweak sector of the Standard Model [Fadin *et al.* (2000)].

Coming back to graviton scattering and introducing the gravitino content of supergravity, at one loop the result is given in Eq. (12.5). This can be compared with the exact amplitude in, *e.g.*, $\mathcal{N}=8$ supergravity (where $N=8$),

$$\mathcal{M}^{(1)}_{4,(N=8)} = \underbrace{\alpha\, t \ln\left(\frac{-s}{-t}\right)\ln\left(\frac{-u}{-t}\right)}_{\text{Double Logs}} - \underbrace{\alpha\frac{(s-u)}{2}\ln\left(\frac{-t}{\lambda^2}\right)\ln\left(\frac{-s}{-u}\right)}_{\text{Eikonal}}$$
$$\underbrace{+\,\alpha\frac{t}{2}\ln\left(\frac{-t}{\lambda^2}\right)\left(\ln\left(\frac{-s}{-t}\right)+\ln\left(\frac{-u}{-t}\right)\right)}_{\text{Trajectory}}. \tag{12.9}$$

There are three contributions: DLs, the eikonal piece and the graviton Regge trajectory. Corrections to this formula at two loops for different supergravities have been given in [Boucher-Veronneau and Dixon (2011)] and at one loop in Einstein–Hilbert gravity ($N = 0$) in [Dunbar and Norridge (1995)]. They all agree with Eq. (12.5) at one–loop order. Its generalization to all orders can be written as the solution of an evolution equation for the partial wave in Eq. (12.7) which is similar to those obtained in QED and QCD, *i.e.*

$$f_\omega^{(N)} = 1 - \alpha t \frac{d}{d\omega} \frac{f_\omega^{(N)}}{\omega} + \alpha t \frac{(N-6)}{2} \frac{f_\omega^{(N)^2}}{\omega^2}. \tag{12.10}$$

The derivative term is due to virtual gravitons with the smallest p_T. Gravitons and gravitinos exchanged in the t–channel with very low energy generate the quadratic contribution. The perturbative solution can be obtained by iteration,

$$\begin{aligned} f_\omega^{(N)} &= 1 + \frac{\alpha t (N-4)}{2w^2} + \frac{\alpha^2 t^2 (N-4)(N-3)}{2w^4} \\ &\quad + \frac{\alpha^3 t^3 (N-4)\left(5N^2 - 26N + 36\right)}{8w^6} + \cdots . \end{aligned} \tag{12.11}$$

Going back to the DL amplitude, this implies,

$$\mathcal{A}_{4,(N)} = \mathcal{A}_4^{\rm Born} \left(\frac{s}{-t}\right)^{\alpha t \ln\left(\frac{-t}{\lambda^2}\right)} \Phi^{(N)}\left(-\alpha t \ln^2\left(\frac{s}{-t}\right)\right), \tag{12.12}$$

where

$$\begin{aligned} \Phi^{(N)}(x) &= 1 - \frac{(N-4)}{2}\frac{x}{2} + \frac{(N-4)}{2}(N-3)\frac{x^2}{4!} \\ &\quad - \frac{(N-4)}{8}(5N^2 - 26N + 36)\frac{x^3}{6!} + \cdots . \end{aligned} \tag{12.13}$$

This formula is, up to two loops for $N = 4, 5, 6, 8$, in agreement with [Boucher-Veronneau and Dixon (2011); Dunbar and Norridge (1995)] and, at three loops, for $N = 8$, with [Henn and Mistlberger (2019)].

It is possible to find the exact solutions to Eq. (12.10) which are compatible with the perturbative expansion in Eq. (12.11). These solutions were studied in [Bartels *et al.* (2014)]. The simplest one corresponds to $N = 4$ since, as it can be seen in Eq. (12.11), only the first term survives. This implies

$$\mathcal{A}_{4,(N=4)} = \mathcal{A}_4^{\rm Born} \left(\frac{s}{-t}\right)^{\alpha t \ln\left(\frac{-t}{\lambda^2}\right)} . \tag{12.14}$$

The next–to–simplest case is $N = 6$ because the quadratic term in the differential equation (12.10) vanishes and one gets

$$\mathcal{A}_{4,(N=6)} = \mathcal{A}_4^{\rm Born} \left(\frac{s}{-t}\right)^{\alpha t \ln\left(\frac{-t}{\lambda^2}\right)} e^{\frac{\alpha t}{2} \ln^2\left(\frac{s}{-t}\right)}. \tag{12.15}$$

The $N = 0, 2$ cases also have a simple representation:

$$\mathcal{A}_{4,(N=2)} = \frac{\mathcal{A}_4^{\text{Born}}}{2} \left(\frac{s}{-t}\right)^{\alpha\, t\, \ln\left(\frac{-t}{\lambda^2}\right)} \left(\left(\frac{s}{-t}\right)^{\sqrt{-\alpha t}} + \left(\frac{s}{-t}\right)^{-\sqrt{-\alpha t}} \right), \tag{12.16}$$

$$\mathcal{A}_{4,(N=0)} = \frac{\mathcal{A}_4^{\text{Born}}}{3} \left(\frac{s}{-t}\right)^{\alpha\, t\, \ln\left(\frac{-t}{\lambda^2}\right)} \left(1 + \left(\frac{s}{-t}\right)^{\sqrt{-3\alpha t}} + \left(\frac{s}{-t}\right)^{-\sqrt{-3\alpha t}} \right). \tag{12.17}$$

The solutions with odd N and $N = 8$ enjoy a richer structure since their partial waves have an infinite number of poles in the complex ω plane placed asymptotically close to the lines with argument $\pm\frac{3}{4}\pi$. For these cases one can write the DL contributions to the amplitude in the form

$$\begin{aligned} r^{(N)}(s,t) &\equiv \frac{\mathcal{A}_{4,(N)}}{\mathcal{A}_4^{\text{Born}}} \left(\frac{s}{-t}\right)^{-\alpha\, t\, \ln\left(\frac{-t}{\lambda^2}\right)} \\ &= \frac{\Gamma\left(\frac{N}{2}-3\right)}{3-\frac{N}{2}} \int_{\delta-i\infty}^{\delta+i\infty} \frac{d\omega}{2\pi i} \left(\frac{s}{-t}\right)^{\omega} \frac{d}{d\omega} \ln\left(e^{-\frac{\omega^2}{4\alpha\, t}} D_{\frac{6-N}{2}}\left(\frac{\omega}{\sqrt{-\alpha\, t}}\right)\right) \end{aligned} \tag{12.18}$$

with the parabolic cylinder function being $D_\nu(x) = \frac{e^{-\frac{x^2}{4}}}{\Gamma(-\nu)} \int_0^\infty \frac{dy}{y^{\nu+1}}\, e^{-\frac{y^2}{2}} e^{-x\, y}$.

It is now possible to numerically compare the behaviour with s of the amplitudes for different N. Taking $b \equiv -\alpha\, t = 1$ one can see the result in Fig. 12.1. The solutions for $N < 4$ grow monotonically with energy. The critical solution with $N = 4$ is constant and equal to one. There are two monotonically decreasing with energy solutions for $N = 5, 6$ and, finally, two oscillatory and decreasing with energy ones for $N = 7, 8$. The high energy behaviour is driven by the poles with the largest real part, whose values can be evaluated numerically. This allows to obtain asymptotic representations for different N:

$$r^{(N=0)}(s,t) \approx s^{\sqrt{-3\alpha\, t}}, \tag{12.19}$$

$$r^{(N=1)}(s,t) \approx s^{1.402\,\sqrt{-\alpha\, t}}, \tag{12.20}$$

$$r^{(N=2)}(s,t) \approx s^{\sqrt{-\alpha\, t}}, \tag{12.21}$$

$$r^{(N=3)}(s,t) \approx s^{0.5508\,\sqrt{-\alpha\, t}}, \tag{12.22}$$

$$r^{(N=4)}(s,t) = 1, \tag{12.23}$$

$$r^{(N=5)}(s,t) \approx s^{-0.762\,\sqrt{-\alpha\, t}}, \tag{12.24}$$

$$r^{(N=6)}(s,t) = e^{\frac{\alpha}{2} t \ln^2 s}, \tag{12.25}$$

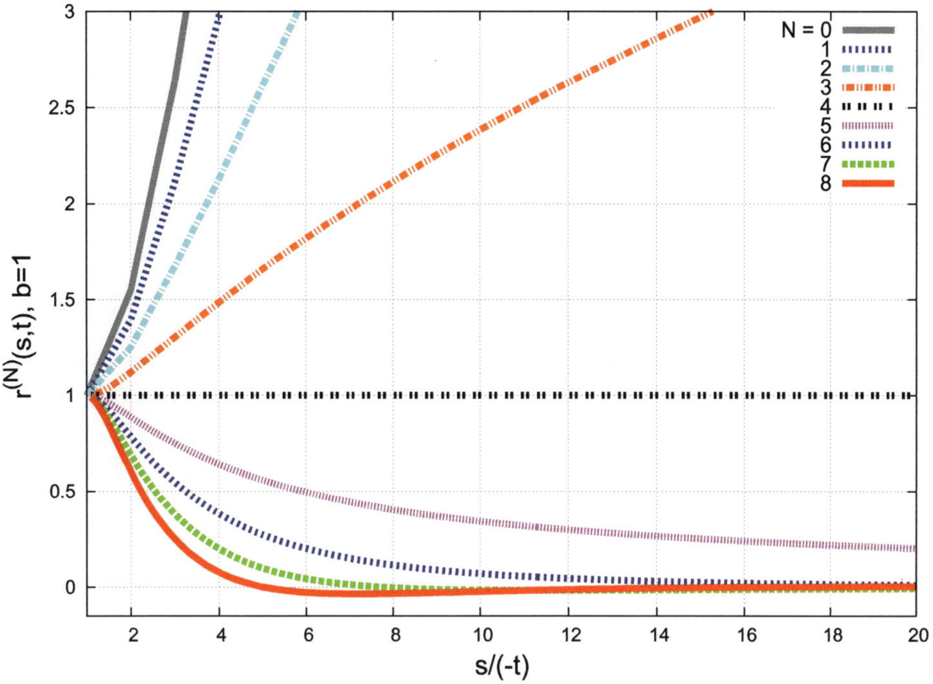

Fig. 12.1 Scattering amplitude for $N = 0, 1, \ldots, 8$.

and

$$r^{(N=7)}(s,t) \approx s^{-2.111\sqrt{-\alpha t}} \cos\left(2.267\sqrt{-\alpha t}\ln s\right), \tag{12.26}$$

$$r^{(N=8)}(s,t) \approx s^{-1.916\sqrt{-\alpha t}} \cos\left(2.816\sqrt{-\alpha t}\ln s\right). \tag{12.27}$$

12.3 Impact parameter representation

In order to gain further understanding for the resummation of DLs, it is useful to work in impact parameter (ρ) representation. This has been done for $\mathcal{N} \geq 4$ supergravities in [Sabio Vera (2019, 2020)] where a novel representation for the coefficients of the perturbative expansion is introduced using mathematical results on exactly solvable recurrences. In this representation it can be shown that the DLs make the gravitational interaction weaker at small ρ and large s. This is so because they introduce corrections to the Born eikonal phase which change the sign of the graviton's deflection angle. As a consequence, there is a constant negative shift in both the eikonal phase and Shapiro's time delay but not large enough to generate causality violation.

The scattering amplitude can be written in the form

$$\mathcal{M}_{4,\mathrm{DL}}^{(N)}(s,t) = \int_{\delta-i\infty}^{\delta+i\infty} \frac{d\,\omega}{2\pi i}\left(\frac{s}{-t}\right)^{\omega}\frac{f_\omega^{(N)}}{\omega} = 1+\sum_{n=1}^{\infty} \mathcal{C}_n^{(N)}\frac{(\alpha\,t)^n}{(2n)!}\ln^{2n}\left(\frac{s}{-t}\right), \tag{12.28}$$

where the coefficients (see Eq. (12.7)) follow the relation (with $\eta_N = (N-6)/2$),

$$\mathcal{C}_n^{(N)} = (\eta_N+2n-1)\,\mathcal{C}_{n-1}^{(N)} + \eta_N\sum_{j=1}^{n-1}\mathcal{C}_j^{(N)}\mathcal{C}_{n-1-j}^{(N)}, \quad \mathcal{C}_0^{(N)} = 1. \tag{12.29}$$

$\mathcal{C}_n^{N=4} = \delta_n^0$ indicating that the DLs are not present in $\mathcal{N}=4$ supergravity which acts as a critical theory marking the transition from poorly behaved amplitudes for $N<4$, blowing up at large s, to finite amplitudes when $N>4$.

The coefficients $\mathcal{C}_n^{(8)}$ have already appeared in many body theory [Ihrig *et al.* (1976)], in QED [Cvitanovic *et al.* (1978)] and in solid state physics [Goodvin *et al.* (2006)]. Eq. (12.29) has been studied in [Martin and Kearney (2010)] in the context of self–convolutive recurrence relations allowing to write

$$\mathcal{C}_n^{(N)} = \int_0^\infty x^n\,\mu^{(N)}(x)\,dx \tag{12.30}$$

with

$$\mu^{(N)}(x) = \frac{\frac{2^{1-\eta_N}}{\eta_N}\Gamma(\eta_N)\sqrt{\frac{2e^x}{\pi x}}}{\Gamma\left(\frac{\eta_N}{2}\right)^2 {}_1F_1\left(\frac{1-\eta_N}{2};\frac{1}{2};\frac{x}{2}\right)^2 + 2\,x\,\Gamma\left(\frac{1+\eta_N}{2}\right)^2 {}_1F_1\left(\frac{2-\eta_N}{2};\frac{3}{2};\frac{x}{2}\right)^2}, \tag{12.31}$$

where Γ is the gamma function and ${}_1F_1$ is the Kummer confluent hypergeometric function. This leads to the novel representation of the scattering amplitude

$$\mathcal{M}_{4,\mathrm{DL}}^{(N)}(s,t) = \int_0^\infty dx\,\mu^{(N)}(x)\cosh\left(\sqrt{\alpha\,t\,x}\ln\left(\frac{s}{-t}\right)\right). \tag{12.32}$$

The $N>4$ cases have been studied in [Sabio Vera (2019, 2020)] and read

$$\mu^{(5)}(x) = \frac{16\sqrt{\frac{e^x}{x}}}{\Gamma\left(-\frac{1}{4}\right)^2 {}_1F_1\left(\frac{3}{4};\frac{1}{2};\frac{x}{2}\right)^2 + 2\,x\,\Gamma\left(\frac{1}{4}\right)^2 {}_1F_1\left(\frac{5}{4};\frac{3}{2};\frac{x}{2}\right)^2}, \tag{12.33}$$

$$\mu^{(6)}(x) = \frac{1}{\sqrt{2\pi x e^x}}, \tag{12.34}$$

$$\mu^{(7)}(x) = \frac{\frac{4\sqrt{2}}{\pi^2 x}}{I_{-\frac{1}{4}}\left(\frac{x}{4}\right)^2 + I_{\frac{1}{4}}\left(\frac{x}{4}\right)^2}, \tag{12.35}$$

$$\mu^{(8)}(x) = \frac{\sqrt{\frac{2e^x}{\pi^3 x}}}{\mathrm{erfi}\left(\sqrt{\frac{x}{2}}\right)^2+1}, \tag{12.36}$$

where $I_a(z)$ is a Bessel function and $\mathrm{erfi}(z)$ the error function.

To work in impact parameter space, ρ, the Fourier transform of the graviton–graviton interaction is introduced (with $q = \sqrt{-t}$), *i.e.*

$$\chi^{(N)}(\rho, s) = \alpha s \int \frac{d^2\vec{q}}{q^2} e^{-i\vec{q}\cdot\vec{\rho}} \mathcal{M}^{(N)}_{4,\mathrm{DL}}(s,t). \tag{12.37}$$

This phase can be exponentiated in a high energy and fixed impact parameter eikonal approach,

$$\begin{aligned}\frac{\alpha s^2}{q^2}\mathcal{M}^{(N)}_{4,\mathrm{DL}}(s,t) &= \frac{s}{4\pi^2}\int d^2\vec{\rho}\, e^{i\vec{q}\cdot\vec{\rho}}\chi^{(N)}(\rho,s)\\ &\simeq -\frac{is}{4\pi^2}\int d^2\vec{\rho}\, e^{i\vec{q}\cdot\vec{\rho}}\left(e^{i\chi^{(N)}(\rho,s)}-1\right).\end{aligned} \tag{12.38}$$

In the forward high energy limit the integral is dominated by the region $\frac{\partial}{\partial\rho}\left(q\rho + \chi^{(N)}\right) = 0$. The corresponding graviton–graviton scattering angle reads

$$\theta^{(N)}(\rho, s) = -\frac{2}{\sqrt{s}}\frac{\partial\chi^{(N)}(\rho,s)}{\partial\rho}. \tag{12.39}$$

The simplest case to study is that of $N = 4$ where an infrared scale λ is needed to define the phase:

$$\begin{aligned}\chi^{(4)}(\rho,s) &= \frac{1}{2s}\int\frac{d^2\vec{q}}{(2\pi)^2}e^{-i\vec{q}\cdot\vec{\rho}}\mathcal{A}_4^{\mathrm{Born}}(s,t) \simeq \frac{\kappa^2 s}{8\pi^2}\int\frac{d^2\vec{q}}{q^2}e^{-i\vec{q}\cdot\vec{\rho}}\,\theta(q-\lambda)\\ &\simeq -\frac{\kappa^2 s}{4\pi}\ln(\rho\lambda) = \chi_{\mathrm{Born}}(\rho,s).\end{aligned} \tag{12.40}$$

The graviton's deflection angle in the forward limit is λ independent,

$$\theta^{(4)}(\rho,s) = \frac{\kappa^2\sqrt{s}}{2\pi\rho} = \theta_{\mathrm{Born}}(\rho,s). \tag{12.41}$$

Its growth for small ρ is characteristic of an attractive interaction. For general N one can write

$$\chi^{(N)}(\rho,s) = \chi_{\mathrm{Born}}(\rho,s) + \chi^{(N)}_{\mathrm{DL}}(\rho,s), \tag{12.42}$$

where (see Eq. (12.28))

$$\begin{aligned}\chi^{(N)}_{\mathrm{DL}}(\rho,s) &= \alpha s\int\frac{d^2\vec{q}}{q^2}e^{-i\vec{q}\cdot\vec{\rho}}\sum_{n=1}^{\infty}\mathcal{C}_n^{(N)}\frac{(\alpha t)^n}{(2n)!}\ln^{2n}\left(\frac{s}{-t}\right)\theta\left(\sqrt{s}-q\right)\\ &= \pi\sum_{m=1}^{\infty}\sum_{n=0}^{m-1}\frac{(-1)^m(\alpha s)^{m+1}\mathcal{C}^{(N)}_{m-n}}{(n!)^2 m^{2(m-n)+1}}\left(\frac{\rho^2}{4\alpha}\right)^n\end{aligned} \tag{12.43}$$

is infrared finite. For $N > 4$ this is a negative function which goes rapidly to zero as ρ grows. The large distance behaviour corresponds to the Born amplitude. This phase is shown in Fig. 12.2 for a large center of mass energy. The DL resummation

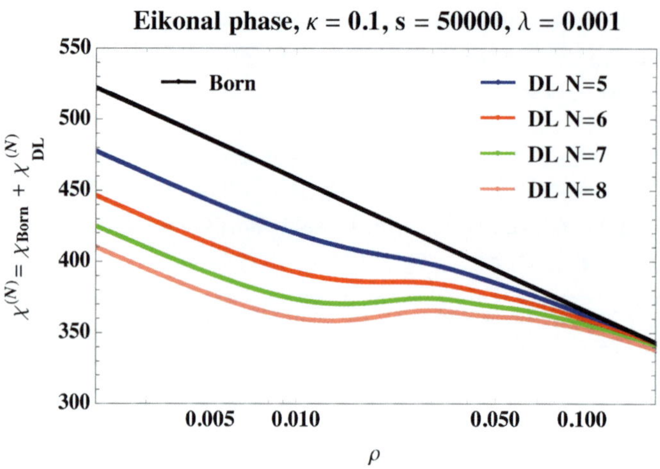

Fig. 12.2 Born eikonal phase plus the DL terms for $s = 50000$ and different supergravities.

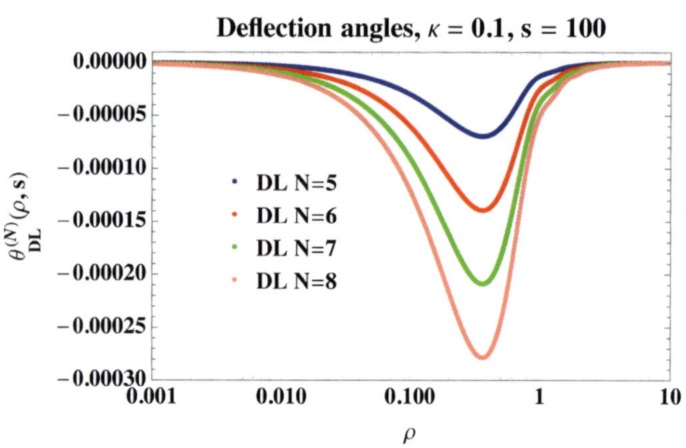

Fig. 12.3 DL deflection angle in different supergravities as a function of the impact parameter.

generates a constant phase shift which is larger as N grows. The deflection angle can be written in the form

$$\theta^{(N)}(\rho, s) = \theta_{\text{Born}}(\rho, s) + \theta_{\text{DL}}^{(N)}(\rho, s), \tag{12.44}$$

$$\theta_{\text{DL}}^{(N)}(\rho, s) = -\theta_{\text{Born}}(\rho, s) \sum_{m=2}^{\infty} \sum_{n=1}^{m-1} \frac{n(-\alpha s)^m \mathcal{C}_{m-n}^{(N)}}{(n!)^2 m^{2(m-n)+1}} \left(\frac{\rho^2}{4\alpha}\right)^n. \tag{12.45}$$

In Fig. 12.3 it can be seen that $\theta_{\text{DL}}^{(N)}$ is negative. This means that the DL resummation has the effect of reducing the Born graviton–graviton deflection angle introducing a screening of the gravitational interaction which is larger for a larger number of gravitinos. This screening is stronger as s grows and it can make the

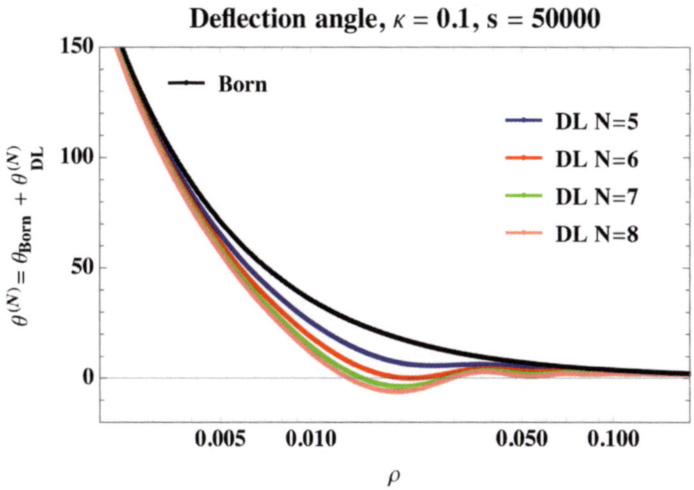

Fig. 12.4 Deflection angle as a function of the impact parameter for different supergravity theories.

graviton's deflection angle negative. This is shown in Fig. 12.4 where $\theta^{(N)} < 0$ for some regions in impact parameter space which are larger as N increases. The s dependence of the impact parameter $\rho_{\max}^{(N)}$ where the DLs contribution is maximal is plotted in Figs. 12.5, 12.6, together with the ratio $-\frac{\theta_{\rm DL}^{(N)}}{\theta_{\rm Born}}$ in the (s, ρ) regions where it is large. The DLs make the Born deflection angle to change sign when $-\frac{\theta_{\rm DL}^{(N)}}{\theta_{\rm Born}} > 1$. The weakening of gravity is larger in Fig. 12.6, for $N = 7, 8$, than in Fig. 12.5, with $N = 5, 6$. When $\rho > \rho_{\max}^{(N)}$ there are oscillations which appear as bands in ρ space.

It is important to stress again that the DLs are in principle subleading with respect to the eikonal contributions (see, *e.g*, [Ciafaloni *et al.* (2019); Di Vecchia *et al.* (2019, 2020)]) since they are enhanced by powers of t instead of powers of s (see Eq. (12.9)). It is nevertheless an interesting sector of the gravitational scattering amplitudes which deserves further study since it is a first example of a full resummation of quantum corrections which are infrared finite. It is already remarkable that they reduce the gravitational interaction at high energies in a region of short distances. The fact that this effect is more acute for more gravitinos in the supergravity theory might be related to the possible lack of ultraviolet divergencies in $\mathcal{N} = 8$ supergravity. This is a point which deserves further investigation.

A negative contribution to the eikonal phase can be interpreted as a negative contribution to the Shapiro's time delay [Shapiro (1964)] experienced by a particle in elastic scattering [Kabat and Ortiz (1992)]. The explicit formula for the delay experienced by the gravitons at Born level would be

$$\Delta_{\rm Born}(\rho, s) = \frac{\partial}{\partial\sqrt{s}}\chi_{\rm Born}(\rho, s) \;=\; -\frac{\kappa^2\sqrt{s}}{2\pi}\ln(\rho\,\lambda), \qquad (12.46)$$

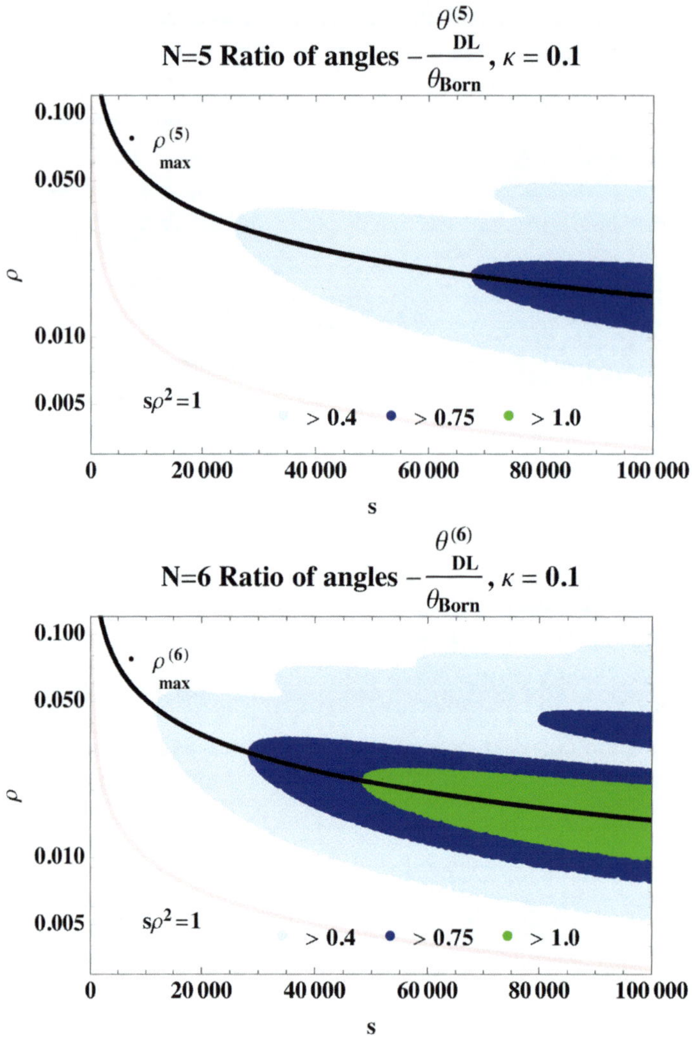

Fig. 12.5 $-\frac{\theta^{(N)}_{\mathrm{DL}}}{\theta_{\mathrm{Born}}}$ as function of s for N=5 (up) and N=6 (bottom) supergravities.

with DL corrections of the form

$$\begin{aligned}\Delta^{(N)}(\rho, s) = \frac{\partial}{\partial\sqrt{s}}\chi^{(N)}(\rho, s) &= \Delta_{\mathrm{Born}}(\rho, s) + \Delta^{(N)}_{\mathrm{DL}}(\rho, s) \\ &= \Delta_{\mathrm{Born}}(\rho, s) + 2\pi\alpha\sqrt{s}\sum_{m=1}^{\infty}\sum_{n=0}^{m-1}\frac{(m+1)(-\alpha s)^m \mathcal{C}^{(N)}_{m-n}}{(n!)^2 m^{2(m-n)+1}}\left(\frac{\rho^2}{4\alpha}\right)^n .\end{aligned} \tag{12.47}$$

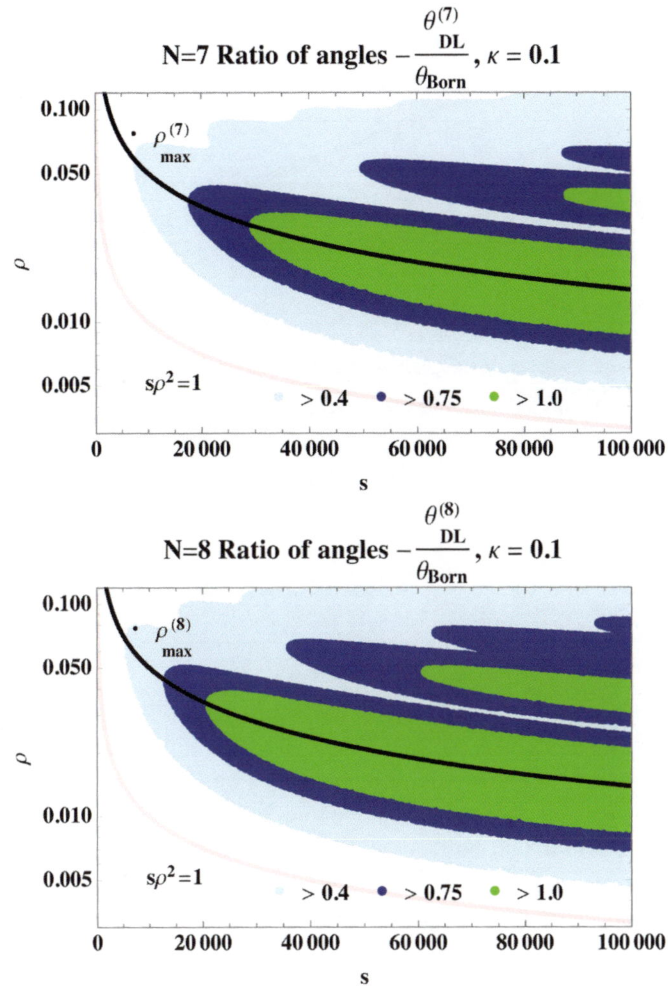

Fig. 12.6 $-\frac{\theta^{(N)}_{\rm DL}}{\theta_{\rm Born}}$ as function of s for N=7 (up) and N=8 (bottom) supergravities.

A numerical representation is shown in Fig. 12.7. A constant negative shift appears which at very small ρ becomes

$$\begin{aligned}\Delta^{(N)}(\rho,s) &\overset{\rho\ll 1}{\simeq} \Delta_{\rm Born}(\rho,s) + 2\pi\alpha\sqrt{s}\sum_{m=1}^{\infty}\frac{(m+1)(-\alpha s)^m \mathcal{C}_m^{(N)}}{m^{2m+1}} \\ &= \frac{\kappa^2\sqrt{s}}{4\pi}\left(-2\ln(\rho\lambda) + \int_0^{\infty}\mu^{(N)}(x)\,dx\sum_{m=1}^{\infty}\left(1+\frac{1}{m}\right)\left(\frac{-\alpha s x}{m^2}\right)^m\right).\end{aligned} \tag{12.48}$$

Quantum corrections contribute to all orders to this expression. In the regions under study the overall time delay is always positive, without challenging the

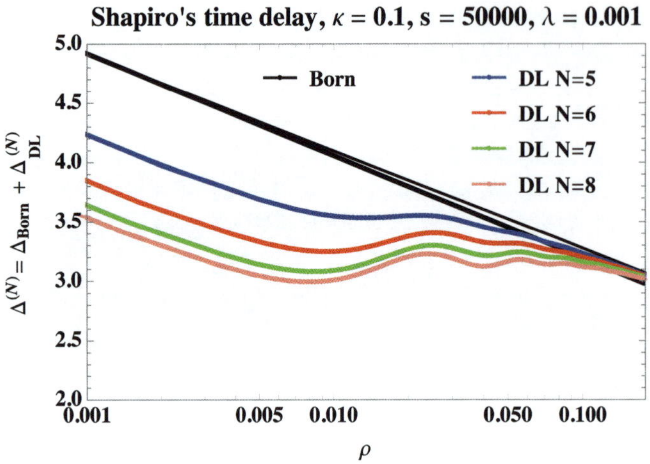

Fig. 12.7 Time delay for graviton–graviton scattering for different supergravities with DL accuracy.

causal structure of the theory (a negative time delay would imply superluminical propagation).

12.4 Connecting with ribbon graphs in $\mathcal{N} = 8$ supergravity

A surprising correspondence between the evolution equation for $f_\omega^{(N=8)}$ and the theory of graphs embedded on orientable surfaces was found in [Sabio Vera (2019)]. A map on an orientable surface is a partition defining vertices, edges and faces on it. Edges can split into two half–edges with a fixed orientation, forming a ribbon graph. If an arrow is assigned to a half–edge it is called rooted. When enumerating equivalence classes of isomorphic rooted maps [Arquès and Béraud (2000)] a relation equivalent to Eq. (12.10) appears.

A graphical bijection between the DL Feynman diagrams and ribbon graphs can be drawn. Similar mappings take place in other theories [Prunotto *et al.* (2018); Gopala *et al.* (2018); Gopala Krishna *et al.* (2018)]. The Born diagram maps to the one root ribbon graph with zero edges,

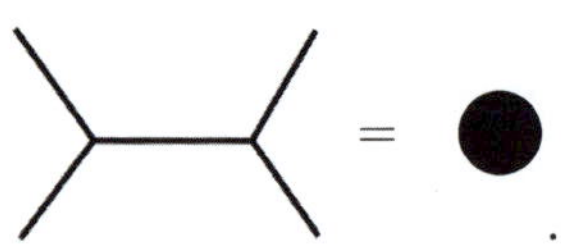
(12.49)

A first set of DLs is generated by attaching soft gravitons and gravitinos to the external lines. The sum of the associated Feynman diagrams is then mapped to the one–edge graph with one vertex,

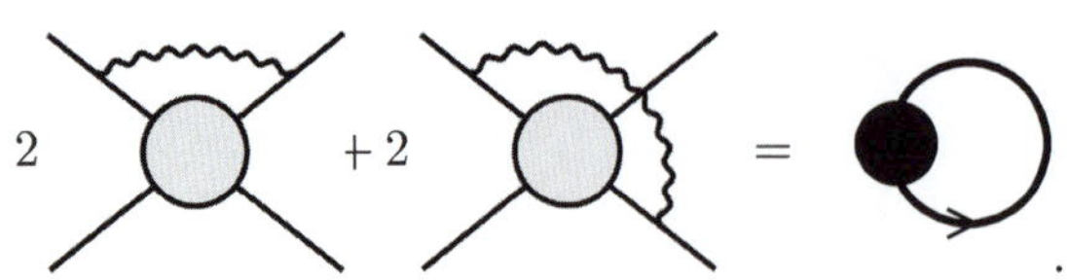
(12.50)

A second set of DLs stems from the exchange in the t–channel of two gravitons or gravitinos. The corresponding Feynman diagram is mapped to the two–vertex 1–rooted ribbon graph:

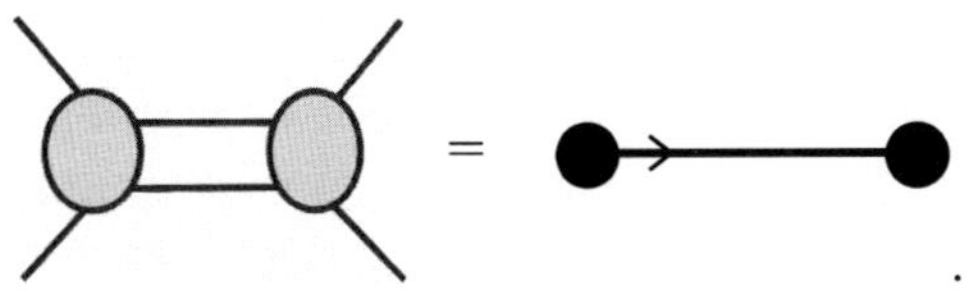

(12.51)

These two configurations generate the second term in the series $(1, 2, 10, 74, \ldots)$ of Eq. (12.7). Further coefficients appear by iteration of these two basic elements. For example, the coefficient 10 stems from

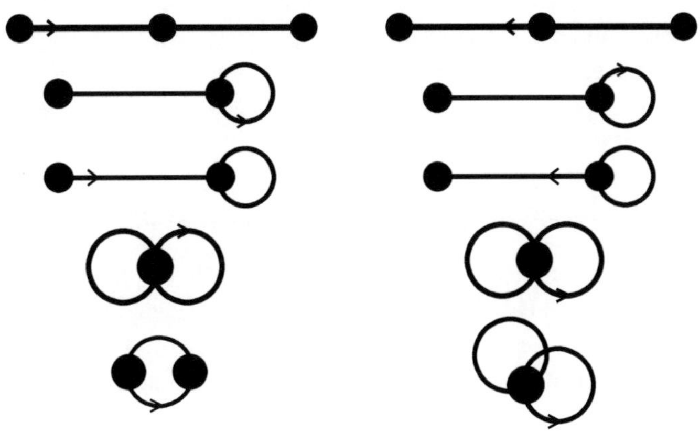

This representation is simpler than the Feynman diagram approach and it will be worth studying further its implications.

12.5 The last scientific discussion with Lev

At the beginning of August 2017 I had my last scientific discussion with Lev. It was at DAMTP, in Cambridge. He had just delivered a seminar on the high energy effective action for gravity and we had a kind of brainstorming session on the blackboard while Elvira, Lev's wife, was making sure all the details of their bus and plane tickets were correct. She always was Lev's connection to planet Earth.

A couple of days before we had been in Southampton visiting Douglas Ross and his wife Jackie and discussing about the meaning of introducing temperature in the high energy limit of Quantum Chromodynamics. That was one of Lev's favourite subjects. He was proud of his work with Hector de Vega where they were able to introduce temperature in the t–channel of a scattering process without breaking the integrability of the BFKL equation. Our discussions with Doug centered around the meaning of this temperature as a parameter to introduce confinement in the whole framework.

During his talk Lev was asked many good questions mainly by Michael Green and Hugh Osborn, who could appreciate the huge amount of work he had devoted to the subject and the implications of his results (a true *tour de force* in Green's words).

Lev was interested in all new developments and wanted to know about the attempts to connect the high energy effective action with the so–called CHY formalism. In the following the first baby steps in this work [Chachamis *et al.* (2018)], which were published with Grigorios Chachamis, Miguel Ángel Vázquez–Mozo (both great physicists with whom it is a pleasure to discus anything, not only science) and Diego Medrano (doctoral student at the time and now fighting in the postdoc jungle) will be briefly described. But before doing so let me write that Lev was very enthusiastic about the possibilities of this line of thinking and we were planning on his next visit to Madrid to work together on this. Needless to write that to make progress in this or any other direction will be much harder without Lev's wisdom and mastery to guide us.

They almost missed the bus to the airport, we had to run along Parker's Piece with all the luggage to catch it. It was there, completely breathless, where we shook hands for the last time.

12.5.1 *CHY scattering equations in Sudakov variables*

The standard Feynman diagram techniques for the evaluation of scattering amplitudes become too complicated when the number of legs or loops grows. The Cachazo–He–Yuan (CHY) approach [Cachazo *et al.* (2014a,b)] offers a promising route to simplify this problem where n–point amplitudes are presented as $(n-3)$–dimensional integrals over the moduli space of n–punctured spheres, which are localized on the solutions to the so-called *scattering equations* (SE).

The bottleneck in this framework is that the number of integrals defining the n–point S-matrix elements grows very rapidly with n. The value of [Chachamis *et al.* (2018)] is to propose the use of Sudakov variables to simplify the problem of finding solutions of the SE since they naturally encode momentum conservation. The SL(2,$\mathbb{C}$) redundancy of the SE is only partially fixed in order to use azimuthal angles (with respect to the axis defined by the incoming particles) to parametrize the position of the punctures on the sphere.

For CHY the set of n on–shell D–dimensional momenta p_i^μ can be mapped into an internal space on the n–punctured Riemann sphere parametrized by $\sigma_i \in \mathbb{CP}^1$ (with $i = 1, \ldots, n$) using

$$p_j^\mu = \oint_{|z-\sigma_j|=\epsilon} \frac{dz}{2\pi i} \frac{v^\mu(z)}{\prod_{k=1}^n (z-\sigma_k)}, \quad v^\mu(z) = \sum_{j=1}^n p_j^\mu \prod_{\substack{k=1 \\ k\neq j}}^n (z-\sigma_k), \tag{12.52}$$

where $v(z)^2 = 0$. This implies the SE ($\sigma_{ij} \equiv \sigma_i - \sigma_j$, $s_{ij} = (p_i + p_j)^2 = 2p_i \cdot p_j$)

$$\mathcal{S}_i(\sigma) \equiv \sum_{j\neq i}^n \frac{s_{ij}}{\sigma_{ij}} = 0. \tag{12.53}$$

There are n equations with n–3 being linearly independent. They have $(n-3)!$ solutions.

CHY have shown [Cachazo *et al.* (2014a,b)] that the tree-level n-point Yang-Mills amplitudes admit the following representation with support on the SE solutions:

$$\mathcal{A}_n = i\, g^{n-2} \int \frac{d^n\sigma}{\mathrm{Vol}[\mathrm{SL}(2,\mathbb{C})]} \sigma_{kl}\sigma_{lm}\sigma_{mk} \prod_{i\neq k,l,m} \delta\left(\sum_{j\neq i}^{n} \frac{2\, p_i \cdot p_j}{\sigma_{ij}}\right) I_L I_R, \tag{12.54}$$

where

$$I_L = \sum_{\beta\in S_n/\mathbb{Z}_n} \frac{\mathrm{Tr}\left(T^{a_{\beta(1)}} T^{a_{\beta(2)}} \ldots T^{a_{\beta(n)}}\right)}{\sigma_{\beta(1)\beta(2)}\sigma_{\beta(2)\beta(3)} \cdots \sigma_{\beta(n)\beta(1)}}, \qquad I_R = \mathrm{Pf}' M_n. \tag{12.55}$$

The latter depends on the reduced Pfaffian of a $2n \times 2n$ antisymmetric matrix which is a function of the momenta and polarizations of external particles. It is remarkable that for amplitudes in Einstein-Hilbert gravity it is enough to make the replacement

$$I_L = \mathrm{Pf}' M_n, \quad I_R = \mathrm{Pf}' M_n, \tag{12.56}$$

exhibiting the double-copy structure of graviton amplitudes.

In the 4–dimensional case there exists an interesting solution to the SE found by Fairlie [Fairlie (2009); Fairlie and Roberts (1972)]. It reads

$$\sigma_j = \frac{p_j^0 + p_j^3}{p_j^1 - i p_j^2} = \frac{p_j^1 + i p_j^2}{p_j^0 - p_j^3}. \tag{12.57}$$

Using the parametrization of on-shell momenta p_j

$$p_j = p_j^{\perp} \left(\cosh Y_j, \cos\phi_j, \sin\phi_j, \sinh Y_j\right), \tag{12.58}$$

with Y_j the rapidity, ϕ_j the azimuthal angle, $p_j^{\perp}$ the modulus of the transverse component of the momentum, and the stereographic coordinates on $\mathbb{S}^2$,

$$p_j = \omega_j \left(1, \frac{\zeta_j + \bar{\zeta}_j}{1+\zeta_j\bar{\zeta}_j}, i\frac{\bar{\zeta}_j - \zeta_j}{1+\zeta_j\bar{\zeta}_j}, \frac{\zeta_j\bar{\zeta}_j - 1}{1+\zeta_j\bar{\zeta}_j}\right), \tag{12.59}$$

one can write Fairlie's solution (12.57) in the form

$$\sigma_j = \zeta_j = e^{Y_j + i\phi_j}. \tag{12.60}$$

p_j is then mapped onto the point $2\sigma_j$ on the projective plane. On the sphere it lies on a circumference of radius $2e^{Y_j}$ parametrized by the azimuthal angle ϕ_j.

The two incoming particles with momenta p and q in a general process in which the particles in the final state have momenta p_i (with $i = 1, \ldots, n-2$) are considered to have their spatial momenta lying along the z axis taking the form

$$p \longrightarrow \frac{\sqrt{s}}{2}(1,0,0,1), \quad q \longrightarrow \frac{\sqrt{s}}{2}(1,0,0,-1). \tag{12.61}$$

The associated punctures solving the SE are located on a circle around the north and south poles of the Riemann sphere, which shrinks to a point when $\epsilon \to 0$, *i.e.*

$$\sigma_p = e^{Y_p + i\phi} = \frac{e^{i\phi}}{\epsilon} \longrightarrow \infty, \quad \sigma_q = -e^{Y_q + i\phi} = -\epsilon\, e^{i\phi} \longrightarrow 0. \tag{12.62}$$

In a general 4–point scattering amplitude the incoming and outgoing momenta given by $\{p, q, p', q'\}$ and constrained by $p+q-p'-q' = 0$ are mapped into the moduli space of spheres with four punctures $\{\sigma_p, \sigma_q, \sigma_{p'}, \sigma_{q'}\} \in \mathbb{CP}^1$. The incoming momenta can be treated as above and the outgoing particles using the Sudakov [Sudakov (1956)] representation as follows,

$$\begin{aligned} q_1 \equiv p - p' &= \alpha\, p + \beta\, q + \mathbf{q}_1, \quad \mathbf{q}_1 = q_1^\perp \left(0, \cos\theta_1, \sin\theta_1, 0\right), \\ p' = p - q_1 &= \left(\frac{\sqrt{s}}{2}\left(1-\alpha-\beta\right), -q_1^\perp \cos\theta_1, -q_1^\perp \sin\theta_1, \frac{\sqrt{s}}{2}(1-\alpha+\beta)\right). \end{aligned}$$

The on-shell condition leads to $|Q_1|^2 \equiv (q_1^\perp)^2 = s(\alpha-1)\beta$ with $Q_j = q_j^\perp e^{i\theta_j}$. A similar parametrisation applies for $q' = q + q_1$ with $|Q_1|^2 \equiv (q_1^\perp)^2 = s\alpha(1+\beta)$. Therefore $\alpha + \beta = 0$.

For the 4-point function, the SE only have one solution, which corresponds to Fairlie's (12.57):

$$\begin{aligned} \sigma_{p'} &\equiv e^{Y_{p'} + i\phi_{p'}} = \frac{Q_1}{\beta\sqrt{s}} = \sqrt{\frac{1-\alpha}{\alpha}}\, e^{i(\theta_1 + \pi)}, \\ \sigma_{q'} &\equiv e^{Y_{q'} + i\phi_{q'}} = \frac{Q_1}{(1-\alpha)\sqrt{s}} = \sqrt{\frac{\alpha}{1-\alpha}}\, e^{i\theta_1}, \end{aligned} \tag{12.63}$$

and the two punctures are located on antipodal points on the sphere as can be seen in Fig. 12.8. Using $s_{ij} = s\widehat{s}_{ij}, Q_i = \sqrt{s}\widehat{Q}_i$, the SE solution now reads

$$\sigma_{p'} = -\frac{\widehat{Q}_1}{\alpha}, \quad \sigma_{q'} = \frac{\widehat{Q}_1}{1-\alpha}. \tag{12.64}$$

In the Sudakov representation the evaluation of scattering amplitudes within the CHY formalism is rather simple. As an example let us consider the 4–point amplitude in a φ^3 scalar theory:

$$\mathcal{A}_4^{\varphi^3} = \int dz_{p'} \left[z_{p'} - \frac{Q_1}{(1-\alpha)\sqrt{s}}\right]^{-2} \frac{Q_1^2}{s^2\alpha^3(\alpha-1)}\delta\left(z_{p'} + \frac{Q_1}{\alpha\sqrt{s}}\right) = \frac{\alpha-1}{s\alpha}. \tag{12.65}$$

Notice that the phase in Q_1 cancels out in the amplitude.

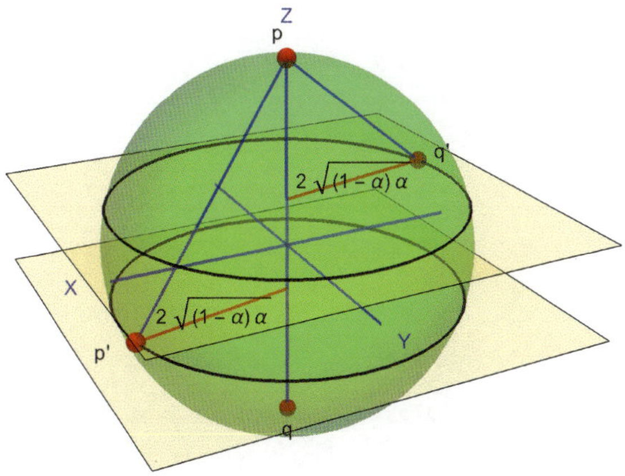

Fig. 12.8 Punctures on the Riemann sphere for 4–particle scattering with momenta $p + q \longrightarrow p' + q'$.

In the case of five particles, $\{p, q, p', k, q'\}$ with $p + q - p' - k - q' = 0$, there exists a second solution to the SE besides Fairlie's. Two sets of Sudakov parameters $\{\alpha_1, \beta_1\}$ and $\{\alpha_2, \beta_2\}$ are introduced, *i.e.*

$$\begin{aligned} q_1 &= p - p' = \alpha_1 p + \beta_1 q + \mathbf{q}_1, \\ q_2 &= q' - q = \alpha_2 p + \beta_2 q + \mathbf{q}_2, \\ k &= q_1 - q_2 = (\alpha_1 - \alpha_2)\, p + (\beta_1 - \beta_2)\, q + \mathbf{q}_1 - \mathbf{q}_2, \end{aligned} \tag{12.66}$$

where $\mathbf{q}_i = q_i^{\perp}\Big(0, \cos\theta_i, \sin\theta_i, 0\Big)$. The on–shell conditions amount to $|Q_1|^2 = s(\alpha_1 - 1)\beta_1$, $|Q_2|^2 = s\alpha_2(1 + \beta_2)$, $|Q_1 - Q_2|^2 = s(\alpha_1 - \alpha_2)(\beta_1 - \beta_2)$ and $Q_1 Q_2^* + Q_1^* Q_2 = s(\alpha_2 - \beta_1 + \alpha_1\beta_2 + \alpha_2\beta_1)$. The punctures associated to Fairlie's solution are then

$$\begin{aligned} \sigma_{p'} &= \frac{Q_1}{\beta_1\sqrt{s}} = \sqrt{\frac{\alpha_1 - 1}{\beta_1}}\, e^{i\theta_1} = e^{Y_{p'} + i\phi_{p'}}, \\ \sigma_{q'} &= \frac{Q_2}{(1 + \beta_2)\sqrt{s}} = \sqrt{\frac{\alpha_2}{1 + \beta_2}}\, e^{i\theta_2} = e^{Y_{q'} + i\phi_{q'}}, \\ \sigma_k &= \frac{Q_1 - Q_2}{(\beta_1 - \beta_2)\sqrt{s}} = \frac{\sqrt{(\alpha_1 - 1)\beta_1}\, e^{i\theta_1} - \sqrt{(1 + \beta_2)\alpha_2}\, e^{i\theta_2}}{\beta_1 - \beta_2} = e^{Y_k + i\phi_k}. \end{aligned} \tag{12.67}$$

As it is shown in Fig. 12.9, on the Riemann sphere they lie onto the equatorial planes on circumferences with radii

$$\begin{aligned} R_{p'} &= 2\sqrt{\frac{(\alpha_1 - 1)\beta_1}{(1 - \alpha_1 - \beta_1)^2}}, \quad R_{q'} = 2\sqrt{\frac{\alpha_2(1 + \beta_2)}{(1 + \alpha_2 + \beta_2)^2}}, \\ R_k &= \frac{2|Q_1 - Q_2|}{\alpha_1 + \beta_1 - \alpha_2 - \beta_2} = 2\sqrt{\frac{(\alpha_1 - \alpha_2)(\beta_1 - \beta_2)}{(\alpha_1 + \beta_1 - \alpha_2 - \beta_2)^2}}. \end{aligned} \tag{12.68}$$

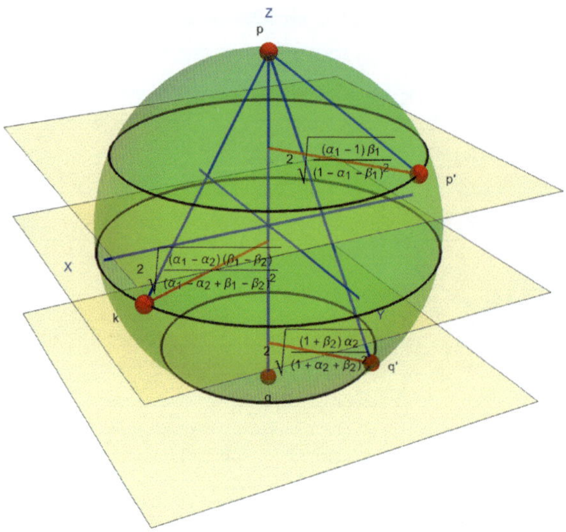

Fig. 12.9 Punctures on the Riemann sphere for the 5–particle amplitude.

It is a simple task to find the second solution of the SE using the Sudakov's variables since it essentially corresponds to the complex conjugate of Fairlie's solution. Both together can then be written as

$$\begin{aligned}
\sigma_{p'}^{(+)} &= \sigma_{p'}^{(-)*} = \frac{\widehat{Q}_1 e^{-i\theta_2}}{\beta_1} = \sqrt{\frac{\alpha_1 - 1}{\beta_1}}\, e^{i(\theta_1-\theta_2+\pi)}, \\
\sigma_{q'}^{(+)} &= \sigma_{q'}^{(-)*} = \frac{\widehat{Q}_2 e^{-i\theta_2}}{1+\beta_2} = \sqrt{\frac{\alpha_2}{1+\beta_2}}, \qquad (12.69) \\
\sigma_{k}^{(+)} &= \sigma_{k}^{(-)*} = \frac{(\widehat{Q}_1 - \widehat{Q}_2) e^{-i\theta_2}}{\beta_1 - \beta_2} = \frac{\sqrt{(\alpha_1-1)\beta_1}\, e^{i(\theta_1-\theta_2)} - \sqrt{\alpha_2(1+\beta_2)}}{\beta_1 - \beta_2}.
\end{aligned}$$

The 5–point amplitude for the φ^3 scalar theory can now be written as the sum over the two solutions:

$$\begin{aligned}
\mathcal{A}_5^{\varphi^3} &= \int dz_{p'} dz_{q'}\, \mathcal{J}^{-1} \delta\big(z_{p'} - \sigma_{p'}\big)\delta\big(z_{q'} - \sigma_{q'}\big) \frac{z_k^2}{z_{q'}^2 z_{q'k}^2 z_{kp'}^2} + \text{c.c.} \\
&= \frac{2}{s^2} \text{Re}\left[\left(\frac{\sigma_{p'}}{\sigma_{q'}}\right) \frac{1}{L\widetilde{L} - R\widetilde{R}}\right] \\
&= \frac{1}{s^2}\left[\frac{1}{\alpha_1+\beta_1} - \frac{1}{\alpha_2+\beta_2} + \frac{1}{(\alpha_1+\beta_1)\beta_1} - \frac{1}{\beta_1\alpha_2} + \frac{1}{\alpha_2(\alpha_2+\beta_2)}\right], \qquad (12.70)
\end{aligned}$$

where

$$\begin{aligned}
L &= \frac{\sigma_{p'k}}{\sigma_{p'q'}}\left[(\alpha_1 - 1)\frac{\sigma_{q'}}{\sigma_{p'}} + \beta_1 \frac{\sigma_{p'}}{\sigma_{q'}}\right], \\
R &= \left(\frac{\sigma_{p'}\sigma_{p'k}}{\sigma_{p'q'}}\right) \frac{(1-\alpha_1+\alpha_2-\beta_1+\beta_2)(\alpha_1+\beta_1)}{(\alpha_1-\alpha_2+\beta_1)\sigma_{p'} - (1+\beta_2)\sigma_{q'}}, \qquad (12.71)
\end{aligned}$$

and the quantities with tilde are defined by

$$\widetilde{\mathcal{O}}\Big(\alpha_1,\alpha_2,\beta_1,\beta_2,\theta_1-\theta_2\Big)=\mathcal{O}\Big(1-\alpha_2,1-\alpha_1,-1-\beta_2,-1-\beta_1,\theta_2-\theta_1\Big). \tag{12.72}$$

This is a first analysis of the use of Sudakov variables in the context of the CHY calculation of scattering amplitudes. In the future it will be interesting to generalize this representation to n-point amplitudes.

The research here presented has been supported by the Spanish Research Agency (Agencia Estatal de Investigación) through the grant IFT Centro de Excelencia Severo Ochoa SEV-2016-0597, and the Spanish Government grants FPA2015-65480-P, FPA2016-78022-P.

Bibliography

Altarelli, G. and Parisi, G. (1977). Asymptotic freedom in parton language, *Nucl. Phys.* **B126**, pp. 298–318, doi:10.1016/0550-3213(77)90384-4.

Arquès, D. and Béraud, J.-F. (2000). Rooted maps on orientable surfaces, Riccatis equation and continued fractions, *Discrete Mathematics* **Elsevier**, pp. 215 (1–3), pp.1–12., doi:10.1016/S0012-365X(99)00197-1.

Balitsky, I. I. and Lipatov, L. N. (1978). The Pomeranchuk singularity in quantum chromodynamics, *Sov. J. Nucl. Phys.* **28**, pp. 822–829, [Yad. Fiz.28,1597(1978)].

Bartels, J., Lipatov, L. N., and Sabio Vera, A. (2014). Double-logarithms in Einstein-Hilbert gravity and supergravity, *JHEP* **07**, p. 056, doi:10.1007/JHEP07(2014)056, arXiv:1208.3423 [hep-th].

Bern, Z. (2002). Perturbative quantum gravity and its relation to gauge theory, *Living Rev. Rel.* **5**, p. 5, doi:10.12942/lrr-2002-5, arXiv:gr-qc/0206071 [gr-qc].

Bern, Z., Carrasco, J. J., Dixon, L. J., Johansson, H., and Roiban, R. (2009). The ultraviolet behavior of N = 8 supergravity at four loops, *Phys. Rev. Lett.* **103**, p. 081301, doi:10.1103/PhysRevLett.103.081301, arXiv:0905.2326 [hep-th].

Bern, Z., Carrasco, J. J. M., Dixon, L. J., Johansson, H., and Roiban, R. (2010). The complete four-loop four-point amplitude in N = 4 Super-Yang-Mills theory, *Phys. Rev.* **D82**, p. 125040, doi:10.1103/PhysRevD.82.125040, arXiv:1008.3327 [hep-th].

Bern, Z., Dixon, L. J., Dunbar, D. C., Perelstein, M., and Rozowsky, J. S. (1998). On the relationship between Yang-Mills theory and gravity and its implication for ultraviolet divergences, *Nucl. Phys.* **B530**, pp. 401–456, doi:10.1016/S0550-3213(98)00420-9, arXiv:hep-th/9802162 [hep-th].

Boucher-Veronneau, C. and Dixon, L. J. (2011). N >- 4 supergravity amplitudes from gauge theory at two loops, *JHEP* **12**, p. 046, doi:10.1007/JHEP12(2011)046, arXiv:1110.1132 [hep-th].

Cachazo, F., He, S., and Yuan, E. Y. (2014a). Scattering of massless particles in arbitrary dimensions, *Phys. Rev. Lett.* **113**, 17, p. 171601, doi:10.1103/PhysRevLett.113.171601, arXiv:1307.2199 [hep-th].

Cachazo, F., He, S., and Yuan, E. Y. (2014b). Scattering of massless particles: Scalars, gluons and gravitons, *JHEP* **07**, p. 033, doi:10.1007/JHEP07(2014)033, arXiv:1309.0885 [hep-th].

Chachamis, G., Medrano Jiménez, D., Sabio Vera, A., and Vázquez-Mozo, M. Á. (2018). Sudakov representation of the Cachazo-He-Yuan scattering equations formalism, *JHEP* **01**, p. 057, doi:10.1007/JHEP01(2018)057, arXiv:1712.04288 [hep-th].

Ciafaloni, M., Colferai, D., and Veneziano, G. (2019). Infrared features of gravitational scattering and radiation in the eikonal approach, *Phys. Rev.* **D99**, 6, p. 066008, doi:10.1103/PhysRevD.99.066008, arXiv:1812.08137 [hep-th].

Cvitanovic, P., Lautrup, B. E., and Pearson, R. B. (1978). The number and weights of Feynman diagrams, *Phys. Rev.* **D18**, p. 1939, doi:10.1103/PhysRevD.18.1939.

Di Vecchia, P., Luna, A., Naculich, S. G., Russo, R., Veneziano, G., and White, C. D. (2019). A tale of two exponentiations in $\mathcal{N} = 8$ supergravity, *Phys. Lett.* **B798**, p. 134927, doi:10.1016/j.physletb.2019.134927, arXiv:1908.05603 [hep-th].

Di Vecchia, P., Naculich, S. G., Russo, R., Veneziano, G., and White, C. D. (2020). A tale of two exponentiations in $\mathcal{N} = 8$ supergravity at subleading level, *JHEP* **03**, p. 173, doi:10.1007/JHEP03(2020)173, arXiv:1911.11716 [hep-th].

Dokshitzer, Y. L. (1977). Calculation of the structure functions for deep inelastic scattering and e+ e− annihilation by perturbation theory in quantum chromodynamics, *Sov. Phys. JETP* **46**, pp. 641–653, [Zh. Eksp. Teor. Fiz.73,1216(1977)].

Dunbar, D. C. and Norridge, P. S. (1995). Calculation of graviton scattering amplitudes using string based methods, *Nucl. Phys.* **B433**, pp. 181–208, doi:10.1016/0550-3213(94)00385-R, arXiv:hep-th/9408014 [hep-th].

Fadin, V. S., Kuraev, E. A., and Lipatov, L. N. (1975). On the Pomeranchuk singularity in asymptotically free theories, *Phys. Lett.* **60B**, pp. 50–52, doi:10.1016/0370-2693(75)90524-9.

Fadin, V. S., Lipatov, L. N., Martin, A. D., and Melles, M. (2000). Resummation of double logarithms in electroweak high-energy processes, *Phys. Rev.* **D61**, p. 094002, doi:10.1103/PhysRevD.61.094002, arXiv:hep-ph/9910338 [hep-ph].

Fairlie, D. B. (2009). A coding of real null four-momenta into world-sheet coordinates, *Adv. Math. Phys.* **2009**, p. 284689, doi:10.1155/2009/284689, arXiv:0805.2263 [hep-th].

Fairlie, D. B. and Roberts, D. E. (1972). Dual models without tachyons — a new approach.

Goodvin, G. L., Berciu, M., and A., S. G. (2006). The Green's function of the Holstein polaron, *Phys. Rev.* **B74**, p. 245104, doi:doi:10.1103/PhysRevB.74.245104.

Gopala, K. K., Labelle, P., and Shramchenko, V. (2018). Enumeration of N-rooted maps using quantum field theory, *Nucl. Phys.* **B936**, pp. 668–689, doi:10.1016/j.nuclphysb.2018.09.017, arXiv:1709.01200 [math-ph].

Gopala Krishna, K., Labelle, P., and Shramchenko, V. (2018). Feynman diagrams, ribbon graphs, and topological recursion of Eynard-Orantin, *JHEP* **06**, p. 162, doi:10.1007/JHEP06(2018)162, arXiv:1802.01773 [hep-th].

Gorshkov, V. G., Gribov, V. N., Lipatov, L. N., and Frolov, G. V. (1966). Double logarithmic asymptotics of quantum electrodynamics, *Phys. Lett.* **22**, pp. 671–673, doi:10.1016/0031-9163(66)90701-3.

Gorshkov, V. G., Gribov, V. N., Lipatov, L. N., and Frolov, G. V. (1968a). Backward electron - positron scattering at high-energies, *Sov. J. Nucl. Phys.* **6**, p. 262, [Yad. Fiz.6,361(1967)].

Gorshkov, V. G., Gribov, V. N., Lipatov, L. N., and Frolov, G. V. (1968b). Doubly logarithmic asymptotic behavior in quantum electrodynamics, *Sov. J. Nucl. Phys.* **6**, p. 95, [Yad. Fiz.6,129(1967)].

Gribov, V. N. and Lipatov, L. N. (1972). Deep inelastic e p scattering in perturbation theory, *Sov. J. Nucl. Phys.* **15**, pp. 438–450, [Yad. Fiz.15,781(1972)].

Grisaru, M. T. and Schnitzer, H. J. (1981). Dynamical calculation of bound state supermultiplets in $N = 8$ supergravity, *Phys. Lett.* **107B**, pp. 196–200, doi: 10.1016/0370-2693(81)90811-X.

Grisaru, M. T., van Nieuwenhuizen, P., and Wu, C. C. (1975). Reggeization and the question of higher loop renormalizability of gravitation, *Phys. Rev.* **D12**, p. 1563, doi:10.1103/PhysRevD.12.1563.

Gubser, S. S., Klebanov, I. R., and Polyakov, A. M. (1998). Gauge theory correlators from noncritical string theory, *Phys. Lett.* **B428**, pp. 105–114, doi:10.1016/S0370-2693(98)00377-3, arXiv:hep-th/9802109 [hep-th].

Henn, J. M. and Mistlberger, B. (2019). Four-graviton scattering to three loops in $\mathcal{N} = 8$ supergravity, *JHEP* **05**, p. 023, doi:10.1007/JHEP05(2019)023, arXiv:1902.07221 [hep-th].

Ihrig, E. I., Rosenstee, G., Chow, H., and Trainor, L. E. H. (1976). Group theory and many body diagrams II, Enumeration methods and number approximations, *Proc. R. Soc. Lond.* **A348**, pp. 339–357.

Kabat, D. N. and Ortiz, M. (1992). Eikonal quantum gravity and Planckian scattering, *Nucl. Phys.* **B388**, pp. 570–592, doi:10.1016/0550-3213(92)90627-N, arXiv:hep-th/9203082 [hep-th].

Kirschner, R. and Lipatov, L. N. (1982a). Double logarithmic asymptotics of quark scattering amplitudes with flavor exchange, *Phys. Rev.* **D26**, pp. 1202–1205, doi: 10.1103/PhysRevD.26.1202.

Kirschner, R. and Lipatov, L. n. (1982b). Doubly logarithmic asymptotic of the quark scattering amplitude with nonvacuum exchange in the T channel, *Sov. Phys. JETP* **56**, pp. 266–273, [Zh. Eksp. Teor. Fiz.83,488(1982)].

Kirschner, R. and Lipatov, L. n. (1983). Double logarithmic asymptotics and Regge singularities of quark amplitudes with flavor exchange, *Nucl. Phys.* **B213**, pp. 122–148, doi:10.1016/0550-3213(83)90178-5.

Kuraev, E. A., Lipatov, L. N., and Fadin, V. S. (1976). Multi-Reggeon processes in the Yang-Mills theory, *Sov. Phys. JETP* **44**, pp. 443–450, [Zh. Eksp. Teor. Fiz.71,840(1976)].

Kuraev, E. A., Lipatov, L. N., and Fadin, V. S. (1977). The Pomeranchuk singularity in nonabelian gauge theories, *Sov. Phys. JETP* **45**, pp. 199–204, [Zh. Eksp. Teor. Fiz.72,377(1977)].

Lipatov, L. N. (1976). Reggeization of the vector meson and the vacuum singularity in nonabelian gauge theories, *Sov. J. Nucl. Phys.* **23**, pp. 338–345, [Yad. Fiz.23,642(1976)].

Lipatov, L. N. (1982a). Graviton Reggeization, *Phys. Lett.* **116B**, pp. 411–413, doi:10.1016/0370-2693(82)90156-3.

Lipatov, L. N. (1982b). Multi-Regge processes in gravitation, *Sov. Phys. JETP* **55**, pp. 582–590, [Zh. Eksp. Teor. Fiz.82,991(1982)].

Lipatov, L. N. (1991). High-energy scattering in QCD and in quantum gravity and two-dimensional field theories, *Nucl. Phys.* **B365**, pp. 614–632, doi:10.1016/0550-3213(91)90512-V.

Lipatov, L. N. (2013). Effective action for the Regge processes in gravity, *Phys. Part. Nucl.* **44**, pp. 391–413, doi:10.1134/S1063779613020214, arXiv:1105.3127 [hep-th].

Maldacena, J. M. (1999). The large N limit of superconformal field theories and supergravity, *Int. J. Theor. Phys.* **38**, pp. 1113–1133, doi:10.1023/A:1026654312961, 10.4310/ATMP.1998.v2.n2.a1, arXiv:hep-th/9711200 [hep-th], [Adv. Theor. Math. Phys.2,231(1998)].

Martin, R. J. and Kearney, M. J. (2010). An exactly solvable self-convolutive recurrence, *Aequationes Math.* **c 80, 3**, pp. 291–318, doi:MR-2739179, arXiv:1103.4936 [math.CO].

Prunotto, A., Alberico, W. M., and Czerski, P. (2018). Feynman diagrams and rooted maps, *Open Phys.* **16**, 1, pp. 149–167, doi:10.1515/phys-2018-0023, arXiv:1312.0934 [nucl-th].

Sabio Vera, A. (2019). Double-logarithms in $\mathcal{N} = 8$ supergravity: impact parameter description & mapping to 1-rooted ribbon graphs, *JHEP* **07**, p. 080, doi:10.1007/JHEP07(2019)080, arXiv:1904.13372 [hep-th].

Sabio Vera, A. (2020). Double logarithms in $\mathcal{N} \geq 4$ supergravity: weak gravity and Shapiro's time delay, *JHEP* **01**, p. 163, doi:10.1007/JHEP01(2020)163, arXiv:1912.00744 [hep-th].

Sabio Vera, A., Serna Campillo, E., and Vazquez-Mozo, M. A. (2012). Graviton emission in Einstein-Hilbert gravity, *JHEP* **03**, p. 005, doi:10.1007/JHEP03(2012)005, arXiv:1112.4494 [hep-th].

Shapiro, I. I. (1964). Fourth test of general relativity, *Phys. Rev. Lett.* **13**, pp. 789–791, doi:10.1103/PhysRevLett.13.789.

Sudakov, V. V. (1956). Vertex parts at very high-energies in quantum electrodynamics, *Sov. Phys. JETP* **3**, pp. 65–71, [Zh. Eksp. Teor. Fiz.30,87(1956)].

Witten, E. (1998). Anti-de sitter space and holography, *Adv. Theor. Math. Phys.* **2**, pp. 253–291, doi:10.4310/ATMP.1998.v2.n2.a2, arXiv:hep-th/9802150 [hep-th].

https://doi.org/10.1142/9789811231124_0013

Chapter 13

$\mathcal{N} = 4$ SYM Quantum Spectral Curve in BFKL Regime

Mikhail Alfimov[*,†,‡,**], Nikolay Gromov[§,¶,††] and Vladimir Kazakov[‖,‡‡]

[*] *National Research University Higher School of Economics, ul. Usacheva, d. 6, Moscow, Russia 119048*
[†] *National Research Nuclear University Moscow Engineering Physical Institute, Kashirskoe sh., d. 31, Moscow, Russia 115409*
[‡] *P.N. Lebedev Physical Institute of the Russian Academy of Sciences, Leninskiy pr., d. 53, Moscow, Russia 119991*
[§] *Mathematics Department, King's College London, The Strand, London WC2R 2LS, UK*
[¶] *St. Petersburg INP, Gatchina, 188 300, St. Petersburg, Russia*
[‖] *Laboratoire de Physique de l'École Normale Supérieure, ENS, Université PSL, CNRS, Sorbonne Université, Université de Paris, F-75005 Paris, France*
[**] *malfimov@hse.ru*
[††] *nikgromov@gmail.com*
[‡‡] *Vladimir.Kazakov@ens.fr*

Abstract

We review the applications of the Quantum Spectral Curve (QSC) method to the Regge (BFKL) limit in $\mathcal{N} = 4$ supersymmetric Yang-Mills theory. QSC, based on quantum integrability of the AdS_5/CFT_4 duality, was initially developed as a tool for the study of the spectrum of anomalous dimensions of local operators in the $\mathcal{N} = 4$ SYM in the planar, $N_c \to \infty$ limit. We explain how to apply the QSC for the BFKL limit, which requires non-trivial analytic continuation in spin S and extends the initial construction for non-local light-ray operators. We give a brief review of high precision non-perturbative numerical solutions and analytic perturbative data resulting from this approach. We also describe as a simple example of the QSC construction the leading order in the BFKL limit. We show that the QSC substantially simplifies in this limit and reduces to the Faddeev-Korchemsky Baxter equation for Q-functions. Finally, we review recent results for the Fishnet CFT, which carries a number of similarities with Lipatov's integrable spin chain for interacting Reggeized gluons.

13.1 Introduction

The Balitsky-Fadin-Kuraev-Lipatov (BFKL) approximation [Kuraev *et al.* (1977); Balitsky and Lipatov (1978)] in Quantum Chromodynamics (QCD) has marked the beginning of a new era in the study of the properties of hadron collisions at high energy. Among various important results following the BFKL approach, the Pomeron spectrum was calculated in the leading [Kuraev *et al.* (1977); Balitsky and Lipatov (1978); Lipatov (1986)], and then, 20 years later, in the next-to-leading [Fadin and Lipatov (1998)] logarithmic approximations (LO and NLO). The complexity of the conventional Feynman perturbation theory increases dramatically at each order and it is unlikely that the NNLO calculation in QCD is reachable with the currently available techniques.

The history of integrability in gauge theories starts from the renowned Lev Lipatov's work [Lipatov (1994)], where the effective Lagrangian for Reggeized gluons in QCD was shown to be integrable. Lipatov was also the first to notice the equivalence of this system to the $SL(2,\mathbb{C})$ integrable quantum spin chain [Lipatov (1993)]. Following the seminal paper [Faddeev and Korchemsky (1995)], the integrability approach to Lipatov's model has been substantially advanced in [Derkachov *et al.* (2001); De Vega and Lipatov (2001); Derkachov *et al.* (2002, 2003)].

The BFKL approximation was then successfully applied to the study of the spectrum of anomalous dimensions of certain operators in $\mathcal{N} = 4$ SYM theory [Kotikov and Lipatov (2000)]. The result of this calculation appears to be slightly less complex than the original Regge limit in QCD, possibly due to the supersymmetry. As it was noticed in [Kotikov and Lipatov (2000, 2003)] the result follows the so-called maximal transcendentality selection principle – only the most complicated in transcedentality parts of the QCD result are present in its

$\mathcal{N} = 4$ SYM counterpart, with exactly the same coefficients.[1] Even in this, maximally super-symmetric theory the complexity of the traditional Feynman perturbation technique increases factorially at each order. However, a powerful alternative method is available in this theory: due to the quantum integrability of planar $\mathcal{N} = 4$ SYM theory it became possible to compute the spectrum of the planar $\mathcal{N} = 4$ SYM theory at any value of the 't Hooft coupling $\lambda = 16\pi^2 g^2 = g_{YM}^2 N_c$! This integrability was discovered at one loop in the seminal Minahan-Zarembo paper [Minahan and Zarembo (2003)],[2] then it was demonstrated at two-loops [Beisert *et al.* (2003)], then at strong coupling in [Bena *et al.* (2004); Kazakov *et al.* (2004); Beisert *et al.* (2006)], via AdS/CFT correspondence, then generalized up to wrapping order in [Janik (2006); Beisert and Roiban (2005); Beisert *et al.* (2007); Ambjorn *et al.* (2006)], and finally to all loops, first in terms of the exact Y-system [Gromov *et al.* (2010)], and then of the TBA equations [Bombardelli *et al.* (2009); Gromov *et al.* (2009); Arutyunov and Frolov (2009)] for the simplest Konishi-like operators. A few years ago, this line of research resulted in the final formulation of the solution of the spectral problem — the Quantum Spectral Curve (QSC) of the AdS_5/CFT_4 duality [Gromov *et al.* (2015a); Gromov and Levkovich-Maslyuk (2016a)].

The QSC equations can be formulated as a finite system of Riemann-Hilbert-type equations on 4+4 Baxter Q-functions. The QSC formalism has led to the tremendous progress in high precision numerical and multi-loop computations in a multitude of particular physical quantities (see the reviews [Gromov (2017); Kazakov (2018)] and citations therein). Notably, the dimension Δ of a twist-2 operator of the type $\mathrm{tr}(Z\nabla_+^S Z)$ was computed for virtually arbitrary coupling and arbitrary complex values of spin S [Gromov *et al.* (2016)] (see the Fig. 13.1 and Fig. 13.2). The particular corner of this picture, when $S \simeq -1 + \mathcal{O}(g^2)$ at weak coupling $g^2 \to 0$, corresponds to the BFKL regime. The loop expansion of S can be written as follows

$$S = -1+g^2\chi_{\mathrm{LO}}(\Delta)+g^4\chi_{\mathrm{NLO}}(\Delta)+g^6\chi_{\mathrm{NNLO}}(\Delta)+g^8\chi_{\mathrm{NNNLO}}(\Delta)+\mathcal{O}(g^8), \quad (13.1)$$

where $\chi_{\mathrm{N}^k\mathrm{LO}}(\Delta)$ is the BFKL kernel eigenvalue at the $(k+1)$-st loop. The BFKL limit of QSC was first explored in [Alfimov *et al.* (2015)] and the LO BFKL spectrum $\chi_{\mathrm{LO}}(\Delta)$ with zero conformal spin was reproduced there. The coefficient $\chi_{\mathrm{NLO}}(\Delta)$ calculated in [Kotikov and Lipatov (2000)] was successfully reproduced by the QSC calculations in [Gromov *et al.* (2015b)]. Then, significant progress was made in the same work [Gromov *et al.* (2015b)], where QSC allowed to derive the NNLO BFKL kernel eigenvalue $\chi_{\mathrm{NNLO}}(\Delta)$, which was later confirmed by

[1] See also [Kotikov *et al.* (2007)].

[2] The first glimpses on such integrability can be found in L.Lipatov's talks [Lipatov (1997, 2001)].

the other method in [Caron-Huot and Herranen (2018)]. It should be noted that all the BFKL kernel eigenvalues up to NNLO included can be written in terms of nested harmonic sums and multiple zeta values. Moreover, the high precision QSC numerical technique [Gromov *et al.* (2016)] is currently adapted in [Alfimov *et al.* (2018)]: for any order $\chi_{\mathrm{N}^k\mathrm{LO}}(\Delta)$[3] can be analyzed by the QSC numerical algorithm. Moreover, the QSC approach allowed to compute anomalous dimensions of the length-2 operators with both non-zero spins, including the conformal one n, $\mathrm{tr}(Z\nabla_+^S\nabla_\perp^n Z)$ [Alfimov *et al.* (2018)].

In this short review, we give a concise formulation of the QSC and its reduction in the case of the BFKL limit. We will briefly review the ideas of numerical and perturbative approaches to the solution of the QSC equations. We also shortly describe the most important numerical and perturbative analytic results for the BFKL limit of $\mathcal{N} = 4$ SYM theory.

The BFKL also gave an inspiration for a number of other research directions over the years. In particular, we will explain here how the BFKL limit influenced the currently being actively developed so-called *fishnet* conformal field theory (FCFT) proposed in [Gurdogan and Kazakov (2016)] as a double scaling limit combining weak coupling and strong *gamma*-deformation of $\mathcal{N} = 4$ SYM theory. Fishnet CFT attracted much of attention in the last few years [Caetano *et al.* (2018); Chicherin *et al.* (2018); Gromov *et al.* (2018); Chicherin *et al.* (2017); Grabner *et al.* (2018); Mamroud and Torrents (2017); Kazakov and Olivucci (2018); Karananas *et al.* (2019); Gromov *et al.* (2019); Derkachov *et al.* (2019); Kazakov *et al.* (2019); Basso *et al.* (2019a); Korchemsky (2019); Dutta Chowdhury *et al.* (2019); Gromov and Sever (2020); Basso *et al.* (2019b); Derkachov and Olivucci (2020); Adamo and Jaitly (2020); de Mello Koch *et al.* (2019); Pittelli and Preti (2019)]. It was generalized to any dimension D [Kazakov and Olivucci (2018)], and for $D = 2$ and a specific representation for conformal spins this model can be identified (for certain physical quantities) with Lipatov's $SL(2,\mathbb{R})$ spin chain for LO approximation of BFKL. As Fishnet CFT has a significant interplay with the initial ideas of BFKL integrability, we briefly review here the construction and integrability properties of fishnet CFT.

13.2 Integrability in $\mathcal{N} = 4$ SYM

In the present Section we briefly review the integrable structure of $\mathcal{N} = 4$ SYM. We connect the BFKL kernel eigenvalues to the dimensions of the twist-2 operators following [Kotikov and Lipatov (2000)], which allows us to apply the QSC to the study of the BFKL spectrum. This step requires certain generalisation of the gluing

[3]In the case of non-zero conformal spin some data connected to the NNNLO BFKL kernel eigenvalues were extracted from QSC in [Alfimov *et al.* (2018)], namely: the intercept function (BFKL kernel eigenvalue at $\Delta = 0$) for different values of conformal spin, the slope-to-intercept function (BFKL kernel eigenvalue derivative with respect to conformal spin) and the curvature function (BFKL kernel eigenvalue 2nd derivative with respect to Δ), both calculated at the special BPS point of dimension 0 and conformal spin 1.

condition in QSC for the local operators, allowing for non-integer spins. As an example we demonstrate how to reproduce the Baxter equation, which initially appeared in the problem of diagonalization of the LO BFKL kernel.

13.2.1 *BFKL spectrum and the maximal transcendentality principle*

Let us start from briefly describing the relation of the BFKL spectrum in $\mathcal{N}=4$ SYM to those in QCD. The key statement governing this connection is the so-called principle of maximal transcendentality [Kotikov and Lipatov (2000, 2003)]. One can find a detailed description of this principle in [Kotikov and Onishchenko (2019)]. Below we demonstrate this principle in the application to the BFKL kernel eigenvalues. In what follows we use the extension of the formula (13.1) for the case of non-zero conformal spin n

$$\begin{aligned}S = &-1 + g^2\chi_{\rm LO}(\Delta,n) + g^4\chi_{\rm NLO}(\Delta,n) + \\ &+ g^6\chi_{\rm NNLO}(\Delta,n) + g^8\chi_{\rm NNNLO}(\Delta,n) + \mathcal{O}(g^{10}).\end{aligned} \tag{13.2}$$

Consider first the LO BFKL Pomeron kernel eigenvalues for QCD and $\mathcal{N}=4$ SYM which can be found in [Kuraev *et al.* (1977); Balitsky and Lipatov (1978)] and [Kotikov and Lipatov (2000)] respectively. We see that the answers are identical and are given by

$$\chi_{\rm LO}(\Delta,n) = -4\left(\psi\left(\frac{1+n+\Delta}{2}\right)+\psi\left(\frac{1+n-\Delta}{2}\right)-2\psi(1)\right), \tag{13.3}$$

where Δ is the dimension of the state and n is the conformal spin. Moreover, in the LO the BFKL kernel eigenvalues coincide for all 4D gauge theories [Kotikov and Lipatov (2000)].

For the next order in perturbation theory the situation is more subtle. The NLO BFKL kernel eigenvalue[4] for QCD was calculated in [Fadin and Lipatov (1998)]

$$\begin{aligned}&\chi^{\rm QCD}_{\rm NLO}(\Delta,n)\\ &= 4\Bigg[-2\Phi\left(n,\frac{1-\Delta}{2}\right)-2\Phi\left(n,\frac{1+\Delta}{2}\right)+6\zeta(3)+\left(\frac{67}{9}-2\zeta(2)-\frac{10n_f}{9N_c}\right)\\ &\times\frac{\chi_{\rm LO}(\Delta,n)}{4}+\psi''\left(\frac{1+n-\Delta}{2}\right)+\psi''\left(\frac{1+n+\Delta}{2}\right)-\frac{1}{2}\left(\frac{11}{3}-\frac{2}{3}\frac{n_f}{N_c}\right)\\ &\times\left(\frac{\chi^2_{\rm LO}(\Delta,n)}{16}-\psi'\left(\frac{1+n+\Delta}{2}\right)+\psi'\left(\frac{1+n-\Delta}{2}\right)\right)+\frac{\pi^2\cos\frac{\pi(1+\Delta)}{2}}{\Delta\sin^2\frac{\pi(1+\Delta)}{2}}\\ &\times\left(\left(3+\left(1+\frac{n_f}{N_c^3}\right)\frac{3\Delta^2-11}{4(\Delta^2-4)}\right)\delta^0_n-\left(1+\frac{n_f}{N_c^3}\right)\frac{\Delta^2-1}{8(\Delta^2-4)}\delta^2_n\right)\Bigg],\end{aligned} \tag{13.4}$$

[4]Our identification of the variable γ from [Kotikov and Lipatov (2000)] with our parameters is: $\gamma=(\Delta-S)/2$, where Δ is the dimension and S is the spin.

where N_c is the number of colors and n_f is the number of flavours of the fermions in the considered version of QCD. The same quantity in $\mathcal{N} = 4$ SYM was first presented in [Kotikov and Lipatov (2000)]. We write it down here from [Kotikov and Lipatov (2003)]

$$\chi_{\mathrm{NLO}}^{\mathcal{N}=4}(\Delta, n) = 4\left[-2\Phi\left(n, \frac{1-\Delta}{2}\right) - 2\Phi\left(n, \frac{1+\Delta}{2}\right) + 6\zeta(3) + \frac{1}{4}\left(\frac{1}{3} - 2\zeta(2)\right)\right.$$
$$\left. \times\, \chi_{\mathrm{LO}}(\Delta, n) + \psi''\left(\frac{1+n-\Delta}{2}\right) + \psi''\left(\frac{1+n+\Delta}{2}\right)\right]. \qquad (13.5)$$

The function $\Phi(n, x)$ in the eigenvalues (13.4) and (13.5) is equal to

$$\Phi(n, x) = \sum_{k=0}^{\infty} \frac{(-1)^{k+1}}{k+x+n/2}\left[\psi'(k+n+1) - \psi'(k+1) + (-1)^{k+1}(\beta'(k+n+1)+\right.$$
$$\left. + \beta'(k+1)) + \frac{1}{k+x+n/2}(\psi(k+n+1) - \psi(k+1))\right] \qquad (13.6)$$

and

$$\beta'(z) = \frac{1}{4}\left[\psi'\left(\frac{z+1}{2}\right) - \psi'\left(\frac{z}{2}\right)\right] = \sum_{k=0}^{+\infty} \frac{(-1)^{k+1}}{(z+k)^2}. \qquad (13.7)$$

Now we are going to explain the maximal transcendentality principle with the example of the NLO BFKL kernel eigenvalue. There is a way to rewrite both $\chi_{\mathrm{LO}}(\Delta, n)$ (13.3) and $\chi_{\mathrm{NLO}}(\Delta, n)$ for both QCD (13.5) and $\mathcal{N} = 4$ SYM (13.5) in terms of nested harmonic sums. In the LO we get the simple result

$$\chi_{\mathrm{LO}}(\Delta, n) = -4\left(S_1\left(\frac{1+\Delta+n}{2} - 1\right) + S_1\left(\frac{1+n-\Delta}{2} - 1\right)\right), \qquad (13.8)$$

from which we see that the LO BFKL kernel eigenvalues in QCD and $\mathcal{N} = 4$ SYM have transcendentality 1, where the transcendentality of a harmonic sum $S_{n_1,\ldots,n_m}$ is defined to be $|n_1| + \cdots + |n_m|$, transcendentality of a product is given by a sum of the transcendentalities of the multipliers. In the same way one defines the transcendentality of the MZV $\zeta_{n_1,\ldots,n_m}$, which is given by the sum $n_1 + \cdots + n_m$. In particular, π and $\log 2$ have transcedentality 1.

To express (13.4) and (13.5) through nested harmonic sums we need to introduce some additional formulas. In [Costa *et al.* (2012)] it was shown, that for $n = 0$ we can represent the NLO BFKL kernel eigenvalue for $\mathcal{N} = 4$ SYM as follows

$$\chi_{\mathrm{NLO}}^{\mathcal{N}=4}(\Delta, 0) = F_2\left(\frac{1+\Delta}{2}\right) + F_2\left(\frac{1-\Delta}{2}\right), \qquad (13.9)$$

where

$$F_2(x) = 4\left(-\frac{3}{2}\zeta(3) + \pi^2\log 2 + \frac{\pi^2}{3}S_1(x-1)\right.$$
$$\left. + \pi^2 S_{-1}(x-1) + 2S_3(x-1) - 4S_{-2,1}(x-1)\right). \tag{13.10}$$

Utilizing (13.9) and (13.10), we are able to rewrite (13.5) in the following way

$$\chi_{\rm NLO}^{\mathcal{N}=4}(\Delta,n) = \frac{1}{2}\left(F_2\left(\frac{1+\Delta+n}{2}\right) + F_2\left(\frac{1-\Delta-n}{2}\right) + F_2\left(\frac{1+\Delta-n}{2}\right) +\right.$$
$$\left. + F_2\left(\frac{1-\Delta+n}{2}\right)\right) + 4\left(R_n\left(\frac{1-\Delta}{2}\right) + R_n\left(\frac{1+\Delta}{2}\right)\right), \tag{13.11}$$

where

$$R_n(\gamma) = -8\left(S_{-2}\left(\gamma+\frac{n}{2}-1\right) + \frac{\pi^2}{12}\right) \times$$
$$\times\left(S_1\left(\gamma+\frac{n}{2}-1\right) - S_1\left(\gamma-\frac{n}{2}-1\right)\right). \tag{13.12}$$

Therefore, from (13.8), (13.9), (13.10), (13.11) and (13.12) we see that all the contributions to (13.11) have transcendentality 3. Furthermore, we see that all the terms of the QCD result with the maximal transcendentality 3 coincide exactly with the result of $\mathcal{N}=4$ SYM!

In the next part we will explain the relation of the BFKL kernel eigenvalues to the spectrum of length-2 (twist-2 in the case of zero conformal spin) operators in $\mathcal{N}=4$ SYM. Then we will explain how to adapt the QSC for the local operators to describe non-integer spin and conformal spin, which would then lead to the BFKL kernel eigenvalue.

13.2.2 *Analytic structure of the twist-2 operator anomalous dimensions*

In this part we describe the analytic structure of the length-2 operators with conformal spin in $\mathcal{N}=4$ SYM. Namely, we restrict ourselves to the states with the quantum numbers $J_1=2$, $J_2=J_3=0$. Let us first consider the case of zero conformal spin $S_2=0$, which are called the twist-2 $\mathfrak{sl}(2)$ operators

$$\mathcal{O} = {\rm tr} Z D_+^S Z + (\text{permutations}). \tag{13.13}$$

In the gauge theory the physical operators should have an even number of derivatives to be conformal primaries. The dimensions of these physical operators Δ were calculated for arbitrary even integer S up to several loops order [Lipatov (2001); Kotikov and Lipatov (2003); Kotikov *et al.* (2003)]. In [Gromov *et al.* (2014a)] it was understood how to describe these operators with integer spins within the QSC approach to the leading order in the perturbation theory. This technique was then generalised to the higher orders in [Marboe and Volin (2015); Marboe *et al.* (2015); Marboe and Velizhanin (2016); Marboe and Volin (2018a,b)].

As we explain below in order to make a connection with the Regge regime one needs to be able to analytically continue in S. In the works [Gromov and Kazakov (2013); Janik (2013)] it was understood how to remove the constraint of integer even spin S from the integrable description, leading to the analytic continuation of the anomalous dimension of the twist-2 $\mathfrak{sl}(2)$ operators. At the 1-loop order one gets the following result [Kotikov and Lipatov (2003); Kotikov and Velizhanin (2005)]

$$\Delta = 2 + S + 8g^2\left(\psi(S+1) - \psi(1)\right) + \mathcal{O}(g^2). \tag{13.14}$$

As we can see the formula above (13.14) is regular for all $S > -1$, but has a simple pole at $S = -1$. In fact as now both S and Δ are allowed to be non-integer we can invert the function for the range where it is regular and instead consider $S(\Delta)$ which is a more convenient function [Kotikov and Lipatov (2003); Brower *et al.* (2007); Costa *et al.* (2012, 2014); Brower *et al.* (2013)]. In particular (13.14) leads to

$$S_{\Delta>1}(\Delta) = \Delta - 2 - 8g^2\left(\psi(\Delta-1) - \psi(1)\right) + \mathcal{O}(g^4). \tag{13.15}$$

Below we describe the method which would allow us to compute this function in $\mathcal{N} = 4$ SYM at finite coupling (see the Fig. 13.1 for $n = 0$). At the moment we only need to know its main features. We see that $S(\Delta)$ is an even function of Δ. At finite g it has a smooth parabolic shape, however, at weak coupling it becomes piece-wise linear: $S = |\Delta| - 2$ for $|\Delta| > 1$ and $S = -1$ for $|\Delta| < 1$. Thus the $g \to 0$ limit breaks the analyticity and the analytic continuation in S does not commute with the $g \to 0$ limit. This implies that even though $S(\Delta)$ is a smooth function at finite g, it generates two different series expansions in g, depending on the range

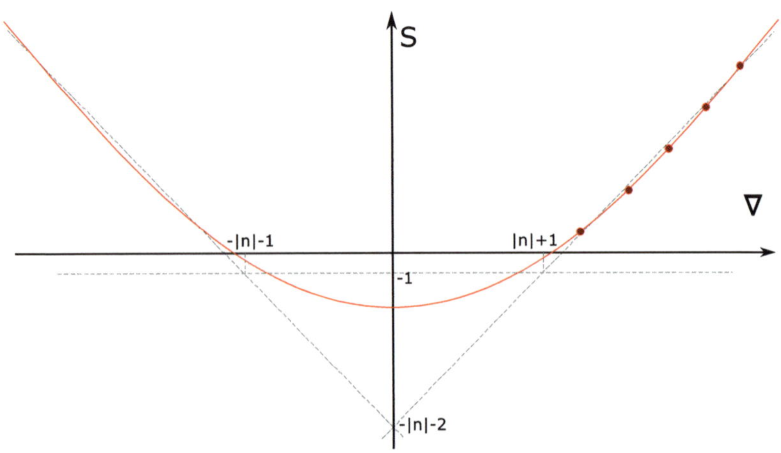

Fig. 13.1 Trajectory of the length-2 operator for conformal spin $n = S_2$ as a function of the full dimension Δ. The dots correspond to the physical operators with $S + n \in 2\mathbb{Z}_{\geq 0}$.

of Δ. Within this picture the BFKL expansion is the series expansion of $S(\Delta)$ for $|\Delta| < 1$ in powers of g^2. Note that all physical operators belong to the opposite region $\Delta > 1$ (the physical operator with the smallest protected dimension 2 is $\text{tr}(Z^2)$), and thus the equation (13.15) is only valid for $\Delta > 1$. At the same time for $|\Delta| < 1$ one can use (13.3) to get

$$S_{|\Delta|<1}(\Delta) = -1 + 4g^2 \left(-\psi\left(\frac{1+\Delta}{2}\right) - \psi\left(\frac{1-\Delta}{2}\right) + 2\psi(1) \right) + \mathcal{O}(g^4). \quad (13.16)$$

To understand how (13.15) and (13.16) are related to each other we have to complexify Δ and S. We explain below how this can be done using QSC at finite g, and the resulting function Re $S(\Delta)$ is presented in the Fig. 13.2. We see that the function $S(\Delta)$ has several sheets, which are connected through a branch cut. When the coupling goes to zero the branch points collide forming the singularity in the function $S(\Delta)$. The domains $|\Delta| < 1$ and $\Delta > 1$ become separated completely by the branch cut dividing the Δ plane into two parts. Thus one can say that (13.15) and (13.16) are the expansions of the two branches of the same function to the left and to the right from the quadratic cut going to infinity along the imaginary axis. This observation resolves the seeming paradox that the two functions are not the same. It also allows to make a prediction: the sum $S_{|\Delta|<1}(\Delta) + S_{\Delta>1}(\Delta)$ should be regular around $\Delta = 1$ at any order in g, as in this combination the branch cut cancels. Indeed, one can check that the simple poles in both functions around $\Delta = 1$ have residues opposite in sign and disappear in the sum at order g^2. At the higher orders the poles at $\Delta = 1$ become more and more severe, but this cancellation also can be verified explicitly to all known orders. This requirement gives non-trivial relations between the functions $S_{\Delta>1}(\Delta)$, obtained perturbatively

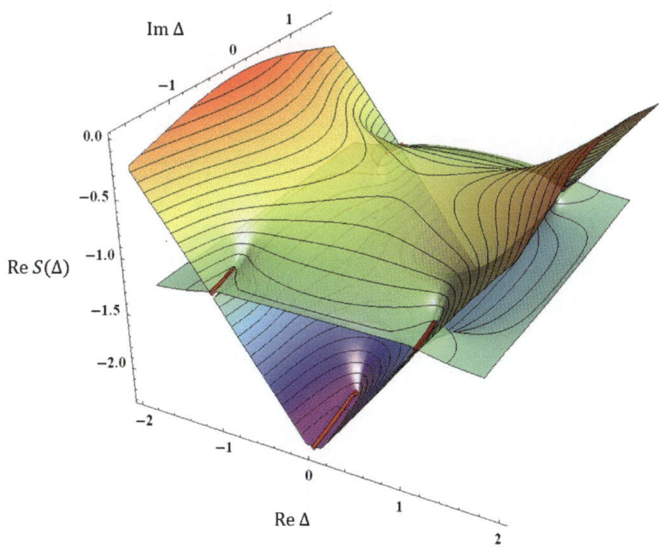

Fig. 13.2 Riemann surface of the function $S(\Delta)$ for twist-2 operators.

as an analytic continuation from the dimensions of physical operators, and the BFKL kernel eigenvalue $S_{|\Delta|<1}(\Delta)$.

Now we are ready to turn to non-zero conformal spin. Namely, non-zero conformal spin adds the derivative in the orthogonal direction to the operators (13.13)

$$\mathcal{O} = \text{tr} Z D_{+}^{S_1} \partial_{\perp}^{S_2} Z + (\text{permutations}). \tag{13.17}$$

The physical states now correspond to non-negative integer S_1 and S_2, whose sum is even. We follow the same strategy for (13.17) as for the case of zero conformal spin. Analogously, having the anomalous dimensions for the physical operators, we can build the analytic continuation in the spins S_1 and S_2 and identify them with the spin S and conformal spin n respectively. This analytic continuation is illustrated with the Fig. 13.1. The physical operators are designated with the dots on the operator trajectory. As in the case of zero conformal spin the exchange of the roles of Δ and $S = S_1$ allows us to reach the BFKL regime.

Having these analytic properties in mind, we understand how to apply the QSC to the study of the BFKL spectrum: we need to analytically continue to non-integer spin S and conformal spin n not only the anomalous dimensions, but the QSC itself: Q-functions, their asymptotics and analytic structure etc. In the next section we are going to start from the brief description of the QSC basics. Then we will use this setup to consider the calculation of the LO BFKL kernel eigenvalue with non-zero conformal spin by the QSC method.

13.2.3 *QSC approach to the BFKL spectrum of $\mathcal{N} = 4$ SYM*

We are going to present here the formulation of the QSC in terms of the Q-system and gluing conditions, the details of which can be found in [Gromov *et al.* (2015a); Gromov (2017); Alfimov *et al.* (2018)]. The algebraic part of the QSC framework is described as follows. For the $\mathcal{N} = 4$ SYM we have the system of 2^8 Q-functions, which are denoted as

$$\mathcal{Q}_{a_1,\ldots,a_n|i_1,\ldots,i_m}(u), \quad 1 \leq n, m \leq 4. \tag{13.18}$$

The Q-functions (13.18) have two groups of the indices: the a's are called "bosonic" and the i's are called "fermionic". They are antisymmetric with respect to the exchange of any pair of indices in these two groups. Not all of the 256 Q-functions in question are independent. They are subject to the set of the so-called Plücker's QQ-relations, which are written in [Gromov *et al.* (2015a)]

$$\begin{aligned} \mathcal{Q}_{A|I}\mathcal{Q}_{Aab|I} &= \mathcal{Q}^{+}_{Aa|I}\mathcal{Q}^{-}_{Ab|I} - \mathcal{Q}^{-}_{Aa|I}\mathcal{Q}^{+}_{Ab|I}, \\ \mathcal{Q}_{A|I}\mathcal{Q}_{A|Iij} &= \mathcal{Q}^{+}_{A|Ii}\mathcal{Q}^{-}_{A|Ij} - \mathcal{Q}^{-}_{Aa|Ii}\mathcal{Q}^{+}_{Ab|Ij}, \\ \mathcal{Q}_{Aa|I}\mathcal{Q}_{A|Ii} &= \mathcal{Q}^{+}_{Aa|Ii}\mathcal{Q}^{-}_{Ab|I} - \mathcal{Q}^{-}_{Aa|Ii}\mathcal{Q}^{+}_{A|I}, \end{aligned} \tag{13.19}$$

where A and I are multi-indices from the set $\{1, 2, 3, 4\}$.

The standard normalization is chosen to be

$$Q_{\emptyset|\emptyset}=1. \tag{13.20}$$

Then the structure of the QQ-relations allows to express the whole Q-system in terms of the "basic" set of 8 Q-functions $Q_{a|\emptyset}$, $a=1,\ldots,4$ and $Q_{\emptyset|i}$, $i=1,\ldots,4$.

Let us now describe two symmetries of the Q-system, which respect the QQ-relations (13.19).

- Imposing the so-called unimodularity condition

$$Q_{1234|1234}=1, \tag{13.21}$$

we are able to introduce the system of Hodge-dual Q-functions

$$Q^{a_1\ldots a_n|i_1\ldots i_m}\equiv(-1)^{nm}\epsilon^{b_{n+1}\ldots b_4 a_1\ldots a_n}\epsilon^{j_{m+1}\ldots j_4 i_1\ldots i_m}Q_{b_{n+1}\ldots b_4|j_{m+1}\ldots j_4}, \tag{13.22}$$

where there is no summation over the repeated indices, and which satisfy the same QQ-relations (13.19). The condition (13.21) allows to write down the following relations between the Q-functions with the lower and upper indices

$$\begin{aligned} Q^{a|i}Q_{b|j}&=-\delta^i_j, & Q^{a|i}Q_{b|i}&=-\delta^a_b,\\ Q^{a|\emptyset}&=(Q^{a|i})^+Q_{\emptyset|i}, & Q^{\emptyset|i}&=(Q^{a|i})^+Q_{a|\emptyset},\\ Q_{a|\emptyset}Q^{a|\emptyset}&=0, & Q_{\emptyset|i}Q^{\emptyset|i}&=0. \end{aligned} \tag{13.23}$$

- Another symmetry is called the H-symmetry. Its general form is given by

$$Q_{A|I}\to\sum_{|B|=|A|,|J|=|I|}(H_b^{[|A|-|I|]})^B_A(H_f^{[|A|-|I|]})^J_I Q_{B|J}, \tag{13.24}$$

where the sum goes over the repeated multi-indices. The definition of H^J_I is the following: $(H_f)^J_I\equiv(H_F(u))^{j_1}_{i_1}(H_F(u))^{j_2}_{i_2}\ldots(H_F(u))^{j_{|I|}}_{i_{|I|}}$ and the same for $(H_b)^B_A$ and $(H_B(u))^{b_k}_{a_k}$, $k=1,\ldots,|A|$ with the $H_{B,F}(u)$ being 4×4 i-periodic matrices. The unimodularity condition leads us to the restriction

$$\det H_B(u)\det H_F(u)=1. \tag{13.25}$$

To proceed with the calculation of the spectrum of $\mathcal{N}=4$ SYM we have to endow the described Q-system with the analytic structure. To start with, we designate the basic Q-functions with this structure as $\mathbf{P}_a$ (same as $Q_{a|\emptyset}$), $\mathbf{P}^a$ ($Q^{a|\emptyset}$), $\mathbf{Q}_i$ ($Q_{\emptyset|i}$) and $\mathbf{Q}^i$ ($Q^{\emptyset|i}$). One can find the asymptotics of the Q-functions in [Gromov (2017)]

$$\mathbf{P}_a\simeq A_a u^{-\tilde{M}_a},\quad \mathbf{P}^a\simeq A^a u^{\tilde{M}_a-1},\quad \mathbf{Q}_i\simeq B_i u^{\hat{M}_i-1},\quad \mathbf{Q}^i\simeq B^i u^{-\hat{M}_i}, \tag{13.26}$$

Fig. 13.3 Analytic structure of the **P**- and **Q**-functions on their defining sheet.

where $\tilde{M}_a$, $a = 1, \dots, 4$ and $\hat{M}_i$, $i = 1, \dots, 4$ are expressed in terms of Cartan charges of PSU(2, 2|4) (i.e. quantum numbers of the states)

$$\tilde{M}_a = \left\{ \frac{J_1 + J_2 - J_3}{2} + 1, \frac{J_1 - J_2 + J_3}{2}, \frac{-J_1 + J_2 + J_3}{2} + 1, \frac{-J_1 - J_2 - J_3}{2} \right\},$$
$$\hat{M}_i = \left\{ \frac{1}{2}(\Delta - S_+ + 2), \frac{1}{2}(\Delta + S_+), \frac{1}{2}(-\Delta - S_- + 2), \frac{1}{2}(-\Delta + S_-) \right\},$$
$$(13.27)$$

where $S_\pm = S_1 \pm S_2$.

As the Q-system is generated by the set of 8 basic Q-functions, we first ascribe them the analytic structure dictated by the classical limit of the Q-functions [Gromov (2017)]. The minimal choice of the cut structure consistent with the asymptotics (13.26) is presented in the Fig. 13.3. Then, according to the (13.19) we can generate two versions of the Q-system: upper half plane and lower half plane analytic (UHPA and LHPA respectively). In what follows we will define by $\mathcal{Q}_{a|i}$ the UHPA solution of the one of the QQ-relations from (13.19)

$$\mathcal{Q}^+_{a|i} - \mathcal{Q}^-_{a|i} = \mathbf{P}_a \mathbf{Q}_i \tag{13.28}$$

with the large u asymptotic

$$\mathcal{Q}_{a|i} \simeq -i \frac{A_a B_i}{-\tilde{M}_a + \hat{M}_i} u^{-\tilde{M}_a + \hat{M}_i}. \tag{13.29}$$

Substitution of the formulas $\mathbf{Q}_i = -\mathcal{Q}^+_{a|i}\mathbf{P}^a$ or $\mathbf{P}_a = -\mathcal{Q}_{a|i}\mathbf{Q}^i$ into (13.28) allows to fix the products of the coefficients of the leading asymptotics of the **P**- and **Q**-functions and for $a_0, i_0 = 1, \dots, 4$ we get

$$A_{a_0} A^{a_0} = i \frac{\prod_{j=1}^4 \left(\tilde{M}_{a_0} - \hat{M}_j \right)}{\prod_{\substack{b=1 \\ b \neq a_0}} \left(\tilde{M}_{a_0} - \tilde{M}_b \right)}, \quad B_{i_0} B^{i_0} = -i \frac{\prod_{a=1}^4 \left(\hat{M}_{i_0} - \tilde{M}_a \right)}{\prod_{\substack{j=1 \\ j \neq i_0}}^4 \left(\hat{M}_{i_0} - \hat{M}_j \right)}, \tag{13.30}$$

where there is no summation over the repeated indices a_0 and i_0.

An important consequence of the formula $\mathbf{Q}_i = -\mathcal{Q}_{a|i}\mathbf{P}^a$ combined with the formula (13.28) is the existence of a 4th order Baxter equation for the functions

$\mathbf{Q}_i$, $i=1,\ldots,4$ (see [Alfimov *et al.* (2015)] for the derivation)

$$\mathbf{Q}_i^{[+4]} - \mathbf{Q}_i^{[+2]}\left[D_1 - \mathbf{P}_a^{[+2]}\mathbf{P}^{a[+4]}D_0\right] + \mathbf{Q}_i\left[D_2 - \mathbf{Q}_a\mathbf{P}^{a[+2]}D_1 + \mathbf{P}_a\mathbf{P}^{a[+4]}D_0\right]$$
$$- \mathbf{Q}_i^{[-2]}\left[\bar{D}_1 + \mathbf{P}_a^{[-2]}\mathbf{P}^{a[-4]}\bar{D}_0\right] + \mathbf{Q}_i^{[-4]} = 0, \quad (13.31)$$

where the functions D_0, D_1 and D_2 are given by

$$D_0 = \begin{vmatrix} \mathbf{P}^{1[+2]} & \mathbf{P}^{2[+2]} & \mathbf{P}^{3[+2]} & \mathbf{P}^{4[+2]} \\ \mathbf{P}^{1} & \mathbf{P}^{2} & \mathbf{P}^{3} & \mathbf{P}^{4} \\ \mathbf{P}^{1[-2]} & \mathbf{P}^{2[-2]} & \mathbf{P}^{3[-2]} & \mathbf{P}^{4[-2]} \\ \mathbf{P}^{1[-4]} & \mathbf{P}^{2[-4]} & \mathbf{P}^{3[-4]} & \mathbf{P}^{4[-4]} \end{vmatrix}, \quad D_1 = \begin{vmatrix} \mathbf{P}^{1[+4]} & \mathbf{P}^{2[+4]} & \mathbf{P}^{3[+4]} & \mathbf{P}^{4[+4]} \\ \mathbf{P}^{1} & \mathbf{P}^{2} & \mathbf{P}^{3} & \mathbf{P}^{4} \\ \mathbf{P}^{1[-2]} & \mathbf{P}^{2[-2]} & \mathbf{P}^{3[-2]} & \mathbf{P}^{4[-2]} \\ \mathbf{P}^{1[-4]} & \mathbf{P}^{2[-4]} & \mathbf{P}^{3[-4]} & \mathbf{P}^{4[-4]} \end{vmatrix},$$

$$D_2 = \begin{vmatrix} \mathbf{P}^{1[+4]} & \mathbf{P}^{2[+4]} & \mathbf{P}^{3[+4]} & \mathbf{P}^{4[+4]} \\ \mathbf{P}^{1[+2]} & \mathbf{P}^{2[+2]} & \mathbf{P}^{3[+2]} & \mathbf{P}^{4[+2]} \\ \mathbf{P}^{1[-2]} & \mathbf{P}^{2[-2]} & \mathbf{P}^{3[-2]} & \mathbf{P}^{4[-2]} \\ \mathbf{P}^{1[-4]} & \mathbf{P}^{2[-4]} & \mathbf{P}^{3[-4]} & \mathbf{P}^{4[-4]} \end{vmatrix}, \quad (13.32)$$

while the bars over $\bar{D}_1$ and $\bar{D}_2$ are understood as the complex conjugation $\bar{f}(u) = \overline{f(\bar{u})}$ of the functions defined above. The same equation for $\mathbf{Q}^i$, $i=1,\ldots,4$ is valid, if we turn the functions $\mathbf{P}^a$ into $\mathbf{P}_a$ for $a=1,\ldots,4$.

One can see from (13.31) that if we are far away (high or low enough) from the real axis in the complex plane, the cut structures (see the Fig. 13.3) of $\mathbf{P}$- and $\mathbf{Q}$-functions do not contradict each other. But as we approach the real axis, this is no longer the case and we have to cross the cut. This makes the QQ-relations ambiguous in the vicinity of the cuts, as the shifts by $\pm i/2$ may cross the cut and one may ask which contour to use to reach $u \pm i/2$ from the point u or, in other words, which branch of the multi-valued function to use.

The way to resolve the analytic continuation ambiguity is to interpret the values of the $\mathbf{Q}$-function in the upper half plane, far enough from the cuts, as a $\mathbf{Q}$-function with the upper index $\mathbf{Q}^i$ and in the lower half plane the same $\mathbf{Q}$-function should obey the QQ-relations as if it was a function with lower index $\mathbf{Q}_i^{\uparrow}$. I.e. in the lower half plane this function satisfies (13.31), whereas in the upper half plane it satisfies the same equation with $\mathbf{P}_a$ interchanged with $\mathbf{P}^a$. The $\uparrow$ is added to indicate that $\mathbf{Q}_i^{\uparrow}$ does not have cuts in the lower half plane. To close the system of equations we notice that there is yet another way to build the set of functions $\mathbf{Q}_i^{\uparrow}$, satisfying (13.31) and having no cuts in the lower half plane. Starting from $\mathbf{Q}^i$, which have no cuts in the upper half plane, and $\mathbf{P}^a$, one can obtain $\mathcal{Q}^{a|i}$ from the finite difference equation and then build $\mathbf{Q}_i^{\downarrow}$ using QQ-relations. Then the complex conjugate of $\mathbf{Q}_i^{\downarrow}$ will satisfy (13.31) and will have no cuts in the lower half plane.

Indeed, due to the property, valid for real S_1, S_2 and Δ we can always assume that

$$\bar{\mathbf{P}}_a = C_a^b\mathbf{P}_b, \quad \bar{\mathbf{P}}^a = -C_b^a\mathbf{P}^b, \quad C = \mathrm{diag}\{1,1,-1,-1\} \quad (13.33)$$

due to the H-symmetry. Thus if we complex conjugate the Baxter equation (13.31), we get the same equation for $\bar{\mathbf{Q}}_j$. This implies that $\mathbf{Q}_i^{\downarrow}(u)$ is a linear combination of $\bar{\mathbf{Q}}_j$ with regular coefficients

$$\mathbf{Q}^i(u) = M^{ij}(u)\bar{\mathbf{Q}}_j(u), \quad \mathbf{Q}_i(u) = \left(M^{-t}\right)_{ij}(u)\bar{\mathbf{Q}}^j(u), \tag{13.34}$$

where $-t$ denotes the inversion and transposition of the matrix. Technically it will be more convenient to work with short cuts $[-2g, 2g]$. We can connect the branch points of $\mathbf{Q}$ the way we like, but this would create an infinite ladder of cuts in the lower half plane. It will also modify the notion of the conjugate function, as the conjugation now will also involve the analytic continuation under the cut. So we conclude that in the short cut conventions the gluing condition reads as

$$\tilde{\mathbf{Q}}^i(u) = M^{ij}(u)\bar{\mathbf{Q}}_j(u), \quad \tilde{\mathbf{Q}}_i(u) = \left(M^{-t}\right)_{ij}(u)\bar{\mathbf{Q}}^j(u), \tag{13.35}$$

which we will use further. Let us now list the properties of $M^{ij}(u)$, which is called the gluing matrix. The details of the derivation of these properties can be found in [Alfimov *et al.* (2018)]. They are:

- $M^{ij}(u)$ is an i-periodic matrix of H-transformation.
- $M^{ij}(u)$ is analytic in the whole complex plane.
- $M^{ij}(u)$ is hermitian as a function

$$\bar{M}^{ij}(u) = M^{ji}(u). \tag{13.36}$$

There exist another additional symmetry of the $\mathbf{P}$-functions. The $\mathbf{P}$-functions of the states with the Cartan charges

$$\begin{aligned}\tilde{M}_a &= \{2, 1, 0, -1\},\\ \hat{M}_i &= \left\{\frac{1}{2}(\Delta - S_+ + 2), \frac{1}{2}(\Delta + S_+), \frac{1}{2}(-\Delta - S_- + 2), \frac{1}{2}(-\Delta + S_-)\right\},\end{aligned} \tag{13.37}$$

where $S_\pm \equiv S_1 \pm S_2$, have the certain parity

$$\mathbf{P}_a(-u) = (-1)^{a+1}\mathbf{P}_a(u), \quad \mathbf{P}^a(-u) = (-1)^a\mathbf{P}^a(u). \tag{13.38}$$

The conjugation symmetry and the symmetry above (13.38) lead to two additional constraints [Alfimov *et al.* (2018)] on the gluing matrix. Let us first concentrate on the physical states, when S_1 and S_2 are integer and have the same parity.[5] As it follows from the power-like large u asymptotics of the $\mathbf{Q}$-functions in this case, the only possible ansatz for the gluing matrix is a constant matrix. In [Gromov *et al.* (2015a); Alfimov *et al.* (2018)] it was shown that for the physical length-2 states

[5]This is dictated by the cyclicity condition for the states in the $\mathfrak{sl}(2)$ Heisenberg spin chain.

with the charges (13.37) from the above-mentioned properties of the gluing matrix and two additional constraints mentioned earlier it follows that

$$M^{ij}\left(e^{i\pi\left(\hat{M}_i-\hat{M}_j\right)}+1\right)=0. \tag{13.39}$$

From (13.39), we immediately see that only when the difference between the charges is an odd integer, M^{ij} is non-zero. It is the case only for

$$\begin{aligned}\hat{M}_1-\hat{M}_2&=-S_1-S_2+1,\\ \hat{M}_3-\hat{M}_4&=-S_1+S_2+1.\end{aligned} \tag{13.40}$$

Therefore, taking into account the hermiticity of the gluing matrix, for integer S_1 and S_2, that have the same parity[6] (i.e. for the physical states) the equations (13.39) lead to the gluing matrix

$$M^{ij}=\begin{pmatrix}0 & M^{12} & 0 & 0\\ \bar{M}^{12} & 0 & 0 & 0\\ 0 & 0 & 0 & M^{34}\\ 0 & 0 & \bar{M}^{34} & 0\end{pmatrix}. \tag{13.41}$$

In the case of at least one of the spins S_1 and S_2 being non-integer, we see that because all differences $\hat{M}_i-\hat{M}_j$ are non-integer the equations (13.39) can have only a zero matrix as a solution, if this matrix is assumed to be constant. This leads us to the conclusion that the gluing matrix cannot be constant anymore for non-integer spins. The minimal way to do this keeping it i-periodic would be to add exponential contributions and get the following gluing matrix

$$M=\begin{pmatrix}M_1^{11} & M_1^{12} & M_1^{13} & M_1^{14}\\ \bar{M}_1^{12} & 0 & 0 & 0\\ \bar{M}_1^{13} & 0 & M_1^{33} & M_1^{34}\\ \bar{M}_1^{14} & 0 & \bar{M}_1^{34} & M_1^{44}\end{pmatrix}+$$

$$+\begin{pmatrix}0 & 0 & M_2^{13} & M_2^{14}\\ 0 & 0 & 0 & 0\\ \bar{M}_2^{13} & 0 & 0 & 0\\ \bar{M}_2^{14} & 0 & 0 & 0\end{pmatrix}e^{2\pi u}+\begin{pmatrix}0 & 0 & M_3^{13} & M_3^{14}\\ 0 & 0 & 0 & 0\\ \bar{M}_3^{13} & 0 & 0 & 0\\ \bar{M}_3^{14} & 0 & 0 & 0\end{pmatrix}e^{-2\pi u}. \tag{13.42}$$

To summarize, the key observation here is that if we want to consider non-physical states with S_1 or S_2 being non-integer, this inevitably requires a modification of the gluing conditions. From now on let us use the notations for the spins coming from BFKL physics $S_1 = S$ and $S_2 = n$.

In the next Subsection we will briefly describe how the QQ-relations together with the gluing condition (13.35) can be used to compute numerically the function $S(\Delta, n)$.

[6]At one loop this is dictated by the cyclicity condition for the states in the $\mathfrak{sl}(2)$ Heisenberg spin chain, appearing in the perturbation theory.

13.2.4 *Numerical solution*

The QQ-system together with the gluing conditions allows for the efficient numerical algorithm, developed in [Gromov *et al.* (2016)]. This method was described in details with the *Mathematica* code attached in a recent review [Gromov (2017)]. Here we describe the main steps very briefly.

First, one notices that the $\mathbf{P}_a$ and $\mathbf{P}^a$ functions, $a = 1, \ldots, 4$, which have only one short cut $[-2g, 2g]$ on the main sheet of the Riemann surface, can be parametrized very efficiently as rapidly converging series expansions in powers of $1/x(u)$, where $x(u) = (u + \sqrt{u-2g}\sqrt{u+2g})/(2g)$. After that one can recover the whole Q-system in terms of these expansion coefficients. For that one first uses

$$\mathcal{Q}^+_{a|i} - \mathcal{Q}^-_{a|i} = -\mathcal{Q}^+_{b|i}\mathbf{P}_a\mathbf{P}^b, \tag{13.43}$$

which follows directly from (13.28) and $\mathbf{Q}_i = -\mathcal{Q}_{a|i}\mathbf{P}^i$, to solve for $\mathcal{Q}_{a|i}$. After that one finds $\mathbf{Q}_i$ and $\tilde{\mathbf{Q}}_i$ from

$$\mathbf{Q}_i = -\mathcal{Q}^+_{a|i}\mathbf{P}^a, \quad \tilde{\mathbf{Q}}_i = -\mathcal{Q}^+_{a|i}\tilde{\mathbf{P}}^a. \tag{13.44}$$

Note that this involves $\tilde{\mathbf{P}}^a$, which is the same series as $\mathbf{P}^a$, but with $x \to 1/x$ for $u \in [-2g, 2g]$. In this way we reconstruct both $\mathbf{Q}_i$ and $\tilde{\mathbf{Q}}_i$ in terms of the expansion coefficients in $\mathbf{P}_a$ and $\mathbf{P}^a$. Finally, one fixes these coefficients from the gluing condition (13.35).

In practice for the numerical purposes one truncates the series expansion in $\mathbf{P}$. In this case the gluing condition cannot be satisfied exactly. The strategy is to minimize the discrepancy in the gluing condition by adjusting the coefficients numerically with a variation of a Newton method.

Application of the numerical procedure allows to calculate the dependence of the intercept function $j = S(\Delta = 0, n = 0) + 2$ for zero conformal spin [Gromov *et al.* (2014b)] on the 't Hooft coupling λ, which is drawn on the Fig. 13.4. By fitting the polynomial dependence of this quantity on the coupling, we obtain several first coefficients

$$\begin{aligned} j(\lambda) &= 1 + 0.07023049277268284\,\lambda - 0.00337167607361\,\lambda^2 \\ &\quad - 0.00064579607573\,\lambda^3 + 0.0002512619258\,\lambda^4 + \mathcal{O}(\lambda^5). \end{aligned} \tag{13.45}$$

Comparing the first two coefficients of (13.45), we find complete agreement with the LO (13.3) and NLO BFKL (13.5) kernel eigenvalues calculated at $\Delta = 0$ and $n = 0$.[7]

Moreover, we can use our numerical algorithm to calculate not only the intercept functions, but also the BFKL kernel eigenvalues at different orders in g. Namely, in [Gromov *et al.* (2015b)] from fitting the numerical values of the spin $S(\Delta, n)$ for $\Delta = 0.45$ and $n = 0$ for different values of the coupling constant one was able to

[7] After renormalizing these results according to the change of the expansions parameter from g^2 to λ.

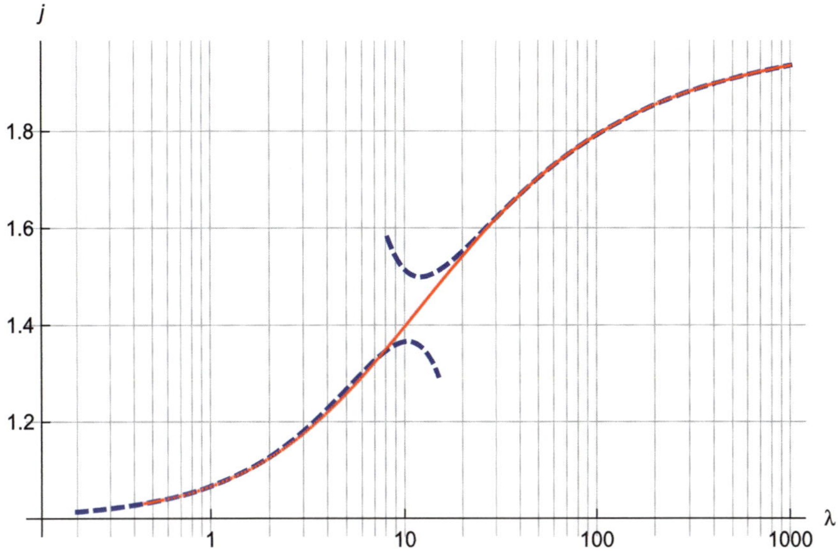

Fig. 13.4 The dependence of the intercept function j on the 't Hooft coupling λ in the logarithmic scale. Data obtained from the numerical procedure are depicted in red. It represents a perfect interpolation between the weak [Kotikov and Lipatov (2003); Brower *et al.* (2007)] and strong [Costa *et al.* (2012); Kotikov and Lipatov (2013); Brower *et al.* (2013); Gromov *et al.* (2014b)] coupling results (blue dashes).

Table 13.1 Numerical values of the BFKL kernel eigenvalue at different orders.

Order	Value	Error
N^2LO	10774.635818847176637957593127192456995929170948057653783424533229	10^{-61}
N^3LO	$-366393.2052053917038937903507478544549935531959333919163403836$	10^{-56}
N^4LO	$1.332736355681126915694044310369828561521940588979476878854 \times 10^7$	10^{-51}
N^5LO	$-4.921740136657916500913955552075070060721450958436559876 \times 10^8$	10^{-47}

extract the numerical values of the coefficients $\chi_{\mathrm{N}^k\mathrm{LO}}(\Delta = 0.45)$ from (13.1), which are listed in the Table 13.1.

The first line of the Table 13.1 contains the numerical value of the NNLO BFKL kernel eigenvalue for $\Delta = 0.45$, which was computed analytically for the first time in [Gromov *et al.* (2015b)] by the QSC method. In that work it was shown that the calculated numerical value in the Table 13.1 coincides with the value of the analytic result for $\Delta = 0.45$ with numerical accuracy 10^{-61}.

In addition, the numerical algorithm can be used to explore the spectrum for non-zero values of conformal spin n (to see the algorithm for different values of Δ, n and g at work one can use the *Mathematica* file `code_for_arxiv.nb` from the [Alfimov *et al.* (2018)] arXiv submission). One can find the plots of the spin $S(0, n)$, which differs from the intercept function $j(n)$ by the additive constant 2, for different (even non-integer) values of the conformal spin n in the Fig. 13.5.

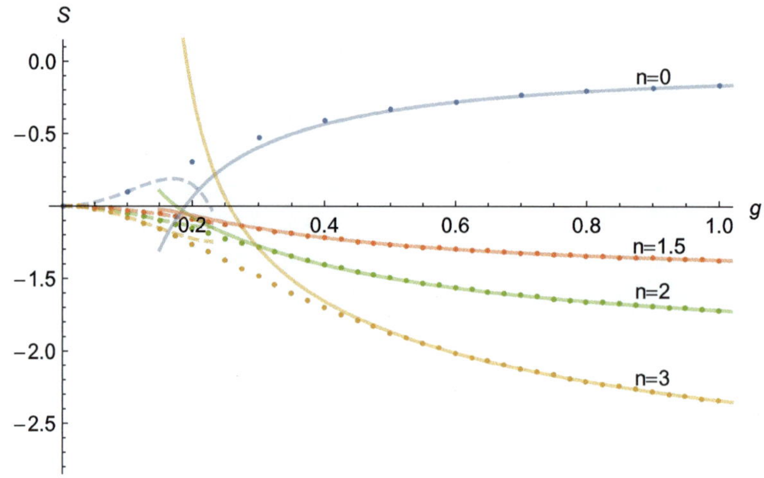

Fig. 13.5 The dependence of the numerical values of the spin $S(0,n)$ on the coupling constant g, calculated for the conformal spins $n = 0$, $n = 3/2$, $n = 2$ and $n = 3$ (dots). The dashed lines represent the weak coupling expansion and the continuous lines represent the strong coupling expansion of the spin obtained in [Alfimov *et al.* (2018)].

To conclude, the numerical method of [Gromov *et al.* (2016)] allowed to obtain the BFKL kernel eigenvalue non-perturbatively with huge precision.

A number of analytic methods was developed to solve the QSC at small g [Marboe and Volin (2015); Marboe *et al.* (2015); Gromov *et al.* (2015b); Marboe and Volin (2018a,b)]. Unfortunately, those methods rely on a particular basis of η-functions, which works very well for the local operators for the BFKL eigenvalues at some integer values of Δ, but are not sufficient in general. In the next section we will use an alternative method to obtain the LO BFKL kernel eigenvalue analytically. Recently, new very promising methods were developed in [Lee and Onishchenko (2018, 2019a,b)], based on the Mellin transformations, which could allow for a systematic analytic calculation of $S(\Delta)$ order by order in g^2 for generic Δ.

13.2.5 *Analytic results from QSC*

Apart from the numerical results mentioned in the previous Section, a number of analytic results related to the BFKL spectrum was obtained recently using the QSC methods in [Gromov *et al.* (2014b); Alfimov *et al.* (2015); Gromov *et al.* (2015b); Gromov and Levkovich-Maslyuk (2016a,b); Alfimov *et al.* (2018)]. To mention a few:

- NNLO BFKL kernel eigenvalue for the conformal spin $n = 0$.
- Intercept function for arbitrary conformal spin n up to NNLO order and partial result at NNNLO order.

- Strong coupling expansion of the intercept function for arbitrary value of the conformal spin n.
- Slope-to-slope function in the BPS point $\Delta=2$, $S=0$ and $n=0$ at all loops.
- Slope-to-intercept $dS(\Delta,n)/dn$ and the curvature $d^2S(\Delta,n)/d\Delta^2$ functions in the BPS point $\Delta=0$, $S=-1$ and $n=1$ at all loops.

Here we demonstrate the power of the QSC method deriving the leading order Faddeev-Korchemsky Baxter equation with non-zero conformal spin for the Lipatov spin chain analytically.

As explained above, we need to study the regime of the anomalous dimensions of the twist-2 operators, when the coupling constant $g^2\to 0$, such that $S\equiv S_1(\Delta,g)$ becomes $-1+\mathcal{O}(g^2)$. According to the discussion in the previous section this is the case for $|\Delta|<1+|n|$. In other words one can say that we keep the ratio $\Lambda\equiv g^2/(S+1)$ finite, whereas the combination $w\equiv S+1=\mathcal{O}(g^2)$ can be used as a small expansion parameter. The second spin $S_2=n$ plays the role of a parameter.

In order to reproduce the Faddeev-Korchemsky Baxter equation we are going to utilise the subset of the QQ-relations known as $\mathbf{P}\mu$-system. The $\mathbf{P}\mu$-system is represented by the functions $\mathbf{P}_a(u)$, $\mathbf{P}^a(u)$, which we introduced before, and of an anti-symmetric matrix $\mu_{ab}(u)$ (see [Gromov *et al.* (2014a, 2015a)] for the detailed description). They satisfy the following equations

$$\tilde{\mu}_{ab}-\mu_{ab}=\mathbf{P}_a\tilde{\mathbf{P}}_b-\mathbf{P}_b\tilde{\mathbf{P}}_a,\qquad \tilde{\mathbf{P}}_a=\mu_{ab}\mathbf{P}^b,$$
$$\tilde{\mu}^{ab}-\mu^{ab}=\mathbf{P}^a\tilde{\mathbf{P}}^b-\mathbf{P}^b\tilde{\mathbf{P}}^a,\qquad \tilde{\mathbf{P}}^a=\mu^{ab}\mathbf{P}_b,$$
$$\mathbf{P}_a\mathbf{P}^a=0,\quad \mu_{ab}\mu^{bc}=\delta_a^c,\qquad \tilde{\mu}_{ab}=\mu_{ab}^{++},\quad \tilde{\mu}^{ab}=\mu^{ab++}. \tag{13.46}$$

Before proceeding we remind of couple of notations. Our notation for the BFKL scaling parameter is $w=S+1$. It is also convenient to use the notation $\Lambda=g^2/w$.

To start solving the $\mathbf{P}\mu$-system in the BFKL regime we have to determine the scaling of the $\mathbf{P}$-, $\mathbf{Q}$- and μ-functions in the small w limit. In what follows we are going to use the arguments from [Alfimov *et al.* (2015)]. Thus as from (13.30) for the length-2 state in question (13.17) in the BFKL limit $A_aA^a=\mathcal{O}(w^0)$ for $a=1,\ldots,4$ and $B_iB^i=\mathcal{O}(w^0)$, these functions can be chosen to scale as w^0

$$\mathbf{P}_a=\mathbf{P}_a^{(0)}+\mathcal{O}(w),\qquad \mathbf{P}^a=\mathbf{P}^{(0)a}+\mathcal{O}(w) \tag{13.47}$$
$$\mathbf{Q}_i=\mathbf{Q}_i^{(0)}+\mathcal{O}(w),\qquad \mathbf{Q}^i=\mathbf{Q}^{(0)i}+\mathcal{O}(w) \tag{13.48}$$

and the μ-functions scale as w^{-2}

$$\mu_{ab}=w^{-2}\left(\mu_{ab}^{(0)}+\mathcal{O}(w)\right),\quad \mu^{ab}=w^{-2}\left(\mu^{(0)ab}+\mathcal{O}(w)\right). \tag{13.49}$$

Additionally, as all the $\mathbf{P}$-functions for the length-2 states being considered possess the certain parity from the $\mathbf{P}\mu$-system equations (13.46) we can conclude that the functions $\mu_{ab}^{+}(u)$ have the certain parity.

Let us restrict ourselves from now on in this Section to the case of integer conformal spin $S_2=n$. We know that the $\mathbf{P}$-functions have only one cut on one of the sheets, therefore they can be written as a series in the Zhukovsky variable $x(u) = (u+\sqrt{u^2 - 4\Lambda w})/(2\sqrt{\Lambda w})$.[8] Then, we are also allowed to apply the certain H-transformation to the **P**-functions, which do not alter their asymptotics and parity. Applying this transformation, we can set the coefficients $A_1 = A_2 = -A^3 = A^4 = 1$ and some other coeficients in the series in $x(u)$ to 0. Thus, we arrive to the formulas in the LO

$$\mathbf{P}_1^{(0)} = \frac{1}{u^2}, \quad \mathbf{P}_2^{(0)} = \frac{1}{u}, \quad \mathbf{P}_3^{(0)} = A_3^{(0)}, \quad \mathbf{P}_4^{(0)} = A_4^{(0)} u + \frac{c_{4,1}^{(1)}}{\Lambda u},$$
$$\mathbf{P}^{(0)1} = A^{1(0)} u + \frac{c^{1,1(1)}}{\Lambda u}, \quad \mathbf{P}^{(0)2} = A^{2(0)}, \quad \mathbf{P}^{(0)3} = -\frac{1}{u}, \quad \mathbf{P}^{(0)4} = \frac{1}{u^2}, \tag{13.50}$$

where

$$A_3^{(0)} = \frac{((\Delta - n)^2 - 1)((\Delta + n)^2 - 9)}{32i},$$
$$A_4^{(0)} = \frac{((\Delta + n)^2 - 1)((\Delta - n)^2 - 25)}{96i} \tag{13.51}$$

originate from the expansion of (13.30) at small w and $c_{4,1}^{(1)}$ and $c^{1,1(1)}$ are some yet unknown coefficients.

The situation with the asymptotics of the μ-functions is more subtle. For non-zero S it appears that the asymptotics of the μ cannot be power-like anymore [Gromov and Sizov (2014)] and the minimal modification of the leading asymptotics of the μ-functions with the lower indices at $u \to \pm\infty$ is

$$\mu_{ab} \sim (u^{-S-1}, u^{-S}, u^{-S+1}, u^{-S+1}, u^{-S+2}, u^{-S+3}) e^{2\pi|u|}, \tag{13.52}$$

while the μ-functions with the upper indices have the same asymptotics but in the reverse order. The fact that the asymptotics should be modified as (13.52) can be derived from the fact that the gluing matrix (13.42) asymptotically receives additional exponential terms. In the work [Alfimov *et al.* (2018)] it was shown, that the correct ansatz for the μ-functions in the LO would be polynomials of the powers consistent with the asymptotics (13.52) for $S = -1 + w$ and w close to 0. We do not write this ansatz explicitly as it is not important for further discussion.

Substituting (13.50) and the ansatz for the μ-functions into (13.46) in the LO and from the analyticity properties of the **P**- and μ-functions we fix the constants $c^{1,1(1)}$ and $c_{4,1}^{(1)}$

$$c^{1,1(1)} = -c_{4,1}^{(1)}, \quad c_{4,1}^{(1)} = -\frac{i\Lambda}{96}\left((\Delta^2 - 1)^2 - 2(\Delta^2 + 1)n^2 + n^4\right) \tag{13.53}$$

and the μ-functions in the LO up to an overall constant (we do not write them explicitly as they are not relevant for further discussion).

[8]Here we rewrote the Zhukovsky variable using $g^2 = \Lambda w$.

After the substitution of the LO **P**-functions (13.50) together with (13.53) into (13.31) in the LO (for the same result with the zero conformal spin see [Alfimov *et al.* (2015)]), one can derive that this 4th order Baxter equation factorizes

$$\left[(u+2i)^2 D + (u-2i)^2 D^{-1} - 2u^2 - \frac{17-(\Delta+n)^2}{4}\right] \times \\ \times \left[D + D^{-1} - 2 - \frac{1-(\Delta-n)^2}{4u^2}\right] \mathbf{Q}_j^{(0)} = 0, \tag{13.54}$$

where $D = e^{i\partial_u}$ is the shift operator (the same equation with n replaced by $-n$ is true for $\mathbf{Q}^{(0)j}$, $j = 1, \ldots, 4$). Thus, two out of four linearly independent solutions of the 4th order Baxter equation can be found by solving the second order Baxter equation

$$\mathbf{Q}^{(0)++} + \mathbf{Q}^{(0)--} + \left(-2 + \frac{(\Delta-n)^2 - 1}{4u^2}\right) \mathbf{Q}^{(0)} = 0. \tag{13.55}$$

Redefining the Q-function to be $Q = \frac{\mathbf{Q}_j}{u^2}$ we exactly reproduce the Baxter equation for the $SL(2,\mathbb{C})$ spin chain [Faddeev and Korchemsky (1995); De Vega and Lipatov (2001); Derkachov *et al.* (2001, 2003, 2002)] from the QSC for $\mathcal{N} = 4$ SYM which is a highly non-trivial test of the all-loop integrability of this theory. This derivation established the desired connection between the integrability in the 4D gauge theory in the BFKL limit and the integrability of the $\mathcal{N} = 4$ SYM. Having this result, one can go further and by using QSC explore many more quantities, such as the NNLO BFKL kernel eigenvalue, numerical twist-2 and length-2 operator trajectories, intercept function, slope, curvature and slope-to-intercept function etc. (see [Gromov *et al.* (2016, 2015b); Alfimov *et al.* (2018)] and references therein).

13.3 Fishnet and BFKL

Above we described how the integrability, discovered in the BFKL regime of the QCD, can be understood as a part of a more general integrable structure of $\mathcal{N} = 4$ SYM. In fact the data coming from the BFKL regime have played an essential role in fixing the dressing phase of the Beisert-Staudacher equations, which eventually resulted in our current understanding of integrability in this model.

Another extremely popular recent topic – the Sachdev-Ye-Kitaev (SYK) model also has many similarities at the technical level with the problem of the BFKL kernel diagonalization [Gross and Rosenhaus (2017); Klebanov and Tarnopolsky (2017); Murugan *et al.* (2017); Kitaev (2015)] (see also a recent review dedicated to this topic [Rosenhaus (2019)]).

In this section, we will describe yet another example of CFT tightly related to the LO BFKL physics – the Fishnet CFT at any dimension (FCFT_D). We will explain its Feynman graph content (in the planar approximation big graphs have the shape of a regular square lattice – at the origin of the name "fishnet") and reveal the origins of its integrability described in terms of the conformal $SO(1,D+1)$ spin chain. We will demonstrate how, in the particular case of $D=2$ and zero conformal spin, the problem of computing the anomalous dimensions of the simplest operators, described by so-called wheel graphs, boils down to the calculation of the spectrum of Lipatov's multi-Reggeon Hamiltonian.

13.3.1 *Fishnet integrable model*

The D–dimensional fishnet biscalar conformal field theory (dubbed usually as fishnet CFT, or FCFT$_D$) is defined by the Lagrangian [Kazakov and Olivucci (2018)]

$$\mathcal{L}_d = N_c\,\mathrm{tr}[\bar{X}\,(-\partial_\mu\partial^\mu)^{\frac{D}{4}-\omega}\,X + \bar{Z}\,(-\partial_\mu\partial^\mu)^{\frac{D}{4}+\omega}\,Z + (4\pi)^{\frac{D}{2}}\xi^2\bar{X}\bar{Z}XZ], \quad (13.56)$$

where X, Z are complex $N_c \times N_c$ matrix fields and $\bar{X} \equiv X^\dagger, \bar{Z} \equiv Z^\dagger$ are their Hermitian conjugates. The differential operator in an arbitrary power is defined in a standard way, as an integral operator. The action (13.56) should be supplemented with certain double-trace counterterms described in [Fokken *et al.* (2014); Sieg and Wilhelm (2016); Grabner *et al.* (2018); Gromov *et al.* (2019)] (see also [Kazakov (2018)] and references therein).

The FCFT$_4$ case of the model (in $D=4$ dimensions), for the particular "isotropic" case $\omega=0$, has a local Lagrangian. It was proposed in [Gurdogan and Kazakov (2016)] as a specific double scaling limit of the γ-deformed $\mathcal{N}=4$ SYM theory, combining strong (imaginary) deformation and weak coupling. The effective coupling constant is $\xi^2 = g^2_{\mathrm{YM}}N_c e^{-i\gamma_3}/(4\pi)^2$, where the 't Hooft coupling $\lambda = g^2_{YM}N_c \to 0$ and the deformations parameter $\gamma_3 \to i\infty$. In this limit, all the fields except for two scalars get decoupled leading to (13.56) with $D=4$ and $\omega=0$, but the model retains the $SU(N_c)$ global symmetry, which is a remnant of the gauge symmetry of the original $\mathcal{N}=4$ SYM theory. In the planar limit $N_c\to\infty$, the FCFT is dominated by so-called fishnet Feynman graphs which have the structure of the regular square lattice. Such graphs appear to be integrable [Zamolodchikov (1980); Chicherin *et al.* (2013); Gromov *et al.* (2018)]. For infinitely long operators at a critical coupling ξ_* those graphs have a continuous limit which is believed to be described by a specific $O(2,4)$ 2D σ-model [Basso and Zhong (2019)]. At strong coupling $\xi\to\infty$ the FQFT$_4$ has a classical description in terms of a dual string-bit *fish-chain* model as it was derived in [Gromov and Sever (2019)]. This exact duality persists at the quantum level too [Gromov and Sever (2020)].

The model has already a rich history of studying various physical quantities, such as explicit computations of spectra of local operators [Gurdogan and Kazakov (2016); Caetano *et al.* (2018)] and 4-point correlation functions [Basso and Dixon (2017); Derkachov *et al.* (2019); Grabner *et al.* (2018); Gromov *et al.* (2019)]. The QSC formalism, adopted for the γ-deformed $\mathcal{N}=4$ SYM theory [Kazakov *et al.* (2016)], appeared in this limit particularly efficient for computations of anomalous dimensions of operators of the type $\mathrm{tr}Z^L(\bar{X}X)^M$ + permutations and later was extended to a much wider class of operators in [Gromov and Sever (2020)]. The FCFT_4 also appears to possess a rich moduli space of flat vacua, which are quantum-mechanically stable in spite of the absence of supersymmetry in the model [Karananas *et al.* (2019)] (in the planar limit).

13.3.2 *Spectrum of the Fishnet model and the LO BFKL*

At the technical level there is a number of places where the study of the spectrum of the fishnet models goes along similar steps as the problem of the BFKL spectrum. The similarity of the integrable structure of both models is discussed below. Here we just consider a simple example of a dimension of operators of the type $\text{tr}\Box^n Z^2 X^2$ in $D = 4$, $w = 0$, where $\Box$ is the 4D Laplace operator. The dimensions Δ of these operators can be obtained by solving the equation [Gromov *et al.* (2019)]

$$\frac{1}{\xi^4} = \frac{\psi^{(1)}\left(\frac{1}{4}(4-\Delta)\right) - \psi^{(1)}\left(\frac{1}{4}(6-\Delta)\right) - \psi^{(1)}\left(\frac{\Delta}{4}\right) + \psi^{(1)}\left(\frac{1}{4}(2+\Delta)\right)}{(\Delta-2)}, \quad (13.57)$$

which is reminiscent of the LO BFKL eigenvalue (13.16), written in the form

$$\frac{1}{g^2} = \frac{4}{S+1}\left(-\psi\left(\frac{1+\Delta}{2}\right) - \psi\left(\frac{1-\Delta}{2}\right) + 2\psi(1)\right) + \mathcal{O}\left(\frac{g^2}{S+1}\right). \quad (13.58)$$

The derivation of [Grabner *et al.* (2020)] from a Baxter equation is also very similar, to what is described in the Sec. 13.2.5.

Whereas the above analogy is demonstrated only at the structural level, below we give a more precise relation between the BFKL Hamiltonian and the graph-building operator of the fishnet theory in $D = 2$ and a particular, spin zero representation for physical spins.

13.3.3 *Graph-building operator and BFKL kernel*

We start from the case of general D, w. Suppose we want to compute the correlator of local operators $\mathcal{O}_L(z) = \text{tr}[Z(z)]^L$ in FCFT_D

$$K(z, y) = \langle \mathcal{O}_L(z)\, \bar{\mathcal{O}}_L(y) \rangle. \quad (13.59)$$

Due to the chiral property of the Lagrangian (13.56) the only planar Feynman graphs for this correlator have a "globe" configuration Fig. 13.6, where "parallels" and "meridians" form a regular square-lattice everywhere except for the "south" and "north" poles. The edges are represented by the propagators in D dimensions given by

$$G_h(x) = (x^2)^{-\frac{D}{4}+\omega}, \quad G_v(x) = (x^2)^{-\frac{D}{4}-\omega} \quad (13.60)$$

in horizontal and vertical directions respectively.

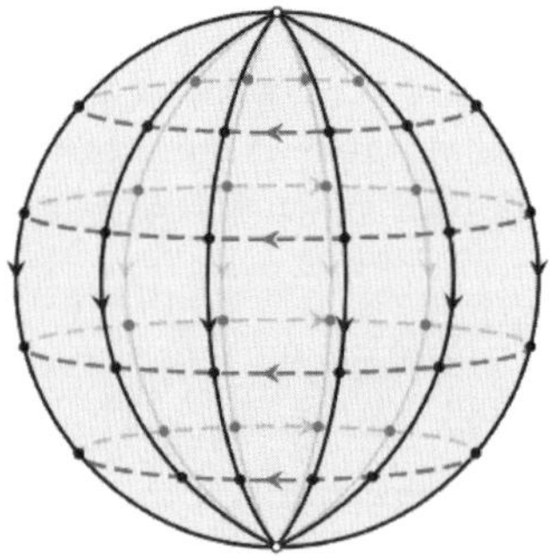

Fig. 13.6 Planar Fishnet graphs of "globe" type arising in the computation of the correlator (13.61). The edges belonging to red "parallels" and the black "meridians" correspond to propagators for two different scalar fields of the Lagrangian (13.56). Due to the chiral properties of vertices of this Lagrangian, no other planar diagrams in each loop order exist.

According to general properties of CFT, such correlator should have the form

$$K(z,y) = \text{const}\ |x-y|^{-2\Delta_{\mathcal{O}_L}}, \tag{13.61}$$

where Δ is the conformal dimension of this operator.

To compute this correlator, it is worth considering a more general operator, of the type $\mathcal{O}(z_1 \dots z_L) = \text{tr}[Z(z_1)\dots Z(z_L)]$ and compute the correlation function

$$K(z_1,\dots,z_L|y_1,\dots,y_L) = \langle \mathcal{O}(z_1 \dots z_L)\ \bar{\mathcal{O}}(y_1 \dots y_L)\rangle. \tag{13.62}$$

The planar Feynman graphs for such correlator are of cylindrical topology. One should open the ends of meridians converging on the "south" and "north" poles on the Fig. 13.6. If we then amputate the propagators converging at the south pole we obtain the cylindrical configuration presented on the Fig. 13.7. Such a graph can be represented at each order of perturbation theory as the corresponding power of the so-called graph-building operator, acting on the space $(\mathbb{R}^D)^{\otimes L}$,

$$\begin{aligned} H_L^{(w)}(z_1,z_2,\dots,z_L|z_1',z_2',\dots,z_L') &= G_v(x_{11'})G_v(x_{22'})\dots G_v(x_{22'}) \\ &\quad \times G_h(x_{1'2'})G_h(x_{2'3'})\dots G_h(x_{L'1'}), \end{aligned} \tag{13.63}$$

where we use the notations $x_{ab} = x_a - x_b$. Then the above correlator (13.62) can be represented as [Gurdogan and Kazakov (2016); Gromov (2017)]

$$K(z_1,\dots,z_L|y_1,\dots,y_L) = \langle z_1 \dots z_L|\ \frac{1}{1-\xi^{2L}H_L^{(s)}}\ |y_1 \dots y_L\rangle, \tag{13.64}$$

where the states $|y_1 \dots y_L\rangle$ are taken in the usual coordinate representation. It is easy to see, already by power counting, or applying the inversion transformation, that each cylindrical graph entering the perturbative expansion in ξ^2 is conformal. Thus the whole correlator will have the coordinate dependence according to the Euclidean D-dimensional conformal symmetry corresponding to the $\mathfrak{so}(1, D+1)$ algebra.

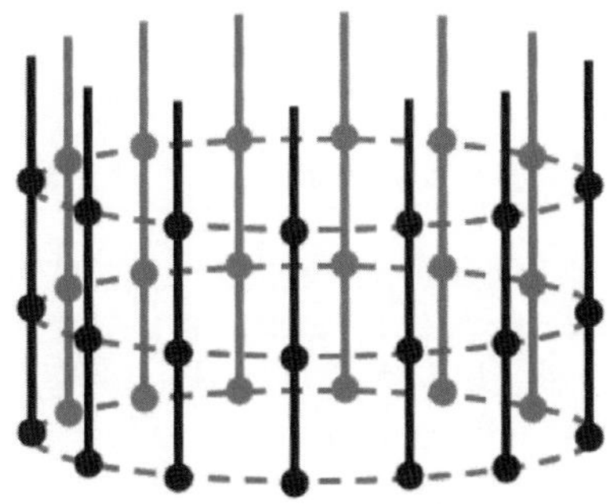

Fig. 13.7 The cylinder fishnet configuration resulting from the opening of the ends of propagators at the south and north poles of the globe (i.e. separating their coordinates) and amputating the propagators at the south pole. Such graphs can be computed by the Bethe-Salpeter procedure, using the integrability of the graph building operator, as described in this Section.

Introducing the 2D momenta conjugated to the coordinates $[p_k, x_k] = i$, we can represent the operator (13.63) in a more compact form

$$\begin{aligned} H_L^{(s)} &= (p_1^2)^{-\frac{D}{4}+\omega}(x_{12})^{-\frac{D}{4}-\omega}(p_2^2)^{-\frac{D}{4}+\omega} \times \\ &\times (x_{23})^{-\frac{D}{4}-\omega}(p_3^2)^{-\frac{D}{4}+\omega}\dots(p_L^2)^{-\frac{D}{4}+\omega}(x_{L1})^{-\frac{D}{4}-\omega}. \end{aligned} \tag{13.65}$$

The operator $H_L^{(w)}$ is known to be a conserved charge from the hierarchy of charges of the $so(1, D+1)$ spin chain encoded into the T-operator [Zamolodchikov (1980); Chicherin *et al.* (2013); Kazakov (2018)]

$$T_L(u) = \mathrm{Tr}_{\mathrm{aux}}(\hat{R}_1^{1'}(u)\hat{R}_2^{2'}(u)\dots\hat{R}_L^{L'}(u)) \tag{13.66}$$

built from L R-matrices acting in the product of principal series representations of $\mathfrak{so}(1, D+\ 1)$ with the weights $(\Delta_1, 0, 0, \dots) \times (\Delta_2, 0, 0, \dots)$

$$\begin{aligned} \hat{R}_2^{2'}(u) &= \langle x_1|R_2^{2'}(u)|x_1'\rangle = \\ &= \frac{1}{(x_{12}^2)^{-u-\frac{D}{2}+\Delta_+}(x_{21'}^2)^{u+\frac{D}{2}+\Delta_-}(x_{1'2'}^2)^{-u+\frac{D}{2}-\Delta_+}(x_{12'}^2)^{u+\frac{D}{2}-\Delta_-}}, \end{aligned} \tag{13.67}$$

where $\Delta_\pm = \frac{\Delta_1 \pm \Delta_2}{2}$. The matrix multiplication in auxiliary space in (13.66) is understood as the integration $\int d^2z\langle x|\dots|z\rangle\langle z|\dots|y\rangle$ and the trace $\mathrm{Tr}_{\mathrm{aux}}$ is understood as $\mathrm{Tr}_{\mathrm{aux}}(\dots) = \int d^2z\langle z|\dots|z\rangle$.

We can also use the momenta p_j to represent the R-matrix in the form

$$R_{ab}(u,\xi) = (x_{ab}^2)^{u+\frac{D}{2}-\Delta_+}(p_a^2)^{u+\Delta_-}(p_b^2)^{u-\Delta_-}(x_{ab}^2)^{u-\frac{D}{2}+\Delta_+}. \tag{13.68}$$

Indeed, if we take $u = -\frac{D}{2} + \Delta_+ + \epsilon$ and $\Delta_+ = \frac{D}{4}$, then the first factor in (13.67) in the limit $\epsilon \to 0$ disappears and the last one effectively becomes

$\frac{1}{(x_{1'2'})^{\frac{D}{2}}} \sim \epsilon^{-1}\delta^{(D)}(x_{1'2'})$, and it is easy to see that

$$T_L(u) \sim \frac{1}{\epsilon^L} H_L^{(w)}, \quad \text{where} \quad \Delta_1 = \frac{D}{4} - w, \quad \Delta_2 = \frac{D}{4} + w. \tag{13.69}$$

An interesting particular case of (13.66), mentioned in [Chicherin *et al.* (2013)] is the limit $u \to 0$, when $\Delta_1 = \Delta_2 \equiv \Delta$, since it is the only way to generate from it the local conserved charge – the spin chain Hamiltonian with nearest neighbour interaction. Indeed, the R-matrix becomes in this limit

$$R_{ab}(u, \xi) = 1 + u h_{ab} + \mathcal{O}(u^2), \tag{13.70}$$

where

$$\begin{aligned} h_{ab}(\Delta) &= 2\log(x_{ab}^2) + (x_{ab}^2)^{\frac{D}{2}-\Delta} \log(p_a^2 p_b^2)(x_{ab}^2)^{-\frac{D}{2}+\Delta} = \\ &= (p_a^2)^{-\frac{D}{2}+\Delta} \log(x_{ab}^2)(p_a^2)^{\frac{D}{2}-\Delta} + (p_b^2)^{-\frac{D}{2}+\Delta} \log(x_{ab}^2)(p_b^2)^{\frac{D}{2}-\Delta} + \log(p_a^2 p_b^2), \end{aligned} \tag{13.71}$$

which gives the Heisenberg Hamiltonian for the non-compact $SO(1, D+1)$ spin chain with nearest neighbour interactions of spins (with conformal spin Δ):

$$\mathcal{H}_\Delta = \sum_{a=1}^{L} h_{ab}(\Delta). \tag{13.72}$$

For the particular dimension $D = 2$ and conformal spin $\Delta = 0$, we obtain from here the famous Lipatov's $SL(2, \mathbb{C})$ spin chain describing the BFKL physics of interacting Reggeized gluons in the LO approximation [Lipatov (1994)]

$$h_{ab}^{BFKL} = 2\log(x_{ab}^2) + (x_{ab}^2)\log(p_a^2 p_b^2)(x_{ab}^2)^{-1}. \tag{13.73}$$

In the particular case of two interacting Reggeized gluons (Pomeron state) the spin chain reduces to the two-spin hamiltonian

$$\mathcal{H}_\Delta = 2h_{12}^{BFKL} \tag{13.74}$$

with the spectrum given by the RHS of (13.58).

In this way, we see that the BFKL physics in the LO approximation emerges as a particular case of D-dimensional Fishnet CFT, as it was first noticed in [Kazakov and Olivucci (2018)].

13.4 Conclusions

In this short review we made an attempt to briefly explain the studies of the BFKL spectrum in $\mathcal{N} = 4$ SYM, which were inspired by the works of Lev Lipatov on the BFKL integrability in the 4D gauge theories without and with supersymmetry. His ideas have led to a very significant progress in understanding the BFKL spectrum of $\mathcal{N} = 4$ SYM, where it was possible to build the bridge between the integrability of $\mathcal{N} = 4$ SYM and the integrability of the BFKL limit in the gauge theory. The usage of the Quantum Spectral Curve method in this limit has led to a series of new analytic and numerical results mentioned in this review.

The key feature utilized to link the results (including the BFKL spectrum) in $\mathcal{N} = 4$ SYM and QCD is the Kotikov-Lipatov's principle of maximal transcendentality [Kotikov and Lipatov (2000, 2001)]. In the Sec. 13.2.1 we illustrated this principle with an example of the NLO BFKL kernel eigenvalue in $\mathcal{N} = 4$ SYM and QCD. In general, this principle goes much beyond the BFKL spectrum. Many more observables in $\mathcal{N} = 4$ SYM admit this principle and reproduce the most transcendental part of the corresponding results in QCD. Another key observation by Lipatov [Kotikov and Lipatov (2000)], which allowed to connect the integrability of $\mathcal{N} = 4$ SYM with the BFKL regime, is the relation between the BFKL regime and the dimensions of the operators in $\mathcal{N} = 4$ SYM explained in the Sec. 13.2.2. This enabled the QSC method to be applied to the BFKL spectrum.

We also described here another special limit of $\mathcal{N} = 4$ SYM similar in spirit to the BFKL limit, leading to the Fishnet CFT, currently actively studied in the literature. We explained here the generalization of the Fishnet CFT from four to any number D of dimensions. It appears that, at $D = 2$ and a special value of spin of the involved fields, the Fishnet CFT reduces to Lipatov's integrable spin chain describing the interacting Reggeized gluons in the LO BFKL approximation of QCD. This remarkable correspondence can open new ways for the study of BFKL physics.

A great deal of the work presented here has been inspired by the ideas of Lev Lipatov. These ideas continue to influence a large community of theoretical physicists working on various non-perturbative aspects of quantum gauge theories. They find applications in a variety of domains of theoretical physics, including the integrable field theories. The BFKL spectrum of $\mathcal{N} = 4$ SYM is only one of many such applications, where Lev Lipatov's ideas gave rise to a great progress and revealed new intriguing problems, which still wait for their solution.

All three of us had the privilege to know Lev Lipatov personally and had the chance to appreciate, in numerous conversations, his outstanding personality and enormous talent. For one of us (V.K.), Lev Lipatov has been a dear friend over many years. For another (N.G.), Lev Nikolaevich was a dedicated and caring teacher who, over many years, greatly shaped his taste in research and was always available with deep insight and advice. And for the latter of us (M.A.), Lev Lipatov, despite being personally acquainted with him for a short time, left a great heritage of inspiring ideas and will always be an outstanding researcher to look up to.

For us, he will always be a great example of selfless devotion to science.

Acknowledgements

M.A. is grateful to Masha N. for her kind support during the work on this text. The work of N.G. was supported by the ERC grant 865075 EXACTC.

Bibliography

Adamo, T. and Jaitly, S. (2020). Twistor fishnets, *J. Phys.* **A53**, 5, p. 055401, doi:10.1088/1751-8121/ab5f88, arXiv:1908.11220 [hep-th].

Alfimov, M., Gromov, N., and Kazakov, V. (2015). QCD Pomeron from AdS/CFT Quantum Spectral Curve, *JHEP* **07**, p. 164, doi:10.1007/JHEP07(2015)164, arXiv:1408.2530 [hep-th].

Alfimov, M., Gromov, N., and Sizov, G. (2018). BFKL spectrum of $\mathcal{N} = 4$: non-zero conformal spin, *JHEP* **07**, p. 181, doi:10.1007/JHEP07(2018)181, arXiv:1802.06908 [hep-th].

Ambjorn, J., Janik, R. A., and Kristjansen, C. (2006). Wrapping interactions and a new source of corrections to the spin-chain/string duality, *Nucl. Phys.* **B736**, pp. 288–301, doi:10.1016/j.nuclphysb.2005.12.007, arXiv:hep-th/0510171 [hep-th].

Arutyunov, G. and Frolov, S. (2009). Thermodynamic Bethe Ansatz for the AdS(5) x S(5) Mirror Model, *JHEP* **05**, p. 068, doi:10.1088/1126-6708/2009/05/068, arXiv:0903.0141 [hep-th].

Balitsky, I. I. and Lipatov, L. N. (1978). The Pomeranchuk singularity in quantum chromodynamics, *Sov. J. Nucl. Phys.* **28**, pp. 822–829, [Yad. Fiz.28,1597(1978)].

Basso, B., Caetano, J., and Fleury, T. (2019a). Hexagons and correlators in the fishnet theory, *JHEP* **11**, p. 172, doi:10.1007/JHEP11(2019)172, arXiv:1812.09794 [hep-th].

Basso, B. and Dixon, L. J. (2017). Gluing ladder Feynman diagrams into fishnets, *Phys. Rev. Lett.* **119**, 7, p. 071601, doi:10.1103/PhysRevLett.119.071601, arXiv:1705.03545 [hep-th].

Basso, B., Ferrando, G., Kazakov, V., and Zhong, D.-l. (2019b). Thermodynamic Bethe ansatz for fishnet CFT, arXiv:1911.10213 [hep-th].

Basso, B. and Zhong, D.-l. (2019). Continuum limit of fishnet graphs and AdS sigma model, *JHEP* **01**, p. 002, doi:10.1007/JHEP01(2019)002, arXiv:1806.04105 [hep-th].

Beisert, N., Eden, B., and Staudacher, M. (2007). Transcendentality and crossing, *J. Stat. Mech.* **0701**, p. P01021, doi:10.1088/1742-5468/2007/01/P01021, arXiv:hep-th/0610251 [hep-th].

Beisert, N., Kazakov, V. A., Sakai, K., and Zarembo, K. (2006). The algebraic curve of classical superstrings on AdS(5) x S**5, *Commun. Math. Phys.* **263**, pp. 659–710, doi:10.1007/s00220-006-1529-4, arXiv:hep-th/0502226 [hep-th].

Beisert, N., Kristjansen, C., and Staudacher, M. (2003). The dilatation operator of conformal $N = 4$ super Yang-Mills theory, *Nucl. Phys.* **B664**, pp. 131–184, doi: 10.1016/S0550-3213(03)00406-1, arXiv:hep-th/0303060 [hep-th].

Beisert, N. and Roiban, R. (2005). Beauty and the twist: The Bethe ansatz for twisted N=4 SYM, *JHEP* **08**, p. 039, doi:10.1088/1126-6708/2005/08/039, arXiv:hep-th/0505187 [hep-th].

Bena, I., Polchinski, J., and Roiban, R. (2004). Hidden symmetries of the AdS(5) x S**5 superstring, *Phys. Rev.* **D69**, p. 046002, doi:10.1103/PhysRevD.69.046002, arXiv:hep-th/0305116 [hep-th].

Bombardelli, D., Fioravanti, D., and Tateo, R. (2009). Thermodynamic Bethe Ansatz for planar AdS/CFT: A Proposal, *J. Phys.* **A42**, p. 375401, doi:10.1088/1751-8113/42/37/375401, arXiv:0902.3930 [hep-th].

Brower, R. C., Costa, M., Djuric, M., Raben, T., and Tan, C.-I. (2013). Conformal Pomeron and Odderon in Strong Coupling, in *International Workshop on Low X Physics*, arXiv:1312.1419 [hep-ph].

Brower, R. C., Polchinski, J., Strassler, M. J., and Tan, C.-I. (2007). The Pomeron and gauge/string duality, *JHEP* **0712**, p. 005, doi:10.1088/1126-6708/2007/12/005, arXiv:hep-th/0603115 [hep-th].

Caetano, J., Gurdogan, O., and Kazakov, V. (2018). Chiral limit of $\mathcal{N} = 4$ SYM and ABJM and integrable Feynman graphs, *JHEP* **03**, p. 077, doi:10.1007/JHEP03(2018)077, arXiv:1612.05895 [hep-th].

Caron-Huot, S. and Herranen, M. (2018). High-energy evolution to three loops, *JHEP* **02**, p. 058, doi:10.1007/JHEP02(2018)058, arXiv:1604.07417 [hep-ph].

Chicherin, D., Derkachov, S., and Isaev, A. P. (2013). Conformal group: R-matrix and star-triangle relation, *JHEP* **04**, p. 020, doi:10.1007/JHEP04(2013)020, arXiv:1206.4150 [math-ph].

Chicherin, D., Kazakov, V., Loebbert, F., Mller, D., and Zhong, D.-l. (2017). Yangian Symmetry for Fishnet Feynman Graphs, *Phys. Rev.* **D96**, 12, p. 121901, doi:10.1103/PhysRevD.96.121901, arXiv:1708.00007 [hep-th].

Chicherin, D., Kazakov, V., Loebbert, F., Mller, D., and Zhong, D.-l. (2018). Yangian Symmetry for Bi-Scalar Loop Amplitudes, *JHEP* **05**, p. 003, doi:10.1007/JHEP05(2018)003, arXiv:1704.01967 [hep-th].

Costa, M. S., Drummond, J., Goncalves, V., and Penedones, J. (2014). The role of leading twist operators in the Regge and Lorentzian OPE limits, *JHEP* **1404**, p. 094, doi:10.1007/JHEP04(2014)094, arXiv:1311.4886 [hep-th].

Costa, M. S., Goncalves, V., and Penedones, J. (2012). Conformal Regge theory, *JHEP* **1212**, p. 091, doi:10.1007/JHEP12(2012)091, arXiv:1209.4355 [hep-th].

de Mello Koch, R., LiMing, W., Van Zyl, H. J. R., and Rodrigues, J. P. (2019). Chaos in the Fishnet, *Phys. Lett.* **B793**, pp. 169–174, doi:10.1016/j.physletb.2019.04.044, arXiv:1902.06409 [hep-th].

De Vega, H. and Lipatov, L. (2001). Interaction of reggeized gluons in the Baxter-Sklyanin representation, *Phys. Rev.* **D64**, p. 114019, doi:10.1103/PhysRevD.64.114019, arXiv:hep-ph/0107225 [hep-ph].

Derkachov, S., Kazakov, V., and Olivucci, E. (2019). Basso-Dixon correlators in two-dimensional fishnet CFT, *JHEP* **04**, p. 032, doi:10.1007/JHEP04(2019)032, arXiv:1811.10623 [hep-th].

Derkachov, S. and Olivucci, E. (2020). Exactly solvable magnet of conformal spins in four dimensions, *Phys. Rev. Lett.* **125**, 3, p. 031603, doi:10.1103/PhysRevLett.125.031603, arXiv:1912.07588 [hep-th].

Derkachov, S. E., Korchemsky, G., Kotanski, J., and Manashov, A. (2002). Noncompact Heisenberg spin magnets from high-energy QCD. 2. Quantization conditions and energy spectrum, *Nucl. Phys.* **B645**, pp. 237–297, doi:10.1016/S0550-3213(02)00842-8, arXiv:hep-th/0204124 [hep-th].

Derkachov, S. E., Korchemsky, G., and Manashov, A. (2001). Noncompact Heisenberg spin magnets from high-energy QCD: 1. Baxter Q operator and separation of variables, *Nucl. Phys.* **B617**, pp. 375–440, doi:10.1016/S0550-3213(01)00457-6, arXiv:hep-th/0107193 [hep-th].

Derkachov, S. E., Korchemsky, G., and Manashov, A. (2003). Noncompact Heisenberg spin magnets from high-energy QCD. 3. Quasiclassical approach, *Nucl. Phys.* **B661**, pp. 533–576, doi:10.1016/S0550-3213(03)00340-7, arXiv:hep-th/0212169 [hep-th].

Dutta Chowdhury, S., Haldar, P., and Sen, K. (2019). On the Regge limit of fishnet correlators, *JHEP* **10**, p. 249, doi:10.1007/JHEP10(2019)249, arXiv:1908.01123 [hep-th].

Faddeev, L. D. and Korchemsky, G. P. (1995). High-energy QCD as a completely integrable model, *Phys. Lett.* **B342**, pp. 311–322, doi:10.1016/0370-2693(94)01363-H, arXiv:hep-th/9404173.

Fadin, V. S. and Lipatov, L. (1998). BFKL pomeron in the next-to-leading approximation, *Phys. Lett.* **B429**, pp. 127–134, doi:10.1016/S0370-2693(98)00473-0, arXiv:hep-ph/9802290 [hep-ph].

Fokken, J., Sieg, C., and Wilhelm, M. (2014). A piece of cake: the ground-state energies in γ_i -deformed $\mathcal{N} = 4$ SYM theory at leading wrapping order, *JHEP* **09**, p. 078, doi:10.1007/JHEP09(2014)078, arXiv:1405.6712 [hep-th].

Grabner, D., Gromov, N., Kazakov, V., and Korchemsky, G. (2018). Strongly γ-Deformed $\mathcal{N} = 4$ Supersymmetric Yang-Mills Theory as an Integrable Conformal Field Theory, *Phys. Rev. Lett.* **120**, 11, p. 111601, doi:10.1103/PhysRevLett.120.111601, arXiv:1711.04786 [hep-th].

Grabner, D., Gromov, N., Kazakov, V., and Korchemsky, G. (2020). In preparation.

Gromov, N. (2017). Introduction to the Spectrum of $N = 4$ SYM and the Quantum Spectral Curve, arXiv:1708.03648 [hep-th].

Gromov, N. and Kazakov, V. (2013). Analytic continuation in spin of the baxter equation solutions for twist-2 operators, *unpublished*.

Gromov, N., Kazakov, V., and Korchemsky, G. (2019). Exact correlation functions in conformal fishnet theory, *JHEP* **08**, p. 123, doi:10.1007/JHEP08(2019)123, arXiv:1808.02688 [hep-th].

Gromov, N., Kazakov, V., Korchemsky, G., Negro, S., and Sizov, G. (2018). Integrability of conformal fishnet theory, *JHEP* **01**, p. 095, doi:10.1007/JHEP01(2018)095, arXiv:1706.04167 [hep-th].

Gromov, N., Kazakov, V., Kozak, A., and Vieira, P. (2010). Exact spectrum of anomalous dimensions of planar $N = 4$ supersymmetric Yang-Mills theory: TBA and excited states, *Lett. Math. Phys.* **91**, pp. 265–287, doi:10.1007/s11005-010-0374-8, arXiv:0902.4458 [hep-th].

Gromov, N., Kazakov, V., Leurent, S., and Volin, D. (2014a). Quantum spectral curve for AdS_5/CFT_4, *Phys. Rev. Lett.* **112**, p. 011602, doi:10.1103/PhysRevLett.112.011602, arXiv:1305.1939 [hep-th].

Gromov, N., Kazakov, V., Leurent, S., and Volin, D. (2015a). Quantum spectral curve for arbitrary state/operator in $\mathrm{AdS}_5/\mathrm{CFT}_4$, *JHEP* **09**, p. 187, doi:10.1007/JHEP09(2015)187, arXiv:1405.4857 [hep-th].

Gromov, N., Kazakov, V., and Vieira, P. (2009). Exact spectrum of anomalous dimensions of planar $N = 4$ supersymmetric Yang-Mills theory, *Phys. Rev. Lett.* **103**, p. 131601, doi:10.1103/PhysRevLett.103.131601, arXiv:0901.3753 [hep-th].

Gromov, N. and Levkovich-Maslyuk, F. (2016a). Quantum spectral curve for a cusped Wilson line in $\mathcal{N} = 4$ SYM, *JHEP* **04**, p. 134, doi:10.1007/JHEP04(2016)134, arXiv:1510.02098 [hep-th].

Gromov, N. and Levkovich-Maslyuk, F. (2016b). Quark-anti-quark potential in $\mathcal{N} = 4$ SYM, *JHEP* **12**, p. 122, doi:10.1007/JHEP12(2016)122, arXiv:1601.05679 [hep-th].

Gromov, N., Levkovich-Maslyuk, F., and Sizov, G. (2015b). Pomeron eigenvalue at three loops in $\mathcal{N} = 4$ supersymmetric Yang-Mills theory, *Phys. Rev. Lett.* **115**, 25, p. 251601, doi:10.1103/PhysRevLett.115.251601, arXiv:1507.04010 [hep-th].

Gromov, N., Levkovich-Maslyuk, F., and Sizov, G. (2016). Quantum spectral curve and the numerical solution of the spectral problem in AdS5/CFT4, *JHEP* **06**, p. 036, doi:10.1007/JHEP06(2016)036, arXiv:1504.06640 [hep-th].

Gromov, N., Levkovich-Maslyuk, F., Sizov, G., and Valatka, S. (2014b). Quantum spectral curve at work: from small spin to strong coupling in $\mathcal{N} = 4$ SYM, *JHEP* **07**, p. 156, doi:10.1007/JHEP07(2014)156, arXiv:1402.0871 [hep-th].

Gromov, N. and Sever, A. (2019). Derivation of the holographic dual of a planar conformal field theory in 4D, *Phys. Rev. Lett.* **123**, 8, p. 081602, doi:10.1103/PhysRevLett.123.081602, arXiv:1903.10508 [hep-th].

Gromov, N. and Sever, A. (2020). The holographic dual of strongly γ-deformed $\mathcal{N} = 4$ SYM theory: derivation, generalization, integrability and discrete reparametrization symmetry, *JHEP* **02**, p. 035, doi:10.1007/JHEP02(2020)035, arXiv:1908.10379 [hep-th].

Gromov, N. and Sizov, G. (2014). Exact slope and interpolating functions in $N = 6$ supersymmetric Chern-Simons theory, *Phys. Rev. Lett.* **113**, 12, p. 121601, doi:10.1103/PhysRevLett.113.121601, arXiv:1403.1894 [hep-th].

Gross, D. J. and Rosenhaus, V. (2017). The bulk dual of SYK: cubic couplings, *JHEP* **05**, p. 092, doi:10.1007/JHEP05(2017)092, arXiv:1702.08016 [hep-th].

Gurdogan, O. and Kazakov, V. (2016). New integrable 4D quantum field theories from strongly deformed planar $\mathcal{N} = 4$ supersymmetric Yang-Mills theory, *Phys. Rev. Lett.* **117**, 20, p. 201602, doi:10.1103/PhysRevLett.117.201602,10.1103/PhysRevLett.117.259903, arXiv:1512.06704 [hep-th], [Addendum: Phys. Rev. Lett.117,no.25,259903(2016)].

Janik, R. A. (2006). The AdS(5) x S**5 superstring worldsheet S-matrix and crossing symmetry, *Phys. Rev.* **D73**, p. 086006, doi:10.1103/PhysRevD.73.086006, arXiv:hep-th/0603038 [hep-th].

Janik, R. A. (2013). Twist-two operators and the BFKL regime — nonstandard solutions of the Baxter equation, *JHEP* **1311**, p. 153, doi:10.1007/JHEP11(2013)153, arXiv:1309.2844 [hep-th].

Karananas, G. K., Kazakov, V., and Shaposhnikov, M. (2019). Spontaneous conformal symmetry breaking in fishnet CFT, arXiv:1908.04302 [hep-th].

Kazakov, V. (2018). *Quantum Spectral Curve of γ-twisted $\mathcal{N} = 4$ SYM Theory and Fishnet CFT*, Vol. 30, p. 1840010, doi:10.1142/9789813233867_0016, arXiv:1802.02160 [hep-th].

Kazakov, V., Leurent, S., and Volin, D. (2016). T-system on T-hook: Grassmannian solution and twisted quantum spectral curve, *JHEP* **12**, p. 044, doi:10.1007/JHEP12(2016)044, arXiv:1510.02100 [hep-th].

Kazakov, V. and Olivucci, E. (2018). Biscalar integrable conformal field theories in any dimension, *Phys. Rev. Lett.* **121**, 13, p. 131601, doi:10.1103/PhysRevLett.121.131601, arXiv:1801.09844 [hep-th].

Kazakov, V., Olivucci, E., and Preti, M. (2019). Generalized fishnets and exact four-point correlators in chiral CFT_4, *JHEP* **06**, p. 078, doi:10.1007/JHEP06(2019)078, arXiv:1901.00011 [hep-th].

Kazakov, V. A., Marshakov, A., Minahan, J. A., and Zarembo, K. (2004). Classical/quantum integrability in AdS/CFT, *JHEP* **05**, p. 024, doi:10.1088/1126-6708/2004/05/024, arXiv:hep-th/0402207 [hep-th].

Kitaev, A. (2015). A simple model of quantum holography, Talks at KITP, http://online.kitp.ucsb.edu/online/entangled15/kitaev/ and http://online.kitp.ucsb.edu/online/entangled15/kitaev2/, April and May, 2015.

Klebanov, I. R. and Tarnopolsky, G. (2017). Uncolored random tensors, melon diagrams, and the Sachdev-Ye-Kitaev models, *Phys. Rev.* **D95**, 4, p. 046004, doi:10.1103/PhysRevD.95.046004, arXiv:1611.08915 [hep-th].

Korchemsky, G. P. (2019). Exact scattering amplitudes in conformal fishnet theory, *JHEP* **08**, p. 028, doi:10.1007/JHEP08(2019)028, arXiv:1812.06997 [hep-th].

Kotikov, A. and Lipatov, L. (2000). NLO corrections to the BFKL equation in QCD and in supersymmetric gauge theories, *Nucl. Phys.* **B582**, pp. 19–43, doi:10.1016/S0550-3213(00)00329-1, arXiv:hep-ph/0004008 [hep-ph].

Kotikov, A. and Lipatov, L. (2001). DGLAP and BFKL evolution equations in the $N = 4$ supersymmetric gauge theory, in *35th Annual Winter School on Nuclear and Particle Physics*, arXiv:hep-ph/0112346.

Kotikov, A. and Lipatov, L. (2003). DGLAP and BFKL equations in the N=4 supersymmetric gauge theory, *Nucl. Phys.* **B661**, pp. 19–61, doi:10.1016/S0550-3213(03)00264-5, arXiv:hep-ph/0208220 [hep-ph].

Kotikov, A. and Onishchenko, A. (2019). DGLAP and BFKL equations in $\mathcal{N} = 4$ SYM: from weak to strong coupling, arXiv:1908.05113 [hep-th].

Kotikov, A. and Velizhanin, V. (2005). Analytic continuation of the Mellin moments of deep inelastic structure functions, arXiv:hep-ph/0501274.

Kotikov, A. V. and Lipatov, L. N. (2013). Pomeron in the $N = 4$ supersymmetric gauge model at strong couplings, *Nucl. Phys.* **B874**, pp. 889–904, doi:10.1016/j.nuclphysb.2013.06.018, arXiv:1301.0882 [hep-th].

Kotikov, A. V., Lipatov, L. N., Rej, A., Staudacher, M., and Velizhanin, V. N. (2007). Dressing and wrapping, *J. Stat. Mech.* **0710**, p. P10003, doi:10.1088/1742-5468/2007/10/P10003, arXiv:0704.3586 [hep-th].

Kotikov, A. V., Lipatov, L. N., and Velizhanin, V. N. (2003). Anomalous dimensions of Wilson operators in $N = 4$ SYM theory, *Phys. Lett.* **B557**, pp. 114–120, doi:10.1016/S0370-2693(03)00184-9, arXiv:hep-ph/0301021 [hep-ph].

Kuraev, E. A., Lipatov, L. N., and Fadin, V. S. (1977). The Pomeranchuk singularity in nonabelian gauge theories, *Sov. Phys. JETP* **45**, pp. 199–204, [Zh. Eksp. Teor. Fiz.72,377(1977)].

Lee, R. and Onishchenko, A. (2019a). Toward an analytic perturbative solution for the ABJM quantum spectral curve, *Teor. Mat. Fiz.* **198**, 2, pp. 292–308, doi:10.1134/S0040577919020077, arXiv:1807.06267 [hep-th].

Lee, R. N. and Onishchenko, A. I. (2018). ABJM quantum spectral curve and Mellin transform, *JHEP* **05**, p. 179, doi:10.1007/JHEP05(2018)179, arXiv:1712.00412 [hep-th].

Lee, R. N. and Onishchenko, A. I. (2019b). ABJM quantum spectral curve at twist 1: algorithmic perturbative solution, *JHEP* **11**, p. 018, doi:10.1007/JHEP11(2019)018, arXiv:1905.03116 [hep-th].

Lipatov, L. (1986). The bare Pomeron in quantum chromodynamics, *Sov. Phys. JETP* **63**, pp. 904–912.

Lipatov, L. N. (1993). High-energy asymptotics of multicolor QCD and two-dimensional conformal field theories, *Phys. Lett.* **B309**, pp. 394–396, doi:10.1016/0370-2693(93)90951-D, [147(1993)].

Lipatov, L. N. (1994). Asymptotic behavior of multicolor QCD at high energies in connection with exactly solvable spin models, *JETP Lett.* **59**, pp. 596–599, arXiv:hep-th/9311037 [hep-th], [Pisma Zh. Eksp. Teor. Fiz. 59,571 (1994)].

Lipatov, L. N. (1997). Evolution equations in QCD, in *Perspectives in hadronic physics. Proceedings, Conference, Trieste, Italy, May 12-16, 1997*, pp. 413–427.

Lipatov, L. N. (2001). Next-to-leading corrections to the BFKL equation and the effective action for high energy processes in QCD, *Nucl. Phys. Proc. Suppl.* **99A**, pp. 175–179, doi:10.1016/S0920-5632(01)01329-9, [175(2001)].

Mamroud, O. and Torrents, G. (2017). RG stability of integrable fishnet models, *JHEP* **06**, p. 012, doi:10.1007/JHEP06(2017)012, arXiv:1703.04152 [hep-th].

Marboe, C. and Velizhanin, V. (2016). Twist-2 at seven loops in planar $\mathcal{N} = 4$ SYM theory: full result and analytic properties, *JHEP* **11**, p. 013, doi:10.1007/JHEP11(2016)013, arXiv:1607.06047 [hep-th].

Marboe, C., Velizhanin, V., and Volin, D. (2015). Six-loop anomalous dimension of twist-two operators in planar $\mathcal{N} = 4$ SYM theory, *JHEP* **07**, p. 084, doi:10.1007/JHEP07(2015)084, arXiv:1412.4762 [hep-th].

Marboe, C. and Volin, D. (2015). Quantum spectral curve as a tool for a perturbative quantum field theory, *Nucl. Phys.* **B899**, pp. 810–847, doi:10.1016/j.nuclphysb.2015.08.021, arXiv:1411.4758 [hep-th].

Marboe, C. and Volin, D. (2018a). The full spectrum of AdS5/CFT4 I: Representation theory and one-loop Q-system, *J. Phys.* **A51**, 16, p. 165401, doi:10.1088/1751-8121/aab34a, arXiv:1701.03704 [hep-th].

Marboe, C. and Volin, D. (2018b). The full spectrum of AdS_5/CFT_4 II: Weak coupling expansion via the quantum spectral curve, arXiv:1812.09238 [hep-th].

Minahan, J. A. and Zarembo, K. (2003). The Bethe-ansatz for N = 4 super Yang-Mills, *JHEP* **03**, p. 013, arXiv:hep-th/0212208.

Murugan, J., Stanford, D., and Witten, E. (2017). More on Supersymmetric and 2d Analogs of the SYK Model, *JHEP* **08**, p. 146, doi:10.1007/JHEP08(2017)146, arXiv:1706.05362 [hep-th].

Pittelli, A. and Preti, M. (2019). Integrable fishnet from γ-deformed $\mathcal{N} = 2$ quivers, *Phys. Lett.* **B798**, p. 134971, doi:10.1016/j.physletb.2019.134971, arXiv:1906.03680 [hep-th].

Rosenhaus, V. (2019). An introduction to the SYK model, *J. Phys. A* **52**, p. 323001, doi:10.1088/1751-8121/ab2ce1, arXiv:1807.03334 [hep-th].

Sieg, C. and Wilhelm, M. (2016). On a CFT limit of planar γ_i-deformed $\mathcal{N} = 4$ SYM theory, *Phys. Lett.* **B756**, pp. 118–120, doi:10.1016/j.physletb.2016.03.004, arXiv:1602.05817 [hep-th].

Zamolodchikov, A. B. (1980). 'Fishnet' diagrams as a completely integrable system, *Phys. Lett.* **97B**, pp. 63–66, doi:10.1016/0370-2693(80)90547-X.

https://doi.org/10.1142/9789811231124_0014

Chapter 14

DGLAP and BFKL Equations in $\mathcal{N} = 4$ SYM: From Weak to Strong Coupling

A. Kotikov* and A. Onishchenko*,†

*Joint Institute for Nuclear Research, Dubna, Russia
†Skobeltsyn Institute of Nuclear Physics, Moscow, Russia

Abstract

DGLAP and BFKL equations are among the cornestones of the contemporary QCD. Moreover, they also played an important role in the recent studies of integrability structure of $\mathcal{N} = 4$ SYM. Lev Lipatov was one of the main contributors to the subject, whose ideas continue to inspire its current development. Here, we review the results obtained along this way together with a brief account of approaches and methods used.

14.1 Introduction

We are very glad that the fate connected us with Lev Lipatov. A.K first saw Lipatov at an international conference in Gatchina in 1989. Two years later he met Lipatov and discussed with him methods for calculation of Feynman diagrams. The first scientific interaction happened in Hamburg in 1999 where A.K had an annual position under the supervision of Professor Bartels supported by the Humbolt Foundation. It was precisely the time when Lev Lipatov also came to Professor Bartels with a scientific visit. Just before that Fadin and Lipatov completed their famous calculation of next-to-leading order (NLO) corrections to the BFKL equation. In this work they used gluon mass as an intermediate regulator. A.K decided to repeat their calculation in the framework of dimensional regularization without using the gluon mass. Learning about this, Lipatov proposed to generalize the calculation given in the article by Fadin and Lipatov to the case of arbitrary values of conformal spin and to the case of supersymmetric generalization of Quantum Chromodynamics. It turned out that the method developed in the previous work [Kazakov and Kotikov (1988)] by A.K and Dmitry Kazakov is well suited for such calculations. A.O first met Lev Nikolaevich in 2000 at the "Quarks" conference held in Pushkin near St. Petersburg. Nevertheless, their first joint paper appeared only in 2004 and was devoted to the reconstruction of three loop anomalous dimensions of twist 2 operators in N=4 SYM based on earlier ideas by A.K and Lev Lipatov, which are now known as maximum transcendentality principle. Right after this work Lev Lipatov invited A.O to Gatchina, where he continued his work under the supervision of Lev Nikolaevich Lipatov.

This paper deals with the review of the known properties of the Balitsky-Fadin-Kuraev-Lipatov (BFKL) [Lipatov (1976); Fadin *et al.* (1975); Kuraev *et al.* (1976, 1977); Balitsky and Lipatov (1978, 1979)] and the Dokshitzer-Gribov-Lipatov-Altarelli-Parisi (DGLAP) [Gribov and Lipatov (1972a,b); Lipatov (1975); Altarelli and Parisi (1977); Dokshitzer (1977)] equations in $\mathcal{N}=4$ Supersymmetric Yang-Mills (SYM) theory [Brink *et al.* (1977); Gliozzi *et al.* (1977)].

Lev Lipatov is one of the main contributors to the discovery and subsequent study of both these well known equations. The BFKL and DGLAP equations resum, respectively, the most important contributions proportional to $\sim \alpha_s \ln(1/x)$ and $\sim \alpha_s \ln(Q^2/\Lambda^2)$ in two different kinematic regions of the Bjorken parameter x and the "mass" Q^2 of the virtual photon for the process of deep inelastic lepton-hadron scattering (DIS) and as such, they are among the main ingredients in the analysis and description of experimental data for lepton-nucleon and nucleon-nucleon scattering.

In the case of supersymmetric theories these equations simplify drastically. Moreover, in the case of $\mathcal{N}=4$ SYM they become related with each other for the nonphysical values of Mellin moments j as it was proposed by Lev Lipatov in [Kotikov and Lipatov (2003)].

The first hint towards the simplification of the BFKL and DGLAP equations in supersymmetric theories came from the study of quasi-partonic operators in $\mathcal{N}=1$ SYM in leading order (LO) approximation [Bukhvostov *et al.* (1985b,a)]. It was shown, that quasi-partonic operators in this theory are unified in supermultiplets and their anomalous dimensions could be described in terms of a single universal anomalous dimension by shifting the argument of the latter by some integer number. Calculations in $\mathcal{N}=4$ SYM, where the coupling constant is not renormalized, gave even more remarkable results. Namely, it turned out, that all twist-2 operators in this theory belong to the same supermultiplet with their anomalous dimension matrix fixed completely by the superconformal invariance. The LO universal anomalous dimension in $\mathcal{N}=4$ SYM was found to be proportional to $\Psi(j-1)-\Psi(1)$, which implies that the DGLAP evolution equations for the matrix elements of quasi-partonic operators are equivalent to the Schrödinger equation for the integrable Heisenberg spin model [Lipatov (1997, 2001)]. In QCD the integrability remains only in a small sector of these operators [Braun *et al.* (1998); Belitsky (1999)] (see also [Ferretti *et al.* (2004); Beisert *et al.* (2005)]). On the other hand, in the case of $\mathcal{N}=4$ SYM the equations for other sets of operators are also integrable [Minahan and Zarembo (2003a); Beisert and Staudacher (2003a); Beisert *et al.* (2003)].

Similar results related to integrability of BFKL and the Bartels-Kwiecinski-Praszalowicz (BKP) [Bartels (1980); Kwiecinski and Praszalowicz (1980)] equations in the multi-colour limit of QCD were obtained earlier in [Lipatov (1994)]. Later it was also shown [Kotikov and Lipatov (2003)], that in the case of $\mathcal{N}=4$ SYM there is a deep relation between the BFKL and DGLAP evolution equations. Namely, the j-plane singularities of the anomalous dimensions of Wilson twist-2 operators in this case can be obtained from the eigenvalues of the BFKL kernel by analytic continuation. The next-to-leading (NLO) calculations in $\mathcal{N}=4$ SYM demonstrated [Kotikov and Lipatov (2003)], that some of these relations are valid also in higher orders of perturbation theory. In particular, the BFKL equation has the property of the hermitian separability and the eigenvalues of the anomalous dimension matrix may be expressed in terms of the universal function $\gamma_{uni}(j)$.

In what follows, in Section 2 we discuss the BFKL equation and its Pomeron solution both at weak and strong coupling values. Section 3 is devoted to the DGLAP equation and related anomalous dimensions of Wilson twist-2 operators. Here, as in the case of the BFKL equation we are not limiting ourselves by the weak coupling regime and discuss also the behaviour of anomalous dimensions at strong coupling. The strong coupling study both in the case of BFKL and DGLAP equations was made possible by an extensive use of both integrability and AdS/CFT correspondence approaches. Finally, we come with a summary in the Conclusion.

14.2 BFKL equation and Pomeron

The behaviour of scattering amplitudes in the high energy or Regge limit may be conveniently described by the positions of singularities of their partial wave amplitudes in the complex angular momentum plane. In particular, the behaviour of the total cross-section for the scattering of colourless particles at high energies is related to the Regge pole with quantum numbers of the vacuum and even parity — the so called Pomeron. Within perturbation theory the latter is given by the bound state of two Reggeized gluons and is described by the now famous BFKL the equation [Lipatov (1976)]. In what follows we are going to describe the properties of the BFKL equation and its Pomeron solution in the case of maximally supersymmetric $\mathcal{N}=4$ Yang-Mills theory.

14.2.1 *BFKL equation and perturbation theory*

In the high-energy limit ($s \gg -t$) the total cross-section $\sigma(s)$ for the scattering of colourless particles A and B takes the following factorized form

$$\sigma(s) = \int \frac{d^2q\, d^2q'}{(2\pi)^2\, q^2\, q'^2} \Phi_A(q)\, \Phi_B(q') \int_{a-i\infty}^{a+i\infty} \frac{d\omega}{2\pi i} \left(\frac{s}{s_0}\right)^{\omega} G_\omega(q,q'), \quad s_0 = |q||q'|, \tag{14.1}$$

where $\Phi_i(q_i)$ are impact factors of the colliding particles with momenta p_A, p_B, $s = 2p_A p_B$ is their squared invariant mass and $G_\omega(q,q')$ is the t-channel partial wave for the Reggeized gluon-gluon scattering. The latter depends on two transverse momenta of the Reggeized gluons in the t channel q and q'[1] and is given by the solution to BFKL equation.

Using the dimensional regularization and the $\overline{MS}$-scheme to deal with ultraviolet and infrared divergences in the intermediate expressions, the BFKL equation for $G_\omega(q,q')$ can be written in the form

$$\omega G_\omega(q,q_1) = \delta^{D-2}(q-q_1) + \int d^{D-2}q_2\, K(q,q_2)\, G_\omega(q_2,q_1), \tag{14.2}$$

where

$$K(q_1,q_2) = 2\,\omega(q_1)\,\delta^{D-2}(q_1-q_2) + K_r(q_1,q_2) \tag{14.3}$$

and the space-time dimension $D = 4 - 2\varepsilon$ for $\varepsilon \to 0$. Here, $\omega(q)$ is the gluon Regge trajectory and the integral kernel $K_r(q_1,q_2)$ is related to the real particle production in t-channel. In $\mathcal{N} = 4$ SYM the description of the BFKL equation follows closely the one in the case of QCD. Moreover, in the leading logarithmic approximation the expressions for the BFKL integral kernel in QCD and $\mathcal{N} = 4$ SYM coincide and the difference starts at the level of NLO corrections. Initially NLO radiative corrections to the BFKL integral kernel at $t = 0$ were calculated in the case of QCD in [Fadin and Lipatov (1998); Ciafaloni and Camici (1998)][2] and later were generalized to the case of $\mathcal{N} = 4$ SYM in [Kotikov and Lipatov (2000)].

As it was shown in [Fadin and Lipatov (1998); Kotikov and Lipatov (2000)], a complete and orthogonal set of eigenfunctions of the homogeneous BFKL equation in LO is given by ($\gamma_{BFKL} = \frac{1}{2} + i\nu$):

$$G_{n,\gamma}(q^2/q'^2,\theta) = \left(\frac{q^2}{q'^2}\right)^{\gamma_{BFKL}-1} e^{in\theta}, \tag{14.4}$$

where the integer conformal spin n and real ν enter the parametrization of conformal weights m and $\tilde{m}$ of the principal series of the unitary Möbius group representation

$$m = \frac{1+n}{2} + i\nu, \quad \tilde{m} = \frac{1-n}{2} + i\nu. \tag{14.5}$$

[1] Here and below we omit arrows in the notation of transverse momenta and simply write q and q' instead of $\vec{q}$ and $\vec{q}'$.

[2] The $t \neq 0$ case can be found in the recent papers [Fadin and Fiore (2008)].

The BFKL kernel in this representation is diagonalized up to the effects related with the running coupling constant $a_s(q^2)$, and up to NLO in the case of QCD we have

$$\omega_{\overline{MS}}^{QCD} = 4a_s(q^2)\Big[\chi(n,\gamma_{BFKL}) + \delta_{\overline{MS}}^{QCD}(n,\gamma_{BFKL})a_s(q^2)\Big], \qquad (14.6)$$

where LO and NLO eigenvalues of the BFKL kernel using formulae of [Kotikov and Lipatov (2000)] are given by

$$\chi(n,\gamma) = 2\Psi(1) - \Psi\left(\gamma+\frac{n}{2}\right) - \Psi\left(1-\gamma+\frac{n}{2}\right) \qquad (14.7)$$

$$\begin{aligned}
\delta_{\overline{MS}}^{QCD}(n,\gamma) &= \left(\frac{67}{9} - 2\zeta(2) - \frac{10}{9}\frac{n_f}{N_c}\right)\chi(n,\gamma) + 6\zeta(3) \\
&\quad + \Psi''\left(\gamma+\frac{n}{2}\right) + \Psi''\left(1-\gamma+\frac{n}{2}\right) - 2\Phi(n,\gamma) - 2\Phi(n,1-\gamma) \\
&\quad - \left(\frac{11}{3} - \frac{2}{3}\frac{n_f}{N_c}\right)\frac{1}{2}\chi^2(n,\gamma) + \frac{\pi^2\cos(\pi\gamma)}{\sin^2(\pi\gamma)(1-2\gamma)} \\
&\quad \times \left\{\left(1+\frac{\tilde{n}_f}{N_c^3}\right)\frac{\gamma(1-\gamma)}{2(3-2\gamma)(1+2\gamma)}\cdot\delta_n^2 \right. \\
&\quad \left. - \left(3 + \left(1+\frac{\tilde{n}_f}{N_c^3}\right)\frac{2+3\gamma(1-\gamma)}{(3-2\gamma)(1+2\gamma)}\right)\cdot\delta_n^0\right\}. \qquad (14.8)
\end{aligned}$$

Here, δ_n^m is the Kronecker symbol, and $\Psi(z)$, $\Psi'(z)$ and $\Psi''(z)$ are the Euler Ψ-function and its derivatives. The function $\Phi(n,\gamma)$ is given by

$$\begin{aligned}
\Phi(n,\gamma) = &\sum_{k=0}^{\infty}\frac{(-1)^{k+1}}{k+\gamma+n/2}[\Psi'(k+n+1) - \Psi'(k+1) \\
&+ (-1)^{k+1}(\beta'(k+n+1) + \beta'(k+1)) \\
&- \frac{1}{k+\gamma+n/2}(\Psi(k+n+1) - \Psi(k+1))] \qquad (14.9)
\end{aligned}$$

and

$$\beta'(z) = \frac{1}{4}\left[\Psi'\left(\frac{z+1}{2}\right) - \Psi'\left(\frac{z}{2}\right)\right].$$

To obtain the corresponding results in $\mathcal{N}=4$ SYM we should account for contributions of scalars and fermions transforming in the adjoint representation of the gauge group. This way in the $\overline{DR}$ scheme [Siegel (1979)] we get

$$\delta_{\overline{DR}}^{N=4}(n,\gamma) = 6\zeta(3) + \Psi''\left(\gamma+\frac{n}{2}\right) + \Psi''\left(1-\gamma+\frac{n}{2}\right) - 2\Phi(n,\gamma) - 2\Phi(n,1-\gamma) - 2\zeta(2)\chi(n,\gamma), \tag{14.10}$$

where the $\overline{DR}$ coupling constant $\hat{a}_s$ is related to the $\overline{MS}$ one a_s as [Altarelli *et al.* (1981); Schuler *et al.* (1987)]

$$\hat{a}_s = a_s + \frac{1}{3}a_s^2. \tag{14.11}$$

It should be noted, that the sum $\Phi(n,\gamma)+\Phi(n,1-\gamma)$ can be further rewritten (see [Kotikov and Lipatov (2003, 2001)]) as a combination of functions with arguments dependent on $\gamma + n/2 \equiv M$ and $1-\gamma+n/2 \equiv \tilde{M}$ only. Indeed, we have

$$\begin{aligned}\Phi(n,\gamma)+\Phi(n,1-\gamma) &= \chi(n,\gamma)\,(\beta'(M)+\beta'(1-\widetilde{M}))\\ &\quad + \Phi_2(M) - \beta'(M)[\Psi(1)-\Psi(M)]\\ &\quad + \Phi_2(1-\widetilde{M}) - \beta'(1-\widetilde{M})[\Psi(1)-\Psi(1-\widetilde{M})],\end{aligned}$$

where $\chi(n,\gamma)$ is given by Eq. (14.7) and

$$\Phi_2(M) = \sum_{k=0}^{\infty}\frac{(\beta'(k+1)+(-1)^k\Psi'(k+1))}{k+M} - \sum_{k=0}^{\infty}\frac{(-1)^k\,(\Psi(k+1)-\Psi(1))}{(k+M)^2}. \tag{14.12}$$

From the latter property we see, that the BFKL equation in $\mathcal{N}=4$ SYM satisfies the property of hermitian separability (see Ref. [Kotikov and Lipatov (2003)] and discussions therein). Moreover, the NLO eigenvalue of the BFKL kernel in $\mathcal{N}=4$ SYM could be further rewritten in terms of analytically continued harmonic sums, see for example appendix C in [Alfimov *et al.* (2018)]. This way we first get the property of maximal transcendentality[3] [Kotikov and Lipatov (2003)] manifest directly[4] for the eigenvalues of the BFKL kernel expressed in terms of harmonic sums with well defined weights, and secondly — the expressions for the Pomeron intercept in $\mathcal{N}=4$ SYM as a function of conformal spin n take the form [Alfimov *et al.* (2018)]:

$$\omega_0^{LO} = 4\mathbb{S}_1\left(\frac{n-1}{2}\right), \tag{14.13}$$

$$\omega_0^{NLO} = 8\mathbb{S}_{2,1}\left(\frac{n-1}{2}\right) + 8\mathbb{S}_3\left(\frac{n-1}{2}\right) + \frac{4\pi^2}{3}\mathbb{S}_1\left(\frac{n-1}{2}\right), \tag{14.14}$$

[3]See the discussion for the case of anomalous dimensions in the next section.

[4]Actually it could be already inferred from the expression in Eq. (14.10), see discussion in [Kotikov and Lipatov (2003)].

where $\omega_0 = \omega(n, 1/2)$ and

$$\mathbb{S}_{i_1,\ldots,i_k}(M) = (-1)^M \sum_{j=1}^{M} (-1)^j \binom{M}{j} \binom{M+j}{j} S_{i_1,\ldots,i_k}(j). \qquad (14.15)$$

are binomial harmonic sums [Vermaseren (1999)].

14.2.2 *AdS/CFT correspondence and the Pomeron at strong coupling*

The AdS/CFT-correspondence [Maldacena (1999); Gubser *et al.* (1998); Witten (1998)] provides us with a unique possibility to study the BFKL equation and its Pomeron solution at strong coupling. In the framework of the AdS/CFT correspondence the BFKL Pomeron is equivalent to the Reggeized graviton [Polchinski and Strassler (2003)], what in particular allows us to construct the Pomeron interaction model as a generally covariant effective theory for the Reggeized gravitons [Lipatov (2013)]. The Pomeron intercept for zero conformal spin $n = 0$ and at the leading order in the inverse coupling constant could be easily obtained using the diffusive approximation for the BFKL equation [Kotikov and Lipatov (2000); Fadin and Fiore (2008)], strong coupling results for the anomalous dimensions [Gubser *et al.* (2002); Frolov and Tseytlin (2002); Roiban *et al.* (2007)] and the mentioned Pomeron-graviton duality [Polchinski and Strassler (2002, 2003)]. This is the approach, which was used in[5] [Kotikov *et al.* (2004)], see also [Brower *et al.* (2007)] for similar results. Moreover, it turns out that the approach of [Kotikov *et al.* (2004)] could be also extended to include higher order corrections in the inverse coupling constant [Kotikov and Lipatov (2013)]. First, one should note that due to the symmetry of the BFKL kernel eigenvalue in $\mathcal{N}=4$ SYM under the substitution $\gamma_{BFKL} \to 1-\gamma_{BFKL}$ the latter is an even function of ν:

$$j - 1 = \omega = \omega_0 + \sum_{m=1}^{\infty} (-1)^m D_m \nu^{2m}, \qquad (14.16)$$

where ($\lambda = g^2 N_c$ is the t'Hooft coupling constant)

$$\omega_0 = 4 \ln 2 \frac{\lambda}{4\pi^2} \left[1 - \bar{c}_1 \frac{\lambda}{16\pi^2} \right] + O(\lambda^3), \qquad (14.17)$$

$$D_m = 2 \left(2^{2m+1} - 1 \right) \zeta_{2m+1} \frac{\lambda}{4\pi^2} + \frac{\delta^{(2m)}(1/2)}{(2m)!} \frac{\lambda^2}{64\pi^4} + O(\lambda^3), \qquad (14.18)$$

and according to [Kotikov and Lipatov (2003)] we have

$$\bar{c}_1 = 2\zeta_2 + \frac{1}{2\ln 2} \left(11\zeta_3 - 32 \mathrm{Ls}_3\left(\frac{\pi}{2}\right) - 14\pi\zeta_2 \right) \approx 7.5812,$$

$$\mathrm{Ls}_3(x) = -\int_0^x \ln^2 \left| 2 \sin\left(\frac{y}{2}\right) \right| dy. \qquad (14.19)$$

[5] See the Erratum to that paper.

Second, the Möbius invariance and hermiticity of the BFKL kernel suggest that the expansion (14.16) should be also valid at strong coupling. Next, to make the connection with the results for anomalous dimensions of local composite operators[6] (string energies in the language of AdS/CFT correspondence) γ at strong coupling we use the relation of the latter to the BFKL anomalous dimension γ_{BFKL} [Fadin and Lipatov (1998); Salam (1998, 1999)]:

$$\gamma = \gamma_{BFKL} + \frac{\omega}{2} = \frac{j}{2} + i\nu. \tag{14.20}$$

Now, due to the energy-momentum conservation $\gamma(j = 2) = 0$ we have

$$\nu(j = 2) = i \tag{14.21}$$

and the small ν expansion for the eigenvalues of the BFKL kernel for Reggeized graviton takes the form

$$j - 2 = \sum_{m=1}^{\infty} D_m((-\nu^2)^m - 1), \tag{14.22}$$

where ν^2 is related to γ according to Eq. (14.20)

$$\nu^2 = -\left(\frac{j}{2} - \gamma\right)^2. \tag{14.23}$$

On the other hand, due to the ADS/CFT correspondence the string energies E in dimensionless units are related to the anomalous dimensions γ of twist-two operators as[7] [Gubser *et al.* (1998); Witten (1998)]

$$E^2 = (j + \Gamma)^2 - 4, \, \Gamma = -2\gamma \tag{14.24}$$

and therefore we can obtain from Eq. (14.23) the relation between the parameter ν of the conformal weights for the principal series of unitary representations of the Möbius group and the string energy E

$$\nu^2 = -\left(\frac{E^2}{4} + 1\right). \tag{14.25}$$

This expression for ν^2 can be inserted in the r.h.s. of Eq. (14.22) and gives us the following expression for the Regge trajectory of the graviton in the AdS space

$$j - 2 = \sum_{m=1}^{\infty} D_m \left[\left(\frac{E^2}{4} + 1\right)^m - 1\right]. \tag{14.26}$$

[6] These are anomalous dimensions of twist 2 operators contributing to the DGLAP equation [Gribov and Lipatov (1972b,a); Lipatov (1975); Altarelli and Parisi (1977); Dokshitzer (1977)].

[7] Note that our expression (14.24) for the string energy E differs from a definition, in which E is equal to the scaling dimension Δ_{sc}. Still the equation (14.24) is correct as it can be written as $E^2 = (\Delta_{sc} - 2)^2 - 4$ and coincides with Eqs. (45) and (3.44) from Refs. [Gubser *et al.* (1998)] and [Witten (1998)], respectively.

Next, we assume that Eq. (14.26) is also valid both at large j and large λ in the region $1 \ll j \ll \sqrt{\lambda}$, where the calculations of string energies at strong coupling were performed [Gromov *et al.* (2011b); Roiban and Tseytlin (2011)]. These energies can be presented in the form[8]

$$\frac{E^2}{4} = \sqrt{\lambda}\frac{S}{2}\left[h_0(\lambda) + h_1(\lambda)\frac{S}{\sqrt{\lambda}} + h_2(\lambda)\frac{S^2}{\lambda}\right] + O(S^{7/2}), \tag{14.27}$$

where

$$h_i(\lambda) = a_{i0} + \frac{a_{i1}}{\sqrt{\lambda}} + \frac{a_{i2}}{\lambda} + \frac{a_{i3}}{\sqrt{\lambda^3}} + \frac{a_{i2}}{\lambda^2}. \tag{14.28}$$

The contribution proportional to S can be extracted from the result of Basso [Basso (2011, 2012)] taking $J_{an} = 2$ according to [Gromov and Valatka (2012)]:

$$h_0(\lambda) = \frac{I_3(\sqrt{\lambda})}{I_2(\sqrt{\lambda})} + \frac{2}{\sqrt{\lambda}} = \frac{I_1(\sqrt{\lambda})}{I_2(\sqrt{\lambda})} - \frac{2}{\sqrt{\lambda}}, \tag{14.29}$$

where $I_k(\sqrt{\lambda})$ is the modified Bessel function. This gives us the following values of coefficients a_{0i}

$$a_{00} = 1, \quad a_{01} = -\frac{1}{2}, \quad a_{02} = a_{03} = \frac{15}{8}, \quad a_{04} = \frac{135}{128}. \tag{14.30}$$

Next, the coefficients a_{10} and a_{20} come from the consideration of the classical energy of the folded spinning string configurations corresponding to the twist-two operators (see, for example, [Roiban and Tseytlin (2011)])

$$a_{10} = \frac{3}{4}, \quad a_{20} = -\frac{3}{16}. \tag{14.31}$$

Finally, the one-loop coefficient a_{11} was found in [Gromov and Valatka (2012)] considering different asymptotic regimes and accounting for the Basso result [Basso (2011)]

$$a_{11} = \frac{3}{16}(1 - \zeta_3). \tag{14.32}$$

Now, comparing both sides of Eq. (14.26) at large j values gives us the desired values of the coefficients D_m (see Appendix A in [Kotikov and Lipatov (2013)]), and the value for the Pomeron intercept j_0 at zero conformal spin in the limit $\lambda \to \infty$ takes

[8]Here we put $S = j - 2$, which in particular is related to the use of the angular momentum $J_{an} = 2$ in calculations of Refs. [Gromov *et al.* (2011b); Roiban and Tseytlin (2011)].

the form[9]

$$j_0 = 2 - \frac{2}{\lambda^{1/2}}\left[1 + \frac{1}{2\lambda^{1/2}} - \frac{1}{8\lambda} - \left(1 + 3\zeta_3\right)\frac{1}{\lambda^{3/2}}\right.$$
$$\left. + \left(2a_{12} - \frac{145}{128} - \frac{9}{2}\zeta_3\right)\frac{1}{\lambda^2} + O\left(\frac{1}{\lambda^{5/2}}\right)\right] \qquad (14.33)$$

with unknown coefficient a_{12} in the $1/\lambda^2$ correction in Eq. (14.33). Later the strong coupling expansion of the Pomeron intercept for zero conformal spin was calculated even further within the quantum spectral curve (QSC) approach with the result given by [Gromov *et al.* (2014b)]:

$$j_0 = = 2 - \frac{2}{\lambda^{1/2}} - \frac{1}{\lambda} + \frac{1}{4\,\lambda^{3/2}} + (6\zeta_3 + 2)\,\frac{1}{\lambda^2} \qquad (14.34)$$
$$+ \left(18\,\zeta_3 + \frac{361}{64}\right)\frac{1}{\lambda^{5/2}} + \left(39\,\zeta_3 + \frac{511}{32}\right)\frac{1}{\lambda^3} + \mathcal{O}\left(\frac{1}{\lambda^{7/2}}\right). \qquad (14.35)$$

14.2.3 *BFKL equation and integrability*

The most intriguing property of the BFKL equation in $\mathcal{N} = 4$ SYM is its all-loop integrability tightly connected to the all-loop integrability of anomalous dimensions of local operators in $\mathcal{N} = 4$ SYM, see [Beisert *et al.* (2012)] for a review. Initially, the integrability of multicolour QCD was discovered in LO by Lev Lipatov precisely in the study of the Regge limit of high-energy scattering [Lipatov (1994); Faddeev and Korchemsky (1995)] and only later in the study of anomalous dimensions of composite operators [Lipatov (1997, 2001)]. Calculations in the maximally supersymmetric $\mathcal{N} = 4$ SYM, where the coupling constant is not renormalized and all twist-2 operators enter in the same multiplet, showed that their anomalous dimension matrix is fixed completely by the super-conformal invariance and its entries are expressed in terms of the universal anomalous dimension, which in LO is proportional to $\Psi(S-1) - \Psi(1)$ (see the discussion in next section). The latter property means, that the evolution equations for the matrix elements of quasi-partonic operators in the multicolour limit $N_c \to \infty$ are equivalent to the Schrödinger equation for an integrable Heisenberg spin model [Lipatov (1997, 2001)]. Subsequent study of these and other operators both at leading and higher orders, starting from the works of [Minahan and Zarembo (2003b); Beisert and Staudacher (2003b)] showed that the discovered integrability property in $\mathcal{N} = 4$ SYM may be extended to all loops. It turned out that a lot of methods developed for two-dimensional integrable systems, such as the sigma-model and the spin-chain S-matrices [Staudacher (2005); Arutyunov *et al.* (2004); Beisert (2008, 2007); Beisert *et al.* (2007a); Janik (2006); Arutyunov and Frolov (2006); Arutyunov *et al.* (2007)],

[9]Using a similar approach, the coefficients $\sim \lambda^{-1}$ and $\sim \lambda^{-3/2}$ were calculated also in the paper [Costa *et al.* (2012)].

Asymptotic Bethe Ansatz (ABA) [Minahan and Zarembo (2003b); Beisert and Staudacher (2003b, 2005); Beisert *et al.* (2007a)], Thermodynamic Bethe Ansatz (TBA) [Gromov *et al.* (2009); Bombardelli *et al.* (2009); Gromov *et al.* (2010a); Arutyunov and Frolov (2009b)] as well as Y and T-systems [Cavaglia *et al.* (2011); Balog and Hegedus (2011); Gromov *et al.* (2011a, 2012)] could be also used for the computation of anomalous dimensions of $\mathcal{N}=4$ SYM operators not only at weak but also strong and in general finite coupling constants. The most advanced framework for such spectral problem calculations at the moment is offered by the quantum spectral curve (QSC) method. The latter is an alternative reformulation of the TBA equations as a finite set of algebraic relations for Baxter type Q-functions together with analyticity and Riemann-Hilbert monodromy conditions for the latter [Gromov *et al.* (2014a, 2015a); Kazakov *et al.* (2016); Marboe and Volin (2017, 2018)]. Within the quantum spectral curve formulation one can relatively easily obtain a numerical solution for any coupling and state [Gromov *et al.* (2016); Hegedus and Konczer (2016)]. Also, the QSC formulation allowed to construct iterative perturbative solutions for these theories at weak coupling up to, in principle, arbitrary loop order [Marboe and Volin (2015, 2018)]. What is more important QSC allows also for the study of the BFKL regime [Alfimov *et al.* (2015); Gromov *et al.* (2015b); Alfimov *et al.* (2018)].

The integrability of the LO BFKL equation (in the leading logarithmic approximation) could be conveniently studied in the impact parameter space $\vec{\rho}$, where the BFKL equation has the Schrödinger like form [Lipatov (1994)]:

$$Ef(\vec{\rho_1},\vec{\rho_2}) = H_{12}f(\vec{\rho_1},\vec{\rho_2}), \tag{14.36}$$

where in the leading logarithmic approximation the Hamiltonian H_{12} takes the holomorphically separable form

$$\begin{aligned} H_{12} &= h_{12} + h_{12}^* \\ h_{12} &= \ln(p_1 p_2) + \frac{1}{p_1}(\ln \rho_{12})p_1 + \frac{1}{p_2}(\ln \rho_{12})p_2 + 2\gamma_E. \end{aligned} \tag{14.37}$$

Here impact parameters are complex coordinates $\rho_k = x_k + iy_k$, $\rho_{12} = \rho_1 - \rho_2$ and p_k, p_k^* play the role of their canonically conjugated momenta. Moreover, it turns out that the Hamiltonian H_{12} coincides with the Hamiltonian of the Heisenberg (XXX) $SL(2,C)$ spin chain [Lipatov (1994)], where $SL(2,C)$ (Möbius) group generators are given by

$$M_k^3 = \rho_k\partial_k, \quad M_k^+ = \partial_k, \quad M_k^- = -\rho_k^2\partial_k \tag{14.38}$$

and $\partial_k = \partial/(\partial\rho_k)$.

The eigenfunctions of the LO BFKL Hamiltonian are easy to find with the use of Möbius invariance. This way we get wave functions labelled by two conformal weights (14.5) [Lipatov (1986)]:

$$f_{m,\tilde{m}}(\vec{\rho}_1, \vec{\rho}_2; \vec{\rho}_0) = \left(\frac{\rho_{12}}{\rho_{10}\rho_{20}}\right)^m \left(\frac{\rho_{12}^*}{\rho_{10}^*\rho_{20}^*}\right)^{\tilde{m}}, \tag{14.39}$$

with eigenvalues

$$E_{m,\tilde{m}} = 4\text{Re}\left[\Psi\left(\frac{1+n}{2} + \imath\nu\right)\right] - 4\Psi(1). \tag{14.40}$$

The Pomeron intercept as a function of conformal spin n is then given by[10]

$$\omega_0 = -\frac{\lambda}{8\pi^2} E_{m,\tilde{m}}\Big|_{\nu=0}. \tag{14.41}$$

Having established a connection of the LO BFKL Hamiltonian with the integrable $SL(2, C)$ spin chain, we may write the generating function for its integrals of motion q_r in terms of the spin chain transfer matrix $T(u)$ [Lipatov (1994)]:

$$T(u) = \text{Tr}\left(L_1(u)L_2(u)\dots L_n(u)\right) = \sum_{r=0}^{n} u^{n-r} q_r, \tag{14.42}$$

where the Lax L_k - operators are given by

$$L_k(u) = \begin{pmatrix} u + \rho_k p_k & p_k \\ -\rho_k^2 p_k & u - \rho_k p_k \end{pmatrix} = \begin{pmatrix} u & 0 \\ 0 & u \end{pmatrix} + \begin{pmatrix} 1 \\ -\rho_k \end{pmatrix} (\rho_k \; 1)\, p_k. \tag{14.43}$$

To study the BFKL equation within the integrability approach beyond LO it is convenient to use the quantum spectral curve method [Gromov *et al.* (2014a, 2015a); Kazakov *et al.* (2016); Marboe and Volin (2017, 2018)]. Besides reproducing the LO BFKL eigenvalues [Alfimov *et al.* (2015)] authors of [Gromov *et al.* (2015b)] were able to deduce the next-to-next-to-leading order (NNLO) corrections for BFKL eigenvalues at zero conformal spin in agreement with [Velizhanin (2015); Caron-Huot and Herranen (2018)]. In addition, the authors of [Alfimov *et al.* (2018)] obtained NNLO corrections for the Pomeron intercept as a function of conformal spin n:

$$\omega_0^{NNLO} = 32\left(\mathbb{S}_{1,4} - \mathbb{S}_{3,2} - \mathbb{S}_{1,2,2} - \mathbb{S}_{2,2,1} - 2\mathbb{S}_{2,3}\right) - \frac{16\pi^2}{3}\mathbb{S}_3 - \frac{32\pi^4}{45}\mathbb{S}_1, \tag{14.44}$$

where the arguments of the binomial sums (14.15) are $(n-1)/2$. There are also partial results for the next-to-next-to-next-to-leading order (NNNLO) intercept [Alfimov *et al.* (2018)].

The range of applicability of the QSC approach is not limited by weak coupling expansions but could be also used to obtain results at strong coupling. This way for

[10] $\lambda = g^2 N_c$ is again the t'Hooft coupling constant.

Pomeron intercept as a function of conformal spin n we get [Alfimov *et al.* (2018)]:

$$j_0 = 2 - n + \frac{(n-1)(n+2)}{\lambda^{1/2}} - \frac{(n-1)(n+2)(2n-1)}{2\lambda} + \frac{(n-1)(n+2)(7n^2-9n-1)}{8\lambda^{3/2}} + \mathcal{O}\left(\frac{1}{\lambda^2}\right). \tag{14.45}$$

There are also analytical results at finite values of the coupling constant for the slope to intercept and curvature functions [Alfimov *et al.* (2018)]. Finally, it should be mentioned that numerically the QSC approach allows the study of the BFKL eigenvalues at arbitrary values of the coupling constant without restrictions to the weak or strong coupling regime considered above, see [Alfimov *et al.* (2018)] for more details.

14.3 Anomalous dimensions and the DGLAP equation

Phenomenologically, Bjorken scaling violation for parton distributions[11] in a framework of QCD is governed by the anomalous dimensions matrix $\gamma_{ab}(j)$ (here $a_s = \alpha_s/(4\pi)$)

$$\gamma_{ab}(j) = \int_0^1 dx\, x^{j-1} W_{b\to a}(x) \;=\; \sum_{k=0}^{\infty} \gamma_{ab}^{(k)}(j) a_s^{k+1}, \tag{14.46}$$

related to Mellin moments of splitting kernels $W_{b\to a}(x)$ entering the DGLAP evolution equations [Gribov and Lipatov (1972b)] for parton distribution densities $f_a(x,Q^2)$ (hereafter $a = \lambda,\, g,\, \phi$ stands for the spinor, vector and scalar particles respectively):

$$\frac{d}{d\ln Q^2} f_a(x,Q^2) = \int_x^1 \frac{dy}{y} \sum_b W_{b\to a}(x/y)\, f_b(y,Q^2),$$

$$\frac{d}{d\ln Q^2} \tilde{f}_a(x,Q^2) = \int_x^1 \frac{dy}{y} \sum_b \tilde{W}_{b\to a}(x/y)\, \tilde{f}_b(y,Q^2). \tag{14.47}$$

In QCD the anomalous dimensions and splitting kernels are completely known up to NNLO of perturbation theory (see [Moch *et al.* (2004); Vogt *et al.* (2004)] and references therein). The similar results in the case of $\mathcal{N}$ extended SYM may be obtained from the corresponding QCD results with the help of Casimir substitutions: $C_A = C_F = N_c$, $T_f n_f = \mathcal{N} N_c/2$. In addition, in the case of $\mathcal{N}=2$ and $\mathcal{N}=4$ SYM we should also account for contributions of scalar particles, see [Kotikov and Lipatov (2003); Kotikov *et al.* (2003)] for the case of $\mathcal{N}=4$ SYM. It turns out, however, that the latter calculations accounting for scalar contributions could be completely abandoned in the case of $\mathcal{N}=4$ SYM.

---

[11] Here we consider only the spin-average case. The results for the spin-dependent case can be found in [Kotikov and Lipatov (2003)].

Indeed, the expressions for the eigenvalues of the anomalous dimension matrix in the $\mathcal{N} = 4$ SYM can be derived directly from the QCD anomalous dimensions without tedious calculations by using a number of plausible arguments. The method elaborated in Ref. [Kotikov and Lipatov (2003)] for this purpose is based on special properties of the integral kernel for the BFKL equation [Lipatov (1976); Fadin and Lipatov (1998); Kotikov and Lipatov (2000)] in this model and a new relation between the BFKL and DGLAP equations (see [Kotikov and Lipatov (2003)]). First, it turns out that the expression for the BFKL kernel in $\mathcal{N} = 4$ SYM possess a property of maximal transcendentality.[12] Next, using the mentioned relation of the BFKL and DGLAP equations in this model [Kotikov and Lipatov (2003)] we may conjecture the similar property of maximal transcendentality for anomalous dimensions [Kotikov and Lipatov (2003)]. Moreover, it turns out that the eigenvalues of the anomalous dimensions matrix in the case of $\mathcal{N} = 4$ SYM are given by the maximal transcendentality part of the eigenvalues for the QCD anomalous dimension matrix. In the NLO approximation this method gives the correct results for anomalous dimension eigenvalues, which were checked by *direct calculations* in Ref. [Kotikov *et al.* (2003)]. Using the results for the NNLO corrections to the anomalous dimensions in QCD [Moch *et al.* (2004); Vogt *et al.* (2004)] and the method of Ref. [Kotikov and Lipatov (2003)] we have also derived the NNLO eigenvalues of the anomalous dimension matrix in $\mathcal{N} = 4$ SYM [Kotikov *et al.* (2004)], which were later confirmed by the integrability predictions in [Staudacher (2005)].

Starting from four loops, i.e. above the existing QCD results, the corresponding results for the anomalous dimensions in $\mathcal{N} = 4$ SYM can be obtained (see [Kotikov *et al.* (2007); Bajnok *et al.* (2009); Lukowski *et al.* (2010)]) for example from the long-range asymptotic Bethe ansatz equations together with some additional terms, the so-called *finite size* or *wrapping corrections*, coming in agreement with the Lüscher approach.[13]

14.3.1 *The maximal transcendentality principle*

As we have already mentioned above in $\mathcal{N} = 4$ SYM there is a way to deduce eigenvalues of the anomalous dimension matrix from QCD results without computing contributions of scalar particles. This possibility is based on the deep relation between the DGLAP and BFKL dynamics in the $\mathcal{N} = 4$ SYM [Kotikov and Lipatov (2000, 2003)].

Indeed, the eigenvalues of the BFKL kernel in $\mathcal{N} = 4$ SYM are analytic functions of the conformal spin n at least in the first two orders of perturbation theory (see

[12] See discussion later in this section.

[13] The three- and four-loop results for the universal anomalous dimension have been also reproduced (see [Kotikov *et al.* (2009); Beccaria *et al.* (2010a)]) by the solution of the so-called asymptotic long-range Baxter equation.

Eqs. (14.6), (14.7) and (14.10)). Moreover, a detailed inspection of equations (14.7) and (14.10) shows that there is no mixing among special functions of different transcendentality levels i.[14] Indeed, in the framework of the $\overline{\text{DR}}$-scheme [Siegel (1979)], all special functions entering the expression for NLO correction contain only sums of the terms $\sim 1/\gamma^i$ $(i = 3)$. More precisely, introducing the transcendentality level i for eigenvalues $\omega(\gamma)$ of the BFKL integral kernel at zero conformal spin in an accordance with the complexity of the terms in the corresponding sums

$$\Psi \sim 1/\gamma, \quad \Psi' \sim \beta' \sim \zeta(2) \sim 1/\gamma^2, \quad \Psi'' \sim \beta'' \sim \Phi \sim \zeta(3) \sim 1/\gamma^3,$$

we see that transcendentality levels of the LO and NLO BFKL kernel eigenvalues are given by $i = 1$ and $i = 3$, respectively.

Now, taking into account the relation between the BFKL and DGLAP equations in $\mathcal{N} = 4$ SYM (see [Kotikov and Lipatov (2000, 2003)]) we may conjecture that similar properties should be also valid for the anomalous dimensions themselves, i.e. the universal anomalous dimensions $\gamma_{uni}^{(0)}(j)$, $\gamma_{uni}^{(1)}(j)$ and $\gamma_{uni}^{(2)}(j)$ are assumed to have transcendentality levels $i = 1$, $i = 3$ and $i = 5$, respectively. An exception could be for the terms appearing at a given order from previous orders of perturbation theory and having a lower transcendentality level. Such contributions could be removed by an approximate finite renormalization of the coupling constant. Moreover, these terms do not appear in the $\overline{\text{DR}}$-scheme.

It is known, that in the LO and NLO approximations (accounting for SUSY relation for QCD colour factors $C_F = C_A = N_c$) the most complicated contributions (with $i = 1$ and $i = 3$, respectively) are the same for all LO and NLO eigenvalues of the QCD anomalous dimensions matrix [Moch *et al.* (2004); Vogt *et al.* (2004)] and also for the LO and NLO scalar-scalar anomalous dimensions in $\mathcal{N} = 4$ SYM [Kotikov *et al.* (2003)]. This property allows one to find the $\mathcal{N} = 4$ SYM universal anomalou dimension $\gamma_{uni}^{(0)}(j)$ and $\gamma_{uni}^{(1)}(j)$ without knowing all elements of the anomalous dimension matrix [Kotikov and Lipatov (2003)], which was verified by the exact calculations in [Kotikov *et al.* (2003)].

Using the above arguments it was concluded in [Kotikov and Lipatov (2003)], that at the NNLO level there is only one possible candidate for $\gamma_{uni}^{(2)}(j)$. Namely, it is the most complicated part of the QCD anomalous dimension matrix (accounting again for SUSY relation for QCD colour factors $C_F = C_A = N_c$). Indeed, after the diagonalization of the QCD anomalous dimension matrix the most complicated parts (having maximum transcendentality level) of its eigenvalues differ from each other by a shift in their arguments (Lorentz spin) and the differences are constructed

[14]Similar arguments were used also in [Fleischer *et al.* (1999); Kotikov *et al.* (2008)] to obtain analytic results for contributions of some complicated massive Feynman diagrams without direct calculations. For the relation of diagrams without sub-divergences and transcendentalities (knots) see [Broadhurst and Kreimer (1995, 1997)].

from less complicated terms. The non-diagonal matrix elements of the anomalous dimensions matrix also contain only less complicated terms (see, for example exact expressions for LO and NLO QCD anomalous dimensions in Refs. [Moch *et al.* (2004); Vogt *et al.* (2004)] and [Kotikov *et al.* (2003)] for the case of $\mathcal{N} = 4$ SYM) and as such they cannot contribute to the most complicated parts of the eigenvalues of the anomalous dimension matrix. Summarizing, the most complicated part of NNLO eigenvalues of QCD anomalous dimensions should coincide (up to colour factors) with the $\mathcal{N} = 4$ SYM universal anomalous dimension $\gamma_{uni}^{(2)}(j)$. All these arguments in general apply to singlet anomalous dimensions. Nevertheless, as was shown in [Kotikov *et al.* (2004)] the universal anomalous dimension in $\mathcal{N} = 4$ SYM could be also obtained from corresponding QCD results for nonsinglet anomalous dimensions available at that moment.

We would like to note that the transcendentality was either used or has been observed also in the study of other quantities. In the same way as above, in [Bianchi *et al.* (2013)] the universal coefficient function of deep inelastic scattering in the framework of $\mathcal{N} = 4$ SYM was obtained from the most complicated part of the corresponding QCD Wilson coefficients [Vermaseren *et al.* (2005)]. For a more recent application of the transcendentality principle see for example the discussion of energy -energy correlations [Dixon *et al.* (2019)] and collinear anomalous dimensions in [Dixon (2018)] in the case of QCD and $\mathcal{N} = 4$ SYM. The maximal transcendentality property was also observed in the study of scattering amplitudes (see [Caron-Huot *et al.* (2019)] and references therein) and formfactors [Loebbert *et al.* (2016)].

14.3.2 *Universal anomalous dimensions in $\mathcal{N} = 4$ SYM*

Let us consider universal[15] anomalous dimensions of the following colour and $SU(4)$ singlet Wilson twist-2 operators

$$\mathcal{O}^{g}_{\mu_1,\ldots,\mu_j} = \hat{S} G^{a}_{\rho\mu_1} \mathcal{D}_{\mu_2} \mathcal{D}_{\mu_3} \ldots \mathcal{D}_{\mu_{j-1}} G^{a}_{\rho\mu_j}, \tag{14.48}$$

$$\mathcal{O}^{\lambda}_{\mu_1,\ldots,\mu_j} = \hat{S} \bar{\lambda}^{a}_{i} \gamma_{\mu_1} \mathcal{D}_{\mu_2} \cdots \mathcal{D}_{\mu_j} \lambda^{a\,i}, \tag{14.49}$$

$$\mathcal{O}^{\phi}_{\mu_1,\ldots,\mu_j} = \hat{S} \bar{\phi}^{a}_{r} \mathcal{D}_{\mu_1} \mathcal{D}_{\mu_2} \cdots \mathcal{D}_{\mu_j} \phi^{a}_{r}, \tag{14.50}$$

where $\mathcal{D}_\mu$ are covariant derivatives. The spinors λ_i and gauge field tensor $G_{\rho\mu}$ describe gluino and gluon field respectively, while ϕ_r stands for complex scalar fields. For all operators in Eqs. (14.48)-(14.50) $\hat{S}$ denotes symmetrization of the tensors in the Lorentz indices $\mu_1, \ldots, \mu_j$ and subtraction of their traces.

[15] The considered operators belong to the same $\mathcal{N} = 4$ SYM superconformal multiplet and as such share the same universal anomalous dimension.

The elements of the LO anomalous dimension matrix in the $\mathcal{N}=4$ SYM are given by (see [Lipatov (2001)]):

$$\begin{aligned}
\gamma_{gg}^{(0)}(j) &= 4\left(\Psi(1)-\Psi(j-1)-\frac{2}{j}+\frac{1}{j+1}-\frac{1}{j+2}\right),\\
\gamma_{\lambda g}^{(0)}(j) &= 8\left(\frac{1}{j}-\frac{2}{j+1}+\frac{2}{j+2}\right), \qquad \gamma_{\varphi g}^{(0)}(j) \;=\; 12\left(\frac{1}{j+1}-\frac{1}{j+2}\right),\\
\gamma_{g\lambda}^{(0)}(j) &= 2\left(\frac{2}{j-1}-\frac{2}{j}+\frac{1}{j+1}\right), \qquad \gamma_{q\varphi}^{(0)}(j) \;=\; \frac{8}{j},\\
\gamma_{\lambda\lambda}^{(0)}(j) &= 4\left(\Psi(1)-\Psi(j)+\frac{1}{j}-\frac{2}{j+1}\right), \quad \gamma_{\varphi\lambda}^{(0)}(j) \;=\; \frac{6}{j+1},\\
\gamma_{\varphi\varphi}^{(0)}(j) &= 4\left(\Psi(1)-\Psi(j+1)\right), \qquad \gamma_{g\varphi}^{(0)}(j) \;=\; 4\left(\frac{1}{j-1}-\frac{1}{j}\right).
\end{aligned} \tag{14.51}$$

This anomalous dimension matrix could be easily diagonalized and we get the simple expression [Lipatov (2001); Kotikov and Lipatov (2003)]:

$$D\Gamma D^{-1} = \begin{vmatrix} -4S_1(j-2) & 0 & 0 \\ 0 & -4S_1(j) & 0 \\ 0 & 0 & -4S_1(j+2) \end{vmatrix},$$

where

$$S_1(j) \equiv \Psi(j+1) - \Psi(1) \equiv \sum_{r=1}^{j} \frac{1}{r}.$$

From this we conclude that at LO the universal anomalous dimension for the considered operator supermultiplet is given by

$$\gamma_{uni}^{(0)}(j) \;=\; -\,4S_1(j-2).$$

The universal anomalous dimensions at NLO and NNLO could be obtained using the method described in subsection 14.3.1. This way up to three loops we get[16] [Kotikov and Lipatov (2003); Kotikov *et al.* (2004)]:

$$\gamma(j) \equiv \gamma_{uni}(j) = \hat{a}\gamma_{uni}^{(0)}(j) + \hat{a}^2\gamma_{uni}^{(1)}(j) + \hat{a}^3\gamma_{uni}^{(2)}(j) + \cdots, \qquad \hat{a} = \frac{\alpha N_c}{4\pi}, \tag{14.52}$$

[16]Note, that in an accordance with Ref. [Fadin and Lipatov (1998)] our normalization of $\gamma(j)$ contains the extra factor $-1/2$ in comparison with the standard normalization (see [Kotikov and Lipatov (2003)]) and differs by sign in comparison with the one from Ref. [Moch *et al.* (2004); Vogt *et al.* (2004)].

where

$$\frac{1}{4}\gamma_{uni}^{(0)}(j+2) = -S_1, \tag{14.53}$$

$$\frac{1}{8}\gamma_{uni}^{(1)}(j+2) = \Big(S_3 + \overline{S}_{-3}\Big) - 2\,\overline{S}_{-2,1} + 2\,S_1\Big(S_2 + \overline{S}_{-2}\Big), \tag{14.54}$$

$$\begin{aligned}\frac{1}{32}\gamma_{uni}^{(2)}(j+2) &= 2\,\overline{S}_{-3}\,S_2 - S_5 - 2\,\overline{S}_{-2}\,S_3 - 3\,\overline{S}_{-5} + 24\,\overline{S}_{-2,1,1,1}\\ &\quad + 6\Big(\overline{S}_{-4,1} + \overline{S}_{-3,2} + \overline{S}_{-2,3}\Big) - 12\Big(\overline{S}_{-3,1,1} + \overline{S}_{-2,1,2} + \overline{S}_{-2,2,1}\Big)\\ &\quad - \Big(S_2 + 2\,S_1^2\Big)\Big(3\,\overline{S}_{-3} + S_3 - 2\,\overline{S}_{-2,1}\Big) - S_1\Big(8\,\overline{S}_{-4} + \overline{S}_{-2}^{2}\\ &\quad + 4\,S_2\,\overline{S}_{-2} + 2\,S_2^2 + 3\,S_4 - 12\,\overline{S}_{-3,1} - 10\,\overline{S}_{-2,2} + 16\,\overline{S}_{-2,1,1}\Big),\end{aligned} \tag{14.55}$$

and $S_a \equiv S_a(j),\ S_{a,b} \equiv S_{a,b}(j),\ S_{a,b,c} \equiv S_{a,b,c}(j)$ are harmonic sums

$$S_{\pm a}(j) = \sum_{r=1}^{j} \frac{(\pm 1)^r}{r^a}, \quad S_{\pm a,\pm b,\cdots}(j) = \sum_{r=1}^{j} \frac{(\pm 1)^r}{r^a} S_{\pm b,\cdots}(r) \tag{14.56}$$

and

$$\overline{S}_{-a,b,c,\cdots}(j) \;=\; (-1)^j\, S_{-a,b,c,\cdots}(j) + S_{-a,b,c,\cdots}(\infty)\Big(1 - (-1)^j\Big). \tag{14.57}$$

The expression (14.57) is defined for positive integer values of arguments (see [Kazakov and Kotikov (1988); Kotikov and Lipatov (2003); Kotikov and Velizhanin (2005)]) but can be easily analytically continued to real and complex j by the method of Refs. [Kazakov and Kotikov (1988); Kotikov and Velizhanin (2005); Kotikov (1994)]. It is interesting to study various limits of the obtained expressions for universal anomalous dimensions.

14.3.2.1 *The limit $j \to 1$*

The limit $j \to 1$ is important for the investigation of the small-x behaviour of parton distributions (see review [Andersson *et al.* (2002)] and references therein). Especially it became popular recently due to new experimental data for high energy processes studied at the LHC. In addition, there are very precise combined experimental data at small x produced by H1 and ZEUS collaborations at HERA [Aaron *et al.* (2010)].

Using asymptotic expressions for harmonic sums at $j = 1+\omega \to 1$ (see [Kotikov and Lipatov (2003); Kotikov *et al.* (2004)]) the universal anomalous dimension $\gamma_{uni}(j)$ in Eq. (14.52) takes the form

$$\gamma_{uni}^{(0)}(1+\omega) = \frac{4}{\omega} + \mathcal{O}\Big(\omega^1\Big), \tag{14.58}$$

$$\gamma_{uni}^{(1)}(1+\omega) = -32\,\zeta_3 + \mathcal{O}\Big(\omega^1\Big), \tag{14.59}$$

$$\gamma_{uni}^{(2)}(1+\omega) = 32\zeta_3\,\frac{1}{\omega^2} - 232\,\zeta_4\,\frac{1}{\omega} - 1120\zeta_5 + 256\zeta_3\zeta_2 + \mathcal{O}\Big(\omega^1\Big) \tag{14.60}$$

in agreement with the predictions coming from the BFKL equation at NLO accuracy [Kotikov and Lipatov (2000)].

14.3.2.2 *The limit $j \to 4$*

This limit corresponds to the Konishi operator [Konishi (1984)], which was used extensively in the integrability [Beisert *et al.* (2012)] and AdS-CFT correspondence [Maldacena (1999); Gubser *et al.* (1998); Witten (1998)] studies recently. The anomalous dimension for the Konishi supermultiplet coincides with our expression (14.52) for $j = 4$

$$\gamma_{uni}(j)\Big|_{j=4} = -6\,\hat{a}_s + 24\,\hat{a}_s^2 - 168\,\hat{a}_s^3, \tag{14.61}$$

which was also confirmed by direct calculations at two [Arutyunov *et al.* (2002); Eden *et al.* (2005b); Eden (2006); Kotikov *et al.* (2003)] and three-loop [Eden *et al.* (2005a)] orders. Now there are results for four, five, six, seven and up to eleven loop corrections to its anomalous dimension [Bajnok and Janik (2009); Fiamberti *et al.* (2008); Velizhanin (2009)], [Bajnok *et al.* (2010); Arutyunov *et al.* (2010); Balog and Hegedus (2010); Eden *et al.* (2012)], [Leurent *et al.* (2012); Marboe *et al.* (2015)], [Bajnok and Janik (2012); Marboe and Velizhanin (2016)] and [Marboe and Volin (2015, 2018)]. Moreover, there are analytical results at strong coupling [Gromov and Valatka (2012); Gromov *et al.* (2011b, 2014b)] and numerical calculations [Gromov *et al.* (2010b); Frolov (2011); Gromov *et al.* (2016); Hegedus and Konczer (2016)] of these anomalous dimensions at any finite coupling.

14.3.2.3 *The limit $j \to \infty$*

This limit is related to the study of the asymptotics of structure functions and cross-sections at $x \to 1$ corresponding to the quasi-elastic kinematics of the deep-inelastic *ep* scattering. In the limit $j \to \infty$ the expressions for anomalous dimensions (14.53)-(14.55) simplify significantly and take the form

$$\gamma_{uni}^{(0)}(j) = -4\Big(\ln j + \gamma_E\Big) + \mathcal{O}\Big(j^{-1}\Big), \tag{14.62}$$

$$\gamma_{uni}^{(1)}(j) = 8\zeta_2\,\Big(\ln j + \gamma_E\Big) + 12\zeta_3 + \mathcal{O}\Big(j^{-1}\Big), \tag{14.63}$$

$$\gamma_{uni}^{(2)}(j) = -88\zeta_4\,\Big(\ln j + \gamma_E\Big) - 16\zeta_2\zeta_3 - 80\zeta_5 + \mathcal{O}\Big(j^{-1}\Big), \tag{14.64}$$

where γ_E is the Euler constant.

The strong coupling expansion in this limit could be again studied with the use of the AdS/CFT correspondence [Maldacena (1999); Gubser *et al.* (1998); Witten (1998)]. The latter provides us with a very interesting strong coupling prediction

[Gubser *et al.* (2002)] (see also [Kruczenski (2002); Makeenko (2003); Axenides *et al.* (2003)]) for the large-j behaviour of the anomalous dimensions of twist-2 operators:

$$\gamma(j) = a(z) \ln j, \quad z = \frac{\alpha N_c}{\pi} = 4\hat{a}_s, \tag{14.65}$$

where (see Ref. [Frolov and Tseytlin (2002)] for asymptotic corrections)

$$\lim_{z\to\infty} a = -z^{1/2} + \frac{3\ln 2}{8\pi} + \mathcal{O}\left(z^{-1/2}\right). \tag{14.66}$$

On the other hand, the results for $\gamma_{uni}(j)$ in Eqs. (14.52) and (14.62)–(14.64) allow us to find three first terms of the small-z expansion of the coefficient $a(z)$

$$\lim_{z\to 0} a = -z + \frac{\pi^2}{12} z^2 - \frac{11}{720}\pi^4 z^3 + \cdots. \tag{14.67}$$

To resum this series Lev Lipatov suggested the following equation for the approximate function $\tilde{a}$ [Kotikov *et al.* (2003)]

$$z = -\tilde{a} + \frac{\pi^2}{12}\tilde{a}^2, \tag{14.68}$$

Solving the latter we may find its weak-coupling expansion

$$\tilde{a} = -z + \frac{\pi^2}{12} z^2 - \frac{1}{72}\pi^4 z^3 + \mathcal{O}(z^4) \tag{14.69}$$

and strong-coupling asymptotics

$$\tilde{a} = -\frac{2\sqrt{3}}{\pi} z^{1/2} + \frac{6}{\pi^2} + \mathcal{O}\left(z^{-1/2}\right) \approx -1.1026\, z^{1/2} + 0.6079 + \mathcal{O}\left(z^{-1/2}\right). \tag{14.70}$$

It is remarkable, that the predictions for the NNLO corrections based on the above simple equation and the NNLO results for universal anomalous dimensions (14.55) and (14.64)) agree with each other with the accuracy $\sim$ 10%. It means, that this extrapolation seems to be good for all values of z.[17]

The limit $j \to \infty$ could be also studied with the help of the Beisert-Eden-Staudacher (BES) integral equation [Beisert *et al.* (2007b)] for some function $f(x, z)$ related to $a(z)$ in (14.65) at $x = 0$, i.e. $f(0, z) = a(z)$. At weak coupling z this equation gives a possibility to easily calculate as many coefficients c_m in the expansion

$$f(0, z) = \sum_{m=0} c_m\, z^m,$$

as needed. The coefficients c_m respect the principle of maximal transcendentality, i.e. $c_m \sim \zeta(2m)$ for $m > 0$ (or products of ζ-functions with the sum of indices equal to $2m$). Moreover, up to 4-loop order these coefficients are in agreement with

[17] Some improvement of (14.68) can be found in [Bern *et al.* (2007b)].

the ones, obtained directly from the calculation of Feynman diagrams [Bern *et al.* (2007a,b)].

The analysis of the strong coupling limit $z \to \infty$ of the BES equation is less straightforward. Still, the Gupser-Klebanov-Polyakov strong coupling asymptotics $\sim z^{1/2}$ (see the r.h.s. of (14.66)) was first reproduced from the BES equation numerically in [Benna *et al.* (2007)] and then analytically in [Kotikov and Lipatov (2007)]. Later the study of the strong coupling limit of the function $f(0,z)$ continued in Refs. [Alday *et al.* (2007); Kostov *et al.* (2008b); Beccaria *et al.* (2007); Casteill and Kristjansen (2007); Basso *et al.* (2008); Kostov *et al.* (2008a)]. In particular, the authors of [Basso *et al.* (2008)] found a way to evaluate the subleading coefficients $\tilde{c}_m$ of the expansion

$$f(0,z) = \sum_{m=0} \tilde{c}_m \, z^{(1-m)/2}.$$

The first three coefficients are in agreement with string side calculations performed in [Gubser *et al.* (2002)], [Frolov and Tseytlin (2002)] and [Roiban *et al.* (2007)], respectively. Moreover, the results of [Basso *et al.* (2008)] are also in agreement with the maximum transcendentality principle: $\tilde{c}_1 \sim \ln 2$ and $\tilde{c}_m \sim \zeta(m)$ for $m > 1$ (or products of ζ-function with the sum of indexes equal to m).

14.3.3 *Anomalous dimensions and integrability*

As we already mentioned in the section devoted to the integrability of the BFKL equation in $\mathcal{N}=4$ SYM the DGLAP equation in this theory is also integrable. Moreover, the integrability property of the DGLAP equation extends in this theory in the planar limit to all loops. First hints that this might be the case came from the calculation of anomalous dimensions of twist 2 operators in $\mathcal{N}=4$ SYM [Lipatov (1997, 2001)]. Nevertheless, the real breakthrough in the study of integrability of $\mathcal{N}=4$ SYM theory started only later with the work of Minahan and Zarembo [Minahan and Zarembo (2003b)]. First, Beisert computed the complete one loop dilatation operator in this theory [Beisert (2004)] and almost immediately Beisert and Staudacher [Beisert and Staudacher (2003b)] showed that the one-loop dilatation operator in $\mathcal{N}=4$ SYM may be identified with the Hamiltonian of the integrable $\mathfrak{psu}(2,2|4)$ super spin chain. This observation together with available techniques for integrable spin chains based on simple Lie algebras [Reshetikhin (1985); Ogievetsky and Wiegmann (1986)] and superalgebras [Saleur (2000)] allowed the authors of [Beisert and Staudacher (2003b)] to write down the one-loop Bethe ansatz equations for the discovered $\mathcal{N}=4$ SYM spin chain. Next, after a lot of preliminary work the authors of [Beisert and Staudacher (2005)] came with the conjecture for the all-loop asymptotic Bethe ansatz equations up to the so called dressing factor [Arutyunov *et al.* (2004)], which was fixed later in [Beisert *et al.* (2007a)] based on the crossing equation proposed in [Janik (2006)]. In what follows we will briefly describe these and subsequent developments in the study of the integrability structure of $\mathcal{N}=4$ SYM.

14.3.3.1 *Asymptotic Bethe-ansatz and Lüscher corrections*

The problem of the calculation of anomalous dimensions could be reduced to the problem of diagonalization of the dilatation operator, which is one of the generators of superconformal symmetry of $\mathcal{N}=4$ SYM. The latter problem in a language of the corresponding $\mathfrak{psu}(2,2|4)$ super spin chain is equivalent to finding a spectrum of the spin chain Hamiltonian upon identification of single trace operators with its states. At one-loop order the Hamiltonian of $\mathfrak{psu}(2,2|4)$ is the one of spin chain with only nearest neighbour interactions. Beyond one loop the spin chain becomes long range with its all-loop asymptotic Bethe ansatz equations given in [Beisert and Staudacher (2005)]. The obtained in [Beisert and Staudacher (2005)] Bethe ansatz is asymptotic in a sense that it is valid only when the length of the spin chain exceeds the interaction order. Another novel feature of the $\mathcal{N}=4$ spin chain is fluctuations of its length, so that the latter is dynamic. The corresponding $\mathcal{N}=4$ spin chain S-matrix [Staudacher (2005); Beisert (2008)] could be decomposed into a product of two $\mathfrak{su}(2|2)$ factors[18]

$$S_{\mathfrak{psu}(2,2|4)}(u_1,u_2)=S_0(u_1,u_2)\cdot\left(S_{\mathfrak{su}(2|2)}(u_1,u_2)\otimes S_{\mathfrak{su}(2|2)}(u_1,u_2)\right)\,, \tag{14.71}$$

where u_1 and u_2 are rapidities of scattering magnons and

$$S_0(u,v)=\frac{x^-(u)-x^+(v)}{x^+(u)-x^-(v)}\frac{1-\frac{g^2}{x^+(u)x^-(v)}}{1-\frac{g^2}{x^-(u)x^+(v)}}\exp(2i\theta(u,v))\,. \tag{14.72}$$

Here $\theta(u,v)$ is the BES dressing phase [Beisert *et al.* (2007a)] and $\mathfrak{su}(2|2)$ S-matrices could be fixed completely from the symmetry requirements [Beisert (2008)]. The variables $x^\pm(u)$ are related to magnon rapidities u through Zhukovsky map

$$x^\pm(u)=x(u^\pm),\qquad u^\pm=u\pm\frac{i}{2},\qquad x(u)=\frac{u}{2}\left(1+\sqrt{1-4\frac{g^2}{u^2}}\right). \tag{14.73}$$

In the case of the $\mathfrak{sl}(2)$ sector corresponding to twist 2 operators relevant to the DGLAP evolution the long-range asymptotic Bethe ansatz equations for the M-magnon state take the form

$$\left(\frac{x_k^+}{x_k^-}\right)^2=\prod_{m=1,m\neq k}^{M}\frac{x_k^- - x_m^+}{x_k^+ - x_m^-}\frac{(1-g^2/x_k^+x_m^-)}{(1-g^2/x_k^-x_m^+)}\exp\left(2\,i\,\theta(u_k,u_j)\right),\quad \prod_{k=1}^{\hat{M}}\frac{x_k^+}{x_k^-}=1. \tag{14.74}$$

The spin length in this case is equal to two and the number of magnons corresponds to the Lorentz spin of twist 2 operators (14.48)-(14.50).

Once we know the solution for the M Bethe roots u_k (magnon rapidities) from the equation above, the asymptotic expression for the anomalous dimension (energy

[18]The choice of a reference state (vacuum) in a spin chain breaks the $\mathfrak{psu}(2,2|4)$ symmetry down to $\mathfrak{su}(2|2)\otimes\mathfrak{su}(2|2)$.

of spin chain state) could be found from the formula

$$\gamma^{ABA}(g) = 2\,g^2 \sum_{k=1}^{M} \left(\frac{i}{x_k^+} - \frac{i}{x_k^-} \right). \tag{14.75}$$

The Bethe ansatz equations in this case can be solved recursively order by order in the coupling constant g at arbitrary integer values of M once the one-loop solution for a given state is known. The latter is known from the exact solution of the Baxter equation [Eden and Staudacher (2006)] and is given by the Hahn polynomial. The equations for quantum corrections to the one-loop Bethe roots are linear and thus numerically solvable with high precision. To find the expression for anomalous dimensions valid for arbitrary spin values M we may use the principle of maximal transcendentality [Kotikov and Lipatov (2003)] and the fact that the function basis for anomalous the dimensions in $\mathcal{N} = 4$ SYM is given by harmonic sums. The reconstruction of the universal anomalous dimension is then reduced to fixing the coefficients in a known basis at a set of fixed integer spin values. Following this procedure for the four-loop universal anomalous dimension we get [Kotikov *et al.* (2007)] $(M = j + 2)$

$$\frac{1}{256}\,\gamma_{uni}^{ABA}(j+2) \;=\; 4\,S_{-7} + 6\,S_7 + \cdots + -\zeta(3)S_1(S_3 - S_{-3} + 2\,S_{-2,1}),$$

where the symbol $\cdots$ stands for a large set of not shown nested harmonic sums of degree seven. The obtained expression for universal anomalous dimension could be analytically continued also for non-integer spin values,[19] in particular to points $M = -r + \omega$ with r being a positive integer. Harmonic sums of degree seven may lead to poles no higher than seventh order in ω. In fact, it is known that none of the sums in the r.h.s. of (14.76) can produce such a high-order pole except for the two sums S_7 and S_{-7}. Their residues at $1/\omega^7$ are of opposite sign. Thus, one immediately sees that the sum of the two residues does *not* cancel. On the other hand, from the BFKL studies [Kotikov and Lipatov (2000, 2003)] we know that in the vicinity of the Pomeron pole at $M = -1+\omega$ the four loop anomalous dimensions behaves as

$$\gamma_{uni}(1+\omega) \;\sim\; 1/\omega^4. \tag{14.76}$$

This latter fact tells us that the asymptotic solution for the four-loop anomalous dimension is not complete and one should account for so called *wrapping corrections.*

The first account for wrapping or finite size corrections in the case of $\mathcal{N} = 4$ SYM was done with the help of the Lüscher type formula [Lüscher (1986)]. The latter consist of two types of contributions: F and μ terms. The F-term corresponds to the interaction of spin chain particles with a virtual particle circulating around a cylinder, while the μ-term corresponds to the splitting of the spin chain particle in two other particles, which then recombine after circulating around a cylinder. To conjecture the Lüscher type formula in the case of $\mathcal{N} = 4$ SYM the authors of [Bajnok and Janik (2009); Bajnok *et al.* (2009)] performed a large volume analysis

[19]An explanation for how this is done may be found in [Kotikov and Velizhanin (2005)].

of the Thermodynamic Bethe Ansatz (TBA) equations for the sinh-Gordon model. The conjectured formula for the F-term contains two contributions: the first one accounts for corrections to ABA and as a consequence to the Bethe roots, while the second one is due to the propagation of virtual particles around the cylinder, which directly changes energy of the spin chain state with all M particles of type a as [Bajnok and Janik (2009); Bajnok *et al.* (2009)]:

$$\Delta E(L) = -\sum_{Q=1}^{\infty} \int_{-\infty}^{\infty} \frac{dq}{2\pi} \mathrm{STr}_{a_1} \left[S_{a_1 a}^{a_2 a}(q,p_1) S_{a_2 a}^{a_3 a}(q,p_2) \dots S_{a_M a}^{a_1 a} \right] e^{\tilde{\epsilon}_{a_1}(q)L}, \quad (14.77)$$

where L is the spin chain length and the sum is all possible virtual particles in the theory Q, their polarizations a_1 and over all possible intermediate states $a_2, \dots, a_M$. The matrix $S_{ba}^{ca}(q,p)$ describes the scattering of the virtual particle of type b and momentum q with the real particle of type a and momentum p, while the exponential factor should be understood as propagator of the virtual particle. The μ-term contribution could be obtained by taking residues at the poles of S-matrices, however in $N = 4$ SYM at weak coupling these contributions are absent. It should be noted, that the same Lüscher type corrections were later much more easily obtained by analyzing the large volume limit of TBA directly for $\mathcal{N} = 4$ SYM, see for example [Gromov *et al.* (2009)].

Accounting for leading wrapping or Lüscher corrections [Bajnok and Janik (2009); Bajnok *et al.* (2009)] the full expression for the four-loop universal anomalous dimension takes the form

$$\gamma_{uni}(j+2) = \gamma_{uni}^{ABA}(j+2) + \gamma_{uni}^{wr}(j+2),$$
$$\frac{1}{256}\gamma_{uni}^{wr}(j+2) = \frac{1}{2} S_1^2 \left[2\,S_{-5} + 2\,S_5 + 4\,(S_{4,1} - S_{3,-2} + S_{-2,-3} - 2\,S_{-2,-2,1}) \right.$$
$$\left. -4\,S_{-2}\zeta(3) - 5\,\zeta(5) \right],$$

which is in full agreement with the BFKL predictions (14.76). Later, using a similar technique and a property of Gribov-Lipatov reciprocity (see [Dokshitzer and Marchesini (2007); Basso and Korchemsky (2007); Beccaria (2007); Beccaria and Forini (2009); Beccaria *et al.* (2009, 2010b)] and references therein), the five-loop corrections of universal anomalous dimensions have been found in [Lukowski *et al.* (2010)].

14.3.3.2 *Thermodynamic Bethe Ansatz and the quantum spectral curve*

The systematic method to account for wrapping or finite size corrections is offered by the Thermodynamic Bethe Ansatz (TBA). First, the idea that the TBA approach could be used for $\mathcal{N} = 4$ SYM appeared in [Ambjorn *et al.* (2006)]. Within the TBA approach the calculation of the spin chain spectrum with account for finite size corrections reduces to the calculation of the partition function of the mirror model. The simplification comes from the fact, that the mirror model is considered at large volume corresponding to large imaginary times in the original model. As a consequence, its spectrum is well under control and may be calculated using the mirror Asymptotic Bethe Ansatz (ABA). Finally, the partition function in the mirror model is calculated using the saddle point approximation resulting in integral equations for Bethe roots and hole densities corresponding to the mentioned saddle point. The mirror ABA equations for primary excitations may be obtained via analytical continuation of the ABA equations in the original theory. In addition one should determine all bound states (string complexes) of the mirror theory and calculate their dispersion relations together with scattering matrices using the bootstrap method, which should allow to write down the generic ABA in the mirror model for all excitations including bound states. First, the mirror ABA for the $\mathcal{N}=4$ spin chain was described in [Arutyunov and Frolov (2007)]. Using the fact that the symmetry structure of the scattering matrix $\mathfrak{su}(2|2)$ (14.71) is the same as that of the Hubbard model one can use the TBA solution of the latter [Takahashi (1999); Essler *et al.* (2005)]. This resulted in string hypotheses formulated in [Arutyunov and Frolov (2009a)]. Finally, the TBA equations for $\mathcal{N}=4$ SYM were written in [Gromov *et al.* (2009); Bombardelli *et al.* (2009); Gromov *et al.* (2010a); Arutyunov and Frolov (2009b)]. The further study of TBA equations continued in the simplified form of Y and T-systems [Cavaglia *et al.* (2011); Balog and Hegedus (2011); Gromov *et al.* (2011a, 2012)].

Eventually, the investigation of the $\mathcal{N}=4$ SYM TBA equations has led to their simplified alternative formulation in terms of the quantum spectral curve (QSC), a finite set of functional equations, so called Q-system, together with their analyticity and Riemann-Hilbert monodromy conditions [Gromov *et al.* (2014a, 2015a); Kazakov *et al.* (2016); Marboe and Volin (2017, 2018)]. The latter is well suited for both numerical solutions for any coupling and state [Gromov *et al.* (2016); Hegedus and Konczer (2016)] as well as for analytical calculations both at weak and strong coupling [Marboe and Volin (2015, 2018); Gromov *et al.* (2014b); Gromov and Levkovich-Maslyuk (2016a,b); Cavagli *et al.* (2018); Marboe *et al.* (2015); Marboe and Velizhanin (2016); Marboe and Widn (2019)].

In the $\mathfrak{sl}(2)$ sector of the $\mathfrak{psu}(2,2|4)$ spin chain relevant for the DGLAP evolution the QSC equations (so called $\mathbf{P}\mu$-system) take the form [Gromov *et al.* (2014a, 2015a); Marboe and Volin (2017, 2018)]:

$$\mu_{ab} - \tilde{\mu}_{ab} = \tilde{\mathbf{P}}_a \mathbf{P}_b - \tilde{P}_b \mathbf{P}_a, \tag{14.78}$$

$$\tilde{\mathbf{P}}_a = (\mu\chi)_a{}^b \mathbf{P}_b, \tag{14.79}$$

$$\tilde{\mu}_{ab} = \mu_{ab}^{[2]}, \tag{14.80}$$

where

$$\chi^{ab} = \begin{pmatrix} 0 & 0 & 0 & -1 \\ 0 & 0 & +1 & 0 \\ 0 & -1 & 0 & 0 \\ +1 & 0 & 0 & 0 \end{pmatrix}, \quad (\mu\chi)_a{}^b \equiv \mu_{ac}\chi^{cb}, \tag{14.81}$$

and the indices a, b take values $1, 2, 3, 4$. μ_{ab} is an antisymmetric matrix constrained by

$$\mu_{12}\mu_{34} - \mu_{13}\mu_{24} + \mu_{14}\mu_{23} = 1, \quad \mu_{14} = \mu_{23}, \tag{14.82}$$

and all functions $\mathbf{P}_a$ and μ_{ab} are functions of the spectral parameter u. The functions $\mathbf{P}_a$ have only one Zhukovsky cut ($u \in [-2g, 2g]$) on the defining (physical) sheet, while the functions μ_{ab} have infinitely many branch points at positions $u = \pm 2g + i\mathbb{Z}$. Next, $f^{[n]}(u) = f(u + \frac{in}{2})$ and $\tilde{f}(u)$ denotes the analytic continuation of the function $f(u)$ around one of its branch points on the real axis located at $-2g$ and $2g$. The boundary conditions for the above Riemann-Hilbert problem are specified by the large u asymptotics of $\mathbf{P}_a$ and μ_{ab} functions:

$$\begin{aligned} &\mathbf{P}_1 \simeq A_1 u^{-\frac{L+2}{2}}, \mathbf{P}_2 \simeq A_2 u^{-\frac{L}{2}}, \mathbf{P}_3 \simeq A_3 u^{\frac{L-2}{2}}, \mathbf{P}_4 \simeq A_4 u^{\frac{L}{2}}, \\ &\mu_{12} \sim u^{\Delta - L},\ \mu_{13} \sim u^{\Delta - 1},\ \mu_{14} \sim u^{\Delta},\ \mu_{24} \sim u^{\Delta + L},\ \mu_{34} \sim u^{\Delta + L} \end{aligned} \tag{14.83}$$

with

$$\begin{aligned} A_1 A_4 &= \frac{[(L - S + 2)^2 - \Delta^2][(L + S)^2 - \Delta^2]}{16 i L (L + 1)}, \\ A_3 &= \frac{[(L + S - 2)^2 - \Delta^2][(L - S)^2 - \Delta^2]}{16 i L (L - 1)}. \end{aligned} \tag{14.84}$$

Here L is the twist, S the spin and Δ is the conformal dimension of the $\mathfrak{sl}(2)$ operator under consideration. Choosing an appropriate ansatz for the $\mathbf{P}_a$ functions it is possible to solve this Riemann-Hilbert problem up to in principle arbitrary order of perturbation theory [Marboe and Volin (2015)]. Moreover, using the computed values of an anomalous dimensions at fixed spin values and the basis of binomial harmonic sums[20] it is possible to reconstruct the universal anomalous dimension at the required order of perturbation theory. This way the universal anomalous dimensions at six and seven loop order were obtained in Refs. [Marboe *et al.* (2015)] and [Marboe and Velizhanin (2016)] correspondingly.

Finally, as we already noted the use of the QSC is not limited by the weak coupling regime and the authors of [Gromov *et al.* (2014b)] were able to obtain analytical results for the anomalous dimensions of the short $\mathfrak{sl}(2)$ operators at the first four orders in the strong coupling expansion.

14.4 Conclusion

We tried to briefly review the present knowledge of the properties of the BFKL and DGLAP equations in the case of $\mathcal{N} = 4$ SYM. A lot of current results in $\mathcal{N} = 4$ SYM were either obtained with the use of the *transcendentality principle* or confirm the latter property of the $\mathcal{N} = 4$ SYM model. Obtained originally as a property of the BFKL kernel eigenvalues at NLO, the maximum transcendentality property

[20]The latter is possible with the use of the Gribov-Lipatov reciprocity.

was later observed also for universal anomalous dimensions up to very high order of perturbation theory, scattering amplitudes, form-factors and correlation functions in $\mathcal{N} = 4$ SYM.

First, the transcendentality principle was used to extract the results for $\mathcal{N} = 4$ SYM anomalous dimensions from the known QCD results at the first three orders of perturbation theory. Next, the results for four and five loop universal anomalous dimensions were obtained from the asymptotic long-range Bethe equations with account for the so-called *wrapping corrections* in the Lüscher approach. Still, the transcendentality property was used again to construct a basis of harmonic sums, in terms of which the result is written. To get the six and seven loop result the quantum spectral curve approach was used instead of asymptotic Bethe equations and the Lüscher corrections.

The AdS/CFT correspondence and the all-loop integrability of the DGLAP and BFKL equations gave us a unique opportunity to study the Pomeron solution and anomalous dimensions of different types of operators at strong and in general at finite coupling. The most suitable approach for later studies at moment is offered by a variant of the Thermodynamic Bethe Ansatz known as the Quantum Spectral Curve approach.

Bibliography

Aaron, F. D. *et al.* (2010). Combined measurement and QCD analysis of the inclusive e+− p scattering cross sections at HERA, *JHEP* **01**, p. 109, doi:10.1007/JHEP01(2010)109, arXiv:0911.0884 [hep-ex].

Alday, L. F., Arutyunov, G., Benna, M. K., Eden, B., and Klebanov, I. R. (2007). On the strong coupling scaling dimension of high spin operators, *JHEP* **04**, p. 082, doi:10.1088/1126-6708/2007/04/082, arXiv:hep-th/0702028 [HEP-TH].

Alfimov, M., Gromov, N., and Kazakov, V. (2015). QCD Pomeron from AdS/CFT quantum spectral curve, *JHEP* **07**, p. 164, doi:10.1007/JHEP07(2015)164, arXiv:arXiv:1408.2530 [hep-th].

Alfimov, M., Gromov, N., and Sizov, G. (2018). BFKL spectrum of $\mathcal{N} = 4$: non-zero conformal spin, *JHEP* **07**, p. 181, doi:10.1007/JHEP07(2018)181, arXiv:1802.06908 [hep-th].

Altarelli, G., Curci, G., Martinelli, G., and Petrarca, S. (1981). QCD Nonleading corrections to weak decays as an application of regularization by dimensional reduction, *Nucl. Phys.* **B187**, pp. 461–513, doi:10.1016/0550-3213(81)90473-9.

Altarelli, G. and Parisi, G. (1977). Asymptotic freedom in parton language, *Nucl. Phys.* **B126**, pp. 298–318, doi:10.1016/0550-3213(77)90384-4.

Ambjorn, J., Janik, R. A., and Kristjansen, C. (2006). Wrapping interactions and a new source of corrections to the spin-chain/string duality, *Nucl. Phys.* **B736**, pp. 288–301, doi:10.1016/j.nuclphysb.2005.12.007, arXiv:hep-th/0510171 [hep-th].

Andersson, B. *et al.* (2002). Small x phenomenology: Summary and status, *Eur. Phys. J.* **C25**, pp. 77–101, doi:10.1007/s10052-002-0998-7, arXiv:hep-ph/0204115 [hep-ph].

Arutyunov, G., Eden, B., Petkou, A. C., and Sokatchev, E. (2002). Exceptional non-renormalization properties and OPE analysis of chiral four point functions in N=4 SYM(4), *Nucl. Phys.* **B620**, pp. 380–404, doi:10.1016/S0550-3213(01)00569-7, arXiv:hep-th/0103230 [hep-th].

Arutyunov, G. and Frolov, S. (2006). On AdS(5) x S**5 String S-matrix, *Phys. Lett.* **B639**, pp. 378–382, doi:10.1016/j.physletb.2006.06.064, arXiv:hep-th/0604043 [hep-th].

Arutyunov, G. and Frolov, S. (2007). On string s-matrix, bound states and TBA, *JHEP* **12**, p. 024, doi:10.1088/1126-6708/2007/12/024, arXiv:0710.1568 [hep-th].

Arutyunov, G. and Frolov, S. (2009a). String hypothesis for the AdS(5) x S**5 mirror, *JHEP* **03**, p. 152, doi:10.1088/1126-6708/2009/03/152, arXiv:0901.1417 [hep-th].

Arutyunov, G. and Frolov, S. (2009b). Thermodynamic Bethe Ansatz for the AdS(5) x S(5) mirror model, *JHEP* **05**, p. 068, doi:10.1088/1126-6708/2009/05/068, arXiv:arXiv:0903.0141 [hep-th].

Arutyunov, G., Frolov, S., and Staudacher, M. (2004). Bethe ansatz for quantum strings, *JHEP* **10**, p. 016, doi:10.1088/1126-6708/2004/10/016, arXiv:hep-th/0406256 [hep-th].

Arutyunov, G., Frolov, S., and Suzuki, R. (2010). Five-loop Konishi from the mirror TBA, *JHEP* **04**, p. 069, doi:10.1007/JHEP04(2010)069, arXiv:1002.1711 [hep-th].

Arutyunov, G., Frolov, S., and Zamaklar, M. (2007). The Zamolodchikov-Faddeev algebra for AdS(5) x S**5 superstring, *JHEP* **04**, p. 002, doi:10.1088/1126-6708/2007/04/002, arXiv:hep-th/0612229 [hep-th].

Axenides, M., Floratos, E., and Kehagias, A. (2003). Scaling violations in Yang-Mills theories and strings in AdS(5), *Nucl. Phys.* **B662**, pp. 170–184, doi:10.1016/S0550-3213(03)00338-9, arXiv:hep-th/0210091 [hep-th].

Bajnok, Z., Hegedus, A., Janik, R. A., and Lukowski, T. (2010). Five loop Konishi from AdS/CFT, *Nucl. Phys.* **B827**, pp. 426–456, doi:10.1016/j.nuclphysb.2009.10.015, arXiv:0906.4062 [hep-th].

Bajnok, Z. and Janik, R. A. (2009). Four-loop perturbative Konishi from strings and finite size effects for multiparticle states, *Nucl. Phys.* **B807**, pp. 625–650, doi:10.1016/j.nuclphysb.2008.08.020, arXiv:0807.0399 [hep-th].

Bajnok, Z. and Janik, R. A. (2012). Six and seven loop Konishi from Luscher corrections, *JHEP* **11**, p. 002, doi:10.1007/JHEP11(2012)002, arXiv:1209.0791 [hep-th].

Bajnok, Z., Janik, R. A., and Lukowski, T. (2009). Four loop twist two, BFKL, wrapping and strings, *Nucl. Phys.* **B816**, pp. 376–398, doi:10.1016/j.nuclphysb.2009.02.005, arXiv:0811.4448 [hep-th].

Balitsky, I. I. and Lipatov, L. N. (1978). The Pomeranchuk Singularity in quantum chromodynamics, *Sov. J. Nucl. Phys.* **28**, pp. 822–829, [Yad. Fiz.28,1597(1978)].

Balitsky, I. I. and Lipatov, L. N. (1979). Calculation of meson meson interaction cross-section in quantum chromodynamics. (in russian), *JETP Lett.* **30**, p. 355, [Pisma Zh. Eksp. Teor. Fiz.30,383(1979)].

Balog, J. and Hegedus, A. (2010). 5-loop Konishi from linearized TBA and the XXX magnet, *JHEP* **06**, p. 080, doi:10.1007/JHEP06(2010)080, arXiv:1002.4142 [hep-th].

Balog, J. and Hegedus, A. (2011). $AdS_5 \times S^5$ mirror TBA equations from Y-system and discontinuity relations, *JHEP* **08**, p. 095, doi:10.1007/JHEP08(2011)095, arXiv:arXiv:1104.4054 [hep-th].

Bartels, J. (1980). High-energy behavior in a Nonabelian Gauge theory (II), *Nucl. Phys.* **B175**, pp. 365–401, doi:10.1016/0550-3213(80)90019-X.

Basso, B. (2011). An exact slope for AdS/CFT, arXiv:1109.3154 [hep-th].

Basso, B. (2012). Scaling dimensions at small spin in N=4 SYM theory, arXiv:1205.0054 [hep-th].

Basso, B. and Korchemsky, G. P. (2007). Anomalous dimensions of high-spin operators beyond the leading order, *Nucl. Phys.* **B775**, pp. 1–30, doi:10.1016/j.nuclphysb.2007.03.044, arXiv:hep-th/0612247 [hep-th].

Basso, B., Korchemsky, G. P., and Kotanski, J. (2008). Cusp anomalous dimension in maximally supersymmetric Yang-Mills theory at strong coupling, *Phys. Rev. Lett.* **100**, p. 091601, doi:10.1103/PhysRevLett.100.091601, arXiv:0708.3933 [hep-th].

Beccaria, M. (2007). Three loop anomalous dimensions of twist-3 gauge operators in N=4 SYM, *JHEP* **09**, p. 023, doi:10.1088/1126-6708/2007/09/023, arXiv:0707.1574 [hep-th].

Beccaria, M., Belitsky, A. V., Kotikov, A. V., and Zieme, S. (2010a). Analytic solution of the multiloop Baxter equation, *Nucl. Phys.* **B827**, pp. 565–606, doi:10.1016/j.nuclphysb.2009.10.030, arXiv:0908.0520 [hep-th].

Beccaria, M., De Angelis, G. F., and Forini, V. (2007). The scaling function at strong coupling from the quantum string Bethe equations, *JHEP* **04**, p. 066, doi:10.1088/1126-6708/2007/04/066, arXiv:hep-th/0703131 [hep-th].

Beccaria, M. and Forini, V. (2009). Four loop reciprocity of twist two operators in N=4 SYM, *JHEP* **03**, p. 111, doi:10.1088/1126-6708/2009/03/111, arXiv:0901.1256 [hep-th].

Beccaria, M., Forini, V., Lukowski, T., and Zieme, S. (2009). Twist-three at five loops, Bethe Ansatz and wrapping, *JHEP* **03**, p. 129, doi:10.1088/1126-6708/2009/03/129, arXiv:0901.4864 [hep-th].

Beccaria, M., Forini, V., and Macorini, G. (2010b). Generalized Gribov-Lipatov reciprocity and AdS/CFT, *Adv. High Energy Phys.* **2010**, p. 753248, doi:10.1155/2010/753248, arXiv:1002.2363 [hep-th].

Beisert, N. (2004). The complete one loop dilatation operator of N=4 super Yang-Mills theory, *Nucl. Phys.* **B676**, pp. 3–42, doi:10.1016/j.nuclphysb.2003.10.019, arXiv:hep-th/0307015 [hep-th].

Beisert, N. (2007). The analytic Bethe Ansatz for a chain with centrally extended $su(2|2)$ symmetry, *J. Stat. Mech.* **0701**, p. P01017, doi:10.1088/1742-5468/2007/01/P01017, arXiv:nlin/0610017 [nlin.SI].

Beisert, N. (2008). The $SU(2|2)$ dynamic S-matrix, *Adv. Theor. Math. Phys.* **12**, pp. 945–979, doi:10.4310/ATMP.2008.v12.n5.a1, arXiv:hep-th/0511082 [hep-th].

Beisert, N., Eden, B., and Staudacher, M. (2007a). Transcendentality and crossing, *J. Stat. Mech.* **0701**, p. P01021, doi:10.1088/1742-5468/2007/01/P01021, arXiv:hep-th/0610251 [hep-th].

Beisert, N., Eden, B., and Staudacher, M. (2007b). Transcendentality and Crossing, *J. Stat. Mech.* **0701**, p. P01021, doi:10.1088/1742-5468/2007/01/P01021, arXiv:hep-th/0610251 [hep-th].

Beisert, N., Ferretti, G., Heise, R., and Zarembo, K. (2005). One-loop QCD spin chain and its spectrum, *Nucl. Phys.* **B717**, pp. 137–189, doi:10.1016/j.nuclphysb.2005.04.004, arXiv:hep-th/0412029 [hep-th].

Beisert, N., Kristjansen, C., and Staudacher, M. (2003). The dilatation operator of conformal N=4 superYang-Mills theory, *Nucl. Phys.* **B664**, pp. 131–184, doi: 10.1016/S0550-3213(03)00406-1, arXiv:hep-th/0303060 [hep-th].

Beisert, N. and Staudacher, M. (2003a). The N=4 SYM integrable super spin chain, *Nucl. Phys.* **B670**, pp. 439–463, doi:10.1016/j.nuclphysb.2003.08.015, arXiv:hep-th/0307042 [hep-th].

Beisert, N. and Staudacher, M. (2003b). The N=4 SYM integrable super spin chain, *Nucl. Phys.* **B670**, pp. 439–463, doi:10.1016/j.nuclphysb.2003.08.015, arXiv:hep-th/0307042 [hep-th].

Beisert, N. and Staudacher, M. (2005). Long-range $psu(2,2|4)$ Bethe Ansatze for gauge theory and strings, *Nucl. Phys.* **B727**, pp. 1–62, doi:10.1016/j.nuclphysb.2005.06.038, arXiv:hep-th/0504190 [hep-th].

Beisert, N. *et al.* (2012). Review of AdS/CFT Integrability: an overview, *Lett. Math. Phys.* **99**, pp. 3–32, doi:10.1007/s11005-011-0529-2, arXiv:1012.3982 [hep-th].

Belitsky, A. V. (1999). Fine structure of spectrum of twist - three operators in QCD, *Phys. Lett.* **B453**, pp. 59–72, doi:10.1016/S0370-2693(99)00326-3, arXiv:hep-ph/9902361 [hep-ph].

Benna, M. K., Benvenuti, S., Klebanov, I. R., and Scardicchio, A. (2007). A Test of the AdS/CFT correspondence using high-spin operators, *Phys. Rev. Lett.* **98**, p. 131603, doi:10.1103/PhysRevLett.98.131603, arXiv:hep-th/0611135 [hep-th].

Bern, Z., Carrasco, J. J. M., Johansson, H., and Kosower, D. A. (2007a). Maximally supersymmetric planar Yang-Mills amplitudes at five loops, *Phys. Rev.* **D76**, p. 125020, doi:10.1103/PhysRevD.76.125020, arXiv:0705.1864 [hep-th].

Bern, Z., Czakon, M., Dixon, L. J., Kosower, D. A., and Smirnov, V. A. (2007b). The four-loop planar amplitude and cusp anomalous dimension in maximally supersymmetric Yang-Mills theory, *Phys. Rev.* **D75**, p. 085010, doi:10.1103/PhysRevD.75.085010, arXiv:hep-th/0610248 [hep-th].

Bianchi, L., Forini, V., and Kotikov, A. V. (2013). On DIS Wilson coefficients in N=4 super Yang-Mills theory, *Phys. Lett.* **B725**, pp. 394–401, doi:10.1016/j.physletb.2013.07.013, arXiv:1304.7252 [hep-th].

Bombardelli, D., Fioravanti, D., and Tateo, R. (2009). Thermodynamic Bethe Ansatz for planar AdS/CFT: A Proposal, *J. Phys.* **A42**, p. 375401, doi:10.1088/1751-8113/42/37/375401, arXiv:arXiv:0902.3930 [hep-th].

Braun, V. M., Derkachov, S. E., and Manashov, A. N. (1998). Integrability of three particle evolution equations in QCD, *Phys. Rev. Lett.* **81**, pp. 2020–2023, doi:10.1103/PhysRevLett.81.2020, arXiv:hep-ph/9805225 [hep-ph].

Brink, L., Schwarz, J. H., and Scherk, J. (1977). Supersymmetric Yang-Mills theories, *Nucl. Phys.* **B121**, pp. 77–92, doi:10.1016/0550-3213(77)90328-5.

Broadhurst, D. J. and Kreimer, D. (1995). Knots and numbers in Phi**4 theory to 7 loops and beyond, *Int. J. Mod. Phys.* **C6**, pp. 519–524, doi:10.1142/S012918319500037X, arXiv:hep-ph/9504352 [hep-ph].

Broadhurst, D. J. and Kreimer, D. (1997). Association of multiple zeta values with positive knots via Feynman diagrams up to 9 loops, *Phys. Lett.* **B393**, pp. 403–412, doi:10.1016/S0370-2693(96)01623-1, arXiv:hep-th/9609128 [hep-th].

Brower, R. C., Polchinski, J., Strassler, M. J., and Tan, C.-I. (2007). The Pomeron and gauge/string duality, *JHEP* **12**, p. 005, doi:10.1088/1126-6708/2007/12/005, arXiv:hep-th/0603115 [hep-th].

Bukhvostov, A. P., Frolov, G. V., Lipatov, L. N., and Kuraev, E. A. (1985a). Evolution equations for quasi-partonic operators, *Nucl. Phys.* **B258**, pp. 601–646, doi:10.1016/0550-3213(85)90628-5.

Bukhvostov, A. P., Kuraev, E. A., Lipatov, L. N., and Frolov, G. V. (1985b). Supersymmetry and anomalous dimensionalities of quasi - parton operators in quantum chromodynamics, *JETP Lett.* **41**, pp. 92–95, [Pisma Zh. Eksp. Teor. Fiz.41,77(1985)].

Caron-Huot, S., Dixon, L. J., Dulat, F., von Hippel, M., McLeod, A. J., and Papathanasiou, G. (2019). Six-gluon amplitudes in planar $\mathcal{N} = 4$ super-Yang-Mills theory at six and seven loops, *Submitted to: J. High Energy Phys.* arXiv:1903.10890 [hep-th].

Caron-Huot, S. and Herranen, M. (2018). High-energy evolution to three loops, *JHEP* **02**, p. 058, doi:10.1007/JHEP02(2018)058, arXiv:1604.07417 [hep-ph].

Casteill, P. Y. and Kristjansen, C. (2007). The strong coupling limit of the scaling function from the quantum string bethe ansatz, *Nucl. Phys.* **B785**, pp. 1–18, doi:10.1016/j.nuclphysb.2007.06.011, arXiv:0705.0890 [hep-th].

Cavaglia, A., Fioravanti, D., and Tateo, R. (2011). Extended Y-system for the AdS_5/CFT_4 correspondence, *Nucl. Phys.* **B843**, pp. 302–343, doi:10.1016/j.nuclphysb.2010.09.015, arXiv:arXiv:1005.3016 [hep-th].

Cavagli, A., Gromov, N., and Levkovich-Maslyuk, F. (2018). Quantum spectral curve and structure constants in N=4 SYM: cusps in the ladder limit, arXiv:arXiv:1802.04237 [hep-th].

Ciafaloni, M. and Camici, G. (1998). Energy scale(s) and next-to-leading BFKL equation, *Phys. Lett.* **B430**, pp. 349–354, doi:10.1016/S0370-2693(98)00551-6, arXiv:hep-ph/9803389 [hep-ph].

Costa, M. S., Goncalves, V., and Penedones, J. (2012). Conformal Regge theory, *JHEP* **12**, p. 091, doi:10.1007/JHEP12(2012)091, arXiv:1209.4355 [hep-th].

Dixon, L. J. (2018). The Principle of Maximal Transcendentality and the Four-Loop Collinear Anomalous Dimension, *JHEP* **01**, p. 075, doi:10.1007/JHEP01(2018)075, arXiv:1712.07274 [hep-th].

Dixon, L. J., Moult, I., and Zhu, H. X. (2019). The collinear limit of the energy-energy correlator, arXiv:1905.01310 [hep-ph].

Dokshitzer, Y. L. (1977). Calculation of the structure functions for deep inelastic scattering and e+ e- annihilation by perturbation theory in quantum chromodynamics. *Sov. Phys. JETP* **46**, pp. 641–653, [Zh. Eksp. Teor. Fiz.73,1216(1977)].

Dokshitzer, Yu. L. and Marchesini, G. (2007). N=4 SUSY Yang-Mills: three loops made simple(r), *Phys. Lett.* **B646**, pp. 189–201, doi:10.1016/j.physletb.2007.01.016, arXiv:hep-th/0612248 [hep-th].

Eden, B. (2006). A Two-loop test for the factorised S-matrix of planar N = 4, *Nucl. Phys.* **B738**, pp. 409–424, doi:10.1016/j.nuclphysb.2006.01.012, arXiv:hep-th/0501234 [hep-th].

Eden, B., Heslop, P., Korchemsky, G. P., Smirnov, V. A., and Sokatchev, E. (2012). Five-loop Konishi in N=4 SYM, *Nucl. Phys.* **B862**, pp. 123–166, doi:10.1016/j.nuclphysb.2012.04.015, arXiv:1202.5733 [hep-th].

Eden, B., Jarczak, C., and Sokatchev, E. (2005a). A Three-loop test of the dilatation operator in N = 4 SYM, *Nucl. Phys.* **B712**, pp. 157–195, doi:10.1016/j.nuclphysb.2005.01.036, arXiv:hep-th/0409009 [hep-th].

Eden, B., Jarczak, C., Sokatchev, E., and Stanev, Ya. S. (2005b). Operator mixing in N = 4 SYM: The Konishi anomaly revisited, *Nucl. Phys.* **B722**, pp. 119–148, doi:10.1016/j.nuclphysb.2005.06.005, arXiv:hep-th/0501077 [hep-th].

Eden, B. and Staudacher, M. (2006). Integrability and transcendentality, *J. Stat. Mech.* **0611**, p. P11014, doi:10.1088/1742-5468/2006/11/P11014, arXiv:hep-th/0603157 [hep-th].

Essler, F. H. L., Frahm, H., Göhmann, F., Klümper, A., and Korepin, V. (2005). *The One-Dimensional Hubbard Model* (Cambridge University Press).

Faddeev, L. D. and Korchemsky, G. P. (1995). High-energy QCD as a completely integrable model, *Phys. Lett.* **B342**, pp. 311–322, doi:10.1016/0370-2693(94)01363-H, arXiv:hep-th/9404173 [hep-th].

Fadin, V. S. and Fiore, R. (2008). The Dipole form of the BFKL kernel in supersymmetric Yang-Mills theories, *Phys. Lett.* **B661**, pp. 139–144, doi:10.1016/j.physletb.2008.01.046, arXiv:0712.3901 [hep-ph].

Fadin, V. S., Kuraev, E. A., and Lipatov, L. N. (1975). On the Pomeranchuk singularity in asymptotically free theories, *Phys. Lett.* **60B**, pp. 50–52, doi:10.1016/0370-2693(75)90524-9.

Fadin, V. S. and Lipatov, L. N. (1998). BFKL pomeron in the next-to-leading approximation, *Phys. Lett.* **B429**, pp. 127–134, doi:10.1016/S0370-2693(98)00473-0, arXiv:hep-ph/9802290 [hep-ph].

Ferretti, G., Heise, R., and Zarembo, K. (2004). New integrable structures in large-N QCD, *Phys. Rev.* **D70**, p. 074024, doi:10.1103/PhysRevD.70.074024, arXiv:hep-th/0404187 [hep-th].

Fiamberti, F., Santambrogio, A., Sieg, C., and Zanon, D. (2008). Wrapping at four loops in N=4 SYM, *Phys. Lett.* **B666**, pp. 100–105, doi:10.1016/j.physletb.2008.06.061, arXiv:0712.3522 [hep-th].

Fleischer, J., Kotikov, A. V., and Veretin, O. L. (1999). Analytic two loop results for selfenergy type and vertex type diagrams with one nonzero mass, *Nucl. Phys.* **B547**, pp. 343–374, doi:10.1016/S0550-3213(99)00078-4, arXiv:hep-ph/9808242 [hep-ph].

Frolov, S. (2011). Konishi operator at intermediate coupling, *J. Phys.* **A44**, p. 065401, doi:10.1088/1751-8113/44/6/065401, arXiv:1006.5032 [hep-th].

Frolov, S. and Tseytlin, A. A. (2002). Semiclassical quantization of rotating superstring in AdS(5) x S**5, *JHEP* **06**, p. 007, doi:10.1088/1126-6708/2002/06/007, arXiv:hep-th/0204226 [hep-th].

Gliozzi, F., Scherk, J., and Olive, D. I. (1977). Supersymmetry, supergravity theories and the dual spinor model, *Nucl. Phys.* **B122**, pp. 253–290, doi:10.1016/0550-3213(77)90206-1.

Gribov, V. N. and Lipatov, L. N. (1972a). Deep inelastic e p scattering in perturbation theory, *Sov. J. Nucl. Phys.* **15**, pp. 438–450, [Yad. Fiz.15,781(1972)].

Gribov, V. N. and Lipatov, L. N. (1972b). e+ e- pair annihilation and deep inelastic e p scattering in perturbation theory, *Sov. J. Nucl. Phys.* **15**, pp. 675–684, [Yad. Fiz.15,1218(1972)].

Gromov, N., Kazakov, V., Kozak, A., and Vieira, P. (2010a). Exact spectrum of anomalous dimensions of planar N = 4 supersymmetric Yang-Mills theory: TBA and excited states, *Lett. Math. Phys.* **91**, pp. 265–287, doi:10.1007/s11005-010-0374-8, arXiv:arXiv:0902.4458 [hep-th].

Gromov, N., Kazakov, V., Leurent, S., and Tsuboi, Z. (2011a). Wronskian Solution for AdS/CFT Y-system, *JHEP* **01**, p. 155, doi:10.1007/JHEP01(2011)155, arXiv:arXiv:1010.2720 [hep-th].

Gromov, N., Kazakov, V., Leurent, S., and Volin, D. (2012). Solving the AdS/CFT Y-system, *JHEP* **07**, p. 023, doi:10.1007/JHEP07(2012)023, arXiv:arXiv:1110.0562 [hep-th].

Gromov, N., Kazakov, V., Leurent, S., and Volin, D. (2014a). Quantum spectral curve for planar $\mathcal{N}$ = super-Yang-Mills theory, *Phys. Rev. Lett.* **112**, 1, p. 011602, doi:10.1103/PhysRevLett.112.011602, arXiv:arXiv:1305.1939 [hep-th].

Gromov, N., Kazakov, V., Leurent, S., and Volin, D. (2015a). Quantum spectral curve for arbitrary state/operator in AdS_5/CFT_4, *JHEP* **09**, p. 187, doi:10.1007/JHEP09(2015)187, arXiv:arXiv:1405.4857 [hep-th].

Gromov, N., Kazakov, V., and Vieira, P. (2009). Exact spectrum of anomalous dimensions of planar N=4 supersymmetric Yang-Mills theory, *Phys. Rev. Lett.* **103**, p. 131601, doi:10.1103/PhysRevLett.103.131601, arXiv:arXiv:0901.3753 [hep-th].

Gromov, N., Kazakov, V., and Vieira, P. (2010b). Exact spectrum of planar $\mathcal{N} = 4$ supersymmetric Yang-Mills theory: Konishi dimension at any coupling, *Phys. Rev. Lett.* **104**, p. 211601, doi:10.1103/PhysRevLett.104.211601, arXiv:0906.4240 [hep-th].

Gromov, N. and Levkovich-Maslyuk, F. (2016a). Quantum spectral curve for a cusped Wilson line in $\mathcal{N} = 4$ SYM, *JHEP* **04**, p. 134, doi:10.1007/JHEP04(2016)134, arXiv:arXiv:1510.02098 [hep-th].

Gromov, N. and Levkovich-Maslyuk, F. (2016b). Quark-anti-quark potential in $\mathcal{N} = 4$ SYM, *JHEP* **12**, p. 122, doi:10.1007/JHEP12(2016)122, arXiv:arXiv:1601.05679 [hep-th].

Gromov, N., Levkovich-Maslyuk, F., and Sizov, G. (2015b). Pomeron eigenvalue at three loops in $\mathcal{N} = 4$ supersymmetric Yang-Mills theory, *Phys. Rev. Lett.* **115**, 25, p. 251601, doi:10.1103/PhysRevLett.115.251601, arXiv:arXiv:1507.04010 [hep-th].

Gromov, N., Levkovich-Maslyuk, F., and Sizov, G. (2016). Quantum spectral curve and the numerical solution of the spectral problem in AdS5/CFT4, *JHEP* **06**, p. 036, doi:10.1007/JHEP06(2016)036, arXiv:arXiv:1504.06640 [hep-th].

Gromov, N., Levkovich-Maslyuk, F., Sizov, G., and Valatka, S. (2014b). Quantum spectral curve at work: from small spin to strong coupling in $\mathcal{N} = 4$ SYM, *JHEP* **07**, p. 156, doi:10.1007/JHEP07(2014)156, arXiv:1402.0871 [hep-th].

Gromov, N., Serban, D., Shenderovich, I., and Volin, D. (2011b). Quantum folded string and integrability: From finite size effects to Konishi dimension, *JHEP* **08**, p. 046, doi:10.1007/JHEP08(2011)046, arXiv:1102.1040 [hep-th].

Gromov, N. and Valatka, S. (2012). Deeper Look into Short Strings, *JHEP* **03**, p. 058, doi:10.1007/JHEP03(2012)058, arXiv:1109.6305 [hep-th].

Gubser, S. S., Klebanov, I. R., and Polyakov, A. M. (1998). Gauge theory correlators from noncritical string theory, *Phys. Lett.* **B428**, pp. 105–114, doi:10.1016/S0370-2693(98) 00377-3, arXiv:hep-th/9802109 [hep-th].

Gubser, S. S., Klebanov, I. R., and Polyakov, A. M. (2002). A Semiclassical limit of the gauge / string correspondence, *Nucl. Phys.* **B636**, pp. 99–114, doi:10.1016/ S0550-3213(02)00373-5, arXiv:hep-th/0204051 [hep-th].

Hegedus, A. and Konczer, J. (2016). Strong coupling results in the AdS_5 /CF T_4 correspondence from the numerical solution of the quantum spectral curve, *JHEP* **08**, p. 061, doi:10.1007/JHEP08(2016)061, arXiv:arXiv:1604.02346 [hep-th].

Janik, R. A. (2006). The AdS(5) x S**5 superstring worldsheet S-matrix and crossing symmetry, *Phys. Rev.* **D73**, p. 086006, doi:10.1103/PhysRevD.73.086006, arXiv:hep-th/0603038 [hep-th].

Kazakov, D. I. and Kotikov, A. V. (1988). Total $\alpha^{-}s$ Correction to deep inelastic scattering cross-section ratio, R = σ^{-}l / σ^{-}t in QCD. calculation of longitudinal structure function, *Nucl. Phys.* **B307**, p. 721, doi:10.1016/0550-3213(88)90106-X, [Erratum: Nucl. Phys.B345,299(1990)].

Kazakov, V., Leurent, S., and Volin, D. (2016). T-system on T-hook: Grassmannian solution and twisted quantum spectral curve, *JHEP* **12**, p. 044, doi:10.1007/ JHEP12(2016)044, arXiv:arXiv:1510.02100 [hep-th].

Konishi, K. (1984). Anomalous supersymmetry transformation of some composite operators in SQCD, *Phys. Lett.* **135B**, pp. 439–444, doi:10.1016/0370-2693(84)90311-3.

Kostov, I., Serban, D., and Volin, D. (2008a). Functional BES equation, *JHEP* **08**, p. 101, doi:10.1088/1126-6708/2008/08/101, arXiv:0801.2542 [hep-th].

Kostov, I., Serban, D., and Volin, D. (2008b). Strong coupling limit of Bethe ansatz equations, *Nucl. Phys.* **B789**, pp. 413–451, doi:10.1016/j.nuclphysb.2007.06.017, arXiv:hep-th/0703031 [HEP-TH].

Kotikov, A., Kuhn, J. H., and Veretin, O. (2008). Two-loop formfactors in theories with mass gap and Z-Boson production, *Nucl. Phys.* **B788**, pp. 47–62, doi:10.1016/j. nuclphysb.2007.07.018, arXiv:hep-ph/0703013 [HEP-PH].

Kotikov, A. V. (1994). Gluon distribution for small x, *Phys. Atom. Nucl.* **57**, pp. 133–144, [Yad. Fiz.57,142(1994)].

Kotikov, A. V. and Lipatov, L. N. (2000). NLO corrections to the BFKL equation in QCD and in supersymmetric gauge theories, *Nucl. Phys.* **B582**, pp. 19–43, doi: 10.1016/S0550-3213(00)00329-1, arXiv:hep-ph/0004008 [hep-ph].

Kotikov, A. V. and Lipatov, L. N. (2001). DGLAP and BFKL evolution equations in the N=4 supersymmetric gauge theory, in *35th Annual Winter School on Nuclear and Particle Physics Repino, Russia, February 19-25, 2001*, arXiv:hep-ph/0112346 [hep-ph].

Kotikov, A. V. and Lipatov, L. N. (2003). DGLAP and BFKL equations in the $N = 4$ supersymmetric gauge theory, *Nucl. Phys.* **B661**, pp. 19–61, doi:10.1016/S0550-3213(03)00264-5, 10.1016/j.nuclphysb.2004.02.032, arXiv:hep-ph/0208220 [hep-ph], [Erratum: Nucl. Phys.B685,405(2004)].

Kotikov, A. V. and Lipatov, L. N. (2007). On the highest transcendentality in N=4 SUSY, *Nucl. Phys.* **B769**, pp. 217–255, doi:10.1016/j.nuclphysb.2007.01.020, arXiv:hep-th/0611204 [hep-th].

Kotikov, A. V. and Lipatov, L. N. (2013). Pomeron in the N=4 supersymmetric gauge model at strong couplings, *Nucl. Phys.* **B874**, pp. 889–904, doi:10.1016/j.nuclphysb.2013.06.018, arXiv:1301.0882 [hep-th].

Kotikov, A. V., Lipatov, L. N., Onishchenko, A. I., and Velizhanin, V. N. (2004). Three loop universal anomalous dimension of the Wilson operators in $N = 4$ SUSY Yang-Mills model, *Phys. Lett.* **B595**, pp. 521–529, doi:10.1016/j.physletb.2004.05.078,10.1016/j.physletb.2005.11.002, arXiv:hep-th/0404092 [hep-th], [Erratum: Phys. Lett.B632,754(2006)].

Kotikov, A. V., Lipatov, L. N., Rej, A., Staudacher, M., and Velizhanin, V. N. (2007). Dressing and wrapping, *J. Stat. Mech.* **0710**, p. P10003, doi:10.1088/1742-5468/2007/10/P10003, arXiv:0704.3586 [hep-th].

Kotikov, A. V., Lipatov, L. N., and Velizhanin, V. N. (2003). Anomalous dimensions of Wilson operators in N=4 SYM theory, *Phys. Lett.* **B557**, pp. 114–120, doi:10.1016/S0370-2693(03)00184-9, arXiv:hep-ph/0301021 [hep-ph].

Kotikov, A. V., Rej, A., and Zieme, S. (2009). Analytic three-loop Solutions for N=4 SYM Twist Operators, *Nucl. Phys.* **B813**, pp. 460–483, doi:10.1016/j.nuclphysb.2008.12.022, arXiv:0810.0691 [hep-th].

Kotikov, A. V. and Velizhanin, V. N. (2005). Analytic continuation of the Mellin moments of deep inelastic structure functions, arXiv:hep-ph/0501274 [hep-ph].

Kruczenski, M. (2002). A Note on twist two operators in N=4 SYM and Wilson loops in Minkowski signature, *JHEP* **12**, p. 024, doi:10.1088/1126-6708/2002/12/024, arXiv:hep-th/0210115 [hep-th].

Kuraev, E. A., Lipatov, L. N., and Fadin, V. S. (1976). Multi — Reggeon processes in the Yang-Mills theory, *Sov. Phys. JETP* **44**, pp. 443–450, [Zh. Eksp. Teor. Fiz.71,840(1976)].

Kuraev, E. A., Lipatov, L. N., and Fadin, V. S. (1977). The Pomeranchuk singularity in nonabelian Gauge theories, *Sov. Phys. JETP* **45**, pp. 199–204, [Zh. Eksp. Teor. Fiz.72,377(1977)].

Kwiecinski, J. and Praszalowicz, M. (1980). Three Gluon integral equation and odd c singlet regge singularities in QCD, *Phys. Lett.* **94B**, pp. 413–416, doi:10.1016/0370-2693(80)90909-0.

Leurent, S., Serban, D., and Volin, D. (2012). Six-loop Konishi anomalous dimension from the Y-system, *Phys. Rev. Lett.* **109**, p. 241601, doi:10.1103/PhysRevLett.109.241601, arXiv:1209.0749 [hep-th].

Lipatov, L. N. (1975). The parton model and perturbation theory, *Sov. J. Nucl. Phys.* **20**, pp. 94–102, [Yad. Fiz.20,181(1974)].

Lipatov, L. N. (1976). Reggeization of the vector meson and the vacuum singularity in nonabelian Gauge theories, *Sov. J. Nucl. Phys.* **23**, pp. 338–345, [Yad. Fiz.23,642(1976)].

Lipatov, L. N. (1986). The Bare Pomeron in Quantum chromodynamics, *Sov. Phys. JETP* **63**, pp. 904–912, [Zh. Eksp. Teor. Fiz.90,1536(1986)].

Lipatov, L. N. (1994). Asymptotic behavior of multicolor QCD at high energies in connection with exactly solvable spin models, *JETP Lett.* **59**, pp. 596–599, arXiv:hep-th/9311037 [hep-th], [Pisma Zh. Eksp. Teor. Fiz.59,571(1994)].

Lipatov, L. N. (1997). Perspectives in Hadronic Physics, in Proc. of the ICTP conf., World Scientific, Singapore, .

Lipatov, L. N. (2001). Next-to-leading corrections to the BFKL equation and the effective action for high energy processes in QCD, *Nucl. Phys. Proc. Suppl.* **99A**, pp. 175–179, doi:10.1016/S0920-5632(01)01329-9, [,175(2001)].

Lipatov, L. N. (2013). Effective action for the Regge processes in gravity, *Phys. Part. Nucl.* **44**, pp. 391–413, doi:10.1134/S1063779613020214, arXiv:1105.3127 [hep-th].

Loebbert, F., Sieg, C., Wilhelm, M., and Yang, G. (2016). Two-loop SL(2) form factors and maximal transcendentality, *JHEP* **12**, p. 090, doi:10.1007/JHEP12(2016)090, arXiv:1610.06567 [hep-th].

Lukowski, T., Rej, A., and Velizhanin, V. N. (2010). Five-loop anomalous dimension of twist-two operators, *Nucl. Phys.* **B831**, pp. 105–132, doi:10.1016/j.nuclphysb.2010.01.008, arXiv:0912.1624 [hep-th].

Lüscher, M. (1986). Volume dependence of the energy spectrum in massive quantum field theories. 1. stable particle states, *Commun. Math. Phys.* **104**, p. 177, doi:10.1007/BF01211589.

Makeenko, Y. (2003). Light cone Wilson loops and the string/gauge correspondence, *JHEP* **01**, p. 007, doi:10.1088/1126-6708/2003/01/007, arXiv:hep-th/0210256 [hep-th].

Maldacena, J. M. (1999). The Large N limit of superconformal field theories and supergravity, *Int. J. Theor. Phys.* **38**, pp. 1113–1133, doi:10.1023/A:1026654312961, 10.4310/ATMP.1998.v2.n2.a1, arXiv:hep-th/9711200 [hep-th], [Adv. Theor. Math. Phys.2,231(1998)].

Marboe, C. and Velizhanin, V. (2016). Twist-2 at seven loops in planar $\mathcal{N} = 4$ SYM theory: full result and analytic properties, *JHEP* **11**, p. 013, doi:10.1007/JHEP11(2016)013, arXiv:1607.06047 [hep-th].

Marboe, C., Velizhanin, V., and Volin, D. (2015). Six-loop anomalous dimension of twist-two operators in planar $\mathcal{N} = 4$ SYM theory, *JHEP* **07**, p. 084, doi:10.1007/JHEP07(2015)084, arXiv:1412.4762 [hep-th].

Marboe, C. and Volin, D. (2015). Quantum spectral curve as a tool for a perturbative quantum field theory, *Nucl. Phys.* **B899**, pp. 810–847, doi:10.1016/j.nuclphysb.2015.08.021, arXiv:arXiv:1411.4758 [hep-th].

Marboe, C. and Volin, D. (2017). The full spectrum of AdS5/CFT4 I: Representation theory and one-loop Q-system, arXiv:arXiv:1701.03704 [hep-th].

Marboe, C. and Volin, D. (2018). The full spectrum of AdS_5/CFT_4 II: Weak coupling expansion via the quantum spectral curve, arXiv:1812.09238 [hep-th].

Marboe, C. and Widn, E. (2019). The fate of the Konishi multiplet in the β-deformed Quantum Spectral Curve, arXiv:1902.01248 [hep-th].

Minahan, J. A. and Zarembo, K. (2003a). The Bethe ansatz for N=4 superYang-Mills, *JHEP* **03**, p. 013, doi:10.1088/1126-6708/2003/03/013, arXiv:hep-th/0212208 [hep-th].

Minahan, J. A. and Zarembo, K. (2003b). The Bethe ansatz for N=4 superYang-Mills, *JHEP* **03**, p. 013, doi:10.1088/1126-6708/2003/03/013, arXiv:hep-th/0212208 [hep-th].

Moch, S., Vermaseren, J. A. M., and Vogt, A. (2004). The Three loop splitting functions in QCD: The Nonsinglet case, *Nucl. Phys.* **B688**, pp. 101–134, doi:10.1016/j.nuclphysb.2004.03.030, arXiv:hep-ph/0403192 [hep-ph].

Ogievetsky, E. and Wiegmann, P. (1986). Factorized S matrix and the Bethe Ansatz for simple lie groups, *Phys. Lett.* **168B**, pp. 360–366, doi:10.1016/0370-2693(86)91644-8.

Polchinski, J. and Strassler, M. J. (2002). Hard scattering and gauge/string duality, *Phys. Rev. Lett.* **88**, p. 031601, doi:10.1103/PhysRevLett.88.031601, arXiv:hep-th/0109174 [hep-th].

Polchinski, J. and Strassler, M. J. (2003). Deep inelastic scattering and gauge/string duality, *JHEP* **05**, p. 012, doi:10.1088/1126-6708/2003/05/012, arXiv:hep-th/0209211 [hep-th].

Reshetikhin, N. Yu. (1985). Integrable models of quantum one-dimensional magnets with O(N) and Sp(2k) symmetry, *Theor. Math. Phys.* **63**, pp. 555–569, doi:10.1007/BF01017501, [Teor. Mat. Fiz.63,347(1985)].

Roiban, R., Tirziu, A., and Tseytlin, A. A. (2007). Two-loop world-sheet corrections in AdS(5) x S**5 superstring, *JHEP* **07**, p. 056, doi:10.1088/1126-6708/2007/07/056, arXiv:0704.3638 [hep-th].

Roiban, R. and Tseytlin, A. A. (2011). Semiclassical string computation of strong-coupling corrections to dimensions of operators in Konishi multiplet, *Nucl. Phys.* **B848**, pp. 251–267, doi:10.1016/j.nuclphysb.2011.02.016, arXiv:1102.1209 [hep-th].

Salam, G. P. (1998). A Resummation of large subleading corrections at small x, *JHEP* **07**, p. 019, doi:10.1088/1126-6708/1998/07/019, arXiv:hep-ph/9806482 [hep-ph].

Salam, G. P. (1999). An Introduction to leading and next-to-leading BFKL, *Acta Phys. Polon.* **B30**, pp. 3679–3705, arXiv:hep-ph/9910492 [hep-ph].

Saleur, H. (2000). The Continuum limit of sl(N/K) integrable superspin chains, *Nucl. Phys.* **B578**, pp. 552–576, doi:10.1016/S0550-3213(00)00002-X, arXiv:solv-int/9905007 [solv-int].

Schuler, G. A., Sakakibara, S., and Korner, J. G. (1987). Use of four-dimensional spin methods in the calculation of radiative QCD corrections, *Phys. Lett.* **B194**, pp. 125–131, doi:10.1016/0370-2693(87)90781-7.

Siegel, W. (1979). Supersymmetric dimensional regularization via dimensional reduction, *Phys. Lett.* **84B**, pp. 193–196, doi:10.1016/0370-2693(79)90282-X.

Staudacher, M. (2005). The Factorized S-matrix of CFT/AdS, *JHEP* **05**, p. 054, doi:10.1088/1126-6708/2005/05/054, arXiv:hep-th/0412188 [hep-th].

Takahashi (1999). *Thermodynamics of One-Dimensional Solvable Models* (Cambridge University Press).

Velizhanin, V. N. (2009). The four-loop anomalous dimension of the Konishi operator in N=4 supersymmetric Yang-Mills theory, *JETP Lett.* **89**, pp. 6–9, doi:10.1134/S0021364009010020, arXiv:0808.3832 [hep-th].

Velizhanin, V. N. (2015). BFKL pomeron in the next-to-next-to-leading approximation in the planar N=4 SYM theory, arXiv:1508.02857 [hep-th].

Vermaseren, J. A. M. (1999). Harmonic sums, Mellin transforms and integrals, *Int. J. Mod. Phys.* **A14**, pp. 2037–2076, doi:10.1142/S0217751X99001032, arXiv:hep-ph/9806280 [hep-ph].

Vermaseren, J. A. M., Vogt, A., and Moch, S. (2005). The third-order QCD corrections to deep-inelastic scattering by photon exchange, *Nucl. Phys.* **B724**, pp. 3–182, doi:10.1016/j.nuclphysb.2005.06.020, arXiv:hep-ph/0504242 [hep-ph].

Vogt, A., Moch, S., and Vermaseren, J. A. M. (2004). The three-loop splitting functions in QCD: The Singlet case, *Nucl. Phys.* **B691**, pp. 129–181, doi:10.1016/j.nuclphysb.2004.04.024, arXiv:hep-ph/0404111 [hep-ph].

Witten, E. (1998). Anti-de Sitter space and holography, *Adv. Theor. Math. Phys.* **2**, pp. 253–291, doi:10.4310/ATMP.1998.v2.n2.a2, arXiv:hep-th/9802150 [hep-th].

https://doi.org/10.1142/9789811231124_0015

Chapter 15

The Regge Limit of N = 4 SUSY Gauge Theories

Jochen Bartels* and Alex Prygarin†

**II. Institut für Theoretische Physik, Universität Hamburg, Luruper Chaussee 149, D-22761 Hamburg, Germany*

†Department of Physics, Ariel University, Ariel,40700, Israel

Both of us have collaborated with Lev Lipatov, and we are deeply grateful that we have had this opportunity. One of us (JB) has met Lev first in 1973 during his first visit to Leningrad, and in the following years he got to know Lev much better: the outstanding depth of his understanding of physics, his friendly personality and his very kind family. Later, with the growing interest of small-x physics at HERA Lev became a regular visitor of Hamburg University and DESY, and these visits have laid the ground for many collaborations with different authors which resulted in numerous and outstanding publications.

The second author (AP) was introduced to the works of Lev Lipatov during his Ph.D. studies under the supervision of Eugene Levin at Tel Aviv. He was deeply impressed by a clear and elegant presentation of rather complicated underlying physics in Lev's papers. The first discussion in person with Lev on BFKL physics during the PNPI Winter School in 2005 was even more inspiring, leaving no doubt regarding the topic to be selected for the Ph.D. thesis. The combination of the very hospitable and vibrating scientific atmosphere of the institutes of Hamburg University and of DESY and the opportunity of having everyday scientific discussions with Lev as a visiting professor at Hamburg determined the choice of my future postdoc position. Beyond the fruitful scientific collaboration it was a big pleasure to have many exciting moments of personal talks on a variety of topics not directly related to physics.

Abstract

In this contribution we summarize some of Lev Lipatov's contributions to the Regge limit of N=4 SUSY gauge theories: the remainder functions of the $2 \to 4$ and the $2 \to 5$ scattering amplitudes. In the last years of his life he had started to generalize these investigations to the $2 \to 6$ amplitude: in the meantime this study has been completed, and we report the main results. In the final part we comment on new and unpublished results on higher order BFKL corrections in which Lev has been involved.

15.1 Introduction

Lev Lipatov has provided many outstanding contributions to QFT. His particular interest has always been devoted to the Regge limit in quantum field theory, in particular in QED, QCD, and Gravity. He immediately realized that the hypothesis of the AdS/CFT correspondence provides an extremely promising framework, and he was one of the first to get involved in the study of the Regge limit of $N = 4$ SUSY theories. In recent years promising new calculation methods have been developed for computing higher order corrections. Lev's contributions are complementary to this since he provides the full information on the structure of the scattering amplitude in the Regge limit. This information is necessary also for the results obtained by the new methods, in order to fully understand the connection with physical cross sections.

Of course, for Lev the structure of the BFKL amplitude [Lipatov (1976); Fadin *et al.* (1975); Kuraev *et al.* (1976, 1977); Balitsky and Lipatov (1978)] (beyond the leading order in perturbation theory) has always been in the focus of his interest. In particular, he has shown [Lipatov (2009)] that in the planar approximation the BFKL kernel belongs to an integrable structure (closed spin chain), and this integrability is expected to hold to all orders. To investigate this structure in more detail one needs to study $2 \to n$ scattering amplitudes in the Regge limit: for $n = 2$ the spin chain consists of two sites (the Regge cut is a bound state two reggeized gluons), for $n = 3$ the spin chain has 3 sites (Regge cuts are bound states of up to three reggeized gluons) and so on. Also, beginning with $n = 3$ higher order corrections to the BFKL kernel may add new structures, e.g. a $3 \to 3$ kernel for the reggeized gluons [Bartels *et al.* (2013)].

Among his main contributions to the investigation of the Regge limit of $N = 4$ SUSY is also the observation that, for $2 \to m$ $(m > 3)$ production amplitudes, the BDS formula [Bern *et al.* (2005)] is incomplete and requires corrections due to Regge cuts [Bartels *et al.* (2009, 2010)]. These corrections are now known as 'remainder functions'. In course of investigating these Regge cut contribution it was realized that in the planar approximation — which is often being used in studying the AdS/CFT correspondence — the structure of Regge pole contributions in $2 \to 2n$ production amplitudes shows peculiar features which differ from those known from signatured amplitudes. In certain kinematic regions the Regge pole contributions become singular, and the singularities are cancelled by the Regge cut contributions which show up in exactly these kinematic regions. Thus, in the planar approximation, Regge cuts are not simply higher order corrections to the Regge poles but contributions which are necessary for obtaining finite and conformal invariant expressions in the Regge limit.

As it has already been said, the interest in BFKL and its generalization has always played a central role in Lev's research. This includes a deeper understanding of higher order corrections for the BFKL kernel and the connection with the graviton in the strong coupling limit.

In this paper we first review the studies of the Regge limit of the $2 \to 4$ and $2 \to 5$ production amplitudes in $N = 4$ SUSY. We also give a brief review of the recent extension to the $2 \to 6$ case. We then present a discussion of higher order corrections for the BFKL kernel which Lev has been pursuing in his last years. For a detailed discussion of the integrability we refer to the contribution of M.Alfimov, N.Gromov and V.Kazakov. The connection between BFKL and the graviton has been discussed by A.Kotikov and A.I.Onishchenko.

15.2 Remainder function of the 6 point amplitude

Starting from the $2 \to 4$ scattering amplitude in the planar approximation in the multi-Regge limit, it was found in [Bartels *et al.* (2009, 2010); Lipatov (2012)] that the BDS-amplitude suggested by Bern, Dixon and Smirnov [Bern *et al.* (2005)] needs corrections in certain kinematic regions. Starting from the kinematic region where all energies are positive, and then analytically continuing into the region where the two produced particles have negative energies, the expression derived from the BDS amplitude is incomplete. First, the Regge pole contribution contains unphysical singularities:

$$s_1^{\omega_1}(-s_2)^{\omega_2}s_3^{\omega_3}\left[e^{i\pi(\omega_a+\omega_b)} - 2ie^{i\pi\omega_2}\frac{\Omega_a\Omega_b}{\Omega_2}\right], \tag{15.1}$$

which we can also write as

$$s_1^{\omega_1}(-s_2)^{\omega_2}s_3^{\omega_3}\left[\cos\pi\omega_{ab} - 2i\left(-\frac{1}{2}\sin\pi(\omega_a+\omega_b) + \cos\pi\omega_2\frac{\Omega_a\Omega_b}{\Omega_2}\right)\right]. \tag{15.2}$$

Here

$$\omega_i = -\frac{\gamma_K}{4} \ln \frac{|q_i|^2}{\lambda^2}, \quad i = 1,2,3$$
$$\gamma_K = 4a, \quad a = \frac{\alpha_s N_c}{2\pi} \tag{15.3}$$

denotes the gluon trajectory function in the t_i-channel, and

$$\Omega_a = -\frac{\gamma_K}{8} \ln \frac{|q_1|^2 |q_2|^2}{|k_a|^2 \lambda^2} \tag{15.4}$$

stands for the production vertex of particle a. In (15.2) it is the round brackets, in particular the terms proportional to $1/\Omega_2$, which are unphysical and have to be removed.

On the other hand it is known from QCD studies that there exists Regge cut contributions. To see this in more detail one writes the scattering amplitude as a sum of five terms which have the maximal number of non-overlapping energy discontinuities and are in agreement with the Steinmann relations.

For the kinematic region where all energies are positive the corresponding energy factors are:

$$\begin{aligned}
RR: &\quad (-s_1)^{\omega_1-\omega_2}(-s_{012})^{\omega_2-\omega_3}(-s)^{\omega_3}\\
LL: &\quad (-s_3)^{\omega_3-\omega_2}(-s_{123})^{\omega_2-\omega_1}(-s)^{\omega_1}\\
RL: &\quad (-s_1)^{\omega_1-\omega_2}(-s_3)^{\omega_3-\omega_2}(-s)^{\omega_2}\\
LR(1): &\quad (-s_2)^{\omega_2-\omega_1}(-s_{012})^{\omega_1-\omega_3}(-s)^{\omega_3}\\
LR(2): &\quad (-s_2)^{\omega_2-\omega_3}(-s_{123})^{\omega_3-\omega_1}(-s)^{\omega_1},
\end{aligned} \tag{15.5}$$

in agreement with the cut structure illustrated in Fig.15.1. The different terms are labeled in the following way: the first letter belongs to the left produced particle a, and 'L' or 'R' indicate whether the uppermost dashed line comes from the left or right hand side.

The two reggeon cut in the t_2 channel is given by the discontinuity in s_2 and therefore can be contained only in the last two terms, $LR(1)$ and $LR(2)$.

Regge theory requires signatured amplitudes, i.e. combinations of uncrossed and crossed amplitudes. As an example, for a signatured amplitude the last term $LR(2)$ in (15.5) becomes

$$|s_2|^{\omega_2-\omega_3}|s_{123}|^{\omega_3-\omega_1}|s|^{\omega_1}\left(e^{-i\pi(\omega_2-\omega_3)}+\tau_2\tau_3\right)\left(e^{-i\pi(\omega_3-\omega_1)}+\tau_1\tau_3\right)\left(e^{-i\pi\omega_1}+\tau_1\right). \tag{15.6}$$

Here the three brackets denote the signature factors. In the planar approximation one does not consider these signatured combinations. Instead, one expands in powers of τ and considers term by term. The region of positive energies corresponds to the term without any factor τ; in the region where the produced particles 'a' and 'b'

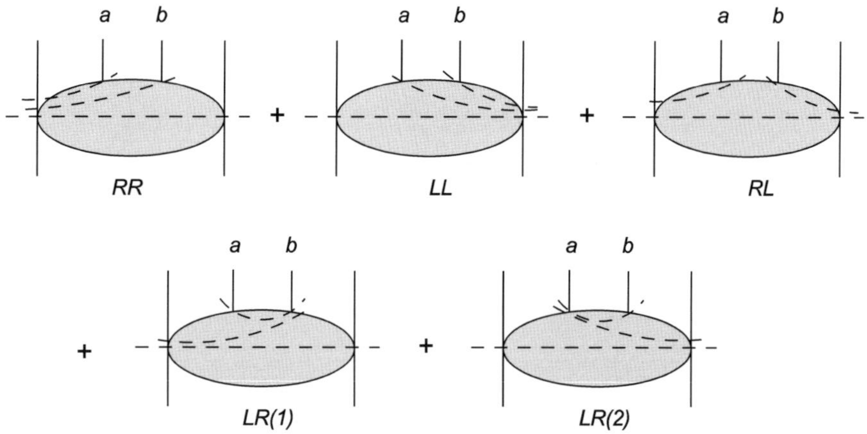

Fig. 15.1 Analytic structure of the 2 → 4 scattering amplitude: The dashed lines indicate the energy variables with a non-vanishing discontinuity.

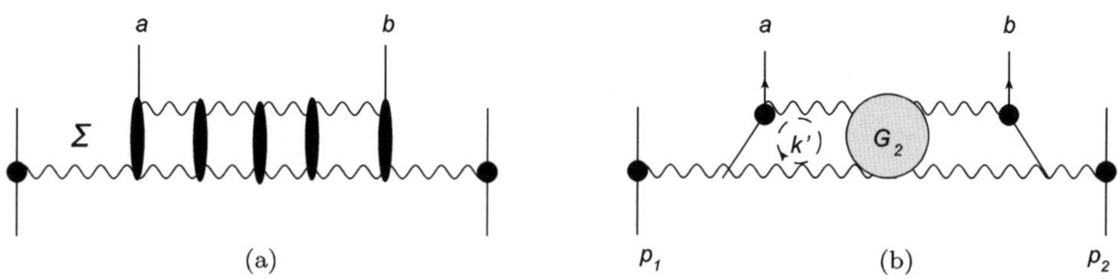

Fig. 15.2 The two reggeon cut in the 2 → 4 scattering amplitude: (a) the unitarity integral and (b) the leading order remainder function f_{ω_2}.

have negative energies both the t_1 and the t_3 channels are 'twisted', i.e. we consider the term proportional to $\tau_1\tau_3$. For the Regge pole contribution, the regions $\tau_1\tau_3$ and $\tau_1\tau_2\tau_3$ are special since they become singular (cf. 15.2)). As to the Regge cut contributions, a detailed analysis of the partial wave expansion of the 2 → 4 scattering amplitude shows that this Regge cut contribution has the following trigonometric factors:

$$LR(1): \quad \frac{W_{\omega_2}}{\Omega_{13}}$$

$$LR(2): \quad \frac{W_{\omega_2}}{\Omega_{31}}. \tag{15.7}$$

Together with the phase factors in (15.5) and going through region by region it is straightforward to see that in all regions except for $\tau_1\tau_3$ (and $\tau_1\tau_2\tau_3$) the Regge cut contributions of $LR(1)$ and $LR(2)$ cancel. In other words, this Regge cut is present only in the kinematic region $\tau_1\tau_3$ and $\tau_1\tau_2\tau_3$, and in the first region the two produced particles a and b are incoming, i.e. have negative energies.

In this region the Regge cut contribution to the scattering amplitude is of the form

$$2is_1^{\omega_1}s_3^{\omega_3}W_{\omega_2} = 2is_1^{\omega_1}s_3^{\omega_3}\int\frac{d\omega_2'}{2\pi i}(-s_2)^{\omega_2'}\tilde{W}(\omega_2'). \tag{15.8}$$

Comparison with (15.2) shows that this Regge cut contribution has the same phase structure as the singular Regge pole term: by re-defining the Regge cut contribution

$$W_{\omega_2} = \delta W_{\omega_2} + W_{\omega_2}^{reg} \tag{15.9}$$

with

$$\delta W_{\omega_2} = (-s_2)^{\omega_2}\left(-\frac{1}{2}\sin\pi(\omega_a+\omega_b) + \cos\pi\omega_2\frac{\Omega_a\Omega_b}{\Omega_2}\right) \tag{15.10}$$

and

$$W_{\omega_2}^{reg} = (-s_2)^{\omega_2}\frac{\pi\delta_{13}}{2} + \int\frac{d\omega_2'}{2\pi i}(-s_2)^{\omega_2'}f_{\omega_2}^{reg} \tag{15.11}$$

the singular pole term cancels. Here we have separated the one loop term of the Regge cut amplitude: the factor $e^{i\pi\delta_{13}}$ is contained in the BDS amplitude. In this sense the BDS amplitude contains the 'beginning of the Regge cut'. The term δW_{ω_2} can be viewed as a subtraction of the ω_2' integral of the Regge cut. It is important to stress that this subtraction not only cancels the singular piece of the Regge pole term in (15.2) but also the regular part $-\frac{1}{2}\sin\pi(\omega_a+\omega_b)$: this is necessary to make the remaining Regge pole term $\cos\pi\omega_{ab}$ finite and conformal invariant.

We stress that the form of the Regge pole contributions as well as the subtractions for the Regge cut is valid to all orders in perturbation theory. It is only the next step, the computation of the Regge cut from unitarity, which as a first approximation is valid to leading order only. However, it is possible to extend this to higher orders.

In leading order unitarity leads to

$$f_{\omega_2}^{reg} = \int\frac{d^2k'}{(2\pi)^3}\frac{d^2k''}{(2\pi)^3}\Phi^a(\mathbf{q}_1,\mathbf{k}',\mathbf{q}_2-\mathbf{k}')\left(\frac{1}{\omega_2'-K_2^{octet}(\mathbf{q}_2,\mathbf{k}',\mathbf{k}'')}-\frac{1}{\omega_2'}\right)$$
$$\cdot\Phi^b(\mathbf{q}_2,\mathbf{k}'',\mathbf{q}_2-\mathbf{k}''), \tag{15.12}$$

where Φ^a denotes the production vertex or impact factor of particle a

$$\Phi^a(\mathbf{q}_1,\mathbf{k}',\mathbf{q}_2-\mathbf{k}') = \frac{\mathbf{q}_1^2}{(\mathbf{q}_1-\mathbf{k}')^2}\frac{(q_1-k')(q_2-k')^*}{q_1q_2^*}$$
$$= \frac{q_1^*}{q_2^*}\frac{q_2^*-k'^*}{q_1^*-k'^*}. \tag{15.13}$$

The Green's function $G_2^{(8_A)}$ of a two gluon system in the octet representation satisfies the equation:

$$\omega G_\omega^{(8_A)}(\mathbf{k},\mathbf{k}',\mathbf{q}) = \frac{\delta^{(2)}(\mathbf{k}-\mathbf{k}')}{\mathbf{k}^2(\mathbf{k}-\mathbf{q})^2} + \frac{1}{\mathbf{k}^2(\mathbf{k}-\mathbf{q})^2}\left(K^{(8_A)}\otimes G_\omega^{(8_A)}\right)(\mathbf{k},\mathbf{k}',\mathbf{q}), \tag{15.14}$$

where the color octet kernel $K^{(8_A)}$ can be decomposed as follows:

$$\begin{aligned}
&K^{(8_A)}(\mathbf{k},\mathbf{k}',\mathbf{q})\\
&= \delta^{(2)}(\mathbf{k}-\mathbf{k}')\left(\omega(-|\mathbf{k}|^2)+\omega(-|\mathbf{q}-\mathbf{k}|^2)\right) + \frac{a}{2}\frac{(q-k)k^*(q-k')^*k' + c.c.}{(\mathbf{k}-\mathbf{k}')^2}\\
&= \delta^{(2)}(\mathbf{k}-\mathbf{k}')\Big[\omega(-|\mathbf{q}]^2) + \frac{1}{2}\left(\omega(-|\mathbf{k}|^2)+\omega(-|\mathbf{q}-\mathbf{k}|^2) - 2\omega(-|q|^2)\right)\Big]\\
&\quad + \frac{1}{2}K^{(1)}(\mathbf{q}-\mathbf{k};\mathbf{q}-\mathbf{k}',\mathbf{k}')\\
&= \delta^{(2)}(\mathbf{k}-\mathbf{k}')\Big[\omega(-|\mathbf{q}]^2) - \frac{a}{2}\ln\frac{\mathbf{k}^2(\mathbf{q}-\mathbf{k})^2}{\mathbf{q}^2\mathbf{q}^2}\Big] + \frac{1}{2}K^{(1)}(\mathbf{q}-\mathbf{k};\mathbf{q}-\mathbf{k}',\mathbf{k}')\\
&= \delta^{(2)}(\mathbf{k}-\mathbf{k}')\omega(-|\mathbf{q}]^2) + K^{octet}(\mathbf{q}-\mathbf{k},\mathbf{k};\mathbf{q}-\mathbf{k}',\mathbf{k}').
\end{aligned} \tag{15.15}$$

Here the kernel K^{octet} is infrared finite. The corresponding Green's function can thus be written as:

$$G_2 = \frac{1}{\omega' - K^{(8_A)}} = \frac{1}{\omega' - \omega(\mathbf{q}) - K^{octet}}. \tag{15.16}$$

When performing the ω_2'-integral, we shift the integration variable

$$\omega_2'' = \omega_2' + \omega_2. \tag{15.17}$$

We thus arrive at the Green's function:

$$G_2^{octet} = \frac{1}{\omega_2'' - K^{octet}}, \tag{15.18}$$

and the integral in (15.11) obtains the prefactor $(-s_2)^{\omega_2}$. In this way, the Regge cut amplitudes are proportional to the BDS amplitude.

It is convenient to use the conformal spin representation:

$$\begin{aligned}
f_{\omega_2} = \frac{g^2N_c}{16\pi^2}\sum_n(-1)^n\int\frac{d\nu}{\nu^2+\frac{n^2}{4}}\left(\frac{q_3^*k_a^*}{k_b^*q_1^*}\right)^{i\nu-\frac{n}{2}}\\
\times\left(\frac{1}{\omega_2'-\omega(\nu,n)}-\frac{1}{\omega_2'}\right)\left(\frac{q_3k_a}{k_bq_1}\right)^{i\nu+\frac{n}{2}},
\end{aligned} \tag{15.19}$$

where

$$\omega(\nu,n) = \frac{g^2N_c}{8\pi^2}\left(2\psi(1) - \mathcal{R}\psi(1+i\nu+\frac{n}{2}) - \mathcal{R}\psi(1+i\nu-\frac{n}{2}\right). \tag{15.20}$$

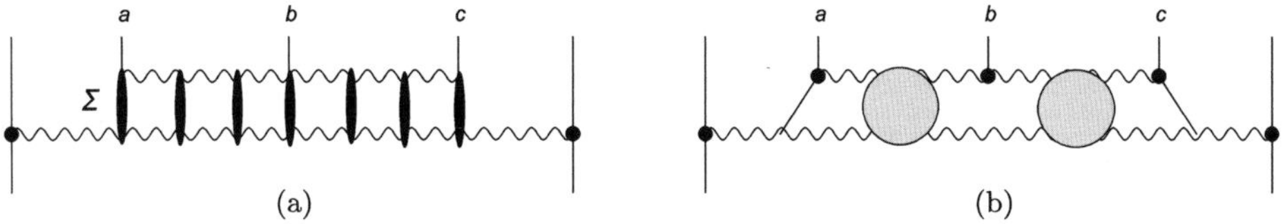

Fig. 15.3 Long Regge cut extending over the t_2 and t_3 channels. (a) unitarity integral and (b) leading order remainder function.

This completes our brief summary of the simplest case, the $2 \to 4$ amplitude in leading order. Several comments are to be made:
1) the impact factor Φ^a in (15.13) in momentum space is not point-like. For this reason, for the attached two gluon state in the color octet representation the bootstrap property does not apply and the two reggeon cut survives.
2) As we have said before, in the planar approximation (large N_c limit) the two gluon Regge cut contributes only in the kinematic regions $\tau_1\tau_3$ and $\tau_1\tau_2\tau_3$. The first one corresponds to the region where the produced particles a and b have negative energies. The absence of the Regge cuts in all other regions can be seen also in another way [Lipatov (2009)]. Consider in Fig.2b the loop on the left hand side of the two reggeon Green's function, and introduce Sudakov variables:

$$k' = \alpha' p_1 + \beta' p_2 + \mathbf{k}'. \tag{15.21}$$

Then the α' integration picks up poles from the propagator of the left impact factor and from the energy factor of the reggeon in the t_1 channel: as long as this reggeon is untwisted, both the energy cut singularity and the pole of the propagator lie on the same side of the α'-integration contour, and the integral vanishes. If on the other hand we twist the t_1 channel, the α'-integration is nonzero.

Since the appearance of [Bartels *et al.* (2009, 2010)] where only the leading log approximation had been used several refinements have been made, e.g. [Lipatov and Prygarin (2011a,b); Bargheer *et al.* (2016, 2020); Bargheer, T. and Bartels, J. (2020)]. First, higher order corrections of the impact factors have been calculated and NLO corrections and NNLO corrections of the BFKL kernel have been obtained. Also, the generalization to the $3 \to 3$ has been found [Bartels *et al.* (2011)].

For the BFKL kernel of the two gluon Regge cut it has been shown [Lipatov (2009)] that it is part of an integrable open spin chain: the two gluon system belongs to a chain consisting two spins, a three reggeon cut (see below) provides the three spin chain etc.

15.3 Remainder function of the 7 point amplitude

The investigation of the $2 \to 4$ amplitude described in the previous section has been extended to the $2 \to 5$ amplitude [Bartels *et al.* (2014, 2015)]. Compared to the $2 \to 4$ scattering amplitude, the new piece of the remainder function is a 'long' Regge cut extending over two t-channels:

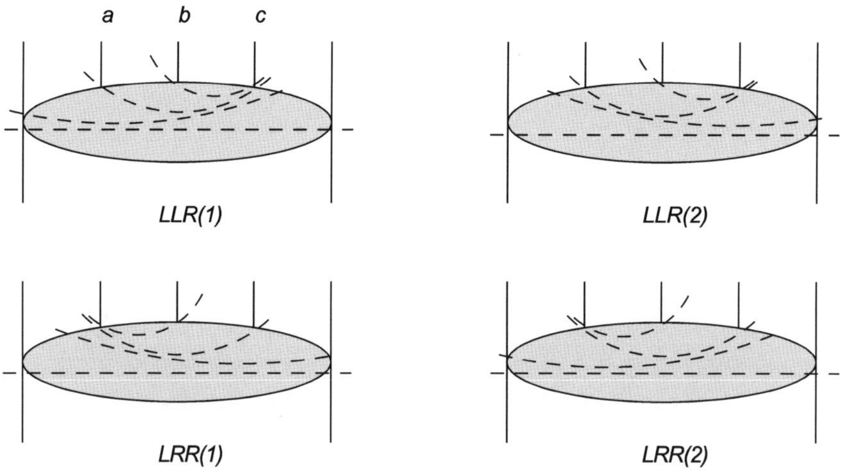

Fig. 15.4 Terms containing the long cut.

Beginning again with the analytic structure, we now have 14 terms. In Fig.15.4 we only show those four terms which contain this long cut: We note the difference between the two terms *LLR* in the upper line and the *LRR* terms in the lower line: as it will become clear later, because of the energy cut structure the production vertex for particle *b* consists of two distinct pieces.

Next the Regge pole terms derived from the BDS formula. As discussed in detail in [Bartels *et al.* (2014)], the signatured $2 \to n+1$ production amplitude can be written in the factorized form:

$$\frac{A_{2\to n+1}^{\tau_i\tau_j\cdots\tau_n}}{\Gamma(t_1)\Gamma(t_n)} = |s_1|^{\omega_1}\xi_1 V^{\tau_1\tau_2;a_1}|s_2|^{\omega_2}\xi_2 V^{\tau_2\tau_3;a_2}|s_3|^{\omega_3}\xi_3 \times \cdots$$
$$\times |s_{n-1}|^{\omega_{n-1}}\xi_{n-1}V^{\tau_{n-1}\tau_n;a_{n-1}}|s_n|^{\omega_n}\xi_n, \tag{15.22}$$

where

$$\xi_i = e^{-i\pi\omega_i} - \tau_i;\ \xi_{ij} = e^{-i\pi\omega_{ij}} + \tau_i\tau_j;\ \xi_{ji} = e^{-i\pi\omega_{ji}} + \tau_i\tau_j \tag{15.23}$$

with

$$\omega_{ij} = \omega_i - \omega_j \tag{15.24}$$

denote the signature factors, and

$$V^{\tau_i\tau_j;a_j} = \frac{\xi_{ij}}{\xi_i}c_R^{ij;a_i} + \frac{\xi_{ji}}{\xi_j}c^{ij;a_i} \tag{15.25}$$

stands for the complex-valued production vertex of the particle a_j.

As it has been said before, the planar approximation is obtained as an expansion in monomials of signatures τ_i, and in this expansion each terms belongs to the planar approximation continued to a particular kinematic region. For this expansion it is convenient to write

$$\tilde{V}^{\tau_1\tau_2;a_1} = \xi_1 V^{\tau_1\tau_2;a_1}\xi_2 \tag{15.26}$$

and

$$\frac{A_{2\to n+1}^{\tau_i\tau_j\cdots\tau_n}}{\Gamma(t_1)|s_1|^{\omega_1}|s_2|^{\omega_2}\cdots|s_n|^{\omega_n}\Gamma(t_n)} = \tilde{V}^{\tau_1\tau_2;a_1}\frac{1}{\xi_2}\tilde{V}^{\tau_2\tau_3;a_2}\frac{1}{\xi_3}\cdots\frac{1}{\xi_{n-1}}\tilde{V}^{\tau_{n-1}\tau_n;a_{n-1}}. \tag{15.27}$$

In order to obtain this representation we observe that the production vertex can be expanded as:

$$\begin{aligned}\tilde{V}^{\tau_1\tau_2;a} &= e^{-i\pi\omega_1}c_R^{12;a_1} + e^{-i\pi\omega_2}c_L^{12;a_1} - \tau_1 e^{-i\pi\omega_1}\left(e^{-i\pi\omega_1}c_R^{12;a_1} + e^{-i\pi\omega_2}c_L^{12;a_1}\right)\\ &\quad - \tau_2 e^{-i\pi\omega_2}\left(e^{-i\pi\omega_1}c_R^{12;a_1} + e^{-i\pi\omega_2}c_L^{12;a_1}\right)\\ &\quad + \tau_1\tau_2\left(e^{-i\pi\omega_2}c_R^{12;a_1} + e^{-i\pi\omega_1}c_L^{12;a_1}\right),\end{aligned} \tag{15.28}$$

and the propagator can be written in the form:

$$\frac{1}{\xi_2} = \frac{1}{e^{-i\pi\omega_2} - \tau_2} = \frac{e^{-i\pi\omega_2} + \tau_2}{-2i\sin(\pi\omega_2)e^{-i\pi\omega_2}}. \tag{15.29}$$

Note the appearance of the nonphysical poles $\sim 1/\sin\pi\omega_2$ which should be cancelled by the Regge cut contributions.

In this way one can determine the Regge pole contributions in all different kinematic regions, labelled by products of τ_i factors. In particular one finds that in all regions containing the product $\tau_1\tau_4$ the Regge pole terms are most singular and contain double poles of the form $\sim 1/(\sin\pi\omega_2\sin\pi\omega_3)$. In particular, in the region $\tau_1\tau_4$ the Regge pole term has the form

$$\begin{aligned}&|s_1|^{\omega_1}(-s_2)^{\omega_2}(-s_3)^{\omega_3}|s_4|^{\omega_4}\\ &\quad\cdot\left[e^{i\pi(\omega_a+\omega_b+\omega_c)} - 2ie^{i\pi(\omega_2+\omega_3)}\frac{\sin(\pi\omega_a)\sin(\pi\omega_b)\sin(\pi\omega_c)}{\sin(\pi\omega_2)\sin(\pi\omega_3)}\right].\end{aligned} \tag{15.30}$$

A complete list of all kinematic regions can be found in [Bartels *et al.* (2014)].

The same result can also be obtained in a different way. Instead of (15.22), we can start from the analytic representation consisting of 14 terms. For each term the Regge pole contribution has a particular form, e.g. for the first term LLR(1) in

Fig.15.4 in the region where all energies are positive:

$$(-s_1)^{\omega_1}(-s_2)^{\omega_2}(-s_3)^{\omega_3}(-s_4)^{\omega_4}F^{pole}_{LLR(1)}$$
$$= (-s_1)^{\omega_1}(-s_2)^{\omega_2}(-s_3)^{\omega_3}(-s_4)^{\omega_4}\frac{\Omega_1\,\Omega_{34}}{\Omega_3\,\Omega_{14}}\,\frac{V_L(a)}{\Omega_{21}}\,\frac{V_L(b)}{\Omega_{32}}\,\frac{V_R(c)}{\Omega_{34}}, \quad (15.31)$$

where

$$\Omega_i = \sin\pi\omega_i,\ \ \Omega_{ij} = \sin\pi(\omega_i-\omega_j),\ \ \omega_{ij} = \omega_i-\omega_j. \quad (15.32)$$

When inserting, for each kinematic region, the corresponding phases of the energy factor, one simply computes the sum of the 14 terms. For the kinematic region $\tau_1\tau_4$ one obtains, after some algebra, the above result (15.30).

Based upon this Regge pole structure it is possible to determine the necessary subtraction terms for the Regge cut contributions, in analogy with the $2 \to 4$ case decribed above. Without going into details which are described in [Bartels *et al.* (2015)] we only list the results for the long cut. Introducing the trigonometric factors of the long cut, analogous to (15.7), we find that the long cut appears in all regions containing the product $\tau_1\tau_4$, i.e. just in those regions where the Regge poles are most singular. From the general analytic representation we find, for the Regge cut contribution in the region $\tau_1\tau_4$,

$$2i|s_1|^{\omega_1}|s_4|^{\omega_4}\left(e^{-i\pi\omega_2}\widetilde{W}_{\omega_2\omega_3;R} + e^{-i\pi\omega_3}\widetilde{W}_{\omega_2\omega_3;L}\right), \quad (15.33)$$

where we have introduced the following combinations of long cut and short cuts:

$$\widetilde{W}_{\omega_2\omega_3;L} = \frac{W_{\omega_2\omega_3;L}}{\Omega_{32}} + \frac{\Omega_a}{\Omega_2}W_{\omega_3} \quad (15.34)$$

$$\widetilde{W}_{\omega_2\omega_3;R} = \frac{W_{\omega_2\omega_3;R}}{\Omega_{23}} + W_{\omega_2}\frac{\Omega_c}{\Omega_3}. \quad (15.35)$$

Here $W_{\omega_2\omega_3;L}$ and $W_{\omega_2\omega_3;R}$ denote the cut terms of the upper and lower lines of Fig.15.4, resp. We define

$$W_{\omega_2\omega_3;L} = \delta W_{\omega_2\omega_3;L} + W^{reg}_{\omega_2\omega_3;L}$$
$$W_{\omega_2\omega_3;R} = \delta W_{\omega_2\omega_3;R} + W^{reg}_{\omega_2\omega_3;R}, \quad (15.36)$$

where the subtraction terms are found to be

$$\Omega_{23}\delta\widetilde{W}_{\omega_2\omega_3;L} = \frac{\Omega_a\Omega_b\Omega_c}{\Omega_3} - \frac{1}{2}\Big[\cos\pi(\omega_3-\omega_b)\cos\pi(\omega_a-\omega_c)$$
$$- \cos\pi(\omega_3-\omega_a-\omega_b-\omega_c)\Big] \quad (15.37)$$

$$\Omega_{32}\delta\widetilde{W}_{\omega_2\omega_3;R} = \frac{\Omega_a\Omega_b\Omega_c}{\Omega_2} - \frac{1}{2}\Big[\cos\pi(\omega_2-\omega_b)\cos\pi(\omega_a-\omega_c)$$
$$- \cos\pi(\omega_2-\omega_a-\omega_b-\omega_c)\Big]. \quad (15.38)$$

With these subtractions we find for the scattering amplitude in the region $\tau_1\tau_4$ the regular expression:

$$|s_1|^{\omega_1}|s_4|^{\omega_4}\left[(-s_2)^{\omega_2}(-s_3)^{\omega_3}e^{i\pi\omega_b}cos\pi\omega_{ac}+2i\left(e^{-i\pi\omega_3}\widetilde{W}^{reg}_{\omega_2\omega_3;L}+e^{-i\pi\omega_2}\widetilde{W}^{reg}_{\omega_2\omega_3;R}\right)\right]. \tag{15.39}$$

So far the expressions are valid to all orders of the coupling constant. However, when computing, from unitarity, the partial waves $W_{\omega_2\omega_3;L}$ and $W_{\omega_2\omega_3;R}$, we will restrict ourselves to the leading approximation. This implies that, in (15.39), we will approximate the phase factors by unity and hence deal with the sum $W^{reg}_{\omega_2\omega_3;L}+W^{reg}_{\omega_2\omega_3;R}$. The energy discontinuity Δ_{123} in leading order is given by:

$$\Delta_{123}=\widetilde{W}^{reg}_{\omega_2\omega_3;L}+\widetilde{W}^{reg}_{\omega_2\omega_3;R}-W^{reg}_{\omega_3}-W^{reg}_{\omega_2}, \tag{15.40}$$

and the unitarity integral which determines the energy discontinuity Δ_{123} in leading order has the form:

$$\Delta_{123}=|s_1|^{\omega_1}|s_2|^{\omega_2}|s_3|^{\omega_3}|s_4|^{\omega_4}\left(f_{\omega_2\omega_3}+\frac{\pi}{2}\delta_{14}-f_{\omega_2}-\frac{\pi}{2}\delta_{13}-f_{\omega_3}-\frac{\pi}{2}\delta_{24}\right), \tag{15.41}$$

where

$$\begin{aligned} f_{\omega_2\omega_3} &= \frac{a}{2}\sum_{n_1,n_2}(-1)^{n_1+n_2}\cdot\int\int\frac{d\nu_1 d\nu_2}{(2\pi)^2}\frac{1}{i\nu_1+\frac{n_1}{2}}\left(\frac{k_a^*q_3^*}{q_1^*k_b^*}\right)^{i\nu_1+\frac{n_1}{2}}\left(\frac{k_a q_3}{q_1 k_b}\right)^{i\nu_1-\frac{n_1}{2}}\\ &\times\left(\frac{s_{12}}{s_{02}}\right)^{\omega(\nu_1,n_1)}\cdot B(\nu_1,\nu_2,n_1.n_2)\left(\frac{s_{23}}{s_{03}}\right)^{\omega(\nu_2,n_2)}\left(\frac{k_b^*q_4^*}{q_2^*k_c^*}\right)^{i\nu_2+\frac{n_2}{2}}\\ &\times\left(\frac{k_b q_4}{q_2 k_c}\right)^{i\nu_2-\frac{n_2}{2}}\frac{1}{i\nu_2-\frac{n_2}{2}}|_{\text{sub}} \end{aligned} \tag{15.42}$$

and

$$\delta_{14}=V_{14}+\pi(\omega_a+\omega_c). \tag{15.43}$$

$f_{\omega_2\omega_3}$ is illustrated in Fig.15.3b. In (15.42) we have carried out the ω-integrals. The impact factors for particles a and c are the same as in the $2\to 4$ amplitude. $B(\nu_1,\nu_2,n_1,n_2)$ denotes the new (leading order) production vertex of particle b (see Fig.15.3b). The subscript $|_{\text{sub}}$ indicates that we have subtracted the one loop contribution.

Combining (15.40) with (15.41) we find

$$W^{reg}_{\omega_2\omega_3;L}+W^{reg}_{\omega_2\omega_3;R}=|s_2|^{\omega_2}|s_4|^{\omega_4}\left(f_{\omega_2\omega_3}+\frac{\pi}{2}\delta_{14}\right), \tag{15.44}$$

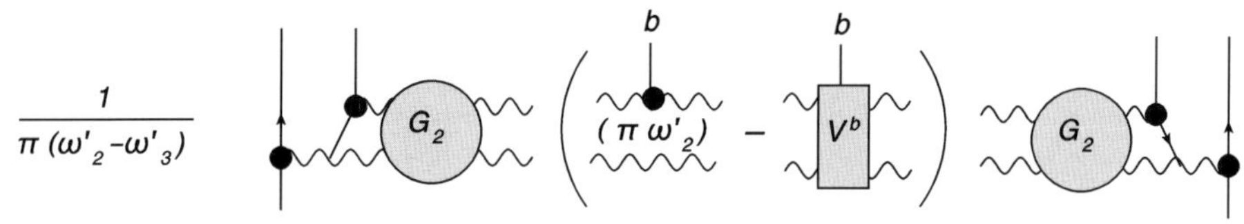

Fig. 15.5 Graphical illustration of the two terms of $\widetilde{W}_{23;R}$.

and the amplitude becomes:

$$|s_1|^{\omega_1}(-s_2)^{\omega_2}(-s_3)^{\omega_3}|s_4|^{\omega_4}\left[e^{i\pi\omega_b}\cos\pi\omega_{ac}+2i\left(f_{\omega_2\omega_3}+\frac{\pi}{2}\delta_{14}\right)\right]. \quad (15.45)$$

Here the energy factors together with the phase $e^{i\pi(\delta_{14}+\omega_b)}$ are contained in the BDS amplitude.

Finally we have to comment on the production vertex of particle b. In (15.42) $B(\nu_1,\nu_2,n_1,n_2)$ denotes the leading order vertex illustrated in Fig.15.3b. However, this is not yet the full LO vertex; to see the full structure we need to compute $W^{reg}_{\omega_2\omega_3;L}$ and $W^{reg}_{\omega_2\omega_3;R}$ separately, not only the sum. As it can be seen from Fig.15.4, in order to discriminate between the first and the second line, we need to compute the double discontinuities $\Delta_{12}\Delta_{123}$ and $\Delta_{23}\Delta_{123}$. This has been done recently in [Bargheer, T. and Bartels, J. (2020)], and we quote the result. Writing down the leading order equations for the double discontinuities and evaluating the corresponding unitarity equations, one finds

$$\widetilde{W}^{\rm reg}_{\omega_2\omega_3;{\rm R}}=\frac{\pi\omega_2' f_{\omega_2\omega_3}-f^{(b)}_{\omega_2\omega_3}}{\pi(\omega_2'-\omega_3')} \quad (15.46)$$

and

$$\widetilde{W}^{\rm reg}_{\omega_2\omega_3;{\rm L}}=\frac{\pi\omega_3' f_{\omega_2\omega_3}-f^{(b)}_{\omega_2\omega_3}}{\pi(\omega_3'-\omega_2')}\,. \quad (15.47)$$

(for simplicity, we have disregarded the one and two loop terms). The first equation is illustrated below:

The new vertex contained in $f^{(b)}_{\omega_2\omega_3}$ can be found in [Bargheer, T. and Bartels, J. (2020)]. It is important to stress that in (15.46) and (15.47) both terms are of the same order in g^2. In the leading term of the scattering amplitude, only the sum $W_{\omega_2\omega_3;L}+W_{\omega_2\omega_3;R}$, (15.44), appears which equals the sum of the two first terms only and cancels the second term. Only if we expand the phases contained in the energy factors also these second terms show up:

$$\begin{aligned}
&2ie^{-i\pi(\omega_2+\omega_3)}\int_2\int_3 s_{12}^{\omega_2'}s_{23}^{\omega_3'}\left(e^{i\pi\omega_2'}\widetilde{W}^{\rm reg}_{\omega_2\omega_3;{\rm R}}+e^{i\pi\omega_3'}\widetilde{W}^{\rm reg}_{\omega_2\omega_3;{\rm L}}\right)\\
&\approx e^{-i\pi\omega_5}e^{i\pi\omega_d}\int_2\int_3 s_{12}^{\omega_2'}s_{23}^{\omega_3'}\left(\widetilde{W}^{reg}_{\omega_2\omega_3;R}+\widetilde{W}^{reg}_{\omega_2\omega_3;L}-i\pi\omega_2'\widetilde{W}^{reg}_{\omega_2\omega_3;R}-i\pi\omega_3'\widetilde{W}^{reg}_{\omega_2\omega_3;L}\right)\\
&\approx 2ie^{-i\pi(\omega_2+\omega_3+\omega_5)}e^{i\pi\omega_d}\int_2\int_3 s_{12}^{\omega_2'}s_{23}^{\omega_3'}\left(f_{\omega_2\omega_3}+if^{(b)}_{\omega_2\omega_3}\right). \quad (15.48)
\end{aligned}$$

The '≈' sign indicates that we have expanded the phase factors, and our equations are valid only up to the first order in $i\pi$. It should be stressed that the second terms in (15.46) and (15.47) still belong to the leading approximation, although in the scattering amplitude they appear proportional to $\sim i\pi$ as 'next-to-leading-order'. Also, the full production vertex for particle b can be written in a factorized and complex valued form.

This completes our summary of the $2 \to 5$ amplitude.

15.4 The 8 point function

One of the last projects Lev was involved in was the generalization to the $2 \to 6$ production process. This process is of special interest since it contains new Regge cut pieces. First, there is a longer cut extending over 3 different t-channels. More important, for the first time, it also contains a Regge cut consisting of three reggeized gluons and the product of two short two-reggeon cuts. We illustrate these new Regge cuts by their unitarity integrals in Figs.15.6 and 15.7a,b:

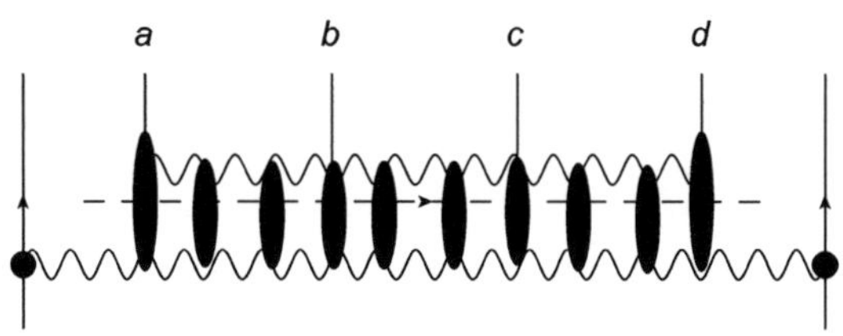

Fig. 15.6 Single discontinuity in s_{1234}.

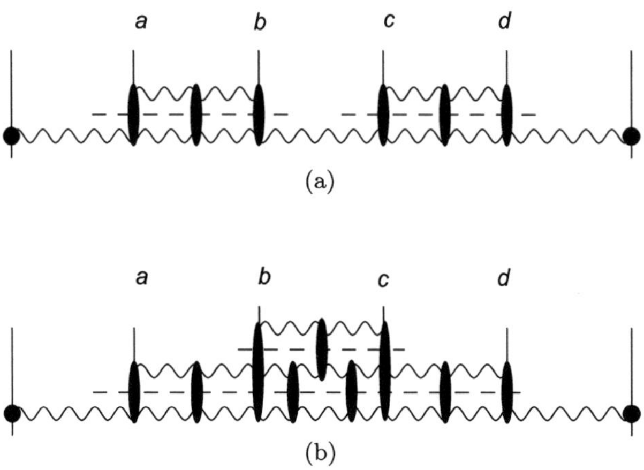

Fig. 15.7 Double discontinuities (a) in s_{12} and s_{34}, (b) in s_{23} and s_{1234}.

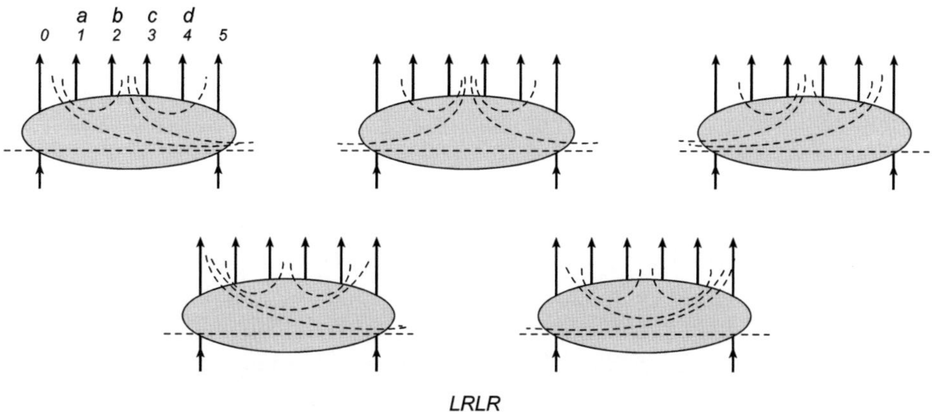

Fig. 15.8 The five terms containing the double cut.

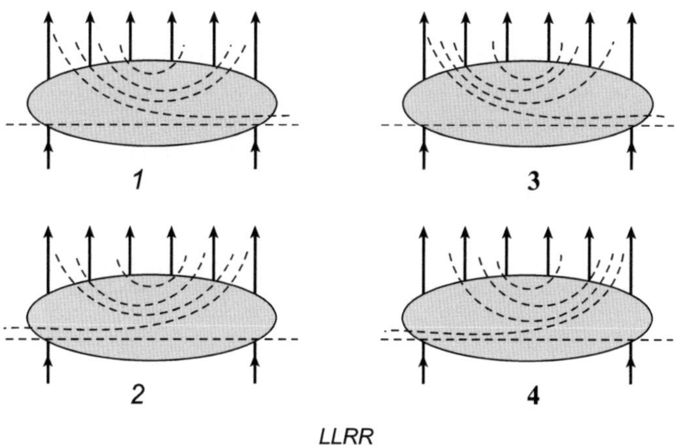

Fig. 15.9 The four terms containing the 3 reggeon cut.

The analytic decomposition of the $2 \to 6$ amplitude now contains 42 terms. In Fig. 15.8 we show those terms which contain the double cuts. For the 3 reggeon cut we have the four terms shown in Fig. 15.9.

The 3 reggeon cut contains a first generalization of the open spin chain mentioned before: whereas the BFKL kernel of the two reggeon cut represents the spin chain of two sites, the kernel of the three reggeon cut has three sites. The diagram in Fig.15.9,b thus provides a new building block of the integrable structure contained in the $2 \to n$ scattering amplitudes in the multi-Regge limit.

Details of this study are given elsewhere [Bartels (2020)], and here we only quote a few results. In the analysis of the $2 \to 4$ and $2 \to 5$ amplitudes, the Regge cut contributions in leading order were proportional to $(i\pi)$. Terms of the order

$(i\pi)^2$ came from the expansion of the phase factors. In contrast, both new Regge cut contributions are proportional to $(i\pi)^2$ and can be obtained only from double discontinuities. This requires an ansatz for the $2 \to 6$ amplitude which takes into account the correct analytic structure in all energy variables, in accordance with the Steinmann relations. For the present case, the scattering amplitude in the multi-Regge limit is written as a sum of 42 terms. In this sense the investigation of the $2 \to 6$ amplitude provides a very critical test of the analytic structure.

As before, in the planar approximation (large-N_c limit) Regge pole and cut singularities appear in specific kinematic regions. As we have already said, this is derived from combining the trigonometric factors of the Regge singularities with the phases of the energy factors. The results are in accordance with the argument given in the appendix of [Lipatov (2009)]. As an example, the very long Regge cut extending over the t_2, t_3, and t_4 channels appears in all those kinematic regions where the t_1 and t_5 channels are twisted; in our notation, all regions containing the product $\tau_1\tau_5$. This includes the region where the four produced particles have negative energies (Fig.15.10a). The three reggeon cut appears in the region where particles a and d have negative energies, particles b and c positive energy (Fig.15.10b): $\tau_1\tau_2\tau_4\tau_5$. Finally, the product of the two Regge cuts appears in all kinematic regions where also the very long cut appears.

Below we list a few results. Beginning with the very long cut, there are four terms which we denote by W_{LL}, W_{LR}, W_{RL}, and W_{RR}. In the leading contribution for the scattering amplitude where all phases are neglected, only the sum appears: $W_{LL} + W_{LR} + W_{RL} + W_{RR}$. The scattering amplitude in the region $\tau_1\tau_5$ then has the form:

$$e^{-i\pi(\omega_2+\omega_3+\omega_4)}\Big[e^{i\pi(\omega_b+\omega_c)}\cos\pi(\omega_a-\omega_d) + 2i\Big[f_{\omega_2\omega_3\omega_4} + \frac{\pi}{2}\delta_{15} \\ -i\left(f_{\omega_2} + \frac{\pi}{2}\delta_{13} - \frac{\pi(\omega_a-\omega_b)}{2}\right)\left(f_{\omega_4} + \frac{\pi}{2}\delta_{35} - \frac{\pi(\omega_d-\omega_c)}{2}\right)\Big]\Big], \tag{15.49}$$

where the long cut is contained in $f_{\omega_2\omega_3\omega_4}$. In addition to the very long cut we also have the double cut which comes as a product of the short cuts in the t_2 and t_4 channels. The remainder function for the very long cut, $f_{\omega_2\omega_3\omega_4}$, has the form

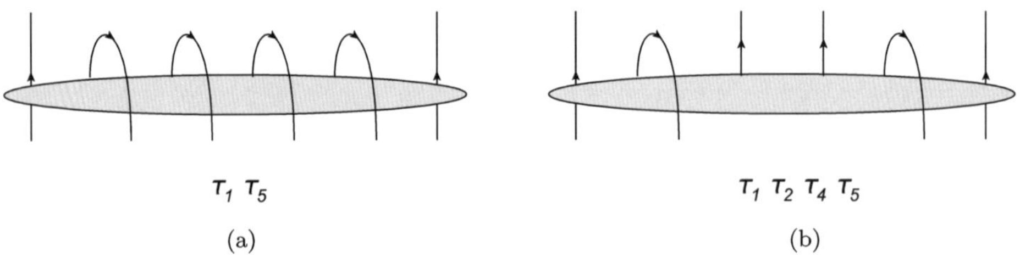

Fig. 15.10 Kinematic regions for (a) the very long cut and (b) the three reggeon cut.

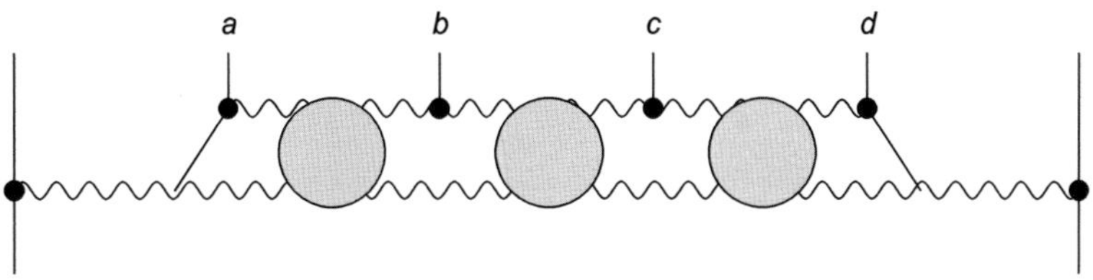

Fig. 15.11 The very long cut in the leading approximation.

(Fig.15.11):

$$f_{\omega_2\omega_3\omega_4} = \frac{a}{2} \sum_{n_2,n_3,n_4} (-1)^{n_2+n_3-n_4} \cdot \int\int \frac{d\nu_2 d\nu_3 d\nu_4}{(2\pi)^3}$$
$$\times \left[\frac{1}{i\nu_2 + \frac{n_2}{2}} \left(\frac{k_a^* q_3^*}{q_1^* k_b^*}\right)^{i\nu_2+\frac{n_2}{2}} \left(\frac{k_a q_3}{q_1 k_b}\right)^{i\nu_2-\frac{n_2}{2}} \cdot \left(\frac{s_{12}}{s_{02}}\right)^{\omega(\nu_2,n_2)} \right.$$
$$\cdot B(\nu_2,\nu_2,n_3.n_3) \left(\frac{k_b^* q_4^*}{q_2^* k_c^*}\right)^{i\nu_3+\frac{n_3}{2}} \left(\frac{k_b q_4}{q_2 k_c}\right)^{i\nu_3-\frac{n_3}{2}} \left(\frac{s_{23}}{s_{03}}\right)^{\omega(\nu_3,n_3)}$$
$$\cdot B(\nu_3,\nu_4,n_3,n_4) \left(\frac{k_c^* q_5^*}{q_3^* k_d^*}\right)^{i\nu_4+\frac{n_4}{2}} \left(\frac{k_c q_5}{q_3 k_d}\right)^{i\nu_4-\frac{n_4}{2}}$$
$$\times \left. \left(\frac{s_{34}}{s_{04}}\right)^{\omega(\nu_4,n_4)} \frac{1}{i\nu_4 - \frac{n_4}{2}} \right] \Bigg|_{\text{sub}}. \tag{15.50}$$

The production vertices for particle a and d are the same as those in the $2 \to 4$ and $2 \to 5$ amplitudes; also the vertex functions for particles b and c are known from before. In (15.49) we have not written the complete answer. In particular, for the terms proportional to $(i\pi)^2$, we have restricted ourselves to the new double cut contribution.

But there are also terms coming from expanding the energy phases in front of the very long cut which, for simplicity, we have not written. As described already for the $2 \to 5$ process, in order to compute these terms of the order $(i\pi)^2$, one needs to compute double discontinuities $\Delta_{12}\Delta_{1234}$, $\Delta_{34}\Delta_{1234}$, and even the triple discontinuity $\Delta_{12}\Delta_{34}\Delta_{1234}$. This then allows to find the four terms W_{LL}, W_{LR}, W_{RL}, and W_{RR} separately, and from this then the terms of the order $(i\pi)^2$. Without going into detail we only mention that the long cut factorizes as illustrated in Fig. 15.12. Details can be found in [Bartels (2020)].

Next the 3 reggeon cut appearing in the region $\tau_1\tau_2\tau_4\tau_5$: The leading order amplitude has the form:

$$e^{-i\pi\omega_3} \left[e^{i\pi(\omega_b+\omega_c-\omega_a-\omega_d)} + 2i \left[f_{\omega_2\omega_3\omega_4} - f_{\omega_2\omega_3} - f_{\omega_3\omega_4} + f_{\omega_3} + \frac{\pi}{2}\delta_{1245} \right.\right.$$
$$\left.\left. -i\left(f_{\omega_2} + \frac{\pi}{2}\delta_{13} - \frac{\pi(\omega_a-\omega_b)}{2}\right)\left(f_{\omega_4} + \frac{\pi}{2}\delta_{35} - \frac{\pi(\omega_d-\omega_c)}{2}\right) + 2if_{\text{3-reggeon}}\right]\right], \tag{15.51}$$

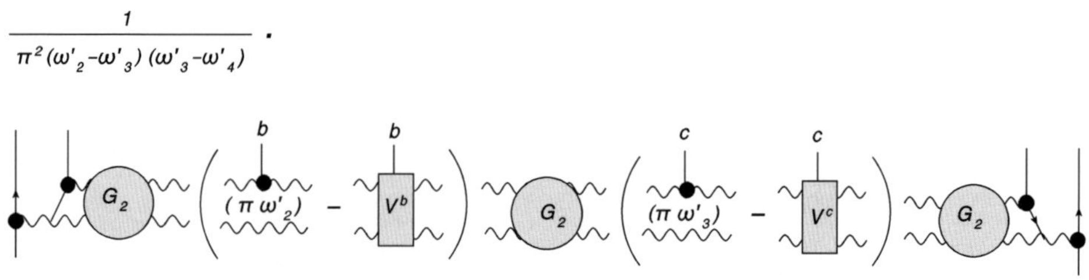

Fig. 15.12 Illustration of $\widetilde{W}_{RR}^{reg}$: Factorization of the production vertices.

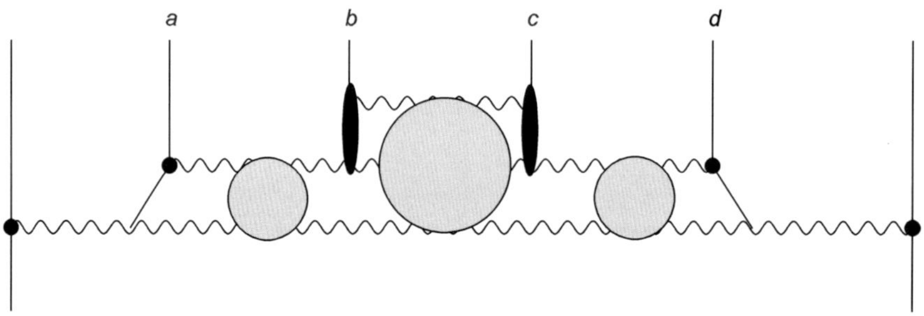

Fig. 15.13 The 3 reggeon cut in the leading approximation.

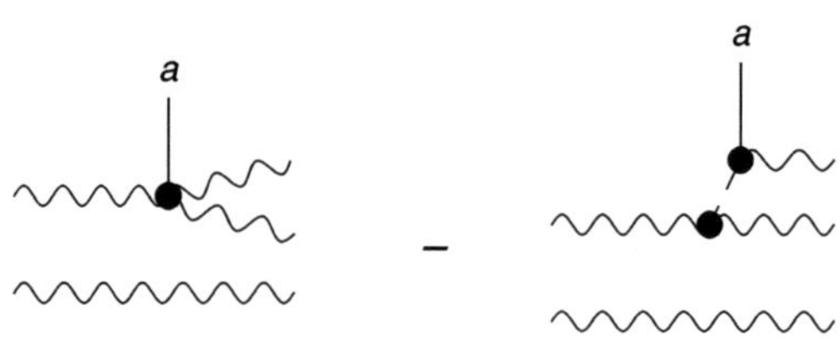

Fig. 15.14 The 2-reggeon → 3-reggeon transition.

where

$$f_{3\text{-reggeon}} = \Phi^d \frac{1}{\omega_2'' - K^{octet}} V^b \frac{1}{\omega_3'' - K^{3;octet}} V^c \frac{1}{\omega_4'' - K^{octet}} \Phi^d|_{\text{more than two loops}}$$
$$+ \text{two loop terms.} \tag{15.52}$$

Here $f_{3\text{-reggeon}}$ is illustrated in Fig.15.13, and we have used the momentum representation. For the produced particles a and d the impact factors are the same as before (e.g. in the long cut or very long cut amplitudes). The transition 2 reggeons → 3 reggeons at the production of particle b is illustrated in Fig.15.14. In particular we note that the production of particle b consists of two terms. In order to obtain this vertex in the conformal (ν, n) representation one has to convolute with BFKL eigenfunctions: from the left with the two reggeon wavefunction, from the right with the three reggeon wavefunction. The result will be published in a separate paper.

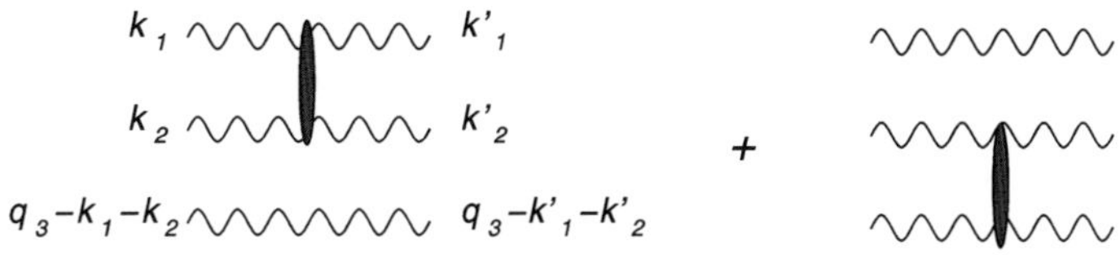

Fig. 15.15 Kernel of the 3-gluon octet state.

For completeness we write the $3 \to 3$ BFKL kernel for the 3 reggeon cut, in momentum space:

$$K^{3-octet}(\mathbf{k}_1, \mathbf{k}_2, \mathbf{q}_3 - \mathbf{k}_1 - \mathbf{k}_2; \mathbf{k}'_1, \mathbf{k}'_2, \mathbf{q}_3 - \mathbf{k}'_1 - \mathbf{k}'_2)$$
$$= -\frac{a}{2} \ln \frac{\mathbf{k}_1^2 (\mathbf{q}_3 - \mathbf{k}_1 - \mathbf{k}_2)^2}{\mathbf{q}_3^2 \mathbf{q}_3^2} + \frac{1}{2}(K^{(1)}(\mathbf{k}_1, \mathbf{k}_2; (\mathbf{k}'_1, \mathbf{k}'_2)$$
$$+ K^{(1)}(\mathbf{k}_2, \mathbf{q}_3 - \mathbf{k}_1 - \mathbf{k}_2; \mathbf{k}'_2, \mathbf{q}_3 - \mathbf{k}'_1 - \mathbf{k}'_2)) \tag{15.53}$$

as a sum of two infrared finite color singlet BFKL kernels $K^{(1)}$. It is this 3-gluon kernel which defines the open string Hamiltonian consisting of three sites.

This completes the summary of the $2 \to 6$ amplitude.

15.5 Higher order amplitudes

From these investigations of the 6, 7, and 8 point amplitudes a few conclusions can already be drawn for the general case, e.g. the $2 \to n + 1$ amplitude.

(1) the longest 2 reggeon cut extending over the t-channels $t_2 ... t_{n-1}$ will appear in all kinematic regions which — in our notation — contains the product $\tau_1 \tau_n$.
(2) The first four reggeon cut will appear in the $2 \to 8$ amplitude. The kinematic regions must contain the products $\tau_1 \tau_2 \tau_3 \tau_5 \tau_6 \tau_7$. In other words, the four reggeon cut needs at least 7 t-channels. From this it is easy to predict where the cut consisting of n reggeized gluons will appear.
(3) Together with the long cuts also products of shorter cuts will be present. As an example, the $2 \to 8$ amplitude contains also the product of three short cuts in the t_2, t_4, and t_6 channels. Also products of longer cuts are expected, e.g. the product of the two long cuts in the $t_2 t_3$ and the $t_5 t_6$ channels.
(4) The leading order BFKL kernel for the n-gluon cut will have the form

$$K^{n-octet}(\mathbf{k}_1, \ldots, \mathbf{k}_n; .\mathbf{k}'_1, \ldots, \mathbf{k}'_{n'}) = -\frac{a}{2} \ln \frac{\mathbf{k}_1^2 \mathbf{k}_n^2}{\mathbf{q}^2 \mathbf{q}^2} + \frac{1}{2} \Big(K^{(1)}(\mathbf{k}_1, \mathbf{k}_2; \mathbf{k}'_1, \mathbf{k}'_2) + \cdots$$
$$+ K^{(1)}(\mathbf{k}_{n-1}, \mathbf{k}_n; \mathbf{k}'_{n-1}, \mathbf{k}'_n) \Big) \tag{15.54}$$

where $\mathbf{q} = \mathbf{k}_1 + \cdots + \mathbf{k}_n$. Beyond leading order, each BFKL kernel for the interaction between two neighboring reggeized gluons will have its higher order corrections, but there are also additional interactions involving, for example, the interactions between three gluons [Bartels *et al.* (2013)].

15.6 Higher order BFKL eigenfunction

We now turn to another aspect of the Regge limit of N=4 SYM theory Lev was involved in, namely the computation of higher order BFKL corrections.

The BFKL equation was originally formulated [Lipatov (1976); Fadin *et al.* (1975); Kuraev *et al.* (1976, 1977); Balitsky and Lipatov (1978)] in the Leading Log Approximation, where each power of the coupling constant is accompanied by a logarithm of the center-of-mass energy. In his inspiring work [Lipatov (1986)] Lev showed that the leading order color singlet BFKL equation possesses a property of Möbius invariance in the impact parameter space, which made it possible to find an analytic solution and laid a path to contemporary integrability approaches exploiting conformal invariance in the dual space and its extensions[1]. It took around a decade for Lev and Victor Fadin [Fadin and Lipatov (1998)] to perform an impressive diagrammatic calculation of the next-to-leading (NLO) order corrections to the BFKL equation in QCD for arbitrary color configuration. Based on this result Lev in collaboration with Anatoly Kotikov [Kotikov and Lipatov (2003)] derived the corresponding NLO eigenvalue for the case of $N = 4$ SYM based on maximal complexity arguments applied to the functions building it.

It is virtually impossible to perform next-to-next-to-leading (NNLO) calculations of the corrections to the BFKL equation using traditional technique of Feynman diagrams in the color singlet case. This fact motivated the community to apply integrability techniques to the BFKL eigenvalue problem in $N = 4$ SYM. The first step in this direction was made by N. Gromov, F. Levkovich-Maslyuk and G. Sizov [Gromov *et al.* (2015)] calculating color singlet NNLO eigenvalue for zero conformal spin $n = 0$ and arbitrary anomalous dimension ν. It was then generalized for $n = 1$ and higher spins by S. Caron-Huot and M. Herranen [Caron-Huot (2018); Caron-Huot and Herranen (2018)]. Another kinematical region of arbitrary n and zero anomalous dimension ν was considered by M. Alfimov, N. Gromov and G. Sizov [Alfimov *et al.* (2018)]. All three calculations were shown to be in agreement while being partially based on each other.

The BFKL equation [Lipatov (1976); Fadin *et al.* (1975); Kuraev *et al.* (1976, 1977); Balitsky and Lipatov (1978)] describes the bound state of two reggeized gluons. Historically the BFKL equation was formulated as the linear Schrödinger equation because of the simplicity and transparency of this representation. The non-linear extension of the BFKL equation is commonly referred to in the literature as the Balitsky hierarchy [Balitsky (1996)] and, more specifically in the color planar limit, the Balitsky-Kovchegov equation [Kovchegov (1999)]. Those are closely

---

[1] See recent review on the modern integrability approaches to the BFKL evolution in Ref. [Alfimov *et al.* (2020)]

related to a more general Bartels-Kwiecinski-Praszalowicz equation [Bartels (1980); Kwiecinski and Praszalowicz (1980)] which describes a complicated interaction between reggeized gluons for an arbitrary color state.

Traditionally a word "non-linearity" in the context of the BFKL approach refers to the above mentioned cases where the number of the reggeized gluons is not conserved due to the interactions between them. In the present study we want to address a different kind of non-linearity, namely, the non-linearity of the eigenvalue of the BFKL equation. In this context by non-linearity we mean a non-separability of the holomorphic and antiholomorphic parts of the BFKL eigenvalue. For many years the BFKL equation was considered only at the leading order level, where a possible non-linearity is not manifested. The holomorphic separability [Lipatov (1990)] appears only at the leading order, while the known higher order corrections in either color singlet or color adjoint state do not have the property of holomorphic separability. The holomorphic separability is related to the integrability of the BFKL equation [Tarasov *et al.* (1983); Lipatov (1994); Faddeev and Korchemsky (1995)] and the holomorphic and antiholomorphic parts originate from different operators acting on the wave function. The operators can be viewed as describing right or left movers in some space and thus it would be good to have those two degrees of freedom separated. Kotikov and Lipatov [Kotikov and Lipatov (2003)] proposed to consider the BFKL equation as the Bethe–Salpeter equation with the kernel having the separation in operators corresponding to the holomorphic and antiholomorphic parts of the BFKL eigenvalue. It is natural to have a bound state of two reggeized gluons to be described by the Bethe–Salpeter equation because it was originally formulated for this kind of problems. We extend this analysis to higher orders making predictions for future calculations. Our arguments are as follows.

Let us consider the BFKL equation in the schematic form of the stationary Schrödinger equation

$$H\psi = \omega\psi, \tag{15.55}$$

where the eigenfunction ψ and the eigenvalue ω can be expanded in powers of the coupling constant $a = \frac{g^2 N_c}{16\pi^2}$ as follows

$$\psi \simeq \psi_0 + a\psi_1 + a^2\psi_2 + \cdots \tag{15.56}$$

and

$$\omega \simeq a(\omega_0 + a\omega_1 + a^2\omega_2 + \cdots). \tag{15.57}$$

It turns out that the leading order eigenfunction ψ_0 is also the eigenfunction of higher order corrections to the BFKL Hamiltonian with eigenvalues $\omega_1, \omega_2, \ldots$ This fact allows to ignore $\psi_1, \psi_2, \ldots$ due to a freedom of redistributing higher order corrections between the eigenfunction and the eigenvalue. Thus higher order corrections to the BFKL eigenvalue $\omega_1, \omega_2, \ldots$ are calculated using $\psi \simeq \psi_0$.

The eigenfunction ψ is a function of the special combination (anharmonic ratio) of the transverse coordinates raised to the power M

$$M = \frac{1}{2} + i\nu + \frac{|n|}{2}, \tag{15.58}$$

where ν is a real number having a meaning of anomalous dimension, while n is a real integer and is called conformal spin. For a clarity of the presentation we do not write the explicit expression for the wave functions and only want to emphasize here that it has also the complex conjugate of the anharmonic ratio of transverse complex coordinates raised to the power

$$1 - \tilde{M} = \frac{1}{2} - i\nu + \frac{|n|}{2}. \tag{15.59}$$

The leading order eigenvalue is rather simple

$$\omega_0 = 4a\left(2\Psi(1) - \Psi(M) - \Psi(1-\tilde{M})\right), \tag{15.60}$$

where $\Psi(z)$ is the digamma function defined as a logarithmic derivative of the Euler Gamma function $\Psi(z) = \frac{d\ln\Gamma(z)}{dz}$. The leading order expression in eq. (15.60) can be written as manifestly Hermitian separable. The Hermitian separability is a restricted case of holomorphic separability, where we use $|n|$ instead of n in eq. (15.58).

The situation is quite different at the next-to-leading (NLO) order [Fadin and Lipatov (1998)]. The NLO expression for the BFKL eigenvalue was derived by Fadin and Lipatov for QCD and then was written for $N = 4$ SYM in a factorized form by Kotikov and Lipatov [Kotikov and Lipatov (2003)][2]

$$\omega_1 = \phi(M) + \phi(1-\tilde{M}) - \frac{\omega_0}{2a}\left(\rho(M) + \rho(1-\tilde{M})\right), \tag{15.61}$$

where

$$\rho(M) = \beta'(M) + \frac{1}{2}\zeta(2) \tag{15.62}$$

for

$$\beta'(z) = \frac{1}{4}\left[\Psi'\left(\frac{z+1}{2}\right) - \Psi'\left(\frac{z}{2}\right)\right] = -\sum_{r=0}^{\infty}\frac{(-1)^r}{(z+r)^2}, \tag{15.63}$$

and the function $\phi(M)$ is defined by

$$\phi(M) = 3\zeta(3) + \Psi''(M) - 2\Phi_2(M) + 2\beta'(M)\left(\Psi(1) - \Psi(M)\right) \tag{15.64}$$

in terms of

$$\Phi_2(M) = \sum_{k=0}^{\infty}\frac{\left(\beta'(k+1) + (-1)^k\Psi'(k+1)\right)}{k+M} - \sum_{k=0}^{\infty}\frac{(-1)^k\left(\Psi(k+1) - \Psi(1)\right)}{(k+M)^2}. \tag{15.65}$$

[2]Please note that we normalized ω_0 of ref. [Kotikov and Lipatov (2003)] by a factor of $4a$ to keep our notation self-consistent.

Note that the last term in eq. (15.61) includes mixed terms of type $\Psi(M)\beta'(1-\tilde{M})$, which break the Hermitian separability. This sort of non-linearity in the functions of holomorphic and anti-holomorphic variables is an alerting feature that suggests that there should be a more transparent way of formulating the BFKL dynamics. Kotikov and Lipatov [Kotikov and Lipatov (2003)] proposed to write the BFKL equation in the Froissart-Gribov pole representation for a complex angular momentum $\omega = 1 + l \to 0$

$$1 = \frac{4a}{\omega}\left(2\Psi(1) - \Psi(M) - \Psi(1-\tilde{M}) + a\left(\phi(M) + \phi(1-\tilde{M})\right)\right) \quad (15.66)$$
$$-2a\left(\beta'(M) + \beta'(1-\tilde{M}) + \zeta(2)\right).$$

This representation corresponds to the Bethe-Salpeter equation with a kernel that respects the Hermitian separability at the level of generating operators. The first term on the right hand side is the leading singularity in the complex ω plane, while the last term corresponds to the zero order expansion in powers of ω. Both of the terms contribute at the NLO level of the BFKL eigenvalue. This can be seen by plugging the expansion of ω in eq. (15.57) into the Froissart-Gribov pole representation of the BFKL equation. The expression in eq. (15.66) was derived for a case of $N = 4$ SYM, but it should be valid for any other theory because the Froissart-Gribov pole representation is based only on general assumptions of analyticity and unitarity consistent with any reasonable theory.

The very special feature of the equality in eq. (15.66) is the fact that all functions in the numerator separately possess the property of the Hermitian separability. This fact allows for a complete separation of operators for left and right movers in modern integrability techniques. In the following we extend this analysis to higher order corrections to the BFKL equation.

Motivated by the NLO expression written by Kotikov and Lipatov [Kotikov and Lipatov (2003)] in eq. (15.66), we write its direct generalization for the NNLO eigenvalue in the form of the Bethe-Salpeter equation as follows.

$$1 = \frac{a(f_0 + af_1 + a^2 f_2)}{\omega} + a(g_0 + ag_1) + \omega a r_0. \quad (15.67)$$

Here we include the first order correction in powers of ω around the pole $\omega \to 0$ (the last term on the rhs) as well higher powers of the coupling constant a in the second term and the numerator of the first term. In our notation each function in eq. (15.67) is symmetrized with respect to $M \leftrightarrow 1-\tilde{M}$, namely

$$f_i = \tilde{f}_i(M) + \tilde{f}_i(1-\tilde{M}) \quad (15.68)$$

and

$$g_i = \tilde{g}_i(M) + \tilde{g}_i(1-\tilde{M}) \quad (15.69)$$

as well as

$$r_0 = \tilde{r}_0(M) + \tilde{r}_0(1-\tilde{M}). \quad (15.70)$$

We believe that the functions $\tilde{f}_i$, $\tilde{g}_i$ and $\tilde{r}_i$ must be real functions of complex arguments.

Plugging the expansion of the BFKL eigenvalue $\omega \simeq a(\omega_0 + a\omega_1 + a^2\omega_2 + ...)$ into eq. (15.67) and solving order by order in powers of the coupling constant a we obtain

$$\omega_0 = f_0 \tag{15.71}$$

and

$$\omega_1 = f_1 + \omega_0 g_0 \tag{15.72}$$

and finally

$$\omega_2 = f_2 + g_0\omega_1 + \omega_0 g_1 + \omega_0^2 r_0. \tag{15.73}$$

Comparing ω_0 and ω_1 with those of eq. (15.60) and eq. (15.61) we find f_1 and g_0 in terms of the known functions $\rho(M)$ and $\phi(M)$ of eq. (15.62) and eq. (15.64) correspondingly.

Thus at the next-to-next-to-leading order in eq. (15.73) we are left with only three unknown functions f_2, g_1 and r_0 to be fixed.

For higher loops we need to include higher powers of ω and a in our expansion. In general, for any order m in the perturbative expansion one can generalize eq. (15.67) as follows

$$1 = \sum_{i=0}^{m} \sum_{k=0}^{m-i} \omega^{i-1} a^{k+1} f_{ik} \tag{15.74}$$

for

$$\omega = \sum_{i=0}^{m} a^{i+1} \omega_i. \tag{15.75}$$

As it was already mentioned, the special feature of the proposed ansatz is that all of the functions f_{ik} have the property of the Hermitian separability and can be written as

$$f_{ik} = \tilde{f_{ik}}\,(M) + \tilde{f_{ik}}(1 - \tilde{M}) \tag{15.76}$$

We believe that the expression in eq. (15.74) should hold for either singlet or adjoint color state in both QCD and $N = 4$ SYM. The functions $f_{i\tilde{k}}$ might differ for different theories, but the general structure should remain the same because it was derived using only unitarity and analyticity assumptions without any reference to a specific dynamics.

15.7 Concluding remarks

In this short article in honor of Lev Lipatov we have briefly summarized two topics related to the Regge limit of N=4 SYM gauge theories. Both topics were among those questions Lev has been thinking about during the last weeks of his life. We expect that the ideas and the results will continue to raise interest and stimulate further investigations.

Bibliography

Alfimov, M., Gromov, N., and Kazakov, V. (2020). $N = 4$ SYM quantum spectral curve in BFKL regime, arXiv:2003.03536 [hep-th].

Alfimov, M., Gromov, N., and Sizov, G. (2018). BFKL spectrum of $\mathcal{N} = 4$: Non-zero conformal spin, *JHEP* **07**, p. 181, doi:10.1007/JHEP07(2018)181, arXiv:1802.06908 [hep-th].

Balitsky, I. (1996). Operator expansion for high-energy scattering, *Nucl. Phys. B* **463**, pp. 99–160, doi:10.1016/0550-3213(95)00638-9, arXiv:hep-ph/9509348.

Balitsky, I. and Lipatov, L. (1978). The Pomeranchuk singularity in quantum chromodynamics, *Sov. J. Nucl. Phys.* **28**, pp. 822–829.

Bargheer, T., Chestnov, V., and Schomerus, V. (2020). The multi-Regge limit from the Wilson loop OPE, *JHEP* **05**, p. 002, doi:10.1007/JHEP05(2020)002, arXiv:1906.00990 [hep-th].

Bargheer, T., Papathanasiou, G., and Schomerus, V. (2016). The two-loop symbol of all multi-Regge regions, *JHEP* **05**, p. 012, doi:10.1007/JHEP05(2016)012, arXiv:1512.07620 [hep-th].

Bargheer, T. and Bartels, J. (2020). Real parts of the $2 \to 5$ amplitude in $N = 4$ SYM in the multi-Regge limit, To be published.

Bartels, J. (1980). High-energy behavior in a nonabelian gauge theory (II): First corrections to $T_{n\to m}$ beyond the leading $\ln s$ approximation, *Nucl. Phys. B* **175**, pp. 365–401, doi:10.1016/0550-3213(80)90019-X.

Bartels, J. (2020). Analytic structure of the 8-point scattering amplitude in multi-Regge kinematics in $N = 4$ SYM: Conformal Regge pole and Regge cut contributions, arXiv:2005.08818 [hep-th].

Bartels, J., Fadin, V., Lipatov, L., and Vacca, G. (2013). NLO corrections to the kernel of the BKP-equations, *Nucl. Phys. B* **867**, pp. 827–854, doi:10.1016/j.nuclphysb.2012.10.024, arXiv:1210.0797 [hep-ph].

Bartels, J., Kormilitzin, A., and Lipatov, L. (2014). Analytic structure of the $n = 7$ scattering amplitude in $\mathcal{N} = 4$ SYM theory in the multi-Regge kinematics: Conformal Regge pole contribution, *Phys. Rev. D* **89**, 6, p. 065002, doi:10.1103/PhysRevD.89.065002, arXiv:1311.2061 [hep-th].

Bartels, J., Kormilitzin, A., and Lipatov, L. N. (2015). Analytic structure of the $n = 7$ scattering amplitude in $\mathcal{N} = 4$ theory in multi-Regge kinematics: Conformal Regge cut contribution, *Phys. Rev. D* **91**, 4, p. 045005, doi:10.1103/PhysRevD.91.045005, arXiv:1411.2294 [hep-th].

Bartels, J., Lipatov, L., and Prygarin, A. (2011). MHV amplitude for $3\to3$ gluon scattering in Regge limit, *Phys. Lett. B* **705**, pp. 507–512, doi:10.1016/j.physletb.2011.09.061, arXiv:1012.3178 [hep-th].

Bartels, J., Lipatov, L., and Sabio Vera, A. (2009). BFKL Pomeron, Reggeized gluons and Bern-Dixon-Smirnov amplitudes, *Phys. Rev. D* **80**, p. 045002, doi:10.1103/PhysRevD.80.045002, arXiv:0802.2065 [hep-th].

Bartels, J., Lipatov, L., and Sabio Vera, A. (2010). N=4 supersymmetric Yang Mills scattering amplitudes at high energies: The Regge cut contribution, *Eur. Phys. J. C* **65**, pp. 587–605, doi:10.1140/epjc/s10052-009-1218-5, arXiv:0807.0894 [hep-th].

Bern, Z., Dixon, L. J., and Smirnov, V. A. (2005). Iteration of planar amplitudes in maximally supersymmetric Yang-Mills theory at three loops and beyond, *Phys. Rev. D* **72**, p. 085001, doi:10.1103/PhysRevD.72.085001, arXiv:hep-th/0505205.

Caron-Huot, S. (2018). Resummation of non-global logarithms and the BFKL equation, *JHEP* **03**, p. 036, doi:10.1007/JHEP03(2018)036, arXiv:1501.03754 [hep-ph].

Caron-Huot, S. and Herranen, M. (2018). High-energy evolution to three loops, *JHEP* **02**, p. 058, doi:10.1007/JHEP02(2018)058, arXiv:1604.07417 [hep-ph].

Faddeev, L. and Korchemsky, G. (1995). High-energy QCD as a completely integrable model, *Phys. Lett. B* **342**, pp. 311–322, doi:10.1016/0370-2693(94)01363-H, arXiv:hep-th/9404173.

Fadin, V. S., Kuraev, E., and Lipatov, L. (1975). On the Pomeranchuk singularity in asymptotically free theories, *Phys. Lett. B* **60**, pp. 50–52, doi:10.1016/0370-2693(75)90524-9.

Fadin, V. S. and Lipatov, L. (1998). BFKL pomeron in the next-to-leading approximation, *Phys. Lett. B* **429**, pp. 127–134, doi:10.1016/S0370-2693(98)00473-0, arXiv:hep-ph/9802290.

Gromov, N., Levkovich-Maslyuk, F., and Sizov, G. (2015). Pomeron eigenvalue at three loops in $\mathcal{N} = 4$ supersymmetric Yang-Mills theory, *Phys. Rev. Lett.* **115**, 25, p. 251601, doi:10.1103/PhysRevLett.115.251601, arXiv:1507.04010 [hep-th].

Kotikov, A. and Lipatov, L. (2003). DGLAP and BFKL equations in the $N = 4$ supersymmetric gauge theory, *Nucl. Phys. B* **661**, pp. 19–61, doi:10.1016/S0550-3213(03)00264-5, arXiv:hep-ph/0208220, [Erratum: Nucl.Phys.B 685, 405–407 (2004)].

Kovchegov, Y. V. (1999). Small x F(2) structure function of a nucleus including multiple pomeron exchanges, *Phys. Rev. D* **60**, p. 034008, doi:10.1103/PhysRevD.60.034008, arXiv:hep-ph/9901281.

Kuraev, E., Lipatov, L., and Fadin, V. S. (1977). The Pomeranchuk singularity in nonabelian gauge theories, *Sov. Phys. JETP* **45**, pp. 199–204.

Kuraev, E. A., Lipatov, L. N., and Fadin, V. S. (1976). Multi-Reggeon processes in the Yang-Mills theory, *Sov. Phys. JETP* **44**, pp. 443–450.

Kwiecinski, J. and Praszalowicz, M. (1980). Three gluon integral equation and odd c singlet Regge singularities in QCD, *Phys. Lett. B* **94**, pp. 413–416, doi:10.1016/0370-2693(80)90909-0.

Lipatov, L. (1976). Reggeization of the vector meson and the vacuum singularity in nonabelian gauge theories, *Sov. J. Nucl. Phys.* **23**, pp. 338–345.

Lipatov, L. (1986). The bare Pomeron in quantum chromodynamics, *Sov. Phys. JETP* **63**, pp. 904–912.

Lipatov, L. (1990). Pomeron and odderon in QCD and a two-dimensional conformal field theory, *Phys. Lett. B* **251**, pp. 284–287, doi:10.1016/0370-2693(90)90937-2.

Lipatov, L. (1994). Asymptotic behavior of multicolor QCD at high energies in connection with exactly solvable spin models, *JETP Lett.* **59**, pp. 596–599, arXiv:hep-th/9311037.

Lipatov, L. (2009). Integrability of scattering amplitudes in N=4 SUSY, *J. Phys. A* **42**, p. 304020, doi:10.1088/1751-8113/42/30/304020, arXiv:0902.1444 [hep-th].

Lipatov, L. (2012). Analytic properties of high energy production amplitudes in $\mathrm{N} = 4$ SUSY, *Theor. Math. Phys.* **170**, pp. 166–180, doi:10.1007/s11232-012-0018-5, arXiv:1008.1015 [hep-th].

Lipatov, L. and Prygarin, A. (2011a). BFKL approach and six-particle MHV amplitude in N=4 super Yang-Mills, *Phys. Rev. D* **83**, p. 125001, doi:10.1103/PhysRevD.83.125001, arXiv:1011.2673 [hep-th].

Lipatov, L. and Prygarin, A. (2011b). Mandelstam cuts and light-like Wilson loops in N=4 SUSY, *Phys. Rev. D* **83**, p. 045020, doi:10.1103/PhysRevD.83.045020, arXiv:1008.1016 [hep-th].

Tarasov, V., Takhtajan, L., and Faddeev, L. (1983). Local Hamiltonians for integrable quantum models on a lattice, *Theor. Math. Phys.* **57**, pp. 1059–1073, doi:10.1007/BF01018648.

https://doi.org/10.1142/9789811231124_0016

Chapter 16

The Discrete BFKL Pomeron and Structure Functions at Low-x

H. Kowalski* and D. A. Ross†

*Deutsches Elektronen-Synchrotron DESY, D-22607 Hamburg, Germany
†School of Physics and Astronomy, University of Southampton, Highfield, Southampton SO17 1BJ, UK

Abstract

This article describes the modification of the perturbative QCD Pomeron, proposed by our friend and collaborator Lev Nikolaevich Lipatov, in order to achieve compatibility with the phenomenologically successful Regge theory of a discrete Pomeron pole, and the application of this modified theory to the description of structure functions at very small values of Bjorken-x, which have been measured with high accuracy at HERA.

16.1 Preface

Around 2005, Lev Nikolayevich Lipatov expressed the view, that the confrontation of the discrete pole solution of BFKL with data may provide new insights into the confinement problem. With time, this motivated us to start an intense collaboration, which was only interrupted by Lev's sudden death. In course of our work, Lev's basic intuition that BFKL equation may offer a new venue to confinement problem started to materialise, as described in this article.

To find solutions to the technical problems we encountered, Lev always went back to the basic properties of the appropriate integro-differential equations. His prowess as a theoretical physicist together with his remarkable capability of algebraic gymnastics coupled with an almost encyclopaedic knowledge of the properties of special functions, was vital to the derivation of our results.

Our discussions often occurred whilst walking to or from the institute at which all three of us were at the same time. This included one particularly fruitful conversation when Lev got us all lost during a "short-cut" through the swampy forest on our way to his own institute at Gatchina, St. Petersburg. Whatever happened, we always enjoyed Lev's gentleness, his broad interest in gluonic physics, his enthusiasm and continuous encouragement.

We were also very grateful to him and Elvira for their tremendous kindness and hospitality whenever we visited St. Petersburg.

16.2 Introduction

Before QCD became accepted as the appropriate Quantum Field Theory of strong interactions, such interactions were described using Regge theory [Regge (1959)] in which amplitudes were determined by poles in the complex angular momentum plane. This theory met with considerable phenomenological success as it was soon noticed that mesons lie on Regge trajectories [Chew and Frautschi (1961, 1962)],[1] whose exchange described very well the total and differential elastic scattering cross-sections, see e.g. [Forshaw and Ross (2011)].

A few years later it was shown that in processes in which the vacuum quantum numbers are exchanged, the amplitudes at high energies should be dominated by a leading pole — the Pomeron [Gribov and Pomeranchuk (1963); Gribov (1968)]. The values of total cross-sections for different processes can be described, over a very large energy range, with only one Regge pole, with an intercept, $\alpha_P = 1.08$. This intercept was found to be the same in various reactions, πp, Kp or pp [Donnachie and Landshoff (1984, 1986, 1992, 1994)].

In the mid-1970s, in a series of experiments at SLAC and DESY, quarks and gluons were discovered and Quantum Chromodynamics was established as a fundamental theory of strong interactions. It then became obvious that if QCD is to be considered as a complete theory of strong interactions, it must reproduce the

[1]Chew and Frautschi noticed in 1961 that the square of the masses of mesons with the same quantum numbers are linearly proportional to their spin. Such trajectories, when continued to the negative values of square-mass, determined the cross-sections for scattering processes in which the corresponding virtual mesons were exchanged.

predictions of Regge theory, at least in the perturbative region. The universality of the rate of rise of hadronic cross-sections at high energies suggested that it could be due to the gluon interactions.

Therefore, in the 1970s, Lipatov, together with his collaborators, studied the gluon-gluon interactions in the limit of very high energies and gluon virtualities which were not too small, and derived the so called BFKL equation [Fadin *et al.* (1975); Kuraev *et al.* (1976); Balitsky and Lipatov (1978)], which describes the Pomeron in QCD. Its solution exhibited a cut (not a pole) in the complex angular momentum plane, when the fixed strong coupling constant, α_S, was used.

In 1986 Lipatov showed that the BFKL amplitude exhibits a pole structure if an infrared cut-off is introduced and a running of α_S is taken into account [Lipatov (1986)]. The properties of the poles were dependent on the cut-off, which was determined by the confinement forces. This creates a possibility to address the confinement problem using QCD in the perturbative region, where it is well understood.

Another motivation to work on the QCD Pomeron is provided by the Pomeron-graviton correspondence, omnipresent in string theories. We recall that these theories were first formulated as a consistent theory of linear Regge trajectories and then applied to the gravitational interactions [Green *et al.* (1988)]. The confinement properties in these theories are today actively studied in string theories by several theory groups [Brower *et al.* (2007); Maldacena (1999); Bern *et al.* (2008, 2010); Ammon and Erdmenger (2015)].

At the beginning of 1990s the HERA collider was built at DESY, which allowed the study of deep inelastic electron scattering on protons at much higher centre-of-mass energies than SLAC. The first measurements of the total virtual γ^* proton cross-sections or F_2 seemed to indicate that BFKL cannot describe the data. However, with the increase of the precision of data and the increase of our understanding of BFKL it became possible, more than two decades later, to provide an excellent description of data at very high energies, where BFKL should be applicable.

In this article we summarize the work initiated by Lipatov and developed in collaboration with the current authors, to derive a mathematically self-consistent discrete solution of BFKL. In this work, it was important to confront the solution, as it was developing, with low-x structure function data.

16.3 The BFKL Pomeron at fixed coupling

In perturbative QCD the Pomeron was described by the amplitude for the scattering at high centre-of-mass energy, $\sqrt{s}$, of two gluons with transverse momenta $\mathbf{k}$ and $\mathbf{k}'$ respectively and momentum transfer $\mathbf{q}$, in the multi-Regge limit, $s \gg k, k', q$. This was calculated in [Fadin *et al.* (1975); Kuraev *et al.* (1976); Balitsky and Lipatov (1978)] for the leading logarithm, $\ln(s/(kk'))$, to all orders in the fixed coupling α_s. The s-dependence of this amplitude is determined by the singularity structure of the Mellin transform, $\mathcal{A}_\omega$, defined by

$$\mathcal{A}_\omega\left(\mathbf{k},\mathbf{k}',\mathbf{q}\right) \equiv \int_0^\infty \frac{ds}{s}\left(\frac{s}{kk'}\right)^{-\omega} \mathcal{A}\left(s,\mathbf{k},\mathbf{k}',\mathbf{q}\right). \tag{16.1}$$

This Mellin transform, was found to obey the Green-function equation

$$\omega \mathcal{A}_\omega\left(\mathbf{k},\mathbf{k}',\mathbf{q}\right) - \int d^2\mathbf{l}\,\mathcal{K}\left(\mathbf{k},\mathbf{l},\mathbf{q}\right)\mathcal{A}_\omega(\mathbf{l},\mathbf{k}',\mathbf{q}) = \delta^2\left(\mathbf{k}-\mathbf{k}'\right). \tag{16.2}$$

Here the kernel, $\mathcal{K}$, has eigenvalues labelled by the continuous variable ν

$$\int d^2\mathbf{l}\,\mathcal{K}\left(\mathbf{k},\mathbf{l},\mathbf{q}\right) f_\nu\left(\mathbf{l},\mathbf{q}\right) = \overline{\alpha}_s\chi_0(\nu) f_\nu(\mathbf{k},\mathbf{q}), \quad \left(\overline{\alpha}_s \equiv \frac{C_A\alpha_s}{\pi}\right) \tag{16.3}$$

with the characteristic function, $\chi_0(\nu)$, given in terms of the digamma function Ψ by

$$\chi_0(\nu) = 2\left[\Psi(1) - \Re e\left\{\Psi\left(\frac{1}{2}+i\nu\right)\right\}\right]. \tag{16.4}$$

The kernel is invariant under Möbius transformations so that the eigenfunctions are representations of the Möbius group. Henceforth we restrict ourselves to the case of zero momentum transfer (the forward scattering amplitude) and suppress $\mathbf{q}$. We also assume azimuthal symmetry so that the amplitude $\mathcal{A}$ is a function only of the magnitudes, k, k' of the gluon transverse momenta. The eigenfunctions are also functions of k and may be written as

$$f_\nu(k) = k^{(2i\nu-1)}, \tag{16.5}$$

with the eigenvalue ω given in leading order by

$$\omega = \overline{\alpha}_s\chi_0(\nu)\,. \tag{16.6}$$

The next order contribution to the amplitude (the summation of the next-to-leading logarithms in all orders of perturbation), were calculated by [Fadin and Lipatov (1998)] and [Ciafaloni and Camici (1998)] and up to this order the eigenvalues have the form

$$\omega = \overline{\alpha}_s\chi_0(\nu) + \overline{\alpha}_s^2\chi_1(\nu)\,. \tag{16.7}$$

These NLO corrections are very large, but [Salam (1998)] has shown that the major part of these corrections are associated with collinear divergences, which can be summed[2] to give

$$\omega = \chi(\overline{\alpha}_s,\nu), \tag{16.8}$$

[2]There are several resummation schemes described in [Salam (1998)], each of which "borrows" different parts of the (uncalculated) NNLO contribution. In our work we chose scheme 3 for convenience. It would be instructive to compare the results with other schemes as an estimate of the theoretical error arising from truncating the characteristic function at NLO.

where the first two terms in the expansion of the function $\chi(\overline{\alpha}_s, \nu)$ in powers of $\overline{\alpha}_s$ yields (16.7).

The solution to the Green-function equation (16.2) (for forward scattering) is then

$$\mathcal{A}_\omega(k,k') \propto \int_{-\infty}^{\infty} d\nu \frac{1}{k\,k'} \frac{1}{(\omega - \chi(\overline{\alpha}_s,\nu))} \left(\frac{k}{k'}\right)^{2i\nu} . \tag{16.9}$$

We see from eq. (16.9) that the Mellin transform of the amplitude is *not* controlled by an isolated pole, as required in Regge theory, but by a cut indicated by an integral over the continuous variable ν. This is a major discrepancy between the prediction of perturbative QCD and that of Regge theory.

One can invert the above Mellin transform to expose the s-dependence of the Pomeron amplitude by integrating over ν. Using the diffusion approximation one can show that it is dominated by the leading singularity

$$\mathcal{A}(s,k,k') \propto \left(\frac{s}{k^2}\right)^{\omega_0} f(s) , \tag{16.10}$$

where, in leading order, $\omega_0 = 4\alpha_s \ln 2$ and the function $f(s)$ has an s-dependence which is weaker than any power of s. Assuming that $\alpha_s \approx 0.2$, which is a realistic value at scales around $Q^2 = 10\,\mathrm{GeV}^2$, one obtains a value $\omega_0 \approx 0.5$. Accounting for the higher order corrections by making use of (16.8) we obtain $\omega_0 \approx 0.3$.

Since the ratio s/k^2 is directly related, by the photon impact factor, to the value of x used to parametrise $F_2(x,Q^2)$ measured in DIS reactions, the value of ω_0 can be easily compared with data. Early measurements of F_2 at HERA led to the observation, that the rise of F_2 with diminishing x can be described by a simple parametrisation; $F_2 \propto (1/x)^\lambda$, where the parameter λ is a function of Q^2 [Bartels and Kowalski (2001)]. The observed values of λ increase with Q^2: at $Q^2 = 0.35\,\mathrm{GeV}^2$ λ is about 0.1 and at $Q^2 = 250\,\mathrm{GeV}^2$ λ is about 0.4.[3] Since the power ω_0 corresponds to λ, and λ values are strongly dependent on Q^2, the HERA measurements were in clear contradiction to the BFKL results at fixed α_S.

In the light of the progress made by Lipatov [Lipatov (1986)] whereby accounting for the running of the coupling led to a series of Regge poles, we investigated the proposition that the Q^2 dependence of λ could be simulated by the Q^2 dependence of the relative strengths of these discrete poles.

16.4 Running coupling

In [Lipatov (1986)] it was shown that a set of discrete poles can be obtained if the spectrum of the kernel $\mathcal{K}(\mathbf{k},\mathbf{k}')$ were discrete owing to the presence of both infrared

[3]For recent evaluation of the parameter λ see [Luszczak and Kowalski (2020)].

and ultraviolet boundary conditions. The ultraviolet boundary conditions follow if one replaces the fixed constant, $\overline{\alpha}_s$, by the running coupling constant, $\overline{\alpha}_s(k)$.[4]

Indeed, since the RHS of eq. (16.8) is constant, it follows that the frequency ν, the argument of the characteristic function, has to depend on k in order to compensate for the k dependence of the running coupling. For sufficiently large k and for positive values of ω, eq. (16.8) no longer has a real solution for ν and the eigenfunctions become exponentially decreasing as opposed to oscillatory. The transition from the real to imaginary values of $\nu(k)$ singles out a special (ω-dependent) value of $k = k_c$, such that the frequency vanishes. The solutions below and above this critical momentum, k_c, must be carefully matched. Lipatov has shown [Lipatov (1986)] that this matching is provided by an Airy function.

For the leading order kernel where the running coupling is taken to be the leading order β-function equation without threshold effects, i.e.

$$\overline{\alpha}_s(k) = \frac{1}{\bar{\beta}_0 \ln(k^2/\Lambda^2_{QCD})},$$

the eigenfunction of the kernel with eigenvalue ω, is given by

$$\begin{aligned} f_\omega(k) &= \frac{1}{k}\sqrt{\ln(k^2/\Lambda^2_{QCD})} \int d\nu k^{2i\nu} \exp\left\{ \frac{-i}{\bar{\beta}_0\omega} \int_0^\nu \chi_0(\mu) d\mu \right\} \\ &= \frac{1}{k}\sqrt{\ln(k^2/\Lambda^2_{QCD})} \int d\nu k^{2i\nu} \left[e^{-2i\Psi(1)\nu} \frac{\Gamma\left(\frac{1}{2}+i\nu\right)}{\Gamma\left(\frac{1}{2}-i\nu\right)} \right]^{1/(\bar{\beta}_0\omega)} . \end{aligned} \tag{16.11}$$

In the semi-classical approximation, the integral over ν in (16.11) is performed by expanding about the ($k-$ and $\omega-$ dependent) saddle-point, $\nu_s^\omega(k)$ given by

$$\chi_0\left(\nu_s^\omega(k)\right) = \bar{\beta}_0\omega \ln\left(\frac{k^2}{\Lambda^2_{QCD}}\right), \tag{16.12}$$

yielding the result (upon integrating by parts)

$$f_\omega(k) \sim \frac{1}{k}\sqrt{\frac{\ln(k^2/\Lambda^2_{QCD})}{\chi_0'\left(\nu_s^\omega(k)\right)}} e^{is_\omega(k)}. \tag{16.13}$$

The phase, $s_\omega(k)$ is given by

$$s_\omega(k) = 2\int_{k_c}^k \nu_s^\omega(l)\frac{dl}{l}. \tag{16.14}$$

For $k > k_c$, $\nu_s^\omega(k)$ is purely imaginary and the oscillatory behaviour of the eigenfunction becomes an evanescent behaviour, diminishing as a negative power of transverse momentum, k. There is also a solution which *increases* as a power k as $k \to \infty$, but this solution is rejected as it is unphysical. The matching of the

[4]In order to maintain the hermiticity of the kernel $\mathcal{K}$, the geometric mean of $\overline{\alpha}_s(k)$ and $\overline{\alpha}_s(k')$ is used.

oscillatory behaviour for $k < k_c$ and the evanescent behaviour for $k > k_c$ provides an ultraviolet boundary condition for the eigenfunctions of the kernel with running coupling.

This result is readily generalized to a general running coupling, $\overline{\alpha}_s(k)$, with mass thresholds as well as taking the β-function beyond leading order, and the resummed NLO characteristic function (16.8). In the case of running coupling, the transverse-momentum dependent saddle-point, $\nu_s^\omega(k)$, obeys the transcendental equation

$$\chi(\overline{\alpha}_s(k), \nu_s^\omega(k)) = \omega \,, \tag{16.15}$$

and the critical transverse momentum, k_c, is the solution to

$$\chi\left(\overline{\alpha}_s(k_c), 0\right) = \omega \,. \tag{16.16}$$

The generalized eigenfunction may be written (in semi-classical approximation)

$$f_\omega(k) \sim \frac{1}{k}\sqrt{\frac{1}{\chi'\left(\overline{\alpha}_s(k), \nu_s^\omega(k)\right)}}e^{is_\omega(k)}, \tag{16.17}$$

with $s_\omega(k)$ given by (16.14).

This semi-classical approximation breaks down for $k \sim k_c$, where $\nu_s^\omega(k)$ and consequently $\chi'\left(\overline{\alpha}_s(k), \nu_s^\omega(k)\right)$ vanishes.

In order to find the eigenfunction in this region, we replace the kernel, $\mathcal{K}$, by the Hermitian differential operator

$$\hat{\Omega}\left(k, \frac{\partial}{\partial \ln(k^2)}\right) \equiv \frac{1}{k}\chi\left(\overline{\alpha}_s(k), i\frac{\partial}{\partial \ln(k^2)}\right)k \,. \tag{16.18}$$

Using (16.16), the equation for the eigenfunction becomes

$$\left[\chi\left(\overline{\alpha}_s(k_c), 0\right) - \hat{\Omega}\left(k, \frac{\partial}{\partial \ln(k^2)}\right)\right] f_\omega(k) = 0 \,. \tag{16.19}$$

In the region $k \sim k_c$ we can expand the operator $\hat{\Omega}$ up to leading order in $\ln(k^2/k_c^2)$ and second order in $(\partial/\partial \ln(k^2)$ to obtain

$$\left[A \ln\left(\frac{k^2}{k_c^2}\right) - B\frac{\partial^2}{\partial\left(\ln(k^2)\right)^2}\right] k f_\omega(k) = 0, \tag{16.20}$$

with

$$A = \frac{\partial}{\partial \ln(k^2)}\left[\chi(\overline{\alpha}_s(k^2), 0)\right]_{|k=k_c},$$

$$B = \frac{1}{2}\frac{\partial^2}{\partial \nu^2}\left[\chi\left(\overline{\alpha}_s(k_c), \nu\right)\right]_{|\nu=0} .$$

We recognize eq. (16.20) as Airy's equation with a solution that decays as $k \to \infty$

$$f_\omega(k) = \frac{1}{k}\mathrm{Ai}\left(\left(\frac{A}{B}\right)^{1/3} \ln\left(\frac{k^2}{k_c^2}\right)\right), \tag{16.21}$$

where $A_i(z)$ is the Airy function which decays as $z \to \infty$.

We note that in the region $k \sim k_c$, the saddle-point frequency, ν_s^ω, is approximated by

$$\nu_s^\omega(k) \approx \sqrt{\frac{A}{B}\ln\left(\frac{k^2}{k_c^2}\right)},$$

and

$$s_\omega(k) \approx \frac{2}{3}\sqrt{\frac{A}{B}}\left(\ln\left(\frac{k^2}{k_c^2}\right)\right)^{3/2},$$

so that we can rewrite (16.21) as

$$f_\omega(k) = \frac{1}{k}\mathrm{Ai}\left(z_\omega(k)\right), \tag{16.22}$$

where the argument, z_ω, of the Airy function is given by

$$\left(-z_\omega(k)\right)^{3/2} = \frac{3}{2}s_\omega(k). \tag{16.23}$$

Note that for $k < k_c$ $z_\omega(k)$ is negative and so we have an oscillatory solution, whereas for $k > k_c$ $z_\omega(k)$ is positive and so we have an evanescent solution,

For $k \ll k_c$ the oscillatory part of the Airy function matches the oscillatory part of the semi-classical result (16.17). We are therefore led to the ansatz for the eigenfunction

$$f_\omega(k) = \frac{1}{k}N_\omega(k)\mathrm{Ai}\left(z_\omega(k)\right). \tag{16.24}$$

Now we expand the operator $\hat{\Omega}\left(k, \partial/\partial\ln(k^2)\right)$ in even powers of $\partial/\partial\ln(k^2)$, maintaining the explicit Hermiticity of the operator

$$\hat{\Omega}\left(k, \partial/\partial\ln(k^2)\right) = \sum_n \sqrt{c^{2n}(k)}\left(i\frac{\partial}{\partial\ln(k^2)}\right)^{2n}\sqrt{c^{2n}(k)}, \tag{16.25}$$

where c^{2n} is the coefficient of ν^{2n} in the expansion of $\chi\left(\overline{\alpha}_s(k), \nu\right)$ in even powers of ν. Inserting the ansatz (16.24) and keeping only the terms in each order that are relevant in the semi-classical approximation, we find that eq. (16.19) is obeyed provided the prefactor, $N_\omega(k)$ is given by[5]

$$N_\omega(k) = \frac{|z_\omega(k)|^{1/4}}{\sqrt{\frac{1}{2}\overline{\alpha}_s(k)\chi'\left(\nu_s^\omega(k)\right)}}. \qquad (16.26)$$

Near $k = k_c$, where the saddle-point frequency $\nu_s^\omega(k) \ll 1$, the prefactor $N_\omega(k)$ becomes constant, reproducing the Airy function solution. Far from this critical value ($k \ll k_c$ or $k \gg k_c$ where $|z_\omega(k)|$ is large) the Airy functions are approximated by the oscillating or decaying behaviour (respectively) and multiplication by the factor $N_\omega(k)$ reproduces the prefactor in eq. (16.17), so that we obtain the eigenfunctions in the semi-classical approximation. The eigenfunctions in these two limits had already been identified in [Lipatov (1986)]. The more general result (16.24) has the advantage of being an analytic function, valid for all values of transverse momentum. Furthermore, since care was taken to preserve explicitly the Hermiticity of the kernel, these eigenfunctions form a complete orthogonal set.

The Airy function $\mathrm{Ai}(z)$ vanishes as $z \to \infty$ so the eigenfunctions (16.24) are well-behaved in the ultraviolet limit. The other Airy function, $\mathrm{Bi}(z)$, diverges as $z \to \infty$ but can nevertheless be used to construct the solution to the Green-function equation, (16.2), with running coupling, so that the Mellin transform of the Pomeron amplitude is

$$\begin{aligned}\mathcal{A}_\omega(k,k') = {} & \frac{\pi}{k\,k'} N_\omega(k) N_\omega(k') \\ & \times \left[\mathrm{Ai}(z_\omega(k))\mathrm{Bi}(z_\omega(k')\theta(k-k') + k \leftrightarrow k'\right]. \end{aligned} \qquad (16.27)$$

We may add to this any term which is a solution to the homogeneous part of the Green-function equation, so that the most general solution is obtained by the transformation

$$\mathrm{Bi}(z_\omega(k)) \to \overline{\mathrm{Bi}}(z_\omega(k)) \equiv \mathrm{Bi}(z_\omega(k)) + \cot\left(\phi(\omega)\right)\mathrm{Ai}(z_\omega(k)). \qquad (16.28)$$

The more general expression for the Green-function in Mellin space is then

$$\begin{aligned}\mathcal{A}_\omega(k,k') = {} & \frac{\pi}{k\,k'} N_\omega(k) N_\omega(k') \\ & \times \left[\mathrm{Ai}(z_\omega(k))\overline{\mathrm{Bi}}(z_\omega(k')\theta(k-k') + k \leftrightarrow k'\right]. \end{aligned} \qquad (16.29)$$

This transformation yields a Green-function with poles at $\omega = \omega_n$ where $\phi(\omega_n) = n\pi$ and these poles determine the s-dependence of the Pomeron amplitude. The amplitude $\mathcal{A}(s,k,k')$, obtained by inverting this Mellin transform, is the sum

[5] Further details can be found in [Kowalski *et al.* (2014)]

of the residues of these poles

$$\mathcal{A}(s,k,k') = \int_{\mathcal{C}} \frac{d\omega}{2\pi i}\left(\frac{s}{k\,k'}\right)^{\omega} \mathcal{A}_\omega(k,k')$$
$$= \sum_n \left(\frac{s}{k\,k'}\right)^{\omega_n} f_{\omega_n}(k) f^*_{\omega_n}(k'), \tag{16.30}$$

where the discrete eigenfunctions are given by

$$f_{\omega_n}(k) = \frac{1}{k}\sqrt{\frac{\pi}{\phi'(\omega_n)}} N_{\omega_n}(k) \mathrm{Ai}\left(z_{\omega_n}(k)\right). \tag{16.31}$$

16.5 The infrared phase

The eigenfunctions (16.31) are still not uniquely determined. They depend on an arbitrary function $\phi(\omega)$. The behaviour of the eigenfunctions as $k \to \infty$ is determined by the matching below and above k_c, but the infrared behaviour as $k \to 0$ is still not specified.

The oscillatory part of the Green-function, $\mathcal{A}_\omega(k,k')$, given by (16.27) with Bi replaced by $\overline{\mathrm{Bi}}$ of (16.28), can be approximated at some small $k = k_0$, where $z(k)$ is large and negative, using the asymptotic properties of Airy functions for large negative argument and the k_0 dependence is given by

$$\mathcal{A}_\omega(k_0,k') \;\sim\; \sin\left(\phi(\omega) + \frac{\pi}{4} + s_\omega(k_0)\right).$$

If we assume that the non-perturbative (infrared) properties of QCD determine the (possibly $\omega-$ dependent) phase, $\eta_{NP}(\omega)$, of the oscillation at the infrared scale, k_0, then the values of $\omega = \omega_n$ which give rise to poles in the Mellin transform obey the relation:

$$\eta_{NP}(\omega_n) = \left(n + \frac{1}{4}\right)\pi + s_\omega(k_0). \tag{16.32}$$

In this way the infrared phase, η_{NP}, provides an infrared boundary condition for the eigenfunctions, determining the positions of the zeroes of $\sin\left(\phi(\omega)\right)$ and thereby restricting the allowed eigenvalues to a discrete set, ω_n.

These discrete eigenfunctions and corresponding eigenvalues are analogous to the wavefunctions and energy levels of a bound-state quantum system (such as the hydrogen atom whose energy levels depend on the principle quantum number n), with the Hamiltonian operator, $\hat{H}$, replaced by the operator $-\hat{\Omega}$ and the energy levels, $E_n \equiv -\omega_n$. Since it is known that for such bound-state systems, the binding energy converges to zero as the principle quantum number becomes large, we expect that

$$\lim_{n\to\infty} \omega_n = 0. \tag{16.33}$$

We can now make an estimate of the n-dependence of ω_n for large n. We can write $s_\omega(k_0)$ in the form

$$s_\omega(k_0) = \overline{\nu} \ln\left(\frac{k_0^2}{k_c^2}\right), \tag{16.34}$$

where $\overline{\nu}$ is the "average" value of the frequency ν between its maximum value $\nu(k_0)$ and zero at $k = k_c$. This quantity does not vary much with n, and for small ω_n we can also assume that the infrared phase, $\eta_{NP}(\omega_n)$, is almost constant. In that case (16.32) leads to the $n-$dependence of k_c given by

$$\ln(k_c(n)) \approx c_0 + c_1 n, \tag{16.35}$$

where c_0 and c_1 are constants.

Now for large values of k_c, the characteristic function χ is well-approximated by its leading order contribution and inserting (16.35) into (16.16) leads to a large-n dependence of ω_n

$$\omega_n = \frac{A}{n+B}, \tag{16.36}$$

which conforms with the expectation (16.33) from the analogy between the BFKL eigenfunctions and the wavefunctions of a quantum bound-state. For the leading order BFKL equation with an ω-independent infrared phase this result was derived briefly in [Lipatov (1986)] and in more detail in [Kowalski *et al.* (2017)].

There remains, nevertheless, a continuum of oscillatory eigenfunctions with *negative* eigenvalues, $f_{-|\omega|}(k)$, since for negative ω there is *no* critical transverse momentum, k_c, at which the oscillatory behaviour is converted into an evanescent one. This continuum of eigenfunctions is given (in the semi-classical approximation) by (16.17) with negative values of ω. This continuum is needed to form a complete set of eigenfunctions. An example of a discrete eigenfunction and a continuum eigenfunction is shown in Fig. 16.1.

With the inclusion of this continuum contribution, the (forward-)Pomeron amplitude is given by

$$\mathcal{A}(s,k,k') = \int_0^\infty d\omega f^*_{-|\omega|}(k) f_{-|\omega|}(k') \left(\frac{s}{k\,k'}\right)^{-|\omega|} + \sum_n f_{\omega_n}(k) f^*_{\omega_n}(k') \left(\frac{s}{k\,k'}\right)^{\omega_n}. \tag{16.37}$$

We see from (16.37) that at sufficiently large energies the contribution from the continuum is suppressed relative to the contribution for the leading Mellin transform poles.

Lipatov worked on a number of models designed to determine the infrared phases. Together with de Vega [de Vega and Lipatov (2013)] they considered the

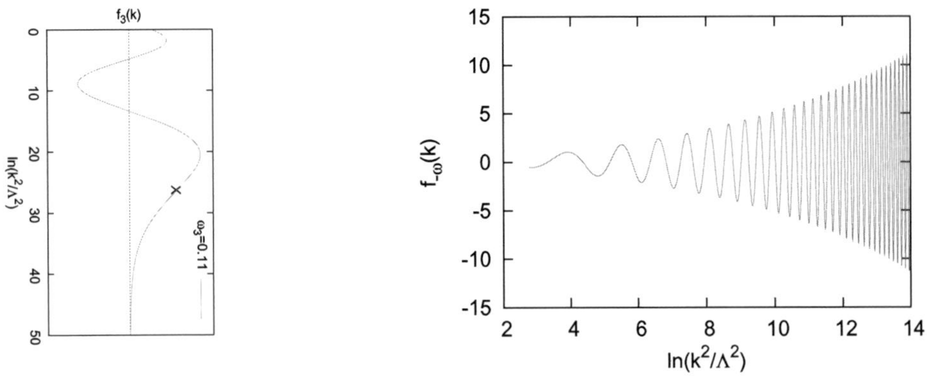

Fig. 16.1 The left-hand plot is the discrete eigenfunction for $n = 3$. The function oscillates for $k < k_c$ (indicated by "X") and then decays exponentially (with $\ln(k^2)$, corresponding to a power suppression with increasing k. The right-hand plot is an example of an eigenfunction for negative ω. In this case there is *no* k_c and the oscillatory behaviour persists for all k.

effect of assuming that in the t-channel, for which the Pomeron is transformed into a superposition of glue-ball states, the gluons are in a plasma of temperature T, which provides an infrared boundary condition for the eigenfunctions of the BFKL kernel. With Levin and Siddikov [Levin *et al.* (2014)] they considered the effect of assuming that the infrared dynamics of QCD provided an effective mass for the gluon.

In our work we tried to determine the infrared phases directly from data by fitting the discrete BFKL amplitude to the data on the proton structure function, $F_2(x, Q^2)$ at very low-x, as described in the next section.

16.6 Application to deep-inelastic scattering

To determine the proton structure function, $F_2(x, Q^2)$, from the forward Pomeron amplitude, we need to make convolutions with the impact factor, $\Phi_\gamma(Q, k)$, for a virtual photon with square-virtuality Q^2 and also with a proton form-factor $\Phi_p(k')$. This is shown schematically in Fig. 16.2. In leading order, the photon impact factor equates the longitudinal component of momentum of the attached gluon with Bjorken-x ($\approx Q^2/s$, for $x \ll 1$). However, as pointed out in [Kwiecinski *et al.* (1997)], this does not apply beyond the leading order and it is necessary to define the longitudinal momentum of the attached gluon to be $\xi\sqrt{s}/2$, and perform a convolution in the quantity, ξ.

The impact factor $\Phi_\gamma(x/\xi, Q.k)$ in [Kwiecinski *et al.* (1997)] is a function of the ratio x/ξ and accounts for the masses of heavy quarks produced by the gluon, which then couple to the virtual photon. Furthermore, the quantity ξ is given by $\xi = (s/k^2)$, whereas the Pomeron amplitude is given in terms of the symmetric quantity $(s/(k\,k'))$. This introduces a correction factor $(k'/k)^\omega$, so that in terms of

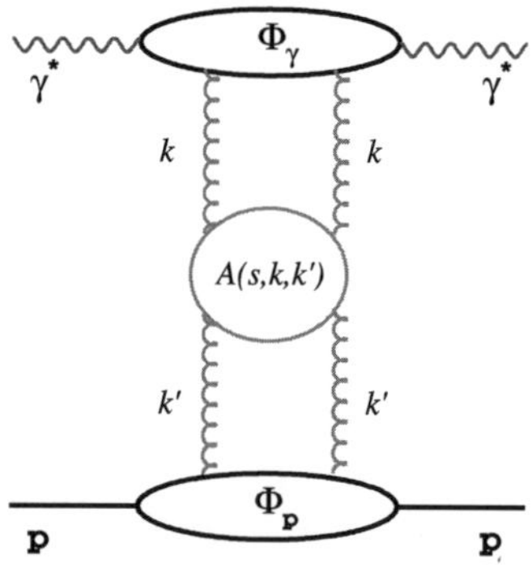

Fig. 16.2 Convolution of the Pomeron amplitude with impact factors.

ξ the Pomeron amplitude (16.37) is modified to

$$\mathcal{A}(\xi k k', k, k') = \int_0^\infty d\omega f_{-|\omega|}(k) f^*_{|\omega|}(k') \left(\xi \frac{k}{k'}\right)^{-|\omega|} + \sum_n f_{\omega_n}(k) f^*_{\omega_n}(k') \left(\xi \frac{k}{k'}\right)^{\omega_n}. \tag{16.38}$$

The proton structure function is then given by

$$F_2(x, Q^2) = \int_x^1 \frac{d\xi}{\xi} \int \frac{dk^2}{k^2} k^2 \int \frac{dk'^2}{k'^2} \times \Phi_\gamma(x/\xi, Q, k) \Phi_p(k') \mathcal{A}(\xi k k', k, k'). \tag{16.39}$$

The photon impact factor can be calculated perturbatively whereas the proton impact factor cannot and has to be modelled. To avoid using such a model, in our first paper [Ellis *et al.* (2008)], we just fitted the overlap coefficients, $a_n = \int dk' f_{\omega_n}(k') \Phi_p(k')$, to HERA F_2 data. Although we obtained a good fit, using only first four eigenfunctions, it was quite clear that this approach cannot be correct, because the effective proton impact factor, resulting from the coefficients a_n, turns out to be highly unstable and becomes negative for sufficiently large values of transverse momentum. Therefore, in all the subsequent papers we assumed a simple model for $\Phi_p(k)$.

The proton impact factor has to be positive, since it gives the probability for the proton to emit a gluon with a given transverse momentum, concentrated around Λ_{QCD}. We have taken the parameterization

$$\Phi_p(k) = a\, k^2 e^{-bk^2}, \tag{16.40}$$

where a and b are fit parameters. We have also investigated other forms of the proton impact factor, e.g. with different powers of k^2 in the prefactor and/or the exponent, but found that the quality of the fit to data has no sensitivity to such alternatives, nor is it sensitive to the precise value of the coefficient b. This is due to the fact that all eigenfunctions have a similar, oscillatory, shape near the infrared boundary and that the period of oscillations is much larger than any physically reasonable support of the proton impact factor. Note that the form (16.40) vanishes as k^2 for small k, as required by colour transparency, and that the coefficient b has the interpretation of the average inverse square transverse momentum of partons inside the proton, and should be of the order of Λ_{QCD}^{-2}.

The fits performed in our subsequent papers [Kowalski *et al.* (2010, 2011, 2013, 2017)] were barely sensitive to the values of the parameter b of the proton impact factor. To be able to describe data, and in particular the dependence of λ on Q^2, it was necessary to assume that the non-pertubative phases, η_{np}, depended on the eigenvalues, i.e. there were a function of n. In our first paper, in which we tried to determine the non-perturbative phases, [Kowalski *et al.* (2010, 2011)], our understanding of the eigenfunction was still limited. We therefore made an oversimplified assumption that the normalization factor, $N_\omega(k)$, of the eigenfunction of eq. (16.24), was independent of k. This dependence on k is very weak and could be neglected in the k-region, corresponding to the Q^2 region accessible by experiments. In [Kowalski *et al.* (2010, 2011, 2013)] we determined the overall constants numerically, from the requirement that the eigenfunctions should be correctly normalized. However, except for the leading one or two eigenfunctions, the dominant k-regions probed by such a normalization program were of order k_c of eq. (16.35), and were much larger than the ones which are experimentally accessible. Therefore, the numerically determined normalisation factors diminished only as $1/\sqrt{n}$, whereas for the correct normalization factors defined by eq. (16.24) diminsh as $1/n$, (see [Kowalski *et al.* (2016)]). Furthermore, the eigenfunctions that were obtained in [Kowalski *et al.* (2010, 2011)] turned out not to be completely orthogonal and the small overlap between nearby eigenfunctions had to be accounted for in the fitting procedure. The unsatisfactory result of this was that, despite the fact that we eventually obtained a good quality fit in [Kowalski *et al.* (2010, 2011)], a very large number of eigenfunctions (>100) was necessary in order to achieve this. The inclusion of the correct transverse-momentum dependence for the normalization factors led to a set of eigenfunctions whose completeness and orthogonality were automatically satisfied (and checked numerically in [Kowalski *et al.* (2016)]) and led to a very good fit using only $O(10)$ eigenfunctions, as shown in [Kowalski *et al.* (2017)].

In [Kowalski *et al.* (2017)] we used the Pomeron amplitude of eq. (16.37) given by the sum over (positive ω) discrete eigenfunctions, eq. (16.24), and the integral over the negative ω contributions to determine the proton structure function, as given by eq. (16.39), to fit the high-precision data from HERA [Abramowicz *et al.* (2015)].

The main results of [Kowalski *et al.* (2017)] are as follows:

(1) In order to avoid the necessity to account for saturation, only measurements for which $Q^2 > 6\,\mathrm{GeV}^2$ were included in the fit. It was found that in order to

be able to obtain a good fit to data, x had to be restricted to $x < 0.001$. This is not surprising as the Pomeron amplitude is calculated to leading and sub-leading powers of $|\ln(x)|$ and so we expect errors of order $1/|\ln(x)|^2$. Therefore x has to be below 0.001 for these errors to be comparable with the experimental accuracy. There are, nevertheless, 51 experimental points with these restrictions on x and Q^2.

(2) The values of the non-perturbative phases, η_n, used in the fit were obtained from the parametrisation

$$\omega_n = \frac{A}{n+B} + C\,, \tag{16.41}$$

due to a one to one correspondence between ω_n and η_n, see eq. (16.32). These phases were defined at the infrared scale k_0, (taken to be 1 GeV — the smallest value for which perturbative QCD can be considered to be reliable).

In addition to the constants A, B, C also a common phase of the negative ω contribution was used as a fit variable. Together with the overall normalization constant the fit used 5 fitting parameters.

Because of the rapidly diminishing value of the normalisation factor $N_{\omega_n}(k)$ with increasing n, it was found that the sum over the pole parts converges to an accuracy within the experimental limits after 10 terms. The parameter A was found to be around 0.5, B around 1.6 and the C parameter was always very small ($C < 0.005$). It is also possible to fit the data assuming $C = 0$, so that the dependence of ω on n is compatible with the expectation (16.33) from the analogy between the BFKL kernel and the Hamiltonian for a bound-state system, but in that case one needs around 20 eigenfunctions.

The ansatz (16.41) is not consistent with the large-n prediction (16.36). Since the parameter C is extremely small the eigenvalues with $\omega < C$ will be so tightly packed as to be indistinguishable from a continuum. We are therefore justified in using (16.41) as a working ansatz, provided we extend the range of the continuum contribution up to C. Since the parameter C is extremely small this extension has no noticeable effect on our fit.

We used in the fits the proton impact factor of eq. (16.40). The fits were not sensitive to the precise value of the parameter b, between $b = 10$ and $b = 20\,\mathrm{GeV}^{-2}$.

The fit is shown in the Fig. 16.3. The fit quality was always very good, $\chi^2/\mathrm{D.o.F} \approx 0.7$, for the fit with $Q^2 > 6\,\mathrm{GeV}^2$ and further decreasing with a growing Q^2 cut.[6] The fits of ref. [Kowalski *et al.* (2017)] were significantly better than those of ref. [Kowalski *et al.* (2010, 2011)].

---

[6]In the fit the statistical and systematic errors were added in quadrature, therefore the best fits can have $\chi^2/\mathrm{D.o.F}$ well below 1.

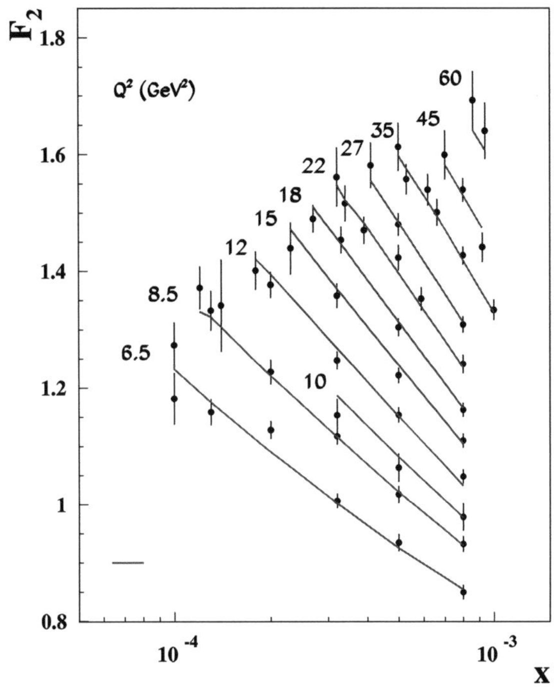

Fig. 16.3 Best fit of the discrete BFKL Pomeron to low-x deep inelastic data.

(3) These high quality fits showed that the contribution of the first eigenfunction was negligible. For this to happen, the infrared-phase of the corresponding eigenfunction, $f_{\omega_1}(k)$, has to be such that the overlap integral, $\int dk/k\ k^{-\omega_1} f_{\omega_1}(k)\Phi_P(k)$, vanishes. This can only happen if the eigenfunction $f_{\omega_1}(k)$ crosses the axis at least once, since the proton impact factor, which is the probability for the proton to emit a gluon with a given transverse momentum, is positive.

It is well-known that the leading eigenfunction of an operator does *not* have nodes throughout its domain. This means that f_{ω_1} is *not* the leading eigenfunction and that there must be at least one eigenfunction with eigenvalue greater than $\omega_1 \approx 0.2$. Since the values of $\ln(k_c)$ have an almost linear dependence of n, (see eq. (16.35)), it is easy to find the k_c value of the zeroth eigenfunction: $k_c^0 \approx 700\,\text{MeV}$.

The existence of the BFKL ground state with support only in the infrared region is one of the main results of our fit.[7] Such a state will not make a significant contribution to the structure function at higher Q^2, because its corresponding transverse momenta are small and so the eigenfunction will have substantially

[7]Similar results were obtained in a numerical analysis of the spectrum of the BFKL Pomeron deformed by the introduction of a Wilsonian infrared regulator, [Bartels *et al.* (2019)].

decayed for values of the transverse momentum comparable with Q. However, at low Q^2, it would dominate F_2 and diffractive amplitudes. At such low values of Q^2, saturation effects will be very important and as was shown in [Golec-Biernat and Wusthoff (1998, 1999)] this can reduce the effective value of ω from around 0.3 to 0.1, thereby giving a contribution consistent with the observed value of λ at low Q^2.

Although the BFKL ground state is essentially non-perturbative ($\alpha_s(k_c^0) \approx 1$) Lipatov was convinced that it should be possible to deduce its properties by extrapolation from the perturbative to non-perturbative region. Such state could provide a bridge between QCD and gravitational theory as in the AdS/CFT-based theoretical studies, gravitational interactions are dual to QCD *with large coupling constant* [Maldacena (1999)]. This connection is currently being investigated by several groups, e.g. [Bern *et al.* (2008, 2010); Ammon and Erdmenger (2015)].

16.7 Communication between small and large scales

Finally we discuss an interesting feature of the discrete QCD Pomeron, its potential sensitivity to new physics at energy scales much higher than the scale of the experiment.

The frequency of the oscillations of an eigenfunction at some transverse-momentum k depends on the location of the critical transverse momentum, k_c, at which the oscillation frequency vanishes. This, in turn, depends on the nature of the running of the coupling and is affected by any threshold for new physics, which opens at a scale below k_c.

For the nth eigenfunction, k_c increases exponentially with n, such that for $n \geq 3$, the value of k_c exceeds the scale at which current experiments are being conducted.[8] Nevertheless a fit of data to the discrete Pomeron can be sensitive to the exact behaviour of the eigenfunctions $f_{\omega_n}(k)$ for $n \geq 3$. These effects, although only present at scales that are much higher than the region of HERA data, can nevertheless affect the quality of the fits to data since BSM effects change the running of the coupling and consequently also substantially change the values of ω_n.

This seems somewhat counter-intuitive. One may ask how it can be possible that a fit to data at relatively low energies can be sensitive to corrections due to loops of particles whose masses are far in excess of those energies. The crucial point is that the above-mentioned large transverse momentum boundary conditions, imposed by the running of the coupling, can occur at very high k scales.[9] For $\omega_n < 0.1$, this is already above the scale at which one might expect BSM physics to occur. The value of the discrete eigenvalues arise from an interplay between these ultraviolet boundary conditions and the infrared boundary condition arising from the imposition of an infrared phase. It is in this sense that in our treatment of the BFKL formalism, there is communication between high and low energy scales.

[8]For $n = 2$, $k_c \approx 3,3\,\text{TeV}$, for $n = 3$, $k_c \approx 260\,\text{TeV}$, for $n = 4$, $k_c \approx 21000\,\text{TeV}$.

[9]Note that this means that formally we determine the eigenvalues at asymptotically large initial energies.

We make the assumption that the allowed eigenvalues can be obtained from the BFKL equation supplemented by an infrared phase condition in a process-independent way, i.e. without needing to impose any infrared or ultraviolet cuts on the integration over transverse momentum. Once these eigenvalues are calculated their corresponding eigenfunctions are convoluted with the necessary impact factors in order to obtain the required amplitudes.

In this way, it turns out that even though these amplitudes, in accordance with kinematical constraints, never involve diffusion into transverse momenta above the threshold for Beyond the Standard Model (BSM) physics, the rapidity dependence of such amplitudes is affected by the substantial changes in the eigenvalues arising as a result of BSM physics.

In [Kowalski *et al.* (2013)] we evaluated the effects of BSM physics on HERA results and found that data are sensitive to such effects. In this evaluation we constructed the Pomeron amplitude, however, from eigenfunctions with the incorrect normalization and therefore we needed to take O(100) eigenfunctions into account. With the better understanding of the properties of the eigenfunctions, developed in our later papers, it is now clear that we need only O(10) eigenfunctions to describe data. This could make a signal from BSM physics somewhat weaker than in [Kowalski *et al.* (2013)], so it is now not clear that we may have such effect in HERA data.

On the other hand, the decoupling of the leading eigenfunction makes the analysis of BSM physics simpler, especially on the future VHEeP [Caldwell and Wing (2015); Caldwell *et al.* (2018)] or LHeC [Abelleira Fernandez *et al.* (2012)] colliders. This is because the ground state is now very well constrained, and the value of ω_2 can be directly measured, at very low x [Kowalski *et al.* (2017)]. This could provide a fascinating connection between the confinement properties and symmetries at very high scales.

Bibliography

Abelleira Fernandez, J. *et al.* (2012). A large hadron electron collider at CERN: Report on the physics and design concepts for machine and detector, *J. Phys. G* **39**, p. 075001, doi:10.1088/0954-3899/39/7/075001, arXiv:1206.2913 [physics.acc-ph].

Abramowicz, H. *et al.* (2015). Combination of measurements of inclusive deep inelastic $e^{\pm}p$ scattering cross sections and QCD analysis of HERA data, *Eur. Phys. J. C* **75**, 12, p. 580, doi:10.1140/epjc/s10052-015-3710-4, arXiv:1506.06042 [hep-ex].

Ammon, M. and Erdmenger, J. (2015). *Gauge/Gravity Duality: Foundations and Applications* (Cambridge University Press, Cambridge), ISBN 978-1-107-01034-5, 978-1-316-23594-2.

Balitsky, I. and Lipatov, L. (1978). The Pomeranchuk singularity in quantum chromodynamics, *Sov. J. Nucl. Phys.* **28**, pp. 822–829.

Bartels, J., Contreras, C., and Vacca, G. P. (2019). A functional RG approach for the BFKL Pomeron, *JHEP* **01**, p. 004, doi:10.1007/JHEP01(2019)004, arXiv:1808.07517 [hep-ph].

Bartels, J. and Kowalski, H. (2001). Diffraction at HERA and the confinement problem, *Eur. Phys. J. C* **19**, pp. 693–708, doi:10.1007/s100520100613, arXiv:hep-ph/0010345.

Bern, Z., Carrasco, J., and Johansson, H. (2008). New relations for gauge-theory amplitudes, *Phys. Rev. D* **78**, p. 085011, doi:10.1103/PhysRevD.78.085011, arXiv:0805.3993 [hep-ph].

Bern, Z., Carrasco, J. J. M., and Johansson, H. (2010). Perturbative quantum gravity as a double copy of gauge theory, *Phys. Rev. Lett.* **105**, p. 061602, doi:10.1103/PhysRevLett.105.061602, arXiv:1004.0476 [hep-th].

Brower, R. C., Polchinski, J., Strassler, M. J., and Tan, C.-I. (2007). The Pomeron and gauge/string duality, *JHEP* **12**, p. 005, doi:10.1088/1126-6708/2007/12/005, arXiv:hep-th/0603115.

Caldwell, A., Ent, R., Levy, A., Newman, P., and Olness, F. (2018). The "DIS and related subjects" strategy document: Fundamental science from Lepton-Hadron scattering, arXiv:1812.08110 [hep-ph].

Caldwell, A. and Wing, M. (2015). VHEeP: A very high energy electron-proton collider based on proton-driven plasma wakefield acceleration, *PoS* **DIS2015**, p. 240, doi:10.22323/1.247.0240, arXiv:1509.00235 [physics.acc-ph].

Chew, G. and Frautschi, S. C. (1961). Principle of equivalence for all strongly interacting particles within the S matrix framework, *Phys. Rev. Lett.* **7**, pp. 394–397, doi:10.1103/PhysRevLett.7.394.

Chew, G. and Frautschi, S. C. (1962). Regge trajectories and the principle of maximum strength for strong interactions, *Phys. Rev. Lett.* **8**, pp. 41–44, doi:10.1103/PhysRevLett.8.41.

Ciafaloni, M. and Camici, G. (1998). Energy scale(s) and next-to-leading BFKL equation, *Phys. Lett. B* **430**, pp. 349–354, doi:10.1016/S0370-2693(98)00551-6, arXiv:hep-ph/9803389.

de Vega, H. and Lipatov, L. (2013). Integrabilty of the BFKL dynamics and Pomeron trajectories in a thermostat, *Phys. Part. Nucl.* **44**, pp. 515–538, doi:10.1134/S1063779613030027.

Donnachie, A. and Landshoff, P. (1984). Elastic scattering and diffraction dissociation, *Nucl. Phys. B* **244**, pp. 322–326, doi:10.1016/0550-3213(84)90315-8.

Donnachie, A. and Landshoff, P. (1986). Dynamics of elastic scattering, *Nucl. Phys. B* **267**, pp. 690–701, doi:10.1016/0550-3213(86)90137-9.

Donnachie, A. and Landshoff, P. (1992). Total cross-sections, *Phys. Lett. B* **296**, pp. 227–232, doi:10.1016/0370-2693(92)90832-O, arXiv:hep-ph/9209205.

Donnachie, A. and Landshoff, P. (1994). Proton structure function at small q**2, *Z. Phys. C* **61**, pp. 139–146, doi:10.1007/BF01641896, arXiv:hep-ph/9305319.

Ellis, J., Kowalski, H., and Ross, D. (2008). Evidence for the discrete asymptotically-free BFKL Pomeron from HERA data, *Phys. Lett. B* **668**, pp. 51–56, doi:10.1016/j.physletb.2008.08.007, arXiv:0803.0258 [hep-ph].

Fadin, V. S., Kuraev, E., and Lipatov, L. (1975). On the Pomeranchuk singularity in asymptotically free theories, *Phys. Lett. B* **60**, pp. 50–52, doi:10.1016/0370-2693(75)90524-9.

Fadin, V. S. and Lipatov, L. (1998). BFKL pomeron in the next-to-leading approximation, *Phys. Lett. B* **429**, pp. 127–134, doi:10.1016/S0370-2693(98)00473-0, arXiv:hep-ph/9802290.

Forshaw, J. R. and Ross, D. (2011). *Quantum Chromodynamics and the Pomeron*, Vol. 9 (Cambridge University Press), ISBN 978-0-511-89326-1, 978-0-521-56880-7.

Golec-Biernat, K. J. and Wusthoff, M. (1998). Saturation effects in deep inelastic scattering at low Q**2 and its implications on diffraction, *Phys. Rev. D* **59**, p. 014017, doi: 10.1103/PhysRevD.59.014017, arXiv:hep-ph/9807513.

Golec-Biernat, K. J. and Wusthoff, M. (1999). Saturation in diffractive deep inelastic scattering, *Phys. Rev. D* **60**, p. 114023, doi:10.1103/PhysRevD.60.114023, arXiv:hep-ph/9903358.

Green, M. B., Schwarz, J., and Witten, E. (1988). *Superstring Theory. Vol. 1: Introduction*, Cambridge Monographs on Mathematical Physics, ISBN 978-0-521-35752-4.

Gribov, V. (1968). A Reggeon diagram technique, *Sov. Phys. JETP* **26**, pp. 414–422.

Gribov, V. and Pomeranchuk, I. (1963). Regge poles and Landau singularities, *Sov. Phys. JETP* **16**, p. 1387, doi:10.1103/PhysRevLett.9.239.

Kowalski, H., Lipatov, L., and Ross, D. (2013). BFKL evolution as a communicator between small and large energy scales, *Phys. Part. Nucl.* **44**, pp. 547–567, doi: 10.1134/S1063779613030052, arXiv:1205.6713 [hep-ph].

Kowalski, H., Lipatov, L., and Ross, D. (2014). The green function for the BFKL Pomeron and the transition to DGLAP evolution, *Eur. Phys. J. C* **74**, 6, p. 2919, doi:10.1140/epjc/s10052-014-2919-y, arXiv:1401.6298 [hep-ph].

Kowalski, H., Lipatov, L., and Ross, D. (2016). The behaviour of the green function for the BFKL Pomeron with running coupling, *Eur. Phys. J. C* **76**, 1, p. 23, doi:10.1140/epjc/s10052-015-3865-z, arXiv:1508.05744 [hep-ph].

Kowalski, H., Lipatov, L., Ross, D., and Schulz, O. (2017). Decoupling of the leading contribution in the discrete BFKL analysis of high-precision HERA data, *Eur. Phys. J. C* **77**, 11, p. 777, doi:10.1140/epjc/s10052-017-5359-7, arXiv:1707.01460 [hep-ph].

Kowalski, H., Lipatov, L., Ross, D., and Watt, G. (2010). Using HERA data to determine the infrared behaviour of the BFKL amplitude, *Eur. Phys. J. C* **70**, pp. 983–998, doi:10.1140/epjc/s10052-010-1500-6, arXiv:1005.0355 [hep-ph].

Kowalski, H., Lipatov, L., Ross, D., and Watt, G. (2011). The new HERA data and the determination of the infrared behaviour of the BFKL amplitude, *Nucl. Phys. A* **854**, pp. 45–56, doi:10.1016/j.nuclphysa.2010.10.002.

Kuraev, E. A., Lipatov, L. N., and Fadin, V. S. (1976). Multi-Reggeon processes in the Yang-Mills theory, *Sov. Phys. JETP* **44**, pp. 443–450.

Kwiecinski, J., Martin, A. D., and Stasto, A. (1997). A unified BFKL and GLAP description of F2 data, *Phys. Rev. D* **56**, pp. 3991–4006, doi:10.1103/PhysRevD.56.3991, arXiv:hep-ph/9703445.

Levin, E., Lipatov, L., and Siddikov, M. (2014). BFKL Pomeron with massive gluons, *Phys. Rev. D* **89**, 7, p. 074002, doi:10.1103/PhysRevD.89.074002, arXiv:1401.4671 [hep-ph].

Lipatov, L. (1986). The bare Pomeron in quantum chromodynamics, *Sov. Phys. JETP* **63**, pp. 904–912.

Luszczak, A. and Kowalski, H. (2020). Investigation of high energy behaviour of HERA data, *Phys. Lett. B* **802**, p. 135199, doi:10.1016/j.physletb.2020.135199, arXiv:1903.09719 [hep-ph].

Maldacena, J. M. (1999). The large N limit of superconformal field theories and supergravity, *Int. J. Theor. Phys.* **38**, pp. 1113–1133, doi:10.1023/A:1026654312961, arXiv:hep-th/9711200.

Regge, T. (1959). Introduction to complex orbital momenta, *Nuovo Cim.* **14**, p. 951, doi: 10.1007/BF02728177.

Salam, G. (1998). A resummation of large subleading corrections at small x, *JHEP* **07**, p. 019, doi:10.1088/1126-6708/1998/07/019, arXiv:hep-ph/9806482.

https://doi.org/10.1142/9789811231124_0017

Chapter 17

BFKL Pomeron and the Survival Factor*

V. A. Khoze[†,‡], A. D. Martin[†] and M. G. Ryskin[†,‡]

[†] *Institute for Particle Physics Phenomenology, University of Durham, Durham, DH1 3LE, UK*

[‡] *Petersburg Nuclear Physics Institute, NRC Kurchatov Institute, Gatchina, St. Petersburg, 188300, Russia*

Abstract

We consider the absorptive corrections and the rapidity gap survival factor which are necessary to provide the unitarization of the BFKL Pomeron. In particular we discuss the role of the enhanced screening diagrams.

---

*A contribution to the memorial volume entitled "From the Past to the Future: The Legacy of Lev Lipatov" edited by J. Bartels *et al.*

17.1 Introduction

Perhaps we may start with a short, but heartfelt, tribute to Lev Lipatov. We were very fortunate to enjoy long-term scientific interactions with Lev; it was a real privilege to collaborate with him on various projects. All his research was at a deep theoretical level and he had a very clear vision of his objectives. His insight was profound and legendary. He seemed materially devoid of ambition and yet scientifically his ambition for the truth was enormous. His deep understanding enabled him to extend our view of nature in many ways. The ten or so years that Lev Lipatov and Victor Fadin spent meticulously extending BFKL to next-to-leading order were a remarkable achievement.

As a colleague and friend Lev was humble and considerate. He was always willing and patient to explain his ideas to those less clever than himself. Indeed, Lev never refused to discuss a topic in depth and took care of the young theorists in the Theory Division of the St. Petersburg Nuclear Physics Institute, creating what became known world-wide as the Lipatov School. We miss him greatly as a person and as a scientist.

An important aspect of Lev Nikolaevich Lipatov's prolific and penetrating research career concerned the application of perturbative Quantum Chromodynamics (QCD) to high-energy interaction amplitudes. In particular how to describe the proton's behaviour in high-energy collisions, a topic extremely important to the analysis of data from the particle physics experiments around the world. In fact, Lipatov was the unique co-author to both of the two relevant theoretical breakthroughs: the DGLAP[Gribov and Lipatov (1972); Lipatov (1975); Altarelli and Parisi (1977); Dokshitzer (1977)] and BFKL[Fadin *et al.* (1975); Kuraev *et al.* (1976, 1977); Balitsky and Lipatov (1978)] formalisms. This memorial article concerns one of the consequences of the BFKL formalism. It deals with the importance of absorptive corrections and informs us about the special behaviour of the survival factors of the Large Rapidity Gaps (LRGs) which can occur in high-energy proton-proton scattering.

In the BFKL formalism, appropriate to the small x domain, high-energy pp scattering is, to leading $(\alpha_s \cdot \ln s)^n$ order, driven by Pomeron exchange, which at the parton level is composed of reggeized two-gluon exchange. A crucial feature of the BFKL Pomeron is the fact that within perturbative QCD it leads to the power growth of high-energy cross section with energy, $\sqrt{s}$. That is, the high-energy (proton-proton) forward scattering amplitude, $A(s,t=0)$, increases with energy as

$$A_{\rm BFKL}(s,t=0) \;=\; i\alpha_s^2\left(\frac{s}{s_0}\right)^{\alpha_P(0)} \qquad (17.1)$$

where α_s is the QCD coupling and the Pomeron intercept $\alpha_P(0)$ is such that

$$\alpha_P(0) \;\equiv\; 1+\omega_0 \;>\; 1. \qquad (17.2)$$

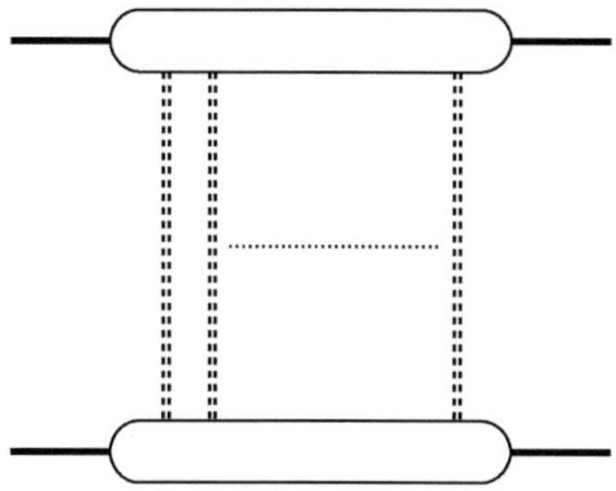

Fig. 17.1 The multi-Pomeron exchange diagram which tames the asymptotic growth of the amplitude. Each double-dotted line corresponds to BFKL Pomeron exchange.

In the leading Log approximation $\omega_0 = (N_c\alpha_s/\pi)4\ln 2$ [Fadin *et al.* (1975); Kuraev *et al.* (1976, 1977); Balitsky and Lipatov (1978)]. If we account for the next-to-leading corrections then $\omega_0 \simeq 0.2 - 0.3$[Fadin and Lipatov (1998); Ciafaloni and Camici (1998); Salam (1998); Ciafaloni and Colferai (1999)].

Recall that this result was obtained for a very small α_s

$$\alpha(s) \ll 1 \quad \text{but} \quad \alpha_s \ln(s/s_0) \sim O(1). \tag{17.3}$$

At larger energies, when the value of $Y = \ln(s/s_0)$ increases, the amplitude (17.1) starts to violate the Froissart bound[Froissart (1961)]. However, the power growth of the amplitude (s^{ω_0+1}) is compensated by *absorptive corrections* described by the multi-Pomeron diagrams shown in Fig. 17.1. Actually the corrections become large already for

$$\omega_0 Y \; > \; 2\ln(1/\alpha_s). \tag{17.4}$$

At a fixed impact parameter, b, the sum of these diagrams transforms the one Pomeron exchange amplitude into

$$A(s,b) = i(1 - e^{-\Omega(s,b)/2}), \tag{17.5}$$

where the opacity $\Omega(s,b)$ is given in terms of the one Pomeron exchange amplitude $\Omega(s,b) = -iA_{\text{BFKL}}(s,b)$. Here $A(s,b)$ is the Fourier transform

$$A(s,b) = \frac{1}{4\pi^2}\int d^2q\; e^{i\vec{q}\cdot\vec{b}}\;\frac{A(s,q^2)}{s}, \tag{17.6}$$

where $q^2 = |t|$. Note that $A(s,b)$ satisfies the unitarity equation

$$2\text{Im}A(Y,b) = |A(Y,b)|^2 + G_{\text{inel}}(Y,b), \tag{17.7}$$

where G_{inel} denotes the total contribution of all the inelastic channels. Here, for convenience, we have replaced the variable s with the rapidity $Y = \ln(s/m_p^2)$ where m_p is the mass of the proton. It is seen from (17.5) that when Ω increases, the amplitude $A(s,b) \to i$, that is $A(s,b)$ tends to the black disk limit.

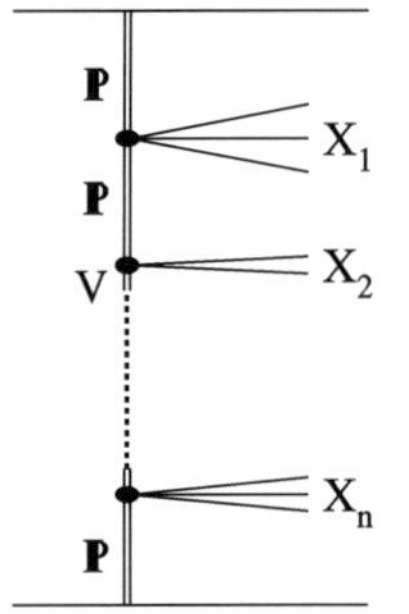

Fig. 17.2 The multi-Reggeon production process of Eq. (17.8).

Note that formally the absorptive effects are of the order of a α_s^2 correction to the elastic amplitude. This correction can be large numerically for $\ln(s/s_0)\omega_0 > 2\ln(1/\alpha_s)$ due to the power growth of the BFKL amplitude proportional to $\alpha_s^2(s/s_0)^{\omega_0}$. However this α_s^2 correction to the amplitude cannot be considered as a next-to-next contribution to the BFKL kernel. This is a completely different correction.

The analogous absorptive corrections described by the multi-Pomeron diagrams are crucially important to restore unitarity for processes with Large Rapidity Gaps (LRG) where some groups of particles are separated from each other by large rapidity intervals. Indeed, it was recognized already in the 1960s[Verdiev *et al.* (1964); Finkelstein and Kajantie (1968)] that the multi-Reggeon reactions,

$$pp \to p + X_1 + X_2 + \cdots + X_n + p\,, \tag{17.8}$$

shown in Fig. 17.2 may cause a problem with unitarity. Being summed over n and integrated over the rapidities of each group, the cross section of such quasi-diffractive production increases faster than a power of s. This was termed in the literature as the Finkelstein-Kajantie (FK) disease, see[Abarbanel *et al.* (1975)] for a review.

Let us explain the problem using the simple example of Central Exclusive Production (CEP) of proton-antiproton pair, as shown in Fig. 17.3(a). Since the proton-proton elastic cross section does not vanish but increases with energy, the corresponding contribution to inelastic cross section reads

$$\begin{aligned}\sigma^{\rm CEP} &= N\int_0^Y dy_1 \int dt_1 dt_2 |A(y_1,t_1)\cdot V\cdot A(Y-y_1,t_2)|^2\\ &\propto \int dy_1 \sigma_{el}(y_1)\sigma_{el}(Y-y_1),\end{aligned} \tag{17.9}$$

where the elastic amplitude $A(y,t)$ is normalized in such a way that $\int dt|A(y,t)|^2 = \sigma_{\rm el}(y)$, and where the upper rapidity $Y = \ln s/m_p^2$. In this example the vertex V describes the central production of the $p\bar{p}$-pair. Thus we find

$$\sigma^{\rm CEP} \propto \ln s \cdot \sigma_{\rm el}^2. \tag{17.10}$$

For a black (or grey) disk of increasing radius $R = c\ln s$, this leads to

$$\sigma^{\mathrm{CEP}} \propto (\ln s)^5 \gg \sigma_{\mathrm{tot}} \propto (\ln s)^2 \tag{17.11}$$

The same result can be obtained in impact parameter, b, space (see[Khoze *et al.* (2018)] for details). Moreover working in b space we have a stronger constraint since for each value of b, that is for each partial wave $l = b\sqrt{s}/2$ of the incoming proton pair, we have the unitarity equation (17.7) and the 'total' cross section, $\sigma(b)_{\mathrm{tot}}$ must be less than the corresponding CEP contribution.

Actually one will face this FK problem in any model where the elastic cross section does not decrease with energy.

17.2 Black disk solution of the FK problem

The only known solution of this multi-Reggeon problem comes from 'black disk' asymptotics of the high-energy cross sections. In such a case the (gap) survival probability, S^2, of the events with LRGs tends to zero as $s \to \infty$ and the value of σ^{CEP} does not exceed σ_{tot}.

In other words besides the contribution of Fig. 17.3(a) we have to consider the diagram of Fig. 17.3(b) where the double dotted line denotes an additional proton-proton interaction. This diagram describes the absorptive correction to the original CEP process and has a negative sign with respect to the amplitude $A^{(a)}$ of Fig. 17.3(a). Therefore to calculate the CEP cross section we have to square the full amplitude

$$|A_{\mathrm{full}}(b)|^2 \;=\; |A^{(a)}(b) - A^{(b)}|^2 \;=\; S^2(b)\cdot|A^{(a)}(b)|^2, \tag{17.12}$$

where

$$S^2(b) = |e^{-\Omega(b)}|, \quad \text{with} \quad \mathrm{Re}\Omega \geq 0, \tag{17.13}$$

and $\Omega(b)$ is the opacity of incoming protons.

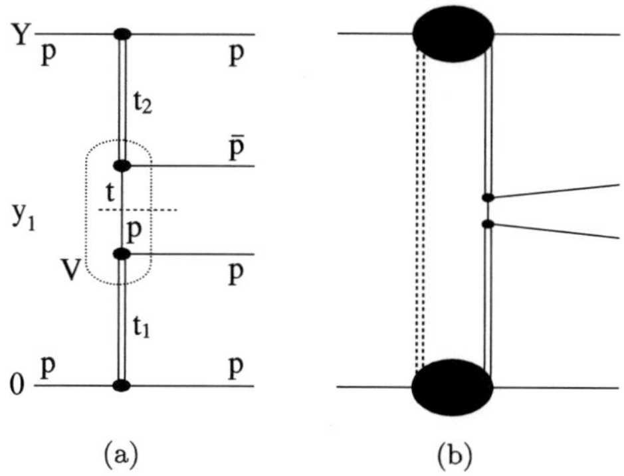

Fig. 17.3 Proton-antiproton pair central exclusive production.

Indeed, in terms of the **S**-matrix, the elastic component $S_l = 1 + iA(b)$; and the unitarity equation (17.7) reflects the probability conservation condition

$$\sum_n \mathbf{S}_l^* |n\rangle\langle n| \mathbf{S}_l \;=\; 1 \tag{17.14}$$

for the partial wave $l = b\sqrt{s}/2$. The solution of unitarity equation (17.7) reads

$$A(b) = i(1 - e^{-\Omega(b)/2}), \tag{17.15}$$

or in terms of the partial wave amplitude with orbital moment $l = b\sqrt{s}/2$

$$a_l \;=\; i(1 - e^{2i\delta_l}) \;=\; i(1 - \eta_l e^{2i\mathrm{Re}\delta_l}), \tag{17.16}$$

where

$$\eta_l = e^{-2\mathrm{Im}\delta_l} \quad \text{with} \quad 0 \le \eta_l \le 1. \tag{17.17}$$

The above discussion shows that $-\Omega(b)/2$ plays the role of $2i\delta_l$. The elastic component of **S** matrix $S_l = \exp(2i\delta_l) = \eta_l \exp(2i\mathrm{Re}\delta_l)$.

The gap survival factor S^2 is the probability to observe a pure CEP event where the LRGs are not populated by secondaries produced in an additional inelastic interaction shown by the dotted line in Fig. 17.3(b). Thus according to (17.17)

$$|S(b)|^2 \;=\; 1 - G_{\mathrm{inel}}(b) \;=\; \eta^2 = \; e^{-\mathrm{Re}\Omega(b)}. \tag{17.18}$$

Equation (17.18) can be rewritten as (see (17.16, 17.17))

$$|S(b)|^2 = |1 + iA(b)|^2 = |S_l|^2. \tag{17.19}$$

In the case of *black disk* asymptotics[1]

$$\mathrm{Re}\Omega(b) \to \infty \quad \text{and} \quad A(b) \to i, \tag{17.20}$$

for $b < R$. That is, we get $S^2(b) \to 0$. The decrease of the gap survival probability S^2 overcompensates the growth of the original CEP cross section (Fig. 17.3(a)), so that finally we have no problem with unitarity.

Recall that this solution of the FK problem was actually realized by Cardy in [Cardy (1974)], where the Reggeon diagrams (generated by Pomerons with intercept $\alpha_P(0) > 1$) were summed up by *assuming analyticity* in the number of Pomerons in a multi-Pomeron vertex. It was shown that the corresponding absorptive corrections (analogous to that shown in Fig. 17.3(b)) suppress not only the growth of a simplest, diagram Fig. 17.3(a), contribution but the growth of cross sections of processes with an arbitrary number of LRG, Fig. 17.2 (for a more detailed discussion see[Khoze *et al.* (2018)]).

[1]Recall that the word 'black' means the *complete* absorption of the incoming state (up to power of s suppressed corrections), i.e. $\mathrm{Re}\Omega(s,b) \to \infty$. 'Black disk' means that in some region of impact parameter space, $b < R$, the whole initial wave function is absorbed. That is, the value of $S(b) = 1 + iA(b) = S_l \to 0$, i.e. $A(b) \to i$.

Note that at the moment we deal with a one-channel eikonal. In other words in Fig. 17.3(b) and in the unitarity equation (17.7) we only account for the pure elastic intermediate states (that is the proton, for the case of pp collisions). In general, there may be $p \to N^*$ excitations shown by the black blobs in Fig. 17.3(b). The possibility of such excitations can be included via the Good-Walker formalism[Good and Walker (1960)] in terms of G-W eigenstates, $|\phi_i\rangle$, which diagonalize the high energy scattering process; that is $\langle\phi_k|A|\phi_i\rangle = A_k\delta_{ki}$. In this case we encounter the FK problem for each state $|\phi_i\rangle$, and then solve it for the individual eigenstates.

Recall that the mean gap survival factor $\langle S^2\rangle$ strongly depends on the particular process. Since the opacity (the optical density) of the proton depends on the impact parameter b we expect a much larger S^2 for the reactions mediated by photons, such as ultraperipheral vector meson production or lepton pair production. These processes occur at a large distances, b, where $\Omega(b)$ is already small. Correspondingly the survival factor $S^2(b) = |\exp(-\Omega(b))|$ of (17.13) is close to 1. On the other hand, the QCD induced exclusive production of a heavy object comes from small distances, and is strongly suppressed by S^2. At a small impact parameter the probability of an additional soft interaction, which fills the rapidity gap, is quite large. That is, there is a small chance for a large rapidity gap to survive.

A more detailed description of the typical values and the behaviour of gap survival factors for different processes can be found in the review given in [Good and Walker (1960)].

17.3 Enhanced diagrams

At first sight it looks sufficient to screen not the whole CEP amplitude, as in Fig. 17.3(b), but just the central vertex V as in Fig. 17.4(a).

Let us consider the enhanced diagram Fig. 17.4(a) in more detail. Note that we have to integrate over the rapidity-positions of the 'effective' triple-Pomeron vertices at y_1 and y_2. Since the original amplitude (shown by the double dotted line) increases with energy, that is with the size of $|y_2 - y_1|$ interval, the main contribution

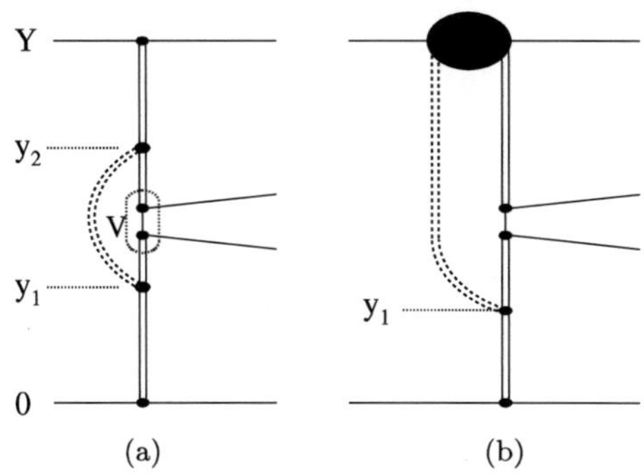

Fig. 17.4 Proton-antiproton pair central exclusive production screened by enhanced (a) and semi-enhanced (b) diagrams.

comes from the configurations where $y_1 \to 0$ and $y_2 \to Y$. In other words, the enhanced diagram,[2] Fig. 17.4(a), acts as the non-enhanced Fig. 17.3(b) graph considered above. The same is valid for the semi-enhanced diagram of Fig. 17.4(b).

Still for the expected value of $\omega_0 \simeq 0.2 - 0.3$ the interval of integration ($|y_2 - y_1| \sim 1/\omega_0 \sim 4$) is not small and the effect of enhanced screening is not negligible. Partly it can be described in terms of the Good-Walker eigenstates $|\phi_i\rangle$, but then we have to account for a large number of eigenstates, up to the mass

$$M_i \sim m_p \cdot \exp(0.5/\omega_0) \sim 5 - 10\,\text{GeV}. \tag{17.21}$$

Note however that the interval of $y_{1,2}$ integration actually depends on the mass of the centrally produced system and the scale corresponding to the subprocess described by the central vertex V. Since we are working in the low-x region for a simplified rough estimate we consider the double logarithmic DGLAP evolution[3] and take $\omega_0 = 0.3$ for the input distribution and account for the $(1-x)^5$ large x behaviour of the gluon at the input scale $Q_0 = 1\,\text{GeV}$. Note that the parton transverse momenta, k_t, are strongly ordered during the DGLAP evolution. The absorptive cross section $\sigma^{\text{abs}} \propto 1/k_t^2$ decreases with k_t. Thus actually only input partons with low transverse momenta ($k_t < Q_0$) participate in an additional interaction and so must be accounted for when we calculate the gap survival factor S^2.

Let us compare the non-enhances diagram Fig. 17.3(b) with the semi-enhaced Fig. 17.4(b). The difference is caused by the following factors:

a) triple-Pomeron coupling g_{3P} instead of the Pomeron-nucleon coupling g_N in the case of Fig. 17.3(b);
b) t-slope B_{enh}, corresponding to the Pomeron loop is about twice smaller than that, B_{el}, for the case of a non-enhanced diagram Fig. 17.3(b);
c) integration over dy_1 (dx_1/x_1).

Note that actually the calculation of the contribution of Fig. 17.3(b) also contains an integration dx_1/x_1 hidden in the convolution of the low scale Q_0 input parton distribution $(1-x)^5 x^{-\omega_0}$ with the Green function given by DGLAP evolution. That is at a large scale μ the lower (shown by the double solid line) propagator in Fig. 17.3(b) within the double logarithmic (DL) approximation reads

$$\int_x^1 \frac{dx_1}{x_1}(1-x_1)^5 x_1^{-\omega_0} \exp(\sqrt{16N_c/\beta_0 \ln(x_1/x)\ln(G)}), \tag{17.22}$$

where $G = \ln(\mu^2/\Lambda^2_{\text{QCD}})/\ln(Q_0^2/\Lambda^2_{\text{QCD}})$.

In the case of a semi-enhanced diagram Fig. 17.4(b) we get a similar expression but multiplied by the factor $x_1^{-\omega_0} = \exp(-y_1\omega_0)$ coming from the screening Pomeron (shown by the double dotted line in Fig. 17.4(b)) which ended now at $y = y_1$

[2]This contribution is *enhanced* by the large intervals of integrations over y_1 and y_2.
[3]The double logarithmic term is exactly the same for BFKL and DGLAP.

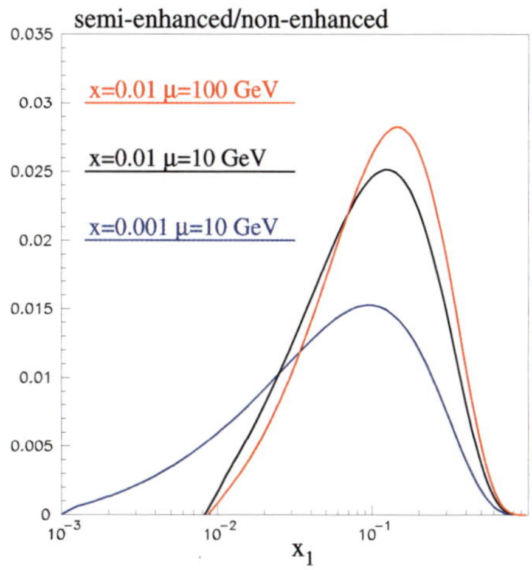

Fig. 17.5 The typical rapidity, y_1, distribution of the semi-enhanced, Fig. 17.4(b), absorptive correction; i.e the x_1 dependence of the integrand of eq. (17.23).

corresponding to the lower triple-Pomeron vertex in Fig. 17.4(b) and not at $y = 0$ as in the case of Fig. 17.3(b). Thus instead of (17.22) we get

$$\int_x^1 \frac{g_{3P}}{\pi g_N} \frac{B_{\rm el}}{B_{\rm enh}} \frac{dx_1}{x_1} (1 - x_1)^5 \exp(\sqrt{16N_c/\beta_0 \ln(x_1/x) \ln(G)}). \tag{17.23}$$

In this way the ratio of the integrals (17.23) and (17.22) will be equal to the ratio of semi-enhanced to non-enhanced contributions. In Fig. 17.5 we show the x_1 dependence of the integrand of (17.23) divided by the integral (17.22) for the case of a typical $x = 0.01$ and scales $\mu = 100$ and $10\,\text{GeV}$, and also for $x = 10^{-3}$ and $\mu = 10\,\text{GeV}$; here we take the one loop running QCD coupling

$$\alpha_s(\mu^2) = 4\pi/(\beta_0 \ln(\mu^2/\Lambda^2) \quad \text{with} \quad \Lambda = 200\text{MeV} \quad \text{and} \quad \beta_0 = 8.333, \tag{17.24}$$

corresponding to 4 light quarks. Also we take $\lambda = g_{3P}/g_N = 0.2$ [Luna *et al.* (2009)] and $B_{\rm el}/B_{\rm enh} = 2$.

Note that in this way we overestimate the enhanced contribution, since:

(a) actually $B_{\rm enh} > B_{\rm el}/2$ due to non-zero slope, $\alpha'_P > 0$, of the effective Pomeron trajectory and non-zero size of the triple-Pomeron vertex;
(b) in the non-enhanced case we only account for the elastic, $p \to p$ intermediate state, and neglect the proton, $p \to N^*$ excitations.

It is seen that for a smaller final x the maximum of the integrand occurs at a smaller x_1 while for a larger final scale μ the maximum moves to a larger x_1. Depending on the mass (i.e. scale) and the momentum fraction carried by the central object the major contribution of the enhanced diagram arises from the region of

$x_1 \sim 0.03 - 0.3$ (i.e. $y_1 \sim 3.5 - 1.2$). For a heavier object the value of y_1 decreases and the enhanced contribution becomes indistinguishable from the non-enhanced absorptive correction of Fig. 17.3(b).[4]

Recall that the major part of proton $p \to N^*$ dissociation was already included into the non-enhanced contribution via the Good-Walker (GW) eigenstates. If the excitations up to $M_X = 3.4\,\text{GeV}$[5] are described within the GW formalism then only the region of $x_1 \leq 0.03$ should be described in terms of enhanced diagrams.[6] That is the absorptive corrections induced by enhanced Pomeron diagrams are practically negligible ($\sim O(0.01)$) for a large mass CEP in comparison with the non-enhanced effect.

17.4 Summary

We emphasize that for the BFKL Pomeron, for which cross sections grow as the power s^ω of energy, black disk absorption is the only cure of the FK disease. Otherwise we have the unphysical situation in which the cross section of events with Large Rapidity Gaps exceed the total cross section. Multi-Pomeron exchange diagrams account for the absorptive effect and compensate the growth of the cross section of the LRG events. The physical meaning of this phenomenon is that simultaneously with the exclusive (LRG) process an additional inelastic interaction takes place, and secondaries from this additional interaction fill the rapidity gaps. Thus instead of an exclusive LRG event we get ordinary multiparticle inelastic production. The probability not to fill the gap is called the gap survival factor S^2.

At asymptotically high energies the value of $S^2 \to 0$. That is the disc becomes black. Any asymptotic behaviour with increasing high-energy cross section which does not lead to complete absorption is not consistent with multi-particle unitarity.

We consider the role of absorptive corrections induced both by the enhanced and the un-enhanced diagrams and show that for the case of central exclusive production of a very heavy object the enhanced contribution acts very similar to that from the non-enhanced diagrams. We emphasize that the value of S^2 is not universal but depends on the nature and the kinematics of a particular process.

Acknowledgements

VAK acknowledges support from a Royal Society of Edinburgh Auber award. MGR thanks the IPPP of Durham University for hospitality.

[4]The fact that enhanced screening becomes negligible for CEP of a heavy object was discussed in [Khoze *et al.* (2006)].

[5]$M_X = 3.4\,\text{GeV}$ is the value used by TOTEM collaboration to separate proton dissociation into low- and high-mass states [Antchev *et al.* (2013)].

[6]In this estimate we assume the mean transverse momentum $p_t \simeq 0.6\,\text{GeV}$; $M_X^2 = m_p^2 + p_t^2/x_1$.

Bibliography

Abarbanel, H. D., Bronzan, J. B., Sugar, R. L., and White, A. R. (1975). Reggeon field theory: Formulation and use, *Phys. Rept.* **21**, pp. 119–182, doi:10.1016/0370-1573(75)90034-4.

Altarelli, G. and Parisi, G. (1977). Asymptotic freedom in parton language, *Nucl. Phys. B* **126**, pp. 298–318, doi:10.1016/0550-3213(77)90384-4.

Antchev, G. *et al.* (2013). Measurement of proton-proton inelastic scattering cross-section at $\sqrt{s} = 7$ TeV, *EPL* **101**, 2, p. 21003, doi:10.1209/0295-5075/101/21003.

Balitsky, I. and Lipatov, L. (1978). The Pomeranchuk singularity in quantum chromodynamics, *Sov. J. Nucl. Phys.* **28**, pp. 822–829.

Cardy, J. L. (1974). General features of the Reggeon calculus with alpha > 1, *Nucl. Phys. B* **75**, p. 413, doi:10.1016/0550-3213(74)90084-4.

Ciafaloni, M. and Camici, G. (1998). Energy scale(s) and next-to-leading BFKL equation, *Phys. Lett. B* **430**, pp. 349–354, doi:10.1016/S0370-2693(98)00551-6, arXiv:hep-ph/9803389.

Ciafaloni, M. and Colferai, D. (1999). The BFKL equation at next-to-leading level and beyond, *Phys. Lett. B* **452**, pp. 372–378, doi:10.1016/S0370-2693(99)00281-6, arXiv:hep-ph/9812366.

Dokshitzer, Y. L. (1977). Calculation of the structure functions for deep inelastic scattering and e+ e− annihilation by perturbation theory in quantum chromodynamics. *Sov. Phys. JETP* **46**, pp. 641–653.

Fadin, V. S., Kuraev, E., and Lipatov, L. (1975). On the Pomeranchuk singularity in asymptotically free theories, *Phys. Lett. B* **60**, pp. 50–52, doi:10.1016/0370-2693(75)90524-9.

Fadin, V. S. and Lipatov, L. (1998). BFKL pomeron in the next-to-leading approximation, *Phys. Lett. B* **429**, pp. 127–134, doi:10.1016/S0370-2693(98)00473-0, arXiv:hep-ph/9802290.

Finkelstein, J. and Kajantie, K. (1968). Multiple Pomeranchuk exchange violates unitarity, *Phys. Lett. B* **26**, pp. 305–307, doi:10.1016/0370-2693(68)90567-4.

Froissart, M. (1961). Asymptotic behavior and subtractions in the Mandelstam representation, *Phys. Rev.* **123**, pp. 1053–1057, doi:10.1103/PhysRev.123.1053.

Good, M. and Walker, W. (1960). Diffraction disssociation of beam particles, *Phys. Rev.* **120**, pp. 1857–1860, doi:10.1103/PhysRev.120.1857.

Gribov, V. and Lipatov, L. (1972). Deep inelastic e p scattering in perturbation theory, *Sov. J. Nucl. Phys.* **15**, pp. 438–450.

Khoze, V., Martin, A., and Ryskin, M. (2006). On the role of hard rescattering in exclusive diffractive Higgs production, *JHEP* **05**, p. 036, doi:10.1088/1126-6708/2006/05/036, arXiv:hep-ph/0602247.

Khoze, V., Martin, A., and Ryskin, M. (2018). Black disk, maximal Odderon and unitarity, *Phys. Lett. B* **780**, pp. 352–356, doi:10.1016/j.physletb.2018.03.025, arXiv:1801.07065 [hep-ph].

Kuraev, E., Lipatov, L., and Fadin, V. S. (1977). The Pomeranchuk singularity in nonabelian gauge theories, *Sov. Phys. JETP* **45**, pp. 199–204.

Kuraev, E. A., Lipatov, L. N., and Fadin, V. S. (1976). Multi-Reggeon processes in the Yang-Mills theory, *Sov. Phys. JETP* **44**, pp. 443–450.

Lipatov, L. (1975). The parton model and perturbation theory, *Sov. J. Nucl. Phys.* **20**, pp. 94–102.

Luna, E., Khoze, V., Martin, A., and Ryskin, M. (2009). Diffractive dissociation re-visited for predictions at the LHC, *Eur. Phys. J. C* **59**, pp. 1–12, doi:10.1140/epjc/s10052-008-0793-1, arXiv:0807.4115 [hep-ph].

Salam, G. (1998). A resummation of large subleading corrections at small x, *JHEP* **07**, p. 019, doi:10.1088/1126-6708/1998/07/019, arXiv:hep-ph/9806482.

Verdiev, L., Kancheli, O., Martinyan, S., Popova, A., and Ter-Martirosyan, K. (1964). Total particle formation cross sections at high energies and the unitarity condition in the s-CHANNEL, in *12th International Conference on High Energy Physics*, Vol. 1, pp. 345–347.

https://doi.org/10.1142/9789811231124_0018

Chapter 18

A Few Topics in BFKL Phenomenology at Hadron Colliders

Agustín Sabio Vera
CERN, Theoretical Physics Department, Geneva, Switzerland
Instituto de Física Teórica UAM/CSIC
Universidad Autónoma de Madrid, Madrid, Spain

Abstract

This is a personal recollection of several results involving the phenomenological study of the multi-Regge limit of scattering amplitudes. None of them would have been possible without the encouragement and constant support from Lev Nikolaevich Lipatov.

18.1 Lev in Madrid

The first time I met Lev N. Lipatov was at a Low x Physics workshop at the end of June of 2001, in Krakow, Poland.[1] At the time I was a postdoctoral fellow at

[1] http://th-www.if.uj.edu.pl/low_x.

the Cavendish Laboratory, in Cambridge, and I thought I knew everything about the high energy limit of Quantum Chromodynamics (QCD). I had been introduced to the concept of the Pomeron in Regge theory by Sandy Donnachie, and had worked during my PhD thesis in Manchester with Jeff Forshaw and Douglas Ross on the understanding of the fundamentals of the Balitsky-Fadin-Kuraev-Lipatov (BFKL) [Balitsky and Lipatov (1978); Kuraev *et al.* (1977, 1976); Lipatov (1976); Fadin *et al.* (1975); Lipatov (1986)] equation at next-to-leading order. Things did not go as planned the day of my talk at that conference because, right after my presentation, Lev stood up and said out loud that all what I had said was completely wrong. Suddenly it felt like a huge black hole opened up under my feet and I got sucked down into it.

After that very rough beginning we became good friends and collaborators for many years. The last time we met and worked together was at the beginning of August 2017, when Lev and his wife Elvira came to Cambridge where I was spending the summer at DAMTP. Already from our first discussions it was clear he was a very special character, full of energy, and an amazing mind. He had an incredible knowledge of all the important works from the old times, before Quantum Field Theory, which he was able to blend with the modern developments in a way I have not witnessed in any other researcher. When working with him you had to be ready to be carried to any uncharted territory. You could start talking about radiative corrections in QED and end up with a blackboard full of results on quantum integrability and tachyon scattering in string theory. When discussing with Lev you always had the impression that something great was about to happen, that a completely new and brilliant idea was about to arise, and this would immediately motivate you to work on the problem at hand as if there was no tomorrow.

We used to meet regularly mainly at four locations: in Hamburg, hosted at DESY by Jochen Bartels; in Geneva, at the CERN Theory Group (then Unit and now Department); in Gatchina, at the Petersburg Nuclear Physics Institute, where he was Director of the Theory Division; and in Madrid, at Autónoma university and the Instituto de Física Teórica, where he was Severo Ochoa (SO(IFT)) Distinguished Professor. During their many visits Lev and Elvira lived in my home town, Alcalá de Henares, and we used to spend endless hours trying to push our projects forward. Funnily enough, we did not publish any papers on QCD but more on supersymmetric and gravitational theories. Nevertheless his encouragement and insight were crucial for the development of the several results which will be described in the following. As a disclaimer, allow me to indicate that this is not a comprehensive review of the subject, it is just a personal recollection of results on the phenomenology of the BFKL formalism which gather together many of the discussions we had.

18.2 Where are the Reggeon fields?

Let us start by motivating the experimental search of BFKL effects at colliders. Right now we are at an interesting period to investigate this subject because we have good control of the theoretical calculations and the Large Hadron Collider (LHC) is operating at full power. This should allow us to determine the region of applicability of asymptotic calculations of scattering amplitudes in the high energy limit.

The study of scattering processes in the high energy limit of QCD offers invaluable information that goes beyond the perturbative aspects of the theory. It touches important issues such as factorization between soft and hard physics and it challenges fixed order calculations. At high energies, the convergence of the perturbative expansion, truncated at a certain order in the strong coupling α_s, is not *a priori* guaranteed. This is because large logarithms of the center-of-mass energy squared, s, appear in Feynman diagrams and it is needed to resum them to all orders. A powerful approach to perform this resummation is the BFKL framework initially developed at the leading logarithmic (LL) approximation [Balitsky and Lipatov (1978); Kuraev *et al.* (1977, 1976); Lipatov (1976); Fadin *et al.* (1975); Lipatov (1986)], where terms of the form $(\alpha_s \log(s))^n$ were resummed. In order to improve those results, the next-to-leading logarithmic (NLL) approximation corrections to the BFKL kernel were calculated [Fadin and Lipatov (1998); Ciafaloni and Camici (1998)], where also terms that behave like $\alpha_s(\alpha_s \log s)^n$ were taken into account. It was soon realised, however, that at NLL the positiveness of cross sections was not always ensured. This is due to the presence of large collinear logarithms that need extra treatment, a step that led to the so-called collinearly improved kernel [Salam (1998); Ciafaloni *et al.* (2003)] allowing for more robust phenomenological studies. Obviously, an important question for collider phenomenology is gauging reliably at which energies the BFKL dynamics becomes relevant and cannot be ignored.

What are the observables where this formalism should apply? In simple terms, in a scattering amplitude we identify and calculate those contributions at each order of perturbation theory which have the largest numerical value when s is big. This shares a common nature with other resummation schemes where a class of terms has to be calculated to all orders. What makes this formalism different is that it is not just a way to calculate some observables more precisely, it also shows that at high energies new degrees of freedom, reggeized particles, play a dominant role. These reggeons are universal in the sense that they drive a large variety of cross sections with very different initial and final state configurations, present in lepton-lepton, lepton-hadron and hadron-hadron colliders. Remarkably, this effective picture also holds for electroweak, supersymmetric and gravitational interactions.

The reggeization picture [Fadin and Fiore (1998)], at LL and NLL accuracies, is robust and has been cross-checked by many different methods, including elegant techniques in string theory [Bajnok *et al.* (2009)]. It is in this linear NLL approximation where we have the opportunity to test the applicability of these emergent degrees of freedom to the LHC phenomenology program since s is large enough to justify the use of reggeized propagators but not so large as to necessarily have to introduce non-linear corrections which go beyond NLL. It is then a timely matter to find out where the window of applicability of the formalism lies. This means to identify observables where this approach is distinct, *i.e.*, quantities where it fits the data and other possible approaches (fixed order or other resummations implemented in general Monte Carlo event generators) fail.

The attempts to search for BFKL effects have had the difficulty of being applied at too low energies or rapidity differences in the final state as to be conclusive. A further drawback is that they dealt with too inclusive observables as to be able to claim that the cross section under study could be described by BFKL dynamics and nothing else. A paradigmatic example is the description of the growth of the proton structure functions at low values of Bjorken x. Indeed it is possible to get a good fit of the combined HERA data for the F_2 and F_L structure functions in the small x region with a NLL BFKL calculation. As an example, see Fig. 18.1, taken from [Hentschinski *et al.* (2013b,a)]. However, it is possible to fit these data with other approaches as well. Moreover, there is model dependence in the BFKL calculation itself, which includes three free parameters for the coupling of the gluon ladder to the hadron, a freezing of the coupling in the infrared and the use of a collinear resummation together with a particular renormalization scheme which allows to reach low values of Q^2 in the fit. All of this pushes us to try to find other, less inclusive, observables to test small x resummations in a more distinct manner.

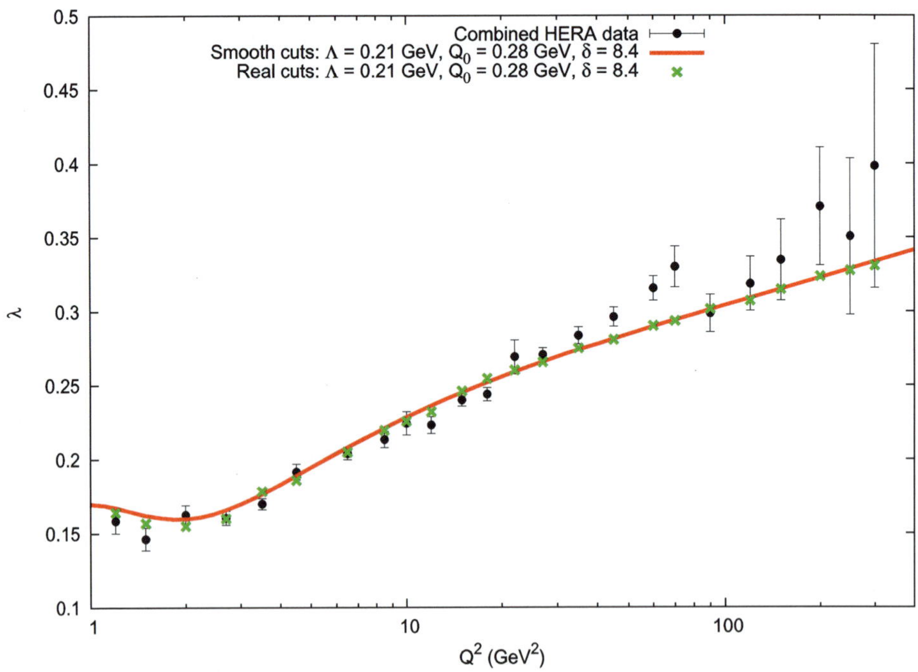

Fig. 18.1 BFKL calculation for $\lambda(Q^2)$ in the parametrization of the structure function $F_2(x, Q^2) = c(Q^2)x^{-\lambda(Q^2)}$ at NLL with collinear improvements and a model for the proton impact factor of the form $\simeq (p^2/Q_0^2)^\delta e^{-p^2/Q_0^2}$, for values of $x < 0.01$.

18.3 More exclusive observables

The LHC energies are higher than those at the Tevatron or HERA and there is enough statistics to allow for the study of more exclusive quantities, fine-tuned with stringent cuts as to be driven by multi-Regge kinematics, which is the underlying principle behind the BFKL approach. The experimental studies should not only consider the usual "growth with energy" signal, associated with the exchange of a hard Pomeron, but also other footprints related to energy flows and azimuthal angle dependences. We will motivate this below.

18.3.1 *Tagging of forward jets hides the growth with energy*

It is only possible to remain in the multi-Regge kinematics if the impact factors associated with each of the incoming protons are strongly peaked at a single hard scale, similar at both hadrons. It turns out that the hard scale is typically associated with the production of forward jets calculated in collinear factorization. This is known as Mueller-Navelet (MN) jet production [Mueller and Navelet (1987)]. MN jets have inclusive final states where two jets with transverse momenta of similar sizes, $k_{A,B}$, are tagged when they are widely separated in rapidity Y.

The further the MN jets are separated in rapidity, the deeper into the $x \to 1$ limit of the collinear parton distribution functions we enter. This is a region characterized by a drop of the cross section due to energy momentum conservation. This hides the BFKL effects in terms of growth of the cross section as can be seen in Fig. 18.2 (obtained in [Caporale *et al.* (2013)]) where the cross section for MN jets

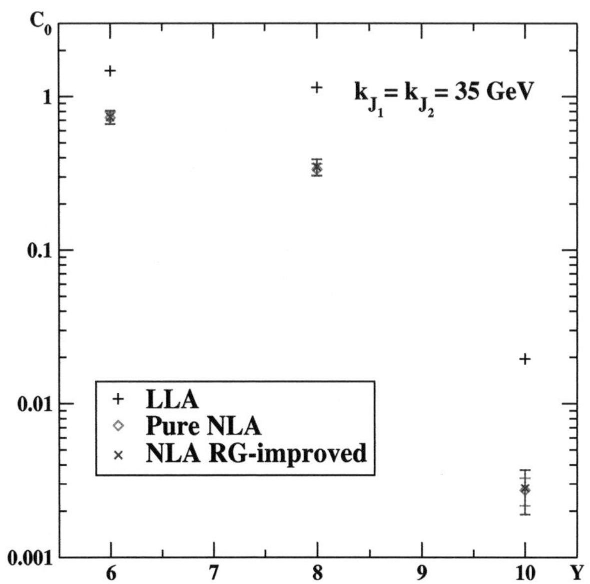

Fig. 18.2 Y dependence of MN cross section C_0 for $|\vec{k}_{J_{1,2}}| = 35\,\text{GeV}$ at $\sqrt{s} = 14\,\text{TeV}$.

with the same lower cut in the p_T of the tagged jets is investigated as a function of their rapidity difference for a LHC run at $\sqrt{s} = 14\,\text{TeV}$. The cross section decreases as Y increases.

18.3.2 *Ratios of azimuthal correlations as a clean signal*

A different signal [Del Duca and Schmidt (1994); Stirling (1994)] which is sensitive to the wealth of the BFKL formalism is to look at the dependence on the relative azimuthal angle, ϕ, between the two MN jets, in the form of n-th moments. The emission of multiple mini-jets manifests as a fast decrease of $<\cos(n\phi)>$ with Y. However, even these differential distributions suffer from a large influence of the collinear region. This is due to the fact that $<\cos(n\phi)>\simeq \exp(\alpha_s Y(\chi_n(1/2) - \chi_0(1/2)))$, where $\chi_n(\gamma)$ is, in Mellin space, the n-th Fourier component of the BFKL kernel. $\gamma \simeq 1/2$ dominates the cross section.

It turns out that the $n = 0$ component is very sensitive to collinear dynamics well beyond the original multi-Regge kinematics. Even though it is possible to include these collinear terms by modifying the original BFKL equation, we believe it is more important to first find the region of applicability of the original formalism by using observables which are far less sensitive to this collinear "contamination". It is only following this philosophy that we will be able to find "distinct" BFKL observables. A relevant step in this direction was taken in [Sabio Vera (2006); Sabio Vera and Schwennsen (2007)] where instead of $<\cos(n\phi)>$ it was proposed to remove the $n = 0$ dependence by studying "conformal ratios"[2] $\mathcal{C}_{m,n} = <\cos(m\phi)>/<\cos(n\phi)>$ which behave like $\simeq \exp(\alpha_s Y(\chi_m(1/2) - \chi_n(1/2)))$. It is important to note that the BFKL kernel for $n \neq 0$ is asymptotically insensitive to collinear regions. In [Sabio Vera (2006); Sabio Vera and Schwennsen (2007)] it was proven that these new ratios are very stable under radiative corrections with the LL result (even with running of the coupling) being very similar to the NLL prediction.

With the advent of the LHC it was soon realized that the NLL predictions for these ratios, including NLO forward jet vertices, were in agreement with the data. Furthermore, these observables are so fine-tuned to the multi-Regge limit that it was difficult for other approaches to describe them. This could be seen in, *e.g.*, the studies presented in [Ciesielski (2014)], see Fig. 18.3, where only a BFKL analysis at NLL is able to fit the large Y tail of the Mueller-Navelet "conformal ratios" proposed in [Sabio Vera (2006); Sabio Vera and Schwennsen (2007)]. In [CMS-Collaboration (2013); Khachatryan *et al.* (2016)] a comparison between data and theory for a number of MN azimuthal ratios is shown. More recent studies can be found in [Duclou *et al.* (2014); Colferai *et al.* (2010)].

---

[2]They capture the $SL(2,C)$ nature of the effective theory of QCD at high energies. When these ratios are calculated at NLL accuracy in the $N = 4$ supersymmetric Yang-Mills model, with four-dimensional conformal invariance, they are in agreement with those obtained in QCD using the Brodsky-Lepage-Mackenzie (BLM) scale-fixing procedure in momentum-subtraction (MOM) renormalization scheme (see [Angioni *et al.* (2011)]).

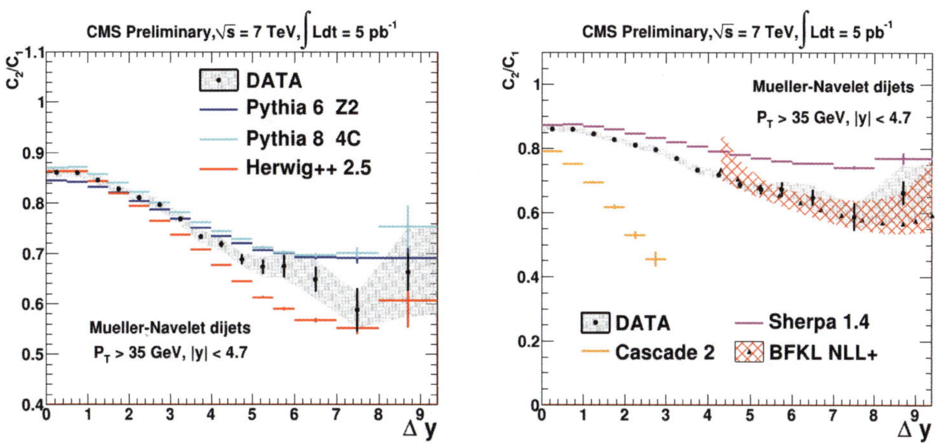

Fig. 18.3 The ratio of average cosines $\mathcal{C}_{2,1} = C_2/C_1$ in bins of $\Delta y = Y$, compared to various Monte Carlo models.

A generalization of the azimuthal ratios has been proposed for processes that have three [Caporale *et al.* (2016c,a)] and four final state jets [Caporale *et al.* (2016b, 2017)]. These are special MN cases since the outermost jets still have a large rapidity distance and other tagged jets are central.

18.4 The Monte Carlo event generator `BFKLex`

For future developments, two main ingredients are needed: new process dependent NLO impact factors and Monte Carlo techniques to control the gluon Green function and to extract its physical content in the most differential form. The former are mandatory to fairly test the theory and correctly control the dependence on the scales appearing in the calculations (running of the coupling and energy scale separating the universal gluon radiation in central regions of rapidity and that stemming from the fragmentation regions). The latter are needed in order to effectively generate differential distributions which are difficult to obtain analytically.

Let us present several observables characterizing multi-jet configurations event by event (averages of transverse momentum, azimuthal angle and ratios of jet rapidities). They can be studied using the Monte Carlo event generator `BFKLex`, developed with Grigorios Chachamis, where higher-order collinear corrections are implemented together with the NLO kernel.

The BFKL formalism allows for the investigation of high-multiplicity final-states even if the scattering energy is not large. If we require two tagged forward/backward jets in the final state the cross sections can be written in the form

$$\sigma(Q_1, Q_2, Y) = \int d^2\vec{k}_A d^2\vec{k}_B \, \phi_A(Q_1, \vec{k}_a) \, \phi_B(Q_2, \vec{k}_b) \, f(\vec{k}_a, \vec{k}_b, Y). \qquad (18.1)$$

The impact factors $\phi_{A,B}$ depend on hard scales $Q_{1,2}$. The gluon Green function f depends on the transverse momenta $\vec{k}_{a,b}$ and the center-of-mass energy in the scattering or, alternatively, on the rapidity difference between the two tagged jets. At LL and NLL f admits the iterative representation [Schmidt (1997); Andersen and Sabio Vera (2003, 2004)]

$$f = e^{\omega(\vec{k}_A)Y}\Bigg\{\delta^{(2)}\left(\vec{k}_A - \vec{k}_B\right) + \sum_{n=1}^{\infty}\prod_{i=1}^{n}\frac{\alpha_s N_c}{\pi}\int d^2\vec{k}_i \frac{\theta\left(k_i^2-\lambda^2\right)}{\pi k_i^2}$$
$$\times \int_0^{y_{i-1}} dy_i e^{\left(\omega\left(\vec{k}_A+\sum_{l=1}^{i}\vec{k}_l\right)-\omega\left(\vec{k}_A+\sum_{l=1}^{i-1}\vec{k}_l\right)\right)y_i}\delta^{(2)}\left(\vec{k}_A+\sum_{l=1}^{n}\vec{k}_l-\vec{k}_B\right)\Bigg\}, \tag{18.2}$$

with $\omega(\vec{q})$ being the gluon Regge trajectory and λ an infrared cut-off. `BFKLex` implements this iteration and has been used for collider phenomenology and other formal studies [Chachamis and Sabio Vera (2013)]. As we have discussed, the BFKL formalism is quite sensitive to collinear configurations. The dominant double-log terms in this region can be resummed [Salam (1998)]. In [Sabio Vera (2005)], it was shown that this can be done in transverse momentum representation using the following prescription in Eq. (18.2):

$$\theta\left(k_i^2-\lambda^2\right) \to \theta\left(k_i^2-\lambda^2\right) + \sum_{n=1}^{\infty}\frac{\left(-\bar{\alpha}_s\right)^n}{2^n n!(n+1)!}\ln^{2n}\left(\frac{\vec{k}_A^2}{\left(\vec{k}_A+\vec{k}_i\right)^2}\right). \tag{18.3}$$

This step improves the convergence of the perturbative expansion and in γ representation corresponds to a J_1-Bessel function originally found in [Sabio Vera (2006)]. It has been implemented in `BFKLex` [Chachamis and Sabio Vera (2016a)] to study MN jet production [Chachamis and Sabio Vera (2016b)]. Three averages in each event have been investigated: of the modulus of the transverse momentum of the final-state jets, of their azimuthal angle and of the rapidity ratio between subsequent jets:

$$\langle p_t\rangle = \frac{1}{N}\sum_{i=1}^{N}|k_i|; \quad \langle\theta\rangle = \frac{1}{N}\sum_{i=1}^{N}\theta_i;$$
$$\langle\mathcal{R}_y\rangle = \frac{1}{N+1}\sum_{i=1}^{N+1}\frac{y_i}{y_{i-1}}. \tag{18.4}$$

Some configurations under study include $k_a = 10\,\text{GeV}$, $k_b = 20\,\text{GeV}$ and $y_a - y_b = 4, 6, 8$. Figure 18.4 shows broad distributions in $\langle p_t\rangle$ with a maximal value at $\langle p_t\rangle \simeq 6\,\text{GeV}$. The cross section receives important contributions from jets with large transverse momentum. The azimuthal angle between the two tagged jets changes randomly event by event. The $\langle\theta\rangle$ per event at which the untagged jets are produced is shown in Fig. 18.4 together with the mean ratios of rapidities $\langle\mathcal{R}_y\rangle$.

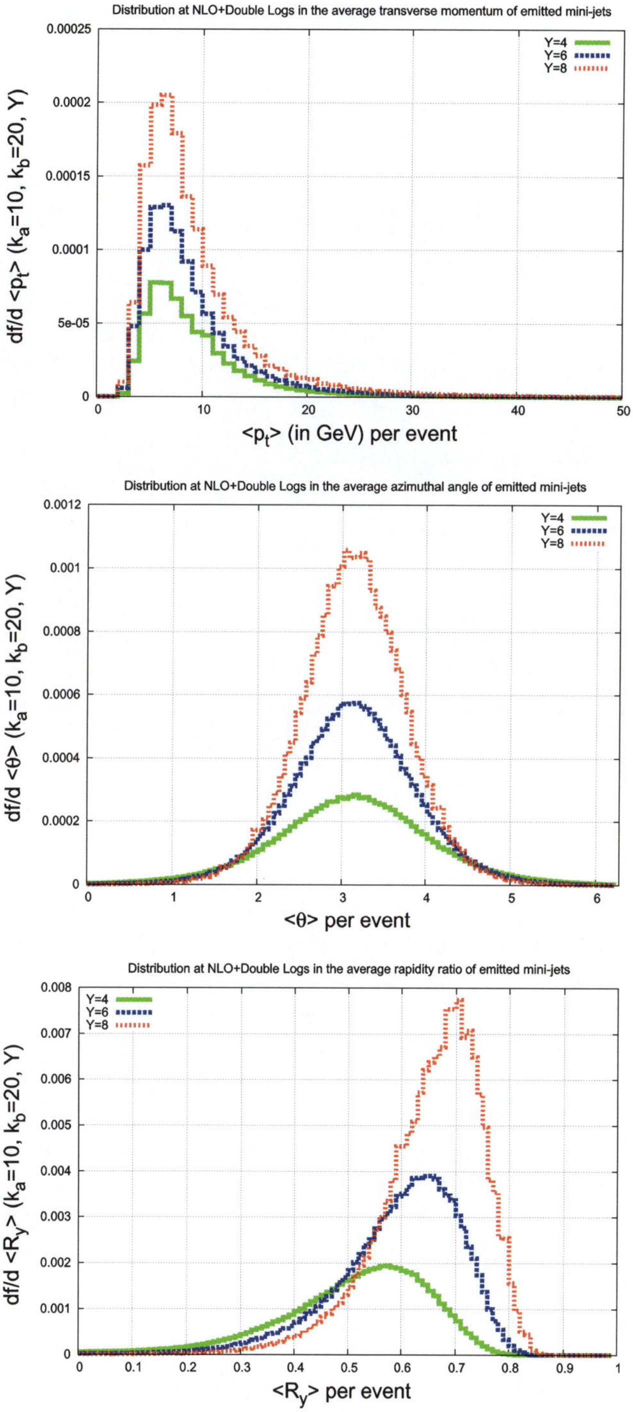

Fig. 18.4 High-multiplicity distributions generated with `BFKLex`.

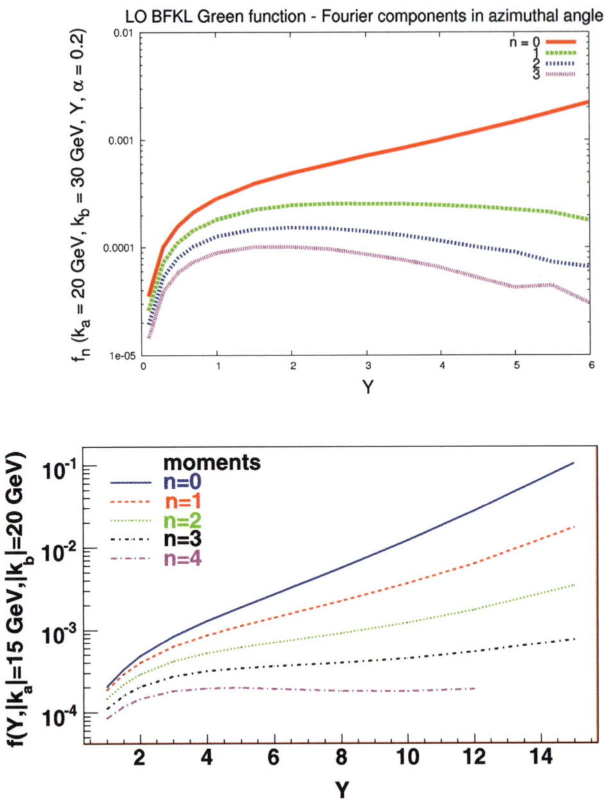

Fig. 18.5 Rapidity dependence of the azimuthal angle Fourier components in the BFKL (top) and CCFM (bottom) Green function.

The distributions have maxima at $\langle \mathcal{R}_y \rangle > 0.5$. As they are quite broad, this means that there are large contributions from preasymptotic configurations.

The value of Monte Carlo studies can be illustrated with another example related to the azimuthal angles. Let us compare the above mentioned n-moments of the BFKL cross section with those obtained from the Catani-Ciafaloni-Fiorani-Marchesini (CCFM) [Catani *et al.* (1990); Marchesini (1995); Ciafaloni (1988); Catani *et al.* (1991)] approach, which, in principle, interpolates the large and small x limits of unintegrated gluon densities (we will see that this is only true for the $n = 0$ projection). This was studied in [Chachamis *et al.* (2011)] and it is shown here in Fig. 18.5. In the top plot we see the BFKL result and at the bottom the CCFM analysis. We can see that the $n = 0$ component has a similar behavior in both cases since it grows with Y. But this is not the case for the $n > 0$ components which in the BFKL case decrease with Y and for CCFM increase. This is a fundamental difference between the Regge limit and approaches based on QCD coherence [Gonzo *et al.* (2019)] which deserves to be explored further in order to disentangle BFKL from other dynamics.

Regarding the impact factors and jet vertices, with Martin Hentschinski, we have developed a method to calculate them based on Lipatov's high energy effective action [Hentschinski and Sabio Vera (2012); Chachamis *et al.* (2013b,a, 2014a)]. Not only have we reproduced the NLO forward jet vertex coupled to an octet [Chachamis *et al.* (2012, 2014b, 2013c)], which is related to forward jet production with associated mini-jet radiation but also when the jet plus proton remnants are attached to a color singlet [Hentschinski *et al.* (2014a,b,c)], which is associated to a gap in rapidity without any hadronic activity in the detectors. For these diffractive events we have the "traditional" problem of the gap survival probability which is difficult to calculate in a reliable manner. `BFKLex` can be very useful in this context of diffractive physics. Further processes at full NLL, like forward production of electroweak bosons, will also require the calculation of NLO impact factors, a problem which can be suitably addressed within the high energy effective action approach.

It is important to put forward this program since there are uncertainties in the BFKL approach itself which need to be fixed. As an example, it is needed to find the dependence of each proposed observable on the renormalization schemes (the above mentioned conformal ratios were shown to be independent of these choices), but also the correct treatment of the running of the coupling must be addressed in an accurate way (we will discuss this further below). Only a fair comparison to experimental data can solve many of these theoretical questions.

If we control some observables which can only be described by BFKL then it makes sense to introduce corrections in the form of hadronization, non-linearities, collinear radiation or study connections with soft-collinear effective theories, in order to extend their range of applicability beyond the multi-Regge kinematics.

18.5 The challenge of the infrared limit

The structure of the BFKL equation changes in the infrared region. A pressing question is to find the optimal treatment of the running of the coupling. This is one of the many attractive features of this formalism since it offers the possibility to study the transition from perturbative to non-perturbative dynamics in a continuous way. Lipatov [Lipatov (1986)] worked on this problem and proved that, if suitable infrared boundary conditions are imposed when treating the running of the coupling, the cut in the complex angular momentum plane transforms into an infinite series of poles. For him this was important since it connected with Gribov's ideas where high energy scattering could be dominated by Regge poles after all.

In [Ross and Sabio Vera (2016)], together with Douglas Ross, we realized that integrating along a contour off the real axis we find a strong dependence of the intercepts and collinear regions on the choice of the boundary conditions. We found that the mean transverse scales dominant in the gluon ladder increase, pushing the gluon evolution towards harder scales. If this is the correct treatment of the running of the coupling, it could have interesting phenomenological consequences worth exploring.

The nutshell of the calculation can be explained using the notation $t_i \equiv \ln(k_i^2/\Lambda_{\rm QCD}^2)$, $\bar{\alpha}_s \equiv \frac{C_A}{\pi}\alpha_s$, $\bar{\alpha}_s(t) = \frac{1}{\bar{\beta}_0 t}$, and a form of running the coupling with a hermitian kernel such that the Green function follows the equation

$$\frac{\partial}{\partial Y}\mathcal{G}(Y,t_1,t_2) = \int \frac{dt}{\sqrt{\bar{\beta}_0 t_1}}\frac{\mathcal{K}(t_1,t)}{\sqrt{\bar{\beta}_0 t}}\mathcal{G}(Y,t,t_2). \tag{18.5}$$

It is convenient to move to Mellin space where the transform can be written in terms of Airy functions [Kowalski *et al.* (2010, 2016)],

$$\mathcal{G}_\omega(t_1,t_2) = \frac{\pi}{4}\frac{\sqrt{t_1 t_2}}{\omega^{1/3}}\left(\frac{\bar{\beta}_0}{14\zeta(3)}\right)^{2/3} Ai\left(z(t_1)\right) Bi\left(z(t_2)\right)\theta\left(t_1-t_2\right) + t_1 \leftrightarrow t_2 \tag{18.6}$$

with $z(t) \equiv \left(\frac{\bar{\beta}_0\omega}{14\zeta(3)}\right)^{1/3}\left(t - \frac{4\ln 2}{\bar{\beta}_0\omega}\right)$. There is some freedom in the way to present this solution. Keeping in mind that the Green function should vanish when $t_{1,2} \to \infty$, which means that we should not tamper with the ultraviolet behaviour, it is possible to add to this function any solution of the homogeneous equation with the same UV behaviour. In this way one can replace the Airy function $Bi(z)$ by $\overline{Bi}(z) \equiv Bi(z) + c(\omega)Ai(z)$. With $c(\omega)$ being

$$c(\omega) = \cot\left(\eta - \frac{2}{3}\sqrt{\frac{\bar{\beta}_0\omega}{14\zeta(3)}}\left(\frac{4\ln 2}{\bar{\beta}_0\omega} - t_0\right)^{3/2}\right), \tag{18.7}$$

an infinite number of Regge poles in the complex ω plane appears when the argument of this function is $-n\pi$. This fixes the phase at the IR fixed point t_0 to be $\frac{\pi}{4}+\eta$. In principle we can only set the value of η using some non-perturbative properties of QCD or, from a more practical point of view, by trying to fit the experimental data. Either way, the interplay between both approaches should teach us some valuable lessons about the confinement region.

Using a quadratic approximation for the kernel simplifies the calculations while keeping the main physical features of the problem. We numerically inverted the Mellin transform and found that the choice for η affects the phase in Eq. (18.7). For example, increasing η reduces the initial rise with energy of the solution (LHS plot in Fig. 18.6). The collinear behaviour is also influenced, specially at small t (RHS plot in Fig. 18.6). This is natural since η has been introduced in order to modify only the IR sector of the theory. The t-profile contains fundamental information about the discrete Pomeron approach since it allows for the study of the diffusion in transverse scales. This is shown in Fig. 18.7. For two large and similar external scales (left) the gluon evolution is governed by a IR/UV symmetric diffusion. This changes when the external scales are smaller (RHS plot) since we find a strong suppression of the diffusion towards the IR. The mean values of the distributions

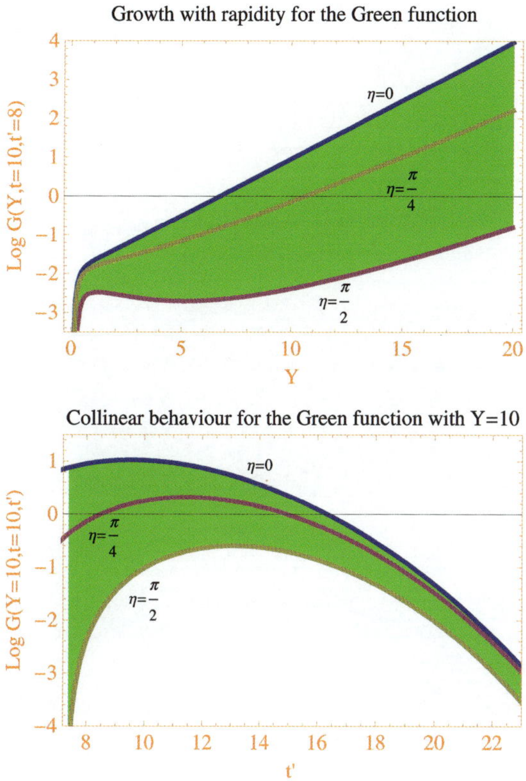

Fig. 18.6 Effect of the non-perturbative phase η on the gluon Green function.

(central line in both plots of Fig. 18.7) are pushed towards the UV. This shows how the IR region in the discrete Pomeron approach has been screened out by the choice of boundary conditions. Since it is difficult to fix the non-perturbative phase from first principles, it will be very interesting to find the best fit to data using this as a free parameter.

18.6 Rapidity veto constraint

In multi-Regge kinematics, one assumes that the cascade of gluons emitted between the two primary gluons have a large relative rapidity. Following the original presentation of this idea by Lipatov in [Lipatov (1998)], in [Schmidt (1999)] it was shown that a significant reduction in the Green function takes place if one considers contributions to the scattering amplitude in which emitted gluons have a minimum rapidity gap, b, relative to the preceding emitted gluon. In [Forshaw *et al.* (1999)] it was proven that the effect of imposing such a restriction reproduces the effect of the collinear corrections to the BFKL equation when $b \approx 2$. This idea has been applied also to non-linear evolution equations [Chachamis *et al.* (2005)]. The mean distance

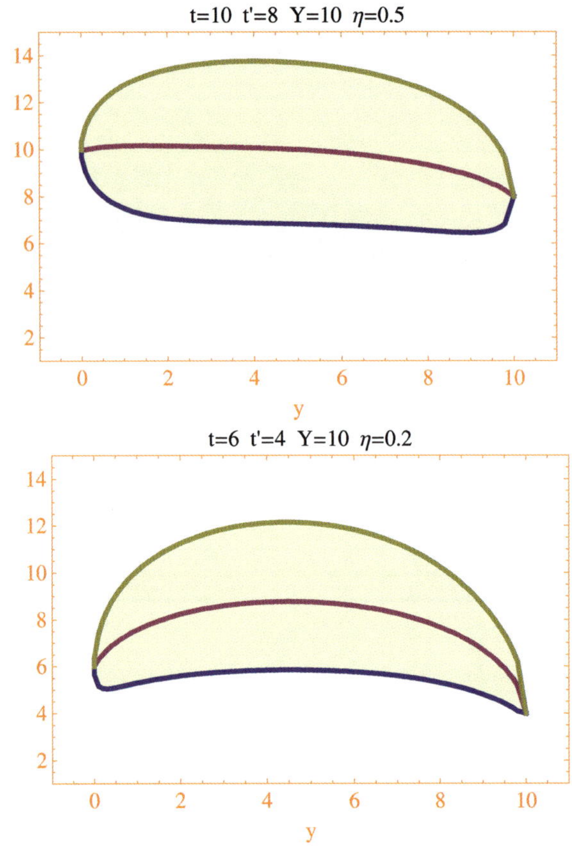

Fig. 18.7 Effect of the non-perturbative phase on the diffusion profile.

in rapidity among emissions in the BFKL ladder, including collinear contributions, has been studied using `BFKLex` in [Chachamis and Sabio Vera (2016b)].

18.6.1 *Effects on the discrete hard Pomeron*

The discrete BFKL spectrum, which we have discussed above, is sensitive to the introduction of this lower cut-off in the relative rapidities of the emitted particles. The eigenvalues associated to each of the discrete eigenfunctions decrease with b. This effect is larger on the lowest eigenfunctions. This limits the growth with energy of cross sections and introduces a fast suppression of the regions with small p_T.

To be more precise, now we do not use the quadratic approximation for the eigenvalues, $\chi(\nu)$, to write, in the semi-classical approximation, the eigenfunctions of the kernel with running coupling, with eigenvalue ω, in the form

$$f_\omega(t) = \frac{|z_\omega(t)|^{1/4}}{\sqrt{\bar{\alpha}_s(t)\chi'\left(\nu_\omega(t)\right)}} Ai\left(z_\omega(t)\right), \tag{18.8}$$

where $\nu_\omega(t) = \chi^{-1}(\bar{\beta}_0 \omega t)$, and $z_\omega(t) = -\left(\frac{3}{2}\int_t^{4\ln 2/\bar{\beta}_0\omega} dt' \nu_\omega(t')\right)^{2/3}$. As explained before, a fixed phase for a negative value of t generates discrete eigenfunctions $f_{\omega_n}(t)$ with n zeroes. The Green function is now

$$\mathcal{G}(Y,t,t') = \sum_n f_{\omega_n}(t) f^*_{\omega_n}(t') e^{\omega_n Y}. \tag{18.9}$$

A rapidity gap veto is then implemented in the form

$$\mathcal{G}(Y,t,t') = \int_C \frac{d\omega}{2\pi i} e^{\omega(Y-b)} \sum_n \frac{f_{\omega_n}(t) f^*_{\omega_n}(t')}{\omega - e^{-b\omega}\omega_n}. \tag{18.10}$$

There is a pole at $\omega = \frac{W(b\,\omega_n)}{b} \equiv \overline{\omega}_n$, where $W(x)$ is the Lambert W-function. Hence

$$\mathcal{G}(Y,t,t') = \sum_n e^{\overline{\omega}_n(Y-b)} \frac{f_{\omega_n}(t) f^*_{\omega_n}(t')}{1 + b\overline{\omega}_n}. \tag{18.11}$$

The gap veto damps the solution in two ways. First, with the shift $Y \to Y - b$. Second, with the replacement $\omega_n \to \overline{\omega}_n$ (see their relative size in Fig. 18.8).

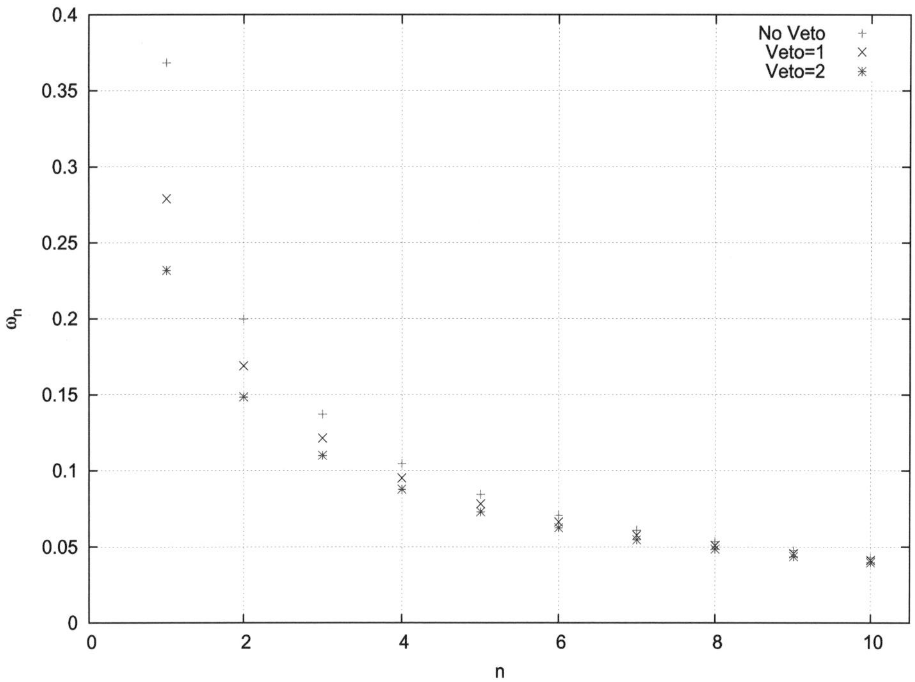

Fig. 18.8 Effect of introducing a rapidity veto in the effective eigenvalues with $b = 1$ (black crosses) and $b = 2$ (blue stars).

18.6.2 *Inclusive dijet hadroproduction*

In [Caporale *et al.* (2018)] we studied ratios of azimuthal angle distributions in MN jets with a rapidity veto constraint. As we have seen, the asymptotic growth with Y considering only the Green function needs a veto of order 2 to generate positive cross sections. From a phenomenological point of view, we found [Caporale *et al.* (2018)] the values of the veto which allow for a good fit of angular distributions at the LHC in a realistic (non asymptotic) set-up.

As we have mentioned, in collisions of two hadrons MN jets have a final state with two tagged jets well separated in rapidity:

$$p\,(p_A) + p\,(p_B) \to J_A\,(k_A) + J_B\,(k_B) + X\,. \tag{18.12}$$

The kinematical configuration is $s \equiv (p_A + p_B)^2 \gg Q^2 \sim \vec{k}_A^2 \sim \vec{k}_B^2 \gg \Lambda_{QCD}^2$, where Q is a hard transverse scale. In Sudakov variables,

$$k_A = x_{J_A} p_A + \frac{\vec{k}_A^2}{x_{J_A} s} p_B + k_{A,\perp}, \quad k_B = x_{J_B} p_B + \frac{\vec{k}_B^2}{x_{J_B} s} p_A + k_{B,\perp},$$

$$k_{A,B,\perp}^2 = -\vec{k}_{A,B}^2, \tag{18.13}$$

with $x_{J_{A,B}}$ being the longitudinal momentum fractions of the jets. The rapidities $y_{A,B}$ of the two tagged jets are

$$y_A = \frac{1}{2}\log\left(\frac{x_{J_A}^2 s}{\vec{k}_A^2}\right), \quad y_B = -\frac{1}{2}\log\left(\frac{x_{J_B}^2 s}{\vec{k}_B^2}\right),$$

$$Y \equiv y_A - y_B = \log\frac{x_{J_A} x_{J_B} s}{|\vec{k}_A||\vec{k}_B|}\,. \tag{18.14}$$

Being a semi-hard process, we need to combine both collinear factorization and BFKL dynamics. We write the cross section as a convolution of the parton distribution functions (PDFs) $f_i\,(x, \mu_F)$ and the partonic cross section $\hat{\sigma}$,

$$\frac{\mathrm{d}\sigma\,(s)}{\mathrm{d}y_A\,\mathrm{d}y_B\,\mathrm{d}^2\vec{k}_A\,\mathrm{d}^2\vec{k}_B} = \sum_{i,j}\int_0^1 f_i\,(x_A, \mu_F)\, f_j\,(x_B, \mu_F)\,\frac{\mathrm{d}\hat{\sigma}_{ij}\,(x_A x_B s, \mu_F)}{\mathrm{d}y_A\,\mathrm{d}y_B\,\mathrm{d}^2\vec{k}_A\,\mathrm{d}^2\vec{k}_B}. \tag{18.15}$$

i, j identify the partons ($i, j = q, \bar{q}, g$), μ_F is the factorization scale and $x_{A,B}$ is the longitudinal momentum fractions of the partons. The partonic cross-section factorizes into jet vertices V and the NLL gluon Green function φ:

$$\begin{aligned}\frac{\mathrm{d}\hat{\sigma}_{ij}\,(x_A x_B s, \mu_F)}{\mathrm{d}y_A\,\mathrm{d}y_B\,\mathrm{d}^2\vec{k}_A\,\mathrm{d}^2\vec{k}_B} &= \frac{x_{J_A} x_{J_B}}{(2\pi)^2}\int\frac{\mathrm{d}^2\vec{q}_A}{\vec{q}_A^{\,2}} V_i\left(\vec{q}_A, x_A, s_0, \vec{k}_A, x_{J_A}, \mu_F, \mu_R\right)\\ &\times\int\frac{\mathrm{d}^2\vec{q}_B}{\vec{q}_B^{\,2}} V_j\left(-\vec{q}_B, x_B, s_0, \vec{k}_B, x_{J_B}, \mu_F, \mu_R\right)\\ &\times\int_C\frac{\mathrm{d}\omega}{2\pi i}\left(\frac{x_A x_B s}{s_0}\right)^{\omega}\varphi_\omega\,(\vec{q}_A, \vec{q}_B).\end{aligned} \tag{18.16}$$

All poles in ω are to the left of the integration contour. An important piece of the computation is the impact factor at NLO, calculated in [Bartels *et al.* (2002, 2003); Caporale *et al.* (2012)] and later in [Hentschinski and Sabio Vera (2012); Chachamis *et al.* (2012, 2014b, 2013c)].

The veto modifies the BFKL kernel only beyond LL accuracy. To introduce some energy-momentum conservation, we expand the Green function in a truncated power series in the coupling to limit the total number of emissions.

We study ratios of azimuthal angle decorrelation [Sabio Vera and Schwennsen (2007)] and compare them to CMS [CMS-Collaboration (2013); Khachatryan *et al.* (2016)] and ATLAS [Aad *et al.* (2011)] data. The CMS cuts are (for $\sqrt{s} = 7$ TeV)

$$\begin{aligned} &35\,\text{GeV} \leq |\vec{k}_A|, |\vec{k}_B| \leq 60\,\text{GeV}, \quad \Big||\vec{k}_A| - |\vec{k}_B|\Big| \geq 2\,\text{GeV}, \\ &0 \leq y_A, |y_B| \leq 4.7\,. \end{aligned} \tag{18.17}$$

We have calculated the quantities

$$\begin{aligned} C_n(Y,b) = \int \mathrm{d}^2\vec{k}_A\,\mathrm{d}^2\vec{k}_B\,\mathrm{d}y_A\,\mathrm{d}y_B\,\delta(y_A - y_B - Y)\cos(n\theta) \\ \times \frac{\mathrm{d}\sigma}{\mathrm{d}y_A\,\mathrm{d}y_B\,\mathrm{d}^2\vec{k}_A\,\mathrm{d}^2\vec{k}_B} \end{aligned} \tag{18.18}$$

and the corresponding ratios $\mathcal{R}^n_m(Y,b) \equiv \frac{C_n(Y,b)}{C_m(Y,b)}$. In Fig. 18.9 the values of the rapidity veto that best fit the LHC data for the correlation functions (left) and the ratios of correlation functions (right) are shown in the $3.25 \leq Y \leq 6.5$ region. The optimal value of the veto grows monotonically with Y. We have a good understanding of the ratios $\mathcal{R}^m_n$ including $m = 0$ or $n = 0$ without having to use the BLM scheme. We conclude that a rough cut-off in the rapidity differences among emitted mini-jets in the final state, $b \gtrsim 1$, is enough to obtain a reasonable global description of many different azimuthal angle correlations in dijet cross sections for current LHC data. This value is far from asymptotic estimates. Introducing jet vertices and parton distribution functions at realistic energies reduces it from $b \gtrsim 2$ to half of this value.

18.7 Forward Drell–Yan production

Forward Drell–Yan (f-DY) production at the LHC was studied in [Brzeminski *et al.* (2017)]. In [Celiberto *et al.* (2018)] we improved their results by using the unintegrated gluon density of [Hentschinski *et al.* (2013b,a)], calculated at NLL with collinear corrections. We found a good description of the f-DY cross section dependence on the invariant mass of the lepton pair both for LHCb and ATLAS data. This has been proposed [Drell and Yan (1970)] as an interesting test of the unintegrated gluon density at small x [Anderson (2010)].

We showed that it is possible to describe HERA F_2 and F_L together with LHC DY data using the NLL BFKL formalism. Our results should be compared

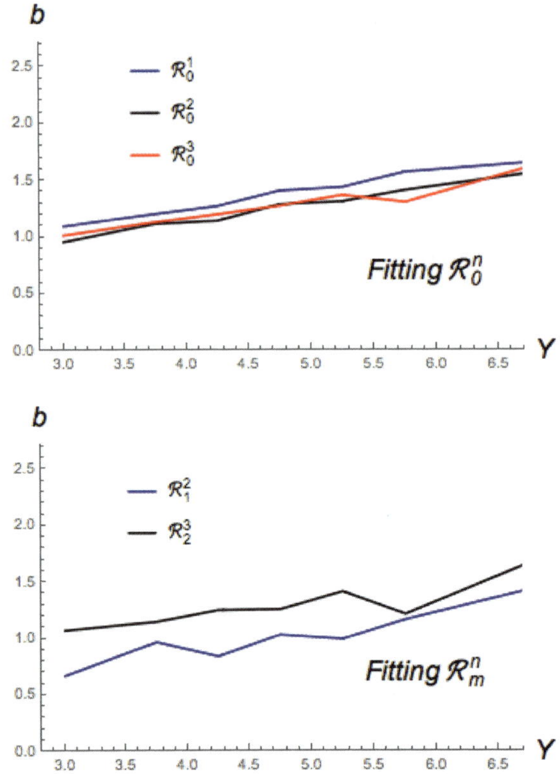

Fig. 18.9 Best-fit values for the rapidity veto b as a function of Y for the $\mathcal{R}_m^n$ ratios.

to [Golec-Biernat *et al.* (2010); Ducati *et al.* (2014); Schaefer and Szczurek (2016); Duclou (2017)]. We studied the production of a lepton-antilepton pair, L^+L^-, in proton-proton collisions:

$$p(P_1) + p(P_2) \to L^+(l^+) + L^-(l^-) + X. \tag{18.19}$$

At LO this is mediated by a virtual photon $\gamma^*(q)$ or a $Z^0(q)$ boson, where the vector boson momentum $q = l^+ + l^-$ carries a q_T transverse component. $M^2 \equiv q^2 > 0$ is the lepton pair invariant mass squared.

Angular distributions of the lepton pair are written in terms of four structure functions, $\mathcal{W}_{[\lambda]}$. The DY formula [Lam and Tung (1978)] factorizes in terms of the lepton pair angular phase space $\mathrm{d}\vartheta^*\,\mathrm{d}\varphi^* \equiv \mathrm{d}\Omega_l^*$ (ϑ^* and φ^* are the polar and azimuthal angles of the lepton momentum vector in the dilepton center-of-mass frame) and the structure functions. The Z^0 contribution can be neglected and the differential cross section is

$$\frac{\mathrm{d}\sigma}{\mathrm{d}M\,\mathrm{d}\Omega_l^*\,\mathrm{d}x_F\,\mathrm{d}q_T} = \frac{\alpha^2\,q_T}{(2\pi\,M)^3}\Big[\left(1-\cos^2\vartheta^*\right)\mathcal{W}_L + \left(1+\cos^2\vartheta^*\right)\mathcal{W}_T + \left(\sin 2\vartheta^*\cos\varphi^*\right)\mathcal{W}_\Delta + \left(\sin^2\vartheta^*\cos 2\varphi^*\right)\mathcal{W}_{\Delta\Delta}\Big]. \tag{18.20}$$

x_F is the longitudinal momentum fraction from the initial-state hadron carried by γ^*, α the electromagnetic coupling, $\mathcal{W}_L$ and $\mathcal{W}_T$ the structure functions for longitudinally and transversely polarized γ^*, $\mathcal{W}_\Delta$ the single-spin-flip structure function, and $\mathcal{W}_{\Delta\Delta}$ the double-spin-flip one. We choose the frame orientation of Gottfried–Jackson [Gottfried and Jackson (1964)]. In the forward region the structure functions can be written in the form

$$\mathcal{W}_{[\lambda]} = \frac{2\pi}{3}\alpha_s(\mu_R)M^2 \int_{x_F}^{1} \frac{\mathrm{d}z}{z^2} f^*\left(\frac{x_F}{z}, \mu_F\right) \int \frac{\mathrm{d}\kappa_T \mathrm{d}\phi_{\kappa_T}}{(\kappa_T^2)^2}\, \mathcal{G}(x_g, \kappa_T^2)\, \Phi_{[\lambda]}(q_T, \vec{\kappa}_T, z). \tag{18.21}$$

z is the longitudinal momentum fraction of the initial-state quark carried by the virtual photon, $\kappa_T \equiv |\vec{\kappa_T}|$ and $q_T \equiv |\vec{q_T}|$. The collinear parton distribution, $f^*(x, \mu_F) = \sum_r f_r(x, \mu_F)$ accounts for incident u, d, s, c and b quarks and antiquarks.

The unintegrated gluon distribution, $\mathcal{G}(x_g, \kappa_T^2)$, generates the small-$x$ gluon evolution and the f-DY impact factors, $\Phi_{[\lambda]}(q_T, \vec{\kappa}_T, z)$, introduce the $\gamma^* \to L^+L^-$ transition. x_g is the gluon longitudinal momentum fraction.

We took the model for the coupling of the Green function to the proton derived in [Hentschinski *et al.* (2013b,a)], also been used in single-bottom quark production [Chachamis *et al.* (2015)], J/Ψ and Υ photoprotoduction [Bautista *et al.* (2016)] and ρ-meson leptoproduction at HERA [Bolognino *et al.* (2018)]. This Green function includes the collinear corrections resummed in the form of a J_1-Bessel function in [Sabio Vera (2005)].

The dependence of the cross-section on the dilepton invariant mass M is

$$\frac{\mathrm{d}\sigma(M)}{\mathrm{d}M} = \int \mathrm{d}\Omega_l^* \int \mathrm{d}x_F \int \mathrm{d}q_T \, \frac{\mathrm{d}\sigma}{\mathrm{d}M\, \mathrm{d}\Omega_l^*\, \mathrm{d}x_F\, \mathrm{d}q_T}\,. \tag{18.22}$$

In terms of $l_T^\pm \equiv |\vec{l}_T^\pm| = (l^\pm)^2 - (l^{z,\pm})^2$, $\eta^\pm = \operatorname{arctanh}\frac{l^{z,\pm}}{l^\pm}$, the LHCb kinematical cuts [LHCb-Collaboration (2012)] (with 5.5 GeV $< M <$ 120 GeV) are

$$2 < \eta^\pm < 4.5, \quad l^\pm > 10\,\text{GeV}, \quad \begin{cases} l_T^\pm > 3\,\text{GeV} & \text{if} \quad M \leq 40\,\text{GeV} \\ l_T^\pm > 15\,\text{GeV} & \text{if} \quad M > 40\,\text{GeV} \end{cases}. \tag{18.23}$$

In [Celiberto *et al.* (2018)] predictions are also given for the ATLAS experimental cuts [Piccaro (2012-08-21); Aad *et al.* (2014)].

Figure 18.10 shows that our calculations reproduce well the LHCb data with the same set of parameters that also fit the HERA data. This is a positive result for the BFKL approach since this global description is a mandatory feature of the framework. The same DY data can, however, be also described by fixed order calculations [Aad *et al.* (2014)] or studies which include saturation corrections [Brzeminski *et al.* (2017)]. In order to disentangle different approaches, future LHC data for f-DY pair production [Akiba *et al.* (2016)] will be very important.

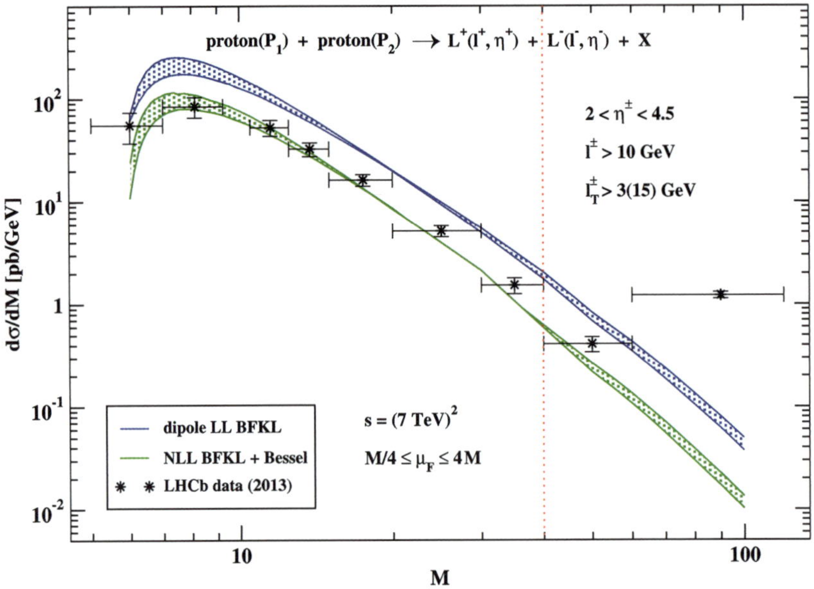

Fig. 18.10 DY cross section as a function of dilepton invariant mass M. Uncertainty bands account for factorization scale shifts $M/4 \leq \mu_F \leq 4M$.

The research here presented has been supported by the Spanish Research Agency (Agencia Estatal de Investigación) through the grant IFT Centro de Excelencia Severo Ochoa SEV-2016-0597, and the Spanish Government grants FPA2015-65480-P, FPA2016-78022-P.

Bibliography

Aad, G. *et al.* (2011). Measurement of dijet production with a veto on additional central jet activity in *pp* collisions at $\sqrt{s} = 7\,\text{TeV}$ using the ATLAS detector, *JHEP* **09**, p. 053, doi:10.1007/JHEP09(2011)053, arXiv:1107.1641 [hep-ex].

Aad, G. *et al.* (2014). Measurement of the low-mass Drell-Yan differential cross section at $\sqrt{s} = 7\,\text{TeV}$ using the ATLAS detector, *JHEP* **06**, p. 112, doi:10.1007/JHEP06(2014)112, arXiv:1404.1212 [hep-ex].

Akiba, K. *et al.* (2016). LHC forward physics, *J. Phys. G* **43**, p. 110201, doi:10.1088/0954-3899/43/11/110201, arXiv:1611.05079 [hep-ph].

Andersen, J. R. and Sabio Vera, A. (2003). Solving the BFKL equation in the next-to-leading approximation, *Phys. Lett. B* **567**, pp. 116–124, doi:10.1016/S0370-2693(03)00871-2, arXiv:hep-ph/0305236 [hep-ph].

Andersen, J. R. and Sabio Vera, A. (2004). The Gluon Green's function in the BFKL approach at next-to-leading logarithmic accuracy, *Nucl. Phys. B* **679**, pp. 345–362, doi:10.1016/j.nuclphysb.2003.11.034, arXiv:hep-ph/0309331 [hep-ph].

Anderson, J. (2010). Constraints on low-x PDFs from Drell Yan processes, and first studies of exclusive dimuon production with the LHCb experiment, in *Proceedings, 40th International Symposium on Multiparticle Dynamics (ISMD 2010): Antwerp, Belgium*, September 21–25, 2010, pp. 227–232, doi:10.5689/UA-PROC-2010-09/28.

Angioni, M., Chachamis, G., Madrigal, J. D., and Sabio Vera, A. (2011). Dijet production at large rapidity separation in $N = 4$ SYM, *Phys. Rev. Lett.* **107**, p. 191601, doi:10.1103/PhysRevLett.107.191601, arXiv:1106.6172 [hep-th].

Bajnok, Z., Janik, R. A., and Lukowski, T. (2009). Four loop twist two, BFKL, wrapping and strings, *Nucl. Phys. B* **816**, pp. 376–398, doi:10.1016/j.nuclphysb.2009.02.005, arXiv:0811.4448 [hep-th].

Balitsky, I. I. and Lipatov, L. N. (1978). The Pomeranchuk singularity in quantum chromodynamics, *Sov. J. Nucl. Phys.* **28**, pp. 822–829, [Yad. Fiz.28,1597(1978)].

Bartels, J., Colferai, D., and Vacca, G. P. (2002). The NLO jet vertex for Mueller-Navelet and forward jets: The Quark part, *Eur. Phys. J. C* **24**, pp. 83–99, doi:10.1007/s100520200919, arXiv:hep-ph/0112283 [hep-ph].

Bartels, J., Colferai, D., and Vacca, G. P. (2003). The NLO jet vertex for Mueller-Navelet and forward jets: The Gluon part, *Eur. Phys. J. C* **29**, pp. 235–249, doi:10.1140/epjc/s2003-01169-5, arXiv:hep-ph/0206290 [hep-ph].

Bautista, I., Fernandez Tellez, A., and Hentschinski, M. (2016). BFKL evolution and the growth with energy of exclusive J/ψ and Υ photoproduction cross sections, *Phys. Rev. D* **94**, 5, p. 054002, doi:10.1103/PhysRevD.94.054002, arXiv:1607.05203 [hep-ph].

Bolognino, A. D., Celiberto, F. G., Ivanov, D. Yu., and Papa, A. (2018). Unintegrated gluon distribution from forward polarized ρ-electroproduction, *Eur. Phys. J. C* **78**, 12, p. 1023, doi:10.1140/epjc/s10052-018-6493-6, arXiv:1808.02395 [hep-ph].

Brzeminski, D., Motyka, L., Sadzikowski, M., and Stebel, T. (2017). Twist decomposition of Drell-Yan structure functions: Phenomenological implications, *JHEP* **01**, p. 005, doi:10.1007/JHEP01(2017)005, arXiv:1611.04449 [hep-ph].

Caporale, F., Celiberto, F. G., Chachamis, G., Gordo Gmez, D., and Sabio Vera, A. (2016a). BFKL azimuthal imprints in inclusive three-jet production at 7 and 13 TeV, *Nucl. Phys. B* **910**, pp. 374–386, doi:10.1016/j.nuclphysb.2016.07.012, arXiv:1603.07785 [hep-ph].

Caporale, F., Celiberto, F. G., Chachamis, G., Gordo Gmez, D., and Sabio Vera, A. (2017). Inclusive four-jet production at 7 and 13 TeV: Azimuthal profile in multi-Regge kinematics, *Eur. Phys. J. C* **77**, 1, p. 5, doi:10.1140/epjc/s10052-016-4557-z, arXiv:1606.00574 [hep-ph].

Caporale, F., Celiberto, F. G., Chachamis, G., Gordo Gmez, D., and Sabio Vera, A. (2018). Inclusive dijet hadroproduction with a rapidity veto constraint, *Nucl. Phys. B* **935**, pp. 412–434, doi:10.1016/j.nuclphysb.2018.09.002, arXiv:1806.06309 [hep-ph].

Caporale, F., Celiberto, F. G., Chachamis, G., and Sabio Vera, A. (2016b). Multi-Regge kinematics and azimuthal angle observables for inclusive four-jet production, *Eur. Phys. J. C* **76**, 3, p. 165, doi:10.1140/epjc/s10052-016-3963-6, arXiv:1512.03364 [hep-ph].

Caporale, F., Chachamis, G., Murdaca, B., and Sabio Vera, A. (2016c). Balitsky-Fadin-Kuraev-Lipatov predictions for inclusive three jet production at the LHC, *Phys. Rev. Lett.* **116**, 1, p. 012001, doi:10.1103/PhysRevLett.116.012001, arXiv:1508.07711 [hep-ph].

Caporale, F., Ivanov, D. Yu., Murdaca, B., Papa, A., and Perri, A. (2012). The next-to-leading order jet vertex for Mueller-Navelet and forward jets revisited, *JHEP* **02**, p. 101, doi:10.1007/JHEP02(2012)101, arXiv:1112.3752 [hep-ph].

Caporale, F., Murdaca, B., Sabio Vera, A., and Salas, C. (2013). Scale choice and collinear contributions to Mueller-Navelet jets at LHC energies, *Nucl. Phys. B* **875**, pp. 134–151, doi:10.1016/j.nuclphysb.2013.07.005, arXiv:1305.4620 [hep-ph].

Catani, S., Fiorani, F., and Marchesini, G. (1990). QCD coherence in initial state radiation, *Phys. Lett. B* **234**, pp. 339–345, doi:10.1016/0370-2693(90)91938-8.

Catani, S., Fiorani, F., Marchesini, G., and Oriani, G. (1991). All loops calculation of the structure function for x —¿ 0 in perturbative QCD, *Nucl. Phys. B* **361**, pp. 645–672, doi:10.1016/0550-3213(91)90598-R.

Celiberto, F. G., Gordo Gmez, D., and Sabio Vera, A. (2018). Forward DrellYan production at the LHC in the BFKL formalism with collinear corrections, *Phys. Lett. B* **786**, pp. 201–206, doi:10.1016/j.physletb.2018.09.045, arXiv:1808.09511 [hep-ph].

Chachamis, G., Deak, M., Hentschinski, M., Rodrigo, G., and Sabio Vera, A. (2015). Single bottom quark production in kT-factorisation, *JHEP* **09**, p. 123, doi:10.1007/JHEP09(2015)123, arXiv:1507.05778 [hep-ph].

Chachamis, G., Deak, M., Vera, A. S., and Stephens, P. (2011). A Comparative study of small x Monte Carlos with and without QCD coherence effects, *Nucl. Phys. B* **849**, pp. 28–44, doi:10.1016/j.nuclphysb.2011.03.011, arXiv:1102.1890 [hep-ph].

Chachamis, G., Hentschinski, M., Madrigal, J. D., and Sabio Vera, A. (2013a). Computing the full two-loop gluon Regge trajectory within Lipatovs high energy effective action, *PoS* **DIS2013**, p. 062, doi:10.22323/1.191.0062, arXiv:1307.7741 [hep-ph].

Chachamis, G., Hentschinski, M., Madrigal, J. D., and Sabio Vera, A. (2014a). High energy factorization at NLO: Lipatov's effective action revisited, *Int. J. Mod. Phys. Conf. Ser.* **25**, p. 1460027, doi:10.1142/S2010194514600271, arXiv:1308.0293 [hep-ph].

Chachamis, G., Hentschinski, M., Madrigal Martinez, J. D., and Sabio Vera, A. (2012). Quark contribution to the gluon Regge trajectory at NLO from the high energy effective action, *Nucl. Phys. B* **861**, pp. 133–144, doi:10.1016/j.nuclphysb.2012.03.015, arXiv:1202.0649 [hep-ph].

Chachamis, G., Hentschinski, M., Madrigal Martinez, J. D., and Sabio Vera, A. (2013b). Gluon Regge trajectory at two loops from Lipatov's high energy effective action, *Nucl. Phys. B* **876**, pp. 453–472, doi:10.1016/j.nuclphysb.2013.08.013, arXiv:1307.2591 [hep-ph].

Chachamis, G., Hentschinski, M., Madrigal Martnez, J. D., and Sabio Vera, A. (2013c). Next-to-leading order corrections to the gluon-induced forward jet vertex from the high energy effective action, *Phys. Rev. D* **87**, 7, p. 076009, doi:10.1103/PhysRevD.87.076009, arXiv:1212.4992 [hep-ph].

Chachamis, G., Hentschinski, M., Madrigal Martnez, J. D., and Sabio Vera, A. (2014b). Forward jet production & quantum corrections to the gluon Regge trajectory from Lipatov‘s high energy effective action, *Phys. Part. Nucl.* **45**, 4, pp. 788–799, doi:10.1134/S1063779614040030, arXiv:1211.2050 [hep-ph].

Chachamis, G., Lublinsky, M., and Sabio Vera, A. (2005). Higher order effects in non linear evolution from a veto in rapidities, *Nucl. Phys. A* **748**, pp. 649–663, doi:10.1016/j.nuclphysa.2004.11.011, arXiv:hep-ph/0408333 [hep-ph].

Chachamis, G. and Sabio Vera, A. (2013). Monte Carlo techniques in small-x physics: Formal studies and phenomenology, *PoS* **DIS2013**, p. 167, doi:10.22323/1.191.0167, arXiv:1307.7750 [hep-ph].

Chachamis, G. and Sabio Vera, A. (2016a). Monte Carlo study of double logarithms in the small x region, *Phys. Rev. D* **93**, 7, p. 074004, doi:10.1103/PhysRevD.93.074004, arXiv:1511.03548 [hep-ph].

Chachamis, G. and Sabio Vera, A. (2016b). The high-energy radiation pattern from BFKLex with double-log collinear contributions, *JHEP* **02**, p. 064, doi:10.1007/JHEP02(2016)064, arXiv:1512.03603 [hep-ph].

Ciafaloni, M. (1988). Coherence effects in initial jets at small q**2 / s, *Nucl. Phys. B* **296**, pp. 49–74, doi:10.1016/0550-3213(88)90380-X.

Ciafaloni, M. and Camici, G. (1998). Energy scale(s) and next-to-leading BFKL equation, *Phys. Lett. B* **430**, pp. 349–354, doi:10.1016/S0370-2693(98)00551-6, arXiv:hep-ph/9803389 [hep-ph].

Ciafaloni, M., Colferai, D., Colferai, D., Salam, G. P., and Stasto, A. M. (2003). Extending QCD perturbation theory to higher energies, *Phys. Lett. B* **576**, pp. 143–151, doi: 10.1016/j.physletb.2003.09.078, arXiv:hep-ph/0305254 [hep-ph].

Ciesielski, R. (2014). Diffraction, forward physics and soft QCD results from CMS, in *Proceedings, 2nd Conference on Large Hadron Collider Physics Conference (LHCP 2014): New York, USA, June 2–7, 2014*, arXiv:1409.5473 [hep-ex], http://www.slac.stanford.edu/econf/C140602.2/papers/1409.5473v1.pdf.

CMS-Collaboration (2013). Azimuthal angle decorrelations of jets widely separated in rapidity in pp collisions at sqrts = 7 TeV, *CMS-PAS-FSQ-12-002*.

Colferai, D., Schwennsen, F., Szymanowski, L., and Wallon, S. (2010). Mueller Navelet jets at LHC — complete NLL BFKL calculation, *JHEP* **12**, p. 026, doi:10.1007/JHEP12(2010)026, arXiv:1002.1365 [hep-ph].

Del Duca, V. and Schmidt, C. R. (1994). Dijet production at large rapidity intervals, *Phys. Rev. D* **49**, pp. 4510–4516, doi:10.1103/PhysRevD.49.4510, arXiv:hep-ph/9311290 [hep-ph].

Drell, S. D. and Yan, T.-M. (1970). Massive lepton pair production in hadron-hadron collisions at high-energies, *Phys. Rev. Lett.* **25**, pp. 316–320, doi:10.1103/PhysRevLett.25.316,10.1103/PhysRevLett.25.902.2, [Erratum: Phys. Rev. Lett.25,902(1970)].

Ducati, M. B. G., Griep, M. T., and Machado, M. V. T. (2014). Study on the low mass Drell-Yan production at the CERN LHC within the dipole formalism, *Phys. Rev. D* **89**, 3, p. 034022, doi:10.1103/PhysRevD.89.034022, arXiv:1307.6882 [hep-ph].

Duclou, B. (2017). Nuclear modification of forward Drell-Yan production at the LHC, *Phys. Rev. D* **96**, 9, p. 094014, doi:10.1103/PhysRevD.96.094014, arXiv:1701.08730 [hep-ph].

Duclou, B., Szymanowski, L., and Wallon, S. (2014). Evidence for high-energy resummation effects in Mueller-Navelet jets at the LHC, *Phys. Rev. Lett.* **112**, p. 082003, doi:10.1103/PhysRevLett.112.082003, arXiv:1309.3229 [hep-ph].

Fadin, V. S. and Fiore, R. (1998). The Generalized nonforward BFKL equation and the 'bootstrap' condition for the gluon Reggeization in the NLLA, *Phys. Lett. B* **440**, pp. 359–366, doi:10.1016/S0370-2693(98)01099-5, arXiv:hep-ph/9807472 [hep-ph].

Fadin, V. S., Kuraev, E. A., and Lipatov, L. N. (1975). On the Pomeranchuk singularity in asymptotically free theories, *Phys. Lett.* **60B**, pp. 50–52, doi:10.1016/0370-2693(75)90524-9.

Fadin, V. S. and Lipatov, L. N. (1998). BFKL pomeron in the next-to-leading approximation, *Phys. Lett. B* **429**, pp. 127–134, doi:10.1016/S0370-2693(98)00473-0, arXiv:hep-ph/9802290 [hep-ph].

Forshaw, J. R., Ross, D. A., and Sabio Vera, A. (1999). Rapidity veto effects in the NLO BFKL equation, *Phys. Lett. B* **455**, pp. 273–282, doi:10.1016/S0370-2693(99)00451-7, arXiv:hep-ph/9903390 [hep-ph].

Golec-Biernat, K., Lewandowska, E., and Stasto, A. M. (2010). Drell-Yan process at forward rapidity at the LHC, *Phys. Rev. D* **82**, p. 094010, doi:10.1103/PhysRevD.82.094010, arXiv:1008.2652 [hep-ph].

Gonzo, R., Mc Loughlin, T., Medrano, D., and Spiering, A. (2019). Asymptotic charges and coherent states in QCD, arXiv:1906.11763 [hep-th].

Gottfried, K. and Jackson, J. D. (1964). On the connection between production mechanism and decay of resonances at high-energies, *Nuovo Cim.* **33**, pp. 309–330, doi:10.1007/BF02750195.

Hentschinski, M., Madrigal Martnez, J. D., Murdaca, B., and Sabio Vera, A. (2014a). The next-to-leading order vertex for a forward jet plus a rapidity gap at high energies, *Phys. Lett. B* **735**, pp. 168–172, doi:10.1016/j.physletb.2014.06.022, arXiv:1404.2937 [hep-ph].

Hentschinski, M., Madrigal Martnez, J. D., Murdaca, B., and Sabio Vera, A. (2014b). The quark induced MuellerTang jet impact factor at next-to-leading order, *Nucl. Phys. B* **887**, pp. 309–337, doi:10.1016/j.nuclphysb.2014.08.010, arXiv:1406.5625 [hep-ph].

Hentschinski, M., Martnez, J. D. M., Murdaca, B., and Sabio Vera, A. (2014c). The gluon-induced MuellerTang jet impact factor at next-to-leading order, *Nucl. Phys. B* **889**, pp. 549–579, doi:10.1016/j.nuclphysb.2014.10.026, arXiv:1409.6704 [hep-ph].

Hentschinski, M. and Sabio Vera, A. (2012). NLO jet vertex from Lipatov's QCD effective action, *Phys. Rev. D* **85**, p. 056006, doi:10.1103/PhysRevD.85.056006, arXiv:1110.6741 [hep-ph].

Hentschinski, M., Sabio Vera, A., and Salas, C. (2013a). F_2 and F_L at small x using a collinearly improved BFKL resummation, *Phys. Rev. D* **87**, 7, p. 076005, doi: 10.1103/PhysRevD.87.076005, arXiv:1301.5283 [hep-ph].

Hentschinski, M., Sabio Vera, A., and Salas, C. (2013b). Hard to soft Pomeron transition in small-x deep inelastic scattering data using optimal renormalization, *Phys. Rev. Lett.* **110**, 4, p. 041601, doi:10.1103/PhysRevLett.110.041601, arXiv:1209.1353 [hep-ph].

Khachatryan, V. *et al.* (2016). Azimuthal decorrelation of jets widely separated in rapidity in pp collisions at $\sqrt{s} = 7\,\text{TeV}$, *JHEP* **08**, p. 139, doi:10.1007/JHEP08(2016)139, arXiv:1601.06713 [hep-ex].

Kowalski, H., Lipatov, L. N., and Ross, D. A. (2016). The behaviour of the green function for the BFKL Pomeron with running coupling, *Eur. Phys. J. C* **76**, 1, p. 23, doi: 10.1140/epjc/s10052-015-3865-z, arXiv:1508.05744 [hep-ph].

Kowalski, H., Lipatov, L. N., Ross, D. A., and Watt, G. (2010). Using HERA data to determine the infrared behaviour of the BFKL amplitude, *Eur. Phys. J. C* **70**, pp. 983–998, doi:10.1140/epjc/s10052-010-1500-6, arXiv:1005.0355 [hep-ph].

Kuraev, E. A., Lipatov, L. N., and Fadin, V. S. (1976). Multi-Reggeon processes in the Yang-Mills theory, *Sov. Phys. JETP* **44**, pp. 443–450, [Zh. Eksp. Teor. Fiz.71,840(1976)].

Kuraev, E. A., Lipatov, L. N., and Fadin, V. S. (1977). The Pomeranchuk singularity in nonabelian gauge theories, *Sov. Phys. JETP* **45**, pp. 199–204, [Zh. Eksp. Teor. Fiz.72,377(1977)].

Lam, C. S. and Tung, W.-K. (1978). A systematic approach to inclusive lepton pair production in hadronic collisions, *Phys. Rev. D* **18**, p. 2447, doi:10.1103/PhysRevD.18.2447.

LHCb-Collaboration (2012). Inclusive low mass Drell-Yan production in the forward region at $\sqrt{s} = 7\,\text{TeV}$, *LHCb-CONF-2012-013, CERN-LHCb-CONF-2012-013.*

Lipatov, L. N. (1976). Reggeization of the vector meson and the vacuum singularity in non-abelian gauge theories, *Sov. J. Nucl. Phys.* **23**, pp. 338–345, [Yad. Fiz.23,642(1976)].

Lipatov, L. N. (1986). The bare Pomeron in quantum chromodynamics, *Sov. Phys. JETP* **63**, pp. 904–912, [Zh. Eksp. Teor. Fiz.90,1536(1986)].

Lipatov, L. N. (1998). Talk presented at the 4th workshop on small-x and diffractive physics, fermi national accelerator laboratory, Sept. 17–20, 1998.

Marchesini, G. (1995). QCD coherence in the structure function and associated distributions at small x, *Nucl. Phys. B* **445**, pp. 49–80, doi:10.1016/0550-3213(95)00149-M, arXiv:hep-ph/9412327 [hep-ph].

Mueller, A. H. and Navelet, H. (1987). An inclusive minijet cross-section and the bare Pomeron in QCD, *Nucl. Phys. B* **282**, pp. 727–744, doi:10.1016/0550-3213(87)90705-X.

Piccaro, E. (2012-08-21). *Measurement of the Low Mass Drell-Yan Cross Section in the Di-Muon Channel in Proton-Proton Collisions at* $s = 7\,TeV$ *with the ATLAS Detector*, Ph.D. thesis, Queen Mary, U. of London.

Ross, D. A. and Sabio Vera, A. (2016). The effect of the infrared phase of the discrete BFKL Pomeron on transverse momentum diffusion, *JHEP* **08**, p. 071, doi:10.1007/JHEP08(2016)071, arXiv:1605.08265 [hep-ph].

Sabio Vera, A. (2005). An 'All-poles' approximation to collinear resummations in the Regge limit of perturbative QCD, *Nucl. Phys. B* **722**, pp. 65–80, doi:10.1016/j.nuclphysb.2005.06.003, arXiv:hep-ph/0505128 [hep-ph].

Sabio Vera, A. (2006). The effect of NLO conformal spins in azimuthal angle decorrelation of jet pairs, *Nucl. Phys. B* **746**, pp. 1–14, doi:10.1016/j.nuclphysb.2006.04.004, arXiv:hep-ph/0602250 [hep-ph].

Sabio Vera, A. and Schwennsen, F. (2007). The azimuthal decorrelation of jets widely separated in rapidity as a test of the BFKL kernel, *Nucl. Phys. B* **776**, pp. 170–186, doi:10.1016/j.nuclphysb.2007.03.050, arXiv:hep-ph/0702158 [HEP-PH].

Salam, G. P. (1998). A Resummation of large subleading corrections at small x, *JHEP* **07**, p. 019, doi:10.1088/1126-6708/1998/07/019, arXiv:hep-ph/9806482 [hep-ph].

Schaefer, W. and Szczurek, A. (2016). Low mass Drell-Yan production of lepton pairs at forward directions at the LHC: A hybrid approach, *Phys. Rev. D* **93**, 7, p. 074014, doi:10.1103/PhysRevD.93.074014, arXiv:1602.06740 [hep-ph].

Schmidt, C. R. (1997). A Monte Carlo solution to the BFKL equation, *Phys. Rev. Lett.* **78**, pp. 4531–4535, doi:10.1103/PhysRevLett.78.4531, arXiv:hep-ph/9612454 [hep-ph].

Schmidt, C. R. (1999). Rapidity separation dependence and the large next-to-leading corrections to the BFKL equation, *Phys. Rev. D* **60**, p. 074003, doi:10.1103/PhysRevD.60.074003, arXiv:hep-ph/9901397 [hep-ph].

Stirling, W. J. (1994). Production of jet pairs at large relative rapidity in hadron hadron collisions as a probe of the perturbative pomeron, *Nucl. Phys. B* **423**, pp. 56–79, doi:10.1016/0550-3213(94)90565-7, arXiv:hep-ph/9401266 [hep-ph].

https://doi.org/10.1142/9789811231124_0019

Chapter 19

Aspects of BFKL Physics at HERA

H. Jung
DESY, Hamburg

Abstract

Aspects of small x physics at HERA are discussed in honor of the work of Lev Lipatov, who influenced and motivated a whole generation of physicists to investigate this new region of phase space. The introduction of transverse momentum dependent parton densities is one of the essential aspects of small x and BFKL physics, and its extension to all flavors with application to various processes marks a new milestone.

19.1 Introduction

It is a funny story how I became interested in small x and BFKL effects and in the work of Lev Lipatov: I started to work on the H1 calorimeter during my first postdoc at RWTH Aachen from 1988 – 1993, and we were searching for an appropriate physics process for the energy calibration. One idea was using the decay electrons from J/ψ production, and I started to write the Monte Carlo event generator EPJPSI [Jung (1992, 1995a)]. With this tool at hand we investigated possibilities to measure the gluon density at small x, even extrapolating to a hypothetical LEP-LHC *ep* collider at high energies at $\sqrt{s} = 1.26$ TeV [Abraham *et al.* (1990)], where clear deviations from DGLAP [Gribov and Lipatov (1972); Lipatov (1975); Altarelli and Parisi (1977); Dokshitzer (1977)] predictions were expected. These studies raised my interest in small x physics and established very close contacts with my Russian colleagues at HERA, resulting in a collaboration, which is still ongoing [Jung *et al.* (2006); Baranov *et al.* (1999, 2001, 2002a,b, 2003)]. In these early days, I never dreamed that I would have a chance to meet Lev in person. Later when he was often at DESY in Hamburg we got to know each other, and the greatest highlight in my scientific career was when Lev came once to my QCD and MC lecture at DESY in 2006 [Bartels and Jung (2006)]. Small x physics fascinated a whole generation of scientists, resulting in several workshops (for example the HERA workshops [Peccei (1987); Buchmuller and Ingelman (1991); Ingelman *et al.* (1996)], the Lund Small x workshops [Andersen *et al.* (2006, 2004); Andersson *et al.* (2002)] and also the HERA-LHC workshop series [Alekhin *et al.* (2005a,b); Ajaltouni *et al.* (2009)]).

19.2 Calculation of small x processes

In this section, I introduce a derivation of the BFKL evolution equation [Kuraev *et al.* (1976, 1977); Balitsky and Lipatov (1978)] in an heuristic approach, based on the formulation of the DGLAP equation as a parton branching evolution.

The DGLAP evolution equation can be written in a form introducing a Sudakov form factor Δ_a [Ellis *et al.* (1996); Hautmann *et al.* (2017)]:

$$x f_a(x,\mu^2) = \Delta_a(\mu^2)\, x f_a(x,\mu_0^2) + \sum_b \int_{\mu_0^2}^{\mu^2} \frac{d\mu'^2}{\mu'^2} \frac{\Delta_a(\mu^2)}{\Delta_a(\mu'^2)} \times \int_x^{z_M} dz\, P_{ab}(\alpha_s(\mu'^2))\, \frac{x}{z} f_b\left(\frac{x}{z}, \mu'^2\right) \qquad (19.1)$$

with

$$\Delta_a(\mu^2) = \exp\left(- \sum_b \int_{\mu_0^2}^{\mu^2} \frac{d\mu'^2}{\mu'^2} \int_0^{z_M} dz\; z\; P_{ba}(\alpha_s(\mu'^2), z) \right), \qquad (19.2)$$

where P_{ab} is the regularized splitting function for a process $b \to a + c$. The Sudakov form factor gives the probability for non-resolvable parton branching between the scales μ^2 and μ_0^2. The parameter z_M separates resolvable from non-resolvable branchings. The Sudakov form factor resums all contributions from non-resolvable branchings as well as all virtual corrections. It is shown explicitly in

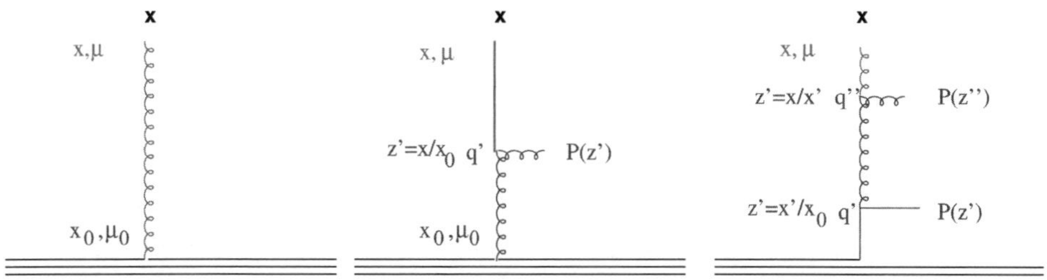

Fig. 19.1 Schematic illustration of the iterative procedure to solve evolution equations: (left) no emission, (middle) one emission, (right) two emissions.

Refs. [Hautmann *et al.* (2017, 2018)], that the formulation of eqs. (19.1) leads exactly to the original DGLAP equation at leading order, next-to-leading and next-to-next-to leading orders in the expansion of the splitting functions, provided z_M is large enough. Away from $x \to 0$, the singularities $1/(1-z)$ in the splitting function for $g \to gg$ and $q \to qg$ are treated with the Sudakov form factor, the resummation of all contributions cancels against the singularity, a feature which we will also use later when discussing BFKL. Equation (19.1) can be solved by iteration, illustrated schematically in Fig. 19.1, leading to an intuitive physical interpretation: the first term of eq. (19.1) gives the probability for no emission between the starting scale μ_0 and the scale μ (left in Fig. 19.1), while the second term describes the branching process (middle and right in Fig. 19.1).

Although the transverse momenta of the emitted- and the propagating-partons are integrated over in DGLAP, for several applications and in particular for the description in terms of parton showers an exclusive formulation (including transverse momenta) of this equation has advantages. The evolution equation can be extended for k_t dependent (unintegrated–uPDFs or Transverse Momentum Dependent–TMD) parton densities $\mathcal{A}(x, \mathbf{k}, t)$ with $\mathbf{k}$ being the two-dimensional (transverse) vector of the propagating parton [Hautmann *et al.* (2018)]:

$$x\mathcal{A}_a(x,\mathbf{k},\mu^2) = \Delta_a(\mu^2)\; x\mathcal{A}_a(x,\mathbf{k},\mu_0^2) + \sum_b \int \frac{d^2\mathbf{q}'}{\pi\mathbf{q}'^2}\, \frac{\Delta_a(\mu^2)}{\Delta_a(\mathbf{q}'^2))} \times \int_x^{z_M} dz\; P_{ab}(\alpha_s)\; \frac{x}{z}\mathcal{A}_b\left(\frac{x}{z}, \mathbf{k} + (1-z)\mathbf{q}', \mathbf{q}'^2\right) \quad . \tag{19.3}$$

In DGLAP we have:

$$\int x\; \mathcal{A}_a(x,\mathbf{k},\mu^2)\; \frac{d^2\mathbf{k}}{\pi} = xf_a(x,\mu^2)\,, \tag{19.4}$$

and the partonic emissions happen in a phase space with increasing scales: $\mu_0 \ll \cdots \ll q' \ll q'' \ll \mu$ for $x_0 \gg \cdots \gg x' \gg x$, as illustrated in Fig. 19.1.

In a formulation, which takes into account color coherence effects, the evolution scale is identified with the angle (angular ordering), while the scale in α_s is the transverse momentum of the emitted parton. The angular ordering condition requires for the rescaled transverse momentum $\tilde{q} = q_t/(1-z)$ that $\tilde{q}'' > z\tilde{q}'$, which for $z \to 0$ leads to very weak constraints [Ciafaloni (1988); Catani *et al.* (1990b,c); Marchesini (1995)].

The BFKL evolution addresses the high energy limit, i.e. $x \to 0$. In this limit, only the gluon contributes and the splitting function becomes $P_{gg} = 6\alpha_s/z$. The longitudinal momenta become small and the transverse momenta of the propagating partons cannot be neglected anymore, leading to the introduction of uPDFs. In analogy to the DGLAP case with the singularity at $1/(1-z)$, the singular behavior at $1/z$ can be regulated by resumming all virtual and non-resolvable branchings, described by the so-called *non-Sudakov* form factor, which includes now only the small x part of P_{gg}:

$$\Delta_{ns} = \exp\left(-\frac{3\alpha_s}{\pi}\int\frac{dq'^2}{q'^2}\int_z^1\frac{dz'}{z'}\Theta(k_t^2-q'^2)\Theta(q'^2-\mu_0^2)\right)$$
$$= \exp\left(-\frac{3\alpha_s}{\pi}\log\frac{1}{z}\log\frac{k_t^2}{\mu_0^2}\right), \qquad (19.5)$$

and the Theta functions limit the available phase space region. While for finite z the transverse momentum of the emitted parton $\mathbf{q}$ is related via $\mu = |\mathbf{q}|/(1-z)$ with the scale μ of the evolution (angular ordering), at small z this reduces to $\mu = |\mathbf{q}|$. The total transverse momentum is calculated via $\mathbf{k} = -\sum_i \mathbf{q}_i$. The angular ordering condition at small z leads to a random walk in k_t, since there is no constraint on q_t coming from $\tilde{q}'' > z\tilde{q}'$ for $z \to 0$.

We can rewrite Δ_{ns} as (using $\bar{\alpha_s} = \frac{3\alpha_s}{\pi}$) :

$$\Delta_{ns} = \exp\left((\log z)^{\bar{\alpha_s}\log\frac{k_t^2}{\mu_0^2}}\right) = z^{\omega},$$

with $\omega = \bar{\alpha_s}\log\frac{k_t^2}{\mu_0^2}$ and obtain the (k_t-dependent) BFKL splitting function as:

$$P_{BFKL} = \frac{6}{z}\Delta_{ns} = z^{-1+\omega}.$$

The BFKL evolution equation in integral form [Kwiecinski *et al.* (1995, 1996a)] can be written as:

$$x\mathcal{A}(x,k_t,q) = x\mathcal{A}_0(x,k_t,q) + \frac{\alpha_s}{2\pi}\int dz\frac{6}{z}\Delta_{ns}\int\frac{dq'^2}{q'^2}\frac{x}{z}\mathcal{A}\left(\frac{x}{z},k_t',q'\right)$$
$$= x\mathcal{A}_0(x,k_t,q) + \bar{\alpha_s}\int\frac{dz}{z}z^{\omega}\int\frac{dq'^2}{q'^2}\frac{x}{z}\mathcal{A}\left(\frac{x}{z},k_t',q'\right). \qquad (19.6)$$

This equation describes parton emissions in the high energy limit, where all recoil effects can be neglected (leading order formulation), and thus is violating (in principle) energy-momentum conservation. Those effects are included as next-to-leading corrections. Recoil effects are also included in the formulation of the CCFM equation, which applies color coherence and includes the $1/(1-z)$ parts of the gluon splitting function, as formulated in Refs. [Ciafaloni (1988); Catani *et al.* (1990b,c); Marchesini (1995); Marchesini and Webber (1991); Webber (1991)]. The CCFM evolution equation reads:

$$x\mathcal{A}(x, \mathbf{k}, \mu^2) = \Delta_a(\mu^2)\; x\mathcal{A}_a(x, \mathbf{k}, \mu_0^2) + \int dz \int \frac{d^2q}{\pi q^2}\Theta(\mu - zq)\frac{\Delta_a(\mu^2)}{\Delta_a(z^2q^2)}\tilde{P}(z, q, k_t)\frac{x}{z}\mathcal{A}_b\left(\frac{x}{z}, \mathbf{k}', \mathbf{q}'^2\right), \tag{19.7}$$

with $\mathbf{k}' = \mathbf{k} + (1 - z)\mathbf{q}'$, and

$$\tilde{P}(z, q, k_t) = \frac{\bar{\alpha_s}(q^2(1 - z^2))}{1 - z} + \frac{\bar{\alpha_s}(k_t^2)}{z}\Delta_{ns}(z, q^2, k_t^2). \tag{19.8}$$

The recently developed Parton Branching (PB) method [Hautmann *et al.* (2017, 2018)] is an application of eq. (19.3) at NLO. Fits to inclusive deep inelastic scattering data over a large range in x and Q^2 leading to the first complete set of TMD distributions at NLO for quarks and gluons are reported in [Bermudez Martinez *et al.* (2019a)].

The parton densities, and also the TMD parton densities are not observables themselves; predictions can be made only after convolution with the appropriate parton cross section. Both BFKL and CCFM gluon densities need to be convoluted with off-shell matrix elements (as required from k_t-factorization [Catani *et al.* (1993, 1991, 1990a)]), while the PB-TMD distributions are convoluted with on-shell ones. In addition, in BFKL and CCFM only gluon densities are calculated, in the PB-TMD calculation the full flavor set of quarks and gluons is determined at NLO. Keeping these differences in mind, it is still of interest to compare the k_t-dependent distributions for gluons (see Fig. 19.2). While both CCFM [Hautmann and Jung (2014)] and PB-TMD [Bermudez Martinez *et al.* (2019a); Hautmann *et al.* (2018)]

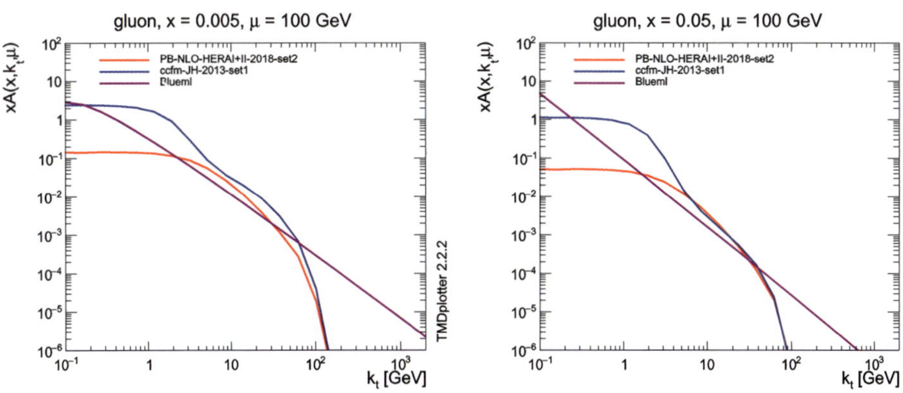

Fig. 19.2 Unintegrated gluon densities obtained in the PB-method [Bermudez Martinez *et al.* (2019a)], CCFM [Hautmann and Jung (2014)] and BKFL [Blumlein (1995)] at different values of x (obtained with TMDlib and TMDplotter [Connor *et al.* (2016); Hautmann *et al.* (2014)]).

distributions fall off quickly at $k_t \sim \mu$, the BFKL distribution [Blumlein (1995)] has a tail to very large values of $k_t > \mu$.

19.3 Measurements with sensitivity to BFKL effects

While parton emissions from the initial state described by DGLAP evolution (i.e. its formulation in terms of exclusive state, like parton showers or with the PB method) are essentially ordered in transverse momentum of the emitted partons (or in its variant with angular ordering), emissions described by BFKL or CCFM evolution at small x can perform a random walk in transverse momentum (only limited by overall kinematic constraints), shown schematically in Fig. 19.3. Partonic emissions in a BFKL evolution will have a different transverse momentum spectrum at a fixed rapidity, compared to those coming from a DGLAP evolution. Differences are expected in particular close to the proton direction, where the transverse momentum coming from a DGLAP evolution is expected to be small.

In the following sections examples of measurements performed at the *ep* collider HERA at DESY are discussed, which probe parton kinematics at small values of x.

19.3.1 *Inclusive cross section measurements*

The inclusive cross section for deep-inelastic scattering was measured already at the start of HERA in 1992 [Abt *et al.* (1993); Derrick *et al.* (1993b)], an example of this very first measurement of $F_2(x, Q^2)$ from ZEUS is shown in Fig. 19.4 (left). The behavior of $F_2(x, Q^2)$ at small values of the momentum fraction x indicates a clear rise, in contrast to some of the predictions.

In Fig. 19.4 (right) the reduced cross section ($\sigma_{red}(x, Q^2) \sim F_2(x, Q^2)$) from a combination of the measurements of H1 and ZEUS after the full HERA running [Abramowicz *et al.* (2015)] is shown, confirming impressively the rise of the cross section for small values of x, but also showing the enormous improvement in

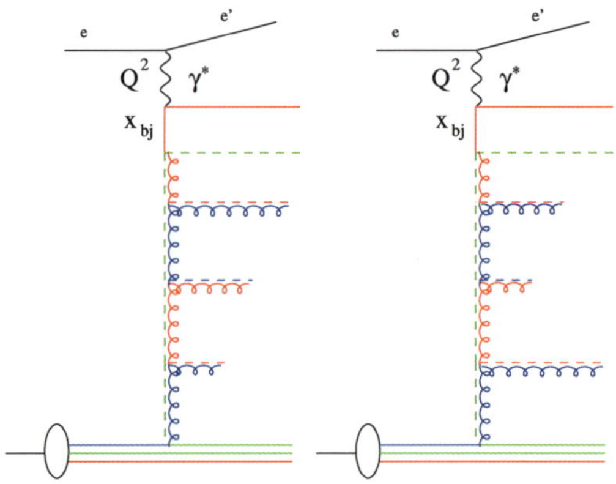

Fig. 19.3 Schematic illustration of a DGLAP-evolution (left) and a BFKL evolution (right). The length of the gluons indicates the size of their transverse momenta.

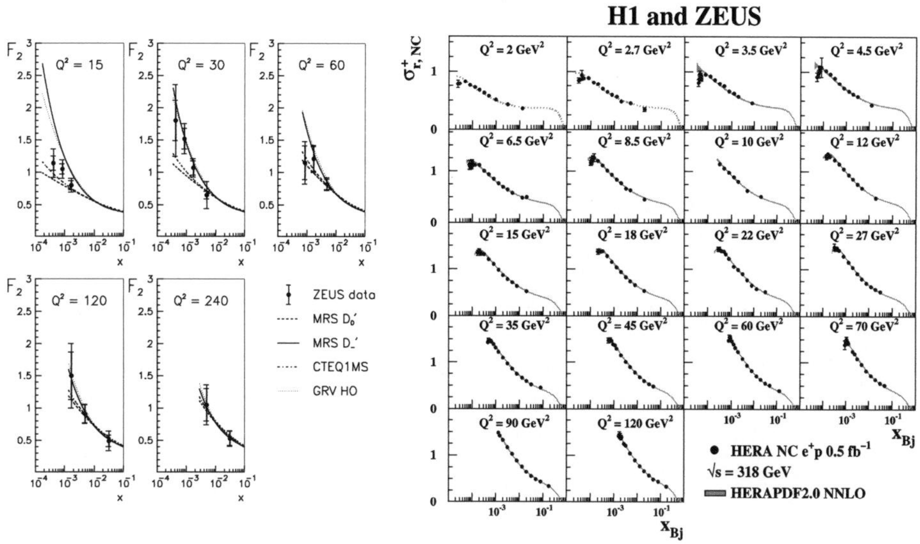

Fig. 19.4 Measurements of deep inelastic scattering cross section: (left) first measurement by of the proton structure function $F_2(x, Q^2)$ [Derrick *et al.* (1993b)], (right) final measurement of the reduced cross section obtained from a combination of the measurements from H1 and ZEUS [Abramowicz *et al.* (2015)]

precision of the measurements from the start of HERA to the final combined results at the end of the HERA running. The rise of F_2 at small x was seen as an indication for BFKL effects (for a discussion see [Bartels (1992); Bartels *et al.* (1991); Bartels (1990); Kwiecinski *et al.* (1991, 1995, 1996b, 1997a,b)]), however, the measurements of the inclusive deep inelastic cross section (σ_{red}) could be well described applying the DGLAP evolution at next-to-leading (NLO) as well as at next-to-next-to-leading order in the expansion of the strong coupling α_s, as shown in Fig. 19.4 (right). The fits performed in [Abramowicz *et al.* (2015)] show a very good χ^2/ndf when measurements with $Q^2 > 3.5\,\text{GeV}^2$ are included.

Recently a dedicated study, including BFKL resummation to the description of the inclusive cross section, proved the need for BFKL effects [Ball *et al.* (2018)]. The effect of small x resummation is shown for the longitudinal structure function $F_L(x, Q^2)$ in Fig. 19.5 (taken from [Ball *et al.* (2018)]). This study showed for the first time, that the measurements can be described better if BFKL effect are included in the calculation.

19.3.2 *Transverse energy flow*

A measurement of the transverse energy flow as a function of the rapidity can give insights in the way transverse momenta of partons along the evolution chain are distributed (as illustrated in Fig. 19.6). In a DGLAP scenario, the transverse momenta of partons close to the proton direction are expected to be small and they increase towards the hard scattering, while in a BFKL scenario the transverse

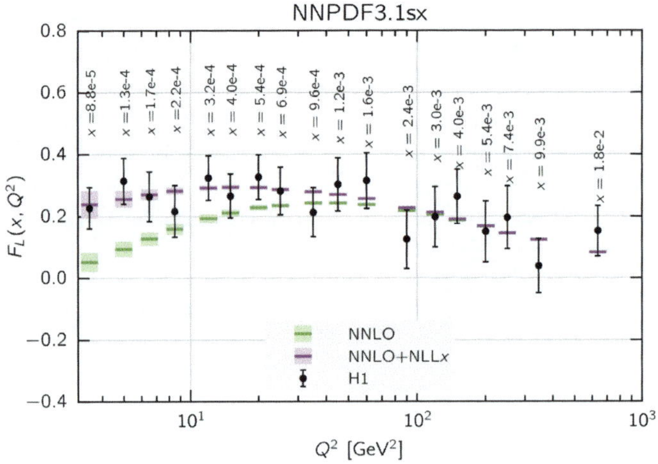

Fig. 19.5 Measurement of the longitudinal structure function $F_L(x, Q^2)$ and comparison with predictions using small x resummation [Ball *et al.* (2018)].

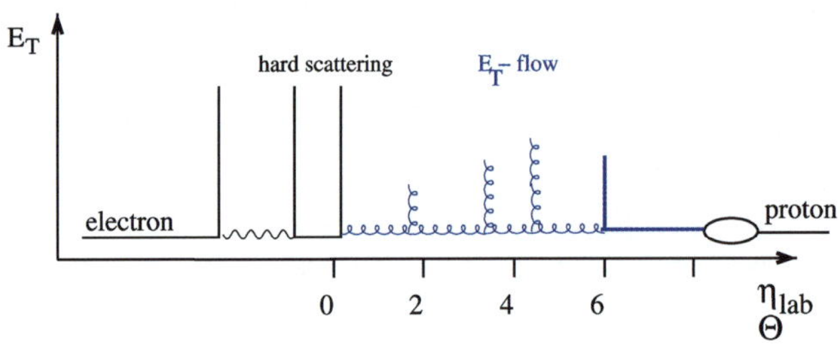

Fig. 19.6 Schematic view of the process contributing to transverse energy flow in deep inelastic scattering.

momenta can be sizable all over the rapidity range, and can form a so-called *Bartels-cigar*[Bartels *et al.* (1996b)], a diffusion pattern of the transverse momenta.

The transverse energy flow in deep inelastic scattering was measured by H1[Abt *et al.* (1994)] and ZEUS[Derrick *et al.* (1993a)]. A significant transverse energy was found away from the hard scattering. An example of one of the first measurements is shown in Fig. 19.7 (right), and compared to predictions based on simulations using parton showers. All predictions[1] fall below the measurements at large rapidities, except the prediction labelled as CDM [Lönnblad (1992)] (which simulates unordered emissions, similar to what is expected from BFKL). These first measurements were also compared to calculations (i.e. in [Golec-Biernat *et al.* (1994)]) obtained with BFKL evolution, which confirmed the increased energy flow in the forward region. Later measurements [Adloff *et al.* (2000)] with much higher

[1]The prediction from HERWIG [Marchesini *et al.* (1992)] had issues with large cluster sizes.

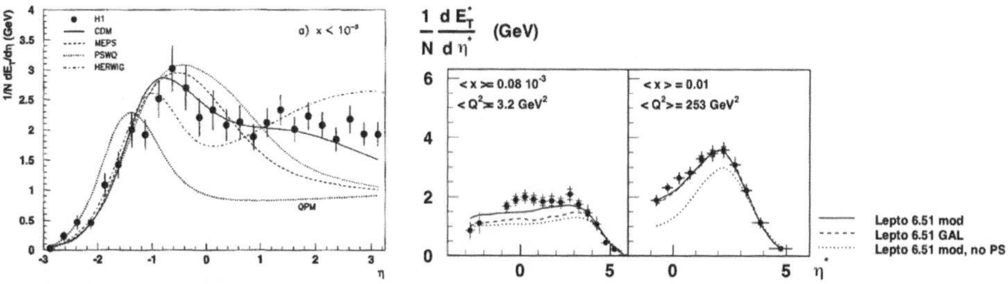

Fig. 19.7 Transverse energy flow in deep inelastic scattering: (left) the distribution as a function of the difference to the direction of the scattered quark is shown (left) for the first measurement [Abt *et al.* (1994)], (right) the distribution with higher statistic, and for different regions of x, is shown [Adloff *et al.* (2000)].

statistics (shown in Fig. 19.7 (left)) and with a finer binning in x and Q^2 confirmed these early findings (note that here the energy flow is plotted in the hadronic center-of-mass system, and the proton direction appears at small η^*). All the different variants of the simulation program LEPTO [Ingelman *et al.* (1997)] (which includes order α_s matrix elements and parton showers) fall below the measurement at $\eta^* \sim 0$ at small x and small Q^2.

19.3.3 *Forward jet and forward π^0 cross section*

Measurements dedicated to investigate BFKL dynamics were proposed in [Mueller (1991b,a)]. The idea is to measure the cross section as a function of x for a jet in the forward region (at large x_{jet}) with a transverse momentum similar to the virtuality of the photon. By this requirement, DGLAP evolution is expected to be suppressed, while the k_t-unordered evolution from BFKL will give significant contributions, at least at small x. The diagram for forward jet production is schematically shown in Fig. 19.8 (left). In Fig. 19.8 (right) calculations for this process are shown: except the predictions labelled as *BFKL LO* [Bartels *et al.* (1997, 1996a)] and ARIADNE (which is the same calculation mentioned in the previous section as CDM) all predictions give a rather small cross section for forward jet production. The forward jet cross section was also of interest for investigations of *hot-spots*, localized regions with high density, inside the proton [Bartels *et al.* (1992)].

Forward jet production was measured by H1 [Aid *et al.* (1995); Adloff *et al.* (1999a); Aktas *et al.* (2006); Aaron *et al.* (2012)] and ZEUS [Breitweg *et al.* (1999, 2000); Chekanov *et al.* (2006, 2007)] collaborations. Measurements from ZEUS are shown in Fig. 19.9 for jets with $p_T > 5\,\text{GeV}$. In Fig. 19.9 (left) it is clearly seen, that the measured cross section increases steeply towards small x, while the predictions obtained from DGLAP parton shower simulations fall below the measurements. Only the prediction from ARIADNE comes closer the the data. In Fig. 19.9 (right) the cross section is shown as a function of E_T^2/Q^2, with E_T being the transverse momentum of the forward jet. For the typical DGLAP region, where E_T^2 is small compared to Q^2, the predictions describe the measurement, while in the region

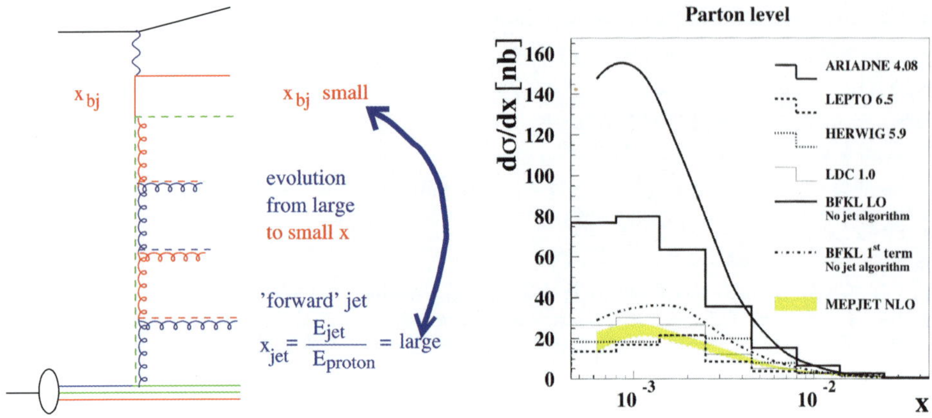

Fig. 19.8 Left: schematic diagram illustrating forward jet production. Right: predictions for forward jet production calculated in different schemes [Breitweg *et al.* (1999)]

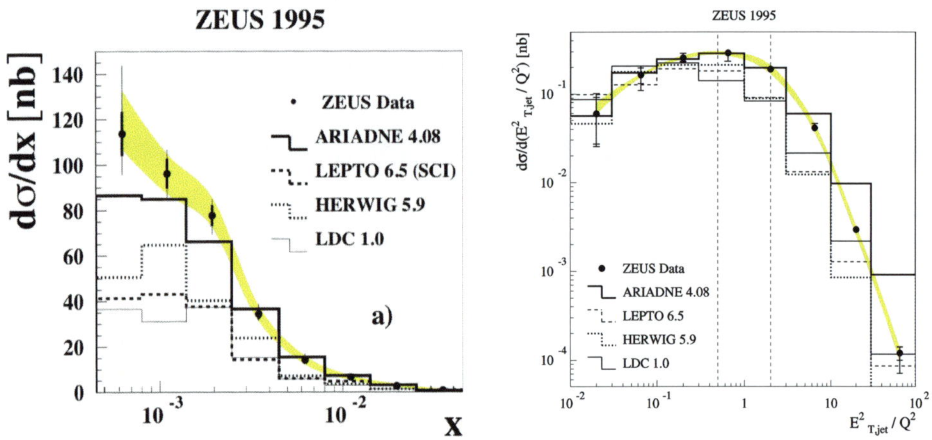

Fig. 19.9 Left: measurement of forward jet production as a function of x [Breitweg *et al.* (1999)]. Right: jet production cross section as a function of E_T^2/Q^2 [Breitweg *et al.* (2000)].

where E_T^2 becomes closer or even larger than Q^2, the measurements are above the predictions, indicating a very different kinematic region.

One of the experimental issues in forward jet production is the requirement of the transverse momentum of the jet, i.e. $p_T > 5\,\mathrm{GeV}$, which limits the kinematic range, but is necessary for a proper jet identification. Identified particles like $\pi^0 \to \gamma\gamma$ have been used by H1 to measure a similar process [Adloff *et al.* (1999b); Aktas *et al.* (2004)], but now being able to access lower x and lower transverse momenta. The measurements from H1 are shown in Fig. 19.10 and compared to predictions using RAPGAP [Jung (1995b)] (which includes order α_s matrix elements, and in addition contributions from resolved virtual photons [Jung *et al.* (1998a,b)]). The model including resolved virtual photons describes the measurement very well, while

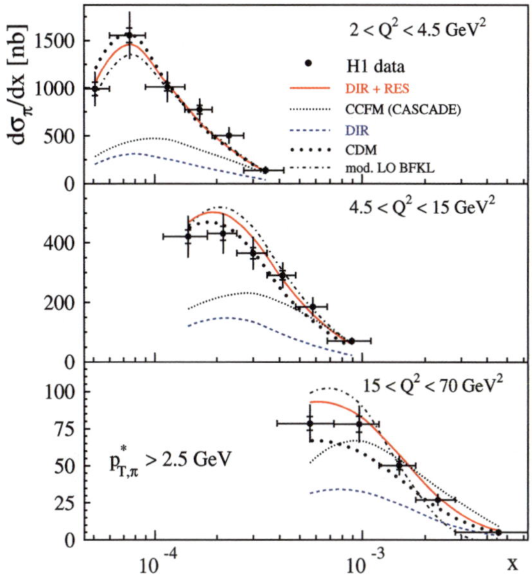

Fig. 19.10 Measurement of forward π^0 production as a function of x and different Q^2 [Aktas *et al.* (2004)].

a prediction based on the CCFM evolution equation [Jung (2002); Jung and Salam (2001)] falls below the measurements.

The forward jet and forward pion cross sections increase towards small x, and only predictions which allow for unordered emissions during the initial state cascade come close to the measurements, while predictions based on DGLAP parton showers, even with order α_s matrix elements are significantly below the data. *Hot-spots* inside the proton could not be identified within the experimental precision reached so far.

19.3.4 *Charged particle spectra*

A new type of measurements, based on inclusive transverse momentum spectra of charged particles was proposed in [Kuhlen (1996)]. The advantage of such measurements is that one is able to probe very small transverse momenta, accessing very low x values. The measurement is performed in the hadronic center-of-mass frame, as indicated in Fig. 19.11. The measurement is shown in Fig. 19.12 (left) for the proton direction and Fig. 19.12 (right) for the photon direction region. In the proton direction region, the predictions based on DGLAP based parton shower simulations (DJANGO [Schuler and Spiesberger (1991)], RAPGAP, HERWIG++ [Bahr *et al.* (2008)]) are significantly below the measurements for transverse momenta $p_T > 1\,\text{GeV}$, while in the photon direction region the predictions (except HERWIG++) come closer to the measurements. The calculation based on CCFM (CASCADE) comes closer to the measurement in the proton direction region at larger transverse momenta.

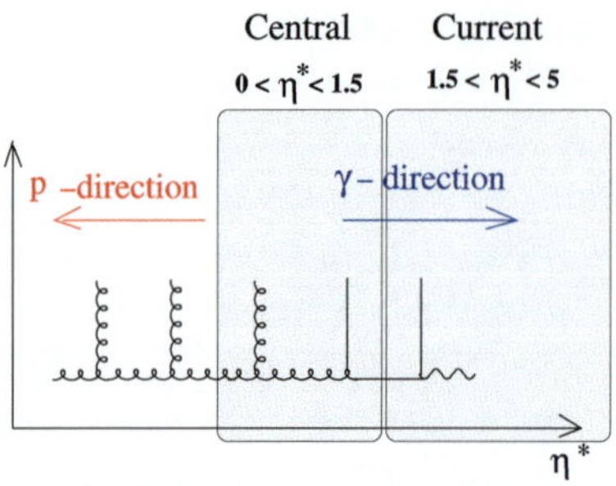

Fig. 19.11 Schematic illustration of the phase space region used for the measurement of charged particle spectra [Alexa *et al.* (2013)]

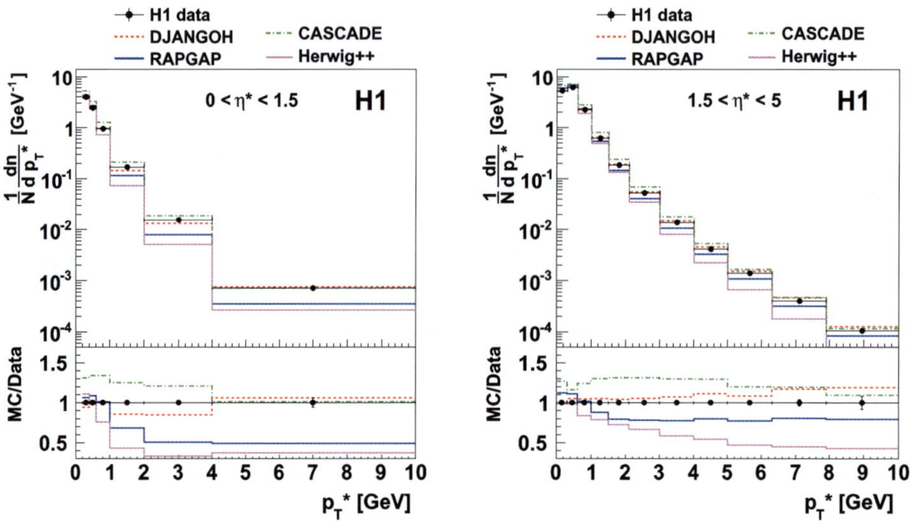

Fig. 19.12 Transverse momentum spectrum of charged particles [Alexa *et al.* (2013)]. Left: the region where the parton evolution dominates (proton direction). Right: the region where the hard process dominates (photon direction)

19.3.5 *Fixed order calculations and small x effects*

Most of the calculations presented so far were based on order α_s matrix elements supplemented with parton showers. Calculations using the complete next-to-leading order matrix elements, matched with parton showers, became available only very recently [Bellm *et al.* (2016)]. However, with advanced techniques of matching and merging of higher order tree-level diagrams, calculations were performed for deep inelastic scattering [Carli *et al.* (2010)] using tree-level diagrams of up to 5 partons.

In Fig. 19.13 a measurement [Adloff *et al.* (2002)] of inclusive jet production as a function of E_T^2/Q^2 is shown for different rapidity regions of the jet, compared to predictions based on different parton multiplicities of the matrix elements. It is

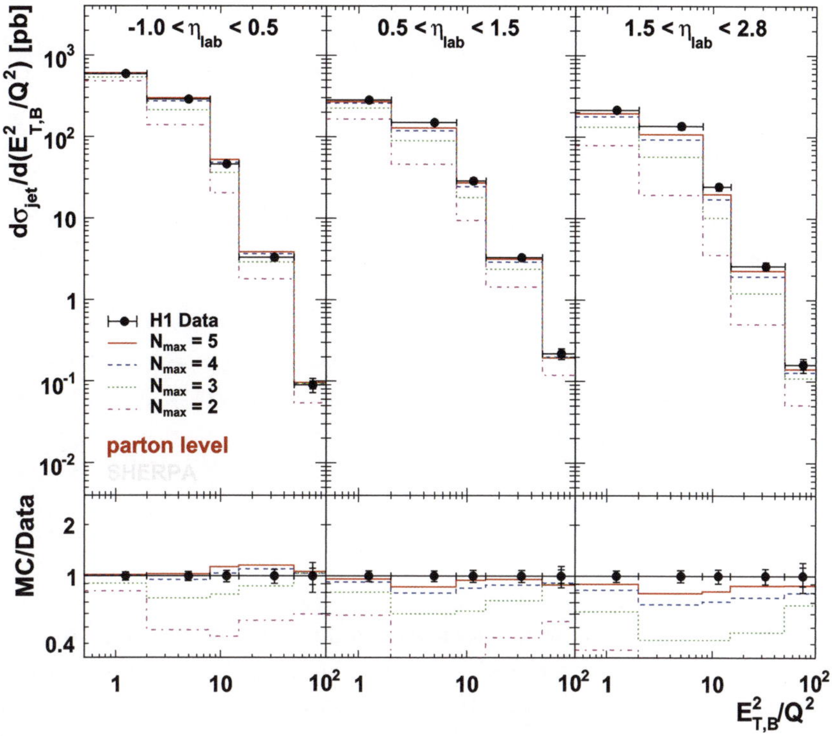

Fig. 19.13 Measurement of inclusive jet production from H1 [Adloff *et al.* (2002)] compared with predictions based on different parton multiplicities of the matrix element calculations [Carli *et al.* (2010)].

interesting to observe, that (at least) some measurements dedicated to observe BFKL effects are well described when higher order matrix elements are included in the calculations. Matrix element calculations cover the full available phase space (since no approximations are applied), but it remains to be seen, whether a fixed order calculation is sufficient to describe the measurements at higher energies, or whether resummation of enhanced contributions to all orders, as performed with BFKL or CCFM calculations, are required.

19.4 The need for full hadronic Monte Carlo event generators for small x physics

Apart from inclusive cross section measurements (as described in section 19.3.1), comparisons with measurements require the simulation of complicated cuts in phase space, and event generators are required. Some event generators provide only the partonic configuration [Nagy and Trocsanyi (2001); Currie *et al.* (2017)], but for many applications a simulation of the complete final state is needed. Most of the event generators applied at HERA were using collinear parton densities and matrix elements up to order α_s to calculate the cross sections supplemented with parton showers based on the DGLAP evolution equation and hadronization (LEPTO, HERWIG and RAPGAP). The package ARIADNE applied color dipoles to simulate the partonic cascade. Only the generators LDCMC [Andersson *et al.* (1996); Gustafson *et al.* (2003)] and CASCADE used directly a BFKL motivated parton shower, in the form of the CCFM formulation. The CASCADE event generator applied unintegrated parton densities obtained from the CCFM evolution with a parton shower, following the uPDF. However, as reported in previous sections, the agreement of predictions obtained from CASCADE with measurements is, in general, not satisfactory; only in dedicated phase space regions the agreement is rather good. One reason is that only gluon initiated processes were simulated, which is appropriate in the asymptotic region of applicability of BFKL, but is clearly not sufficient for the kinematic region at HERA.

A significant step forward has been made recently with the development and determination of the so-called Parton-Branching TMD distributions [Bermudez Martinez *et al.* (2019a); Hautmann *et al.* (2018)], which are uPDFs for all flavors and obtained with NLO evolution. These uPDFs can be applied in a new version of CASCADE for detailed comparison with measurements, and impressive agreement with measurements at the LHC [Bermudez Martinez *et al.* (2019b)] have been obtained. The new TMD (or uPDF) distributions are a significant extension of the early HERA unintegrated gluon densities. With this development, a new and promising attempt to understand small x measurements from HERA has been started [Monfared *et al.* (2019)]. In the last years a major step forward has been achieved with the automated calculation package KaTie [van Hameren (2019, 2018)] for off-shell matrix elements for all flavors in k_t-factorization, also applicable to deep-inelastic scattering.

19.5 Outlook

The study and investigations of small x effects at HERA, which were triggered by the pioneering work of Lev, have led to a very deep understanding of parton dynamics at large energies. Although, a final confirmation for the need of BFKL resummation could only be given recently for inclusive deep inelastic cross section at small Q^2, the studies of final states in deep inelastic scattering, which aimed to confirm deviations from predictions of collinear factorization, were extremely important for the investigations of new processes at large energies.

Recently a first complete set of unintegrated (Transverse Momentum Dependent) parton densities including all flavors was determined and can be used in event simulation with its unique application to parton showers. This development has great impact on present measurements (not only at small x) at the LHC but also influences the preparation for measurements at new colliders like the LHeC, the EIC or FCC, and thus the basic ideas of Lev which influenced so much a whole generation of physicists at HERA, are alive also in the studies for future colliders.

Bibliography

Aaron, F. D. *et al.* (2012). Measurement of the azimuthal correlation between the most forward jet and the scattered positron in Deep-Inelastic Scattering at HERA, *Eur. Phys. J.* **C72**, p. 1910, doi:10.1140/epjc/s10052-012-1910-8, arXiv:1111.4227 [hep-ex].

Abraham, K. J., Jung, H., Schuler, G. A., and de Troconiz, J. F. (1990). Gluon extraction from charm and bottom production at LEP / LHC, in *ECFA Large Hadron Collider Workshop, Aachen, Germany, 4-9 Oct 1990: Proceedings.2.*, pp. 899–916, http://www-library.desy.de/cgi-bin/showprep.pl?DESY-90-160.

Abramowicz, H. *et al.* (2015). Combination of measurements of inclusive deep inelastic $e^{\pm}p$ scattering cross sections and QCD analysis of HERA data, *Eur. Phys. J.* **C75**, 12, p. 580, doi:10.1140/epjc/s10052-015-3710-4, arXiv:1506.06042 [hep-ex].

Abt, I. *et al.* (1993). Measurement of the proton structure function $F_2(x, Q^2)$ in the low x region at HERA, *Nucl. Phys.* **B407**, pp. 515–538, doi:10.1016/0550-3213(93)90090-C.

Abt, I. *et al.* (1994). Energy flow and charged particle spectrum in deep inelastic scattering at HERA, *Z. Phys.* **C63**, pp. 377–390, doi:10.1007/BF01580319.

Adloff, C. *et al.* (1999a). Forward jet and particle production at HERA, *Nucl. Phys.* **B538**, pp. 3–22, doi:10.1016/S0550-3213(98)00745-7, arXiv:hep-ex/9809028 [hep-ex].

Adloff, C. *et al.* (1999b). Forward π^0 meson production at HERA, *Phys. Lett.* **B462**, pp. 440–452, doi:10.1016/S0370-2693(99)00906-5, arXiv:hep-ex/9907030 [hep-ex].

Adloff, C. *et al.* (2000). Measurements of transverse energy flow in deep inelastic-scattering at HERA, *Eur. Phys. J.* **C12**, pp. 595–607, doi:10.1007/s100520000287, arXiv:hep-ex/9907027.

Adloff, C. *et al.* (2002). Measurement of inclusive jet cross-sections in deep inelastic ep scattering at HERA, *Phys. Lett.* **B542**, pp. 193–206, doi:10.1016/S0370-2693(02)02375-4, arXiv:hep-ex/0206029 [hep-ex].

Aid, S. *et al.* (1995). Transverse energy and forward jet production in the low x regime at HERA, *Phys. Lett.* **B356**, pp. 118–128, doi:10.1016/0370-2693(95)00804-T, arXiv:hep-ex/9506012.

Ajaltouni, Z. J. *et al.* (2009). Proceedings of the workshop: HERA and the LHC workshop series on the implications of HERA for LHC physics, arXiv:0903.3861 [hep-ph].

Aktas, A. *et al.* (2004). Forward π^0 production and associated transverse energy flow in deep-inelastic scattering at HERA, *Eur. Phys. J.* **C36**, pp. 441–452, doi:10.1140/epjc/s2004-01983-1, arXiv:hep-ex/0404009 [hep-ex].

Aktas, A. *et al.* (2006). Forward jet production in deep inelastic scattering at HERA, *Eur. Phys. J.* **C46**, pp. 27–42, doi:10.1140/epjc/s2005-02471-x, arXiv:hep-ex/0508055 [hep-ex].

Alekhin, S. *et al.* (2005a). HERA and the LHC - A workshop on the implications of HERA for LHC physics: Proceedings Part A, arXiv:hep-ph/0601012.

Alekhin, S. *et al.* (2005b). HERA and the LHC — A workshop on the implications of HERA for LHC physics: Proceedings Part B, arXiv:hep-ph/0601013.

Alexa, C. *et al.* (2013). Measurement of charged particle spectra in Deep-Inelastic *ep* Scattering at HERA, *Eur. Phys. J.* **C73**, 4, p. 2406, doi:10.1140/epjc/s10052-013-2406-x, arXiv:1302.1321 [hep-ex].

Altarelli, G. and Parisi, G. (1977). Asymptotic freedom in parton language, *Nucl. Phys.* **B126**, p. 298, doi:10.1016/0550-3213(77)90384-4.

Andersen, J. R. *et al.* (2004). Small x phenomenology: summary and status 2002, *Eur. Phys. J.* **C35**, pp. 67–98, doi:10.1140/epjc/s2004-01780-x, arXiv:hep-ph/0312333.

Andersen, J. R. *et al.* (2006). Small x phenomenology: summary of the 3rd Lund Small x workshop in 2004, *Eur. Phys. J.* **C48**, pp. 53–105, doi:10.1140/epjc/s2006-02615-6, arXiv:hep-ph/0604189.

Andersson, B., Gustafson, G., and Samuelsson, J. (1996). Discrete QCD: a new approximation for QCD cascades, *Nucl. Phys.* **B463**, pp. 217–237, doi:10.1016/0550-3213(96)00022-3.

Andersson, B. *et al.* (2002). Small x phenomenology: summary and status, *Eur. Phys. J.* **C25**, pp. 77–101, doi:10.1007/s10052-002-0998-7, arXiv:hep-ph/0204115.

Bahr, M., Gieseke, S., Gigg, M., Grellscheid, D., Hamilton, K., *et al.* (2008). Herwig++: physics and manual, *Eur.Phys.J.* **C58**, pp. 639–707, doi:10.1140/epjc/s10052-008-0798-9, arXiv:0803.0883 [hep-ph].

Balitsky, I. I. and Lipatov, L. N. (1978). The Pomeranchuk singularity in Quantum Chromodynamics, *Sov. J. Nucl. Phys.* **28**, pp. 822–829.

Ball, R. D., Bertone, V., Bonvini, M., Marzani, S., Rojo, J., and Rottoli, L. (2018). Parton distributions with small-x resummation: evidence for BFKL dynamics in HERA data, *Eur. Phys. J.* **C78**, 4, p. 321, doi:10.1140/epjc/s10052-018-5774-4, arXiv:1710.05935 [hep-ph].

Baranov, S. P., Jung, H., Jonsson, L., Padhi, S., and Zotov, N. P. (2002a). A phenomenological interpretation of open charm production at HERA in terms of the semi-hard approach, *Eur. Phys. J.* **C24**, pp. 425–437, doi:10.1007/s10052-002-0957-3, arXiv:hep-ph/0203025.

Baranov, S. P., Jung, H., Jonsson, L., Padhi, S., and Zotov, N. P. (2002b). Associated D* and dijet production at HERA as a test of the k(T)-factorization approach, *Acta Phys. Polon.* **B33**, pp. 3257–3262.

Baranov, S. P., Jung, H., Jonsson, L., Padhi, S., and Zotov, N. P. (2003). Heavy quark production at HERA with BFKL and CCFM dynamics, in *Multiparticle dynamics: Proceedings, 32nd International Symposium, ISMD 2002, Alushta, Crimea, Ukraine, September 7-13, 2002*, pp. 195–198, doi:10.1142/9789812704962_0038, arXiv:hep-ph/0301017 [hep-ph].

Baranov, S. P., Jung, H., and Zotov, N. P. (1999). Charm production in the semihard approach of QCD and the unintegrated gluon distribution, in *Monte Carlo generators for HERA physics. Proceedings, Workshop, Hamburg, Germany, 1998-1999*, pp. 484–491, arXiv:hep-ph/9910210 [hep-ph].

Baranov, S. P., Jung, H., and Zotov, N. P. (2001). Pomeron intercept from BFKL gluon dynamics in deep inelastic charm production at HERA, *Nucl. Phys. Proc. Suppl.* **99A**, pp. 192–195, arXiv:hep-ph/0011127.

Bartels, J. (1990). The small x limit and the pomeron in QCD, *Nucl. Phys. Proc. Suppl.* **12**, pp. 201–218, doi:10.1016/0920-5632(90)90191-V.

Bartels, J. (1992). Deep inelastic structure functions at very small x, *Nucl. Phys.* **A546**, pp. 107c–129c, doi:10.1016/0375-9474(92)90504-D.

Bartels, J., de Roeck, A., and Loewe, M. (1992). Measurement of hot spots inside the proton at HERA and LEP/LHC, *Z. Phys.* **C54**, pp. 635–642, doi:10.1007/BF01559492.

Bartels, J., Del Duca, V., De Roeck, A., Graudenz, D., and Wusthoff, M. (1996a). Associated jet production at HERA, *Phys. Lett.* **B384**, pp. 300–306, doi:10.1016/0370-2693(96)00786-1, arXiv:hep-ph/9604272.

Bartels, J., Del Duca, V., and Wusthoff, M. (1997). Azimuthal dependence of forward-jet production in DIS in the high-energy limit, *Z. Phys.* **C76**, pp. 75–79, doi:10.1007/s002880050529, arXiv:hep-ph/9610450.

Bartels, J. and Jung, H. (2006). QCD and collider physics II, https://www.desy.de/~jung/qcd_collider_physics_sose_2006/welcome.html.

Bartels, J., Lotter, H., and Vogt, M. (1996b). A numerical estimate of the small-k_T region in the BFKL Pomeron, *Phys. Lett.* **B373**, pp. 215–222, doi:10.1016/0370-2693(96)00115-3, arXiv:hep-ph/9511399.

Bartels, J., Schuler, G. A., and Blumlein, J. (1991). A numerical study of the small x behavior of deep inelastic structure functions in QCD, *Z. Phys.* **C50**, pp. 91–102, doi:10.1007/BF01558562.

Bellm, J. *et al.* (2016). Herwig 7.0/Herwig++ 3.0 release note, *Eur. Phys. J.* **C76**, 4, p. 196, doi:10.1140/epjc/s10052-016-4018-8, arXiv:1512.01178 [hep-ph].

Bermudez Martinez, A., Connor, P., Hautmann, F., Jung, H., Lelek, A., Radescu, V., and lebk, R. (2019a). Collinear and TMD parton densities from fits to precision DIS measurements in the parton branching method, *Phys. Rev.* **D99**, 7, p. 074008, doi:10.1103/PhysRevD.99.074008, arXiv:1804.11152 [hep-ph].

Bermudez Martinez, A. *et al.* (2019b). Production of Z-bosons in the parton branching method, *Phys. Rev.* **D100**, 7, p. 074027, doi:10.1103/PhysRevD.100.074027, arXiv:1906.00919 [hep-ph].

Blumlein, J. (1995). On the k_T dependent gluon density in hadrons and in the photon, in *'95 QCD and high-energy hadronic interactions. Proceedings, 30th Rencontres de Moriond, Moriond Particle Physics Meetings, Hadronic Session, Le Arcs, France, March 19-25, 1995*, pp. 191–197, arXiv:hep-ph/9506446 [hep-ph], http://www-library.desy.de/cgi-bin/showprep.pl?desy95-125.

Breitweg, J. *et al.* (1999). Forward jet production in deep inelastic scattering at HERA, *Eur. Phys. J.* **C6**, pp. 239–252, doi:10.1007/s100529801018, arXiv:hep-ex/9805016 [hep-ex].

Breitweg, J. *et al.* (2000). Measurement of the $E^2_{(T,jet)}/Q^2$ dependence of forward jet production at HERA, *Phys. Lett.* **B474**, pp. 223–233, doi:10.1016/S0370-2693(99)01478-1, arXiv:hep-ex/9910043 [hep-ex].

Buchmuller, W. and Ingelman, G. (eds.) (1991). *Workshop on Physics at HERA*, CNUM: C91-10-29 (DESY), http://inspirehep.net/search?p=773__w%3AC91/10/29+or+773__w%3AC91-10-29+and+980__a%3AConferencePaper.

Carli, T., Gehrmann, T., and Hoeche, S. (2010). Hadronic final states in deep-inelastic scattering with Sherpa, *Eur. Phys. J.* **C67**, pp. 73–97, doi:10.1140/epjc/s10052-010-1261-2, arXiv:0912.3715 [hep-ph].

Catani, S., Ciafaloni, M., and Hautmann, F. (1990a). Gluon contributions to small x heavy flavor production, *Phys. Lett.* **B242**, p. 97, doi:10.1016/0370-2693(90)91601-7.

Catani, S., Ciafaloni, M., and Hautmann, F. (1991). High-energy factorization and small x heavy flavor production, *Nucl. Phys.* **B366**, pp. 135–188, doi:10.1016/0550-3213(91)90055-3.

Catani, S., Ciafaloni, M., and Hautmann, F. (1993). High-energy factorization in QCD and minimal subtraction scheme, *Phys. Lett.* **B307**, pp. 147–153, doi:10.1016/0370-2693(93)90204-U.

Catani, S., Fiorani, F., and Marchesini, G. (1990b). QCD coherence in initial state radiation, *Phys. Lett.* **B234**, p. 339, doi:10.1016/0370-2693(90)91938-8.

Catani, S., Fiorani, F., and Marchesini, G. (1990c). Small x behavior of initial state radiation in perturbative QCD, *Nucl. Phys.* **B336**, p. 18, doi:10.1016/0550-3213(90)90342-B.

Chekanov, S. *et al.* (2006). Forward jet production in deep inelastic ep scattering and low-x parton dynamics at HERA, *Phys. Lett.* **B632**, pp. 13–26, doi:10.1016/j.physletb.2005.09.066, arXiv:hep-ex/0502029 [hep-ex].

Chekanov, S. *et al.* (2007). Forward-jet production in deep inelastic ep scattering at HERA, *Eur. Phys. J.* **C52**, pp. 515–530, doi:10.1140/epjc/s10052-007-0418-0, arXiv:0707.3093 [hep-ex].

Ciafaloni, M. (1988). Coherence effects in initial jets at small Q^2/s. *Nucl. Phys.* **B296**, p. 49, doi:10.1016/0550-3213(88)90380-X.

Connor, P., Jung, H., Hautmann, F., and Scheller, J. (2016). TMDlib 1.0.8 and TMDplotter 2.1.1, *PoS* **DIS2016**, p. 039, doi:10.3204/PUBDB-2016-05953.

Currie, J., Gehrmann, T., Huss, A., and Niehues, J. (2017). NNLO QCD corrections to jet production in deep inelastic scattering, *JHEP* **07**, p. 018, doi:10.1007/JHEP07(2017) 018, arXiv:1703.05977 [hep-ph].

Derrick, M. *et al.* (1993a). Hadronic energy distributions in deep inelastic electron - proton scattering, *Z. Phys.* **C59**, pp. 231–242, doi:10.1007/BF01566687.

Derrick, M. *et al.* (1993b). Measurement of the proton structure function F_2 in *ep* scattering at HERA, *Phys. Lett.* **B316**, pp. 412–426, doi:10.1016/0370-2693(93)90347-K.

Dokshitzer, Y. L. (1977). Calculation of the structure functions for Deep Inelastic Scattering and e^+e^- annihilation by perturbation theory in Quantum Chromodynamics. *Sov. Phys. JETP* **46**, pp. 641–653, [Zh. Eksp. Teor. Fiz.73,1216(1977)].

Ellis, R. K., Stirling, W. J., and Webber, B. R. (1996). QCD and collider physics, *Camb. Monogr. Part. Phys. Nucl. Phys. Cosmol.* **8**, pp. 1–435.

Golec-Biernat, K. J., Kwiecinski, J., Martin, A. D., and Sutton, P. J. (1994). Transverse energy flow at HERA, *Phys. Lett.* **B335**, pp. 220–225, doi:10.1016/0370-2693(94) 91417-6, arXiv:hep-ph/9405400.

Gribov, V. N. and Lipatov, L. N. (1972). Deep inelastic *ep* scattering in perturbation theory, *Sov. J. Nucl. Phys.* **15**, pp. 438–450, [Yad. Fiz.15,781(1972)].

Gustafson, G., Lönnblad, L., and Miu, G. (2003). Hadronic collisions in the linked dipole chain model, *Phys.Rev.* **D67**, p. 034020, doi:10.1103/PhysRevD.67.034020, arXiv:hep-ph/0209186 [hep-ph].

Hautmann, F. and Jung, H. (2014). Transverse momentum dependent gluon density from DIS precision data, *Nuclear Physics B* **883**, p. 1, arXiv:1312.7875 [hep-ph].

Hautmann, F., Jung, H., Krämer, M., Mulders, P., Nocera, E., *et al.* (2014). TMDlib and TMDplotter: library and plotting tools for transverse-momentum-dependent parton distributions, *Eur. Phys. J.* **C 74**, 12, p. 3220, doi:10.1140/epjc/s10052-014-3220-9, arXiv:1408.3015 [hep-ph].

Hautmann, F., Jung, H., Lelek, A., Radescu, V., and Zlebcik, R. (2017). Soft-gluon resolution scale in QCD evolution equations, *Phys. Lett.* **B772**, p. 446, doi:10.1016/ j.physletb.2017.07.005, arXiv:1704.01757 [hep-ph].

Hautmann, F., Jung, H., Lelek, A., Radescu, V., and Zlebcik, R. (2018). Collinear and TMD quark and gluon densities from Parton Branching solution of QCD evolution equations, *JHEP* **01**, p. 070, doi:10.1007/JHEP01(2018)070, arXiv:1708.03279 [hep-ph].

Ingelman, G., De Roeck, A., and Klanner, R. (eds.) (1996). *Future physics at HERA. Proceedings, Workshop, Hamburg, Germany, September 25, 1995-May 31, 1996. Vol. 1, 2*, http://www.desy.de/~heraws96/.

Ingelman, G., Edin, A., and Rathsman, J. (1997). LEPTO 6.5: A Monte Carlo generator for deep inelastic lepton - nucleon scattering, *Comput.Phys.Commun.* **101**, pp. 108–134, doi:10.1016/S0010-4655(96)00157-9, arXiv:hep-ph/9605286 [hep-ph].

Jung, H. (1992). The Monte Carlo generator EPJPSI for J/ψ mesons in high-energy electron proton collisions, in *Workshop on Physics at HERA Hamburg, Germany, October 29-30, 1991*, pp. 1488–1496, [,1488(1992)].

Jung, H. (1995a). Epjpsi: A monte carlo generator for j/ψ mesons in high energy γp, ep and $p\bar{p}$ collisions, version 3.3, http://www.desy.de/~jung/epjpsi.html.

Jung, H. (1995b). Hard diffractive scattering in high-energy e p collisions and the Monte Carlo generator RAPGAP, *Comp. Phys. Commun.* **86**, pp. 147–161, doi:10.1016/ 0010-4655(94)00150-Z.

Jung, H. (2002). The CCFM Monte Carlo generator CASCADE, *Comput. Phys. Commun.* **143**, p. 100, doi:10.1016/S0010-4655(01)00438-6, arXiv:hep-ph/0109102.

Jung, H., Jonsson, L., and Kuster, H. (1998a). Resolved photon processes in DIS and small x dynamics, arXiv:hep-ph/9805396.

Jung, H., Jonsson, L., and Kuster, H. (1998b). Resolved virtual photons in the small x domain, arXiv:hep-ph/9811368.

Jung, H., Kotikov, A. V., Lipatov, A. V., and Zotov, N. P. (2006). Towards precision determination of unintegrated parton density functions, arXiv:hep-ph/0611093.

Jung, H. and Salam, G. P. (2001). Hadronic final state predictions from CCFM: The hadron- level Monte Carlo generator CASCADE, *Eur. Phys. J.* **C19**, pp. 351–360, doi:10.1007/s100520100604, arXiv:hep-ph/0012143.

Kuhlen, M. (1996). A New method to probe the low x parton dynamics at HERA, *Phys. Lett.* **B382**, pp. 441–446, doi:10.1016/0370-2693(96)00750-2, arXiv:hep-ph/9606246 [hep-ph].

Kuraev, E. A., Lipatov, L. N., and Fadin, V. S. (1976). Multi - Reggeon processes in the Yang-Mills theory, *Sov. Phys. JETP* **44**, pp. 443–450.

Kuraev, E. A., Lipatov, L. N., and Fadin, V. S. (1977). The Pomeranchuk singularity in nonabelian gauge theories, *Sov. Phys. JETP* **45**, pp. 199–204.

Kwiecinski, J., Martin, A. D., and Stasto, A. M. (1997a). A unified BFKL and GLAP description of F_2 data, *Phys. Rev.* **D56**, pp. 3991–4006, doi:10.1103/PhysRevD.56. 3991, arXiv:hep-ph/9703445.

Kwiecinski, J., Martin, A. D., and Stasto, A. M. (1997b). A unified BFKL/DGLAP description of deep inelastic scattering, arXiv:hep-ph/9706455.

Kwiecinski, J., Martin, A. D., and Sutton, P. (1996a). Constraints on gluon evolution at small x, *Z.Phys.* **C71**, pp. 585–594, doi:10.1007/s002880050206, arXiv:hep-ph/9602320 [hep-ph].

Kwiecinski, J., Martin, A. D., and Sutton, P. J. (1991). The Small x behavior of the gluon, *Phys. Lett.* **B264**, pp. 199–203, doi:10.1016/0370-2693(91)90727-8.

Kwiecinski, J., Martin, A. D., and Sutton, P. J. (1995). The gluon distribution at small x obtained from a unified evolution equation, *Phys. Rev.* **D52**, pp. 1445–1458, doi: 10.1103/PhysRevD.52.1445, arXiv:hep-ph/9503266.

Kwiecinski, J., Martin, A. D., and Sutton, P. J. (1996b). The description of F_2 at small x incorporating angular ordering, *Phys. Rev.* **D53**, pp. 6094–6099, doi: 10.1103/PhysRevD.53.6094, arXiv:hep-ph/9511263.

Lipatov, L. N. (1975). The parton model and perturbation theory, *Sov. J. Nucl. Phys.* **20**, pp. 94–102, [Yad. Fiz.20,181(1974)].

Lönnblad, L. (1992). ARIADNE version 4: A Program for simulation of QCD cascades implementing the color dipole model, *Comput. Phys. Commun.* **71**, pp. 15–31, doi: 10.1016/0010-4655(92)90068-A.

Marchesini, G. (1995). QCD coherence in the structure function and associated distributions at small x, *Nucl. Phys.* **B445**, pp. 49–80, doi:10.1016/0550-3213(95)00149-M, arXiv:hep-ph/9412327.

Marchesini, G. and Webber, B. R. (1991). Simulation of QCD initial state radiation at small x, *Nucl. Phys.* **B349**, pp. 617–634, doi:10.1016/0550-3213(91)90389-F.

Marchesini, G. *et al.* (1992). HERWIG: A Monte Carlo event generator for simulating hadron emission reactions with interfering gluons. Version 5.1 - April 1991, *Comput. Phys. Commun.* **67**, pp. 465–508, doi:10.1016/0010-4655(92)90055-4.

Monfared, S. T., Hautmann, F., Jung, H., and Schmitz, M. (2019). Extending parton branching TMDs to small x, arXiv:1908.01621 [hep-ph].

Mueller, A. H. (1991a). Jets at LEP and HERA, *J. Phys.* **G17**, pp. 1443–1454, doi:10. 1088/0954-3899/17/10/001.

Mueller, A. H. (1991b). Parton distributions at very small x values, *Nucl. Phys. Proc. Suppl.* **18C**, pp. 125–132, doi:10.1016/0920-5632(91)90101-J.

Nagy, Z. and Trocsanyi, Z. (2001). Multijet cross-sections in deep inelastic scattering at next-to-leading order, *Phys. Rev. Lett.* **87**, p. 082001, doi:10.1103/PhysRevLett.87.082001, arXiv:hep-ph/0104315 [hep-ph].

Peccei, R. (ed.) (1987). *DESY Workshop 1987: Physics at HERA*, CNUM: C87-10-12 (DESY), http://inspirehep.net/search?p=773__w%3AC87/10/12+or+773__w%3AC87-10-12+and+980__a%3AConferencePaper.

Schuler, G. A. and Spiesberger, H. (1991). DJANGO: The interface for the event generators HERACLES and LEPTO, in *Workshop on Physics at HERA Hamburg, Germany, October 29-30, 1991*, pp. 1419–1432.

van Hameren, A. (2018). KaTie : For parton-level event generation with k_T-dependent initial states, *Comput. Phys. Commun.* **224**, pp. 371–380, doi:10.1016/j.cpc.2017.11.005, arXiv:1611.00680 [hep-ph].

van Hameren, A. (2019). A note on QED gauge invariance of off-shell amplitudes, arXiv:1902.01791 [hep-ph].

Webber, B. R. (1991). Simulation of final states in small x processes, *Nucl. Phys. Proc. Suppl.* **18C**, pp. 38–48, doi:10.1016/0920-5632(91)90093-T.

https://doi.org/10.1142/9789811231124_0020

Chapter 20

Aspects of BFKL Physics at LEP, Tevatron and LHC

Victor T. Kim

Petersburg Nuclear Physics Institute, Russia

20.1 Introduction

Quantum Chromodynamics (QCD), being an essential ingredient of the Standard Model, is well tested in hard processes when the transferred momentum is of the order of the total collision energy (Bjorken limit: $Q^2 \sim s \to \infty$). The cornerstones of perturbative QCD in this hard kinematic regime (QCD-improved parton model): factorization of inclusive hard processes and the Gribov–Lipatov–Altarelli–Parisi–Dokshitzer (DGLAP) evolution equation [Gribov and Lipatov (1972a,b); Lipatov (1975); G. Altarelli, G. Parisi (1977); Dokshitzer (1977)] provide a basis for the successful QCD-improved parton model. The factorization theorem for inclusive hard processes with hadron(s) ensures that the inclusive cross section factorizes into partonic subprocess cross section and parton distribution function(s). The DGLAP evolution equation governs the $\log Q^2$-dependence (at $Q^2 \to \infty$) of the inclusive hard-process cross-sections at the fixed scaling variable $x = Q^2/s$.

Another important kinematic domain at high-energy is expected to be described by the BFKL (Balitsky–Fadin–Kuraev–Lipatov) evolution [Fadin *et al.* (1975); Kuraev *et al.* (1976, 1977); Balitsky and Lipatov (1978)], in the so-called QCD Gribov-Regge limit, whereby at fixed $Q^2 \gg \Lambda^2_{QCD}$, $s \to \infty$. In the Gribov-Regge limit, the BFKL evolution in the leading logarithmic (LL) approximation governs the $\log(1/x)$ evolution (at $x \to 0$). Note that the BFKL evolution in the next-to-leading logarithmic (NLL) approximation [Fadin and Lipatov (1998); Ciafaloni and Camici (1998); Brodsky *et al.* (1999); Fadin *et al.* (2002)], unlike the LL BFKL [Fadin *et al.* (1975); Kuraev *et al.* (1976, 1977); Balitsky and Lipatov (1978)], partly includes the leading order (LO) DGLAP evolution with the running coupling constant, $\alpha_S(Q^2) = 4\pi/\beta_0 \log(Q^2/\Lambda^2_{QCD})$.

Therefore, the LL BFKL and especially the NLL BFKL are anticipated to be important theoretical tools for exploring the high-energy limit of QCD. In particular, this importance arises since the highest eigenvalue, ω^{max}, of the BFKL equation is related to the intercept of the Pomeron, which in turn governs the high-energy asymptotics of the total cross-sections: $\sigma \sim (s/s_0)^{\alpha_{IP}-1} = (s/s_0)^{\omega^{max}}$, where the Regge parameter s_0 defines the approach to the asymptotic regime. The BFKL Pomeron intercept in the LL turns out to be rather large: $\alpha_{IP} - 1 = \omega^{max}_{LO} = 12\log 2\,(\alpha_S/\pi) \simeq 0.54$ for $\alpha_S = 0.2$. The NLL corrections to the BFKL have been calculated in [Fadin and Lipatov (1998); Ciafaloni and Camici (1998)], but the results in the $\overline{\text{MS}}$ scheme have a strong renormalization scale dependence. This scale ambiguity dependence is, in particular, due to large running coupling corrections. The naive application of the BLM approach [Brodsky *et al.* (1983)] for the reduction of the scale ambiguity for the NLL BFKL result in the $\overline{\text{MS}}$ scheme does not lead to a sensible result.

In [Brodsky *et al.* (1999)] an approach has been formulated based on the asymptotic conformal properties of the theory generalizing the Brodsky–Lepage–Mackenzie (BLM) optimal scale setting procedure [Brodsky *et al.* (1983)] for non-Abelian cases, i.e., for the processes which contain in the leading order non-Abelian interactions. The BLM optimal scale setting resums the conformal-violating β_0-terms into the running coupling in all orders of perturbation theory, thus preserving the conformal properties of the theory. It turns out [Brodsky *et al.* (1999)] that the BLM procedure within the physical non-Abelian renormalization schemes [Celmaster and Gonsalves (1979a,b)] eliminates the renormalization scale ambiguity of the NLL BFKL [Fadin and Lipatov (1998); Ciafaloni and Camici (1998)]. (Hereafter, we will call the generalized BLM approach of [Brodsky *et al.* (1999)] as BFKLP).

Strictly speaking, the integral kernel [Fadin and Lipatov (1998); Ciafaloni and Camici (1998)] of the BFKL equation at NLL is not conformally invariant and, hence, one should use a more accurate method for its solution [Lipatov (1986)]. But in the BFKLP approach [Brodsky *et al.* (1999)], the dependence of the eigenvalue

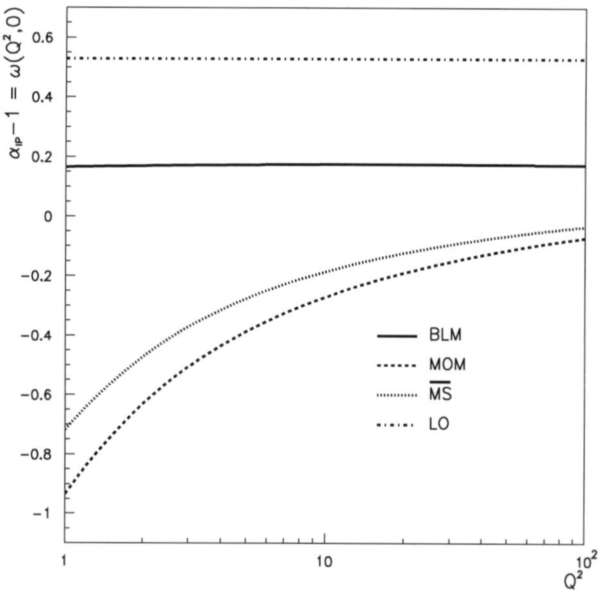

Fig. 20.1 The intercept of the BFKL Pomeron in NLL $\alpha_{IP} - 1 = \omega^{max}_{NLL}$, improved by BFKLP [Brodsky *et al.* (1999)], the generalized BLM approach for the non-Abelian case

of the kernel from the gluon virtuality is extremely weak and, therefore, ω_{NLL} coincides basically with the eigenvalue [Fadin *et al.* (2002)](arXiv:hep-ph/0207296).

Therefore, one of the striking features of the NLL BFKL analysis by BFKLP [Brodsky *et al.* (1999)] is that the NLL value for the intercept of the BFKL Pomeron, improved by the generalized BLM procedure, has a very weak dependence on the gluon virtuality Q^2: $\alpha_{IP} - 1 = \omega^{max}_{NLL} \simeq 0.13 - 0.18$ at $Q^2 = 1 - 100\,\text{GeV}^2$ (Fig. 20.1). This agrees with the conventional Regge theory where one expects a universal intercept of the Pomeron without any Q^2-dependence. The minor Q^2-dependence obtained leads to approximate conformal invariance.

Note, that the BFKLP approach generalizes the BLM procedure for non-Abelian cases, when the processes contain non-Abelian contributions at the LO, while the recent approaches in [Mikhailov (2007); Kataev and Mikhailov (2012, 2015)] and in [Brodsky and Wu (2012); Brodsky and Di Giustino (2012); Mojaza *et al.* (2013); Brodsky *et al.* (2014)] generalize it for higher orders.

It is important that the BFKLP approach allows to use in the NLL BFKL many of celebrated features of LL BFKL, such as a conformal invariance [Lipatov (1986)], using the fact that a source of the conformal invariance violation is the running coupling.

I was priveleged to have many conversations and discussions for almost three decades with Lev N. Lipatov. I got interested in his work in the middle of 80s when I have been learning QCD phenomenology as an undergraduate student of Anatoli Efremov at JINR, Dubna. At that time, the DGLAP evolution equations for hard processes in QCD were usually cited as the Altarelli-Parisi equations. But Anatoli

Efremov cited them then as the Lipatov-Altarell-Parisi ones. After checking the reference [Lipatov (1975)], which he provided, I have learned myself that it was indeed correct.

When I moved to the Petersburg Nuclear Physics Institute (PNPI), Gatchina in the end of 80s I could see Lev Lipatov often at seminars and discussions, but I never thought then that I could work myself with such a brilliant theorist. It is interesting that our first close personal encounter happened at a soccer field, when at some conference in Crimea we played by chance in the same team. Since then we were always playing, as we joked, in the same team. When in the beginning of the 90s our St. Petersburg Winter Schools were restarted, we organized together our traditional snow football games. Since 2000 I helped him to organize the biannual conferences "Hadron Structure and Quantum Chromodynamics" (HSQCD) and the annual St. Petersburg Winter Schools.

From the early 80s, the small-x and QCD Pomeron physics were one of the main topics discussed at Gatchina due to works by Eugene Levin, Mikhail Ryskin and Leonid Gribov, based on the BFKL approach. Attending in 1993–94 the CTEQ Schools in the USA I observed a spiking interest in small-x physics and in the Balitsky-Fadin-Kuraev-Lipatov (BFKL) evolution equations for semi-hard processes in QCD, because of the newest HERA data with fastly growing cross sections for deep inelastic scattering and by the building of the SSC and the LHC colliders. Since then I started to work with Grigorii Pivovarov on BFKL physics, and we had often discussions with Lev Lipatov.

The high energy physics community in the 90s was anticipating the forthcoming results of the BFKL equation in the NLL approximation. I still remember when in February of 1998 I met Lev Lipatov near our Institute, and he told me that he and Victor Fadin finally completed the derivation of the NLL BFKL evolution equation and just posted it on the arXive. When I asked him about the value of the BFKL Pomeron intercept, he smiled and mentioned that they got a very intriguing result for it, so I should see the paper. Shortly afterwards Gianni Camici and Marcelo Ciafaloni confirmed the original NLL BFKL results. The result was indeed rather striking and even puzzling, and it got immediate attention of a broad audience. However, the initial great expectations were changed to some dissapointment because the rather large NLL correction to the BFKL Pomeron intercept value qualitatively changed the high energy behaviour of the cross sections. A naive application of NLL BFKL seemed to yield a decreasing cross section with energy, which could lead to unacceptable physical consequences.

During meetings and discussions throughout Spring and Summer of 1998, Lev Lipatov, Stan Brodsky, Victor Fadin, Grigorii Pivovarov and myself developed a new approach. It is from one side a generalized Brodsky-Lepage-McKenzie (BLM) method for coupling constant resummation in non-Abelian cases with physical renormalization schemes, and from another side it provided a reasonable way for NLL BFKL physical applications [Brodsky *et al.* (1999)]. The obtained results yieded a visible relief for Lev Lipatov, though he had never doubted that the problem can be solved in one way or another. The results have been presented for the first time at the Small-x Physics Workshop at Fermilab in September 1998 and triggered many papers on this subject. After completion of our next paper [Brodsky *et al.* (2002)] on application of NLL BFKL to photon-photon scattering, Lev Lipatov confessed to me with a smile that for the first time in his career there was a comparison of theoretical results with experimental data in his article.

Lev Lipatov was an outspoken leader in theoretical high energy physics for several decades [Belavin *et al.* (2017)], and we will miss him a lot.

20.2 BFKL physics at LEP

As a phenomenological application of the NLL BFKL improved by BFKLP [Brodsky *et al.* (1999)], the generalized BLM procedure, with its effective resummation of the conformal-violating β_0-terms into the running coupling in all orders of the perturbation theory, one can consider scattering highly virtual photons [Brodsky *et al.* (2002)]. The LL BFKL application for highly virtual photon scattering was considered in [Bartels *et al.* (1996); Brodsky *et al.* (1997a,b); Bialas *et al.* (1998); Boonekamp *et al.* (1999)].

Photon–photon collisions, particularly processes with high virtualities, play a special role in QCD, since their analysis is under much better control than the calculations of lepton–hadron and hadron–hadron processes, which require the input of non-perturbative hadronic structure functions or wave functions. In addition, unitarization (screening) corrections due to multiple Pomeron exchange should be less important for the scattering of highly virtual photons (γ^*) than for hadronic collisions or real photon (γ) scattering [Goncalves *et al.* (2011)].

The high-energy asymptotic behaviour of the $\gamma\gamma$ total cross section in QED can be calculated [Frolov *et al.* (1970); Gribov *et al.* (1971); Cheng and Wu (1970)] by an all-orders resummation of the leading logarithmic terms: $\sigma \sim \alpha^4 s^\omega$, $\omega = \frac{11}{32}\pi\alpha^2 \simeq 6 \times 10^{-5}$ (Fig. 20.2). However, the slowly rising asymptotic behaviour of

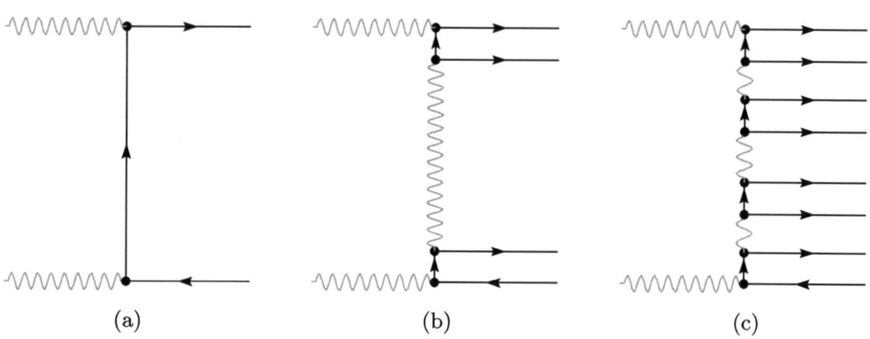

Fig. 20.2 Photon–Photon collisions in QED: (a) electron-box diagram: $\sigma \sim \alpha^2(\log s)/s$; (b) one-photon exchange diagram: $\sigma \sim \alpha^4 s^0$; (c) a typical higher-order diagram; its resummation leads to $\sigma \sim \alpha^4 s^\omega$, $\omega = \frac{11}{32}\pi\alpha^2$ [Gribov *et al.* (1971)]

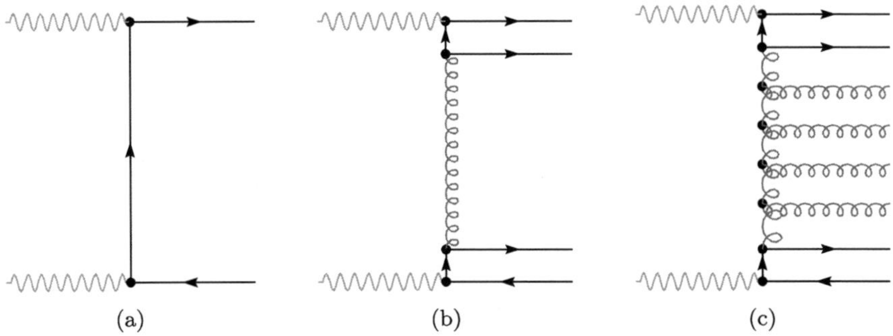

Fig. 20.3 High-energy photon-photon collisions in QCD: (a) quark-box diagram: $\sigma \sim \alpha^2(\log s)/s$; (b) one-gluon exchange diagram: $\sigma \sim \alpha^2\alpha_S^2 s^0$; (c) a typical higher-order diagram; its resummation leads to $\sigma \sim \alpha^2\alpha_S^2 s^\omega$, $\omega_{LL} = 12\ \ln 2\,(\alpha_S/\pi) \simeq 0.55$[Fadin *et al.* (1975); Kuraev *et al.* (1976, 1977); Balitsky and Lipatov (1978)] and $\omega_{NLL} = 0.13\text{-}0.18$ [Brodsky *et al.* (1999)]

the QED cross section is not apparent since large contributions come from other sources, such as the cut of the fermion-box contribution: $\sigma \sim \alpha^2(\log s)/s$ [Budnev *et al.* (1975)] (which although subleading in energy dependence, dominates the rising contributions by powers of the QED coupling constant) and QCD-driven processes (Fig. 20.3).

Fig. 20.4 compares the LL and BLM scale-fixed NLL BFKL predictions $\sigma \sim \alpha^2\alpha_S^2 s^\omega$ [Brodsky *et al.* (2002)] with CERN LEP2 data from OPAL [G. Abbiendi *et al.* (2002)] and L3 [P. Achard *et al.* (2002)]. The spread in the curves reflects the uncertainty in the choice of the Gribov-Regge scale parameter, which defines the beginning of the asymptotic regime: $s_0 = Q^2$ to $4Q^2$ for LL and NLL BFKL, where Q^2 is the mean virtuality of the colliding photons. One can see from Fig. 20.4 that the agreement of the NLL BFKL predictions [Brodsky *et al.* (2002)], improved by BFKLP [Brodsky *et al.* (1999)], with the data is quite good. The sensitivity of the NLL BFKL results [Brodsky *et al.* (2002)] to the Gribov-Regge parameter s_0 is much smaller than in the case of the LL BFKL. The variation of the predictions in the value of s_0 reflects uncertainties from uncalculated subleading terms. The parametric variation of the LL BFKL predictions is so large that it can be, in fact, neither ruled out nor confirmed at the energy range of CERN LEP2.

The next-to-leader order (NLO) impact factor of the virtual photon, which is needed for complete NLL BFKL calculations, has became recently known [Bartels and Kyrieleis (2004); Balitsky and Chirilli (2011, 2013)]. However, it is still difficult to apply it for full NLL calculation of highly virtual photon cross section. So, currently one can use the leading order (LO) impact factor of [Gribov *et al.* (1971)], assuming that the main energy-dependent NLL corrections come from the NLL BFKL subprocess rather than from the photon impact factors [Brodsky *et al.* (2002); Caporale *et al.* (2008)].

The double-logarithmic DGLAP asymptotics related with $\log(Q_A^2/Q_B^2)$-terms for the total photon–photon cross section was considered in [Boonekamp *et al.* (1999)] and found to be small for the CERN LEP2 kinematical region. The point

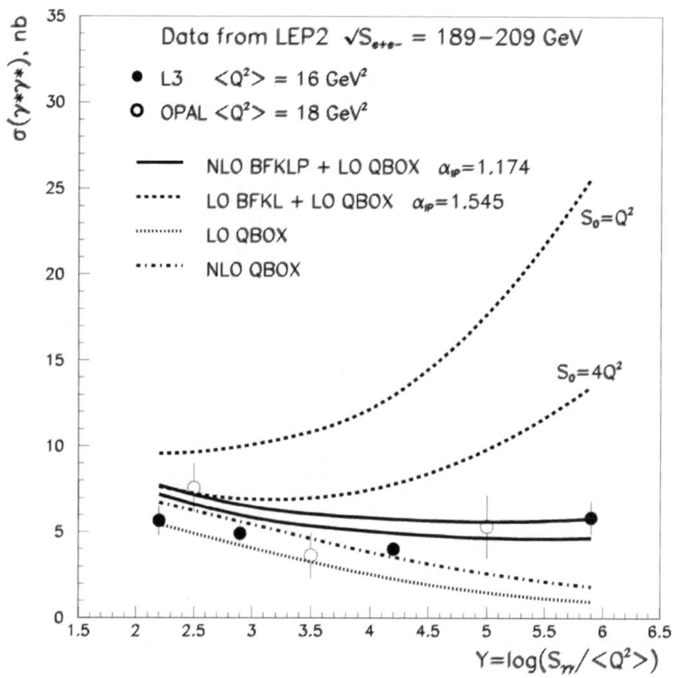

Fig. 20.4 The energy dependence of the total cross section for highly virtual photon–photon collisions predicted by the generalized BLM scale-fixed NLL BFKL [Brodsky *et al.* (2002)] compared with OPAL [G. Abbiendi *et al.* (2002)] and L3 [P. Achard *et al.* (2002)] data from LEP2 at CERN. The (solid) dashed curves correspond to the (N)LL BFKL predictions for two different choices of the Regge scale: $s_0 = Q^2$ for upper curves and $s_0 = 4Q^2$ for lower curves

is that most of the CERN LEP2 data [G. Abbiendi *et al.* (2002); P. Achard *et al.* (2002)] are collected at the approximately equal virtualities of the colliding photons: $1/2 < Q_A^2/Q_B^2 < 2$. It should be stressed that the soft Pomeron contribution to the $\gamma^\star\gamma^\star$ total cross section, if estimated within the vector-dominance model, is proportional to $\sigma_{\gamma^\star\gamma^\star} \sim (m_V^2/Q^2)^4\sigma_{\gamma\gamma}$ and therefore suppressed for such highly virtual photons as those under consideration.

We also note that for consistency, quark and anti-quark exchange contributions should be added to the NLL BFKL predictions [Brodsky *et al.* (2002); Caporale *et al.* (2008)] improved by BFKLP [Brodsky *et al.* (1999)]. This addition corresponds to the secondary Reggeon exchange which should not play an important role at very high energies. However, at not very high energies such as at LEP2, it can yield a noticeable contribution [Bartels and Lublinsky (2003)].

The NLO quark-box contribution [Cacciari *et al.* (2001)], which is leading in coupling constant counting and subleading in energy dependence, underestimates the L3 data point at $Y \equiv \log(s_{\gamma\gamma}/\langle Q^2\rangle) = 6$ by 4 standard deviations. Indeed, the NLO quark-box contribution, calculated in massless approximation, can be scaled down from general considerations with the quark masses. For example, at leading order, the inclusion of masses to the quark-box diagram reduces its contribution by

10–15% [Cacciari *et al.* (2001)]. Also, the one–gluon exchange added to the (N)LO quark-box contribution is not sufficient to describe the data at $Y = 6$ within (3) 4 standard deviations.

NLL BFKL calculations for virtual gamma-gamma scattering were performed by several groups [Bartels and Kyrieleis (2004)], [Balitsky and Chirilli (2011, 2013); Chirilli and Kovchegov (2014)], [Caporale *et al.* (2008); Ivanov *et al.* (2014)]. However, the latter and former results are in disagreement with each other. According to the results of [Ivanov *et al.* (2014)] there are negligeable BFKL contributions at the LEP2 data.

To summarize this section, highly virtual photon–photon collisions provide a very unique opportunity to test high-energy asymptotics of QCD. The NLL BFKL predictions with LO impact factors for the $\gamma^*\gamma^*$ total cross section improved by the BFKLP approach [Brodsky *et al.* (1999)] show good agreement with the data from OPAL [G. Abbiendi *et al.* (2002)] and L3 [P. Achard *et al.* (2002)] at CERN LEP2. However, the apparent disagreement between the full calculations of [Balitsky and Chirilli (2011, 2013); Chirilli and Kovchegov (2014)] and [Caporale *et al.* (2008); Ivanov *et al.* (2014)] prevents decisive conclusions. It is quite important to resolve the disgareement in order to have the complete NLL BFKL calculations for highly virtual photon collisions.

20.3 BFKL physics at the TEVATRON and the LHC

Experimental tests of BFKL predictions and finding of good observables for BFKL effects are very important for studying of the new high-energy domain of QCD.

The first LL BFKL phenomenology calculations were provided for heavy quark production in photon collisions [Balitsky and Lipatov (1978)], heavy lepton-pair hadro-production [Levin and Ryskin (1980)] , deep inelastic lepton-nucleon scattering [Gribov *et al.* (1981b)], and high p_T hadron production in hadronic collisions [Ryskin (1980); Gribov *et al.* (1981a, 1983a)] (for a review see [Gribov *et al.* (1983b)]).

20.3.1 *Jets with large rapidity separation: Ratios*

The proposal to study in rapidity the most forward and most backward jets (Mueller-Navelet (MN) jets [Mueller and Navelet (1987)]) in hadronic collisions was very promising, since the BFKL effects could be very pronounced with increasing rapidity separation between Mueller-Navelet jets. Mueller-Navelet dijets is a special case of inclusive dijets [Kim and Pivovarov (1996a)], the latter are more suitable for jet tagging in limited rapidity coverage detectors. However, the K-factor for Mueller-Navelet dijet production, which is ratio of the Mueller-Navelet dijet cross section to the Born cross section of two jet production, can not be experimentally measured. This is because the Born cross section is a purely theoretical quantity:

e.g., one cannot forbid by experimental conditions the virtual corrections to Born process.

So, to avoid the above problem, other jet observables were suggested : dijet azimuthal angle decorrelations [Del Duca and Schmidt (1994a); Stirling (1994); Kim and Pivovarov (1996a); Sabio Vera (2006); Sabio Vera and Schwennsen (2007)], single jet ratio [Kim and Pivovarov (1998)] and dijet ratio at different collision energies [Orr and Stirling (1998b)]. Also, in [Gavrilov *et al.* (2008, 2013)] it was proposed to use a dijet "K-factor" (a measurable analog of a dijet K-factor): instead of the Born dijet cross section one can use exactly two-jet cross section, that is the cross section of events where there are only two jets above p_{Tmin} (veto on extra jets with $p_T > p_{Tmin}$).

At the LL BFKL approximation, various approaches were developed to calculate (multi-)jet production in hadronic collisions. Among them there are approaches with effective Feynman-rules for scattering amplitudes [Antonov *et al.* (2005)] and for cross sections [Kim and Pivovarov (1996b)]. Also, several BFKL Monte Carlo (MC) event generators appeared: [Schmidt (1997); Orr and Stirling (1998a)], HEJ+ARIADNE [Andersen *et al.* (2011)], BFKLex[Chachamis and Sabio Vera (2016a)], and the CASCADE generator [Jung and Salam (2001); Jung *et al.* (2010)] based on CCFM-evoluton and k_T-factorization.

At the NLL BFKL approximation, several groups made calculations for jet production [Ostrovsky (2000)], [Bartels *et al.* (2006); Hentschinski and Sabio Vera (2012); Chachamis *et al.* (2013, 2014); Caporale *et al.* (2013b)] [Ivanov and Papa (2012); Caporale *et al.* (2013a)], [Colferai *et al.* (2010); Ducloué *et al.* (2012); Ducloue *et al.* (2013); Duclou*et al.* (2014)].

At the Fermilab Tevatron, the D∅ Collaboration has measured Mueller-Navelet dijets azimuthal angle decorrelations [S. Abachi *et al.* (1996)] and ratio of Mueller-Navelet dijet cross sections at 1800 GeV and 630 GeV collisions energies [S. Abbott *et al.* (2000)]. Unfortunately, those two measurements for Mueller-Navelet dijets are not consistent with each other. The former measurement [S. Abachi *et al.* (1996)] shows no clear indication of BFKL evolution manifestation: PYTHIA 6 predictions with LL DGLAP parton shower overestimate the data for azimuthal decorrelations of Mueller-Navelet dijets, while PYTHIA 6 with the switched-on color coherence option can describe the data reasonably well. At the same time the latter measurement [S. Abbott *et al.* (2000)] for the dijet ratio for the two energies shows an overwhelming BFKL-like contribution, while it can not be described by the PYTHIA nor the HERWIG generators.

The dijet "K-factor" as a function of rapidity separation was measured at LHC Run I at 7 TeV by ATLAS (inverse "K-factor") [G. Aad *et al.* (2011)] and CMS [S. Chatrchyan *et al.* (2012)]. The ATLAS measurement was more limited than the CMS one in the range of the rapidity separation and the p_{Tmin} (ATLAS: $\Delta y < 6$ and $p_{Tmin} > 70$ GeV versus CMS: $\Delta y < 9.4$ and $p_{Tmin} > 35$ GeV).

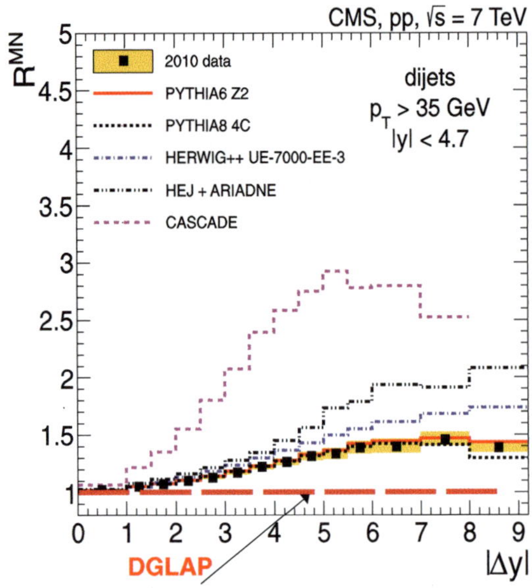

Fig. 20.5 Ratio of Mueller-Navelet (R^{MN}) dijet production cross section over two-jet cross section (dijet "K-factor") as a function of rapidity separation between jets at the LHC: CMS data at 7 TeV versus prediction of DGLAP event generators with LL parton shower improved by beyond-DGLAP effects such as color coherence. The dashed line shows the expectation from pure DGLAP evolution without color coherence effects

Figure 20.5 shows a comparison of the CMS data with various Monte Carlo (MC) event generators. The LL BFKL based MC generators CASCADE [Jung *et al.* (2010)] and HEJ+ARIADNE [Andersen *et al.* (2011)] overestimate the CMS data. The MC generators based on LL BFKL, which is designed for infinite energy, inherits its drawbacks at finite energy. So, this is not a big surprise that LL BFKL based MC generators do not describe the data.

The LL DGLAP based MC generator: HERWIG++ [Bahr *et al.* (2008)] overshoots the data, while the LL DGLAP PYTHIA [Sjöstrand *et al.* (2006)] and 8 [Sjöstrand *et al.* (2008)] provide a good agreement with the data. However, both the HERWIG++ and the PYTHIA generators contain beyond DGLAP contributions such as color coherence, polar angle ordering, color dipole approximation, etc. Those contributions are small corrections in the DGLAP kinematical domain and often improve the quality of the data description. However, those corrections are no longer valid in the BFKL domain, and one has to use a proper tool: BFKL approach. As one can see in Fig. 20.5, instead of being constant, both the LL DGLAP based MC generators HERWIG++ and PYTHIA yield a noticeable rise with increasing rapidity separation. Different implementations of those corrections provide a quite different behavior for HERWIG++ and PYTHIA. Unfortunately, there is neither a NLL BFKL generator nor a pure DGLAP generator to confront the data. So, one can check whether the description of the data by PYTHIA generators is accidental or real by testing other observables.

20.3.2 *Jets with large rapidity separation: Azimuthal angle decorrelations*

According to [Del Duca and Schmidt (1994b); Stirling (1994)], the normalized cross-section for MN jets with $p_T > p_{Tmin}$ can be written as a Fourier series:

$$\frac{1}{\sigma}\frac{\mathrm{d}\sigma}{\mathrm{d}(\Delta\phi)}(\Delta y, p_{Tmin}) = \frac{1}{2\pi}\left[1 + 2\sum_{n=1}^{\infty} C_n(\Delta y, p_{Tmin})\cos\left(n(\pi - \Delta\phi)\right)\right],$$

where C_n, the Fourier coefficients, are the average cosines of the decorrelation angle, $(\pi - \Delta\phi)$, $C_n = \langle\cos\left(n(\pi - \Delta\phi)\right)\rangle$, where $\Delta\phi = \phi_1 - \phi_2$ is the difference between the azimuthal angles ϕ_1 and ϕ_2 of MN jets.

The average cosines C_1, C_2 (Fig. 20.6) and C_3 were measured by the CMS Collaboration [Khachatryan *et al.* (2016)] for jets with $p_T > 35\,\text{GeV}$ and $|y| < 4.7$

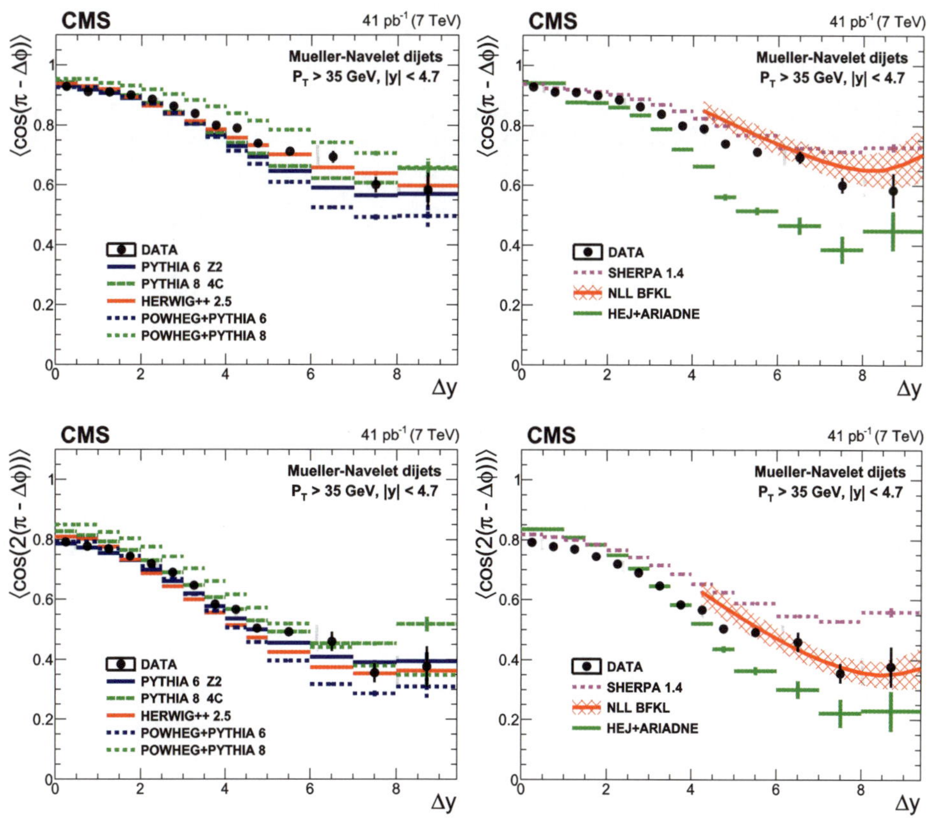

Fig. 20.6 Measured average cosines C_1 (*top* row), and C_2 (*middle* row) as a function of the rapidity separation Δy compared with the LL DGLAP MC generators and the NLO generator POWHEG (*left* column). Comparisons with the tree approximation MC generator SHERPA, the LL BFKL-LL BFKL-motivated HEJ+ARIADNE and analytical NLL BFKL parton level calculations $(4.0 < \Delta y < 9.4)$ (*right* column). The *shaded band* indicates the total systematic uncertainty of the data.

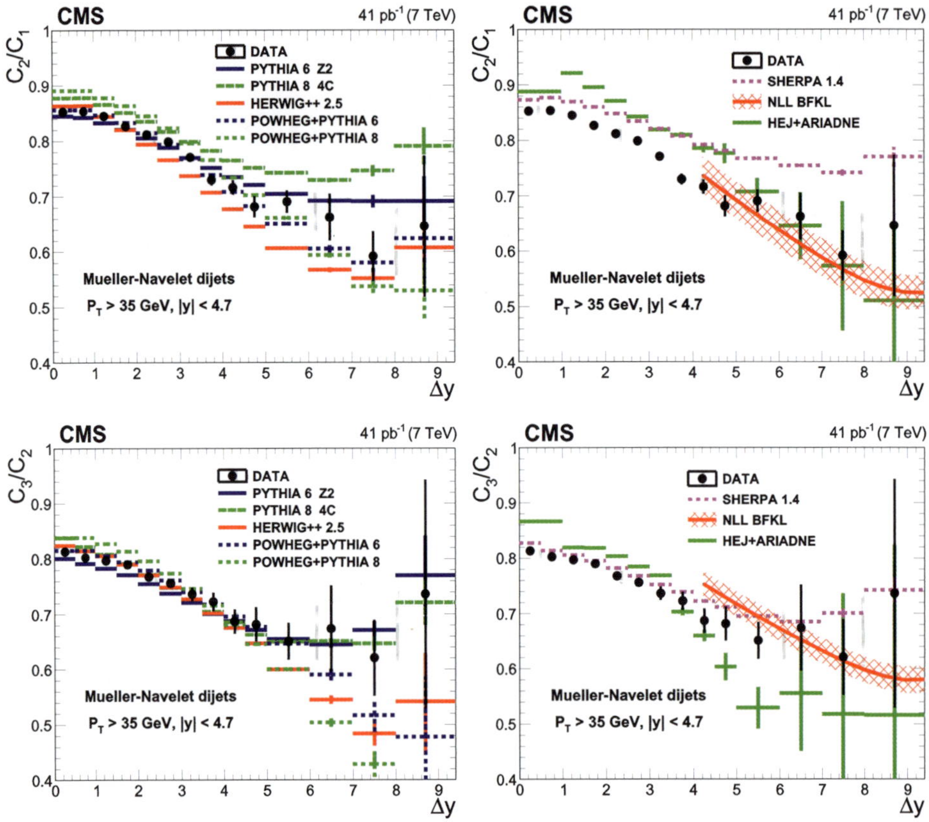

Fig. 20.7 Measured average cosine ratios C_2/C_1 (*top*) and C_3/C_2 (*bottom*) as a function of the rapidity separation Δy compared with the LL DGLAP MC generators and the NLO generator POWHEG (*left* column). Comparisons with the tree approximation MC generator SHERPA, the LL BFKL-LL BFKL-motivated HEJ+ARIADNE and analytical NLL BFKL parton level calculations (*right* column). The *shaded band* indicates the total systematic uncertainty of the data.

for pp-collisions at $\sqrt{s} = 7\,\text{TeV}$. The data and event selection are the same as in [S. Chatrchyan *et al.* (2012)]. Also, in this paper the average cosine ratios C_2/C_1 and C_3/C_2 (Fig. 20.7) were measured as proposed in [Sabio Vera and Schwennsen (2007)]. The measured observables were compared with predictions from the LL DGALP MC generators (PYTHIA6 tune Z2 [Sjöstrand *et al.* (2006); Field (2010); Chatrchyan *et al.* (2011)], PYTHIA8 tune 4C [Sjöstrand *et al.* (2008); Corke and Sjöstrand (2011)] and HERWIG++ tune UE-7000-EE-3 [Bahr *et al.* (2008); Gieseke *et al.* (2011)]), the tree approximation MC generator SHERPA [Freeman *et al.* (2001)], the next-to-leading-order (NLO) matrix element MC generator POWHEG [Nason (2004); Frixione *et al.* (2007); Alioli *et al.* (2010)] interfaced with the LL DGLAP-based PYTHIA6 and PYTHIA8 generators, LL BFKL-motivated HEJ+ARIADNE [Andersen *et al.* (2011)] and NLL BFKL parton level analytical calculations [Duclou*et al.* (2014)].

The predictions of LL DGLAP MC generator HERWIG++ show reasonable agreement with the data for average cosines, while for their ratios it cannot describe the data for $\Delta y \geq 5$. On the contrary, the LL DGLAP MC generators PYTHIA6 and PYTHIA8 have a reasonable description for the average cosine ratios from $\Delta y \geq 5$, while it cannot describe the average cosines. The results of the NLO matrix element MC generator POWHEG interfaced with the LL DGLAP-based parton shower of PYTHIA6 and PYTHIA8 show no better description of the measured observables than standalone results of PYTHIA6 and PYTHIA8. The tree approximation MC generator SHERPA also provides the same level of MC data description as PYTHIA8. The LL BFKL-motivated MC generator HEJ+ARIADNE provides a stronger decorrelation than observed in the data. An analytical NLL BFKL calculation provides reasonable agreement with data for $\Delta y \geq 4$.

All these facts (sensitivity to an implementation of color-coherence effects in different MC generators and reasonable data description by analytical NLL BFKL calculations) indicate that, given a kinematic region (jets with $p_T > 35\,\text{GeV}$ and $\Delta y \geq 5$ at collision energy 7 TeV), the data may be beyond the domain described by the DGLAP approximation. Indeed, recent CMS data [Khachatryan *et al.* (2016)] for azimuthal angle decorrelations of dijets as a function of rapidity separation between jets show better overall description by HERWIG++ than PYTHIA. However, the best description is provided by analytical NLL BFKL calculations [Duclou*et al.* (2014)], improved by BFKLP [Brodsky *et al.* (1999)]. Furher progress for experimental tests of BFKL predictions will be related with LHC Run II data, development of BFKL Monte Carlo generators with NLL corrections, and with finding new observables.

Thre are a few recent suggestions for new observables for BFKL effects in jet physics at LHC, e.g., three-jet observables [Caporale *et al.* (2016b, 2017b, 2016a, 2017a)], correlations of average p_T and azimutal angles of the jets additional to Mueller-Navelet dijjets [Chachamis and Sabio Vera (2016b)], and also dihadron production [Celiberto *et al.* (2017)].

20.3.3 *Jets with large rapidity gaps*

Jet production with rapidity gaps has been calcuated in the LL BFKL approach for non-forward scattering amplitude in [Mueller and Tang (1992)] for hadronic collisions and in [Bartels *et al.* (1995)] for deep inelastic processes. Calculations with NLL elements were performed in [Enberg *et al.* (2002); Chevallier *et al.* (2009); Hatta and Ueda (2009); Kepka *et al.* (2011); Hentschinski *et al.* (2014a,b)].

The D0 [Abachi *et al.* (1996); Abbott *et al.* (1998)] and CDF [Abe *et al.* (1998a,b)] Collaborations observed dijet production with rapidity gaps between jets at Tevatron energies 1800 and 630 GeV. The CMS Collaboration perfomed a search for BFKL effects in dijet production with rapidity gaps at LHC with 7 TeV [Sirunyan *et al.* (2018)] measuring jet p_T and rapidity gap-size dependences. None of the theoretical models can describe all the data features, however many of these features are reasonably described by a NLL BFKL based approach improved by soft color interaction model [Ekstedt *et al.* (2017)].

20.4 Summary and outlooks

The BFKLP approach [Brodsky *et al.* (1999)], the generalized BLM scale setting for non-Abelian case, opens new windows for applications of NLL BFKL evolution to the high-energy phenomenology.

Hiighly virtual photon–photon collisions provide a very unique opportunity to test high-energy asymptotics of QCD. The NLL BFKL predictions [Brodsky *et al.* (2002); Caporale *et al.* (2008)] for the $\gamma^*\gamma^*$ total cross section with LO photon impact factors and with the renormalization scale fixed by the BFKLP approach [Brodsky *et al.* (1999)] show a good agreement with the data of LEP2. However, the apparent disagreement between the full NLL BFKL calculations of [Balitsky and Chirilli (2011, 2013); Chirilli and Kovchegov (2014)] and [Caporale *et al.* (2008); Ivanov *et al.* (2014)] is preventing from reaching decisive conclusions. It is quite important to resolve the disagreement in order to have complete NLL BFKL calculations for highly virtual photon collisions.

The LHC Run I data at 7 TeV [Khachatryan *et al.* (2016)] on azimuthal dijet decorrelation with wide rapidity separation between jets are in good agreement with analytical NLL BFKL calculations [Duclou*et al.* (2014)] improved by BFKLP [Brodsky *et al.* (1999)].

Therefore, at present, there are some evidences for BFKL evolution manifestations at LEP2, Tevatron and LHC. Futher progress for experimental tests of BFKL predictions for jet production will be related with the recently completed LHC Run II and with finding new observables for BFKL evolution manifestation. The progress also requires the development of BFKL Monte Carlo generators with NLL corrections.

Bibliography

Aad, G. *et al.* (2011). Measurement of dijet production with a veto on additional central jet activity in pp collisions at $\sqrt{s} = 7$ TeV using the ATLAS detector, *JHEP* **1109**, p. 053, doi:10.1007/JHEP09(2011)053, arXiv:1107.1641 [hep-ex].

Abachi, S. *et al.* (1996). Jet production via strongly-interacting color-singlet exchange in $p\bar{p}$ collisions, *Phys. Rev. Lett.* **76**, pp. 734–739, doi:10.1103/PhysRevLett.76.734, arXiv:hep-ex/9509013 [hep-ex].

Abachi, S. *et al.* (1996). The Azimuthal decorrelation of jets widely separated in rapidity, *Phys. Rev. Lett.* **77**, pp. 595–600, doi:10.1103/PhysRevLett.77.595, arXiv:hep-ex/9603010 [hep-ex].

Abbiendi, G. *et al.* (2002). Measurement of the hadronic cross-section for the scattering of two virtual photons at LEP, *Eur. Phys. J. C* **24**, pp. 17–31, doi:10.1007/s100520200940, arXiv:hep-ex/0110006 [hep-ex].

Abbott, B. *et al.* (1998). Probing hard color-singlet exchange in $p\bar{p}$ collisions at $\sqrt{s} =$ 630 GeV and 1800 GeV, *Phys. Lett.* **B440**, pp. 189–202, doi:10.1016/S0370-2693(98)01238-6, arXiv:hep-ex/9809016 [hep-ex].

Abbott, S. *et al.* (2000). Probing BFKL dynamics in the dijet cross section at large rapidity intervals in $p\bar{p}$ collisions at $\sqrt{s} = 1800\,\text{GeV}$ and 630-GeV, *Phys. Rev. Lett.* **84**, pp. 5722–5727, doi:10.1103/PhysRevLett.84.5722, arXiv:hep-ex/9912032 [hep-ex].

Abe, F. *et al.* (1998a). Dijet production by color — singlet exchange at the Fermilab Tevatron, *Phys. Rev. Lett.* **80**, pp. 1156–1161, doi:10.1103/PhysRevLett.80.1156.

Abe, F. *et al.* (1998b). Events with a rapidity gap between jets in $\bar{p}p$ collisions at $\sqrt{s} = 630\,\text{GeV}$, *Phys. Rev. Lett.* **81**, pp. 5278–5283, doi:10.1103/PhysRevLett.81.5278.

Achard, P. *et al.* (2002). Double tag events in two photon collisions at LEP, *Phys. Lett. B* **531**, pp. 39–51, doi:10.1016/S0370-2693(02)01340-0, arXiv:hep-ex/0111012 [hep-ex].

Alioli, S., Nason, P., Oleari, C., and Re, E. (2010). A general framework for implementing NLO calculations in shower Monte Carlo programs: The Powheg box, *JHEP* **06**, p. 043, doi:10.1007/JHEP06(2010)043, arXiv:1002.2581 [hep-ph].

Altarelli, G. and Parisi, G. (1977). Asymptotic freedom in parton language, *Nucl. Phys. B* **126**, p. 298.

Andersen, J. R., Lonnblad, L., and Smillie, J. M. (2011). A parton shower for high energy jets, *JHEP* **07**, p. 110, doi:10.1007/JHEP07(2011)110, arXiv:1104.1316 [hep-ph].

Antonov, E. N., Lipatov, L. N., Kuraev, E. A., and Cherednikov, I. O. (2005). Feynman rules for effective Regge action, *Nucl. Phys.* **B721**, pp. 111–135, doi:10.1016/j.nuclphysb.2005.05.013,10.1016/j.nuclphysb.2005.013, arXiv:hep-ph/0411185 [hep-ph].

Bahr, M. *et al.* (2008). Herwig++ physics and manual, *Eur. Phys. J. C* **58**, pp. 639–707, doi:10.1140/epjc/s10052-008-0798-9, arXiv:0803.0883 [hep-ph].

Balitsky, I. and Chirilli, G. A. (2011). Photon impact factor in the next-to-leading order, *Phys. Rev. D* **83**, p. 031502, doi:10.1103/PhysRevD.83.031502, arXiv:1009.4729 [hep-ph].

Balitsky, I. and Chirilli, G. A. (2013). Photon impact factor and k_T-factorization for DIS in the next-to-leading order, *Phys. Rev. D* **87**, 1, p. 014013, doi:10.1103/PhysRevD.87.014013, arXiv:1207.3844 [hep-ph].

Balitsky, I. I. and Lipatov, L. N. (1978). The Pomeranchuk singularity in quantum chromodynamics, *Sov. J. Nucl. Phys.* **28**, pp. 822–829, [Yad. Fiz. 28,1597 (1978)].

Bartels, J., De Roeck, A., and Lotter, H. (1996). The gamma* gamma* total cross-section and the BFKL pomeron at e+ e− colliders, *Phys. Lett. B* **389**, pp. 742–748, doi:10.1016/S0370-2693(96)01344-5, arXiv:hep-ph/9608401 [hep-ph].

Bartels, J., Forshaw, J. R., Lotter, H., Lipatov, L. N., Ryskin, M. G., and Wusthoff, M. (1995). How does the BFKL pomeron couple to quarks? *Phys. Lett.* **B348**, pp. 589–596, doi:10.1016/0370-2693(95)00201-U, arXiv:hep-ph/9501204 [hep-ph].

Bartels, J. and Kyrieleis, A. (2004). NLO corrections to the gamma* impact factor: First numerical results for the real corrections to gamma*(L), *Phys. Rev. D* **70**, p. 114003, doi:10.1103/PhysRevD.70.114003, arXiv:hep-ph/0407051 [hep-ph].

Bartels, J. and Lublinsky, M. (2003). Quark anti-quark exchange in gamma* gamma* scattering, *JHEP* **09**, p. 076, doi:10.1088/1126-6708/2003/09/076, arXiv:hep-ph/0308181 [hep-ph].

Bartels, J., Sabio Vera, A., and Schwennsen, F. (2006). NLO inclusive jet production in k_T-factorization, *JHEP* **11**, p. 051, doi:10.1088/1126-6708/2006/11/051, arXiv:hep-ph/0608154 [hep-ph].

Belavin, A. A., Vysotsky, M. I., Gershtein, S. S., Zakharov, V. I., Ioffe, B. L., Kazakov, D. I., Kim, V. T., Matveev, V. A., Polyakov, A. M., Rubakov, V. A., Slavnov, A. A., and Fadin, V. S. (2017). In memory of Lev Nikolaevich Lipatov, *Phys. Usp.* **60**, p. 1306, doi:10.3367/UFNe.2017.11.038249.

Bialas, A., Czyz, W., and Florkowski, W. (1998). Total gamma* gamma* cross-section and the BFKL pomeron, *Eur. Phys. J. C* **2**, pp. 683–689, doi:10.1007/s100520050171, arXiv:hep-ph/9705470 [hep-ph].

Boonekamp, M., De Roeck, A., Royon, C., and Wallon, S. (1999). Gamma* gamma* total cross-section in the dipole picture of BFKL dynamics, *Nucl. Phys.* **B555**, pp. 540–564, doi:10.1016/S0550-3213(99)00345-4, arXiv:hep-ph/9812523 [hep-ph].

Brodsky, S., Fadin, V., Kim, V., Lipatov, L., and Pivovarov, G. (1999). The QCD pomeron with optimal renormalization, *JETP Lett.* **70**, pp. 155–160, doi:10.1134/1.568145, arXiv:hep-ph/9901229 [hep-ph].

Brodsky, S. J. and Di Giustino, L. (2012). Setting the renormalization scale in QCD: The principle of maximum conformality, *Phys. Rev. D* **86**, p. 085026, doi:10.1103/PhysRevD.86.085026, arXiv:1107.0338 [hep-ph].

Brodsky, S. J., Fadin, V. S., Kim, V. T., Lipatov, L. N., and Pivovarov, G. B. (2002). High-energy QCD asymptotics of photon-photon collisions, *JETP Lett.* **76**, pp. 249–252, doi:10.1134/1.1520615, arXiv:hep-ph/0207297 [hep-ph], [Pisma Zh. Eksp. Teor. Fiz. 76, 306 (2002)].

Brodsky, S. J., Hautmann, F., and Soper, D. E. (1997a). Probing the QCD pomeron in e+ e− collisions, *Phys. Rev. Lett.* **78**, pp. 803–806, doi:10.1103/PhysRevLett.78.803, arXiv:hep-ph/9610260 [hep-ph], [Erratum: Phys. Rev. Lett.79,3544(1997)].

Brodsky, S. J., Hautmann, F., and Soper, D. E. (1997b). Virtual photon scattering at high-energies as a probe of the short distance pomeron, *Phys. Rev. D* **56**, pp. 6957–6979, doi:10.1103/PhysRevD.56.6957, arXiv:hep-ph/9706427 [hep-ph].

Brodsky, S. J., Lepage, G. P., and Mackenzie, P. B. (1983). On the elimination of scale ambiguities in perturbative quantum chromodynamics, *Phys. Rev. D* **28**, p. 228, doi:10.1103/PhysRevD.28.228.

Brodsky, S. J., Mojaza, M., and Wu, X.-G. (2014). Systematic scale-setting to all orders: The principle of maximum conformality and commensurate scale relations, *Phys. Rev. D* **89**, 1, p. 014027, doi:10.1103/PhysRevD.89.014027, arXiv:1304.4631 [hep-ph].

Brodsky, S. J. and Wu, X.-G. (2012). Scale setting using the extended renormalization group and the principle of maximum conformality: The QCD coupling constant at four loops, *Phys. Rev. D* **85**, p. 034038, doi:10.1103/PhysRevD.85.034038,10.1103/PhysRevD.86.079903, arXiv:1111.6175 [hep-ph], [Erratum: Phys. Rev. D 86, 079903 (2012)].

Budnev, V. M., Ginzburg, I. F., Meledin, G. V., and Serbo, V. G. (1975). The Two photon particle production mechanism. Physical problems. Applications. Equivalent photon approximation, *Phys. Rept. C* **15**, pp. 181–281, doi:10.1016/0370-1573(75)90009-5.

Cacciari, M., Del Duca, V., Frixione, S., and Trocsanyi, Z. (2001). QCD radiative corrections to gamma* gamma* —> hadrons, *JHEP* **02**, p. 029, doi:10.1088/1126-6708/2001/02/029, arXiv:hep-ph/0011368 [hep-ph].

Caporale, F., Celiberto, F. G., Chachamis, G., Gomez, D. G., and Sabio Vera, A. (2017a). Stability of azimuthal-angle observables under higher order corrections in inclusive three-jet production, *Phys. Rev.* **D95**, 7, p. 074007, doi:10.1103/PhysRevD.95.074007, arXiv:1612.05428 [hep-ph].

Caporale, F., Celiberto, F. G., Chachamis, G., Gordo Gmez, D., and Sabio Vera, A. (2016a). BFKL azimuthal imprints in inclusive three-jet production at 7 and 13 TeV, *Nucl. Phys.* **B910**, pp. 374–386, doi:10.1016/j.nuclphysb.2016.07.012, arXiv:1603.07785 [hep-ph].

Caporale, F., Celiberto, F. G., Chachamis, G., Gordo Gmez, D., and Sabio Vera, A. (2017b). Inclusive four-jet production at 7 and 13 TeV: Azimuthal profile in multi-Regge kinematics, *Eur. Phys. J.* **C77**, 1, p. 5, doi:10.1140/epjc/s10052-016-4557-z, arXiv:1606.00574 [hep-ph].

Caporale, F., Chachamis, G., Murdaca, B., and Sabio Vera, A. (2016b). Balitsky-Fadin-Kuraev-Lipatov predictions for inclusive three jet production at the LHC, *Phys. Rev. Lett.* **116**, 1, p. 012001, doi:10.1103/PhysRevLett.116.012001, arXiv:1508.07711 [hep-ph].

Caporale, F., Ivanov, D. Yu., Murdaca, B., and Papa, A. (2013a). Mueller-Navelet small-cone jets at LHC in next-to-leading BFKL, *Nucl. Phys. B* **877**, pp. 73–94, doi: 10.1016/j.nuclphysb.2013.09.013, arXiv:1211.7225 [hep-ph].

Caporale, F., Ivanov, D. Yu., and Papa, A. (2008). BFKL resummation effects in the gamma* gamma* total hadronic cross section, *Eur. Phys. J. C* **58**, pp. 1–7, doi: 10.1140/epjc/s10052-008-0732-1, arXiv:0807.3231 [hep-ph].

Caporale, F., Murdaca, B., Sabio Vera, A., and Salas, C. (2013b). Scale choice and collinear contributions to Mueller-Navelet jets at LHC energies, *Nucl. Phys.* **B875**, pp. 134–151, doi:10.1016/j.nuclphysb.2013.07.005, arXiv:1305.4620 [hep-ph].

Celiberto, F. G., Ivanov, D. Yu., Murdaca, B., and Papa, A. (2017). Dihadron production at the LHC: full next-to-leading BFKL calculation, *Eur. Phys. J.* **C77**, 6, p. 382, doi:10.1140/epjc/s10052-017-4949-8, arXiv:1701.05077 [hep-ph].

Celmaster, W. and Gonsalves, R. J. (1979a). QCD perturbation expansions in a coupling constant renormalized by momentum space subtraction, *Phys. Rev. Lett.* **42**, p. 1435, doi:10.1103/PhysRevLett.42.1435.

Celmaster, W. and Gonsalves, R. J. (1979b). The renormalization prescription dependence of the QCD coupling constant, *Phys. Rev. D* **20**, p. 1420, doi:10.1103/PhysRevD.20.1420.

Chachamis, G., Hentschinski, M., Madrigal Martnez, J. D., and Sabio Vera, A. (2013). Next-to-leading order corrections to the gluon-induced forward jet vertex from the high energy effective action, *Phys. Rev.* **D87**, 7, p. 076009, doi:10.1103/PhysRevD.87.076009, arXiv:1212.4992 [hep-ph].

Chachamis, G., Hentschinski, M., Madrigal Martnez, J. D., and Sabio Vera, A. (2014). Forward jet production & quantum corrections to the gluon Regge trajectory from Lipatov's high energy effective action, *Phys. Part. Nucl.* **45**, 4, pp. 788–799, doi: 10.1134/S1063779614040030, arXiv:1211.2050 [hep-ph].

Chachamis, G. and Sabio Vera, A. (2016a). Phenomenology and formal studies on small-x physics by using Monte Carlo techniques, *Nucl. Part. Phys. Proc.* **273–275**, pp. 2767–2769, doi:10.1016/j.nuclphysbps.2015.10.057.

Chachamis, G. and Sabio Vera, A. (2016b). The high-energy radiation pattern from BFKLex with double-log collinear contributions, *JHEP* **02**, p. 064, doi:10.1007/JHEP02(2016)064, arXiv:1512.03603 [hep-ph].

Chatrchyan, S. *et al.* (2011). Measurement of the underlying event activity at the LHC with $\sqrt{s} = 7\,\text{TeV}$ and comparison with $\sqrt{s} = 0.9$ TeV, *JHEP* **09**, p. 109, doi: 10.1007/JHEP09(2011)109, arXiv:1107.0330 [hep-ex].

Chatrchyan, S. *et al.* (2012). Ratios of dijet production cross sections as a function of the absolute difference in rapidity between jets in proton-proton collisions at $\sqrt{s} = 7\,\text{TeV}$, *Eur. Phys. J. C* **72**, p. 2216, doi:10.1140/epjc/s10052-012-2216-6, arXiv:1204.0696 [hep-ex].

Cheng, H. and Wu, T. T. (1970). Logarithmic factors in the high-energy behavior of quantum electrodynamics, *Phys. Rev. D* **1**, pp. 2775–2794, doi:10.1103/PhysRevD.1.2775.

Chevallier, F., Kepka, O., Marquet, C., and Royon, C. (2009). Gaps between jets at hadron colliders in the next-to-leading BFKL framework, *Phys. Rev.* **D79**, p. 094019, doi: 10.1103/PhysRevD.79.094019, arXiv:0903.4598 [hep-ph].

Chirilli, G. A. and Kovchegov, Y. V. (2014). $\gamma^*\gamma^*$ Cross section at NLO and properties of the BFKL evolution at higher orders, *JHEP* **05**, p. 099, doi:10.1007/JHEP05(2014)099,10.1007/JHEP08(2015)075, arXiv:1403.3384 [hep-ph], [Erratum: JHEP08,075(2015)].

Ciafaloni, M. and Camici, G. (1998). Energy scale(s) and next-to-leading BFKL equation, *Phys. Lett. B* **430**, pp. 349–354, doi:10.1016/S0370-2693(98)00551-6, arXiv:hep-ph/9803389 [hep-ph].

Colferai, D., Schwennsen, F., Szymanowski, L., and Wallon, S. (2010). Mueller Navelet jets at LHC — complete NLL BFKL calculation, *JHEP* **12**, p. 026, doi:10.1007/JHEP12(2010)026, arXiv:1002.1365 [hep-ph].

Corke, R. and Sjöstrand, T. (2011). Interleaved parton showers and tuning prospects, *JHEP* **3**, p. 32, doi:10.1007/JHEP03(2011)032, arXiv:1011.1759 [hep-ph].

Del Duca, V. and Schmidt, C. R. (1994a). Dijet production at large rapidity intervals, *Phys. Rev. D* **49**, pp. 4510–4516, doi:10.1103/PhysRevD.49.4510, arXiv:hep-ph/9311290 [hep-ph].

Del Duca, V. and Schmidt, C. R. (1994b). Mini-jet corrections to Higgs production, *Phys. Rev. D* **49**, pp. 177–182, doi:10.1103/PhysRevD.49.177, arXiv:hep-ph/9305346 [hep-ph].

Dokshitzer, Y. L. (1977). Calculation of the structure functions for deep inelastic scattering and e+ e− annihilation by perturbation theory in quantum chromodynamics, *Sov. Phys. JETP* **46**, pp. 641–653, [Zh. Eksp. Teor. Fiz. 73,1216 (1977)].

Ducloué, B., Szymanowski, L., and Wallon, S. (2012). Mueller-Navelet jets at LHC: the first complete NLL BFKL study, *PoS* **QNP2012**, p. 165, arXiv:1208.6111 [hep-ph].

Ducloue, B., Szymanowski, L., and Wallon, S. (2013). Confronting Mueller-Navelet jets in NLL BFKL with LHC experiments at 7 TeV, *JHEP* **05**, p. 096, doi:10.1007/JHEP05(2013)096, arXiv:1302.7012 [hep-ph].

Duclou, B., Szymanowski, L., and Wallon, S. (2014). Evidence for high-energy resummation effects in Mueller-Navelet jets at the LHC, *Phys. Rev. Lett.* **112**, p. 082003, doi:10.1103/PhysRevLett.112.082003, arXiv:1309.3229 [hep-ph].

Ekstedt, A., Enberg, R., and Ingelman, G. (2017). Hard color singlet BFKL exchange and gaps between jets at the LHC, arXiv:1703.10919 [hep-ph].

Enberg, R., Ingelman, G., and Motyka, L. (2002). Hard color singlet exchange and gaps between jets at the Tevatron, *Phys. Lett.* **B524**, pp. 273–282, doi:10.1016/S0370-2693(01)01379-X, arXiv:hep-ph/0111090 [hep-ph].

Fadin, V. and Lipatov, L. (1998). BFKL pomeron in the next-to-leading approximation, *Phys. Lett. B* **429**, pp. 127–134, doi:10.1016/S0370-2693(98)00473-0, arXiv:hep-ph/9802290 [hep-ph].

Fadin, V. S., Kim, V. T., Lipatov, L. N., and Pivovarov, G. B. (2002). The BFKL pomeron within physical renormalization schemes and scales, in *35th Annual Winter School on Nuclear and Particle Physics*, arXiv:hep-ph/0207296.

Fadin, V. S., Kuraev, E. A., and Lipatov, L. N. (1975). On the Pomeranchuk singularity in asymptotically free theories, *Phys. Lett. B* **60**, pp. 50–52, doi:10.1016/0370-2693(75)90524-9.

Field, R. (2010). Early LHC underlying event data — findings and surprises, in *21st Hadron Collider Physics Symposium*, arXiv:1010.3558 [hep-ph].

Freeman, P. E., Doe, S., and Siemiginowska, A. (2001). Sherpa: A mission-independent data analysis application, *Proc. SPIE Int. Soc. Opt. Eng.* **4477**, p. 76, doi:10.1117/12.447161, arXiv:astro-ph/0108426.

Frixione, S., Nason, P., and Oleari, C. (2007). Matching NLO QCD computations with parton shower simulations: The POWHEG method, *JHEP* **11**, p. 070, doi:10.1088/1126-6708/2007/11/070, arXiv:0709.2092 [hep-ph].

Frolov, G. V., Gribov, V. N., and Lipatov, L. N. (1970). On Regge poles in quantum electrodynamics, *Phys. Lett. B* **31**, p. 34, doi:10.1016/0370-2693(70)90013-4.

Gavrilov, V., Kim, V., Krokhotin, A., Pivovarov, G., and Safronov, G. (2008). Asymptotic effects in jet production at high energies, in *15th International Seminar on High Energy Physics*, arXiv:0901.1985 [hep-ph].

Gavrilov, V. B., Kim, V. T., Murzin, V. A., Oreshkin, V. A., Pivovarov, G. B., Pozdnyakov, I. Yu., and Safronov, G. B. (2013). Forward dijets with wide rapidity separation in *pp*-collisions at LHC and Tevatron: dijet ratios and azimuthal decorrelations, *Nucl. Phys. Proc. Suppl.* **245**, pp. 153–156, doi:10.1016/j.nuclphysbps.2013.10.029.

Gieseke, S. *et al.* (2011). Herwig++ 2.5 Release note, arXiv:1102.1672 [hep-ph].

Goncalves, V. P., Kugeratski, M. S., Cazaroto, E. R., Carvalho, F., and Navarra, F. S. (2011). Non-linear QCD dynamics in two-photon interactions at high energies, *Eur. Phys. J.* **C71**, p. 1779, doi:10.1140/epjc/s10052-011-1779-y, arXiv:1009.1112 [hep-ph].

Gribov, L. V., Levin, E. M., and Ryskin, M. G. (1981a). High p_T hadrons in the pionization region in QCD, *Phys. Lett. B* **100**, pp. 173–176, doi:10.1016/0370-2693(81)90768-1.

Gribov, L. V., Levin, E. M., and Ryskin, M. G. (1981b). Singlet structure function at small x: Unitarization of gluon ladders, *Nucl. Phys. B* **188**, pp. 555–576, doi:10.1016/0550-3213(81)90007-9.

Gribov, L. V., Levin, E. M., and Ryskin, M. G. (1983a). Large — Et processes as a main source of hadrons at very high energies, *Phys. Lett. B* **121**, pp. 65–71, doi:10.1016/0370-2693(83)90204-6.

Gribov, L. V., Levin, E. M., and Ryskin, M. G. (1983b). Semihard processes in QCD, *Phys. Rept. C* **100**, pp. 1–150, doi:10.1016/0370-1573(83)90022-4.

Gribov, V. N. and Lipatov, L. N. (1972a). Deep inelastic e p scattering in perturbation theory, *Sov. J. Nucl. Phys.* **15**, pp. 438–450, [Yad. Fiz. 15, 781 (1972)].

Gribov, V. N. and Lipatov, L. N. (1972b). e+ e- pair annihilation and deep inelastic e p scattering in perturbation theory, *Sov. J. Nucl. Phys.* **15**, pp. 675–684, [Yad. Fiz. 15, 1218 (1972)].

Gribov, V. N., Lipatov, L. N., and Frolov, G. V. (1971). The leading singularity in the j plane in quantum electrodynamics, *Sov. J. Nucl. Phys.* **12**, p. 543, [Yad. Fiz. 12, 994(1970)].

Hatta, Y. and Ueda, T. (2009). Jet energy flow at the LHC, *Phys. Rev.* **D80**, p. 074018, doi:10.1103/PhysRevD.80.074018, arXiv:0909.0056 [hep-ph].

Hentschinski, M., Madrigal Martinez, J. D., Murdaca, B., and Sabio Vera, A. (2014a). The gluon-induced Mueller-Tang jet impact factor at next-to-leading order, *Nucl. Phys.* **B889**, pp. 549–579, doi:10.1016/j.nuclphysb.2014.10.026, arXiv:1409.6704 [hep-ph].

Hentschinski, M., Madrigal Martinez, J. D., Murdaca, B., and Sabio Vera, A. (2014b). The quark induced Mueller-Tang jet impact factor at next-to-leading order, *Nucl. Phys.* **B887**, pp. 309–337, doi:10.1016/j.nuclphysb.2014.08.010, arXiv:1406.5625 [hep-ph].

Hentschinski, M. and Sabio Vera, A. (2012). NLO jet vertex from Lipatov's QCD effective action, *Phys. Rev.* **D85**, p. 056006, doi:10.1103/PhysRevD.85.056006, arXiv:1110.6741 [hep-ph].

Ivanov, D. Yu., Murdaca, B., and Papa, A. (2014). The $\gamma^*\gamma^*$ total cross section in next-to-leading order BFKL and LEP2 data, *JHEP* **10**, p. 058, doi:10.1007/JHEP10(2014)058, arXiv:1407.8447 [hep-ph].

Ivanov, D. Yu. and Papa, A. (2012). The next-to-leading order forward jet vertex in the small-cone approximation, *JHEP* **05**, p. 086, doi:10.1007/JHEP05(2012)086, arXiv:1202.1082 [hep-ph].

Jung, H., Baranov, S., Deak, M., Grebenyuk, A., Hautmann, F., *et al.* (2010). The CCFM Monte Carlo generator CASCADE version 2.2.03, *Eur. Phys. J. C* **70**, pp. 1237–1249, doi:10.1140/epjc/s10052-010-1507-z, arXiv:1008.0152 [hep-ph].

Jung, H. and Salam, G. P. (2001). Hadronic final state predictions from CCFM: The Hadron level Monte Carlo generator CASCADE, *Eur. Phys. J. C* **19**, pp. 351–360, doi:10.1007/s100520100604, arXiv:hep-ph/0012143 [hep-ph].

Kataev, A. L. and Mikhailov, S. V. (2012). New perturbation theory representation of the conformal symmetry breaking effects in gauge quantum field theory models, *Theor. Math. Phys.* **170**, pp. 139–150, doi:10.1007/s11232-012-0016-7, arXiv:1011.5248 [hep-ph], [Teor. Mat. Fiz. 170, 174 (2012)].

Kataev, A. L. and Mikhailov, S. V. (2015). Generalization of the Brodsky-Lepage-Mackenzie optimization within the ?-expansion and the principle of maximal conformality, *Phys. Rev. D* **91**, 1, p. 014007, doi:10.1103/PhysRevD.91.014007, arXiv:1408.0122 [hep-ph].

Kepka, O., Marquet, C., and Royon, C. (2011). Gaps between jets in hadronic collisions, *Phys. Rev.* **D83**, p. 034036, doi:10.1103/PhysRevD.83.034036, arXiv:1012.3849 [hep-ph].

Khachatryan, V. *et al.* (2016). Azimuthal decorrelation of jets widely separated in rapidity in pp collisions at $\sqrt{s} = 7\,\text{TeV}$, *JHEP* **08**, p. 139, doi:10.1007/JHEP08(2016)139, arXiv:1601.06713 [hep-ex].

Kim, V. T. and Pivovarov, G. B. (1996a). BFKL QCD pomeron in high-energy hadron collisions: Inclusive dijet production, *Phys. Rev. D* **53**, pp. 6–10, doi:10.1103/PhysRevD.53.6, arXiv:hep-ph/9506381 [hep-ph].

Kim, V. T. and Pivovarov, G. B. (1996b). New effective Feynman — like rules for the multi-Regge QCD asymptotics of inclusive multi-jet production, *Phys. Rev.* **D54**, pp. R725–R729, doi:10.1103/PhysRevD.54.R725, arXiv:hep-ph/9601335 [hep-ph].

Kim, V. T. and Pivovarov, G. B. (1998). A BFKL pomeron manifestation in inclusive single jet production at high-energy hadron collisions, *Phys. Rev. D* **57**, pp. 1341–1344, doi:10.1103/PhysRevD.57.1341, arXiv:hep-ph/9709433 [hep-ph].

Kuraev, E. A., Lipatov, L. N., and Fadin, V. S. (1976). Multi-Reggeon processes in the Yang-Mills theory, *Sov. Phys. JETP* **44**, pp. 443–450, [Zh. Eksp. Teor. Fiz. 71, 840 (1976)].

Kuraev, E. A., Lipatov, L. N., and Fadin, V. S. (1977). The Pomeranchuk singularity in nonabelian gauge theories, *Sov. Phys. JETP* **45**, pp. 199–204, [Zh. Eksp. Teor. Fiz. 72, 377 (1977)].

Levin, E. M. and Ryskin, M. G. (1980). Quantum chromodynamics prediction for heavy lepton pair production in pionization region, *Yad. Fiz.* **32**, pp. 802–816.

Lipatov, L. N. (1975). The parton model and perturbation theory, *Sov. J. Nucl. Phys.* **20**, pp. 94–102, [Yad. Fiz. 20,181 (1974)].

Lipatov, L. N. (1986). The bare Pomeron in quantum chromodynamics, *Sov. Phys. JETP* **63**, pp. 904–912, [Zh. Eksp. Teor. Fiz. 90,1536 (1986)].

Mikhailov, S. V. (2007). Generalization of BLM procedure and its scales in any order of pQCD: A Practical approach, *JHEP* **06**, p. 009, doi:10.1088/1126-6708/2007/06/009, arXiv:hep-ph/0411397 [hep-ph].

Mojaza, M., Brodsky, S. J., and Wu, X.-G. (2013). Systematic all-orders method to eliminate renormalization-scale and scheme ambiguities in perturbative QCD, *Phys. Rev. Lett.* **110**, p. 192001, doi:10.1103/PhysRevLett.110.192001, arXiv:1212.0049 [hep-ph].

Mueller, A. H. and Navelet, H. (1987). An inclusive minijet cross-section and the bare Pomeron in QCD, *Nucl. Phys. B* **282**, p. 727, doi:10.1016/0550-3213(87)90705-X.

Mueller, A. H. and Tang, W.-K. (1992). High-energy parton-parton elastic scattering in QCD, *Phys. Lett.* **B284**, pp. 123–126, doi:10.1016/0370-2693(92)91936-4.

Nason, P. (2004). A New method for combining NLO QCD with shower Monte Carlo algorithms, *JHEP* **11**, p. 040, doi:10.1088/1126-6708/2004/11/040, arXiv:hep-ph/0409146 [hep-ph].

Orr, L. H. and Stirling, W. J. (1998a). A BFKL Monte Carlo approach to jet production at hadron hadron and lepton — hadron colliders, in *High-energy physics. Proceedings, 29th International Conference, ICHEP'98, Vancouver, Canada, July 23–29, 1998. Vol. 1, 2*, pp. 755–759, arXiv:hep-ph/9811423 [hep-ph].

Orr, L. H. and Stirling, W. J. (1998b). The Collision energy dependence of dijet cross-sections as a probe of BFKL physics, *Phys. Lett. B* **429**, pp. 135–144, doi:10.1016/S0370-2693(98)00471-7, arXiv:hep-ph/9801304 [hep-ph].

Ostrovsky, D. (2000). NLO correction to one particle inclusive production at high-energies, *Phys. Rev. D* **62**, p. 054028, doi:10.1103/PhysRevD.62.054028, arXiv:hep-ph/9912258 [hep-ph].

Ryskin, M. G. (1980). High transverse momentum hadron production in the pionization region and vacuum singularity in QCD, *Yad. Fiz.* **32**, pp. 259–274.

Sabio Vera, A. (2006). The effect of NLO conformal spins in azimuthal angle decorrelation of jet pairs, *Nucl. Phys. B* **746**, pp. 1–14, doi:10.1016/j.nuclphysb.2006.04.004, arXiv:hep-ph/0602250 [hep-ph].

Sabio Vera, A. and Schwennsen, F. (2007). The Azimuthal decorrelation of jets widely separated in rapidity as a test of the BFKL kernel, *Nucl. Phys. B* **776**, pp. 170–186, doi:10.1016/j.nuclphysb.2007.03.050, arXiv:hep-ph/0702158 [HEP-PH].

Schmidt, C. R. (1997). A Monte Carlo solution to the BFKL equation, *Phys. Rev. Lett.* **78**, pp. 4531–4535, doi:10.1103/PhysRevLett.78.4531, arXiv:hep-ph/9612454 [hep-ph].

Sirunyan, A. M. *et al.* (2018). Study of dijet events with a large rapidity gap between the two leading jets in pp collisions at $\sqrt{s} = 7$ TeV, *Eur. Phys. J.* **C78**, 3, p. 242, doi:10.1140/epjc/s10052-018-5691-6, arXiv:1710.02586 [hep-ex].

Sjöstrand, T., Mrenna, S., and Skands, P. Z. (2006). PYTHIA 6.4 Physics and Manual, *JHEP* **05**, p. 026, doi:10.1088/1126-6708/2006/05/026, arXiv:hep-ph/0603175 [hep-ph].

Sjöstrand, T., Mrenna, S., and Skands, P. Z. (2008). A brief introduction to PYTHIA 8.1, *Comput. Phys. Commun.* **178**, pp. 852–867, doi:10.1016/j.cpc.2008.01.036, arXiv:0710.3820 [hep-ph].

Stirling, W. J. (1994). Production of jet pairs at large relative rapidity in hadron hadron collisions as a probe of the perturbative pomeron, *Nucl. Phys. B* **423**, pp. 56–79, doi:10.1016/0550-3213(94)90565-7, arXiv:hep-ph/9401266 [hep-ph].

https://doi.org/10.1142/9789811231124_0021

Chapter 21

Lipatov's Legacy and the Future of Deep Inelastic Scattering

Paul Newman

University of Birmingham, UK

21.1 Introduction

Around 1993-4, as a PhD student looking for an analysis topic using the first data from the world's only ever lepton-hadron collider facility, HERA, I became intrigued by the first observations [Derrick *et al.* (1993b); Ahmed *et al.* (1994)] of 'diffractive' deep inelastic scattering (DDIS, $ep \to eXp$) events in which the proton appeared to stay intact, despite the very large momentum transfer from the electron. Hearing about the mysterious 'Pomeron' that was believed to mediate the process, I was completely hooked! In those early days, the questions we asked were relatively simple. Was this the 'soft' Pomeron [Feinberg and Pomerančuk (1956)] that historically described elastic, total and soft diffractive cross sections, with some degree of universality [Kaidalov (1979); Goulianos (1983)], or was it a first sighting of a 'hard' Pomeron, emerging as a high energy 'Regge' limit of QCD according to the work of Lev Nikolaevich Lipatov and others [Lipatov (1986, 1997)]. Searching for evidence for this QCD BFKL Pomeron was a major theme throughout the lifetime of HERA and in some sense I am still working on it.

Applying the QCD Pomeron to DDIS is only a small part of a much richer phenomenology that arises as a consequence of the BFKL equation [Kuraev *et al.* (1976, 1977); Balitsky and Lipatov (1978)], a differential equation in $\ln 1/x$ for an 'unintegrated' gluon density at asymptotically large energies. The unintegrated gluon density can be convoluted with off-mass-shell parton-level cross sections to make predictions for observables in the k_T factorisation scheme [Gribov *et al.* (1983); Levin *et al.* (1991); Catani *et al.* (1991); Collins and Ellis (1991)]. The corresponding Pomeron emerges starting from a basic two-gluon exchange [Low (1975); Nussinov (1975)], developing into a 'ladder' with rungs corresponding via the optical theorem to x-ordered successive emissions of partons. As well as DDIS, the BFKL Pomeron can be applied to predict energy dependences of inclusive deep inelastic scattering (DIS) $ep \to eX$ and quasi-elastic vector meson production $ep \to eVp$. Meanwhile, the properties of the hadronic final state of the inclusive DIS process are driven by the details of the parton cascade between the proton and the virtual exchange photon. Often at HERA, we (almost certainly naiively) framed physics questions in terms of which of the two major evolution schemes was the dominant feature of the data; BFKL with its $\ln 1/x$ resummation and x ordering of partons, or DGLAP [Gribov and Lipatov (1972a,b); Lipatov (1975); Altarelli and Parisi (1977); Dokshitzer (1977)] with $\ln Q^2$ resummations and transverse momentum ordering. Whilst it is undoubtedly BFKL for which Lipatov is best known and most closely associated, it is a measure of his stature and contribution that he is the 'L' in both evolution schemes; the common link that joins the field together.

Whilst DGLAP evolution is fully accepted and widely used, BFKL is not yet well established in existing data, neither from HERA nor at the proton-proton energy frontier at the LHC. As a basic prediciton of QCD, it must ultimately exist; the question is where one needs to look. Large logarithms of $1/x$ play the central role and, through the kinematic correlation in electron-proton scattering $x = Q^2/sy$ where the kinematic variables have their usual meanings, large $1/x$ implies large centre of mass energy s. Given the dominance of the gluon density in low x proton structure, it seems apparent that low x DIS and BFKL physics are intimately linked. Searching for BFKL physics therefore sits, and will continue to sit, as a basic experimental question at the energy frontier of QCD.

This article describes low x physics aspects of future proposed DIS facilities. After a brief introduction to the currently proposed future faciltities, inclusive DIS, exclusive vector meson production, DDIS and final state studies are all reviewed, emphasising the role that might be played by Lipatov's work.

21.2 Possible futures for deep inelastic scattering

An overview of past, present and future electron-proton scattering facilties is shown in the plane of energy versus luminosity in Fig. 21.1. The only collider facility realised to date is HERA. Its large centre of mass energy opened up the field of low x physics. There are at present three proposals under discussion for future DIS collider facilities. The EIC and LHeC / FCC-eh both offer three orders of magnitude increases in luminosity over HERA, but have rather different physics focuses, commensuarate with their centre of mass energies. Whilst the EIC will operate with polarised protons and heavy ions to investigate three-dimensional proton structure and nuclear effects, the LHeC and FCC-eh offer new phase space in *ep* physics towards higher $\sqrt{s}$ or equivalently low x. A third potential future facility, VHEeP is not shown in Fig. 21.1, but would extend the $\sqrt{s}$ or $1/x$ reach still further, albeit at low luminosities. These three proposed facilities are briefly described in the following. Their comparisons and complementarity are expanded in more detail in [Caldwell *et al.* (2018)].

21.2.1 *Electron-Ion Collider (EIC)*

Designs for a future high-luminosity polarised electron ion collider (EIC) have evolved in the past few years at Jefferson Laboratory and Brookhaven National

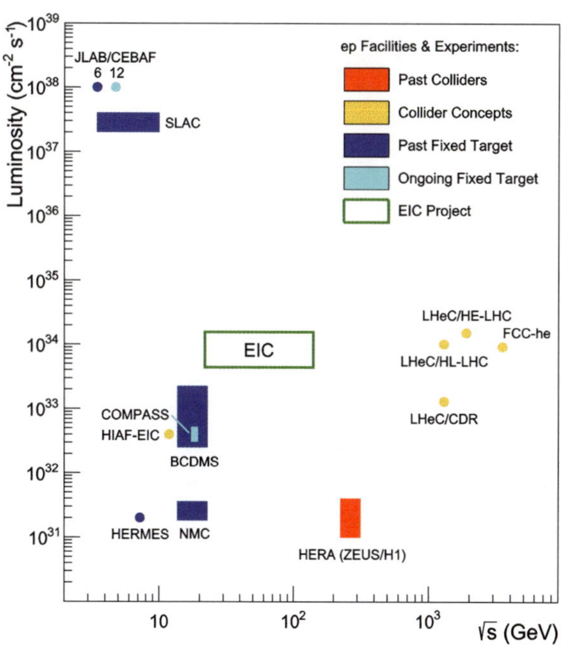

Fig. 21.1 Illustration [Ullrich (2019)] of the past, present and possible future landscape of DIS experiments in terms of *ep* centre of mass energy $\sqrt{s}$ versus instantaneous luminosity. VHEeP is not shown, but would sit in an energy range $1 < \sqrt{s} < 9\,\text{TeV}$ at luminosities up to the order of $10^{28}\,\text{cm}^{-2}\text{s}^{-1}$.

Laboratory (BNL) in the United States [Accardi *et al.* (2016)]. The EIC project has recently received a major boost by attracting Department of Energy CD0 'mission need' approval, together with a site selection at BNL. It is therefore the most advanced of the future faciltities in terms of approval status and might be ready for data taking on a timescale of around 2030 at a cost of \$1.6 – 2.6 billion. The EIC will be realised by adding an electron storage ring to the existing RHIC heavy ion collider complex at BNL to enable electron-ion collisions. The forseen energy range is $20 < \sqrt{s} < 140\,\text{GeV}$ at luminosities of order $10^{34}\,\text{cm}^{-2}\text{s}^{-1}$. The EIC design incorporates a wide range of beam energies and polarisations for protons, deuterons and ions, enabling an unprecedented nucleon and nuclear femtography program.

Due to its high luminosity and favourable kinematics, the EIC is expected to provide much-needed *ep* scattering data that are sensitive to parton densities at large x. Since the EIC centre of mass energy is smaller than that of HERA, it is unlikely that its inclusive data will shed fresh light on low x energy-driven effects such as BFKL dynamics. However, it remains plausible that its much higher luminosity than that of HERA and extensive programme of investigating exclusive processes may reveal something that has previously been missed. The facility may also be sensitive to density-driven (as distinct from energy-driven) low x effects such as parton saturation [Mueller (1990)], due to its extensive plans for electron-ion scattering, the density of partons being enhanced over the case for electron-proton scattering by a factor $A^{1/3}$.

21.2.2 *Large Hadron-Electron Collider (LHeC) and Future Circular Electron-Hadron Collider (FCC-eh)*

The LHeC [Abelleira Fernandez *et al.* (2012); LHeC and FCC-he Study Group (2020)] is a proposal to add an intense 60 GeV electron beam to the LHC proton (and heavy ion) accelerator complex, enabling *ep* or *eA* collisions to take place at the same time as the ongoing *pp*, *pA* and *AA* programme. The electron beam is based on a racetrack energy-recovery linear accelerator (ERL) configuration, in which beams circulate three times through two 10 GeV linear accelerator structures before being brought into collision with one of the LHC proton beams and subsequently winding back down through the same structures, enabling energy recovery. The first step towards the LHeC is to build a lower energy prototype ('PERLE'), which is currently planned for Orsay [Angal-Kalinin *et al.* (2018)]. The LHeC reaches previously unexplored energies, (nominally $\sqrt{s} = 1300\,\text{GeV}$) at high luminosity, $10^{34}\,\text{cm}^{-2}\text{s}^{-1}$. Its future prospects depend on the overall strategy for paricle physics in Europe and worldwide. With a price tag of the order of CHF 1 Billion, its cost is an order of magnitude lower than currently discussed future energy frontier e^+e^- and *pp* collider projects. The LHeC physics programme is very broad; highlights include complementary characterisation of the Higgs boson to that at the LHC, reaching a new level of precision in proton PDF deteminations which, particularly at high x, will empower the LHC to extend its search programme to new particles with larger masses, as well as a major extension in kinematic range for the exploration of low x effects, as discussed here.

The FCC project [Abada *et al.* (2019)] is investigating the possibility and physics capabilities of adding a new 100 km circumference tunnel to the CERN infrastructure, housing electron or hadron beams. The hadron-hadron collider design (FCC-hh) could be complemented by an electron accelerator of similar design to that of the LHeC, enabling very high energy ep and eA collisions. The FCC-eh would increase the energy reach for electron-proton collisions to $\sqrt{s} = 3500\,\text{GeV}$, again with luminosity of order $1.5 \times 10^{34}\,\text{cm}^{-2}\text{s}^{-1}$. Timescales for realisation are long and depend critically on the future direction of CERN's programme.

Integrated luminosities of order 1 ab^{-1} and 3 ab^{-1} are envisioned for the LHeC and FCC-eh, respectively. However, a low x programme can be realised with much smaller amounts of data. As described in [LHeC and FCC-he Study Group (2020)], the LHeC programme might be staged, with a first component overlapping with LHC operation and an extended programme taking place in standalone mode. An initial luminosity of $50\,\text{fb}^{-1}$, as envisioned for a three year 'LHeC First Run' programme during LHC run 5, would already be more than sufficient for the investigation of the observables discussed in this document.

21.2.3 *Very-High Energy Electron-Proton Collider (VHEeP)*

The VHEeP design [Caldwell and Wing (2016)] aims to exploit the highly encouraging progress made by the CERN AWAKE project [Edli *et al.* (2018)], in plasma wakefield acceleration. The existing 7 TeV per proton LHC bunches[1] are used as the driver for the protons to create a wakefield which then accelerates electrons to 3 TeV. The electrons then collide with the other LHC proton beam. The potential energy reach of the VHEeP design thus extends to $\sqrt{s} = 9000\,\text{GeV}$, the highest values among current proposals. If sufficient beam stability can be achieved, luminosities of up to $10^{29}\,\text{cm}^{-2}\text{s}^{-1}$ have been predicted. Although this is only 1% of the maximum achieved by HERA, VHEeP would still be able to measure the ultra-high-energy behavior of hadronic cross sections in regions where new QCD dynamics must exist.

21.3 Phenomenology and background from HERA

Much of low x physics, including the phenomenological implications of the BFKL Pomeron, is concerned with energy dependences. These are conveniently expressed in terms of Regge phenomenology, which predates QCD and applies at asymptotically large $\sqrt{s}$. In terms of a 'Pomeron' trajectory $\alpha(t) = \alpha(0) + \alpha' t$ with intercept $\alpha(0)$, elastic cross sections behave as $d\sigma_{\text{el}}/dt \propto s^{2\alpha(t)-2}$ at fixed t and, via

[1] A similar scheme, using the CERN SPS accelerator as the proton driver, known as PEPIC, has also been suggested.

the optical theorem, total cross sections follow $\sigma_{\text{tot}} \propto s^{\alpha(0)-1}$. In the context of DIS at modest Q^2, it is natural to think of the cross section for the inclusive neutral current DIS process $ep \to eX$ as factorising into a flux of virtual photons from the electron and a total $\gamma^* p$ cross section. The $\gamma^* p$ centre of mass energy W is related to the familiar kinematic variables of DIS via $W^2 = Q^2/x$. In Regge language, the total cross section at fixed Q^2 is thus expected to follow a power-law behaviour $\sigma_{\text{tot}}^{\gamma^* p} \propto x^{-(\alpha(0)-1)}$ and may be discussed in terms of an effective Pomeron intercept. Similarly, the forward (i.e. $t \to 0$) quasi-elastic cross section for exclusive vector meson production follows $\mathrm{d}\sigma_{\text{EL}}^{\gamma^* p}/dt \propto x^{-2(\alpha(0)-1)}$.

The 'soft' Pomeron intercept, governing for example the slowly rising total pp and $\bar{p}p$ high energy cross sections, has been found to be a little larger than unity; for example the famous Dannachie and Landshoff simple pole analysis finds $\alpha(0) = 1.08 - 1.10$ [Donnachie and Landshoff (1992); Cudell *et al.* (1997)].[2] The solution to the BFKL evolution equation is a gluon density that grows with $1/x$ as a power, $x^{-\omega_{I\!P}}$, such that $1 + \omega_{I\!P}$ is the hard Pomeron intercept. In the leading logarithmic approximation, $\omega_{I\!P} = 4\bar{\alpha}_s \ln 2$ and $\bar{\alpha}_s = 3\alpha_s/\pi$ for the three colours of QCD, such that $\alpha(0) \simeq 1.5$, a much larger value than that for the soft Pomeron. By analogy with the soft Pomeron, the BFKL Pomeron then provides a contribution to elastic, total and diffractive dissociation cross sections which can be characterised in the language of Regge phenomenology in terms of steeply rising cross sections with increasing energy. In particular, elastic cross sections at vanishing t are predicted to follow $\mathrm{d}\sigma_{\text{EL}}^{\gamma^* p}/\mathrm{d}t \propto \left(W^2\right)^{2\omega_{I\!P}}$. For a detailed discussion, see e.g. [Forshaw and Ross (1997)].

When the first results on inclusive DIS emerged from HERA [Abt *et al.* (1993); Derrick *et al.* (1993a)] at intermediate $Q^2 \sim 10 - 100\,\text{GeV}^2$, steep rises with decreasing x were observed for the first time, not explainable by the soft Pomeron. Whilst not entirely unexpected from more traditional approaches [De Rujula *et al.* (1974)], this led to speculation that the hard Pomeron was at work. However, matters became still more ingtriguing when Lipatov and co-authors published the masterful next-to-leading-logarithm (NLL) corrections to the BFKL equation, indicating strongly downward changes in $\omega_{I\!P}$ [Fadin and Lipatov (1998); Ciafaloni and Camici (1998)]. As discussed elsewhere in this book [Stasto, A. M. (2020)], there has since been a long process of understanding and ultimately resumming these corrections, leading to a picture where $\omega_{I\!P}$ is well controlled and typically takes a value of around 0.3 for α_s values of relevance to low x physics in DIS.

Detailed data from HERA on power law fits are shown in Fig. 21.2. Figure 21.2(left) corresponds to inclusive data. The data exhibit clear transitions from a relatively soft energy dependence at low scales Q^2 to a stronger dependence at larger scales. As these more precise data emerged and were interpreted in more

[2]Results from elastic and total cross sections at the LHC suggest a breakdown of this behaviour, possibly due to multi-Regge exchanges as discussed for example in [Khoze, D. and Ryskin, M and Tasevsky, M. (2020)].

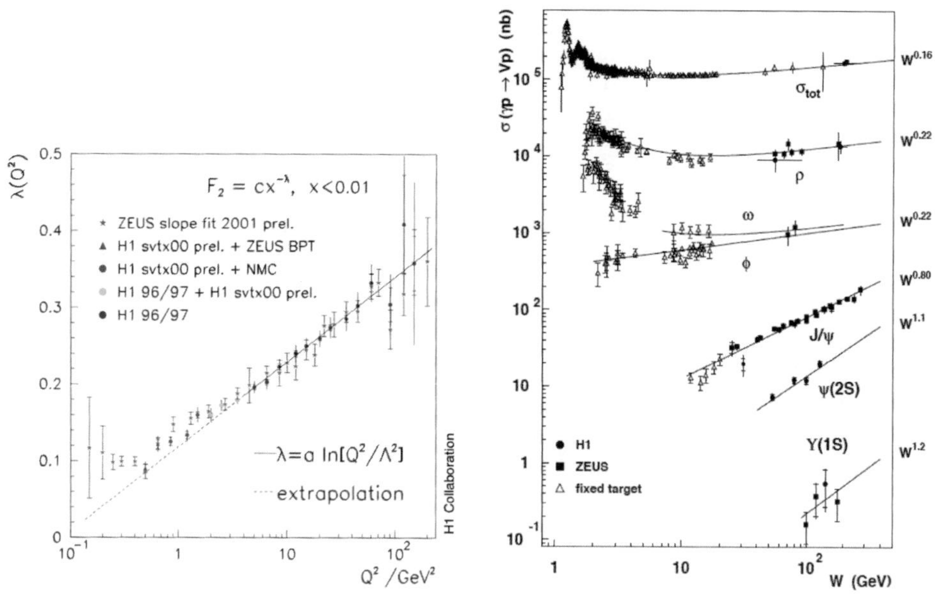

Fig. 21.2 Energy dependences for a subset of HERA data. (left) The Q^2 dependence of λ when inclusive DIS data are fitted to the form $F_2 \propto \sigma_{\rm tot}(\gamma^* p \to X) \propto x^{-\lambda}$ at each fixed Q^2 [Newman (2004)]. (right) W dependences for quasi-elastic vector meson photoproduction $\gamma p \to Vp$ for different vector meson species, V. The results of power law fits of the form $\sigma \propto W^{\delta}$ are indicateed in each case [Levy (2009)].

detail and subjected to fits to extract proton PDFs, it became well understood that a slowly rising behaviour at low Q^2 naturally changes into a faster-rising behaviour with increasing Q^2 through standard DGLAP evolution. This conclusion was further refined with increasingly precise HERA data over many years [Klein and Yoshida (2008)], culminating in the final combined H1 and ZEUS inclusive DIS measurements and their detailed interpretation in terms of proton structure [Abramowicz *et al.* (2015)].

Although the truly elastic $\gamma^* p$ cross section is not directly accessible experimentally, HERA provided extensive data [Newman and Wing (2014)] on exclusive vector meson production $\gamma^{(*)} p \to Vp$ for real and virtual photons, with vector meson species including $V = \rho, \phi, J/\psi, \psi(2S)$ and Υ. These processes can be interpreted in terms of the elastic scattering of a $q\bar{q}$ dipole produced from a photon splitting, with subsequent formation of the vector meson. The closely related Deeply Virtual Compton Scattering process (DVCS, $\gamma^{(*)} p \to \gamma p$) was also studied [Aaron *et al.* (2008, 2009); Chekanov *et al.* (2009a)]. All of the measurements conformed to power law W dependences, with an effective Pomeron intercept that evolved with scale from a soft behaviour e.g. in ρ photoproduction to a much harder value where large Q^2 or vector meson mass scales were extant. This tendency is illustrated for the photoproduction $Q^2 \to 0$ limit in Fig. 21.2(right). It is also clearly observed as Q^2 increases in ρ electroproduction data [Chekanov *et al.* (2007a); Aaron *et al.* (2010)].

Following the development of a framework [Bonvini *et al.* (2016)] to incorporate $\ln 1/x$ resummation [Altarelli *et al.* (2000, 2002)] at NLL accuracy, it became possible to include BFKL effects in otherwise standard DGLAP fits to inclusive *ep* data. A highly revealing fit to HERA data came from the recent study in the NNPDF framework [Forte *et al.* (2002)], including next-to-next-lo-leading order (NNLO) DGLAP evolution with NLL $\ln 1/x$ resummation [Ball *et al.* (2018)]. The substantial improvement in quality of fit to the data in [Abramowicz *et al.* (2015)] when the BFKL contribution is included originates from the small x, small Q^2 kinematic region, providing perhaps the most compelling evidence for BFKL dynamics in HERA data.

The interpretation of the novel effects that seem to be present in the lowest x HERA data is complicated by successful alternative descriptions, in which the $\gamma^* p$ cross section contains a low x saturation behaviour [Golec-Biernat and Wusthoff (1998)], which in the perturbative region might be interpreted as the onset of gluon recombination effects [Mueller (1990)]. This approach, based on density-driven low x effects in contrast to the energy-driven effects of BFKL, is usually implemented in dipole models, which can be applied deep into the non-perturbative Q^2 region. The full story of low x physics at HERA is thus rather complicated and uncertain, possibly including a delicate interplay of effects which may not be fully describable with any single approximation to QCD. A fully detailed interpretation will surely not be possible without data from a future higher energy *ep* facility.

21.4 Inclusive measurements

21.4.1 *Inclusive measurements at VHEeP*

The VHEeP concept is very well suited to the study of inclusive cross sections at low x, since the centre of mass energies are the largest among current proposals and the low luminosity is not a major problem at modest Q^2. With the anticipated beam energies, values of W in the range $10^3 - 10^4$ GeV would be reachable, corresponding to x of $10^{-7} - 10^{-8}$ for $Q^2 = 1\,\text{GeV}^2$.

Figure 21.3 [Caldwell and Wing (2016)] shows an initial study that illustrates the power of a machine such as VHEeP for the exploration of low x physics well beyond the kinematic region studied to date. Existing measurements of the inclusive neutral current *ep* cross section are shown in the form of the total $\gamma^* p$ cross section as a function of centre of mass energy W. Fits are shown for a range of scales Q^2 and two assumptions on the small x behavior. In one case, the cross section is assumed to continue to follow the simple power-law behavior $\sigma^{\gamma p} \propto x^{-\lambda(Q^2)}$ of Fig. 21.2(a). In the second case, double asymptotic scaling predictions [Ball and Forte (1994)] are invoked, such that $\sigma^{\gamma p} \propto e^{B(Q^2)\sqrt{log(1/x)}}$. Whilst the HERA data are compatible with both of these assumptions, the curves diverge dramatically in the TeV energy range where VHEeP becomes sensitive. The double asymptotic scaling prediction is derived using DGLAP evolution assuming a soft input behaviour at small x, though other approaches including for example the BFKL resummation of

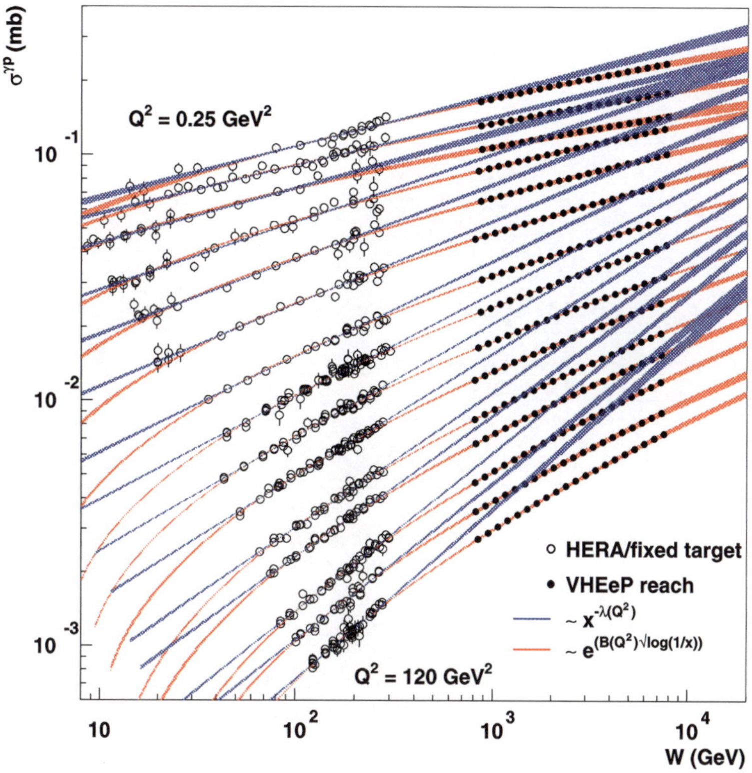

Fig. 21.3 Measurements (open points) of the total $\gamma^* p$ cross section as a function of centre of mass energy W for $0.25 < Q^2 < 120\,\text{GeV}^2$ from HERA [Abramowicz *et al.* (2015)] and fixed-target experiments [Adams *et al.* (1996); Arneodo *et al.* (1997)]. The blue and red lines show different fits to the data as described in the text. The kinematic reach of VHEeP is illustrated by the closed points. The expected uncertainties per data point are yet to be evaluated, but at modest Q^2 thhey might be expected to reach the 1% level achieved at HERA. Figure taken from [Caldwell and Wing (2016)].

[Ball *et al.* (2018)] or high density saturation [Golec-Biernat and Wusthoff (1998)] would have the same property of damping the growth with increasing W (or decreasing x) in the very high energy region, whilst maintaining an acceptable description of HERA data. If the luminosities achievable are sufficient to reach even a fraction of the Q^2 range shown, VHEeP would be an excellent facility to investigate the new low x effects implied by Lipatov's work in the context of inclusive data.

21.4.2 *Inclusive measurements at LHeC and FCC-eh*

The physics capabilities of the LHeC and FCC-eh concepts have been explored in considerable detail. The kinematic plane in (x, Q^2) for inclusive scattering is illustrated in Fig. 21.4, and placed in the context of other ep scattering facilities. Figure 21.4(left) shows the full kinematic plane, illustrating the extension towards lower x at fixed Q^2 by around an order of magnitude for LHeC over HERA and

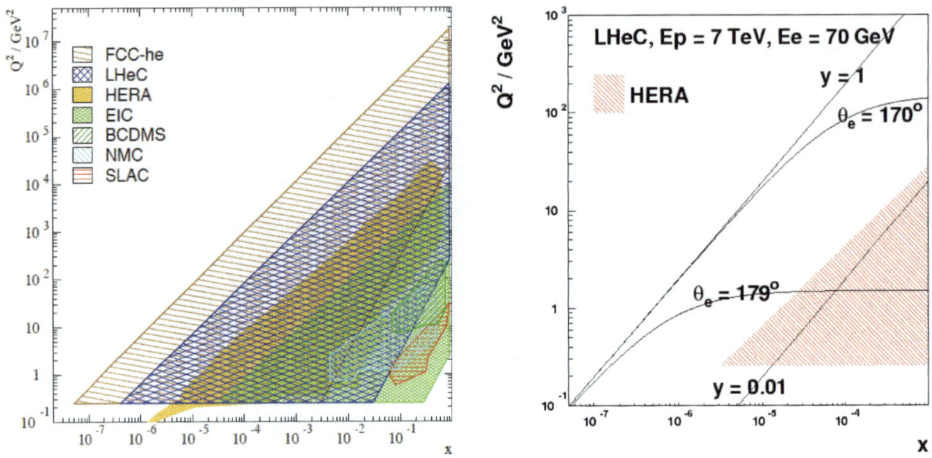

Fig. 21.4 (left) Kinematic plane showing acceptances as a function of x and Q^2 for various past and future DIS facilities, from [LHeC and FCC-he Study Group (2020)]. (right) Zoomed version of the kinematic plane for kinematics similar to those of the LHeC, restricted to $x < 10^{-3}$, from [Abelleira Fernandez *et al.* (2012)]. Lines of constant electron scattering angle $\theta_e = 1°$ and $10°$ are indicated, as is the approximate coverage of HERA.

a further order of magnitude for FCC-eh. This gives the potential to approach $x \sim 10^{-6}$ and $x \sim 10^{-7}$ at LHeC and FCC-eh, respectively, whilst remaining in the perturbative ($Q^2 > 1\,\text{GeV}^2$) regime, a highly attractive perspective for investigating low x effects. The EIC is also shown in this figure. It has a role to play in bridging the gap between HERA and the fixed target facilities and providing information from a single facility in the high x region, potentially without complications from nuclear binding effects.

Figure 21.4(right) illustrates the importance, in the LHeC context, of ensuring that the detector acceptance for the scattered electron extends as close to the beamline as possible. A zoom is shown of the low x, low Q^2 corner of the kinematic plane ($x < 10^{-3}$), with lines of constant scattered electron angle superimposed. If the instrumentation required to detect and reconstruct the scattered electron stops at $10°$ from the beamline, only a small fraction of the low x corner of the kinematic plane, mostly with $Q^2 > 100\,\text{GeV}^2$, is accessible. On the other hand, if scattered electrons can be efficiently and precisely reconstructed to within $1°$ to the beamline, the full kinematic plane with $Q^2 > 1\,\text{GeV}^2$ becomes accessible, including the lowest x values. The need for near-hermetic acceptance, in the forward as well as the backward direction, is a recurring theme in LHeC physics.

One way to illustrate the power of the LHeC to constrain low x physics is by quantifying the improvements in precision on PDFs that it makes possible, assuming (bravely!) that $\ln 1/x$ effects can be ignored and that standard collinear evolution techniques remain applicable. In order to study this, LHeC 'pseudodata' are produced, simulating the sort of measurement that could be be made in terms of kinematic range and statistical and systematic precision. An integrated luminosity

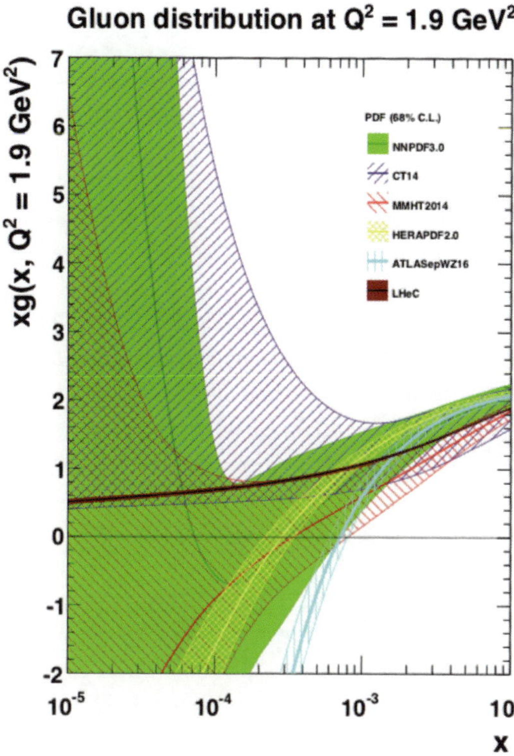

Fig. 21.5 Results of various NLO DGLAP QCD fits to extract the gluon density of the proton, shown in the form of uncertainty bands as a function of x at a scale $Q^2 = 1.9\,\mathrm{GeV}^2$. The global fits NNPDF3.0 [Ball *et al.* (2015)], CT14 [Dulat *et al.* (2016)] and MMHT2014 [Harland-Lang *et al.* (2015)] are shown, as well as fits that use HERA data only (HERAPDF2.0 [Abramowicz *et al.* (2015)]) and HERA plus selected LHC data (ATLASSepWZ16 [Aaboud *et al.* (2017)]). The brown band labelled LHeC shows the precision achieved when LHeC pseudodata are included in addition to HERA data, as described in detail in [LHeC and FCC-he Study Group (2020)].

of 1 ab^{-1} is considered, though the results, especially for the low x region, are hardly affected if the 'LHeC first run' luminosity of $50\,\mathrm{fb}^{-1}$ is assumed instead. The pseudodata are included along with existing data in NLO and NNLO DGLAP-based fits, in the xFitter [Alekhin *et al.* (2015)] framework in a fitting procedure similar to that of HERAPDF2.0 [Abramowicz *et al.* (2015)]. One example result from this approach is shown in Fig. 21.5 [LHeC and FCC-he Study Group (2020)]. The current knowledge of the low x gluon density at a typical starting scale for DGLAP evolution in the fits ($Q^2 = 1.9\,\mathrm{GeV}^2$) is shown via error bands from various groups. Taking the error band from the NNPDF study suggests that there is some current knowledge of the gluon density down to $x \sim 10^{-4}$. However, when comparing with other fit results, there are differences between approaches for all values below $x \sim 10^{-3}$ despite them having similar input data. The error bars also allow the possibility of a negative gluon density, with associated conceptual difficulties. In contrast, the fit in which LHeC pseudodata are included in addition to HERA data retains high precision on the gluon density throughout the region shown ($x > 10^{-5}$). As described in more detail in [LHeC and FCC-he Study Group (2020)], the relative precision obtained in the fits including LHeC pseudodata is typically at the 1% on the gluon density down to $x = 10^{-5}$, with a further order of magnitude extension at FCC-eh. A full flavour decomposition for the quarks is also obtained at similar levels of precision. It is worth noting in passing that the projected impact of the 'ultimate' LHC *pp* data has also been investigated. Whilst the HL-LHC has some impact at intermediate and large x [Abdul Khalek *et al.* (2018)], it doesn't extend into the low x region of interest here.

The studies of PDF precision in the context of NLO or NNLO DGLAP QCD fits show that the LHeC offers a transformational level of precision at low x. However, it is also important to establish whether the kinematic region of sensitivity is sufficient to fully reveal and establish BFKL effects. This question is addressed in Figs. 21.6 and 21.7 [Ball *et al.* (2018)]. In Fig. 21.6 predictions for both of the inclusive DIS structure functions[3] F_2 and F_L are shown in the LHeC and FCC-eh kinematic regions based on extrapolations of the fits to HERA data with and without additional $\ln 1/x$ effects implemented through the HELL package [Bonvini *et al.* (2016)] (see section 21.3). The large uncertainty bands are simply a consequence of the lack of current constaints in the low x region beyond the sensitivity of HERA. Whilst the impact of the BFKL terms is subtle in HERA

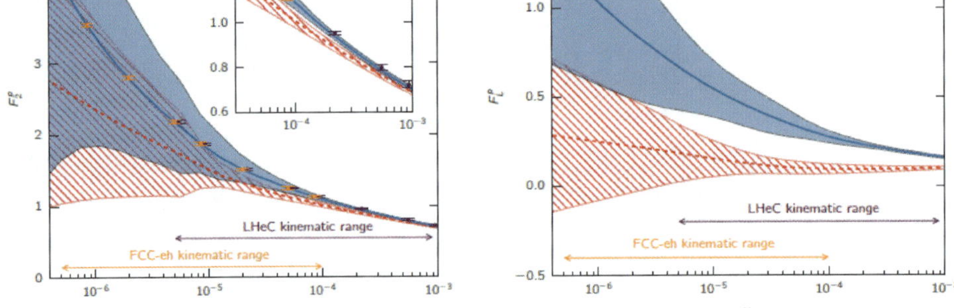

Fig. 21.6 Extrapolations of predictions for F_2 (left) and F_L (right) at $Q^2 = 5\,\text{GeV}^2$ according to fits to HERA data in the NNPDF framework both with ('NNLO + NLLx') and without ('NNLO') including NLL BFKL $\ln 1/x$ resummation in the framework of [Bonvini *et al.* (2016)]. The error bands reflect the uncertainties on the basis of currently available data from HERA. The kinematic coverages at the LHeC and FCC-eh are indicated. The data points in the insert for F_2 show selected LHeC pseudodata [LHeC and FCC-he Study Group (2020)]. Figures taken from [Ball *et al.* (2018)].

[3]See e.g. [Abramowicz *et al.* (2015)] for full formal definitions.

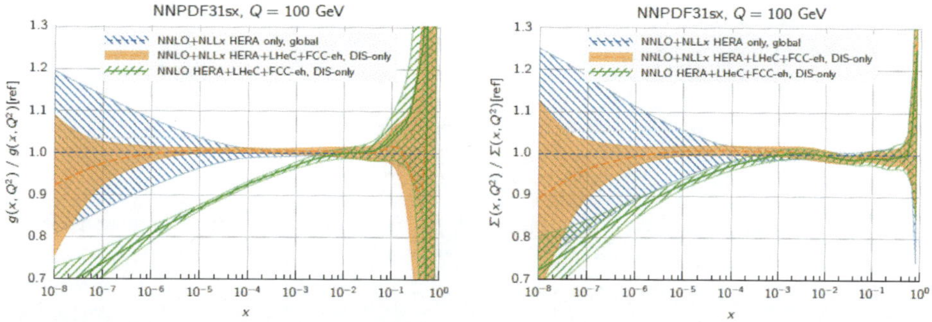

Fig. 21.7 Gluon (left) and singlet quark (right) distributions obtained in fits in the NNPDF framework relative to that in NNPDF3.1 [Ball *et al.* (2015)] at a scale of $Q^2 = 10^4$ GeV^2, extending to very low x. The error bands show the relative precision obtained when including only HERA data in the fit, or when additionally including LHeC and FCC-eh pseudodata, as indicated in the key. The central values correspond to fits in which no BFKL resummation is included ('NNLO') or in which NLL resummation is included in the framework of [Bonvini *et al.* (2016)] ('NNLO + NLLx'). The kinematic coverages at the LHeC and FCC-eh are indicated. Figures taken from [Ball *et al.* (2018)].

data, only being visible from careful consideration of the change in the χ^2 with the inclusion or removal of selected low x, low Q^2 data points, the effects become much more dramatic in the new low x region opened up by the proposed machines, as can be inferred from comparisons of the central values of the predictions. Due to its close relationship to the gluon density, the impact of including the NLL $\ln 1/x$ resummation is particularly pronounced for the structure function F_L, reaching a factor of four at $x = 10^{-6}$. As described in [LHeC and FCC-he Study Group (2020)], F_L can be measured at small x at the 10% level of precision at the LHeC from the change in the cross section at fixed x and Q^2 while $\sqrt{s}$ varies by taking data at three electron beam energies. Further immprovements are possible at FCC-eh. Figure 21.6(right) suggests that such a precision is comfortably sufficient to identify and characterise low x effects beyond NNLO DGLAP. As can be seen in Fig. 21.6(left), the differences between the predictions for F_2 with and without $\ln 1/x$ resummation are also substantial, and much larger than the expected 1% level of experimental precision.

The comparisons in Fig. 21.6 have been developed further to investigate the impact of $\ln 1/x$ resummation on the proton parton densities in the LHeC and FCC-eh kinematic range. Figure 21.7 shows the gluon and singlet quark distributions at an example $Q^2 = 10^4$ GeV^2 as a function of x relative to a reference provided by the NNPDF 3.1 extrapolation to very low x including the NLL BFKL resummation. Comparison of the error bands for the 'NNLO + NLLx' predictions with and without pseudodata from the new proposed colliders illustrates once again that the sensitivity of LHeC and FCC-eh extends to around $x = 10^{-6} - 10^{-7}$. Comparison of the central values for 'NNLO + NLLx' and 'NNLO' illustrates the large effect of including BFKL effects on the extracted parton densities The size of the BFKL effect is slightly larger for the gluon density than for the quarks, reaching almost 30% at $x = 10^{-7}$.

21.5 Exclusive and semi-inclusive measurements

21.5.1 *Exclusive vector meson production*

The lowest order underlying parton level process leading to the quasi-elastic, or exclusive, production of vector mesons is illustrated in Fig. 21.8(left). As discussed in section 21.3, the predicted cross section follows a power law $d\sigma_{\rm el}/dt \propto s^{2\alpha(0)-2}$ at vanishing t and if the two gluon exchange shown is replaced by the full BFKL Pomeron, $\alpha(0) = 1 + \omega_{I\!P}$.

Among the possible vector meson studies, J/ψ photoproduction has perhaps a special role. In models built using QCD starting from the low x gluon density [Martin *et al.* (1997)], the relevant scale of $\bar{Q}^2 = (Q^2 + M_V^2)/4 \sim 3\,\text{GeV}^2$ lies in the perturbative domain, whilst effective x values $\bar{x} = (Q^2 + M_V^2)/(Q^2 + W^2)$ as low as 10^{-5} are probed at the LHeC, with further extensions towards lower x at FCC-eh and VHEeP. As well as being studied in detail at HERA, this process has been the subject of substantial investigation in ultra-peripheral collisions at the LHC, where an incoming proton or lead ion provides the source of photons in place of the electron at HERA. As shown in Fig. 21.9(left), the LHC measurements[4] confirm the HERA power law energy dependence and, particularly for the favourable forward acceptance of LHCb, extend the kinematic range to higher W. The energy dependence of the J/ψ production process is fully compatible with a power law behaviour and corresponds to $\alpha(\langle t\rangle) \simeq 1.2$ when averaged over t. It has been interpreted starting from a knowledge of the proton gluon density by a number

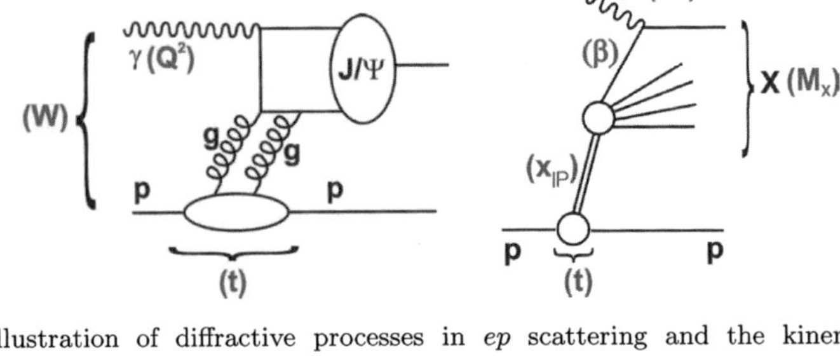

Fig. 21.8 Illustration of diffractive processes in ep scattering and the kinematic variables used to describe them. (left) Lowest order QCD diagram for exclusive or 'quasi-elastic' J/ψ photoproduction ($\gamma p \to J/\psi p$), where the incoming photon couples to the electron. (right) Inclusive diffractive deep inelastic scattering ($ep \to eXp$), shown mediated by a generic Pomeron.

[4]The CMS collaboration has made similar measurements for exclusive ρ [Sirunyan *et al.* (2019a)] and Υ [Sirunyan *et al.* (2019b)] meson production.

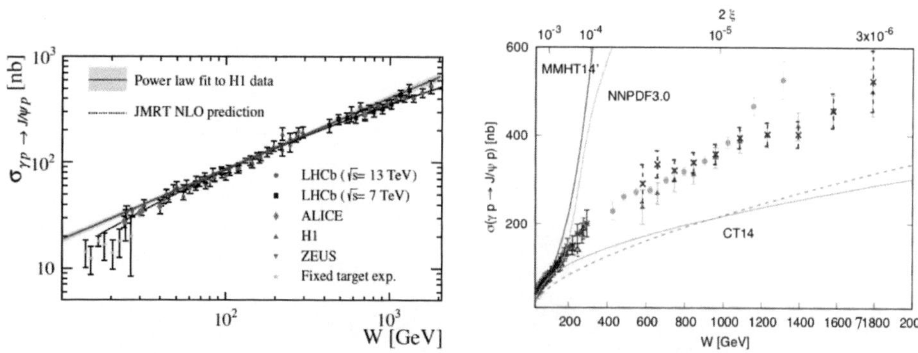

Fig. 21.9 (left) Current data on the total cross section for quasi-elastic J/ψ photoproduction ($\gamma p \to J/\psi p$) as a function of γp centre of mass energy, including measurements from H1 [Alexa *et al.* (2013); Aktas *et al.* (2006a)] and ZEUS [Chekanov *et al.* (2002)] at HERA, as well as ALICE [Abelev *et al.* (2014)] and LHCb [Aaij *et al.* (2014, 2018)] at the LHC (figure from [Aaij *et al.* (2018)]). (right) HERA J/ψ photoproduction data compared with predictions from a k_T factorisation model [Flett *et al.* (2019)] assuming different input gluon distributions [Harland-Lang *et al.* (2015); Dulat *et al.* (2016); Ball *et al.* (2015)].

of groups [Martin *et al.* (2000); Frankfurt *et al.* (2001); Martin *et al.* (2008)] with the conclusion that the data can be described using the gluon density extracted from inclusive DIS, but with scale uncertainties that are hard to control. A recent approach [Flett *et al.* (2019)] attempts to tackle these issues in the framework of k_T factorisation. As shown in Fig. 21.9(right), in this approach the J/ψ data potentially powerfully distinguish between different proton PDF sets. From an alternative viewpoint, this strong variation of the predictions with input PDF set suggests that unintegrated gluons may ultimately be the appropriate building blocks for the study of low x final states including vector meson production.

Independently of the detailed conclusions from HERA and LHC data, it is clear that high energy exclusive vector meson photoproduction and leptoproduction data may play a crucial role in our future understanding of low x physics. They are therefore an important target for the future *ep* facilities currently under discussion. The limited centre of mass reach of the EIC and its ability to polarise both beams imply that its vector meson and DVCS programme will focus on spatial imaging, correlations and spin structure [Aschenauer *et al.* (2019)] rather than on low x aspects. The potential ultra-low-x reach of the VHEeP facility has exciting potential and the low luminosities should not be a strong impediment to the study of at least ρ, ϕ and J/ψ production. However detailed studies have yet to be carried out. On the other hand, the LHeC's capabilities have been investigated in some detail, in particular for J/ψ photoproduction. As shown in Fig. 21.10(left), a very precise measuerment is possible in the decay channel $J/\psi \to \mu^+\mu^-$ with a modest luminosity of $2\,\text{fb}^{-1}$, provided the muon detector acceptance extends to within 1° of the beamline. Whilst the W range is a substantial improvement over that available at HERA, it does not extend beyond the upper limits reached by LHCb. However, the power of a future *ep* machine may lie more in well controlled proton tagging

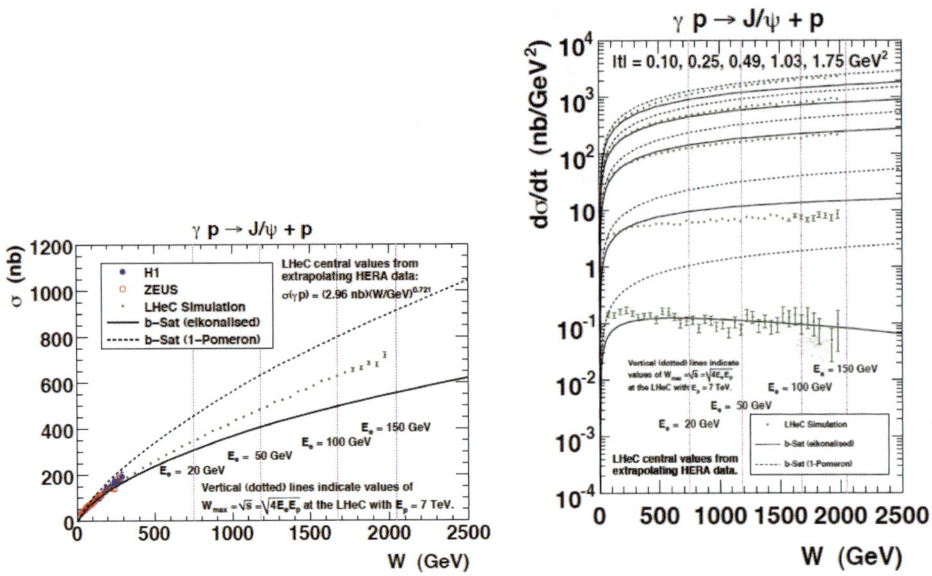

Fig. 21.10 Pseudodata based on LHeC kinematics for quasi-elastic J/ψ photoproduction ($\gamma p \to J/\psi p$). The cross section is shown as a function of W in the left panel, whilst the right panel shows W dependences differentially in t at the t values indicated, running from top to bottom. Limits imposed by different electron beam energies are indicated. An integrated luminosity of $2\,\text{fb}^{-1}$ is assumed. The full and dotted lines correspond to two different versions of the b-Sat dipole model [Kowalski *et al.* (2006)], in which low x density-driven saturation is included as a function of impact parameter ('eikonalised'), or is omitted ('1 Pomeron'). The data shown in the left panel are taken from HERA [Aktas *et al.* (2006a); Chekanov *et al.* (2002)]. The figures are taken from [Abelleira Fernandez *et al.* (2012)].

and precise transverse momentum meassurements, for example using the Roman pot technique that was already successfully applied at HERA [Aaron *et al.* (2012a)]. This becomes clear from simulations of differential measurements as a function of W and t, as shown in Fig. 21.10(right). The delicate interplay between energy-driven BFKL effects and density driven saturation effects will require this sort of detail, and possibly the additional inclusion of coherent vector meson production from nuclei $eA \to eVA$, as explored further in [LHeC and FCC-he Study Group (2020)].

Although not yet studied in the context of the proposed high energy colliders, it is worth noting that quasi-elastic processes at very large $|t|$ offer a highly promising environment in which to search for BFKL effects. The large $|t|$ ensures that the exchange is genuinely hard, whilst still satisfying the Regge limit $s \gg |t|$. One such process is the production of large rapidity gaps between high transverse momentum jets in photoproduction, where the HERA data [Adloff *et al.* (2002); Chekanov *et al.* (2007b)] have been compared with leading log BFKL calculations [Mueller and Tang (1992); Enberg *et al.* (2002)] with some degree of success, the achievable precision being limited by the poorly constrained rapidity gap survival probability. A related process is diffractive vector meson (particularly J/ψ) production at large

$|t|$, where again the data [Aktas *et al.* (2003); Chekanov *et al.* (2010)] and BFKL-based predictions [Forshaw and Ryskin (1995); Bartels *et al.* (1996b); Forshaw and Poludniowski (2003)] are broadly in agreement. Whilst the calculations are not yet at a stage where they can accurately reproduce the data, the shere existence of such processes is highly suggestive of the presence of BFKL dynamics and further high luminosity data at the LHeC, FCC-eh or VHEeP could have a major impact.

21.5.2 *Inclusive diffractive DIS*

The inclusive diffractive process in DIS ($ep \to eXp$) is illustrated in Fig. 21.8(right). The proton remains intact, tranferring a fraction $x_{I\!P} \equiv \xi$ of its longitudinal momentum through a colourless exchange to create a multi-particle system X with a continuous mass distribution that can be viewed as the products of the dissociation of the virtual photon, correspondingly with overall 1^{--} quantum numbers. The fraction of the proton longitudinal momentum loss that is carried by the struck quark is denoted β, such that $\beta x_{I\!P} = x$. One important observation from HERA was that this DDIS process comprises a large ($\sim 10\,\%$) fraction of diffractive events, to first approximation independently of Q^2 [Derrick *et al.* (1993b); Ahmed *et al.* (1994)]. In a long series of papers (see [Newman and Wing (2014)] for a review), the HERA experiments determined the deep inelastic structure of the t-channel exchange in these events in the form of diffractive parton densities, as well as studying the energy dependence in the form of the variation with $x_{I\!P}$ at fixed β and Q^2. The diffractive parton densities are often interpreted [Ingelman and Schlein (1985)] in terms of a structure for the Pomeron. They were found to be strongly gluon dominated, encouraging interpretations in which DDIS is closely related to the gluon dynamics also discussed in the inclusive case. Interestingly, the final results from HERA on the effective Pomeron intercept describing DDIS are well aligned at around $\alpha(0) \sim 1.12^{+0.03}_{-0.01}$ [Aktas *et al.* (2006c); Chekanov *et al.* (2009b); Aaron *et al.* (2012b)], more compatible with a soft Pomeron than a BFKL Pomeron, so the picture for the dominant process is more one of the virtual photon probing a soft Pomeron exchange than of the progressively harder processes that are observed at large scales in the vecor meson case (section 21.5.1). The picture is therefore a rather complex one, whose thorough understanding requires the study of DDIS at future higher energy ep colliders. Mapping in detail the evolution with energy has considerable promise to unravel the delicate interplay between higher twist effects at low Q^2, the onset of BFKL effects at low β or x and eventually non-linear dynamics at high densities.

An investigation of the potential studies of inclusive diffraction that would be possible at the EIC and LHeC is presented in [Armesto *et al.* (2019)]. Although that work is performed at NLO, similar analyses in the HERA context have recently been extended to NNLO [Khanpour (2019)]. In Fig. 21.11, the accessible kinematic range in (x, Q^2) is shown for HERA, EIC, LHeC and FCC-eh. For the LHeC design the range in x is increased by a factor ~ 20 over HERA and the maximum available Q^2

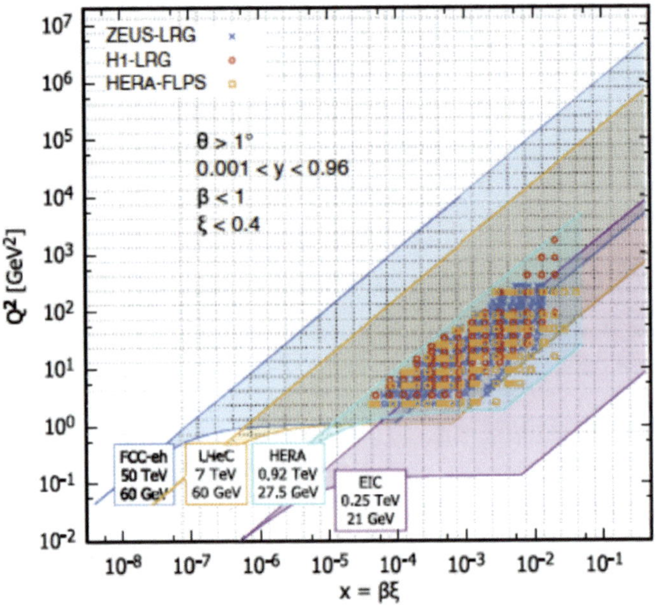

Fig. 21.11 Kinematic plane indicating the acceptance for inclusive diffraction in (x, Q^2) for the EIC (magenta region), the LHeC (orange region) and the FCC-eh (dark blue region) as compared with the HERA data (light blue region, [Chekanov *et al.* (2009b); Aaron *et al.* (2012b,a)]). The acceptance limit for the scattered electron is assumed to be 1°, and ξ values up to 0.4 are included. Figure taken from [Armesto *et al.* (2019)].

by a factor ~ 100. The FCC-eh machine would further increase this range by roughly one order of magnitude in both x and Q^2. The EIC coverage is complementary, adding sensitivity at higher x than HERA, a region which is currently unexplored due to the lack of DDIS measurements in fixed target ep scattering experiments. The LHeC (FCC-eh) machine probes very small values of ξ, reaching 10^{-4} (10^{-5}) for $\beta \sim 10^{-2}$. At $\xi = 0.1$, the LHeC and FCC-eh kinematic ranges extend down to $\beta \sim 10^{-5}$ and $\beta \sim 10^{-6}$, respectively, offering the enticing prospect of studying diffractive DIS in a wider β, Q^2 kinematic range than has been possible in (x, Q^2) for inclusive DIS at HERA. These kinematic coverages are clearly very well suited to the study of new low x effects.

Based on the studies in [Armesto *et al.* (2019)], the experimental precision with which the diffractive gluon density could be extracted when LHeC and FCC-eh data become available is illustrated in Fig. 21.12. The predicted fractional uncertainties are shown for the LHeC and FCC-eh at two different scales Q^2 as a function of parton momentum fraction z. The different bands show the variation with the upper cut on the available ξ range, from 0.01 to 0.32. In the best constrained region of $z \simeq 0.1$, the precision reaches the 1% level, with only a small dependence on the available ξ range. The low x DPDF determination accuracy improves with respect to HERA by a factor of 5–7 for the LHeC and 10–15 for the FCC-eh and completely new kinematic regimes are accessed.

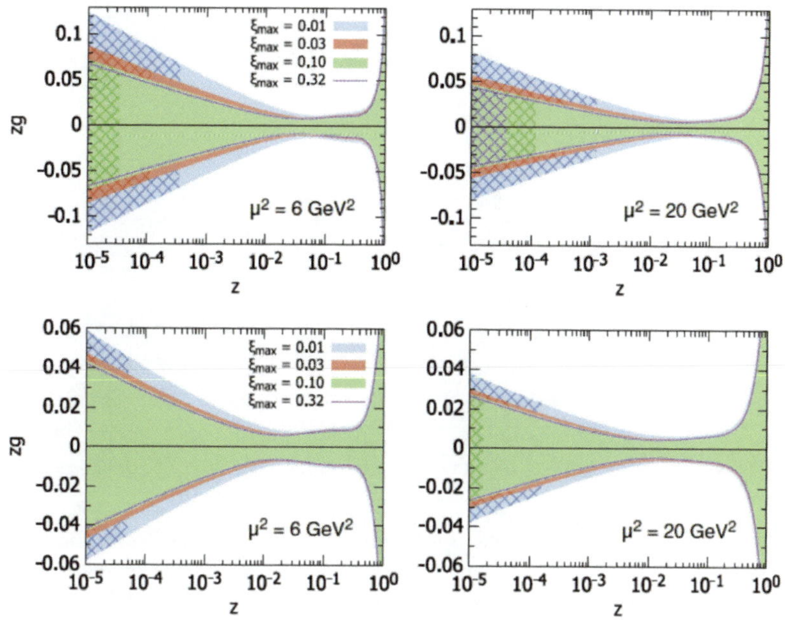

Fig. 21.12 Relative experimental uncertainties on the diffractive gluon density for LHeC kinematics (upper panel) and FCC-eh kinematics (lower panel). Two different choices of scale, $\mu^2 = 6$ and $20\,\mathrm{GeV}^2$, are considered. The blue, red and green bands and the magenta line correspond to different maximal values of $\xi = 0.01, 0.03, 0.1, 0.32$, respectively. The cross-hatched areas show regions which are not directly accessible in data and require extrapolations. Theoretical uncertainties, for example due to parameterisation bias or variations in scales and heavy quark masses, are not included. Figure taken from [Armesto *et al.* (2019)].

The hadronic final state in DDIS may offer further insights into low x effects. For example, diffractive charm production [Aktas *et al.* (2007); Chekanov *et al.* (2003)] has been successfully modelled in terms of a gluon pair or ladder coupling to a basic $\gamma^* \to q\bar{q}$ or $\gamma^* \to q\bar{q}g$ fluctuation, using k_T factorisation and an unintegrated gluon density in the 'BJKLW' model [Bartels *et al.* (1996c,a, 1999, 2002)]. The prospects at future facilities have yet to be explored in detail. The much larger photon dissociation masses that become accessible kinematically at the LHeC or FCC-eh offer the interesting possibility of accessing very low z values, though this region also enhances the $q\bar{q}g$ and higher multiplicity photon fluctuations that are difficult to handle theoretically.

21.6 Inclusive hadronic final state measurements

Much of the search programme for BFKL effects in HERA data was focused on hadronic final state measurements. A large part of this work involved looking for deviations from the transverse momentum ordering of successively emitted partons along the parton cascade between the photon and the proton that is implied by DGLAP evolution. If BFKL effects are the dominant feature of the data, one expects instead to find strong x (or energy) ordering, as illustrated in Fig. 21.13(left).

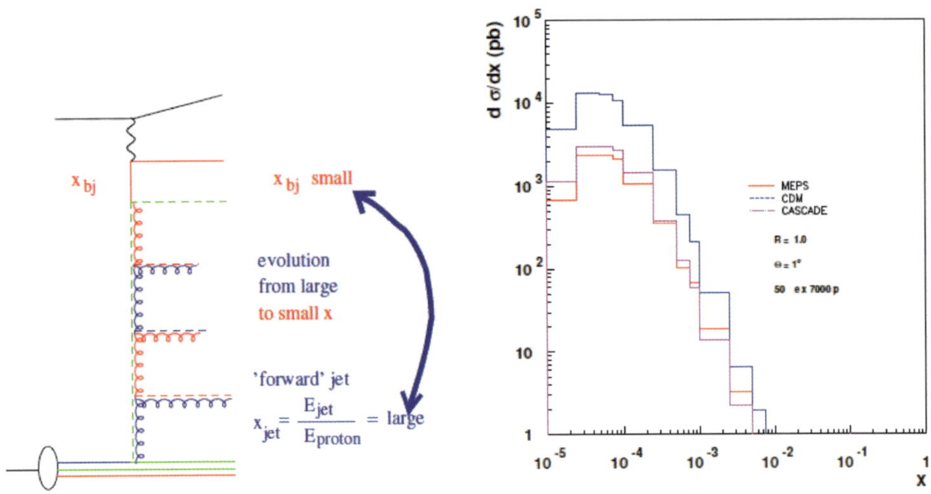

Fig. 21.13 (left) Illustration of a parton cascade in a low x ep process in which BFKL evolution dominates (the lengths of the emitted partons is representative of their energies). (right) Simulated cross sections for forward jet production with $E_T > 10\,\text{GeV}$ in the angular range $3 < \theta < 20°$ relative to the outgoing proton beam as a function of Bjorken x in the LHeC context. Jets are found using the SIScone algorithm [Salam and Soyez (2007)] with $R = 0.5$. The predictions of three Monte Carlo models with differening parton cascade dynamics as described in the text are compared. Figure taken from [Abelleira Fernandez *et al.* (2012)].

Observables that are sensitive to these differences include azimuthal decorrelations between jets, Mueller-Navelet jet configurations and forward particle or energy flows, as summarised for example in [Jung, H. (2020)].

The potential to reveal BFKL effects in the hadronic final state has yet to be investigated in detail for future *ep* colliders, though clearly the enhanced reach towards low x will be a major benefit. A flavour of the potential to observe these effects is shown in Fig. 21.13(right). Following the approach in [Aktas *et al.* (2006b)], simulated cross sections are shown for forward jets in the range of angles relative to the outgoing proton direction $1° < \theta < 20°$, with transverse energy $E_T > 10\,\text{GeV}$, as a function of Bjorken-x. The curve labelled 'MEPS' corresponds to a standard (RAPGAP [Jung (1995)]) Monte Carlo implementation of LO matrix elements interfaced to DGLAP parton showers, implying transverse momentum ordering along the parton cascade. On the other hand, the 'CDM' curve is obtained from the DJANGO / ARIADNE programmes [Charchula *et al.* (1994); Lonnblad (1992)] and includes the BFKL-like feature of energy ordering. The CASCADE programme [Jung and Salam (2001); Jung (2002)] uses an unintegrated gluon density as input and implements the angular parton cascade ordering of the CCFM formalism [Ciafaloni (1988); Catani *et al.* (1990a,b); Marchesini (1995)], in which both $\ln 1/x$ and $\ln Q^2$ terms are resummed in the relevant limits. The CDM model predicts almost a factor of five enhancement of forward jet production over the other models. With LHeC energies and sufficiently forward instrumentation, observables such as forward jets thus appear likely to be highly powerful in distinguishing between different types of parton evolution dynamics and ultimately characterising the transition to a BFKL-dominated domain.

21.7 Conclusion

Lev Lipatov is unquestionably a giant of low x physics. Whilst his work already underlies a wide range of theoretical aspects of the field, the impact of his best-known contributions in the BFKL context only just start to be seen in experimental data, presumably due to the need for high energies to approach the rather ill-defined 'asymptotic' kinematic region. Here only a flavour has been given of the possible future directions of the subject. As we now look towards the development of new DIS facilities, with increased centre-of-mass energies a major theme, it is inevitable that Lipatov's legacy will become more and more apparent over the coming decades.

Bibliography

Aaboud, M. *et al.* (2017). Precision measurement and interpretation of inclusive W^+, W^- and Z/γ^* production cross sections with the ATLAS detector, *Eur. Phys. J.* **C77**, 6, p. 367, doi:10.1140/epjc/s10052-017-4911-9, arXiv:1612.03016 [hep-ex].

Aaij, R. *et al.* (2014). Updated measurements of exclusive J/ψ and ψ(2S) production cross-sections in pp collisions at $\sqrt{s} = 7\,\text{TeV}$, *J. Phys.* **G41**, p. 055002, doi:10.1088/0954-3899/41/5/055002, arXiv:1401.3288 [hep-ex].

Aaij, R. *et al.* (2018). Central exclusive production of J/ψ and $\psi(2S)$ mesons in pp collisions at $\sqrt{s} = 13\,\text{TeV}$, *JHEP* **10**, p. 167, doi:10.1007/JHEP10(2018)167, arXiv:1806.04079 [hep-ex].

Aaron, F. D. *et al.* (2008). Measurement of deeply virtual Compton scattering and its t-dependence at HERA, *Phys. Lett.* **B659**, pp. 796–806, doi:10.1016/j.physletb.2007.11.093, arXiv:0709.4114 [hep-ex].

Aaron, F. D. *et al.* (2009). Deeply virtual compton scattering and its beam charge asymmetry in e+− collisions at HERA, *Phys. Lett.* **B681**, pp. 391–399, doi:10.1016/j.physletb.2009.10.035, arXiv:0907.5289 [hep-ex].

Aaron, F. D. *et al.* (2010). Diffractive electroproduction of rho and phi mesons at HERA, *JHEP* **05**, p. 032, doi:10.1007/JHEP05(2010)032, arXiv:0910.5831 [hep-ex].

Aaron, F. D. *et al.* (2012a). Combined inclusive diffractive cross sections measured with forward proton spectrometers in deep inelastic *ep* scattering at HERA, *Eur. Phys. J.* **C72**, p. 2175, doi:10.1140/epjc/s10052-012-2175-y, arXiv:1207.4864 [hep-ex].

Aaron, F. D. *et al.* (2012b). Inclusive measurement of diffractive deep-inelastic scattering at HERA, *Eur. Phys. J.* **C72**, p. 2074, doi:10.1140/epjc/s10052-012-2074-2, arXiv:1203.4495 [hep-ex].

Abada, A. *et al.* (2019). FCC physics opportunities, *Eur. Phys. J.* **C79**, 6, p. 474, doi:10.1140/epjc/s10052-019-6904-3.

Abdul Khalek, R., Bailey, S., Gao, J., Harland-Lang, L., and Rojo, J. (2018). Towards ultimate parton distributions at the high-luminosity LHC, *Eur. Phys. J.* **C78**, 11, p. 962, doi:10.1140/epjc/s10052-018-6448-y, arXiv:1810.03639 [hep-ph].

Abelev, B. B. *et al.* (2014). Exclusive J/ψ photoproduction off protons in ultra-peripheral p-Pb collisions at $\sqrt{s_{\text{NN}}} = 5.02\,\text{TeV}$, *Phys. Rev. Lett.* **113**, 23, p. 232504, doi:10.1103/PhysRevLett.113.232504, arXiv:1406.7819 [nucl-ex].

Abelleira Fernandez, J. L. *et al.* (2012). A large hadron electron collider at CERN: report on the physics and design concepts for machine and detector, *J. Phys.* **G39**, p. 075001, doi:10.1088/0954-3899/39/7/075001, arXiv:1206.2913 [physics.acc-ph].

Abramowicz, H. *et al.* (2015). Combination of measurements of inclusive deep inelastic $e^{\pm}p$ scattering cross sections and QCD analysis of HERA data, *Eur. Phys. J.* **C75**, 12, p. 580, doi:10.1140/epjc/s10052-015-3710-4, arXiv:1506.06042 [hep-ex].

Abt, I. *et al.* (1993). Measurement of the proton structure function F2 (x, Q**2) in the low x region at HERA, *Nucl. Phys.* **B407**, pp. 515–538, doi:10.1016/0550-3213(93) 90090-C.

Accardi, A. *et al.* (2016). Electron ion collider: The next QCD frontier, *Eur. Phys. J.* **A52**, 9, p. 268, doi:10.1140/epja/i2016-16268-9, arXiv:1212.1701 [nucl-ex].

Adams, M. R. *et al.* (1996). Proton and deuteron structure functions in muon scattering at 470-GeV, *Phys. Rev.* **D54**, pp. 3006–3056, doi:10.1103/PhysRevD.54.3006.

Adloff, C. *et al.* (2002). Energy flow and rapidity gaps between jets in photoproduction at HERA, *Eur. Phys. J.* **C24**, pp. 517–527, doi:10.1007/s10052-002-0988-9, arXiv:hep-ex/0203011 [hep-ex].

Ahmed, T. *et al.* (1994). Deep inelastic scattering events with a large rapidity gap at HERA, *Nucl. Phys.* **B429**, pp. 477–502, doi:10.1016/0550-3213(94)90151-1.

Aktas, A. *et al.* (2003). Diffractive photoproduction of J/ψ mesons with large momentum transfer at HERA, *Phys. Lett.* **B568**, pp. 205–218, doi:10.1016/j.physletb.2003.06. 056, arXiv:hep-ex/0306013 [hep-ex].

Aktas, A. *et al.* (2006a). Elastic J/psi production at HERA, *Eur. Phys. J.* **C46**, pp. 585–603, doi:10.1140/epjc/s2006-02519-5, arXiv:hep-ex/0510016 [hep-ex].

Aktas, A. *et al.* (2006b). Forward jet production in deep inelastic scattering at HERA, *Eur. Phys. J.* **C46**, pp. 27–42, doi:10.1140/epjc/s2005-02471-x, arXiv:hep-ex/0508055 [hep-ex].

Aktas, A. *et al.* (2006c). Measurement and QCD analysis of the diffractive deep-inelastic scattering cross-section at HERA, *Eur. Phys. J.* **C48**, pp. 715–748, doi:10.1140/ epjc/s10052-006-0035-3, arXiv:hep-ex/0606004 [hep-ex].

Aktas, A. *et al.* (2007). Diffractive open charm production in deep-inelastic scattering and photoproduction at HERA, *Eur. Phys. J.* **C50**, pp. 1–20, doi:10.1140/epjc/ s10052-006-0206-2, arXiv:hep-ex/0610076 [hep-ex].

Alekhin, S. *et al.* (2015). HERAFitter, *Eur. Phys. J.* **C75**, 7, p. 304, doi:10.1140/epjc/ s10052-015-3480-z, arXiv:1410.4412 [hep-ph].

Alexa, C. *et al.* (2013). Elastic and proton-dissociative photoproduction of J/psi mesons at HERA, *Eur. Phys. J.* **C73**, 6, p. 2466, doi:10.1140/epjc/s10052-013-2466-y, arXiv:1304.5162 [hep-ex].

Altarelli, G., Ball, R. D., and Forte, S. (2000). Resummation of singlet parton evolution at small x, *Nucl. Phys.* **B575**, pp. 313–329, doi:10.1016/S0550-3213(00)00032-8, arXiv:hep-ph/9911273 [hep-ph].

Altarelli, G., Ball, R. D., and Forte, S. (2002). Factorization and resummation of small x scaling violations with running coupling, *Nucl. Phys.* **B621**, pp. 359–387, doi: 10.1016/S0550-3213(01)00563-6, arXiv:hep-ph/0109178 [hep-ph].

Altarelli, G. and Parisi, G. (1977). Asymptotic freedom in parton language, *Nucl. Phys.* **B126**, pp. 298–318, doi:10.1016/0550-3213(77)90384-4.

Angal-Kalinin, D. *et al.* (2018). PERLE. Powerful energy recovery linac for experiments. Conceptual design report, *J. Phys.* **G45**, 6, p. 065003, doi:10.1088/1361-6471/ aaa171, arXiv:1705.08783 [physics.acc-ph].

Armesto, N., Newman, P. R., Somiski, W., and Stato, A. M. (2019). Inclusive diffraction in future electron-proton and electron-ion colliders, *Phys. Rev.* **D100**, 7, p. 074022, doi:10.1103/PhysRevD.100.074022, arXiv:1901.09076 [hep-ph].

Arneodo, M. *et al.* (1997). Measurement of the proton and deuteron structure functions, F2(p) and F2(d), and of the ratio sigma-L / sigma-T, *Nucl. Phys.* **B483**, pp. 3–43, doi:10.1016/S0550-3213(96)00538-X, arXiv:hep-ph/9610231 [hep-ph].

Aschenauer, E. C., Fazio, S., Lee, J. H., Mantysaari, H., Page, B. S., Schenke, B., Ullrich, T., Venugopalan, R., and Zurita, P. (2019). The electronion collider: assessing the energy dependence of key measurements, *Rept. Prog. Phys.* **82**, 2, p. 024301, doi: 10.1088/1361-6633/aaf216, arXiv:1708.01527 [nucl-ex].

Balitsky, I. I. and Lipatov, L. N. (1978). The Pomeranchuk singularity in quantum chromodynamics, *Sov. J. Nucl. Phys.* **28**, pp. 822–829, [Yad. Fiz.28,1597(1978)].

Ball, R. D., Bertone, V., Bonvini, M., Marzani, S., Rojo, J., and Rottoli, L. (2018). Parton distributions with small-x resummation: evidence for BFKL dynamics in HERA data, *Eur. Phys. J.* **C78**, 4, p. 321, doi:10.1140/epjc/s10052-018-5774-4, arXiv:1710.05935 [hep-ph].

Ball, R. D. and Forte, S. (1994). Double asymptotic scaling at HERA, *Phys. Lett.* **B335**, pp. 77–86, doi:10.1016/0370-2693(94)91561-X, arXiv:hep-ph/9405320 [hep-ph].

Ball, R. D. *et al.* (2015). Parton distributions for the LHC Run II, *JHEP* **04**, p. 040, doi:10.1007/JHEP04(2015)040, arXiv:1410.8849 [hep-ph].

Bartels, J., Ewerz, C., Lotter, H., and Wusthoff, M. (1996a). Azimuthal distribution of quark - anti-quark jets in DIS diffractive dissociation, *Phys. Lett.* **B386**, pp. 389–396, doi:10.1016/0370-2693(96)81071-9, arXiv:hep-ph/9605356 [hep-ph].

Bartels, J., Forshaw, J. R., Lotter, H., and Wusthoff, M. (1996b). Diffractive production of vector mesons at large t, *Phys. Lett.* **B375**, pp. 301–309, doi:10.1016/0370-2693(96)00203-1, arXiv:hep-ph/9601201 [hep-ph].

Bartels, J., Jung, H., and Kyrieleis, A. (2002). Massive c anti-c g: Calculation in diffractive DIS and diffractive D* production at HERA, *Eur. Phys. J.* **C24**, pp. 555–560, doi: 10.1007/s10052-002-0991-1, arXiv:hep-ph/0204269 [hep-ph].

Bartels, J., Jung, H., and Wusthoff, M. (1999). Quark - anti-quark gluon jets in DIS diffractive dissociation, *Eur. Phys. J.* **C11**, pp. 111–125, doi:10.1007/s100520050618,10.1007/s100529900081, arXiv:hep-ph/9903265 [hep-ph].

Bartels, J., Lotter, H., and Wsthoff, M. (1996c). Quark-antiquark production in DIS diffractive dissociation, *Phys. Lett.* **B379**, pp. 239–248, doi:10.1016/0370-2693(96)00412-1,10.1016/0370-2693(96)00840-4, arXiv:hep-ph/9602363 [hep-ph], [Erratum: Phys. Lett.B382,449(1996)].

Bonvini, M., Marzani, S., and Peraro, T. (2016). Small-x resummation from HELL, *Eur. Phys. J.* **C76**, 11, p. 597, doi:10.1140/epjc/s10052-016-4445-6, arXiv:1607.02153 [hep-ph].

Caldwell, A., Ent, R., Levy, A., Newman, P., and Olness, F. (2018). The "DIS and related subjects" strategy document: fundamental science from Lepton-Hadron scattering, arXiv:1812.08110 [hep-ph].

Caldwell, A. and Wing, M. (2016). VHEeP: A very high energy electronproton collider, *Eur. Phys. J.* **C76**, 8, p. 463, doi:10.1140/epjc/s10052-016-4316-1, arXiv:1606.00783 [hep-ex].

Catani, S., Ciafaloni, M., and Hautmann, F. (1991). High-energy factorization and small x heavy flavor production, *Nucl. Phys.* **B366**, pp. 135–188, doi:10.1016/0550-3213(91)90055-3.

Catani, S., Fiorani, F., and Marchesini, G. (1990a). QCD coherence in initial state radiation, *Phys. Lett.* **B234**, pp. 339–345, doi:10.1016/0370-2693(90)91938-8.

Catani, S., Fiorani, F., and Marchesini, G. (1990b). Small x behavior of initial state radiation in perturbative QCD, *Nucl. Phys.* **B336**, pp. 18–85, doi:10.1016/0550-3213(90)90342-B.

Charchula, K., Schuler, G. A., and Spiesberger, H. (1994). Combined QED and QCD radiative effects in deep inelastic lepton — proton scattering: The Monte Carlo generator DJANGO6, *Comput. Phys. Commun.* **81**, pp. 381–402, doi:10.1016/0010-4655(94)90086-8.

Chekanov, S. *et al.* (2002). Exclusive photoproduction of J/psi mesons at HERA, *Eur. Phys. J.* **C24**, pp. 345–360, doi:10.1007/s10052-002-0953-7, arXiv:hep-ex/0201043 [hep-ex].

Chekanov, S. *et al.* (2003). Measurement of the open charm contribution to the diffractive proton structure function, *Nucl. Phys.* **B672**, pp. 3–35, doi:10.1016/j.nuclphysb.2003.09.001, arXiv:hep-ex/0307068 [hep-ex].

Chekanov, S. *et al.* (2007a). Exclusive rho0 production in deep inelastic scattering at HERA, *PMC Phys.* **A1**, p. 6, doi:10.1186/1754-0410-1-6, arXiv:0708.1478 [hep-ex].

Chekanov, S. *et al.* (2007b). Photoproduction of events with rapidity gaps between jets at HERA, *Eur. Phys. J.* **C50**, pp. 283–297, doi:10.1140/epjc/s10052-007-0241-7, arXiv:hep-ex/0612008 [hep-ex].

Chekanov, S. *et al.* (2009a). A Measurement of the Q**2, W and t dependences of deeply virtual Compton scattering at HERA, *JHEP* **05**, p. 108, doi:10.1088/1126-6708/2009/05/108, arXiv:0812.2517 [hep-ex].

Chekanov, S. *et al.* (2009b). Deep inelastic scattering with leading protons or large rapidity gaps at HERA, *Nucl. Phys.* **B816**, pp. 1–61, doi:10.1016/j.nuclphysb.2009.03.003, arXiv:0812.2003 [hep-ex].

Chekanov, S. *et al.* (2010). Measurement of J/psi photoproduction at large momentum transfer at HERA, *JHEP* **05**, p. 085, doi:10.1007/JHEP05(2010)085, arXiv:0910.1235 [hep-ex].

Ciafaloni, M. (1988). Coherence effects in initial jets at small q**2/s, *Nucl. Phys.* **B296**, pp. 49–74, doi:10.1016/0550-3213(88)90380-X.

Ciafaloni, M. and Camici, G. (1998). Energy scale(s) and next-to-leading BFKL equation, *Phys. Lett.* **B430**, pp. 349–354, doi:10.1016/S0370-2693(98)00551-6, arXiv:hep-ph/9803389 [hep-ph].

Collins, J. C. and Ellis, R. K. (1991). Heavy quark production in very high-energy hadron collisions, *Nucl. Phys.* **B360**, pp. 3–30, doi:10.1016/0550-3213(91)90288-9.

Cudell, J. R., Kang, K., and Kim, S. K. (1997). Simple pole fits to p p and anti-p p total cross-sections and real parts, *Phys. Lett.* **B395**, pp. 311–331, doi:10.1016/S0370-2693(97)00046-4, arXiv:hep-ph/9601336 [hep-ph].

De Rujula, A., Glashow, S. L., Politzer, H. D., Treiman, S. B., Wilczek, F., and Zee, A. (1974). Possible nonRegge behavior of electroproduction structure functions, *Phys. Rev.* **D10**, p. 1649, doi:10.1103/PhysRevD.10.1649.

Derrick, M. *et al.* (1993a). Measurement of the proton structure function F2 in e p scattering at HERA, *Phys. Lett.* **B316**, pp. 412–426, doi:10.1016/0370-2693(93)90347-K.

Derrick, M. *et al.* (1993b). Observation of events with a large rapidity gap in deep inelastic scattering at HERA, *Phys. Lett.* **B315**, pp. 481–493, doi:10.1016/0370-2693(93)91645-4.

Dokshitzer, Y. L. (1977). Calculation of the structure functions for deep inelastic scattering and e+ e− annihilation by perturbation theory in quantum chromodynamics, *Sov. Phys. JETP* **46**, pp. 641–653, [Zh. Eksp. Teor. Fiz.73,1216(1977)].

Donnachie, A. and Landshoff, P. V. (1992). Total cross-sections, *Phys. Lett.* **B296**, pp. 227–232, doi:10.1016/0370-2693(92)90832-O, arXiv:hep-ph/9209205 [hep-ph].

Dulat, S., Hou, T.-J., Gao, J., Guzzi, M., Huston, J., Nadolsky, P., Pumplin, J., Schmidt, C., Stump, D., and Yuan, C. P. (2016). New parton distribution functions from a global analysis of quantum chromodynamics, *Phys. Rev.* **D93**, 3, p. 033006, doi: 10.1103/PhysRevD.93.033006, arXiv:1506.07443 [hep-ph].

Edli, E. *et al.* (2018). Acceleration of electrons in the plasma wakefield of a proton bunch, *Nature* **561**, p. 363.

Enberg, R., Ingelman, G., and Motyka, L. (2002). Hard color singlet exchange and gaps between jets at the Tevatron, *Phys. Lett.* **B524**, pp. 273–282, doi:10.1016/S0370-2693(01)01379-X, arXiv:hep-ph/0111090 [hep-ph].

Fadin, V. S. and Lipatov, L. N. (1998). BFKL pomeron in the next-to-leading approximation, *Phys. Lett.* **B429**, pp. 127–134, doi:10.1016/S0370-2693(98)00473-0, arXiv:hep-ph/9802290 [hep-ph].

Feinberg, E. L. and Pomerančuk, I. (1956). High energy inelastic diffraction phenomena, *Il Nuovo Cimento* **3**, 4, pp. 652–671, doi:10.1007/BF02746068.

Flett, C. A., Jones, S. P., Martin, A. D., Ryskin, M. G., and Teubner, T. (2019). Towards a determination of the low x gluon via exclusive J/ψ production, *PoS* **DIS2019**, p. 053, doi:10.22323/1.352.0053, arXiv:1907.06471 [hep-ph].

Forshaw, J. and Ross, D. (1997). *Quantum Chromodynamics and the Pomeron (Cambridge Lecture Notes in Physics)* (Cambridge University Press), doi:10.1017/CBO9780511524387.

Forshaw, J. R. and Poludniowski, G. (2003). Vector meson photoproduction at high t and comparison to HERA data, *Eur. Phys. J.* **C26**, pp. 411–415, doi:10.1140/epjc/s2002-01078-1, arXiv:hep-ph/0107068 [hep-ph].

Forshaw, J. R. and Ryskin, M. G. (1995). Diffractive vector meson production at large momentum transfer, *Z. Phys.* **C68**, pp. 137–148, doi:10.1007/BF01579812, arXiv:hep-ph/9501376 [hep-ph].

Forte, S., Garrido, L., Latorre, J. I., and Piccione, A. (2002). Neural network parametrization of deep inelastic structure functions, *JHEP* **05**, p. 062, doi:10.1088/1126-6708/2002/05/062, arXiv:hep-ph/0204232 [hep-ph].

Frankfurt, L., McDermott, M., and Strikman, M. (2001). A Fresh look at diffractive J/psi photoproduction at HERA, with predictions for THERA, *JHEP* **03**, p. 045, doi:10.1088/1126-6708/2001/03/045, arXiv:hep-ph/0009086 [hep-ph].

Golec-Biernat, K. J. and Wusthoff, M. (1998). Saturation effects in deep inelastic scattering at low Q**2 and its implications on diffraction, *Phys. Rev.* **D59**, p. 014017, doi:10.1103/PhysRevD.59.014017, arXiv:hep-ph/9807513 [hep-ph].

Goulianos, K. A. (1983). Diffractive interactions of hadrons at high-energies, *Phys. Rept.* **101**, p. 169, doi:10.1016/0370-1573(83)90010-8.

Gribov, L. V., Levin, E. M., and Ryskin, M. G. (1983). Semihard processes in QCD, *Phys. Rept.* **100**, pp. 1–150, doi:10.1016/0370-1573(83)90022-4.

Gribov, V. N. and Lipatov, L. N. (1972a). Deep inelastic e p scattering in perturbation theory, *Sov. J. Nucl. Phys.* **15**, pp. 438–450, [Yad. Fiz.15,781(1972)].

Gribov, V. N. and Lipatov, L. N. (1972b). e+ e- pair annihilation and deep inelastic e p scattering in perturbation theory, *Sov. J. Nucl. Phys.* **15**, pp. 675–684, [Yad. Fiz.15,1218(1972)].

Harland-Lang, L. A., Martin, A. D., Motylinski, P., and Thorne, R. S. (2015). Parton distributions in the LHC era: MMHT 2014 PDFs, *Eur. Phys. J.* **C75**, 5, p. 204, doi:10.1140/epjc/s10052-015-3397-6, arXiv:1412.3989 [hep-ph].

Ingelman, G. and Schlein, P. E. (1985). Jet structure in high mass diffractive scattering, *Phys. Lett.* **152B**, pp. 256–260, doi:10.1016/0370-2693(85)91181-5.

Jung, H. (1995). Hard diffractive scattering in high-energy e p collisions and the Monte Carlo generator RAPGAP, *Comput. Phys. Commun.* **86**, pp. 147–161, doi:10.1016/0010-4655(94)00150-Z.

Jung, H. (2002). The CCFM Monte Carlo generator CASCADE, *Comput. Phys. Commun.* **143**, pp. 100–111, doi:10.1016/S0010-4655(01)00438-6, arXiv:hep-ph/0109102 [hep-ph].

Jung, H. and Salam, G. P. (2001). Hadronic final state predictions from CCFM: The Hadron level Monte Carlo generator CASCADE, *Eur. Phys. J.* **C19**, pp. 351–360, doi:10.1007/s100520100604, arXiv:hep-ph/0012143 [hep-ph].

Jung, H. (2020). Aspects of BFKL Physics at HERA, To appear elsewhere in this volume.

Kaidalov, A. B. (1979). Diffractive production mechanisms, *Phys. Rept.* **50**, pp. 157–226, doi:10.1016/0370-1573(79)90043-7.

Khanpour, H. (2019). Phenomenology of diffractive DIS in the framework of fracture functions and determination of diffractive parton distribution functions, *Phys. Rev.* **D99**, 5, p. 054007, doi:10.1103/PhysRevD.99.054007, arXiv:1902.10734 [hep-ph].

Khoze, D. and Ryskin, M and Tasevsky, M. (2020). High energy soft QCD and diffraction, chapter to appear in the 2020 PDG review, in litt.

Klein, M. and Yoshida, R. (2008). Collider Physics at HERA, *Prog. Part. Nucl. Phys.* **61**, pp. 343–393, doi:10.1016/j.ppnp.2008.05.002, arXiv:0805.3334 [hep-ex].

Kowalski, H., Motyka, L., and Watt, G. (2006). Exclusive diffractive processes at HERA within the dipole picture, *Phys. Rev.* **D74**, p. 074016, doi:10.1103/PhysRevD.74.074016, arXiv:hep-ph/0606272 [hep-ph].

Kuraev, E. A., Lipatov, L. N., and Fadin, V. S. (1976). Multi-Reggeon processes in the Yang-Mills theory, *Sov. Phys. JETP* **44**, pp. 443–450, [Zh. Eksp. Teor. Fiz.71,840(1976)].

Kuraev, E. A., Lipatov, L. N., and Fadin, V. S. (1977). The Pomeranchuk singularity in nonabelian gauge theories, *Sov. Phys. JETP* **45**, pp. 199–204, [Zh. Eksp. Teor. Fiz.72,377(1977)].

Levin, E. M., Ryskin, M. G., Shabelski, Yu. M., and Shuvaev, A. G. (1991). Heavy quark production in semihard nucleon interactions, *Sov. J. Nucl. Phys.* **53**, p. 657, [Yad. Fiz.53,1059(1991)].

Levy, A. (2009). Electroproduction of vector mesons, in *Proceedings, 17th International Workshop on Deep-Inelastic Scattering and Related Subjects (DIS 2009): Madrid, Spain, April 26-30, 2009*, p. 177, arXiv:0907.2178 [hep-ex].

LHeC and FCC-he Study Group (2020). The large hadron-electron collider at the LHeC, Extensive update to the LHeC CDR, CERN-Acc-Note-2020-0002, doi:10.1088/1361-6471/abf3ba arXiv:2007.14491 [hep-ex].

Lipatov, L. N. (1975). The parton model and perturbation theory, *Sov. J. Nucl. Phys.* **20**, pp. 94–102, [Yad. Fiz.20,181(1974)].

Lipatov, L. N. (1986). The bare Pomeron in quantum chromodynamics, *Sov. Phys. JETP* **63**, pp. 904–912, [Zh. Eksp. Teor. Fiz.90,1536(1986)].

Lipatov, L. N. (1997). Small x physics in perturbative QCD, *Phys. Rept.* **286**, pp. 131–198, doi:10.1016/S0370-1573(96)00045-2, arXiv:hep-ph/9610276 [hep-ph].

Lonnblad, L. (1992). ARIADNE version 4: A Program for simulation of QCD cascades implementing the color dipole model, *Comput. Phys. Commun.* **71**, pp. 15–31, doi:10.1016/0010-4655(92)90068-A.

Low, F. E. (1975). A model of the bare Pomeron, *Phys. Rev.* **D12**, pp. 163–173, doi:10.1103/PhysRevD.12.163.

Marchesini, G. (1995). QCD coherence in the structure function and associated distributions at small x, *Nucl. Phys.* **B445**, pp. 49–80, doi:10.1016/0550-3213(95)00149-M, arXiv:hep-ph/9412327 [hep-ph].

Martin, A. D., Nockles, C., Ryskin, M. G., and Teubner, T. (2008). Small x gluon from exclusive J/psi production, *Phys. Lett.* **B662**, pp. 252–258, doi:10.1016/j.physletb.2008.02.067, arXiv:0709.4406 [hep-ph].

Martin, A. D., Ryskin, M. G., and Teubner, T. (1997). The QCD description of diffractive rho meson electroproduction, *Phys. Rev.* **D55**, pp. 4329–4337, doi:10.1103/PhysRevD.55.4329, arXiv:hep-ph/9609448 [hep-ph].

Martin, A. D., Ryskin, M. G., and Teubner, T. (2000). Q**2 dependence of diffractive vector meson electroproduction, *Phys. Rev.* **D62**, p. 014022, doi:10.1103/PhysRevD.62.014022, arXiv:hep-ph/9912551 [hep-ph].

Mueller, A. H. (1990). Small x behavior and parton saturation: A QCD model, *Nucl. Phys.* **B335**, pp. 115–137, doi:10.1016/0550-3213(90)90173-B.

Mueller, A. H. and Tang, W.-K. (1992). High-energy parton-parton elastic scattering in QCD, *Phys. Lett.* **B284**, pp. 123–126, doi:10.1016/0370-2693(92)91936-4.

Newman, P. (2004). Deep inelastic lepton nucleon scattering at HERA, *Int. J. Mod. Phys.* **A19**, pp. 1061–1073, doi:10.1142/S0217751X0401897X, arXiv:hep-ex/0312018 [hep-ex].

Newman, P. and Wing, M. (2014). The hadronic final state at HERA, *Rev. Mod. Phys.* **86**, 3, p. 1037, doi:10.1103/RevModPhys.86.1037, arXiv:1308.3368 [hep-ex].

Nussinov, S. (1975). Colored quark version of some hadronic puzzles, *Phys. Rev. Lett.* **34**, pp. 1286–1289, doi:10.1103/PhysRevLett.34.1286.

Salam, G. P. and Soyez, G. (2007). A practical seedless infrared-safe cone jet algorithm, *JHEP* **05**, p. 086, doi:10.1088/1126-6708/2007/05/086, arXiv:0704.0292 [hep-ph].

Sirunyan, A. M. *et al.* (2019a). Measurement of exclusive $\rho(770)^0$ photoproduction in ultraperipheral pPb collisions at $\sqrt{s_{\mathrm{NN}}} = 5.02\,\mathrm{TeV}$, *Eur. Phys. J.* **C79**, 8, p. 702, doi:10.1140/epjc/s10052-019-7202-9, arXiv:1902.01339 [hep-ex].

Sirunyan, A. M. *et al.* (2019b). Measurement of exclusive Υ photoproduction from protons in pPb collisions at $\sqrt{s_{\mathrm{NN}}} = 5.02\,\mathrm{TeV}$, *Eur. Phys. J.* **C79**, 3, p. 277, doi:10.1140/epjc/s10052-019-6774-8, arXiv:1809.11080 [hep-ex].

Stasto, A. M. (2020). Contribution to Lev Lipatov's Memorial Book, To appear elsewhere in this volume.

Ullrich, T. (2019). Luminosity v Energy plot (after M. Klein). Private Communication.